Methods in Enzymology

Volume 370
RNA POLYMERASES AND ASSOCIATED FACTORS
Part C

METHODS IN ENZYMOLOGY

EDITORS-IN-CHIEF

John N. Abelson   Melvin I. Simon

DIVISION OF BIOLOGY
CALIFORNIA INSTITUTE OF TECHNOLOGY
PASADENA, CALIFORNIA

FOUNDING EDITORS

Sidney P. Colowick and Nathan O. Kaplan

Methods in Enzymology

Volume 370

RNA Polymerases and Associated Factors

Part C

EDITED BY

Sankar Adhya

NATIONAL CANCER INSTITUTE
NATIONAL INSTITUTES OF HEALTH
BETHESDA, MARYLAND

Susan Garges

NATIONAL INSTITUTE OF ALLERGY AND INFECTIOUS DISEASES
NATIONAL INSTITUTES OF HEALTH
BETHESDA, MARYLAND

ELSEVIER
ACADEMIC
PRESS

AMSTERDAM • BOSTON • HEIDELBERG • LONDON
NEW YORK • OXFORD • PARIS • SAN DIEGO
SAN FRANCISCO • SINGAPORE • SYDNEY • TOKYO
Academic Press is an imprint of Elsevier

Elsevier Academic Press
525 B Street, Suite 1900, San Diego, California 92101-4495, USA
84 Theobald's Road, London WC1X 8RR, UK
http://www.academicpress.com

International Standard Book Number: 0-12-182273-7

PRINTED IN THE UNITED STATES OF AMERICA
03 04 05 06 07 08   9 8 7 6 5 4 3 2 1

Table of Contents

Section I. RNA Polymerase Structure and Properties

Section II. Analyses of Promoter and Transcription Patterns

Section III. Polymerase Associated Factors

Section IV. Transcription Initiation

Contributors to Volume 370

Article numbers are in parentheses and following the names of contributors. Affiliations listed are current.

ANNIE-CLAUDE ALBERT (11), *Biology Department, Washington University in St. Louis, One Brookings Drive, St. Louis, Missouri 63130*

J.-F. ALLEMAND (49), *Laboratoire de Physique Statistique, CNRS UMR 8550, Ecole Normale Superiure, 24 rue Lhomond, 75231 Paris, France*

JENNIFER R. ANTHONY (5), *Department of Bacteriology, University of Wisconsin – Madison, 1550 Linden Drive, Madison, Wisconsin 53706*

LARRY C. ANTHONY (16), *McArdle Laboratory of Cancer Research, University of Wisconsin, 1400 University Avenue, Madison, Wisconsin 53706*

JAE-BUM BAE (7), *Laboratory of Molecular Microbiology, School of Biological Sciences, Seoul National University, Seoul 151-742, Korea*

VLADIMIR B. BAJIC (21), *Laboratories for Information Technology, 21, Heng Mui Keng Terrace, Singapore 119613*

RAJIV P. BANDWAR (55), *Department Biochemistry, UMDNJ, Robert Wood Johnson Medical School, 675 Hoes Lane, Piscataway, New Jersey 08854*

PETER B. BECKER (42), *Adolf-Butenandt-Institut, Molekularbiologie, Ludwig-Maximilians-Universitat, Schillerstrasse 44, D80336 Munchen, Germany*

VEIT BERGENDAHL (17), *McArdle Laboratory of Cancer Research, University of Wisconsin, 1400 University Avenue, Madison, Wisconsin 53706*

SUKESH R. BHAUMIK (38), *Department of Biochemistry and Molecular Biology, Southern Illinois University, Carbondale, Illinois 62901*

VLADIMIR BONDARENKO (29), *Department of Biochemistry and Molecular Biology, Wayne State University School of Medicine, Detroit, Michigan 48201*

PATRICIA BORDES (53), *Department of Biological Sciences, National College of Science, Technology, and Medicine, SAFB, Imperial College Road, London SW7 2AZ, United Kingdom*

ROBERT BRITTON (23), *Department of Microbiology and Molecular Genetics, Michigan State University, East Lansing, Michigan 48824*

VLADIMIR BRUSIC (21), *Laboratories for Information Technology, 21, Heng Mui Keng Terrace, Singapore 119613*

MARTIN BUCK (3, 53), *Department of Biological Sciences, Wolfson Laboratories, Imperial College of London, Rm. 313, London SW7 2AY, United Kingdom*

RICHARD BURGESS (16, 17), *McArdle Laboratory of Cancer Research, University of Wisconsin, 1400 University Avenue, Madison, Wisconsin 53706*

PATRICIA C. BURROWS (3), *Department of Biological Sciences, Wolfson Laboratories, Imperial College of London, Rm. 313, London SW7 2AY, United Kingdom*

JULIO CABRERA (1), *Laboratory of Molecular Biology, National Cancer Institute, National Institutes of Health, Bldg. 37, Rm. 5144, Bethesda, Maryland 20892*

WENDY CANNON (53), *Department of Biological Sciences, National College of Science, Technology, and Medicine, SAFB, Imperial College Road, London SW7 2AZ, United Kingdom*

MICHAEL W. CAPP (45), *Department of Chemistry, University of Wisconsin, 433 Babcock Drive, Madison, Wisconsin 53706-1544*

FREDERIC COIN (58), *Institut de Genetique et de Biologie Moleculaire et Cellulaire, Dept. of Transcription, CNRS/INSERM/ ULP, B.P. 162, 67404 Illkirch Cedex, C.U. de Strasbourg, France*

RONALD C. CONAWAY (59), *Dept. of Biochemistry and Molecular Biology, Kansas University Medical Center, Kansas City, Missouri 66160*

BENOIT COULOMBE (57), *Laboratory of Gene Transcription, Clinical Research Institute of Montreal, 110 Pine Avenue West, Montreal, Quebec H2W 1R7, Canada*

V. CROQUETTE (49), *Laboratoire de Physique Statistique, CNRS UMR 8550, Ecole Normale Superiure, 24 rue Lhomond, 75231 Paris, France*

MICHAEL E. DAHMUS (13), *Section of Molecular and Cell Biology, Division of Biological Science, University of California, Davis, Davis, California 95616-5224*

SETH A. DARST (4), *The Rockefeller University, 1230 York Avenue, Box 224, New York, New York 10021*

DIPAK DASGUPTA (50), *Biophysics Division, Saha Institute of Nuclear Physics, 37, Belgachia Road, Calcutta 700037, India*

MARGARET A. DAUGHERTY (31), *Department of Biochemistry, University of Vermont College of Medicine, Burlington, Vermont 05405*

ELENA K. DAVYDOVA (8), *Department of Molecular Genetics and Cell Biology, University of Chicago, 920 E. 58th Street, Chicago, Illinois 60637*

PIETER DEHASETH (47), *Department of Biochemistry, Case Western Reserve University, Cleveland, Ohio 44106-4935*

BORRIES DEMELER (43), *Center for Analytical Ultracentrifugation of Macromolecular Assemblies, University of Texas Health Science Center, San Antonio, Texas*

JONATHAN A. DODD (10), *Department of Biological Chemistry, University of California – Irvine, 240 D Medical Sciences I, Irvine, California 92697-1700*

TIMOTHY J. DONOHUE (5), *Department of Bacteriology, University of Wisconsin – Madison, 1550 Linden Drive, Madison, Wisconsin 53706*

DAVID DUNLAP (32), *DIBIT 3A3, san Raffaele Scientific Institute, via Olgettina, 58, Milan 20132, Italy*

ARIK DVIR (59), *Department of Biological Sciences, Oakland University, Rochester, Michigan 48309*

KEITH EARLEY (11), *Biology Department, Washington University in St. Louis, One Brookings Drive, St. Louis, Missouri 63130*

R. H. EBRIGHT (49), *Howard Hughes Medical Institute, Waksman Institute, Department of Chemistry, Rutgers University, Piscataway, New York 08854*

JEAN MARC EGLY (58), *Institut de Genetique et de Biologie Moleculaire et Cellulaire, Department of Transcription, CNRS/IN-SERM/ULP, B.P. 162, 67404 Illkirch Cedex, C.U. de Strasbourg, France*

ROBERT D. FINN (3), *The Wellcome Sanger Institute, Wellcome Trust Genome Campus, Hinxton, Cambridge CB10 1SA, United Kingdom*

LAURA FINZI (32), *Dip Biol "Luigi Gorini," U Degli Studi Milano, Via Celoria 26, Milan I-20133 Italy*

KATHERINE M. FOLEY (16), *McArdle Laboratory of Cancer Research, University of Wisconsin, 1400 University Avenue, Madison, Wisconsin 53706*

DIANE FORGET (57), *Laboratory of Gene Transcription, Clinical Research Institute of Montreal, 110 Pine Avenue West, Montreal, Quebec H2W 1R7, Canada*

MICHAEL G. FRIED (31), *Department of Biochemistry and Molecular Biology, The Pennsylvania State University College of Medicine, 500 University Drive, Hershey, Pennsylvania 17033*

SUSAN GARGES (25), *Laboratory of Molecular Biology, CCR, National Cancer Institute, National Institutes of Health, Bldg. 37, Rm. 5138, Bethesda, Maryland 20892*

S. GILFILLAN (40), *GSF-National Research Center for Environment and Health, Department of Gene Expression, Institute of Molecular Immunology, Marchionini-str. 25, Munich D-81377, Germany*

JAMES A. GOODRICH (56), *Department of Chemistry and Biochemistry, University of Colorado at Boulder, Boulder, Colorado 80309-0215*

RICHARD L. GOURSE (51), *Department of Bacteriology, University of Wisconsin – Madison, 1550 Linden Drive, Madison, Wisconsin 53706-1567*

BRENT GOWEN (3), *627 Gower Point Road, Gibson's Landing, British Columbia V0N 1V8, Canada*

HEATHER A. GREEN (5), *Department of Bacteriology, University of Wisconsin – Madison, 1550 Linden Drive, Madison, Wisconsin 53706*

MICHAEL R. GREEN (36, 38), *Program in Gene Function and Expression, 364 Plantation Street, Worcester, Massachusetts 01605-2377*

JACK GREENBLATT (34), *Banting and Best Department of Medical Research, University of Toronto, Toronto, Ontario M5G 1L6, Canada*

CAROL A. GROSS (18), *Departments of Stomatology and Microbiology and Immunalogy, University of California, San Francisco, California 94143-0512*

TANJA M. GRUBER (18), *Departments of Stomatology and Microbiology and Immunalogy, University of California, San Francisco, California 94143-0512*

MI-YOUNG HAHN (7), *Laboratory of Molecular Microbiology, School of Biological Sciences, Seoul National University, Seoul 151-742, Korea*

MITSUHIRO HAMADA (14), *Department of Integrated Biosciences, University of Tokyo, Tokyo, Japan*

JOHN D. HELMANN (2), *Department of Microbiology, Cornell University, Ithaca, New York 14853*

SUSANNE HOHEISEL (34), *Department of Molecular and Cell Biology, Division of Biochemistry and Molecular Biology, 401 Barker Hall, Berkeley, Californaia 94720-3202*

YING HUANG (14), *Laboratory of Molecular Growth Regulation, National Institute of Child Health and Human Development, National Institutes of Health, Bethesda, Maryland 20892-2753*

DIANE IMBURGIO (19), *Department of Molecular Biology and Biochemistry, Rutgers, The State University of New Jersey, 190 Frelinghuysen Road, Piscataway, New Jersey 08854*

DING JUN JIN (1, 15, 25, 26), *Laboratory of Molecular Biology, National Cancer Institute, National Institutes of Health, Bldg. 37, Rm. 5144, Bethesda, Maryland 20892*

SUSAN JONES (53), *Department of Biological Sciences, National College of Science, Technology, and Medicine, SAFB, Imperial College Road, London SW7 2AZ, United Kingdom*

CAROLINE M. KANE (34), *Department of Molecular and Cell Biology, Division of Biochemistry and Molecular Biology, 401 Barker Hall, Berkeley, California 94720-3202*

CHANGWON KANG (54), *Department of Biological Sciences, Korea Advanced Institute of Science and Technology, 373-1 Guseong-dong, Yuseong-gu, Daejon 305-701, Republic of Korea*

MIKHAIL KASHLEV (12), *NCI Center for Cancer Research, National Cancer Institute, Frederick Cancer Research and Development Center, Bldg. 539, Rm. 222, Frederick, Maryland 21702*

KRYSTYNA M. KAZMIERCZAK (8), *Department of Molecular Genetics and Cell Biology, University of Chicago, 920 E. 58th Street, Chicago, Illinois 60637*

PATRICIA J. KILEY (27), *Department of Biomolecular Chemistry, University of Wisconsin, 574 Medical Science Center, 1300 University Avenue, Madison, Wisconsin 53706*

MARIA L. KIREEVA (12), *NCI Center for Cancer Research, National Cancer Institute, Frederick Cancer Research and Development Center, Bldg. 539, Rm. 222, Frederick, Maryland 21702*

BRUNO P. KLAHOLZ (3), *Institut de Genetique et de Biologie Moleulaire et Cellelaire, Dept. de Biologie et Genomique Structurales, 1, rue Laurent Fries, 67404 Illkirch, France*

MICHAEL S. KOBOR (34), *Department of Molecular and Cell Biology, Division of Biochemistry and Molecular Biology, 401 Barker Hall, Berkeley, California 94720-3202*

NATALIA KOMISSAROVA (12), *NCI Center for Cancer Research, National Cancer Institute, Frederick Cancer Research and Development Center, Bldg. 539, Rm. 222, Frederick, Maryland 21702*

E. KREMMER (40), *GSF-National Research Center for Environment and Health, Department of Gene Expression, Institute of Molecular Immunology, Marchionini-str. 25, Munich D-81377, Germany*

JENNIFER F. KUGEL (56), *Department of Chemistry and Biochemistry, University of Colorado at Boulder, Boulder, Colorado 80319-0215*

KONSTANTIN KUZNEDELOV (9), *Limnological Institute of the Russian Academy of Sciences, Irkutsk, Russia*

LESTER J. LAMBERT (43), *Laboratory of Molecular Biophysics, The Rockefeller University, New York, New York 10021*

DALE E. A. LEWIS (52), *Laboratory of Molecular Biology, National Cancer Institute, 37 Convent Drive, Rm. 5138, Bethesda, Maryland 20892-4255*

PATRICK S. LIN (13), *Section of Molecular and Cell Biology, Division of Biological Science, University of California, Davis, Davis, California 95616-5224*

YE V. LIU (29), *Department of Biochemistry and Molecular Biology, Wayne State University School of Medicine, Detroit, Michigan 48201*

LUCYNA LUBKOWSKA (12), *NCI Center for Cancer Research, National Cancer Institute, Frederick Cancer Research and Development Center, Bldg. 539, Rm. 222, Frederick, Maryland 21702*

BETINA MACHO (35), *Institut de Genetique et de Biologie Moleculaire et Cellulaire, 1, rue Laurent Fries, 67404 Illkirch, Strasbourg, France*

RICHARD MARAIA (14), *Laboratory of Molecular Growth Regulation, National Institute of Child Health and Human Development, National Institutes of Health, Bethesda, Maryland 20892-2753*

ROBERT G. MARTIN (24), *Laboratory of Molecular Biology, NIDDK, National Institutes of Health, Bethesda, Maryland 20892-0560*

SHOKO MASUDA (4), *The Rockefeller University, 1230 York Avenue, Box 224, New York, New York 10021*

KALAI MATHEE (22), *Department of Biological Sciences, Florida International University, Miami, Florida 33199*

WALTER MESSER (30), *Max-Planck-Institute for Molecular Genitics, D-14195 Berlin, Germany*

M. MIESTERERNST (40), *GSF-National Research Center for Environment and Health, Department of Gene Expression, Institute of Molecular Immunology, Marchionini-str. 25, Munich D-81377, Germany*

LEONID MINAKHIN (9), *Waksman Institute of Microbiology, Rutgers, The State University of New Jersey, 190 Frelinghuysen Road, Piscataway, New Jersey 08854*

AARON P. MITCHELL (41), *Department of Microbiology and Institute of Cancer Research, Columbia University, 701 West 168th Street, New York, New York 10032*

KATSUHIKO S. MURAKAMI (4), *The Rockefeller University, 1230 York Avenue, Box 224, New York, New York 10021*

HEATH D. MURRAY (51), *Department of Bacteriology, University of Wisconsin – Madison, 1550 Linden Drive, Madison, Wisconsin 53706-1567*

YOSHIHIRO NAKATANI (37), *Dana-Farber Cancer Institute and Harvard Medical School, Boston, Maryland 02115*

GIRI NARASIMHAN (22), *School of Computer Science, ECS 389, Florida International University, Miami, Florida 33199*

SERGEI NECHAEV (19, 53), *CMG, University of California San Diego, 9500 Gilman Drive, La Jolla, California 92093*

YUR A. NEDIALKOV (44), *Department of Biochemistry and Molecular Biology, Michigan State University, East Lansing, Michigan 48824-1319*

ALEXANDER J. NINFA (29), *Department of Biochemistry, University of Michigan Medical School, Ann Arbor, Michigan 48109-0606*

MASAYASU NOMURA (10), *Department of Biological Chemistry, University of California – Irvine, 240 D Medical Sciences I, Irvine, California 92697-1700*

MICHAEL C. O'NEILL (20), *Department of Biological Sciences, University of Maryland, Baltimore County (UMBC), 1000 Hilltop Circle, Baltimore, Maryland 21250*

VASILY OGRYZKO (37), *Laboratoire Oncogenese, Differenciation et Transduction du Signal, CNRS UPR 9079, Institut Andre Lwoff, Villejuif, France*

ELENA V. ORLOVA (3), *Department of Crystallography, Birbeck College, Malet Street, London WC1E 7HX, United Kingdom*

JOO-HONG PARK (7), *Laboratory of Molecular Microbiology, School of Biological Sciences, Seoul National University, Seoul 151-742, Korea*

SMITA S. PATEL (55), *Department Biochemistry, UMDNJ, Robert Wood Johnson Medical School, 675 Hoes Lane, Piscataway, New Jersey 08854*

ERIK PIERSTORFF (34), *Department of Molecular and Cell Biology, Division of Biochemistry and Molecular Biology, 401 Barker Hall, Berkeley, California 94720-3202*

CRAIG S. PIKAARD (11), *Biology Department, Washington University in St. Louis, One Brookings Drive, St. Louis, Missouri 63130*

APARNA RAVAL (33), *Experimental Immunology Branch, National Cancer Institute, National Institutes of Health, Bldg. 10, Rm. 4B36, Bethesda, Maryland 20892-1360*

PAMPA RAY (3), *Department of Biological Sciences, Wolfson Laboratories, Imperial College of London, Rm. 313, London SW7 2AY, United Kingdom*

M. THOMAS RECORD JR. (45), *Department of Chemistry, University of Wisconsin, 433 Babcock Drive, Madison, Wisconsin 53706-1544*

JOSEPH C. REESE (36), *Deptartment of Biochemistry and Molecular Biology, Pennsylvania State University, 203 Althouse Laboratory, University Park, Pennsylvania 16802*

JOHN N. REEVE (6), *Department of Microbiology, Ohio State University, Columbus, Ohio 43210*

A. REVYAKIN (49), *Howard Hughes Medical Institute, Waksman Institute, Department of Chemistry, Rutgers University, Piscataway, New York 08854*

VIKAS RISHI (39), *Laboratory of Metabolism, National Cancer Institute, National Institutes of Health, Bldg. 37, Rm. 2D24, Bethesda, Maryland 20892*

JUNG-HYE ROE (7), *Laboratory of Molecular Microbiology, School of Biological Sciences, Seoul National University, Seoul 151-742, Korea*

JUDAH L. ROSNER (24), *Laboratory of Molecular Biology, NIDDK, National Institutes of Health, Bethesda, Maryland 20892-0560*

LUCIA B. ROTHMAN-DENES (8), *Department of Molecular Genetics and Cell Biology, University of Chicago, 920 E. 58th Street, Chicago, Illinois 60637*

SIDDHARTHA ROY (46, 48), *Department of Biophysics, Bose Institute, P-1/12, C.I.T., Scheme VII M, Calcutta 700 054, India*

RUTH M. SAECKER (45), *Department of Chemistry, University of Wisconsin, 433 Babcock Drive, Madison, Wisconsin 53706-1544*

JULIO SAEZ-VASQUEZ (11), *Biology Department, Washington University in St. Louis, One Brookings Drive, St. Louis, Missouri 63130*

RAPHAEL SANDALTZOPOULOS (42), *Department of Molecular Biology, Democritus University of Thrace, University Hospital at Dragana, G-68100 Alexandroupolis, Greece*

PAOLA SASSONE-CORSI (35), *Institut de Genetique et de Biologie Moleculaire et Cellulaire, 1, Rue Laurent Fries, 67404 Illkirch, Strasbourg, France*

VIRGIL SCHIRF (43), *Center for Analytical Ultracentrifugation of Macromolecular Assemblies, University of Texas Health Science Center, San Antonio, Texas*

DAVID A. SCHNEIDER (51), *Department of Bacteriology, University of Wisconsin – Madison, 1550 Linden Drive, Madison, Wisconsin 53706-1567*

RANJAN SEN (50), *Laboratory of Transcription Biology, Center for DNA Fingerprinting and Diagnosis, ECIL Road, Nacharam, Hyderabad 500076, India*

STEVE D. SEREDICK (28), *Department of Microbiology and Immunology, University of British Columbia, 6174 University Boulevard, Vancouver, British Columbia V6T 123, Canada*

KONSTANTIN SEVERINOV (9, 19, 53), *Department of Molecular Biology and Biochemistry, Rutgers, The State University of New Jersey, 190 Frelinghuysen Road, Piscataway, New Jersey 08854*

MITSUHIRO SHIMIZU (41), *Deptartment of Chemistry, Meisei University, Tokyo 191-8506, Japan*

INKYUNG SHIN (54), *Department of Biological Sciences, Korea Advanced Institute of Science and Technology, 373-1 Guseong-dong, Yuseong-gu, Daejon 305-701, Republic of Korea*

DEAN D. SHOOLTZ (44), *Department of Biochemistry and Molecular Biology, Michigan State University, East Lansing, Michigan 48824-1319*

DINAH S. SINGER (33), *Experimental Immunology Branch, National Cancer Institute, National Institutes of Health, Bldg. 10, Rm. 4B36, Bethesda, Maryland 20892-1360*

GEORGE B. SPIEGELMAN (28), *Department of Microbiology and Immunology, University of British Columbia, 6174 University Boulevard, Vancouver, British Columbia V6T 123, Canada*

G. STELZER (40), *GSF-National Research Center for Environment and Health, Department of Gene Expression, Institute of Molecular Immunology, Marchionini-str. 25, Munich D-81377, Germany*

T. R. STRICK (49), *Cold Spring Harbor Laboratory, Cold Spring Harbor, New York 11724*

VASILY M. STUDITSKY (29), *Department of Biochemistry and Molecular Biology, Wayne State University School of Medicine, Detroit, Michigan 48201*

MAXIM V. SUKHODOLETS (25), *Developmental Genetics Section, Laboratory of Molecular Biology, CCR, National Cancer Institute, National Institutes of Health, Bldg. 37, Rm. 5138 Bethesda, Maryland 20892*

VICTORIA SUTTON (27), *Department of Biomolecular Chemistry, University of Wisconsin, 574 Medical Science Center, 1300 University Avenue, Madison, Wisconsin 53706*

NANCY E. THOMPSON (16), *McArdle Laboratory of Cancer Research, University of Wisconsin, 1400 University Avenue, Madison, Wisconsin 53706*

PRASAD TONGAONKAR (10), *Department of Biological Chemistry, University of California – Irvine, 240 D Medical Sciences I, Irvine, California 92697-1700*

STEVEN J. TRIEZENBERG (44), *Department of Biochemistry and Molecular Biology, Michigan State University, East Lansing, Michigan 48824-1319*

OLEG V. TSODIKOV (45), *Harvard Medical School, Boston, Massachusetts 02115*

LAURA TSUJIKAWA (47), *Department of Biochemistry, Case Western Reserve University, Cleveland, Ohio 44106-4935*

BARB M. TURNER (28), *Department of Microbiology and Immunology, University of British Columbia, 6174 University Boulevard, Vancouver, British Columbia V6T 123, Canada*

MARIN VAN HEEL (3), *Department of Biological Sciences, Wolfson Laboratories, Imperial College of London, Rm. 313, London SW7 2AY, United Kingdom*

CHARLES VINSON (39), *Laboratory of Metabolism, National Cancer Institute, National Institutes of Health, Building 37, Rm. 2D24, Bethesda, Maryland 20892*

CHRISTOPH WEIGEL (30), *Max-Planck-Institute for Molecular Genitics, D-14195 Berlin, Germany*

JOCELYN D. WEISSMAN (33), *Experimental Immunology Branch, National Cancer Institute, National Institutes of Health, Bldg. 10, Rm. 4B36, Bethesda, Maryland 20892-1360*

JOAN WELIKY CONAWAY (59), *Stowers Institute of Medical Research, 1000 E. 50th Street, Kansas City, Missouri 64110*

MILTON H. WERNER (43), *Laboratory of Molecular Biophysics, The Rockefeller University, New York, New York 10021*

SIVA R. WIGNESHWERARAJ (53), *Department of Biological Sciences, National College of Science, Technology, and Medicine, SAFB, Imperial College Road, London SW7 2AZ, United Kingdom*

YUNWEI XIE (6), *Department of Microbiology, Ohio State University, Columbus, Ohio 43210*

WENXUE YANG (26), *Laboratory of Molecular Biology, National Cancer Institute, National Institutes of Health, Bldg. 37, Rm. 5144, Bethesda, Maryland 20892*

HUIJUN ZHI (15, 26), *Laboratory of Molecular Biology, National Cancer Institute, National Institutes of Health, Bldg. 37, Rm. 5144, Bethesda, Maryland 20892*

Preface

It has been just seven years since Volumes 273 and 274 of *Methods in Enzymology*, which covered RNA polymerase and associated activities, were published. Since then, there has been an explosion in the amount of information that has come out on RNA polymerase and transcription, driven by the intensification in technology. In Volumes 370 and 371, we try to take up where Volumes 273 and 274 left off, but we include and emphasize what we feel is an important aspect of RNA polymerase that was only touched on in the previous volumes. It is quite clear now that RNA polymerase does not act alone. Its associated factors are key in determining initiation, elongation, and termination occurrences. The reliance on the associated factors is often so great that it is difficult to determine whether a given protein is an RNA polymerase-associated factor or is, in fact, a subunit of the enzyme.

The chapters in these volumes describe the RNA polymerase enzymes and the associated factors, and their effects on transcription initiation, elongation, and termination. The chapters expose both prokaryotic and eukaryotic systems, but they are purposely kept undivided irrespective of the origins of the RNA polymerases. We predict that as more is learned about the prokaryotic and eukaryotic systems of transcription and gene regulation, we will realize how similarly they behave rather than how differently.

SANKAR ADHYA
SUSAN GARGES

METHODS IN ENZYMOLOGY

VOLUME 335. Flavonoids and Other Polyphenols
Edited by LESTER PACKER

VOLUME 336. Microbial Growth in Biofilms (Part A: Developmental and Molecular Biological Aspects)
Edited by RON J. DOYLE

VOLUME 337. Microbial Growth in Biofilms (Part B: Special Environments and Physicochemical Aspects)
Edited by RON J. DOYLE

VOLUME 338. Nuclear Magnetic Resonance of Biological Macromolecules (Part A)
Edited by THOMAS L. JAMES, VOLKER DÖTSCH, AND ULI SCHMITZ

VOLUME 339. Nuclear Magnetic Resonance of Biological Macromolecules (Part B)
Edited by THOMAS L. JAMES, VOLKER DÖTSCH, AND ULI SCHMITZ

VOLUME 340. Drug–Nucleic Acid Interactions
Edited by JONATHAN B. CHAIRES AND MICHAEL J. WARING

VOLUME 341. Ribonucleases (Part A)
Edited by ALLEN W. NICHOLSON

VOLUME 342. Ribonucleases (Part B)
Edited by ALLEN W. NICHOLSON

VOLUME 343. G Protein Pathways (Part A: Receptors)
Edited by RAVI IYENGAR AND JOHN D. HILDEBRANDT

VOLUME 344. G Protein Pathways (Part B: G Proteins and Their Regulators)
Edited by RAVI IYENGAR AND JOHN D. HILDEBRANDT

VOLUME 345. G Protein Pathways (Part C: Effector Mechanisms)
Edited by RAVI IYENGAR AND JOHN D. HILDEBRANDT

VOLUME 346. Gene Therapy Methods
Edited by M. IAN PHILLIPS

VOLUME 347. Protein Sensors and Reactive Oxygen Species (Part A: Selenoproteins and Thioredoxin)
Edited by HELMUT SIES AND LESTER PACKER

VOLUME 348. Protein Sensors and Reactive Oxygen Species (Part B: Thiol Enzymes and Proteins)
Edited by HELMUT SIES AND LESTER PACKER

VOLUME 349. Superoxide Dismutase
Edited by LESTER PACKER

VOLUME 350. Guide to Yeast Genetics and Molecular and Cell Biology (Part B)
Edited by CHRISTINE GUTHRIE AND GERALD R. FINK

VOLUME 351. Guide to Yeast Genetics and Molecular and Cell Biology (Part C)
Edited by CHRISTINE GUTHRIE AND GERALD R. FINK

Section I

RNA Polymerase Structure and Properties

[1] Construction, Purification, and Characterization of *Escherichia coli* RNA Polymerases Tagged with Different Fluorescent Proteins

By Julio E. Cabrera and Ding Jun Jin

The green fluorescent protein (GFP) from *Aequorea victoria* and its color variants, the cyan fluorescent protein (CFP) and the yellow fluorescent protein (YFP), have been widely used in cellular and molecular biology studies.[1–3] These proteins are intrinsically fluorescent in a wide range of hosts and, in most of the cases, remain fluorescent when fused to a target protein. Some of these fusion proteins have both the physical properties of the fluorescent proteins (i.e., fluorescence) and the biochemical properties of the target protein. This article describes a method used to generate *Escherichia coli* strains with all their RNA polymerase (RNAP) molecules labeled with GFP, CFP, or YFP. It also describes the physical and biochemical properties of purified RNAP molecules tagged with different fluorescent proteins.

Construction of *E. coli* Strains with Chromosomal Gene Fusions to the *rpoC* Gene

Escherichia coli core RNAP is composed of four subunits, $\alpha_2\beta\beta'$. The β and β' subunits are respectively encoded by the *rpoB* and *rpoC* genes in the same operon. We fused *gfp*, *cfp*, or *yfp* genes at the 3' end terminus of the *rpoC* gene in the chromosome. The choice of fusing the C terminus of the β' subunit with a fluorescent protein was based on the previous finding that the C terminus of the β' subunit can be fused to a 200 amino acid fragment of the β galactosidase without loss of β' function.[4] A schematic of the construction of gene fusions is represented in Fig. 1.

[1] R. Y. Tsien, *Annu. Rev. Biochem.* **67,** 509 (1998).
[2] M. Chalfie and S. Kain (eds.), "Green Fluorescent Protein, Properties, Applications and Protocols." Wiley-Liss, New York, 1998.
[3] M. C. Southward and M. Surette, *Mol. Microbiol.* **45,** 1191 (2002).
[4] G. C. Rowland and R. E. Glass, *Mol. Microbiol.* **17,** 401 (1995).

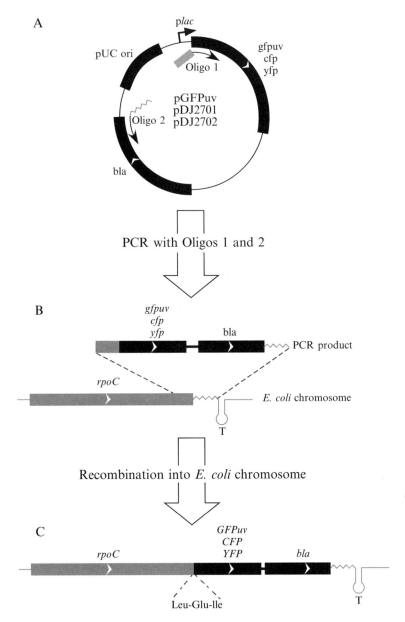

FIG. 1. A schematic for the construction of chromosomal gene fusions to the *rpoC* gene. Regions representing sequences of the *E. coli* chromosome and sequences with homology to it are shaded in gray. The drawing is not to scale. (A) Plasmid DNA templates and relative orientation of the oligonucleotides used in the PCR reaction. (B) Relative location of the

Generating Linear DNA Fragments That Carry gfpuv, cfp, or yfp Genes and the bla Gene Flanked by Chromosomal Sequences Near the End of rpoC

We used the polymerase chain reaction (PCR) to generate linear fragments to be used for recombination in the next step. The DNA templates for these PCR reactions are pGFPuv (Clontech), pDJ2701 (which harbors the *cfp* gene), or pDJ2702 (which harbors the *yfp* gene). *Note.* The GFPuv protein is a mutant GFP optimized for higher bacterial expression and maximal fluorescence when excited by UV light. The plasmids pDJ2701 and pDJ2702 are pGFPuv derivatives. They have been constructed by replacing the *gfpuv* gene by the *cfp* or *yfp* genes. The *cfp* and *yfp* genes were obtained from the pDH3 and pDH5 plasmids (University of Washington, Seattle, Yeast Resource Center),[5] respectively. A schematic map of these plasmids is shown in Fig. 1A. The relative positions of the two oligos nucleotides (oligo 1 and oligo 2) for these PCR reactions are also shown in Fig. 1A. The sequence of oligo 1 is 5′ <u>CCAy GCC TGG CAG AAC TGC TGA ACG CAG GTC TGG GCG GTT CTG ATA ACG AGC TAG AAA TAA TGA GTA AAG GAG AAG AAC TTT TCA CTG G</u> 3′. The sequence of oligo 2 is 5′ <u>CCC CCC ATA AAA AAA CCC GCC GAA GCG GGT TTT TAC GTT ATT TGC GGA TTA</u> TGG TCT GAC AGT TAC CAA TGC 3′. For oligo 1, the first 50 nucleotides (nt) of the 5′ end (underlined) are identical to the last 50 nt coding sequence of *rpoC* immediately before the stop codon, followed by an 9 nt (bold) encoding for a three amino acid linker, Leu-Glu-Ile, that replaces the stop codon of the *rpoC* gene. Downstream of the three amino acid linker, the oligo contains sequences encoding the first 29 nt coding sequences of the *gfpuv*, *cfp*, or *yfp* genes (the 5′ end sequences of the three genes are the same). For oligo 2, the first 51 nt of the 5′ end (underlined) are identical to the bases located 4 bp downstream from the stop codon of the *rpoC* gene, and the remaining 21 nt are identical to the last coding sequences, including the stop codon of the *bla* gene encoding ampicillin resistance (Amp[r]). The expected PCR products are DNA fragments of ∼ 2230 bp. The DNA fragments

[5] D. W. Hailey, T. N. Davis, and E. G. Muller, *Methods Enzymol.* **351,** 34 (2002).

homolog regions between the linear DNA fragment and the chromosomal *rpoC* gene. The 5′ end of the linear DNA fragment, represented as a solid gray bar (▬), is homolog to the 50 bp upstream of the stop codon of the *rpoC* gene. The 3′ end of the linear DNA fragment, represented as a gray wavy line (∿∿∿), is homolog to sequences located 4 bp downstream of the *rpoC* gene stop codon. (C) Gene fusion at the *rpoC* locus. A linker of three codons (Leu-Glu-Ile) replaces the stop codon and fuses the *rpoC* gene to genes encoding the fluorescent proteins.

are purified. The overall homology of this DNA fragment relative to the end of the *rpoC* gene in the chromosome is shown in Fig. 1B.

Recombining the Linear DNA Fragment into the E. coli *Chromosome*

We used the recombination method described by Yu and collaborators,[6] which is particularly useful for this purpose, and selected the recombinant candidates on LB + Amp plates (Fig. 1C). The Amp[r] colonies obtained are checked to ensure they are recombinants because minute amounts of intact plasmid templates could be present in the linear DNA preparation and thereby transform into the recipient cells, resulting in Amp[r] colonies. A rapid method used to check this is to analyze total proteins from the Amp[r] colonies by Western blot followed by immunostaining with either an anti-GFP antibody or a monoclonal anti-β' antibody. Proteins from recombinant Amp[r] cells show a reactive band corresponding to a fusion protein with a molecular weight higher than that of the β' polypeptide. To confirm that the fusion is correct, we sequenced the *rpoC* regions flanking the insertion points, including the *gfpuv*, *cfp*, or *yfp* genes. *rpoC–gfpuv* and other fusions can be transferred to other strains' backgrounds by phage P1 transduction using the *bla* gene (Amp[r]) as a selection marker (the linkage between *rpoC* fusion and *bla* is almost 100%). *Note.* Cells expressing β' fusion proteins have a reduced ability to grow at temperatures higher than 37°. This defect is probably caused by a misfolding of the fluorescent proteins, which alters the conformation of the entire fusion protein at temperature higher than 37°.[7–9] Thus, strains containing *rpoC* fusions should be grown at 32°.

Purification of RNAP Tagged with Green, Yellow, or Cyan Fluorescent Proteins

RNAP fused with different fluorescent proteins are purified essentially as described for the purification of RNAP.[10] The only modification is that before Mono Q chromatography, the ammonium sulfate precipitates of RNAP fusion proteins are resuspended in 0.2 *M* NaCl in TGED instead of TGED only. This is done because the NaCl apparently facilitates the

[6] D. Yu, H. M. Ellis, E. C. Lee, N. A. Jenkins, N. G. Copeland, and D. L. Court, *Proc. Natl. Acad. Sci. USA* **97,** 5978 (2000).

[7] P. A. Levin, I. G. Kurtser, and A. D. Grossman, *Proc. Natl. Acad. Sci. USA* **96,** 9642 (1999).

[8] K. R. Siemering, R. Golbick, R. Sever, and J. Haseloff, *Curr. Biol.* **6,** 1653 (1996).

[9] G. S. Gordon, D. Sitnikov, C. D. Webb, A. Teleman, A. Straight, R. Losick, A. W. Murray, and A. Wright, *Cell* **90,** 1113 (1997).

[10] M. V. Sukhodolets and D. J. Jin, *J. Biol. Chem.* **273,** 7018 (1998).

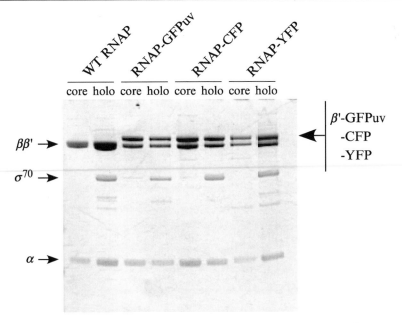

FIG. 2. SDS–PAGE of wild-type (WT) RNAP and RNAP tagged with green, yellow, or cyan fluorescent proteins. Positions of the bands corresponding to the different subunits are indicated.

solubilization of RNAP fluorescent fusion proteins. It has been reported that the GFP dimerizes in solutions of ionic strengths less than 100 mM^{11}; it is possible that at a very low ionic strength the highly concentrated RNAP fluorescent fusion proteins become insoluble. A typical yield for core and holo RNAP fusion proteins is about 15 mg from about 100 g of wet cell paste, similar to that of RNAP preparations. When purified, each RNAP fusion protein has the characteristic color of the corresponding fluorescent protein. Figure 2 shows a SDS–PAGE analysis of purified wild-type and different RNAP fluorescent fusion proteins. As expected, β' subunits from RNAP fusion proteins, due to fused fluorescent proteins, exhibit a higher molecular weight than that of wild-type RNAP.

Physical and Biochemical Properties of RNAP Tagged with Green, Yellow, or Cyan Fluorescent Proteins

We studied purified RNAP fusion proteins by fluorescence spectroscopy. Each tagged RNAP has emission and excitation spectra that are coincident with spectra of the corresponding isolated fluorescent protein.[1]

[11] F. Yang, L. G. Moss, and G. N. Phillips, *Nature Biotechnol.* **14,** 1246 (1996).

For example, peak values in excitation and emission spectra of the RNAP–GFPuv protein fusion are 396 and 507 nm, respectively (Fig. 3A). Peak values in excitation spectra of the RNAP–CFP protein fusion are between 436 and 450 nm, whereas the peak value in emission spectra of the fusion protein is 474 nm (Fig. 3B). Peak values in excitation and emission spectra of the RNAP–YFP protein fusion are 517 and 530 nm, respectively (Fig. 3C).

RNAP fluorescent fusion proteins are fully active in RNA synthesis. The fact that cells carrying chromosomal fusions between the *rpoC* gene

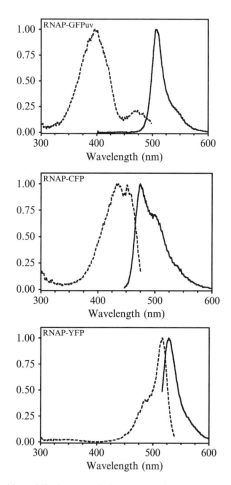

Fig. 3. Excitation (dotted line) and emission spectra (continuous line) of RNAP enzymes labeled with green, yellow, or cyan fluorescent proteins. Data were collected with a Quanta Master fluorometer (Photon Technology International).

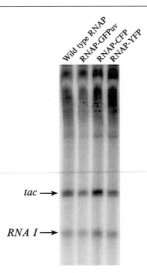

FIG. 4. *In vitro* transcription assays with wild-type RNAP or RNAP tagged with green, yellow, or cyan fluorescent proteins. *In vitro* transcription reactions were carried out as described previously[12] in a transcription buffer containing 50 mM KCl. The DNA template was plasmid pDJ631 containing the *tac* promoter, and RNAs were labeled with (α-^{32}P) UTP. Reactions were analyzed in an 8% sequencing gel. An autoradiograph of the gel is shown. Positions of transcripts synthesized from *tac* and *RNAI* promoters are indicated.

and the genes encoding fluorescent proteins are viable demonstrates that RNAP fusion proteins are functional *in vivo*. *In vitro*, transcriptional activities of purified RNAP fluorescent fusion proteins are comparable to those of the purified wild-type RNAP (Fig. 4). Furthermore, RNAP fluorescent fusion proteins are as active as the wild-type RNAP in *in vitro* transcription assays at 37° and 42° (data not shown). These results are consistent with the notion that the folding step of fluorescent proteins, rather than the function of RNAP fluorescent fusion proteins, is temperature sensitive *in vivo*.[7–9]

Potential Uses of RNAP Tagged with Green, Yellow, or Cyan Fluorescent Proteins

RNAP fused with fluorescent proteins will provide a powerful new tool to study RNAP functions and transcriptions both *in vivo* and *in vitro*. The following examples are potential uses of RNAP fusion proteins. *In vivo*, the distribution of RNAP inside cells under different physiological

[12] Y. N. Zhou and D. J. Jin, *Proc. Natl. Acad. Sci. USA* **95**, 2908 (1998).

conditions could be visualized using a fluorescence microscope.[13] Also, the locations of RNAP and another transcription factor(s) could be addressed simultaneously by imaging cells expressing both RNAP tagged with CFP (or YFP) and a transcription factor(s) tagged with YFP (or CFP), respectively. Many transcription factors are interesting candidates for these experiments: the σ factors, RapA, NusA, and SspA, just to mention a few. In addition, analyses of fluorescence resonance energy transfer (FRET) between the tagged RNAP and a tagged transcription factor could be used to analyze the physical interaction between RNAP and the transcription factor.

In vitro, the movements of single RNAP molecules could be followed or visualized under different transcription conditions using these RNAP fluorescent fusion proteins. In addition, the FRET experiments mentioned earlier could also be used to study the interactions of RNAP and another transcription factor(s) and to follow the association and/or dissociation of a different transcription factor(s) during the transcription cycle under defined conditions.

Acknowledgments

We thank Drs. Wenxue Yang and Huijun Zhi for their assistance in the purification of RNAPs.

[13] J. E. Cabrera and D. J. Jin, *Mol. Microbiol.* (2003, in press).

[2] Purification of *Bacillus subtilis* RNA Polymerase and Associated Factors

By JOHN D. HELMANN

Introduction

Bacterial RNA polymerase (RNAP) has been purified using a wide variety of techniques. In addition to continuing interest in its fundamental enzymatic properties, purified RNAP is used as a tool for the investigation of transcriptional control mechanisms and for the preparation of specific RNA transcripts. This article focuses on techniques developed for the purification of RNAP and its associated factors from the model gram-positive bacterium *Bacillus subtilis*.

Bacillus subtilis RNAP has a complex subunit structure. The minimal catalytic moiety, consisting of the $\beta\beta'\alpha_2$ complex, is associated with two ω

subunits (of uncertain function) to generate the core enzyme (E = $\beta\beta'\alpha_2\omega^1$ ω^2). The core enzyme interacts with a 21-kDa auxiliary specificity factor known as δ and a variable population of specificity (σ) factors.[1–3] Altogether, the B. subtilis genome encodes at least 17 σ factors.[2] The primary σ, σ^A, is the most abundant σ during logarithmic growth and the only essential σ factor in B. subtilis. The remaining σ factors are present at much lower amounts during growth, but some become quite abundant during sporulation. The identity and relative proportions of the various σ factors vary depending on growth stage and whether the cells have initiated sporulation.[4,5]

As purified, B. subtilis RNAP is a complex mixture of core (E), core-δ (Eδ), core-σ (Eσ), and core-σ-δ (E$\delta\sigma$).[6] Only the forms of RNAP associated with a σ factor (Eσ or E$\sigma\delta$) can recognize promoter sites. During chromatography on DNA–cellulose[6] or heparin–agarose,[7] those fractions enriched in δ tend to elute at lower salt concentrations. Conversely, fractions containing the σ^A-saturated holoenzyme are often depleted of δ. This is consistent with the observation that σ^A and δ bind to the core with negative cooperativity, although binding does not appear to be mutually exclusive.[8] δ has dramatic effects on the transcriptional properties of the B. subtilis enzyme and, therefore, its presence should be monitored closely.[9–14] For consistency, it is possible to add stoichiometric amounts of purified δ to RNAP preparations or, alternatively, to start with strains (rpoE mutants) deficient in δ.

This article describes methods used for the purification of E, Eδ, Eσ^A, and E$\delta\sigma^A$. Methods are also summarized for the production and purification of individual σ factors and the δ subunit, important reagents for the reconstitution of defined holoenzyme species. For some applications, it may be desirable to purify individual RNAP subunits or associated factors.

[1] W. G. Haldenwang, Microbiol. Rev. 59, 1 (1995).
[2] J. D. Helmann and C. P. Moran, Jr., in "Bacillus subtilis and Its Relatives: From Genes to Cells" (A. L. Sonenshein and R. Losick, eds.), p. 289. ASM Press, Washington, DC, 2002.
[3] J. D. Helmann, Adv. Microbiol. Phys. 46, 47 (2002).
[4] M. Fujita, Genes Cells 5, 79 (2000).
[5] L. Kroos, B. Zhang, H. Ichikawa, and Y. T. Yu, Mol. Microbiol. 31, 1285 (1999).
[6] R. H. Doi, in "The Molecular Biology of the Bacilli" (D. A. Dubnau, ed.), p. 7. Academic Press, New York, 1982.
[7] J. L. Wiggs, M. Z. Gilman, and M. J. Chamberlin, Proc. Natl. Acad. Sci. USA 78, 2762 (1981).
[8] E. I. Hyde, M. D. Hilton, and H. R. Whiteley, J. Biol. Chem. 261, 16565 (1986).
[9] E. C. Achberger and H. R. Whiteley, J. Biol. Chem. 256, 7424 (1981).
[10] K. F. Dobinson and G. B. Spiegelman, Biochemistry 26, 8206 (1987).
[11] Y. L. Juang and J. D. Helmann, J. Mol. Biol. 239, 1 (1994).
[12] Y. L. Juang and J. D. Helmann, Biochemistry 34, 8465 (1995).
[13] F. J. Lopez de Saro, A. Y. Woody, and J. D. Helmann, J. Mol. Biol. 252, 189 (1995).
[14] Y. F. Chen and J. D. Helmann, J. Mol. Biol. 267, 47 (1997).

TABLE I
RNA POLYMERASE SUBUNITS AND INITIATION FACTORS

Protein	Gene	Function(s)	Amino acid	Reference
β	rpoB	Core RNAP	1193	—
β'	rpoC	Core RNAP	1199	—
α	rpoA	Core RNAP	314	15,16
δ	rpoE	Increases selectivity, stimulates recycling	173	13
ω^1	ykzG	Unknown	69	—
ω^2	yloH	Chaperone ?	67	—
σ^A	sigA	Primary σ	371	36,39,47
σ^B	sigB	General stress response	264	44
σ^D	sigD	Flagella, chemotaxis, autolysins	254	34
σ^E	sigE	Sporulation—early mother cell	239	24
σ^F	sigF	Sporulation—late mother cell	255	16a,40,46
σ^G	sigG	Sporulation—late forespore	260	16b, 42
σ^H	sigH	Sporulation, competence	218	36
σ^I	ykoZ	Heat stress	251	—
σ^K	sigK	Sporulation, late forespore		—
σ^L	sigL	Levanase, amino acid catabolism	436	—
σ^M	sigM	Salt resistance	163	—
σ^V	sigV	Unknown	166	—
σ^W	sigW	Antimicrobial resistance	187	16b,42
σ^X	sigX	Cell surface properties	194	40
σ^Y	sigY	Unknown	178	—
σ^Z	sigZ	Unknown	176	—
σ^{ylaC}	ylaC	Unknown	173	—

For example, the purified *B. subtilis* α subunit binds DNA at upstream promoter (UP) element sequences,[15] and RNAP can be denatured and then reconstituted with truncated or mutant α subunit to test for contacts between activator proteins and the α carboxyl-terminal domain (CTD).[16] Where procedures have been described for overproduction and purification of RNAP subunits, they are referenced in Table I.

General Considerations

Prior to embarking on a purification of RNAP, one must first determine the quantity and purity of the material desired. Because most isolation procedures lead to a complex mixture of enzyme forms, individual

[15] K. Fredrick, T. Caramori, Y. F. Chen, A. Galizzi, and J. D. Helmann, *Proc. Natl. Acad. Sci. USA* **92,** 2582 (1995).
[16] M. Mencia, M. Monsalve, F. Rojo, and M. Salas, *Proc. Natl. Acad. Sci. USA* **93,** 6616 (1996).
[16a] L. Duncan, S. Alper, and R. Losick, *J. Mol. Biol.* **260,** 147 (1996).
[16b] E. M. Kellner, A. Decatur, and C. P. Moran, Jr., *Mol. Microbiol.* **21,** 913 (1996).

holoenzymes are generally reconstituted by the addition of a molar excess of the desired σ factor (see later). Alternatively, procedures have been developed for the reconstitution of bacterial RNAP from individually overproduced subunits, but this approach has not been widely adopted.[17,18] Moreover, *Escherichia coli* RNAP reconstituted from overproduced β, β', α, and σ subunits lacks the ω polypeptide and other less abundant factors that often copurify with RNAP (e.g., GreA and GreB). As a result, the reconstituted enzyme is not necessarily a good model for the enzyme present in cells.

In growing cells, RNAP is about 1% by weight of soluble protein and yields of >10 mg of partially purified enzyme are typical from 50 g (wet weight) of cells. The use of strains containing a histidine-tagged β' subunit allows recovery of >1 mg of high-purity enzyme starting with 10 g of cell paste (~2 liters of culture volume).[19,20] In a typical *in vitro* transcription reaction, RNAP is added at concentrations of 40–50 μg/ml, so a purification yielding a milligram of enzyme is sufficient for many reactions. However, because RNAP can be stored for months or years with little loss in activity, it may be worth preparing a larger amount of enzyme to have as a laboratory stock.

For most investigators, the desired end product is likely to be a holoenzyme: either the σ^A holoenzyme or one containing an alternate σ factor. Conditions have been described for the isolation of RNAP fractions enriched in various holoenzymes beginning with cells in various growth stages.[6,7,21] For example, RNAP fractions containing σ^A, σ^H, σ^B, and σ^D are all purified from late logarithmic or early stationary phase cells, whereas fractions containing sporulation σ factors are isolated from sporulating cells. The multiple holoenzymes present in the cell can be separated by chromatography on DNA–cellulose[6] or heparin–agarose.[7] This approach could be combined with the initial isolation of His-tagged RNAP populations from sporulating cells.[21] However, the resulting fractions will have variable and usually substoichiometric amounts of σ factors. For biochemical studies, it is most convenient to reconstitute the desired σ-saturated holoenzyme starting with purified RNAP (see later).

[17] N. Fujita and A. Ishihama, *Methods Enzymol.* **273**, 121.

[18] H. Tang, Y. Kim, K. Severinov, A. Goldfarb, and R. H. Ebright, *Methods Enzymol.* **273**, 130.

[19] Y. Qi and F. M. Hulett, *Mol. Microbiol.* **28**, 1187 (1998).

[20] L. Anthony, I. Artsimovitch, V. Svetlov, R. Landick, and R. R. Burgess, *Protein Exp. Purif.* **19**, 350 (2000).

[21] M. Fujita and Y. Sadaie, *Gene* **221**, 185 (1998).

Buffers and Reagents

Many of the buffers used during the purification of RNAP and its associated factors are based on TGMED [10 mM Tris, pH 8.0, 10% (v/v) glycerol, 10 mM MgCl$_2$, 0.1 mM EDTA, 0.1 mM dithiothreitol) (DTT)] or derivatives of TGMED lacking one or more component (as indicated by the abbreviation). Protease inhibitors, such as phenylmethylsulfonyl fluoride (PMSF), should be included in the early stages of RNAP isolation and all procedures should be performed as rapidly as practical. Much valuable background information is contained in previous reviews by Chamberlin et al.[22] and Doi.[6]

Method 1. Purification of His-Tagged RNAP Core Enzyme

Probably the simplest purification procedures take advantage of strains engineered to have a His-tagged β-prime subunit. Such strains have been developed in both the Hulett[19] and Fujita[21] laboratories. Strain MH5636 encodes a hexa-histidine tag on the carboxyl-terminal end of the β-prime subunit.[19] Purification of the core enzyme starting from this strain has been described in detail by Anthony et al.[20] Briefly, cells are grown to late logarithmic phase, harvested, and lysed by sonication in buffer P (300 mM NaCl, 50 mM Na$_2$HPO$_3$, 3 mM 2-mercaptoethanol, 5% glycerol, and a protease inhibitor mix). After clarification, a slurry of Ni–NTA agarose (1/10th volume) is added and mixed for 30 min at 8° for 30 min. The mixture is collected in a column and washed extensively (20 volumes) with buffer P + 60 mM imidazole. RNAP is then eluted with buffer P + 400 mM imidazole and assayed by SDS–PAGE for the presence of RNAP. At this stage the partially purified RNAP contains significant amounts of σ^A and can be used for in vitro transcription reactions.

For purification of the core enzyme, RNAP-containing fractions are pooled and further purified by sizing gel chromatography (FPLC Superdex-200) followed by an FPLC Mono Q ion-exchange column. The RNAP core enzyme elutes in the first peak (0.38 M NaCl) from Mono Q. A yield of >1 mg of 98% pure core enzyme is obtained starting with 10 g (wet weight) of cell paste. Inspection of the published SDS–PAGE gel[20] indicates that this fraction contains at least some δ. Immunoblot analysis indicates that the resulting enzyme is free from detectable σ^A, NusA, or GreA.[20] However, when this enzyme was used for in vitro transcription

[22] M. Chamberlin, R. Kingston, M. Gilman, J. Wiggs, and A. DeVera, Methods Enzymol. 101, 540.

reactions, numerous specific products appeared to result from the presence of low levels of σ factors.[23]

Method 2. Purification of RNAP Using Polymin P, DNA–Cellulose, and A1.5M Chromatography

Because RNAP has a high affinity for DNA, most purification procedures begin with one or more steps designed to remove nucleic acids. Three common approaches include (i) PEG/dextran phase partitioning, (ii) polyethyleneimine (polymin P) precipitation, or (iii) heparin–agarose affinity chromatography. A method for purification of the σ^E holoenzyme using PEG/dextran phase partitioning has been described previously.[24] Here, a method for the large-scale preparation of RNAP beginning with polymin P precipitation is presented.

This procedure, based loosely on the procedure of Burgess and Jendrisak,[25] was developed to allow the preparation of high-purity RNA polymerase in quantities of 80 to 100 mg.[7,26,27] The essential steps in this procedure are the use of polymin P to precipitate DNA (and RNA polymerase) from the crude extract, elution of RNAP with elevated salt, and desalting and affinity chromatography using DNA–cellulose, followed by sizing chromatography. DNA cellulose is prepared as described[28] or can be obtained commercially. Unless noted otherwise, all fractionation steps should be performed at 4°. The purification procedure presented begins with ∼500 g of cell paste (equivalent to the yield from a 50-liter fermenter culture grown in rich medium to late logarithmic phase) and is similar to that described by Helmann *et al.*[27] This procedure can also be scaled down by ∼10-fold if desired.

Lysis

Cell paste (stored frozen at −80° prior to use) is broken into small pieces using a hammer and is suspended by the gradual addition of 2 liters of lysis buffer 1 [50 mM Tris, pH 8.0, 10 mM MgCl$_2$, 2 mM EDTA, 0.1 mM DTT, 1 mM 2-mercaptoethanol, 233 mM NaCl, 10% (v/v) glycerol] using a Waring blender. Lysis is achieved by at least two passages through a

[23] M. Cao, P. A. Kobel, M. M. Morshedi, M. F. Wu, C. Paddon, and J. D. Helmann. *J. Mol. Biol.* **316,** 443 (2002).

[24] K. M. Tatti and C. P. Moran, Jr., *Methods Enzymol.* **273,** 149.

[25] R. R. Burgess and J. J. Jendrisak, *Biochemistry* **14,** 4634 (1975).

[26] J. Jaehning, J. Wiggs, and M. Chamberlin, *Proc. Natl. Acad. Sci. USA* **76,** 5470 (1979).

[27] J. D. Helmann, F. R. Masiarz, and M. J. Chamberlin, *J. Bacteriol.* **170,** 1560 (1988).

[28] B. Alberts and G. Herrick, *Methods Enzymol.* **21,** 198.

Manton–Gaulin homogenizer at approximately 10,000 psi. For smaller preparations (e.g., 50 g of cell paste), a French pressure cell or sonication can be used for lysis. The efficiency of cell lysis can be improved optionally by the addition of lysozyme (0.1 mg/ml final) to weaken the cell wall. However, it is important that lysozyme be added after the cells are suspended uniformly in lysis buffer or the increased viscosity resulting from released DNA can make it difficult to complete the resuspension process. To inhibit proteolysis, freshly prepared PMSF (100 mM in absolute ethanol) is added to 1 mM during the lysis procedure.

Immediately following lysis, the cell extract is diluted twofold with 2 liters of TGMED + 0.2 M NaCl. This buffer is prepared in advance and stored overnight at 4°. Alternatively, some or all of this buffer can be frozen at −20° (e.g., using ice cube trays). This has the advantage that the addition of frozen buffer between passages through the Manton–Gaulin homogenizer (or French pressure cell) helps maintain a low temperature. A thermometer can be used to monitor the temperature of the lysate, which should not rise above 15°. The lysate is clarified by centrifugation at 10,000 g for 45 min in a prechilled rotor in a refrigerated centrifuge (4°). For smaller volumes, the length of the centrifugation can be shortened.

Polymin P Fractionation

The clarified lysate (∼3.5 liter) is poured carefully into a clean beaker, avoiding the carryover of pellet material. Polymin P precipitation is achieved by the addition, with stirring, of a 10% polyethyleneimine solution (polymin P; BASF) to a final concentration of 0.5% over a period of 10 to 20 min. For details on the preparation of polymin P stocks, the reader is referred to Jendrisak and Burgess.[29] The solution is stirred for an additional 30 min at 4°, and the precipitate is collected by centrifugation for 15 min at 10,000 g in a refrigerated (4°) centrifuge. The polymin P supernatant is stored at 4°, and purification continues with the polymin P pellet, which contains both chromosomal DNA and associated proteins (including RNAP).

The polymin P pellet is washed by suspension in 1.6 liters of TMED + 0.4 M NH$_4$Cl. This can be achieved by scraping the pellet with a spatula into a cuisinart or blender and then adding wash buffer prior to mixing. Care should be taken to avoid excessive foaming. Alternatively, the wash buffer can added directly to the centrifuge bottles and the pellet suspended using a spatula. Centrifugation is repeated as described earlier, and the supernatant fraction (0.4 M wash) is stored for later analysis. To elute RNAP, the pellets are suspended thoroughly in 1.0 liter of TMED +

---

[29] J. J. Jendrisak and R. R. Burgess, *Biochemistry* **14,** 4639 (1975).

1.0 M NH$_4$Cl, incubated at 4° for 5 min., and centrifuged again. The supernatant fraction is the 1.0 M eluate. Note that this procedure uses NH$_4$Cl for the elution of the polymin P pellet, as suggested in the original description of the procedure,[30] rather than NaCl as in Burgess and Jendrisak.[25] This substitution increases the efficiency of the subsequent ammonium sulfate fractionation greatly.[27]

Ammonium Sulfate Fractionation

Proteins are precipitated from the 1.0 M eluate by the gradual addition of solid ammonium sulfate (0.40 g/ml of eluate) while the pH is maintained between 7 and 8 by the addition of 5 N NH$_4$OH. After stirring at 4° for 30 min, the precipitate is collected by centrifugation at 16,000 g for 1 h. The pellet is suspended in a minimal volume (ca. 150 ml) of TGED + 50 mM NaCl and is desalted rapidly by passage over a 600-ml P-6 desalting column (48 × 4 cm; Bio-Rad) equilibrated previously in the same buffer. Note that Mg(II) is omitted to increase the stability of the DNA–cellulose during the subsequent chromatography step. Prior to running the DNA–cellulose affinity column, fractions from the initial stages of purification are analyzed by SDS–PAGE to determine if polymin P precipitation and elution led to a good recovery of RNAP. The β and β' subunits, diagnostic of fractions containing RNAP, should be enriched in the desalted ammonium sulfate pellet and the 1.0 M eluate and not present in significant amounts in the polymin P supernatant or the 0.4 M wash.

DNA–Cellulose Chromatography

Protein containing fractions from the P-6 column are pooled and applied to a 500-ml DNA–cellulose column (40 × 4 cm) equilibrated in TGED + 50 mM NaCl. The column is washed overnight with at least 1 liter of the same buffer, and then RNAP is eluted with a 1.5-liter linear gradient from 0.05 to 1.0 M NaCl in TGED. MgCl$_2$ is added to the fractions to a final concentration of 10 mM to help stabilize RNAP.[6] Fractions containing RNAP, as judged either by an enzymatic assay or by SDS–PAGE (to monitor fractions containing the large β and β' subunits) are pooled and precipitated with ammonium sulfate (0.42 g/ml).

A1.5M Sizing Chromatography

The ammonium sulfate pellet is suspended in a minimal volume (ca. 10 ml) of TGMED + 0.5 M NaCl and chromatographed over a 500-ml Bio-Gel A1.5M column (110 × 2.5 cm; Bio-Rad) equilibrated in the same

[30] W. Zillig, K. Zechel, and H. Halbwachs, *Hoppe-Seyler's Z. Physiol. Chem.* **351,** 221 (1970).

buffer at a flow rate of ca. 25 ml/h. Alternatively, the A1.5M column can be substituted with Sephacryl S-300 as suggested by Hager *et al.*[31] RNAP-containing fractions are dialyzed into RNAP storage buffer [TMED, 50% (v/v) glycerol, 100 mM NaCl] and stored at $-20°$. These fractions retain activity for at least several years. At this stage, the purified RNAP contains all the core subunits (including δ) and a substoichiometric complement of σ^A. Other σ factors are also present at much lower levels. For example, preparations from late logarithmic phase cells contain σ^D at levels of about 1–2% saturation relative to core.[27]

Separation of Eσ^A from core (Eδ containing a small amount of E$\delta\sigma^D$) can be achieved using heparin–agarose chromatography and a shallow gradient[7] or using FPLC MonoS chromatography (see later). The aforementioned purification procedure can be modified by substituting heparin–agarose chromatography in place of steps 2 and 3 (polymin P precipitation/ammonium sulfate precipitation). For an example of this approach, see Cummings and Haldenwang.[32] Heparin-containing columns are available commercially or may be prepared using published procedures.[33]

Method 3. Purification of RNAP Using Heparin–Sepharose, Sephacryl S-300, and Mono-Q

Procedures beginning with direct loading of crude extracts onto heparin columns have been highly successful for RNAP (and for many other DNA-binding proteins). Heparin is an anionic (sulfated) polysaccharide that acts as both an ion exchange and an affinity column for DNA-binding proteins. Heparin columns have the advantage that RNAP will bind even in the presence of large amounts of DNA. As a result, crude extracts can be loaded directly onto heparin-containing columns and, after extensive washing to remove nucleic acids, the bound proteins can be eluted relatively free of contaminating DNA or RNA.[22,33] We have used this procedure to prepare RNAP from strains lacking δ[13] or σ^D.[34] However, it should be applicable to most *B. subtilis* strains. Note that an FPLC Superose 6 column can be used in place of Sephacryl S-300.

Eight liters of *B. subtilis* RP17 (*rpoE*::ΔSspI-BglII::*cat*) cells is grown to late logarithmic phase in 20 g/liter tryptone, 10 g/liter yeast extract, 3 g/liter disodium phosphate, 30 g/liter glucose, and 5 μg/ml chloramphenicol to yield ~35 g of cells.[13] The cells are harvested, resuspended in lysis

[31] D. A. Hager, D. J. Jin, and R. R. Burgess, *Biochemistry* **29**, 7890 (1990).
[32] C. W. Cummings and W. G. Haldenwang, *J. Bacteriol.* **170**, 5863 (1988).
[33] B. L. Davison, T. Leighton, and J. C. Rabinowitz, *J. Biol. Chem.* **254**, 9220 (1979).
[34] Y. F. Chen and J. D. Helmann, *J. Mol. Biol.* **249**, 743 (1995).

buffer 2 (50 mM Tris HCl, pH 8.0, 10 mM MgCl$_2$, 2 mM EDTA, 100 mM NaCl, 0.1 mM DTT, 5% glycerol, 1 mM PMSF), and lysed by three passages through a French pressure cell at 20,000 psi. After centrifugation at 23,000 g for 10 min, the supernatant fraction is loaded on a 60-ml heparin–sepharose CL-2B (Pharmacia) column equilibrated in TMED with 5% glycerol and 0.1 M NaCl. After washing with 10 column volumes of the same buffer, proteins are eluted with a 400-ml gradient of 0.1 to 1.0 M NaCl. RNAP-containing fractions are pooled and precipitated with ammonium sulfate (65% saturation), and the precipitate is collected by centrifugation (17,000 g). RNAP-containing fractions are resuspended in a minimal volume of TMED and loaded on a Sephacryl S-300 column in TMED with 5% glycerol and 0.5 M NaCl. The peak fractions are pooled, diluted threefold with TMED with 5% glycerol, and loaded on to an FPLC Mono Q HR5/5 column. After elution with a NaCl gradient, RNA polymerase-containing fractions (∼ 3 mg) are dialyzed against storage buffer 2 (50 mM Tris acetate, pH 8.0, 10 mM MgCl$_2$, 0.1 mM EDTA, 0.1 mM DTT, 50% glycerol) and stored at −20°.

Method 4. Core RNA Polymerase Containing δ (Eδ)

Eδ is prepared by dialysis of RNAP (fractions from either A1.5M in Method 2 or Mono Q in Method 3) into HMGED buffer [50 mM Na·HEPES, pH 8.0, 10 mM MgCl$_2$, 50% (v/v) glycerol, 0.1 mM EDTA, and 1 mM DTT] and application to an FPLC Mono S column. Core enzyme containing approximately stoichiometric amounts of δ protein flows through this column and is dialyzed into storage buffer.[11]

Method 5. Core RNA Polymerase Lacking δ (E) and Purified δ Protein

Separation of core RNAP from σ factors can also be achieved by chromatography on Bio-Rex-70 as originally reported for the E. coli core.[35] Often, this procedure needs to be repeated to remove trace amounts of the σ factor that remain bound during the first passage over Bio-Rex-70. This procedure has been used to prepare the B. subtilis core enzyme and removed all detectable σ^A and σ^H.[36]

We have used Bio-Rex-70 to prepare core RNAP starting with Eδ (Method 4). The Mono S flow-through fractions (2 ml at ca. 0.7 mg/ml) containing Eδ are loaded on a 1-ml Bio-Rex-70 column equilibrated in TGED buffer. Purified δ is recovered in the flow-through fraction by

[35] P. A. Lowe, D. A. Hager, and R. R. Burgess, Biochemistry 18, 344 (1979).
[36] M. Fujita and Y. Sadaie, J. Biochem. 124, 89 (1998).

washing the column with 15 ml of TGED buffer. Core RNA polymerase (E) is recovered by elution with TGED containing 1 M NaCl. All fractions are dialyzed into RNAP storage buffer and stored at $-20°$.[11]

Method 6. Overexpression and Purification of δ

δ is a highly charged polypeptide (net charge at neutral pH of -47) with a structured amino-terminal domain and a largely unstructured, polyanionic carboxyl-terminal domain.[13] Perhaps because of these unusual properties, δ is highly soluble and does not form inclusion bodies upon overproduction in *E. coli*. The purification developed for δ takes advantage of the high net negative charge density.[13]

Three liters of *E. coli* BL21/DE3 (pFL31) is grown in 2xYT medium containing 0.2% (w/v) glucose and 100 μg/ml ampicillin; at an OD_{600} of 0.6, Isopropyl-β-D-thiogalactoside (IPTG) is added to 0.8 mM and cell growth continues for 1.5 h. Cells (ca. 9 g) are harvested, resuspended in 20 ml of lysis buffer 2 (100 mM Bis–Tris, pH 6.0, 2 mM EDTA, 0.1 mM DTT, 1 mM 2-mercaptoethanol, 100 mM NaCl), and lysed by three passages through a French pressure cell. PMSF is added to a final concentration of 1 mM, and the lysate is clarified by centrifugation at 30,000 g for 30 min. δ is found in the supernatant fraction.

The supernatant fraction is diluted to a final volume of 200 ml with buffer A (50 mM Bis–Tris, pH 6.0, 1 mM EDTA) plus 100 mM NaCl and is applied to a 11-ml QAE–Sepharose column. At this pH, δ is still negatively charged and binds tightly to QAE, whereas other less acidic proteins (p$I > 6.0$) tend to flow through. After washing with 100 ml of buffer A plus 100 mM NaCl, the bound proteins are eluted with a 200-ml gradient from 0.1 to 1.0 M NaCl in buffer A. Using this approach, aproximately 120 mg of δ is isolated in a broad peak containing nucleic acids as the major contaminant (as judged by A_{260}/A_{280} ratio). Initial attempts to concentrate δ by ammonium sulfate precipitation were unsuccessful until we elected to lower the pH to near the pI of δ (pH 3.6). Subsequent experiments established that δ precipitates at pH 3.6 even in the absence of ammonium sulfate.

To concentrate δ by isoelectric point precipitation, δ-containing fractions are pooled and sodium acetate buffer (pH 3.6) is added to 50 mM. The abundant precipitate is collected by centrifugation at 12,000 g for 5 min. δ is dissolved in 50 mM TrisHCl (pH 8.0), 1 mM EDTA at about 6 mg/ml and is stored at $-20°$. As a final purification step, δ is fractionated on an FPLC Mono Q column with a gradient of 0 to 1.0 M NaCl in 50 mM Tris–HCl (pH 8.0), 1 mM EDTA. δ elutes between 0.5 and 0.8 M NaCl as a single sharp peak.[13]

Method 7. Purification of σ Factors

In general, σ factors can be overproduced to high levels in *E. coli* using T7 RNAP-driven expression systems. For an excellent review of the T7 system, readers are advised to consult Studier *et al.*[37] Because σ factors are often toxic in *E. coli*, and many T7 expression systems are leaky, care must be taken to avoid selection against the overproduction plasmid. To avoid plasmid loss, the overproduction plasmid is transformed into *E. coli* BL21/DE3, and transformants are selected on ampicillin plates containing 0.4% glucose. In some cases, it is necessary to use strains also carrying either pLysS or pLysE plasmids.[37]

Most σ factors form inclusion bodies after overproduction.[38] Washing of inclusion bodies with Triton X-100 improves the purity of the recovered σ factors. While we have often used guanidine–HCl to denature the inclusion bodies, we have shifted to the anionic detergent Sarkosyl, as described for *E. coli* σ factors.[38] Sarkosyl has the advantage that it tends to help avoid protein loss due to aggregation during refolding. In general, renatured σ factors can often be purified using cation exchange (e.g., DEAE, QAE–Sephadex, or a Mono Q column)[34,36,39,40] followed by a sizing column (e.g., Sephadex 75). High-purity preparations can usually be achieved with one or two column chromatography steps. However, at least one σ factor does not bind to cation-exchange columns and is purified instead using heparin–sepharose.[41] It is also possible to purify the overproduced σ factors starting with the soluble fraction, particularly if steps are taken to limit inclusion body formation. These can include growth at 30° rather than 37°. Examples of this approach, for σ^A and σ^F, are described in Lord *et al.*[42] Alternatively, procedures for the purification of σ factors as His-tagged or glutathione–sulfuryl transferase (GST) fusions[43] or self-cleavable intein tags[44] have been described. Other approaches include the use of an affinity column containing an immobilized anti-σ factor[45] or coexpression of a σ factor and anti-σ factor to form a more soluble complex.[46]

[37] F. W. Studier, A. H. Rosenberg, J. J. Dunn, and J. W. Dubendorff, *Methods Enzymol.* **185,** 60.
[38] R. R. Burgess, *Methods Enzymol.* **273,** 145.
[39] Y. L. Juang and J. D. Helmann, *J. Mol. Biol.* **235,** 1470 (1994).
[40] X. Huang, A. Decatur, A. Sorokin, and J. D. Helmann, *J. Bacteriol.* **179,** 2915 (1997).
[41] X. Huang, K. L. Fredrick, and J. D. Helmann, *J. Bacteriol.* **180,** 3765 (1998).
[42] M. Lord, D. Barilla, and M. D. Yudkin, *J. Bacteriol.* **181,** 2346 (1999).
[43] A. J. Dombroski, *Methods Enzymol.* **273,** 134.
[44] O. Delumeau, R. J. Lewis, and M. D. Yudkin, *J. Bacteriol.* **184,** 5583 (2002).
[45] T. Magnin, M. Lord, and M. D. Yudkin, *J. Bacteriol.* **179,** 3922 (1997).
[46] E. A. Campbell and S. A. Darst, *J. Mol. Biol.* **300,** 17 (2000).

Purification of σ^A

σ^A is purified from *E. coli* BL21/DE3 containing plasmid pLC2, a pBSK[+] (Stratagene) derivative that supports T7-driven overproduction of σ^A.[39] The procedure described here is very similar to that first reported by Chang and Doi.[47] Cultures (50 ml) are grown at 37° in 2xYT medium containing 0.4% glucose and 100 μg/ml ampicillin. At an OD_{595} of 0.7, IPTG is added to 0.4 mM and growth continued for 2 h. Cell pellets are suspended in 1.2 ml of lysis buffer 3 (lysis buffer 1 without $MgCl_2$), hen egg white lysozyme is added to 130 μg/ml, and the suspension is incubated on ice for 10 min. Sodium deoxycholate is added to 0.05% (w/v), and the cell suspension is disrupted by pulsed sonication for 2 min. Lysates are diluted with 1.2 ml of TGED containing 200 mM NaCl and clarified by centrifugation for 30 min in a microcentrifuge (4°). The cell debris is washed three times by suspension in 5 ml of TGEDX [TGD containing 0.5% (w/v) Triton-X-100 and 10 mM EDTA] followed by centrifugation for 20 min.

The σ^A protein is solubilized from the washed cell debris with 0.8 ml of 6 M guanidine hydrochloride. Insoluble material is separated by centrifugation, and the soluble portion is diluted by a gradual addition of 55 ml of TGED buffer. One milliliter of a DE-52 suspension in TGED + 100 mM NaCl is added, mixed thoroughly, and allowed to settle for 30 min. min. The supernatant fraction is decanted and the settled DE-52 is washed twice with 2-ml portions of TGED buffer containing 200 mM NaCl. The σ^A protein is then eluted twice with 2-ml portions of TGED buffer containing 330 mM NaCl and dialyzed into RNAP storage buffer. If desired, σ^A can be purified further by application to an FPLC Mono Q (HR5/5) column and elution with a 0.25 to 0.4 M NaCl gradient in TGED buffer. The σ^A protein elutes at 330 mM NaCl. This procedure was used for the purification of σ^A and many altered function σ^A mutants.[39]

Purification of σ^D

Eight liters of BL21/DE3 (pYFC11)[34] is grown and induced as described for σ^A. After centrifugation, cells are suspended in 60 ml of lysis buffer 3 and lysed, and the cell pellet containing inclusion bodies is recovered by centrifugation. σ^D is renatured using 6 M guanidine hydrochloride as described for σ^A, batch adsorbed to 80 ml QAE–Sepharose, and then poured into a column. The column is washed with TED + 0.15 M NaCl and σ^D eluted with 0.40 M NaCl. Using this procedure, 160 mg of σ^D is recovered with peak fractions containing 8 mg/ml protein.[35]

[47] B. Y. Chang and R. H. Doi, *J. Bacteriol.* **172,** 3257 (1990).

At this high concentration, σ^D tends to aggregate slowly. To avoid this problem, the protein is stored in aliquots at $-80°$, and samples are removed for further purification by chromatography on an FPLC Superdex-75 column in TED buffer containing 0.15 M NaCl.

Purification of σ^X

One liter of *E. coli* BL21/DE3(pXH10)[40] is grown and induced as for σ^D. Cells are suspended in 20 ml lysis buffer 3 and lysed by sonication, and the inclusion bodies are recovered by centrifugation. Inclusion bodies are washed and denatured with TGED + 6 M guanidine hydrochloride, and then σ^X is renatured by dilution with TGED + 0.01% Triton X-100. Renatured σ^X is recovered using a 5-ml QAE–Sepharose column equilibrated with the same buffer. After washing with 50 ml TGED + 0.15 M NaCl + 0.01% Triton X-100, σ^X is eluted with TGEDG + 0.4 M NaCl + 0.01% Triton X-100. σ^X can be purified further using an FPLC Superdex-75 column in TGED buffer containing 0.2 M NaCl followed by dialysis into TGED + 0.1 M NaCl + 0.01% (v/v) Triton X-100 + 50% (v/v) glycerol and storage at $-80°$.

Purification of σ^W

One liter of *E. coli* BL21/DE3 (pKF86)[41] is used for the production of σ^W as described for σ^D. Cells are suspended in 20 ml of lysis buffer 3 and lysed by sonication, and the inclusion bodies are recovered by centrifugation. The inclusion bodies are washed twice with 100 ml TGEDX and dissolved in 20 ml TGEDX plus 0.4% (w/v) Sarkosyl. The dissolved σ^W is diluted gradually to 200 ml with TGEDX to allow renaturation of σ^W, and the diluted σ^W is dialyzed twice for 8 h against 10 volumes of TGEDX at $4°$. The dialyzed material is centrifuged to remove any precipitate and loaded onto a 20-ml heparin–Sepharose CL-6B column equilibrated with TGEDX. After washing with 200 ml of TGEDX and 0.2 M NaCl, σ^W is eluted with TGEDX and 0.5 M NaCl. The peak fraction of σ^W is purified further by chromatography on an FPLC Superdex-75 column in TGEDX plus 0.2 M NaCl buffer, and the peak fractions are stored frozen at $-80°$.

Method 8. Reconstitution of Holoenzyme

For many biochemical and *in vitro* transcription studies, it is desirable to have a homogeneous population of holoenzyme. To reconstitute holoenzymes, one can start with a purified RNAP fraction such as those resulting from Methods 1 through 3. Because the various σ factors bind competitively to a common core enzyme, the desired holoenzyme can be generated most simply by adding a large molar excess of the desired σ and incubating for 10 to 20 min. However, this does not remove the other σ factors that were originally present in the sample.

To separate the desired holoenzyme from excess σ (and other displaced σ factors), sizing gel chromatography (e.g., FPLC Superose-6 or Superdex-200) is convenient.[14] An alternative approach, although not one yet reported in the literature, is to bind a His-tagged RNAP to a Ni-NTA affinity column and wash the column thoroughly with the σ factor of interest. Subsequent elution is expected to yield an RNAP fraction highly enriched for the holoenzyme of interest. It may also be possible to use a His-tagged σ factor to bind an excess of RNAP (not containing a His tag). Other holoenzymes should flow through the column, whereas the core enzyme could potentially bind to the immobilized σ to form the holoenzyme. Regardless of the approach used to generate the holoenzyme, the presence or absence of the δ subunit should also be monitored closely. For convenience, we routinely include an approximately fivefold molar excess of δ in our *in vitro* transcription reactions.

Acknowledgments

Our work on *Bacillus subtilis* RNAP is supported by the National Institutes of Health (Grant GM-47446).

[3] Determination of *Escherichia coli* RNA Polymerase Structure by Single Particle Cryoelectron Microscopy

By Pampa Ray, Bruno P. Klaholz, Robert D. Finn,
Elena V. Orlova, Patricia C. Burrows, Brent Gowen,
Martin Buck, and Marin van Heel

Transmission electron microscopy (TEM) is an important technique for studying large biological macromolecular complexes. TEM provides projection images of the macromolecules, and elaborate three-dimensional (3D) image processing then yields structures of the complexes. This approach can be applied to a wide range of different molecules of sizes typically greater than 200 kDa. The quality of the resulting reconstructions depends on many factors, including the sample quality; the size of the complexes; their internal ("point group") symmetry; the type of electron microscope used and the imaging parameters applied; and, to a very large extent, on the computing procedures applied. The resolution obtained in a reconstruction indicates the size of the smallest individual detail that may be reliably interpreted in a structure. Using single particle analysis, the best resolutions achieved are for a highly symmetrical icosahedral viral structure[1,2] of about 7, 5.9,[3] and even 4.5 Å.[4] For asymmetrical molecules, resolution levels of ~7.5 Å have been achieved.[5] Although better structural details can be obtained using the established techniques of X-ray crystallography and nuclear magnetic resonance (NMR) spectroscopy, electron microscopy offers unique opportunities to study large, uncrystallized protein complexes, which are available only in small quantities and at low concentrations.

Cryoelectron microscopy (cryo-EM) provides 3D images of frozen-hydrated particles or "solution" structures of the investigated molecules with minimal structural distortions. Although there is no real upper limit to the size of the molecular complexes that can be examined, a lower practical limit of 150–200 kDa arises because a minimum contrast is required for recognizing the individual molecules in the electron micrographs prior to all image processing. A limiting factor for cryo-EM is the low homogeneity of samples, a restriction that also affects X-ray crystallography and NMR spectroscopy. Classification approaches[3] can help sorting of heterogeneous populations of single particles according to their composition or conformational states.

The multisubunit RNA polymerase (RNAP) enzyme from *Escherichia coli* has undoubtedly been the most extensively structurally studied bacterial polymerase.[6,7] The structural organization of the subunits was first visualized in 1989 by electron crystallography,[8] but for many years the structural interpretation of the enzyme had been restricted by the low-resolution levels of the reconstructions.[9] More flexible advanced single-particle structure analysis methods[3] have now opened new alleys of structure determination for the *E. coli* core RNA polymerase, a relatively

[1] B. Böttcher, S. A. Wynne, and R. A. Crowther, *Nature* **386,** 88 (1997).

[2] J. F. Conway, N. Cheng, A. Zlotnick, P. T. Wingfield, S. J. Stahl, and A. C. Steven, *Nature* **386,** 91 (1997).

[3] M. van Heel, B. Gowen, R. Matadeen, E. V. Orlova, R. Finn, T. Pape, D. Cohen, H. Stark, R. Schmidt, M. Schatz, and A. Patwardhan, *Q. Rev. Biophys.* **33,** 307 (2000).

[4] A. Patwardhan *et al.*, unpublished results

[5] R. Matadeen, A. Patwardhan, B. Gowen, E. V. Orlova, T. Pape, F. Mueller, R. Brimacombe, and M. van Heel, *Structure* **7,** 1575 (1999).

[6] S. A. Darst, *Curr. Opin. Struct. Biol.* **11,** 155 (2001).

[7] R. D. Finn, E. V. Orlova, M. van Heel, and M. Buck, "Signals, Switches, Regulons and Cascades: Control of Bacterial Gene Expression." Cambridge Univ. Press. Cambridge, 2002.

[8] S. A. Darst, E. W. Kubalek, and R. D. Kornberg, *Nature* **340,** 730 (1989).

[9] S. A. Darst, A. Polyakov, C. Richter, and G. Zhang, *J. Struct. Biol.* **124,** 115 (1998).

small particle (~380 kDa) with no internal symmetry.[10] The combination of cryo-EM and X-ray crystallography has helped the interpretation of the single-particle electron density map, leading to a molecular model of the enzyme.[11] The approach has allowed the identification of large sequence insertions (so-called dispensable regions) and carboxy-terminal domains of the α subunits (α-CTD). Comparison of the low-resolution *E. coli* RNAP cryo-EM structure with X-ray crystal structures of other bacterial RNAPs,[12,13] and of the *Saccharomyces cerevisiae* RNAP II,[14] demonstrates a striking conservation between bacterial and eukaryotic enzymes.[15–18] The comparisons help establish the principal features of the multisubunit RNAP enzyme that have been retained through evolution and that common mechanisms of transcription are used among prokaryotes and eukaryotes.

Purification of *Escherichia coli* Core RNA Polymerase

A 200-liter culture of *E. coli* strain MRE600 is grown in Luria Bertoni media to midlog phase in a fermentor at 37° with constant agitation to maintain an aerobic growing environment. The cells are harvested and immediately flash frozen in liquid nitrogen. Frozen cells (250 g) are resuspended in a lysis buffer solution consisting of 10 mM Tris, pH 8.0, 5% glycerol, 1 mM dithiothreitol (DTT), 0.1 mM EDTA (TGED), 200 mM NaCl containing a cocktail of protease inhibitors (Roche). Cells are lysed using a French press under high pressure (1400 kg/cm^3). The lysate is centrifuged at 4° for 45 min at 18,000 rpm in a Beckman JA-20 rotor. Polymin P is added to the supernatant from a stock of 10% (w/v) in H_2O to a final concentration of 0.35% under constant stirring at 4° and allowed to stir for 30 min. The precipitate is then pelleted by centrifugation at 18,000 rpm for 30 min at 4° in a JA-20 rotor. The supernatant is discarded, and the pellet is resuspended in TGED buffer containing 0.4 M NaCl, which

[10] R. D. Finn, E. V. Orlova, B. Gowen, M. Buck, and M. van Heel, *EMBO J.* **19,** 6833 (2000).
[11] R. D. Finn, E. V. Orlova, B. Klaholz, B. Gowen, M. van Heel, and M. Buck, Manuscript in preparation.
[12] G. Zhang, E. A. Campbell, L. Minakhin, C. Richter, K. Severinov, and S. A. Darst, *Cell* **98,** 811 (1999).
[13] D. G. Vassylyev, S. Sekine, O. Laptenko, J. Lee, M. N. Vassylyeva, S. Borukhov, and S. Yokoyama, *Nature* **417,** 712 (2002).
[14] P. Cramer, D. A. Bushnell, and R. D. Kornberg, *Science* **292,** 1863 (2001).
[15] R. H. Ebright, *J. Mol. Biol.* **304,** 687 (2000).
[16] E. P. Geiduschek and M. S. Bartlett, *Nature Struct. Biol.* **7,** 437 (2000).
[17] L. Minakhin, S. Bhagat, A. Brunning, E. A. Campbell, S. A. Darst, R. H. Ebright, and K. Severinov, *Proc. Natl. Acad. Sci. USA* **98,** 892 (2001).
[18] P. Cramer, *Curr. Opin. Struct. Biol.* **12,** 89 (2002).

selectively solubilizes many of the unwanted contaminants. Core RNAP is solubilized between 0.5 and 0.7 M NaCl.[19] The supernatant containing RNAP is separated from the remaining polymin P precipitate by centrifugation. Solid ammonium sulfate is added to the supernatant to 50% (w/v) concentration over a 30-min period at 4°. After centrifugation in a JA-20 rotor at 18,000 rpm for 40 min at 4°, the ammonium sulfate precipitate is resuspended in TGED buffer containing 200 mM NaCl and dialyzed against 1 liter of this buffer over 16 h with two changes to remove the ammonium sulfate.

Prior to ion-exchange chromatography using FPLC (Amersham Pharmacia), the protein is dialyzed against buffer A$^{Q\text{-seph}}$ containing 20 mM imidazole, 200 mM NaCl, 1 mM DTT, 0.1 mM EDTA, 5% (v/v) glycerol, pH 7.0. The dialysate is centrifuged at 18,000 rpm in a JA-20 rotor for 45 min at 4° and loaded at a rate of 1 ml/min onto a 10-ml Q-Sepharose column (Pharmacia) preequilibrated with buffer A$^{Q\text{-seph}}$. The protein is eluted with a salt gradient from 200 mM to 1 M NaCl at a rate of 2 ml/min. Fractions containing core RNAP are analyzed using SDS–PAGE and dialyzed against TGED buffer containing 200 mM NaCl. After centrifugation of the dialysate, 1 M MgCl$_2$ is added to the protein solution to a final concentration of 10 mM and loaded onto a 5-ml Hi-Trap heparin column (Pharmacia) at 1 ml/min equilibrated previously, with TGED buffer, pH 8.0, containing 10 mM MgCl$_2$ and 200 mM NaCl (buffer A^{Hep}). After a 10 column volume wash with buffer A^{Hep}, the protein is eluted with a staggered gradient of buffer B^{Hep} (TGED containing 1 M NaCl and 10 mM MgCl$_2$) at 1 ml/min. Core polymerase elutes at 0.6 M NaCl. Separation of the core enzyme from the holoenzyme is achieved using a BioRex-70 column. The protein is dialyzed into TGED containing 200 mM NaCl and, after centrifugation at 18,000 rpm using a JA-20 rotor for 45 min at 4°, is loaded at a flow rate of 0.2 ml/min onto a 3-ml column containing Bio-Rad BioRex-70 resin, equilibrated previously with TGED buffer, pH 8.0, containing 150 mM NaCl (buffer A^{BioRex}). The σ^{70} holoenzyme elutes in the flow through and wash. The core enzyme is eluted with a NaCl gradient of 0% buffer B^{BioRex} (TGED with 1 M NaCl) to 50% buffer B^{BioRex} over 40 column volumes at a rate of 1 ml/min.

An alternative method for purifying core RNAP involves using a 5-ml Hi-Trap heparin column with buffer A^{Hep} as the starting buffer at a flow rate of 2 ml/min. Core polymerase elutes with a salt gradient at 0.6 M NaCl. Fractions containing core polymerase are dialyzed against 20 mM Tris, pH 8.3, 200 mM KCl, 0.1 mM DTT, 0.1 mM EDTA, 5% glycerol.

---

[19] R. R. Burgess and Jendrisak, *Biochemistry* **14,** 4634 (1975).

Protein (1.5–1.8 ml) is loaded onto a 120-ml Sephacryl 300 column (Pharmacia) at a flow rate of 0.8 ml/min. Core RNAP elutes after 40 min over 12 ml. This is dialyzed into TGED containing 200 mM NaCl. A solution of 1 M MgCl$_2$ is added to the dialysate to 10 mM concentration and rechromatographed on a 1-ml Hi-Trap heparin column, which concentrates the protein. The protein elutes at 0.6 M NaCl. The protein is dialyzed into TGED, pH 8.0, 200 mM NaCl containing 50% glycerol for storage. Concentration is determined using bovine serum albumin (BSA) as a standard and reagents from Bio-Rad,[20] measuring absorbance at 595 nm.

Verification of RNA Polymerase Composition and Purity

The purity, presence, and integrity of α, β, and β' and ω subunits are verified by SDS–PAGE analysis on an 18% acrylamide gel and compared to commercially available RNAP samples (Epicentre Technologies). The presence of the ω subunit is confirmed by loading 4 μg of core polymerase onto the aforementioned gel and following the manufacturer's protocol for blotting using a Bio-Rad Sequi-Blot PVDF membrane for protein sequencing. Bands corresponding to 14 and 30 kDa are analyzed and cut out for N-terminal sequencing and identification as the ω subunit and *E. coli* ProQ, the major contaminant.

Characterization of Core RNA Polymerase

The core RNAP is characterized by *in vitro* reconstitution of the RNA polymerase holoenzyme as outlined in Buck *et al.*[21]

Core RNA Polymerase Catalytic Activity Assay

This assay[22] measures the catalytic activity of the core RNAP by its ability to nonspecifically bind and transcribe from a poly(dA-dT) template (Sigma). The reaction mixture contains 3.4 μl of ATP (30 mg/ml), 40 μl of poly(dA-dT) (0.5 mg/ml), 4.5 μl of DTT (0.1 M), 390.1 μl of H$_2$O, and 562 μl of 2× master mix (consisting of 142 μM Tris–HCl, pH 8.0, 357 μM EDTA, 15 mM MgCl$_2$, 1.5 mM KPO$_4$, 1.78 mg/ml of BSA).

Of the reaction mixture, 25 μl is mixed with 2 μl of core RNA polymerase (final concentrations ranging from 2.5 to 250 nM) and 10 μCi of

[20] M. Bradford, *Anal Biochem* **72,** 248 (1976).
[21] M. Buck, S. R. Wigneshwereraj, S. Nechaev, P. Bordes, S. Jones, W. Cannon, and K. Severinov, *Methods Enzymol.* **370,** [53] 2003 (this volume).
[22] S. Borukhov and A. Goldfarb, *Methods Enzymol.* **274,** 315 (1996).

α-P^{32}-UTP and incubated at 37° for 10 min; 1 μl of unlabeled UTP (1.5 mg/ml) is added and the mix is incubated at 37° for an additional 10 min. The reactions are stopped by placing them on ice. The transcription products are analyzed by spotting 2 μl of each reaction mixture on a strip of a polyethyleneimine-cellulose TLC plate, 8 mm up from the bottom. After air drying, the bottom 5 mm of the plate is placed in 2 *M* HCl and left until the HCl chromatographs to the top of the plate. The activity of the polymerase enzyme is quantified by a phosphoimager, which indicates the presence of incorporated ^{32}P-UTP at the origin.

Electron Microscopy

Preparation of Holey Carbon Grids

Glass slides are sonicated twice in acetone for 10 s and allowed to air dry. This procedure is repeated twice. The slides are transferred to a 0.5% aqueous solution of Osvan (alkylbenzyldimethyl-ammoniumchloride) for 30 min, which renders the slide hydrophobic, rinsed in distilled water, and then dried. The dry slides are placed onto a metal block (precooled to −20°) to allow water droplets to condense on the surface for 15–20 s, depending on the ambient humidity and the desired hole size, which is more or less proportional to the amount of time the slide is left on the metal block. The optimum number of holes per grid square in a 300 mesh size grid (Agar Scientific) is ~100. About 2 ml of 0.5% Triafol (cellulose acetobutyrate) is pipetted over the slide on the surface containing the condensation droplets and excess liquid is removed by blotting. This yields a plastic film containing holes where the Triafol cannot penetrate through the water droplets. The slide is then submerged in Pellex (dioctyl sulfosuccinate sodium salt) for a minimum of an hour to render the film hydrophilic. The film is removed from the slide by slowly entering the slide at an acute angle into a beaker of distilled water. The water penetrates between the glass slide and the Triafol film and the film floats off onto the water meniscus. Three hundred square mesh-size dry copper grids, cleaned previously by sonication in acetone for 2 min, are placed onto the floating film, transferred onto blotting paper, and air dried. The grids are placed in a carbon coater. Thin carbon films of equal thickness are required to maintain a thin layer of vitrified ice within the holes and therefore subsequently provide good contrast in the images of the molecules. The Triafol film is dissolved from the grids by placing them in ethyl acetate. The holey carbon grids are glow discharged in an inert atmosphere of argon to render the carbon hydrophilic to facilitate the entry of the samples into the holes.[23]

Vitrification of Protein Samples

Vitrification of the protein solution is achieved when a layer of sample is less than 2 μm in thickness and the rate of cooling of the sample is greater than $10^{4\circ}$/s.[23–26] A 3-μl aliquot of purified RNAP (0.3 mg/ml) is applied to a holey carbon grid for a minute to allow the sample to pass into the holes. Although the presence of NaCl and glycerol in the protein solution is necessary to maintain RNAP in a monodispersed state, the presence of these buffer components causes phase separation upon cooling and diminishes image contrast. Therefore the concentration of these components is reduced by floating the holey carbon grid on a drop (~40 μl) of distilled water.[27] This method is applicable only in situations where dilution of the divalent ion concentration will not affect the stability of the protein complexed to its associated factors. After the application and washing of the sample on the grid, the grid is secured with a pair of forceps and is placed in a guillotine-like device (similar to the one depicted in the review by Dubochet *et al.*[23]). Prior to vitrification, the grid is blotted nearly dry by pressing a piece of Whatman filter paper directly against the grid for 2–3 s on the same side to which the sample was applied. The guillotine is then released, and the grid is plunged rapidly into liquid ethane. Excess drying of the grid prior to vitrification should be avoided because it forms empty holes and leads to increased solute concentrations, which may distort the structure.[28]

Cryo-EM of RNA Polymerase

The main difference between cryo-EM and conventional TEM is that in cryo-EM, the sample is always kept below the devitrification temperature (~$-170°$)[23] so as to prevent conversion of water to a crystalline state. Grids are transferred to a FEG CM-200 Philips transmission electron microscope using a Gatan side-entry cryoholder. The EM grids are searched for suitable specimen areas at low magnification × 3800 and low irradiation level (less than 0.05 e$^-$/Å^2). Once good areas are found, the electron beam is switched to high magnification and the beam is moved by ~3 μm to an adjacent area for focusing ("low-dose" mode). The micrographs are recorded

[23] J. Dubochet, M. Adrian, J. J. Chang, J. C. Homo, J. Lepault, A. W. McDowall, and P. Schultz, *Q. Rev. Biophys.* **21,** 11056 (1988).
[24] W. Chiu, K. H. Downing, J. Dubochet, R. M. Glaeser, G. H. Heide, E. Knapek, M. K. Lamvik, J. Lepault, J. D. Robertson, E. Zeitler, and F. Zemlin, *J. Microsc.* **141,** 385 (1986).
[25] J. Berriman and N. Unwin, *Ultramicroscopy* **56,** 241 (1994).
[26] T. S. Baker, N. H. Olson, and S. D. Fuller, *Microbiol. Mol. Biol. Rev.* **63,** 862 (1999).
[27] M. Cyrklaff, N. Roos, H. Gross, and J. Dubochet, *J. Microsc.* **175,** 135 (1994).
[28] E. M. Mandelkow and E. Mandelkow, *J. Mol. Biol.* **181,** 123 (1985).

under an electron dose of $\sim$10 e$^-$/Å^2 at a nominal magnification of $\times$50,000 (calibrated magnification: $\times$48,600) using an acceleration voltage of 200 KV and an exposure time of 1 s on Kodak S0163 film (yielding micrographs with an optical density between 0.4 and 1.2). The micrographs are collected using a range of defocus values ($\sim$1.5–4 μm underfocus). Kodak S0163 negatives are developed in a D19 Kodak developer for 12 min and fixed for 5–10 min before being rinsed in tap water and dried.

Image Processing

Structure determination of individual macromolecular complexes by single-particle cryo-EM techniques involves several image processing steps, applied to the data in two and three dimensions. Biological samples are very sensitive to radiation damage even at liquid nitrogen (or liquid helium) temperatures. The overall electron dose to which the specimens are exposed must therefore be kept at a minimum. At the same time, a minimum electron dose has to be maintained to provide sufficient signal to find the molecular image in the micrographs. For single particles, this dose is typically $\sim$10 e$^-$/Å^2. Therefore, an inherent problem of cryo-EM is the low signal-to-noise ratio (SNR) of the molecular images in the low-dose electron micrographs (see Fig. 1A).

The following section presents an overview of the main image processing steps partitioned into the preprocessing phase and the structure determination phase. Preprocessing covers the digitization of the micrographs, particle selection, correction of the contrast transfer function (CTF), and band-pass filtering of the selected particle images. The structure determination phase encompasses image alignments, multivariate statistical analysis, including automatic classification, Euler angle assignments by angular reconstitution (AR), 3D reconstruction, and interpretation of the final maps. Finally, assessment of the resolution of the final 3D reconstruction is described.

Preprocessing

Digitization of Images. Good micrographs, selected for the absence of astigmatism and drift using an optical diffractometer, are digitized using a high-resolution Image-Science patchwork densitometer.[3] The images are digitized with a 5.0-μm pixel size, corresponding to a pixel size of 1.05 Å pixel size at the specimen level as can be derived by dividing the densitometer pixel size by the overall electron microscopical magnification.

For this RNA polymerase project, the images are coarsened by a factor of "2," i.e., 2×2 pixels from the digitized images, a total of four pixels, are

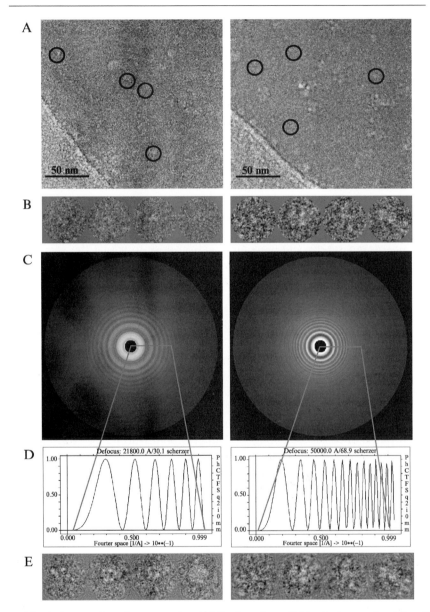

FIG. 1. An overview of preprocessing. (A) Patches of two digitized micrographs of *E. coli* RNA polymerase holoenzyme (in this case with a fragment of the sigma factor[29]) taken at calculated defocus values of 2.18 μm (left) and 5 μm (right) with a 10 μm step size. The original micrographs were taken from two different regions of the holey carbon grid. Individual particles, which were selected from the micrographs, are ringed. (B) Four

merged into a single pixel. The final pixel size of the digitized data set is thus 2.1 Å. Under optimal conditions, the best achievable resolution is equal to *three* times this pixel size[30] and thus for this RNAP data set, 3D reconstructions could yield a resolution of, at best, around 6 Å.

Particles Selection. RNA polymerase, like all biological molecules, has poor electron-scattering properties, leading to low contrast molecular images in the micrographs. As the contrast in the low-frequency image components is typically much stronger than in very noisy high-frequency components, the visibility of the particles can be increased by low-pass filtering the electron images. Such low-pass filtered images are used for interactive or semiautomatic particle selection only. Selected particles are extracted into individual frames from the original images, rather than low-pass filtered ones, with a linear size of approximately twice the maximum particle diameter.

Correction of the Contrast Transfer Function. Images obtained from weak-phase biological specimens are influenced strongly by the characteristics of the imaging process, such as the accelerating voltage, the spherical aberration of the objective lens, and the exact focus setting for each area of the sample or even for each individual macromolecule. The combined influence of these imaging parameters, some of which are constant (e.g., spherical aberration, acceleration voltage) but some of which differ from image to image (defocus), can be expressed in a contrast transfer function (CTF). The CTF is a function of spatial frequency, meaning that it affects the different sinusoidal components of the images differently. If the electron microscopical imaging system is rotationally symmetric, i.e., free of such aberrations as astigmatism, the CTF is also rotationally symmetric and real (as opposed to complex).[3] The net effect of differences in defocus within a data set is that in some molecular images, the details corresponding to, say, 8 Å are depicted as dark, whereas in other images they are

[29] M. Buck, M. T. Gallegos, D. J. Studholme, Y. Guo, and J. D. Gralla, *J. Bacteriol.* **182,** 4129 (2000).

[30] E. V. Orlova, P. Dube, J. R. Harris, E. Beckman, F. Zemlin, J. Markl, and M. van Heel, *J. Mol. Biol.* **271,** 417 (1997).

individual particles selected from the corresponding micrograph in A and excised into smaller images (150 × 150 pixels). (C) Power spectrums derived from all individual particle images from the corresponding micrographs in A. (D) Squares of theoretical CTF curves. The lines from C correspond to the same distance in Fourier space. Resolution derived from measurement of the radii of the individual Thon rings in C corresponds to the zero points in the plots in D when the defocus is estimated correctly. This defocus value is used for CTF correction. (E) The four particles in B after application of CTF correction, band-pass filtering (to remove frequencies below 0.005 and above 0.9), and normalization.

depicted as light, or even not depicted at all if the CTF has a "zero cross-ing" in that frequency range. In principle, one would like to know the CTF parameters associated with each individual particle. This information is needed in order to apply reliable CTF correction, which is crucial for a high-resolution structure determination. Without such a CTF correc-tion, the high-frequency image components from different particles will have wildly fluctuating phases, which will therefore cause these image components to fully disappear from the averaged final results.

Spherical aberration and acceleration voltage are parameters that can be considered constant during the experiment and the same may even apply to some uncorrected astigmatic component in the imaging system; the parameter one needs to concentrate on here is the defocus value. First of all, one cannot at any time focus the instrument precisely on the desired specimen area, as the electron exposure required for this operation is an order of magnitude more than the sensitive biological specimen can toler-ate. Thus low-dose techniques are applied in which one focuses on adjacent specimen areas for imaging rather than on the area of interest itself.[31] The electron microscopy specimens are normally not really flat, and local fluctuations of the grid topology can lead to significant defocus differ-ences within a single micrograph. The molecules may also lie at different heights within the layer of vitreous ice; this type of unintended defocus dif-ference is especially hard to account for.[3] Apart from such unintended or unavoidable defocus differences, one typically (and deliberately) collects the micrographs over a range of different defocus values (e.g., 1.5–4 μm), the reason being that apart from contrast reversals, the zeroes of the CTF represent missing information. The collection of images over a range of defocus values ensures that the collected information covers all relevant frequency ranges, without information gaps.

Defocus parameters are measured based on the position of the maxima and minima in the power spectrum, which are used for fitting theoretical (squared) CTF curves (Fig. 1D). Differences between maxima and minima within the power spectra (here based on individual single-particle images) are amplified by removing the very low-frequency component background of the rotationally averaged sums of Fourier transforms. Defocus estimates of RNAP data have been performed using either a semiautomated "CTF wizard" based method[32] or IMAGIC-5 CTF software.[3] CTF correction is achieved by pure phase flipping on individual particle images using the "transfer" option of IMAGIC-5[3]; this approach ensures maintaining the continuity of the measured experimental amplitudes precisely.

[31] P. N. Unwin and R. Henderson, *J. Mol. Biol.* **94,** 425, (1975).
[32] R. D. Finn *et al.*, unpublished.

Band-Pass Filtering. Prior to alignment, CTF-corrected images are band-pass filtered and normalized.[33] The band-pass filter is a combination of low-pass and high-pass filters. This suppresses some very low frequencies that are not correlated to the actual structure and some high frequencies, which, certainly in the early phases of the analysis, contain predominantly noise. Typically for these RNA polymerase data sets, a low-frequency cut-off of 0.01 (about ~400 Å for a 2.1-Å pixel size) and a high-frequency cut-off of 0.9 (5.6 Å) are used. To apply such filtering, the particle images are Fourier transformed, multiplied by the band-pass filter function, and Fourier transformed back to real space to yield the filtered images.

Structure Determination

Reference-Free Alignment and Multireference Alignment. All the individual molecules embedded in vitreous ice are in arbitrary orientations and positions. The aim of the alignment procedure is to ensure that all the individual images in any one particular orientation are brought into register so that they can subsequently be analyzed and averaged.

In order to center and eventually position the particle images rotationally, two types of alignments can be performed. Cross-correlation functions are used to find the best translational and rotational alignments of the particle images. The reference-free alignment by classification uses the rotationally averaged total sum of all particle images as a reference for a translation-only alignment of the particles.[34] This reference-free alignment (RFA) method avoids bias resulting from using specific reference images. The RFA is the initial step for any new structure determination where no similar 3D structure is already available. This translational alignment (in x and y directions) results in centering of the images so that an automatic multivariate statistical classification can be run that provides the first characteristic views (classes) of the molecule. The resulting class averages can then be used as a first set of references for multireference alignment (MRA). Whereas the RFA starts with a translational alignment with respect to a single amorphous, rotationally symmetric "blob," MRA uses whole series of reference images with an—in the course of the interative refinement scheme—increasing level of image detail.[3] For the later rounds of MRA during the structure refinements, a larger number of reference images are used, which are generated by reprojections of the previous 3D result in directions ~15° apart. Whereas analysis of the RNAP core structure is initiated from scratch with a RFA step for determining the 3D

---

[33] M. van Heel and M. Stöffler-Meilicke, *EMBO J.* **4,** 2389 (1985).
[34] P. Dube, P. Tavares, R. Lurz, and M. van Heel, *EMBO J.* **15,** 1303 (1993).

structure of RNAP holoenzyme complexes, the 3D RNAP core structure is used to generate the first set of reference images for MRA.

Multivariate Statistical Analysis and Classification. The overall statistical analysis of data sets comprises of two main steps: data compression by multivariate statistical analysis (MSA) of the whole data set (i.e., all centered/aligned particle images), followed by hierarchical ascendant classification into classes and summing into classums.[35–38] The classification allows similar images and/or classes to be merged with the constraint of a minimal increase in intraclass variance. Depending on the original SNR values in the raw molecular images, the number of images per classum differs: for a negatively stained specimen, 3–5 particle images can be sufficient, whereas cryo-EM micrographs typically require 10–20 class members, depending on the defocus value (see Figs. 1A and 1B for the effect of different defocus values on the contrast). Using MSA analysis, it is possible to group images (aligned previously by MRA or RFA) of particles that have the same orientation and to obtain averages ("classums") of these images with enhanced SNRs.[3] The next steps are the assignment of Euler angles to each characteristic view, i.e., determination of the orientation of each classum relative to that of all other characteristic views, and finally the calculation 3D reconstruction of the complex.

Euler Angle Assignment. To be able to perform a 3D reconstruction based on 2D projection images of a 3D object, one needs to obtain relative Euler angle orientations of the projection images, i.e., the classums in this case. The basis for the Euler angle assignment is the common line projection theorem: two 2D projections of a 3D object share a common line projection.[3,39] In theory, three different projections of an asymmetric object are sufficient to find their relative Euler angle orientations.[39,40] In practice, however, many projection images are required to achieve a good 3D reconstruction with an isotropic resolution. Hence an object can be reconstructed from its projection images as long as we have enough different views available to fill the information volume.

Common line projections are determined in real space with the angular reconstitution program,[39] as implemented in the IMAGIC-5 software system.[41] First, all possible line projections ("sinograms") are determined for each classums and then all possible correlation functions between the

[35] M. van Heel, *Ultramicroscopy* **13**, 16 (1984).
[36] M. van Heel, *Optik* **82**, 114 (1989).
[37] L. Borland and M. van Heel, *J. Opt. Soc. Am.* **A7**, 601 (1990).
[38] G. Golub and C. van Loan, "Matrix Computations," 3rd Ed. The John Hopkins Univ. Press, London, 1996.
[39] M. van Heel, *Ultramicroscopy* **21**, 111 (1987).
[40] A. B. Goncharov and M. S. Gelfand, *Ultramicroscopy* **25**, 317 (1988).

sinogram of each new classum and a set of sinograms derived from well established views are determined ("sinogram correlation functions").[3] What follows finally is a 'brute-force' search over all possible Euler orientations of the new classum in order to find its best possible orientation.

Three-Dimensional Reconstruction. Three-dimensional reconstructions based on the angles found by angular reconstitution are achieved using the exact-filtered back-projection algorithm,[42,43] an approach that not only reduces the overweighting of low frequencies generated by overlapping of the central sections in Fourier space in the proximity of the origin of the coordinates,[3] but also takes into account the nonhomogeneous distribution of projections over the unit sphere covering all possible particle orientations. Reprojection of the 3D reconstruction, using the Euler directions found for each individual classum, allows one to verify the correctness of the angle assignment: a high correlation coefficient is expected between a classum and its corresponding reprojection. In order to obtain an isotropic resolution for the final 3D reconstruction, it is important to ensure a full distribution of views without gaps or "missing cones."

Refinements. Multiple iterations of MRA, MSA, Euler-angle assignment, and 3D reconstructions (see Fig. 2) are required to refine the RNAP structures. At the end of each refinement cycle, the most recent 3D model is masked in order to remove noise outside the model and is reprojected to generate a large number of reference images for the next round of refinements. The same 3D reconstruction is also used for generating a small set of anchor reprojections (typically 30 images).[3,44,45] These reprojections have a reduced level of noise compared to the classums and are fully consistent with each other. Therefore they are used as an "anchor set" for refining the angular orientations of the set of classums. The various refinement scheme details are discussed in van Heel *et al.*[3]

Resolution Estimation and Filtering of the Final Three-Dimensional Reconstruction

Upon convergence of the structure, resolution of the final 3D reconstruction is assessed by calculating the Fourier shell correlation (FSC)[42] between two independent 3D reconstructions obtained from the set of

---

[41] M. van Heel, G. Harauz, E. V. Orlova, R. Schmidt, and M. Schatz, *J. Struct. Biol.* **116,** 17 (1996).

[42] G. Harauz and M. van Heel, *Optik* **73,** 146 (1986).

[43] M. Radermacher, *J. Elect. Microsc. Tech.* **9,** 359 (1988).

[44] M. Schatz, E. V. Orlova, P. Dube, J. Jäger, and M. van Heel, *J. Struct. Biol.* **114,** 28 (1995).

[45] I. Serysheva, E. V. Orlova, M. Sherman, W. Chiu, S. Hamilton, and M. van Heel, *Nature Struct. Biol.* **2,** 18 (1995).

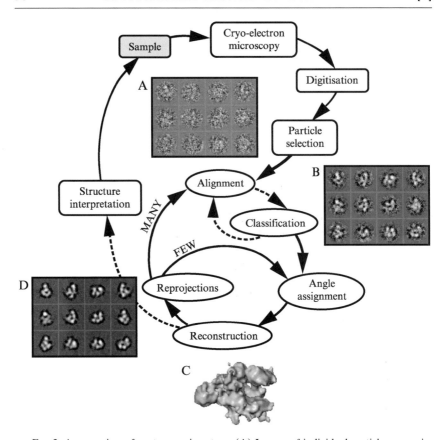

Fig. 2. An overview of postprocessing steps. (A) Images of individual particles are noisy and have to be averaged. To average like images, the images are aligned and classified into groups of approximately 17 like images, which are averaged together to produce classes (B). The relative orientation of each class with respect to one another is determined (angle assignment). The classes are then used to reconstruct a three-dimensional volume (C). This structure is reprojected (D) either to refine the alignment or for angular assignment. The whole procedure is applied in an iterative manner until a stable structure is achieved. Adapted from Finn et al.[7] with permission.

classums data split into two subsets. The different ways of determining resolution are discussed in detail elsewhere.[46] Here it suffices to say that the 3σ threshold curve indicates at which resolution one has collected information significantly above the random noise fluctuations. Finally, it should be noted that reliable visualization of the 3D maps may require additional

[46] M. van Heel and M. Schatz, in preparation (2002).

band-pass filtering of the 3D reconstruction; low frequencies heavily affect the interpretation, and frequencies higher than the real maximal resolution would only add noise. This issue is discussed elsewhere in detail.[47]

Interpretation of RNA Polymerase Three-Dimensional Cryo-EM Maps

The following sections describe the interpretation of the core *E. coli* RNAP cryo-EM map derived from single particle analysis with help of the crystal structures of core RNAP from *Thermus aquaticus*[12] and RNAP II from *Saccharomyces cerevisiae*.[14]

Fitting of the RNAP Crystal Structure from T. aquaticus

The fitting of the RNA polymerase crystal structure from *T. aquaticus*[12] into the density map of the *E. coli* RNAP is achieved in two steps: (i) a global fitting of the structure as a rigid body, first interactively and eventually by reciprocal space fitting, and (ii) an interactive fitting of the individual subunits and their domains. Real space refinement trials with the program ESSENS (part of the RAVE suite, http://xray.bmc.uu.se/usf/rave.html) for fitting of RNA polymerase domains are not appropriate because of extremely long calculation times. A molecular replacement solution is obtained by interactive fitting using "O" software,[48] whereas reciprocal space fitting, achieved using the program X-PLOR,[49] is used to further refine the fit. For the latter procedure, the electron density map is Fourier transformed into a P1 cell (using the SFTOOL subroutine of the CCP4 program suite http://www.ccp4.ac.uk/main.html). The resulting structure factors are used as a target for rigid body positional refinement (X-PLOR) of the *T. aquaticus* RNAP coordinates.[12] The subsequent global fit allows the identification of structural domains that required interactive fitting, achieved by small translational and rotational movements respecting the molecular connectivity.[7] Least-square refinement of the crystal structure of the holoenzyme from *T. thermophilus*[13] to the *T. aquaticus* core RNAP crystal structure using the program "O" reveals their high similarity. Thus the *T. thermophilus* structure provides no significant additional information for interpretation of the *E. coli* core RNAP map.

[47] M. van Heel, *Curr. Op. in Struct. Biol.* **10,** 259 (2000).
[48] T. A. Jones, J. Y. Zou, S. W. Cowan, and M. Kjeldgaard, *Acta Crystallogr. A* **47,** 110 (1991).
[49] T. Brünger, "XPLOR Manual," Version 3.1, Yale Univ. Press, New Haven, 1993.

Homology Modeling of the E. coli RNA Polymerase

In order to generate a molecular model of the *E. coli* RNA polymerase, two steps follow the fitting of the *T. aquaticus* RNA polymerase crystal structure: (i) amino acid sequence alignments and (ii) homology modeling. Multiple sequence alignments between *E. coli* RNA polymerase and *T. aquaticus* and *S. cerevisiae* RNA polymerase are performed using the ClustalX program.[50] For every RNAP subunit, sequence alignments are guided by secondary structure profiles and verified interactively. Final sequence alignments are used to carry out homology modeling using MODELLER6a.[51] For each RNAP subunit, 10 models are made to identify poorly modeled regions, which arise from the sequence present in the *E. coli* RNAP but absent in *T. aquaticus* RNAP. These insertions correspond to the dispensable regions (DRs) in the β and β' subunits of the *E. coli* enzyme and cannot be modeled based on the *T. aquaticus* or *T. thermophilus* RNAP crystal structures.[12,13] The obtained model is fitted interactively to the map using individual RNAP subunits and some of their domains as described earlier. The global fit of the *E. coli* homology model is shown in Fig. 3 (top). Locations of the α-carboxy-terminal domains (α-CTDs) relative to their respective N-terminal domains based on their crystal structures[52,53] are shown in Fig. 3 (bottom). This is consistent with the various modes of interaction of the α-CTD with activators presented at different promoters.[54]

The additional density remaining after fitting the *E. coli* homology model to the cryo-EM electron density map corresponds to the DRs. These can be described by certain subunits of the yeast RNAP II when the coordinates of the *S. cerevisiae* RNAP II[14] are superimposed to the *E. coli* RNAP homology model using the LSQ option in the program "O." The overall architecture of the *E. coli* and *S. cerevisiae* enzymes is the same. At the subunit level, the α_1-NTD is structurally homologous to the Rpb3 and Rpb11 subunits of the *S. cerevisiae* RNAP II enzyme. DR2, which is located close to the C terminus of the β subunit, in the vicinity of the α_1-NTD of *E. coli* RNAP, appears to be structurally homologous to the *S. cerevisiae* RNAP II subunits Rpb10 and Rpb12. Finally, the interaction of DR1 (β subunit) with DR (β' subunit) of *E. coli* RNAP is similar to that between Rpb9 and a fragment of Rpb1 in *S. cerevisiae*.[11]

[50] A. Aiyar, *Methods Mol. Biol.* **132,** 221 (2000).

[51] A. Sali, F. Yuan, H. van Vlijmen, and M. Karplus, *Proteins*, **23,** 318 (1995).

[52] Y. H. Jeon, T. Negishi, M. Shirakawa, T. Yamazaki, N. Fujita, A. Ishihama, and Y. Kyogoku, *Science* **270,** 1495 (1995).

[53] G. Zhang and S. A. Darst, *Science* **281,** 262 (1998).

[54] G. S. Lloyd, W. Niu, J. Tebbutt, R. H. Ebright, and S. J. Busby, *Genes. Dev.* **16,** 2557 (2002).

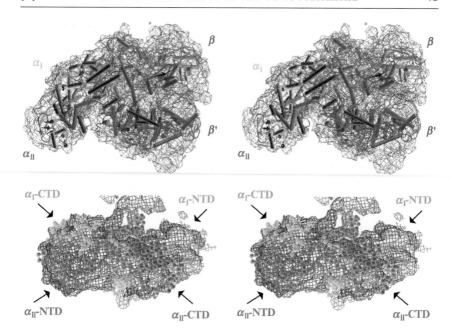

Fig. 3. (Top) Stereo view of the global fit of the *E. coli* homology model of RNAP into the three-dimensional map derived from cryo-EM studies. This orientation of RNAP—looking into the active site of the enzyme—illustrates its typical "crab claw" shape. The Cα backbone is represented as cylinders for α helices and arrows for β sheets. (Bottom) Stereo views of the refined fit illustrate localization of the α_I and α_{II} C-terminal domains relative to the corresponding N-terminal domains based on their individual crystal structures. Relative to the global view above, these domains have been rotated anticlockwise by 90° about an axis perpendicular to the plane of the paper and flipped 180° around a horizontal axis in the plane of the paper. (See color insert.)

Concluding Remarks

Single particle cryoelectron microscopy has allowed determination of the structure of a bacterial mesophile RNAP. When compared with the crystal structures of RNAP from bacterial thermophiles, the medium-resolution electron density map highlights features that are unique to the *E. coli* enzyme and has allowed the localization of domains that otherwise appear disordered in crystal structures of bacterial RNAP. Promoter specificity of RNAP is conferred by the additional sigma factor in the RNAP holoenzyme. Comparison of the different functional states of RNAP in *E. coli* should prove valuable in examining the structural basis of the transcription cycle. Single particle cryo-EM studies can be extended to encompass the interaction of RNA polymerase with transcription factors in ternary complexes. The technique allows the study of different complexes in different functional states without the necessity to grow crystals first. Atomic fittings and homology modeling based on cryo-EM structural studies can be used to ascertain the functional role of such transcription complexes.

Acknowledgments

This work was supported by the BBSRC (UK) and was conducted in the Centre for Biomolecular Electron Microscopy. The CM-200/FEG microscope was funded by a BBSRC/ HEFCE Joint Research Equipment Initiative grant.

[4] Crystallographic Analysis of *Thermus aquaticus* RNA Polymerase Holoenzyme and a Holoenzyme/Promoter DNA Complex

By Katsuhiko S. Murakami, Shoko Masuda, and Seth A. Darst

RNA in all cellular organisms is synthesized by a complex molecular machine, the DNA-dependent RNA polymerase (RNAP). The catalytically competent bacterial core RNAP (subunit composition $\alpha_2\beta\beta'\omega$) is conserved evolutionarily in sequence, structure, and function from bacteria to humans.[1-4] In both prokaryotes and eukaryotes, promoter-specific initiation of transcription requires protein factors in addition to the catalytic core RNAP, ranging from an $\sim$750-kD collection of more than a dozen basal initiation factors for messenger RNA (mRNA) synthesis in eukaryotes[5] to a single polypeptide in bacteria, the σ subunit, which binds core RNAP to form the holoenzyme.[6] The holoenzyme locates specific DNA sequences called promoters, forms the open complex by melting the DNA surrounding the transcription start site, and initiates the synthesis of an RNA chain.[7]

To understand the molecular mechanism of transcription initiation, it is necessary to solve the structures of the holoenzyme and holoenzyme–DNA

[1] S. A. Darst, *Curr. Opin. Struct. Biol.* **11**, 155 (2001).

[2] P. Cramer, *Curr. Opin. Struct. Biol.* **12**, 89 (2001).

[3] J. Archambault and J. D. Friesen, *Microbiol. Rev.* **57**, 703 (1993).

[4] R. H. Ebright, *J. Mol. Biol.* **293**, 199 (2000).

[5] N. A. Woychik and M. Hampsey, *Cell* **108**, 453 (2002).

[6] C. Gross, C. Chan, A. Dombroski, T. Gruber, M. Sharp, J. Tupy, and B. Young, *Cold Spring Harb. Symp. Quant. Biol.* **63**, 141 (1998).

[7] P. L. deHaseth, M. L. Zupancic, and M. T. J. Record, *J. Bacteriol.* **180**, 3019 (1998).

complex. We have reported structures of the bacterial RNAP holoenzyme and a holoenzyme–promoter DNA complex at 4 and 6.5 Å resolution, respectively.[8,9] These structures shed light on the first steps of RNA synthesis. This article reports on details of the purification, crystallization, and structure determination of these macromolecular complexes.

Purification

Core Enzyme

By far the best-characterized cellular RNAP is that from *Escherichia coli*. However, despite years of effort by many groups, three-dimensional crystals of *E. coli* RNAP suitable for X-ray analysis have never been obtained. We expanded our investigations to thermophilic eubacteria, focusing on *Thermus aquaticus* (*Taq*) because of the relative ease with which large amounts of biomass could be obtained. The core RNAP isolated from *Taq* comprised four distinct polypeptides, β', β, α, and ω, of about 170, 125, 35, and 11 kDa, respectively.[10] The preparative procedure for *Taq* core RNAP was similar to the preparation of *E. coli* core RNAP.[11] A schematic outlining the purification procedure is shown in Fig. 1.

Polymin-P Precipitation. Approximately 100 g of frozen cell paste is thawed and mixed with 70 ml of lysis buffer [50 mM Tris–HCl (pH 8.0 at 4°), 2 mM EDTA, 233 mM NaCl, 5% glycerol, 1 mM dithiothreitol (DTT), 1 mM β-mercaptoethanol] using a Waring blender. During mixing, 3.5 ml of a 100× protease inhibitor cocktail [696 mg phenylmethylsulfonyl fluoride (PMSF), 1.248 g benzamidine, 20 mg chymostain, 20 mg leupeptin, 4 mg pepstain, 40 mg aprotinin in 40 ml cold ethanol], an additional 35 ml of lysis buffer, 8% Na-deoxycholate (final 0.2%), and 100 ml of TGED buffer [10 mM Tris–HCl (pH 8.0 at 4°), 0.1 mM EDTA, 5% glycerol, 1 mM DTT] are added. The cells are lysed using a continuous flow French press (~15,000 psi), centrifuged for 30 min at 30,000 g, and the supernatant collected. The supernatant is diluted with TGED buffer until the conductivity is equal to TGED + 0.15 M NaCl and stirred gently while slowly adding a 10% Polymin-P, pH 7.9, solution to a final concentration of 0.6%. The mixture is centrifuged for 30 min at 30,000 g. The pellet is

[8] K. S. Murakami, S. Masuda, and S. A. Darst, *Science* **296,** 1280 (2002).

[9] K. S. Murakami, S. Masuda, E. A. Campbell, O. Muzzin, and S. A. Darst, *Science* **296,** 1285 (2002).

[10] G. Zhang, E. A. Campbell, L. Minakhin, C. Richter, K. Severinov, and S. A. Darst, *Cell* **98,** 811 (1999).

[11] A. Polyakov, E. Severinova, and S. A. Darst, *Cell* **83,** 365 (1995).

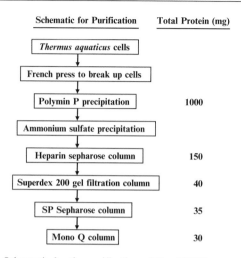

FIG. 1. Schematic for the purification of *Taq* RNAP core enzyme.

washed twice by resuspending in 400 ml of TGED + 0.3 M NaCl and is recovered by centrifugation. RNAP is eluted from the pellet by resuspending in 200 ml of TGED + 1 M NaCl buffer. Solid ammonium sulfate is added to a final saturation of 60% (100% saturation = 4 M) and pelleted by centrifugation.

Heparin–Sepharose Column Chromatography. Prior to loading the sample, the heparin-Sepharose (heparin-Sepharose FF, Amersham Biosciences) column must be regenerated to obtain the maximum RNAP yield. The column is washed with 2 column volumes (CV) of regeneration buffer (50 mM Tris–HCl, pH 7.5, 1 mM EDTA, 2 M KCl, 6 M Urea) and 2 CV of acetate buffer (60 mM sodium acetate, 0.5 M NaCl, pH 5.5) at least twice. The column is equilibrated with TGED + 0.2 M NaCl. The ammonium sulfate pellet is suspended in TGED buffer, and the solution is diluted with TGED until the conductivity is equal to TGED + 0.2 M NaCl. During dilution, the solution may become cloudy, presumably due to precipitate protein and residual Polymin-P. Apply the whole sample to the column without clarification. The protein sample is loaded onto a 50-ml column (2.5 × 10 cm) at 5 ml min^{-1}. The column is washed with TGED + 0.2 M NaCl until A_{280} reaches background and then RNAP is eluted with TGED + 0.6 M NaCl. The sample is precipitated with ammonium sulfate (final concentration of 60% saturation) and pelleted by centrifugation. Approximately 1 g of total protein is loaded onto the heparin column for each run.

Superdex 200 Gel Filtration Column Chromatography. The ammonium sulfate pellet is suspended in TGED buffer and fractionated over a Superdex 200 (Amersham Biosciences) gel filtration column (120 ml, 1.6×60 cm) equilibrated with TGED + 0.5 M NaCl.

SP Column Chromatography. Pooled fractions containing RNAP are diluted with TGED buffer until the conductivity is equal to TGED + 0.1 M NaCl and then loaded onto an SP-Sepharose (Amersham Biosciences) ion-exchange column (20 ml, 1.25×20 cm) equilibrated with TGED + 0.1 M NaCl. The protein is eluted with a gradient from 0.1 to 0.5 M NaCl. The RNAP peak elutes at around 0.3 M NaCl.

Mono Q Column Chromatography. Fractions containing RNAP are diluted with TGED buffer until the conductivity is equal to TGED + 0.15 M NaCl and then loaded onto a Mono Q (Amersham Biosciences) ion-exchange column (8 ml, 1×10 cm) equilibrated with TGED + 0.15 M NaCl. The protein is eluted with a gradient from 0.15 to 0.5 M NaCl. The RNAP peak elutes at around 0.3 M NaCl. Peak fractions containing RNAP are pooled, and glycerol is added to a final concentration of 15% (v/v). The samples are then aliquoted, frozen in liquid nitrogen, and stored at $-80°$ until use. One hundred grams of wet cell paste typically yields 30 mg of core RNAP, which is more than 99% pure as judged from overloaded, Coomassie-stained SDS gels (Fig. 2A, lane 1).

σ^A

The gene encoding for wild-type *Taq* σ^A or a region 1.1 deletion mutant (*Taq* σ^A[92-438], or $\Delta 1.1 \sigma^A$)[12,13] is PCR subcloned into the *Nde*I/*Bpu*1102I sites of the pET21a expression vector (Novagen). *E. coli* strain BL21(DE3)pLysS transformed with the expression plasmid is grown at $37°$ in 2 liter of LB medium supplemented with ampicillin (100 μg/ml) and chloramphenicol (34 μg/ml) to an A_{600} of $\sim$0.5. Expression is induced by adding 1 mM isopropyl-β-D-thiogalactopyranoside (IPTG) for 3 h. Cells are harvested and stored frozen at $-80°$.

Frozen cells are suspended in 50 ml of TGED + 0.5 M NaCl, lysed with a continuous flow French press or by sonication, and centrifuged at 20,000 g for 10 min at $4°$. The supernatant is heat treated at $65°$ for 30 min, and σ^A is recovered from the supernatant by centrifugation. The supernatant is diluted until the conductivity is equal to TGED + 0.1 M NaCl and is then applied to a cation-exchange column [HiTrap SP-sepharose

[12] L. Minakhin, S. Nechaev, E. A. Campbell, and K. Severinov, *J. Bacteriol.* **183,** 71 (2001).
[13] E. A. Campbell, O. Muzzin, M. Chlenov, J. L. Sun, C. A. Olson, O. Weinman, M. L. Trester-Zedlitz, and S. A. Darst, *Mol. Cell* **9,** 527 (2002).

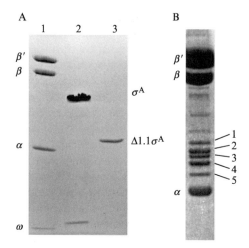

FIG. 2. (A) SDS–PAGE analysis of *Taq* core RNAP (lane 1) and recombinant σ^A (lane 2, wild-type σ^A; lane 3, $\Delta 1.1\sigma^A$) stained with Coomassie blue. (B) *In situ* proteolysis. The *Taq* holoenzyme was incubated for several months in hanging drop crystallization trials and then analyzed by SDS–PAGE and Coomassie blue staining. Core enzyme subunits and the major, stable proteolytic fragments of σ^A are labeled. N-terminal cleavage sites of these fragments are (band, residue) 1, 35; 2, 49; 3, 58; 4, 71; and 5, 92.

(Amersham Biosciences) or an equivalent] equilibrated with TGED + 0.1 *M* NaCl. Proteins are eluted with a linear gradient of 0.1 to 0.8 *M* NaCl in TGED buffer. The protein is purified further by gel filtration over Superdex 200 (Amersham Biosciences) equilibrated with TGED + 0.5 *M* NaCl. The wild-type σ^A (50 kDa) shows slower mobility than expected based on the molecular weight, but $\Delta 1.1\sigma^A$ (40 kDa) runs with the expected mobility on a denaturing gel (Fig. 2A, lanes 2 and 3). Peak fractions containing σ^A are pooled, and glycerol is added to a final concentration of 15%, aliquoted, frozen in liquid nitrogen, and stored at $-80°$ until use.

DNA

Lyophilized, tritylated, single-stranded oligonucleotides (Oligos Etc.) are detritylated and purified as described.[14] Dried oligonucleotides are dissolved in annealing buffer (5 m*M* sodium cacodylate, pH 7.4, 0.5 m*M* EDTA, 0.2 *M* NaCl) to a concentration of ~2 m*M*. Equimolar amounts of complementary oligonucleotides are annealed at 1 m*M* concentration

[14] A. Aggarwal, *Methods* **1**, 83 (1990).

by heating to $90°$ for 5 min and cooling to $25°$ at a rate of $0.01°/s$. Annealed oligonucleotides are stored at $-20°$.

Crystallization

Holoenzyme

The transcriptionally active *Taq* RNAP holoenzyme is reconstituted *in vitro* by combining native *Taq* core RNAP and recombinant *Taq* σ^A (molar ratio 1:1.2) in reconstitution buffer (10 mM Tris–HCl, pH 8.0, 0.2 M NaCl, 1 mM DTT, 0.1 mM EDTA, 15% glycerol) and is then incubated at $30°$ for 30 min. Initial crystallization trials yield small, needle-like crystals. Examination of the protein present in the crystals by SDS–PAGE reveals *in situ* degradation of σ^A by unknown contaminating proteases. N-terminal sequencing shows that the σ^A polypeptide is cleaved at its N terminus at residues 35, 49, 58, 71, and 92 (Fig. 2B). The shortest σ^A fragment (residues 92–438), which lacks the entire σ region 1.1 ($\Delta1.1\sigma^A$), is then cloned, expressed, and purified from *E. coli* (described earlier) and used for reconstitution and crystallization. Crystals are grown by vapor diffusion with microseeding. To prepare crystal seeds, the $\Delta1.1\sigma^A$–holoenzyme solution (15 mg/ml) is mixed with the same volume of crystallization solution containing 0.1 M HEPES–NaOH, pH 8.0, 3 M sodium formate, and the mixture is incubated as a hanging drop over the crystallization solution. Needle-like crystals grow in a week. These can be stabilized and stored in crystallization solution. Fresh $\Delta1.1\sigma^A$–holoenzyme solution (15 mg/ml) is mixed with the same volume of crystallization solution containing crystal seeds, and the mixture is incubated over the crystallization solution containing 2.4–2.6 M sodium formate. Plate-like crystals with typical dimensions of $0.4 \times 0.1 \times 0.02$ mm grow in 2 months at $22°$. The crystals belong to the primitive space group $P2_1$ (a = 155.0 Å, b = 271.2 Å, c = 155.3 Å, $\beta = 91.4°$), with two 430-kDa $\Delta1.1\sigma^A$–holoenzyme molecules per asymmetric unit and a solvent content of 65%. For cryocrystallography, crystals are presoaked in crystallization solution, followed by four 10-min soaks in crystallization solution containing increasing concentrations of sodium formate to a final concentration of 6 M. Crystals are then flash frozen by dunking in liquid ethane and held at liquid nitrogen temperature.

Holoenzyme/Fork–Junction DNA Complex

The holoenzyme/fork–junction DNA complex is prepared by mixing holoenzyme and annealed fork–junction DNA (molar ratio 1:1.5) in 10 mM Tris–HCl (pH 8.0 at $4°$), 0.1 M NaCl, 20 mM MgCl$_2$, 1 mM DTT,

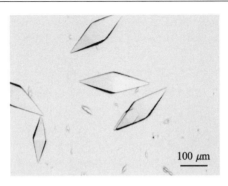

FIG. 3. Crystals of the *Taq* holoenzyme/fork–junction DNA complex.

0.1 m*M* EDTA, and incubated for 30 min at room temperature. Crystals are grown by vapor diffusion. The protein/DNA solution (1 μl at a final RNAP holoenzyme concentration of 15 mg/ml) is mixed with an equal volume of 1.6–1.8 *M* $(NH_4)_2SO_4$ (pH adjusted to 8.0 with NaOH), 0.1 *M* Tris–HCl (pH 8.0 at 22°), 20 m*M* $MgCl_2$, and incubated as a hanging drop over the same solution at 4°. Crystals grow in about 3 weeks to typical dimensions of 0.3 × 0.1 × 0.1 mm (Fig. 3). The crystals belong to the space group P$4_3$22 (unit cell parameters a = b = 182 Å, c = 522 Å), with one 450-kDa protein/DNA complex per asymmetric unit and a solvent content of 75%. Unfrozen crystals mounted in a capillary do not show any diffraction. For cryocrystallography, the crystals are presoaked in 1.8 *M* $(NH_4)_2SO_4$, 0.1 *M* Tris–HCl, pH 8.0, 20 m*M* $MgCl_2$, then soaked directly in 6 *M* sodium formate, 0.1 *M* Tris–HCl, pH 8.0, 20 m*M* $MgCl_2$, for 10 min before flash freezing by dunking in liquid ethane at liquid nitrogen temperature. Crystals larger than 0.4 mm in length are damaged during cryoprotectant soaking. The frozen crystals diffract to 7 Å resolution from an in-house X-ray generator. Spots can sometimes be observed, in one direction, to 4.5 Å resolution at synchrotron beam lines.

Data Collection

Holoenzyme

Because of the thin, plate-like shape (0.4 × 0.1 × 0.02 mm) of the crystal, most holoenzyme crystals are somewhat bent during freezing, resulting in poor diffraction with elongated spots. We attempted annealing the frozen crystals[15] but with no improvement. Third-generation synchrotron beam line SBC-19ID at APS produced an extremely intense X-ray beam, allowing a fairly small beam (0.1×0.1 mm) to be used without intensity reduction. Searching with the small beam over the surface of the crystal, a relatively flat region could be found that gave sharp diffraction spots at all orientations of the crystal. Diffraction to 4 Å resolution was obtained. Diffraction data were processed using DENZO and SCALEPACK.[16]

Holoenzyme/Fork–Junction DNA Complex

Crystals were aligned with the long c axis approximately along the spindle axis to minimize reflection overlaps. Low-resolution data were extremely important in obtaining a continuous electron density map for model fitting described later. Using a small beam stop and a large crystal-to-detector distance (900 mm) at CHESS-F1 beam line allowed collection of data down to 80 Å resolution. Low-resolution data sets were collected from 80 to 9 Å, and high-resolution data sets were collected from 10 to 6.5 Å; these data sets were combined to obtain 80 to 6.5 Å.

Phasing, Model Building, and Refinement

Figure 4 shows strategy used to solve the structures of the holoenzyme and holoenzyme/fork–junction DNA complex (RF). At resolutions less than about 3.5 Å, it would normally not be possible to construct an accurate model *de novo*. However, the availability of atomic structures of both core RNAP[10] and domain structures of σ^A facilitated the construction of a model for RF at 6.5 Å resolution and holoenzyme at 4 Å resolution.[13] We attempted to obtain phases from RF crystals, as the crystallization was highly reproducible, the crystals were isomorphous, and only one macromolecular complex was contained in the asymmetric unit (as opposed to a dimer in the asymmetric unit of the holoenzyme crystals). Because the crystals diffracted in the worst direction to only around 6.5 Å resolution and contained a large molecular mass ($\sim$450 kDa) in the asymmetric unit, we used heavy metal clusters for multiple isomorphous replacement (MIR) phasing. Soaks were done with tungsten clusters[17] (W_{18}, $K_6P_2W_{18}O_{62}$; W_{12}, $K_5HCoW_{12}O_{40}$) and tantalum bromide[18] (Ta_6Br_{14}). The W_{18} cluster disrupted the crystals, but the W_{12} cluster showed excellent phasing, which was the milestone in determining the

[15] F. A. Samatey, K. Imada, F. Vonderviszt, Y. Shirakihara, and K. Namba, *J. Struct. Biol.* **132,** 106 (2002).

[16] Z. Otwinowski and W. Minor, *Methods Enzymol.* **276,** 307 (1997).

[17] N. Casan-Pastor, P. Gomez-Romero, G. B. Jameson, and L. C. W. Baker, *J. Am. Chem. Soc.* **113,** 5685 (1991).

[18] G. Schneider and Y. Lindqvist, *Acta Cryst.* **D50,** 186 (1994).

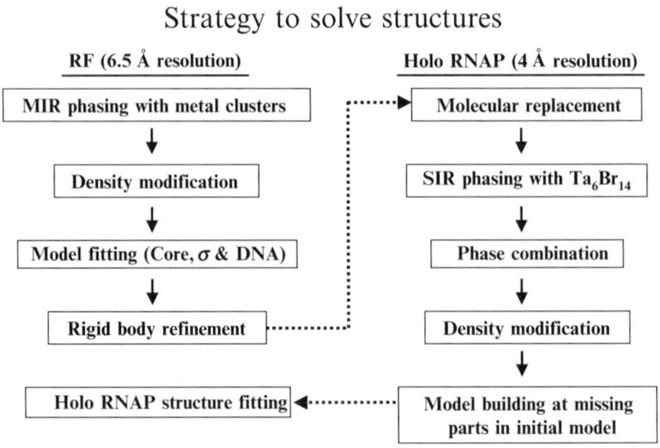

Strategy to solve structures

FIG. 4. Summary of structure determination procedures for holoenzyme and holoenzyme/fork–junction DNA complex.

structures. The locations of three W_{12}-binding sites were derived using SHELX-97.[19] Initial phases were calculated using MLPHARE.[20] Spherically averaged models[21] for the structure of the W and Ta clusters were used for heavy atom refinement and phasing calculations. The four-Gaussian fit to the scattering curve yielded the following coefficients: W_{12} (useful to 7 Å resolution): $c = -239.701$, $a_1 = 422.612$, $b_1 = 287.463$, $a_2 = 312.267$, $b_2 = 287.532$, $a_3 = 215.955$, $b_3 = 286.933$, $a_4 = 162.945$, $b_4 = 287.325$; Ta_6Br_{14} (useful to 4.5 Å resolution): $c = 17976.4$, $a_1 = 27.1519$, $b_1 = 673.365$, $a_2 = 6696.77$, $b_2 = 22.1632$, $a_3 = 4994.74$, $b_3 = 22.1958$, $a_4 = -29267.1$, $b_4 = 6.99315$. This treatment improved phases and map quality. The resulting MIR map was not interpretable, but density modification using SOLOMON[22] improved the density map dramatically. A systematic variation of the solvent content and truncation factors used as inputs to SOLOMON was necessary to determine the optimal solvent-flattening procedure.[23] For each set of these parameters, the electron density map was inspected visually (e.g., four helix bundle of the α subunit N-terminal domain or

[19] G. M. Sheldrick, *in* "Proc. CCP4 Study Weekend" (W. Wolf, P. R. Evans, and A. G. W. Leslie, eds.), p. 23. SERC Daresbury Laboratory, Warrington, UK, 1991.

[20] Z. Otwinowski, *in* "Proc. of the CCP4 Study Weekend" (W. Wolf, P. R. Evans, and A. G. W. Leslie, eds.), p. 80 SERC Daresbury Laboratory, Warrington, UK, 1991.

[21] J. Fu, A. L. Gnatt, D. A. Bushnell, G. J. Jensen, N. E. Thompson, R. R. Burgess, P. R. David, and R. D. Kornberg, *Cell* **98**, 799 (1999).

[22] J. P. Abrahams and A. G. W. Leslie, *Acta Cryst.* **D52**, 30 (1996).

[23] W. M. Clemons, Jr., D. E. Brodersenl, J. P. McCutcheon, J. L. C. May, A. P. Carter, R. J. Morgan-Warren, B. T. Wimberly, and V. Ramakrishnan, *J. Mol. Biol.* **310**, 827 (2001).

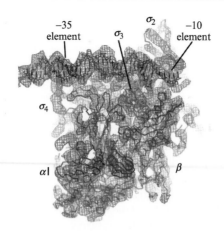

FIG. 5. *Taq* RNAP holoenzyme/fork–junction DNA complex. The α-carbon backbones of the holoenzyme subunits and the fork–junction DNA are shown. The three structural domains of σ^A (σ_2, σ_3, and σ_4) are labeled. The direction of transcription (downstream) would be to the right. The experimental electron density map, calculated using F_o coefficients, is shown (net, contoured at 1.5σ) and was computed using MIR phases, followed by density modification. The entire complex is viewed from in front of the β subunit (β' is obscured behind).

the DNA double helix was easily recognized in the maps) and optimal density modification conditions were noted. Solvent flattening using a solvent content of 55% produced the best density map, even though the calculated solvent content was 75%. The protein phases were recalculated with the heavy atom parameters refined against the solvent-flattened phase.[24]

At 6.5 Å resolution, the major and minor grooves of the double-stranded DNA, as well as the $3'$ single-stranded tail of the fork–junction nontemplate strand, were readily apparent in the electron density maps. Moreover, the protein/solvent boundary was clearly discernible, and rod-like densities were interpretable as protein α helices. The core RNAP structure was divided into five "modules".[8] Using the program O,[25] the modules were fit manually as rigid bodies into the density, mainly by aligning α helices with the rod-like densities (Fig. 5).

After fitting the core RNAP modules, the three structural domains of σ^A (σ_2, σ_3, and σ_4[13]) were readily placed into the density map, and the DNA model was then added and adjusted manually to fit the density. Crystallographic rigid body refinement was carried out using CNS.[26] The model

[24] M. A. Rould, J. J. Perona, and T. A. Steitz, *Acta Cryst.* **A48,** 751 (1992).

[25] T. A. Jones, J.-Y. Zou, S. Cowan, and M. Kjeldgaard, *Acta Cryst.* **A47,** 110 (1991).

[26] P. D. Adams, N. S. Pannu, R. J. Read, and A. T. Brunger, *Proc. Natl. Acad. Sci. USA* **94,** 5018 (1997).

of holoenzyme in RF was used for the molecular replacement search against 4-Å resolution data from the *Taq* RNAP holoenzyme crystals. Using diffraction data from 15 to 5 Å resolution, rotation and translation searches were performed using CNS (correlation coefficient of first peak, 0.111; second peak, 0.100; third peak, 0.035). After a solution for one molecule in the asymmetric was obtained, its position was fixed and another search was performed (correlation coefficient of first peak, 0.217; second peak, 0.091). The molecular replacement phases were improved by phase combination with single isomorphous replacement (SIR) phases from a Ta_6Br_{14} derivative. The electron density map was improved further by density modification and noncrystallographic symmetry averaging using CNS. Comparison of the molecular replacement solution with the resulting electron density map revealed significant deviations, indicating that model bias was removed effectively by the inclusion of the Ta-cluster phases. The structure was divided into modules and rigid body refinements were performed by CNS. Adjustments to the model were then made by hand to better fit the electron density map. Connecting segments of σ^A that were not modeled in the original lower resolution RF complex (most significantly the extended polypeptide chain and α helix comprising σ conserved region 3.2), and segments of the RNAP subunits that were disordered and not modeled in the core RNAP structure,[10] were modeled as polyalanine. Several solvent-exposed loops within the RNAP subunits were missing electron density, presumably because of disorder within the crystals, and were removed. A Zn^{2+} ion was placed according to the ethyl mercury thiosalicylate Fourier difference peak from the RF complex. Structural calculations were performed using CNS with very tight noncrystallographic symmetry restraints (NCS restraint weight = 1000) between the two molecules in the asymmetric unit at all times. Map inspection and model building were done using O.[25] The current, unrefined model has an R factor of 0.345 (R_{free} = 0.397). Eventually, the final holoenzyme model derived from the 4-Å resolution holoenzyme crystals was fit into the RF electron density map with the DNA model, conformations of domains were altered to fit the density map, and a final round of rigid body refinement was executed to obtain the final RF structure. The final round of rigid body refinement yielded a crystallographic R factor of 0.448 (R_{free} = 0.452), using diffraction data from 15 to 6.5 Å resolution.

Concluding Remarks

Some general conclusions for structure determination of large macromolecules can be drawn from this work. The interaction of RNAP holoenzyme with promoter DNA initiates a series of structural transitions from the initial closed promoter complex to the transcription competent open complex. The reaction is in rapid equilibrium with several intermediate complexes and this equilibrium depends on many factors, including temperature, Mg^{2+} concentration, and DNA sequence. This presents challenges for structure determination, where large quantities of a homogeneous complex must be isolated. We used fork–junction DNA in order to "freeze" the kinetics of RNAP–promoter DNA interaction and prepare a homogeneous complex for crystallization.

With thin crystals of large asymmetric units, the small number of unit cells in the crystal results in weak diffraction. This can be overcome by the use of intense beams from undulator insertion device beam lines at third-generation synchrotrons, where the beam size is matched carefully to the crystal dimensions so that excess background scattering is minimized. Also, the beam stop size is matched carefully to the beam to collect low-resolution diffraction in order to obtain a continuous electron density map.

A sufficient heavy atom signal can be obtained even for very large structures with low-resolution data using heavy metal clusters. The metal cluster phasing (using the spherically averaged model as described in Fu et al.[21]) and solvent density modification generate excellent phases with excellent low-resolution maps and can be used to determine sites of additional derivatives using difference Fourier techniques.

The careful density modification protocol is essential in obtaining an interpretable electron density map. In our practice, it was necessary to vary the input parameters in SOLOMON iteratively in order to obtain the best map with the most continuous chain density and to minimize truncation of weak density. A visual inspection of many maps may be necessary to obtain the best combination of parameters. While laborious and time-consuming, the final results can be well worth the effort.

Acknowledgments

We thank M. Pope for the gift of the W clusters and G. Schneider for the gift of Ta_6Br_{14}. We are indebted to A. Joachimiak, S. L. Ginell, and N. Duke at the Advanced Photon Source Structural Biology Center, D. Thiel and staff at the Cornell High Energy Synchrotron Source, and M. Becker and L. Berman at NSLS X25 for support during data collection. We thank D. Jeruzalmi, S. Nair, and H. Yamaguchi for invaluable advice. Figure 5 was made using the program O. K.S.M. was supported by a Norman and Rosita Winston Postdoctoral Fellowship and a Human Frontiers Sciences Program Postdoctoral Fellowship. Work in the laboratory of S.A.D. is supported by the National Institutes of Health (GM53759 and GM61898).

[5] Purification of *Rhodobacter sphaeroides* RNA Polymerase and Its Sigma Factors

By JENNIFER R. ANTHONY, HEATHER A. GREEN, and TIMOTHY J. DONOHUE

Like other facultative bacteria, *Rhodobacter sphaeroides* has a variety of metabolic lifestyles. In addition to generating energy by respiration or photosynthesis, this α-proteobacterium uses catabolic, anabolic, or energetic pathways of environmental, agricultural, or industrial interest.[1] The *R. sphaeroides* basal transcription apparatus includes the typical bacterial RNA polymerase subunits ($\alpha_2\beta\beta'\omega$) and 17 sigma factors.[2] *R. sphaeroides* sigma factors include a homolog of the *Escherichia coli* housekeeping sigma factor (hereafter referred to as σ^{70} or RpoD),[3] two members of the heat shock family of alternate sigma factors (RpoH$_I$ and RpoH$_{II}$),[4,5] nine proteins that are related to sigma factors in the extracytoplasmic stress family, ECF (including one, σ^E or RpoE, which is currently being analyzed),[2,6–8] and four members of the RpoN family.[2,9] When one considers the metabolic activities that allow *R. sphaeroides* to generate energy, it is not surprising that alternate sigma factors and regulatory strategies are present that are not found in well-studied enteric or gram-positive bacteria. There is also experimental evidence[10] and inference[8] that some of the unusual signal transduction pathways from *R. sphaeroides* are present in proteobacteria of environmental, agricultural, or medical interest.

This article presents methods used to obtain *R. sphaeroides* core RNA polymerase and holoenzyme, as well as approaches used to purify active sigma factors or their cognate antisigma factors. It also outlines assays to monitor the ability of RNA polymerase to transcribe target genes, either

---

[1] R. E. Blankenship, M. T. Madigan, and C. E. Bauer, Kluwer Academic, Boston, 1995.

[2] C. MacKenzie, M. Choudhary, F. W. Larimer, P. F. Predki, S. Stilwagen, J. P. Armitage, R. D. Barber, T. J. Donohue, J. P. Hosler, J. Newman, J. P. Shapliegh, R. E. Sockett, J. Zeilstra-Ryalls, and S. Kaplan, *Photosynth. Res.* **70**, 19 (2001).

[3] T. M. Gruber and D. A. Bryant, *J. Bacteriol.* **179**, 1734 (1997).

[4] R. K. Karls, J. Brooks, P. Rossmeissl, J. Luedke, and T. J. Donohue, *J. Bacteriol.* **180**, 10 (1998).

[5] H. A. Green and T. J. Donohue, in preparation.

[6] S. Nechaev and K. Severinov, *J. Mol. Biol.* **289**, 815 (1999).

[7] J. D. Newman, M. J. Falkowski, B. A. Schilke, L. C. Anthony, and T. J. Donohue, *J. Mol. Biol.* **294**, 307 (1999).

[8] J. Newman, J. Anthony, and T. J. Donohue, *J. Mol. Biol.* **313**, 485 (2001).

[9] S. Poggio, A. Osorio, G. Dreyfus, and L. Camarena, *Mol. Microbiol.* **46**, 75 (2002).

alone or in the presence of specific transcription factors.[10,11] Many of these methods are modifications of techniques used to monitor the activity of analogous proteins from other well-studied prokaryotes. The systems described here should be useful for analyzing many transcriptional regulatory networks in R. sphaeroides and other bacteria.

I. General Considerations

When sonication is used to lyse cells, a VWR Scientific sonicator with a microtip is used. SDS–PAGE is commonly used to assay for the presence of RNA polymerase subunits.[12] All dialysis steps are performed at $4°$.

II. Purification of RNA Polymerase

For many experiments, samples containing a mixture of RNA polymerase holoenzymes are sufficient. Such samples can be obtained rapidly and in high yield using heparin–agarose chromatography (Section II, A).[13] Crude RNA polymerase holoenzymes can be used to confirm the location of putative promoters, to assess promoter strength, or to determine if a potential transcription factor alters promoter activity, especially if the test promoter is recognized by the housekeeping sigma factor.[10,11] When it is necessary to reconstitute holoenzyme preparations with a specific sigma factor,[7,8,14] it is preferable to use core RNA polymerase holoenzyme that is free of endogenous sigma factors. We present one approach to obtain R. sphaeroides core RNA polymerase that uses affinity purification on monoclonal antibody resin (Section II, B).[15] Other methods can be used to separate most of the σ^{70}-containing holoenzyme from core RNA polymerase or to provide holoenzymes that are enriched for alternate sigma factors (Section II, C).[16]

For RNA polymerase purification from small (<10 liter) quantities of cells, we routinely use highly aerated, exponential phase aerobic cultures of R. sphaeroides to minimize complications due to the large amount of photosynthetic membrane that is made in the absence of O_2.[17] Larger amounts of cells are typically grown at $30°$ in a fermentor, maintaining

[10] J. C. Comolli and T. J. Donohue, Mol. Microbiol. **45,** 755 (2002).

[11] J. Comolli, A. Carl, C. Hall, and T. J. Donohue, J. Bacteriol. **184,** 390 (2002).

[12] R. K. Karls, D. J. Jin, and T. J. Donohue, J. Bacteriol. **175,** 7629 (1993).

[13] C. Gross, F. Engbaek, T. Flammang, and R. Burgess, J. Bacteriol. **128,** 382 (1976).

[14] J. T. Newlands, W. Ross, K. K. Gosink, and R. L. Gourse, J. Mol. Biol. **220,** 569 (1991).

[15] N. E. Thompson, D. A. Hager, and R. R. Burgess, Biochemistry **31,** 7003 (1992).

[16] D. A. Hager, D. J. Jin, and R. R. Burgess, Biochemistry **29,** 7890 (1990).

[17] J. Zeilstra-Ryalls, M. Gomelsky, J. M. Eraso, A. Yeliseev, J. O'Gara, and S. Kaplan, J. Bacteriol. **180,** 2801 (1998).

the dissolved O_2 content by vigorous sparging and stirring. In either case, late exponential phase cells (density $\sim 2 \times 10^9$ cells/ml) are harvested by centrifugation (8000 g, 10 min at 4°), the pellets frozen under liquid N_2, and stored at $-80°$ until use.

A. Crude R. sphaeroides RNA Polymerase

This is a modification of a previously published protocol that allows one to rapidly obtain a mixture of RNA polymerase holoenzymes from relatively small quantities of cells.[11,13]

Approximately 2 g (wet weight) of freshly harvested or frozen ($-80°$) R. sphaeroides cells are resuspended in 6 ml of 10 mM Tris–Cl, pH 8.0; 50 mM NaCl; 10 mM MgCl$_2$; 1 mM EDTA; 0.3 mM dithiothreitol (DTT); 7.5% glycerol; 0.5 mM phenylmethylsulfonyl fluoride (PMSF); and 1.5 mg/ml lysozyme. The sample is incubated on ice for 10 min before sonicating for five 1-min intervals at 50% duty cycle. The sample is centrifuged at 10,000 g for 40 min, and the supernatant is mixed for 60 min with 3 volumes of 50% heparin–agarose equilibrated in heparin–agarose buffer (10 mM Tris–Cl, pH 8.0; 10 mM MgCl$_2$; 1 mM EDTA; and 0.3 mM DTT). This slurry is poured into a column and washed with 5 volumes of heparin–agarose buffer containing 7.5% glycerol and 50 mM NaCl. RNA polymerase is eluted in five $\sim$5-ml fractions with heparin–agarose buffer containing 7.5% glycerol and 0.6 M NaCl. Samples containing RNA polymerase subunits are pooled and concentrated $\sim$5-fold using a Centricon YM10 (Millipore Corporation, Billerica, MA) concentrator prior to storage at $-20°$ where the activity is stable for months. Typically, approximately 10 μg of RNA polymerase is obtained from 2 g of cells; this is sufficient for 50–100 in vitro transcription assays assuming that a final concentration of 50 nM RNA polymerase is used in a typical 20-μl assay.

B. Affinity Purification of R. sphaeroides Core RNA Polymerase

In this procedure, R. sphaeroides core RNA polymerase is purified via a combination of affinity chromatography using the polyol-responsive 4RA2 monoclonal antibody (Dr. Richard Burgess laboratory, UW–Madison)[18] and anion-exchange chromatography.

Aerobically grown R. sphaeroides cells ($\sim$55 g wet weight) are resuspended in 60 ml of TE buffer (10 mM Tris–Cl, pH 8.0 and 0.1 mM EDTA) containing 1 M NaCl, fresh 1 mM DTT, and 100 μg/ml of lysozyme. Deoxycholic acid is added to 0.2%, and the cells are incubated on ice for 20 min. After cells are lysed by sonication (six 1-min bursts at 50% duty

---

[18] S. A. Lesley and R. R. Burgess, Biochemistry 28, 7728 (1989).

cycle), the extract is centrifuged at 10,000 g for 25 min. The supernatant is treated with 0.2% poly(ethyleneimine) (polymin P; P-3143, Sigma-Aldrich Corp., St. Louis, MO) and incubated on ice for 15 min. After centrifuging at 7000 g for 15 min, proteins in the supernatant are precipitated by adding ammonium sulfate to 60% saturation. Samples are centrifuged at 10,000 g for 20 min and the precipitate is resuspended in sufficient TE buffer ($\sim$80 ml) to achieve a conductivity similar to 0.15 M NaCl. Samples are loaded at a flow rate of $\sim$2 ml/min onto an $\sim$3-ml 4RA2 antibody column[18] preequilibrated with TE buffer containing 0.15 M NaCl. The flow-through is collected, and proteins bound to the 4RA2 antibody column are eluted by adding TE buffer containing 0.75 M NaCl and 40% polyethylene glycol. SDS–PAGE analysis of fractions indicates that various RNA polymerase holoenzymes are bound to the 4RA2 antibody column, while the desired core enzyme is present in the flow-through. The 4RA2 column is regenerated by washing with 2 column volumes of TE buffer containing 2 M KSCN followed by 2 column volumes of TE buffer. The previous flow-through fractions, containing core RNA polymerase subunits, are pooled and reapplied to the 4RA2 antibody column; the column is washed with 3 column volumes of TE buffer containing 0.15 M NaCl. The column is washed with 3 column volumes of TE buffer containing 0.5 M NaCl. Core RNA polymerase is eluted in ten 1-ml fractions by adding TE buffer containing 0.75 M NaCl and 40% propylene glycol. After SDS–PAGE identifies fractions containing RNA polymerase subunits, the appropriate samples are pooled and diluted with sufficient TGED (10 mM Tris–HCl, pH 7.9; 10% glycerol; 0.1 mM EDTA; and fresh 0.1 mM DTT) to achieve a conductivity similar to a 0.15 M NaCl solution. The sample is filtered through a 0.45-μm membrane and applied to a Mono Q anion-exchange column (Amersham Pharmacia, Piscataway, NJ) equilibrated with TGED and 50 mM NaCl at a flow rate of 1 ml/min.[16] Core RNA polymerase is eluted over 45 min with a linear gradient of 300–500 mM NaCl (core RNA polymerase elutes at $\sim$425 mM NaCl). After pooling fractions that contain RNA polymerase, the sample is dialyzed against 1 liter of TGED containing 50% glycerol and 100 mM NaCl. The material is separated into 0.3-ml aliquots and frozen in liquid N_2 prior to storage at -80°. We typically obtain 2.5 mg of high-purity core RNA polymerase from 55 g wet weight of cells.

C. Highly Purified R. sphaeroides RNA Polymerase

This method is a modification[12] of the traditional method for isolating *E. coli* RNA polymerase[19] that provides enzyme with high *in vitro* transcription activity. After cell lysis, all steps are performed on ice or at 4°.

Typically, ~200 g wet weight of cells (from ~20 liters of cells) are resuspended in 320 ml (1.6 volume) of ice-cold 0.5 M Tris–HCl (pH 7.9), 10% glycerol, 2 mM EDTA (pH 7.9), 0.233 M NaCl, and fresh 0.1 mM DTT. To this slurry, lysozyme is added to 1.5 mg/ml, and the sample is incubated for 30 min and sonicated at 50% duty cycle until no longer viscous (approximately ten 1-min pulses). After adding PMSF to 0.1 mM and incubation for 45 min on ice, the suspension is placed in a prechilled blender along with 150 ml ice-cold TGED (Section II,B) containing 0.2 M NaCl. The mixture is blended at high speed for 30 s to shear DNA. The lysate is centrifuged at 10,000 g for 45 min prior to treatment with polymin P to precipitate DNA-binding proteins.

Before processing the entire sample, we add 0, 10, 20, 30, 35, 40, 50, or 60 μl of 10% (v/v) polymin P (pH 7.9) to 1.5-ml centrifuge tubes containing 1 ml of the lysate. The samples are mixed in a vortex and centrifuged in a bench-top microfuge for 5 min, and the supernatant is observed to note the minimum concentration of polymin P that gives a clear supernatant (typically ~0.2–0.3% polymin P). Using this information, the appropriate amount of polymin P is added to the remaining lysate (e.g., if clarification occurred at 0.3% polymin P, add 3 ml of 10% polymin P for every 100 ml of sample) while blending at low speed over 5 min. After centrifugation at 9000 g for 10 min, the precipitate is resuspended in 400 ml TGED containing 0.5 M NaCl, blended at low speed for 5 min, and centrifuged (9000 g for 10 min). To dissociate RNAP from DNA, the resulting pellet is resuspended in 400 ml TGED containing 1.0 M NaCl, blended, and centrifuged (9000 g for 10 min). To the supernatant, ammonium sulfate is added to 35% saturation with stirring and incubated for >1 h (at this point, the sample can be left refrigerated overnight). The sample is centrifuged at 10,000 g for 30 min, and the pellet is suspended in sufficient TGED + 0.25 M NaCl so that the conductivity approximates that of TGED with 0.25 M NaCl. A final centrifugation at 10,000 g for 10 min removes insoluble material.

To enrich for DNA-binding proteins, the sample is mixed with 30–40 ml of single-stranded (ss)DNA cellulose (Worthington Biochemicals, 3 mg of DNA bound per gram of cellulose) that was equilibrated previously with 5 volumes of TGED plus 0.25 M NaCl, centrifuged at 50 g for 5 min, min, and washed twice with 50 ml TGED + 0.25 M NaCl. The resulting slurry is poured into a column and washed with ~100 ml of TGED + 0.25 M NaCl (flow rate 50 ml/h). Proteins are eluted in two steps. The first elution is performed with 50 ml TGED + 0.4 M NaCl, collecting 4-ml fractions. Next, 4-ml fractions are collected during a 50-ml elution with TGED

[19] R. R. Burgess and J. J. Jendrisak, *Biochemistry* **14**, 4634 (1975).

containing 1.0 M NaCl (flow rate ~35–40 ml/h). Spotting 1-μl samples on nitrocellulose and assaying for protein by Amido black staining[12] identifies fractions enriched for protein, while SDS–PAGE monitors the polypeptide composition of these fractions.

The ssDNA cellulose eluate can be used for *in vitro* transcription assays, but chromatography on Sephacryl-300 increases the enzyme purity and specific activity.[12] Fractions from the ssDNA column that contain RNA polymerase subunits are pooled, and proteins are precipitated by adding ammonium sulfate to 35% saturation. The proteins are resuspended in 0.5 ml TGED plus 0.5 M NaCl before being applied to a 100 ml Sephacryl S-300 column (1.6 × 100 cm) that is equilibrated with TGED plus 0.5 M NaCl. The column is run at ~16 ml/h, collecting 1.0-ml fractions (RNA polymerase generally elutes after ~60 ml). Amido black staining and SDS–PAGE analysis are used to identify fractions enriched in RNA polymerase; these samples are pooled and dialyzed against two changes of 250 volumes of TGED + 0.5 M NaCl. We typically obtain ~17 mg of RNA polymerase from 200 g wet weight of aerobically grown cells; enzyme activity is stable for months at $-80°$.

This preparation contains a mixture of core RNA polymerase, $E\sigma^{70}$, and holoenzymes containing alternate sigma factors. We were initially surprised at the relative abundance of several potential sigma factors in these *R. sphaeroides* RNA polymerase samples,[4,12] as comparable preparations from *E. coli* do not contain as many of these proteins.[16] However, the polypeptide complexity of these preparations probably reflects the presence of 16 alternate sigma factors in *R. sphaeroides*,[2] some of which are present in reasonably high abundance.[4]

III. Purification of Recombinant *R. sphaeroides* Sigma Factors

Sigma factors are commonly obtained by overexpression in *E. coli*. The following procedures describe purification of these recombinant proteins from either the soluble fraction or from insoluble inclusion bodies.

A. Purification of Recombinant R. sphaeroides *RpoD*

When RpoD is overexpressed in *E. coli*, it forms insoluble aggregates. To obtain RpoD, these aggregates are isolated and the protein is refolded and purified by anion-exchange chromatography.

One liter of *E. coli* M15pREP4 (Qiagen, Valencia, CA) cells containing a plasmid-encoded N-terminal His$_6$-tagged *R. sphaeroides* RpoD protein is treated with 1 mM isopropylthio-β-D-galactoide (IPTG) for 3 h to induce sigma factor expression. Cells (~2 g wet weight) are centrifuged at

8000 g for 15 min, and the resulting pellet is resuspended in 10 ml phosphate-buffered saline (PBS, pH 7.4; 140 mM NaCl; 2.7 mM KCl; 10 mM Na$_2$HPO$_4$; and 1.8 mM KH$_2$PO$_4$) containing 100 μg/ml lysozyme. Following incubation on ice for 30 min, the cells are lysed by sonicating five times for 1 min using a 50% duty cycle. The sample is centrifuged at 14,000 g for 20 min, and the supernatant is discarded. The insoluble material is washed with 5 ml PBS, centrifuged at 14,000 g for 20 min, and resuspended in 10 ml of PBS containing 6 M guanidine–HCl. The sample is dialyzed (Spectra/Por2, Spectrum Laboratories Inc., Rancho Dominguez, CA) into refolding buffer (50 mM Tris–HCl, pH 8.0; 0.5 mM EDTA; 5% glycerol; 50 mM NaCl; and 0.1 mM DTT) for 24 h with stirring, the sample is removed and centrifuged at 15,000 g for 15 min to remove any insoluble protein. The supernatant is applied to a Mono Q anion-exchange column (Amersham Pharmacia) preequilibrated with TGED and 50 mM NaCl at a flow rate of 1 ml/min. RpoD is eluted over 30 min with a linear gradient of 50–500 mM NaCl (RpoD elutes at ~320 mM NaCl). Peak fractions are analyzed by SDS–PAGE and those containing RpoD are pooled for dialysis against 1 liter of storage buffer (50 mM Tris–HCl, pH 7.9; 0.5 mM EDTA; 0.1 mM DTT; 50% glycerol; and 100 mM NaCl) for 12 h and then stored at $-80°$. We routinely obtain ~1 mg of His$_6$-RpoD per liter of cells.

B. Purification of Recombinant R. sphaeroides RpoH$_I$ and RpoH$_{II}$

Each of these heat shock sigma factors is purified as recombinant N-terminal His$_6$-tagged proteins using pET-15b plasmids from *E. coli* BL21 (DE3) (Novagen, Madison, WI). The purification procedures vary based on the behavior of the recombinant proteins; His$_6$-RpoH$_I$ is soluble, whereas most of the His$_6$-RpoH$_{II}$ is insoluble when overexpressed.[5]

The expression of His$_6$-RpoH$_I$ is induced by treating a 100-ml culture (OD$_{600}$ ~0.4 to 0.6) grown at 37° with 1 mM IPTG for 3 h. The cells are centrifuged (5000 g for 10 min), resuspended in 4 ml binding buffer (50 mM NaH$_2$PO$_4$, pH 8.0; 300 mM NaCl; and 10 mM imidazole) containing 0.5 mg/ml lysozyme, incubated on ice for 30 min, and lysed by sonication at 50% duty cycle until no longer viscous (~six 20-s pulses). The lysate is centrifuged at 10,000 g for 20 min and the supernatant is filtered through a 0.45-μm membrane. The supernatant is mixed for 1 h on a rotary shaker with 2 ml Ni^{2+}-NTA resin (Qiagen), that was equilibrated in binding buffer. This slurry is poured into a 10-ml disposable column and washed with 12 ml of wash buffer (50 mM NaH$_2$PO$_4$, pH 8.0; 300 mM NaCl; and 20 mM imidazole). His$_6$-RpoH$_I$ is eluted with four 1-ml samples of elution buffer (50 mM NaH$_2$PO$_4$, pH 8.0; 300 mM NaCl; and 250 mM imidazole).

Eluates containing His$_6$-RpoH$_I$ are pooled and dialyzed against storage buffer (50 mM Tris–HCl, pH 7.9; 0.5 mM EDTA; and 50 mM NaCl containing 50% glycerol). These samples are stable for months if stored at $-80°$. We typically obtain 5–10 mg of recombinant His$_6$-RpoH$_I$ from 100 ml of *E. coli* cells.

Active His$_6$-RpoH$_{II}$ is obtained from the soluble fraction of cells that are grown at 20°. A 100-ml culture is grown to an OD$_{600}$ of ~0.4 at 20° and treated with 1 mM IPTG for 5–6 h to induce sigma factor expression. The cells are centrifuged at 5000 g for 10 min, resuspended in 5 ml binding buffer (as described earlier) containing 0.1% Triton X-100 and 1 mg/ml lysozyme, incubated on ice for 30 min, sonicated for seven 20-s pulses at 50% duty cycle, and centrifuged at 10,000 g for 30 min. The supernatant is added to 1 ml Ni^{2+}-NTA resin (Qiagen) equilibrated in binding buffer, mixed on a rotary mixer for 1 h, and loaded into a disposable 10-ml column. The column is washed with 8 ml of wash buffer (as described earlier) containing 0.1% Triton X-100 prior to eluting the His$_6$-RpoH$_{II}$ with four 0.5-ml volumes of elution buffer (as described earlier) containing 0.1% Triton X-100. Most of the His$_6$-RpoH$_{II}$ elutes in the second fraction. The eluate is dialyzed against the same storage buffer for His$_6$-RpoH$_I$ containing 40% glycerol. The protein is stable for months when stored at $-80°$. We routinely obtain 0.3–0.5 mg of recombinant His$_6$-RpoH$_{II}$ per 100 ml of *E. coli* cells.

C. Purification of Recombinant R. sphaeroides RpoE

Rhodobacter sphaeroides RpoE is insoluble when overexpressed in *E. coli*, and attempts to purify any soluble recombinant protein have proved unsuccessful.[7,8] However, denatured RpoE can be purified and refolded into active protein.

One liter of *E. coli* M15pREP4 (Qiagen) cells containing a pQE plasmid-encoded N-terminal His$_6$-tagged *R. sphaeroides* RpoE protein is treated with 1 mM IPTG for 3 h to induce sigma factor expression. Cells are centrifuged at 8000 g for 15 min, and the pellets (~2 g wet weight) are resuspended in 10 ml PBS containing 100 μg/ml lysozyme. The sample is then incubated on ice for 30 min. Following lysis by sonication for five 1-min intervals at 50% duty cycle, the sample is centrifuged at 14,000 g for 20 min. The pellet is washed with 5 ml PBS and centrifuged at 14,000 g for 20 min. The resulting pellet is dissolved in 10 ml of PBS containing 8 M urea and mixed with ~2 ml Ni^{2+}-NTA agarose (Qiagen) equilibrated in PBS containing 8M urea. The slurry is incubated at 4° for 1 h on a rotary shaker and then poured into a 5-ml gravity flow column. The column is washed with five column volumes of PBS containing 8 M urea and 10 mM

imidazole before eluting RpoE with five 1-ml aliquots of PBS containing 8 M urea and 300 mM imidazole. Samples containing RpoE are pooled and diluted with FoldIt buffer #16 (55 mM Tris–HCl, pH 8.2; 264 mM NaCl; 11 mM KCl; 0.055% PEG 3350; 550 mM guanidine–HCl; 2.2 mM MgCl$_2$; 2.2 mM CaCl$_2$; 440 mM sucrose; and 550 mM L-arginine; Hampton Research, Laguna Niguel, CA) to a protein concentration of 0.4–0.5 mg/ml and incubated on a rotary shaker at 4° for 12 h. The sample is then centrifuged at 15,000 g for 20 min to remove precipitate, and the supernatant is dialyzed against 50 mM Tris–HCl, pH 7.9; 0.5 mM EDTA; 0.1 mM DTT; 50% glycerol; and 100 mM NaCl for 12 h prior to storage at −80°. We typically obtain 1.5–3.0 mg of His$_6$-RpoE per liter of cells.

D. Purification of R. sphaeroides RpoE:ChrR Complex

A soluble RpoE–ChrR complex is obtained by affinity chromatography from cells containing an intact *rpoEchrR* operon behind an inducible promoter.[8]

One liter of *E. coli* M15pREP4 (Qiagen) cells containing a pQE plasmid-encoded N-terminal His$_6$-tagged *R. sphaeroides* RpoE gene cloned immediately upstream of *chrR* is treated with 1 mM IPTG for 3 h. Cells are centrifuged at 8000 g for 15 min at 4°, and the pellets (~2 g wet weight) are resuspended in 10 ml PBS containing 100 μg/ml lysozyme and incubated on ice for 30 min. Following lysis by sonication for five 1-min intervals at 50% duty cycle, cells are centrifuged at 14,000 g for 20 min. The supernatant is applied to ~2 ml of Ni^{2+}-NTA agarose (Qiagen) preequilibrated with PBS. The slurry is incubated for 1 h on a rotary shaker and is then poured into a 5-ml gravity flow column. The column is washed with 5 column volumes of PBS buffer containing 10 mM imidazole, and the RpoE:ChrR complex is eluted with five 1-ml aliquots of PBS containing 300 mM imidazole. Eluates containing the RpoE:ChrR complex are pooled and dialyzed against 1 liter of 50 mM Tris–HCl, pH 7.9; 0.5 mM EDTA; 0.1 mM DTT; 50% glycerol; and 100 mM NaCl prior to storage at −80°. We typically obtain 10–12 mg of the His$_6$-σ^E–ChrR complex per liter of cells.

E. Purification of MBP–ChrR

When overexpressing ChrR, the majority of the protein is insoluble and inactive.[7] By placing a maltose-binding protein (MBP) domain on the N terminus of ChrR, the majority of the protein remains soluble when overexpressed and an active protein can be purified by affinity chromatography.[8] While the MBP protein can be removed by thrombin cleavage after purification, this protease also cleaves ChrR, resulting in a truncated protein that is not active *in vitro*.[8]

One liter of *E. coli* cells containing a plasmid-encoded MBP-tagged (New England Biolabs, Beverly, MA) *chrR* gene is treated with 1 mM IPTG for 3 h to induce MBP–ChrR expression. Cells are centrifuged at 8000 g for 15 min at 4°, and the pellets (~2 g wet weight) are resuspended in 10 ml PBS (Section III,A) containing 100 μg/ml lysozyme. The sample is incubated on ice for 30 min. Following lysis by sonication for five 1-min intervals at 50% duty cycle, cells are centrifuged at 14,000 g for 20 min at 4°, and the resulting supernatant is mixed with 40 ml of PBS containing 0.15 M KCl and fresh 2 mM DTT. This sample is mixed with ~4 ml of amylose resin (New England Biolabs) that is equilibrated with PBS containing 0.15 M NaCl and 2 mM DTT. The slurry is incubated for 1 h at 4° on a rotary shaker and poured into a 5-ml gravity flow column. The column is washed with 5 column volumes of PBS buffer with 0.15 M KCl and 2 mM DTT, and MBP–ChrR is eluted with seven 1-ml aliquots of PBS buffer with 0.15 M KCl, 2 mM DTT, and 14 mM maltose. Eluates containing MBP–ChrR are pooled, immediately concentrated ~2-fold using a Centricon YM10 membrane (Millipore Corporation), and dialyzed against 1 liter of 25 mM Tris–HCl, pH 7.5; 300 mM KCl; 50 % glycerol; and 2 mM DTT at 4° for 12 h. The protein remains active as an antisigma factor if it is stored at −80°. We typically obtain 25–50 mg of MBP–ChrR per liter of cells.

IV. Transcription Templates

We routinely use circular plasmids that contain a test promoter cloned upstream of transcription terminators as templates (Fig. 1) because they give rise to high levels of promoter-specific transcripts with several *R. sphaeroides* RNA polymerase holoenzymes.[4,5,7,8,10–12,20] The use of circular transcription templates also prevents complications associated with the production of transcripts that arise from initiation by RNA polymerase at the ends of linear DNA fragments. Because the plasmid templates also contain the σ^{70}-dependent *oriV* promoter,[21] the abundance of RNA-1 serves as an internal control when testing the function of an activator or when using a reconstituted holoenzyme in which a specific sigma factor is added to core RNA polymerase. Two similarly sized transcripts are often observed due to termination at adjacent sites by *R. sphaeroides* RNA polymerase *in vitro* (Fig. 1).[4,5,7,8,10–12,20]

V. Assay Conditions

Conditions used for *in vitro* transcription assays with *R. sphaeroides* proteins and promoters are similar to those used with other bacterial systems. However, the assay conditions are modified slightly depending

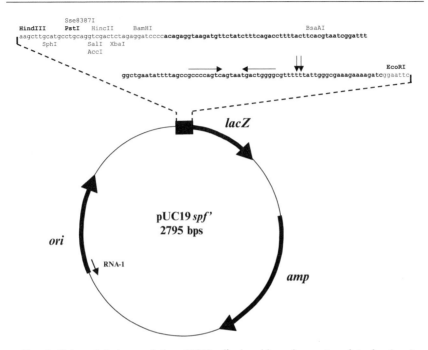

```
                Sse8387I
HindIII         PstI    HincII   BamHI                                      BsaAI
 aagcttgcatgcctgcaggtcgactctagaggatcccc acagaggtaagatgttctatctttcagacctttacttcacgtaatcggattt
    SphI            SalI  XbaI
                    AccI


                                      ─────→   ←─────    ⇊
                                                                                EcoRI
             ggctgaatattttagccgccccagtcagtaatgactggggcgtttttattgggcgaaagaaaagatcggaattc
```

FIG. 1. Relevant features of the pUC19spf' plasmid used as a template for *in vitro* transcription reactions.[18] The ampicillin resistance gene (*amp*), the ColE1 origin of DNA replication (*ori*), the site of a region containing an $E\sigma^{70}$-dependent promoter that encodes the RNA primer used to initiate plasmid replication (RNA-1), and the region of the *spf* gene (sequences indicated in bold) that was cloned within multiple cloning site in *lacZ*α of pUC19 are shown. The region of the *spf* gene includes the Spot 42 transcription terminator (indicated by the inverted arrows). When test promoters are cloned upstream of this region (using the indicated unique restriction sites), transcripts should terminate 103 or 104 nucleotides downstream of the *Bam*HI site, depending on the precise position of termination (vertical arrows). The control RNA-1 transcript can be 102–108 nucleotides long depending on the site of transcription termination.

on the source of *R. sphaeroides* RNA polymerase. When using either a crude (Sections II, A and II, C) or a highly purified (Section II, B) enzyme from *R. sphaeroides*, a final concentration of 25–100 nM RNA polymerase is used in a total volume of 20 μl. When RNA polymerase holoenzymes are reconstituted by adding recombinant sigma factors (Sections III, A–D) to

[20] R. K. Karls, J. R. Wolf, and T. J. Donohue, *Mol. Microbiol.* **34,** 822 (1999).
[21] J. W. Erickson and C. A. Gross, *Genes Dev.* **3,** 1462 (1989).

core RNA polymerase (Sections II, B and II, C), we initially determine the amount of sigma factor that is required to produce a maximal amount of test transcript in the presence of 25–100 nM core enzyme. Depending on the activity of the individual recombinant sigma factor, it is typically present in a ~1- to 10-fold excess over core RNA polymerase.

Summary

This article summarized methods to obtain RNA polymerase and sigma factors that can be used to analyze the *in vitro* control of gene expression by the facultative phototroph *R. sphaeroides*. While not a topic of this article, these purified components also allow one to analyze *R. sphaeroides* promoters that use activators to stimulate transcription.[10,11,20] We expect that these approaches will be increasingly useful as investigators continue to dissect the number of unusual signal transduction pathways that control gene expression in this and other related species.[17, 22–28]

Acknowledgments

The development of these *in vitro* technologies was supported by NIH Grant GM37509 to TJD. The authors thank Jason Hickman and Christine Tavano for their comments on this manuscript.

[22] J. H. Zeilstra-Ryalls, M. Gomelsky, A. A. Yeliseev, J. M. Eraso, and S. Kaplan, *Methods Enzymol.* **297,** 151 (1998).
[23] J. Pemberton, I. Horne, and A. McEwan, *Microbiology* **144,** 267 (1998).
[24] M. Gomelsky and G. Klug, *Trends Biochem. Sci.* **27,** 497 (2002).
[25] T. J. Donohue, *Proc. Natl. Acad. Sci. USA* **94,** 4821 (1997).
[26] C. E. Bauer, S. Elsen, and T. H. Bird, *Annu. Rev. Microbiol.* **53,** 495 (1999).
[27] S. Masuda, C. Dong, D. Swem, A. T. Setterdahl, D. B. Knaff, and C. E. Bauer, *Proc. Natl. Acad. Sci. USA* **99,** 7078 (2002).
[28] S. Masuda and C. Bauer, *Cell* **110,** 613 (2002).

[6] *In Vitro* Transcription Assays Using Components from *Methanothermobacter thermautotrophicus*

By YUNWEI XIE and JOHN N. REEVE

Eukarya, Bacteria, and *Archaea* have diverged to form the three primary domains of life.[1] Although *Archaea* are prokaryotes, archaeal RNA polymerases (RNAP) are most similar to eukaryotic RNA polymerase II,[2,3] but transcribe both protein and stable RNA encoding genes. A functional archaeal RNA polymerase has been reconstituted from 12 individual recombinant subunits,[3] and archaeal RNAP subunits were used as surrogates in determining the RNA polymerase II structure.[4,5] However, the largest subunit of the archaeal RNAP does not contain a heptapeptide repeat-containing C-terminal domain (CTD), consistent with this regulatory domain having evolved later during eukaryotic divergence.[6] In addition to RNAP, transcription initiation in *Archaea* requires at least two general transcription factors, archaeal homologs of the eukaryotic TATA box-binding protein (TBP) and transcription factor TFIIB, the latter known as TFB in *Archaea*.[2] *Archaea* also contain a protein homologous to the N-terminal, zinc ribbon-containing domain of the α subunit of eukaryotic TFIIE, and addition of this protein, designated TFE, stimulates transcription *in vitro* from some but not all archaeal promoters.[7,8] Based on the available genome sequences, *Archaea* do not have homologs of the β subunit of TFIIE, nor of TFIIA, TFIIF, or TFIIH and although some *Archaea* have histones,[9] they do not appear to have their DNA packaged into regular chromatin. There is also no evidence for archaeal homologs of the multisubunit eukaryotic histone modification, chromatin remodeling, and transcription activation complexes. In contrast, the archaeal transcription regulators investigated so far appear to function in a manner analogous to bacterial repressors. They bind to the promoter region,

[1] M. L. Wheelis, O. Kandler, and C. R. Woese, *Proc. Natl. Acad. Sci. USA* **89**, 2930 (1992).
[2] J. Soppa, *Adv. Appl. Microbiol.* **50**, 171 (2001).
[3] F. Werner and R. O. J. Weinzierl, *Mol. Cell* **10**, 635 (2002).
[4] P. Cramer, D. A. Bushnell, J. Fu, A. L. Gnatt, B. Maier-Davis, N. E. Thompson, R. R. Burgess, A. M. Edwards, P. R. David, and R. D. Kornberg, *Science* **288**, 640 (2000).
[5] F. Todone, P. Brick, F. Werner, R. O. J. Weinzierl, and S. Onesti, *Mol. Cell* **8**, 1137 (2001).
[6] J. W. Stiller and B. D. Hall, *Proc. Natl. Acad. Sci USA* **99**, 6091 (2002).
[7] B. L. Hanzelka, T. J. Darcy, and J. N. Reeve, *J. Bacteriol.* **183**, 1813 (2001).
[8] S. D. Bell, A. B. Brinkman, J. van der Oost, and S. P. Jackson, *EMBO Rep.* **2**, 133 (2001).
[9] K. Sandman and J. N. Reeve, *Adv. Appl. Microbiol.* **50**, 75 (2001).

preventing TBP and/or TFB binding to the TATA box/BRE region or they block RNAP access to the site of transcription initiation.[10]

This article describes the purification and assay of RNAP from the thermophilic, anaerobic *Archaeon, Methanothermobacter thermautotrophicus* (formerly *Methanobacterium thermoautotrophicum* strain ΔH^{11}), and its use with recombinant *M. thermautotrophicus* TBP, TFB, and TFE to obtain promoter-dependent transcription *in vitro*.[12,13]

Growth of *Methanothermobacter thermautotrophicus*

Methanothermobacter thermautotrophicus is a thermophilic, obligately anaerobic autotroph.[11] For growth it requires a highly reduced environment, H_2 plus CO_2 supplied at an 8:1 (v/v) ratio, and a buffered salt solution. Sufficient cell mass ($\sim$20 g wet weight) for RNAP purification (see later) can be obtained from one 20-liter culture, inoculated with the cells from a 1- to 2-L culture, or the combined cell mass from several such smaller cultures (see Fig. 1). The *M. thermautotrophicus* salts solution contains (per L H_2O), 4 g $NaHCO_3$, 0.3 g KH_2PO_4, 1 g NH_4Cl, 0.6 g NaCl, 100 mg $MgCl_2 \cdot 6H_2O$, 60 mg $CaCl_2 \cdot 2H_2O$, a 10-ml trace element (TES) solution, 0.5 g cysteine, 0.62 g Na-thiosulfate, and 1 ml resazurin (2 mg/ml). To make 1 liter of TES, 12.8 g nitrilotriacetic acid is dissolved in 200 ml H_2O, the pH is adjusted to 6.5, and the following salts are added and dissolved: 50 mg $AlCl_3 \cdot 6H_2O$, 100 mg $CaCl_2 \cdot 2H_2O$, 100 mg $CoCl_2 \cdot 6H_2O$, 25 mg $CuCl_2 \cdot 2H_2O$, 1.35 g $FeCl_3 \cdot 6H_2O$, 10 mg H_3BO_3, 100 mg $MnCl_2 \cdot 4H_2O$, 1 g NaCl, 24 mg $Na_2\ MoO_4 \cdot 2H_2O$, 26 mg $Na_2SeO\ _4 \cdot 6H_2O$, 120 mg $NiCl_2 \cdot 6H_2O$, and 100 mg $ZnCl_2$. This solution can be sterilized by autoclaving at 121° for 20 min and it is then stable for an extended period if stored at 4°. The growth salts solution is sterilized *in situ* inside a sealed fermentor vessel (2 or 20 liter), allowed to cool to 65°, and then reduced by sparging with an 8:1 mixture of H_2:CO_2 (v/v). The resazurin redox indicator is initially pink but with reduction the medium becomes colorless and can then be inoculated using an $\sim$10% volume inoculum. Provided that strictly anaerobic conditions are maintained, aliquots of fully grown *M. thermautotrophicus* cultures can be stored for at least 3 months at room

[10] S. D. Bell and S. P. Jackson, *Curr. Opin. Microbiol.* **4,** 208 (2001).

[11] A. Wasserfallen, J. Nölling, P. Pfister, J. Reeve, and E. Conway de Macario, *Int. J. System. Evol. Microbiol.* **50,** 43 (2000).

[12] T. J. Darcy, in *"In Vitro* Analysis of Transcription from the Thermophilic Archaeon *Methanobacterium thermautotrophicum* Strain ΔH," Ph.D. dissertation, Ohio State University, 1999.

[13] T. J. Darcy, W. Hausner, D. E. Awery, A. Edwards, M. Thomm, and J. N. Reeve, *J. Bacteriol.* **181,** 4424 (1999).

A    B

FIG. 1. Examples of fermentation equipment used to grow (A) small (1–2 liter) and (B) large ($\sim$ 20 liter) cultures of *M. thermautotrophicus* at 65° under anaerobic conditions using a 8:1 gas mixture of H_2:CO_2. The rate of growth is dependent on the rate of gas dissolution[16] with methanogenesis adding methane to the exhaust gas.

temperature for future use as inocula. The culture is incubated at 65° with continuous H_2:CO_2 sparging at a flow rate of $\sim$200 ml/min and impellor mixing at 600 rpm until the OD_{600} reaches $\sim$1. The cells should be harvested and maintained under anaerobic conditions. The resulting cell paste is transferred into an anaerobic work chamber for immediate use or is frozen rapidly by immersion in liquid N_2 and stored at $-70°$ in an airtight container.

Purification of *M. thermautotrophicus* RNA Polymerase

Stock solutions A (1 M KCl, 10 mM $MgCl_2$, 50 mM Tris–HCl), B (50 mM KCl, 10 mM $MgCl_2$, 50 mM Tris–HCl), and C (10 mM $MgCl_2$, 50 mM Tris–HCl) are adjusted to pH 8, and glycerol [20% (v/v) final concentration] and resazurin (100 μl of 2 mg/ml) are added. After filtration, autoclaving at 121° for 20 min, and cooling, these solutions are transferred into the anaerobic chamber, 174 mg Na-thiosulfate is added per liter to each solution, and the solution colors should change with reduction from blue to pink to colorless.

[16] J. N. Reeve, J. Nölling, R. M. Morgan, and D. R. Smith, *J. Bacteriol.* **179**, 5975 (1997).

Resuspend ~20 g *M. thermautotrophicus* cell paste in 40 ml of solution B, rupture the cells by passage through a French pressure cell at 18,000 psi, and collect the resulting cell lysate into a centrifuge tube preflushed with N_2 that contains 200 μl of cysteine (150 mg/ml) and 200 μl of Na-thiosulfate (186 mg/ml). Seal the tube and centrifuge at 10,000 g for 90 min at 4°. The supernatant is collected, and all subsequent chromatography steps are undertaken at room temperature inside an anaerobic chamber containing an atmosphere of 95% N_2:5% H_2. The supernatant is loaded at 2 ml/min onto a 200-ml bed volume DEAE cellulose column (Whatman, Fairfield, NJ) preequilibrated with solution B, and after washing with 300 ml of solution B, bound protein is eluted using a 50 to 525 mM KCl gradient (800 ml) generated by mixing solutions A and B. The presence of RNAP activity in each fraction (9 ml) collected is determined by assaying [α-^{32}P]UTP incorporation into trichloroacetic acid (TCA)-precipitable material using poly(dA-dT) as the DNA template. To do so, an aliquot (10 μl) of each fraction is mixed with 90 μl of 20 mM Tris–HCl (pH 8), 40 mM KCl, 10 mM MgCl$_2$, 1 mM ATP, 0.1 mM UTP, 0.7 μCi [α-^{32}P]UTP (~3 kCi/mM), and 9 μg poly(dA-dT) (ICN, Costa Mesa, CA) and the mixtures are incubated at 58° for 30 min. The reaction is stopped by the addition of 900 μl ice-cold 5% TCA containing 165 mM NaCl and, after a 5-min incubation on ice, the resulting precipitates are collected by filtration onto glass microfiber filters (934-AH, Whatman). Each filter is washed three times using 10 ml of cold 5% TCA and once with 10 ml of cold 95% ethanol; after drying, the amount of radioactive material bound to the filters is determined by scintillation counting.

Fractions containing RNAP activity are pooled and solution C is added to reduce the KCl concentration to 70 mM. Load this solution (1 ml/min) onto a 20-ml bed volume heparin–Sepharose (Amersham Pharmacia, Piscataway, NJ) column preequilibrated with solution B. Wash with 60 ml of solution B, and elute bound proteins using a 200-ml gradient of 50 mM to 1 M KCl made by mixing solutions A and B. Collect 4-ml fractions and identify the fractions containing RNAP activity as described earlier. Pool these fractions and add solution C to reduce the KCl concentration to 70 mM. Load this solution onto a 1-ml bed volume Mono Q column (Amersham Pharmacia) preequilibrated with solution B, wash with 3 ml of solution B, and elute bound proteins using a 15-ml gradient of 50 mM to 1 M KCl made by mixing solutions A and B. Collect 0.5-ml fractions and assay for RNAP activity. Equilibrate a Hi-Load 16/60 Superdex 200 gel filtration column (120-ml bed volume, 60 cm height; Amersham Pharmacia) for 20 h by pumping a 27:73 mixture of solutions A and B at 0.35 ml/min through the column. Then inject a 1-ml aliquot of the pooled RNAP-containing fractions from the Mono Q column and elute using the

27:73 mixture of solutions A and B. Collect 50 fractions (2 ml) and identify the RNAP-containing fractions. Pool these fractions (usually fractions 30 and 31), add 2 μl 1 M dithiothreitol (DTT) per fraction, aliquot (~1 ml) into 1.5-ml microfuge tubes, freeze in liquid nitrogen, and store at $-70°$ for future use. Such fractions retain full RNAP activity when stored at $-70°$ for at least 1 year. Repeat the Superdex 200 gel filtration column chromatography, each time injecting a 1-ml aliquot of the RNAP-containing pooled material from the Mono Q column until all this material has fractionated.

Preparation of Recombinant *M. thermautotrophicus* TBP, TFB, and TFE

Promoter-specific transcription initiation by *M. thermautotrophicus* RNAP requires the additional presence of at least two general transcription factors, TBP and TFB.[12,13] TFE addition further stimulates transcription *in vitro* from some but not all *M. thermautotrophicus* promoters.[7] Genes MTH1627, MTH0885, and MTH1669 which encode TBP, TFB, and TFE, respectively, have been amplified from the *M. thermautotrophicus* genome[14] and cloned into expression vectors. They direct the synthesis of soluble (his)$_6$-tagged recombinant versions of these transcription factors in *Escherichia coli*, which function *in vitro*. These proteins can be purified easily by standard Ni-NTA affinity-imidazole elution chromatography.[7,12,13] Isopropyl-β-D-thiogalactoside-inducible expression plasmids encoding TBP, TFB, and TFE, designated pTD105, pTD103, and pTrc1669, respectively, are available at O.S.U. on request. These transcription factor preparations should be dialyzed against 50 mM Tris–HCl, pH 8, 300 mM KCl, 10 mM MgCl$_2$, 1 mM DTT, and 20% (v/v) glycerol to remove all traces of imidazole and to substitute KCl for NaCl and their concentrations determined by Bradford assays; they can be stored as aliquots frozen at $-70°$.

Promoter-Specific *in Vitro* Transcription

Promoter-specific transcription initiation is obtained *in vitro* using native *M. thermautotrophicus* RNAP and recombinant *M. thermautotrophicus* TBP, TFB, and TFE, purified as described earlier, from some but not all *M. thermautotrophicus* promoters.[7] The basis for the inactivity *in vitro* of some promoters known to be active *in vivo* is currently under investigation. Templates carrying the promoter for the archaeal histone-encoding gene *hmtB*[15] can be used to establish the *in vitro* system as they consistently direct abundant transcription *in vitro*. *Dde*I digestion of

[14] D. R. Smith *et al.*, *J. Bacteriol.* **179**, 7135 (1997).

plasmid pRT7412[15] DNA (available on request) generates a convenient linear template from which the *hmtB* promoter directs the synthesis of a 193 nucleotide runoff transcript. *In vitro* transcription reaction mixtures (50 μl) containing 20 mM Tris–HCl, pH 8, 120 mM KCl, 10 mM MgCl$_2$, 2 mM DTT, 100 ng template DNA, 50 ng TBP, 300 ng TFB, 5 μl RNAP, 30 μM ATP, 30 μM CTP, 30 μM GTP, 2 μM UTP, and $\sim$2 μCi of [α-^{32}P]UTP (3 kCi/mM) are incubated at 58°. Although the concentrations of the individual reaction components can be varied slightly, this *in vitro* transcription system is particularly sensitive to the KCl concentration. Nonspecific (promoter-independent) transcription increases substantially at KCl concentrations below 100 mM, and all transcription is inhibited above 150 mM KCl.[12] Based on runoff transcript accumulation, the rate of transcription remains linear for at least 30 min at 58°. Transcription is terminated by adding 30 μl of 95% formamide, 20 mM EDTA, 0.05% bromphenol blue, and 0.05% xylene cyanol and placing the reaction mixture at 95° for 3 min. The radioactively labeled transcripts synthesized can then be detected and quantitated by autoradiography or phosphorimaging after their separation by electrophoresis through a denaturing polyacryamide gel. Addition of TFE to the transcription reaction mixture does not stimulate *hmtB* promoter function *in vitro*, but has been shown to result in a one- to three-fold stimulation in transcription *in vitro* from several other *M. thermautotrophicus* promoters.[7] As the *M. thermautotrophicus* genome sequence is available,[14] template DNAs can be generated with any promoter of interest directly by polymerase chain reaction (PCR) amplification from *M. thermautotrophicus* genomic DNA. Such PCR-generated DNA molecules should be purified using a Qiaquick PCR cleanup kit (Qiagen, Valencia, CA) before being used as templates to direct *in vitro* transcription.

Purification and Use of Stalled Transcription Ternary Complexes to Assay Transcription Elongation

Templates have been constructed on which the *hmtB* promoter directs transcription initiation at the start of a 24-bp sequence that can be transcribed in the absence of UTP. Such U-less cassette templates can be used to investigate the consequences of the addition of inhibitors or changes in reaction conditions on transcription elongation without concern for spurious concomitant effects on transcription initiation (Fig. 2A). After transcription of the first 24 nucleotides in the absence of UTP but in the presence of ATP, GTP, and [^{32}P] CTP, a stable ternary complex is generated that contains the template DNA, a ^{32}P-labeled stalled transcript, and

---

[15] R. Tabassum, K. M. Sandman, and J. N. Reeve, *J. Bacteriol.* **174,** 7890 (1992).

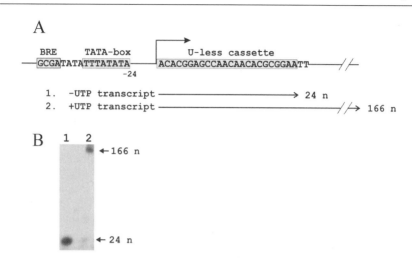

Fig. 2. U-less template and stalled-transcript elongation. (A) A U-less cassette template DNA containing the TFB-responsive element (BRE) and TATA box from the *hmtB* promoter. Transcription initiated at the site indicated by the arrow results in a 24 nucleotide stalled transcript in the absence of UTP and in a 116 nucleotide runoff transcript in the presence of UTP. (B) Ternary complexes that contained the template DNA, *M. thermautotrophicus* RNAP, and ^{32}P-labeled stalled 24 nucleotide transcript were purified by centrifugation through a Sephadex G-50 spin column. An aliquot of the 24 nucleotide transcript was isolated and subjected to electrophoresis in lane 1. The remaining complexes were incubated in a transcription reaction mixture that contained all four unlabeled rNTPs, and an aliquot of the resulting runoff transcripts was subjected to electrophoresis in lane 2. Radioactively labeled transcripts were then visualized by autoradiography.

RNAP. Such complexes can be separated from the transcription reaction mixture by centrifugation through a Sephadex G-50 spin column and then added back to a complete *in vitro* transcription reaction mixture that contains all four unlabeled ribonucleotide triphosphates to obtain stalled transcript elongation. The stalled, ^{32}P-labeled transcripts are extended, resulting in full-length ^{32}P-labeled runoff transcripts, but all newly initiated transcripts are unlabeled. The effects of a potential inhibitor or DNA-binding protein on transcription elongation, as opposed to initiation, are then determined easily by measuring the effects of their addition on the accumulation of ^{32}P-labeled full-length transcripts.

Acknowledgments

Research in the authors' laboratory is supported by grants from the Department of Energy (DE-FGO2-87ER13731) and the National Institutes of Health (GM53185).

[7] Isolation and Characterization of *Streptomyces coelicolor* RNA Polymerase, Its Sigma, and Antisigma Factors

By Mi-Young Hahn, Jae-Bum Bae, Joo-Hong Park, and Jung-Hye Roe

Streptomyces coelicolor is a model organism for gram-positive soil bacteria of high G+C content that undergoes a complex life cycle of mycelial growth and spore formation and produces a variety of antibiotics and other drugs during the differentiation process. Its complete genome sequence information has been reported,[1] revealing the presence of 7825 protein-coding genes in a 8,667,507-bp linear chromosome whose G+C content is 72.1%. The number of protein-coding genes far exceeds that (6023) of yeast *Saccharomyces cerevisiae*, suggesting the richness of new protein families as well as more members of known families. About 12.3% of proteins are predicted to have regulatory functions, and among them RNA polymerase sigma factors constitute one major group. It is predicted that there are 64 genes for the chromosome-encoding sigma factors of the σ^{70} family. Among them, 50 genes encode sigma factors of the extracytoplasmic function (ECF) subfamily.[2] Two additional ECF sigma genes are predicted to reside in plasmid SCP1. Only a handful of sigma factors have been characterized for their function; σ^{HrdB} as the principal sigma factor, σ^F, σ^{WhiG}, σ^U and σ^{BldN} for differentiation, σ^E for cell wall homeostasis, σ^R and possibly σ^T for redox regulation, and σ^H and σ^B for a general stress response.[3–5]

The phylogenetic relationship among these sigma factors is presented in Fig. 1 using the PHYDIT program developed by J. Chun as described later. The non-ECF sigmas include four σ^{HrdB}-type factors, nine σ^B-type factors, and σ^{WhiG}, a sporulation-specific sigma factor. The ECF subfamily includes σ^R, σ^U, σ^T, and σ^{BldN}. The majority of these sigma factors are thought to be controlled by their cognate antisigma factors, whose coding genes in many cases are located immediately downstream or near the sigma factor gene.

---

[1] S. D. Bentley *et al.*, *Nature* **417**, 143 (2002).

[2] M. A. Lonetto, K. L. Brown, K. E. Rudd, and M. J. Buttner, *Proc. Natl. Acad. Sci. USA* **91**, 7573 (1994).

[3] M. S. Paget, H. J. Hong, M. J. Bibb, and M. J. Buttner, *Soc. Gen. Microbiol. Symp.* **61**, 106 (2002).

[4] J. Helmann, *Adv. Micro. Physiol.* **46**, 47 (2002).

[5] Y. H. Cho, E. J. Lee, B. E. Ahn, and J. H. Roe, *Mol. Microbiol.* **42**, 205 (2001).

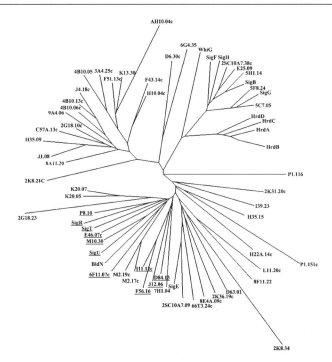

FIG. 1. Phylogenetic relationship of 66 sigma factors in *Strepomyces coelicolor*. The phylogenetic relationship of sigma factors is analyzed by the PHYDIT program as described in the text. Each sigma factor is indicated by a cosmid and gene number as appears in the *S. coelicolor* genome database at the Sanger Institute. Several characterized sigma factors are indicated by their given names; SigB(sig8, SCF55.24), SigE(SCE94.07), SigF(2SCD60.01c), SigG(SC4G10.20c), SigH(sig1, 2SC7G11.05c), SigR(SC7E4.13), SigT(SCH24.14c), SigU(SCE59.13c), and BldN (SCE68.21). Eleven σ^R-like sigma factors whose coding genes are neighbored by open reading frames containing conserved HX_3CX_2C motif are underlined.

One interesting feature in the phylogenetic tree is the clustering of 18 ECF members represented by σ^R. As in the *sigR-rsrA* operon, all the sigma factor genes are neighbored by open reading frames (ORFs), which most likely encode antisigma factors. RsrA, the anti-σ^R factor, contains a motif HX_3CX_2C, which is critical for its function as a redox-sensitive antisigma factor.[6,7] This motif is conserved in the neighboring genes in 11 out of the 18 members of this subgroup and is not found near other sigma factors (Fig. 1).

[6] J. G. Kang, M. S. Paget, Y. J. Seok, M. Y. Hahn, J. B. Bae, J. S. Hahn, C. Kleanthous, M. J. Buttner, and J. H. Roe, *EMBO J.* **18**, 4292 (1999).
[7] M. S. Paget, J. B. Bae, M. Y. Hahn, W. Li, C. Kleanthous, J. H. Roe, and M. J. Buttner, *Mol. Microbiol.* **39**, 1036 (2001).

This article describes methods for preparing *S. coelicolor* holo and core RNA polymerases, isolation of sigma factors, and characterization of sigma and antisigma interaction using the example of σ^R and RsrA.

Phylogenetic Sequence Analysis of Multiple Sigma Factors in *Streptomyces coelicolor* A3(2)

For phylogenetic analysis of sigma factors in *S. coelicolor*, nucleotide sequences of 66 genes for sigma factors (64 from chromosome, 2 from plasmid SCP1) were acquired from the genome database at Sanger Institute (http://www.sanger.ac.uk/Projects/S_coelicolor/) and NCBI databases (http://www.ncbi.nlm.nih.gov/Database/) and were analyzed using the PHYDIT program developed by Dr. Jongsik Chun (http://plaza.snu.ac.kr/~jchun/phydit/). Amino acids converted from the nucleotide sequences are aligned by Clustal X in the PHYDIT analysis menu, and the result is saved as a GDE file, which is then imported into PHYDIT. The phylogenetic relationships are calculated using the Jukes and Cantor distance model. The tree is generated using the neighbor-joining method by the TreeView program in the PHYDIT package. The *Escherichia coli rpoD* gene was used as an outgroup to generate the tree (Fig. 1).

Purification of Core and Holo RNA Polymerases

Preparation of Cell Extracts from Streptomyces coelicolor *Mycelia*

Streptomyces coelicolor A3(2) strain M145 is grown in seed medium (3% dextrose, 1.7% soytone, 0.3% peptone, 0.4% CaCO$_3$) by inoculating with a spore suspension (10^8 spores/100 ml broth) and incubation at 30° for 2 days.[8] The freshly grown seed culture (200 ml) is inoculated into 4 liter YEME medium (1% glucose, 0.5% Bacto-peptone, 0.3% malt extract, 0.3% yeast extract) containing 5 mM MgCl$_2$·6H$_2$O and 10.3% sucrose in a 5-liter fermenter, aerated at a 0.5 air/medium (v/v) ratio, and agitated at 250 rpm at 30°. Mycelia cells are harvested 12 h after inoculation of seed culture by centrifugation at 6000 g for 10 min. Cells are washed twice with a washing buffer [10 mM Tris–HCl, 1 mM EDTA, 150 mM NaCl, 0.1 mM dithiothreitol (DTT), 1 mM phenylmethylsulfonyl fluoride (PMSF) and stored at −70° until use.

---

[8] T. Kieser, M. J. Bibb, M. J. Buttner, K. F. Chater, and D. A. Hopwood, "Practical *Streptomyces* Genetics." The John Innes Foundation, Norwich, 2000.

RNA polymerase is purified from cell pellets according to the procedures developed to purify *E. coli* RNAP with slight modifications.[8–11] All purification steps are carried out at 4°. The protease inhibitor PMSF is added at 1 mM final concentration to all the buffers used throughout enzyme purification. *S. coelicolor* A3(2) cells frozen at −70° are disrupted with an ultrasonicator. Approximately 50 g (wet weight) of cells is resuspended in 100 ml of lysis buffer [20 mM Tris–HCl, pH 7.9, 10% (v/v) glycerol, 5 mM EDTA, 0.1 mM DTT, 10 mM MgCl$_2$, 1 mM PMSF, 0.15 M NaCl]. The cell suspension is treated with 100 mg of lysozyme for 30 min on ice and is disrupted by sonication with a flat tip (60-W, 10-mm-diameter probe; Ultrasonics Ltd.) at 40% of the amplitude until the viscosity of the lysate is reduced greatly. The homogenates are clarified by centrifugation at 16,000 g for 30 min.

Polyethyleneimine (PEI) Fractionation

To fractionate nucleic acid-binding proteins, a 5% (v/v) solution of PEI (pH 7.9, polymin P, Sigma) is added slowly with thorough stirring to the final concentration of approximately 0.3%. Following 5 min of continuous stirring, the mixture is centrifuged at 6000 g for 5 min and the supernatant is discarded. The drained pellet, containing nucleic acids and their binding proteins, is crushed with a glass homogenizer and is resuspended in 200 ml of TGED buffer (10 mM Tris–HCl, pH 7.9 at 4°, 0.1 mM EDTA, 0.1 mM DTT, 10% glycerol) containing 0.5 M NaCl. The buffer removes extractable protein while the RNA polymerase still remains in the pellet. The suspension is centrifuged at 6000 g for 5 min and the supernatant is discarded. The washed pellet is again crushed and resuspended in 100 ml of TGED buffer containing 0.9–1 M NaCl to release the bound RNA polymerase. The suspension is centrifuged at 6000 g for 10 min and the supernatant is collected (0.9–1 M NaCl eluate).

In order to remove residual PEI and concentrate RNA polymerase before loading onto a column, 200 ml of 100% saturated ammonium sulfate solution, pH 7.9, is added to 100 ml of the 0.9–1 M NaCl eluate with stirring. Following 30 min of stirring, the mixture is centrifuged for 30 min at 16,000 g and the drained pellet is dissolved in 10 ml of TED (TGED minus glycerol). The dissolved sample is centrifuged for 10 min at 16,000 g.

[9] D. A. Hager, D. J. Jin, and R. R. Burgess, *Biochemistry* **29,** 7890 (1990).
[10] N. Fujita, T. Nomura, and A. Ishihama, *J. Biol. Chem.* **262,** 1855 (1987).
[11] M. J. Buttner and N. L. Brown, *J. Mol. Biol.* **185,** 177 (1985).

Heparin–Sepharose Column

The supernatant is then applied to a 40-ml heparin–Sepharose column (Pharmacia) preequilibrated with TGED buffer and washed with 50 ml of TGED buffer containing 0.2 M NaCl. The proteins are eluted with a 100-ml linear salt gradient of 0.2 to 0.7 M NaCl in TGED buffer at a flow rate of 0.5 ml/min. Fractions (3 ml) are collected into tubes, and a small amount of each fraction is analyzed by SDS–PAGE as well as by an *in vitro* transcription assay (see later). The $E\sigma^{HrdB}$ activity is eluted at approximately 0.6 M NaCl, whereas $E\sigma^R$ activity is eluted at a slightly higher salt concentration. Following SDS–PAGE, fractions containing α, β, and β' subunits are pooled and concentrated by ammonium sulfate precipitation to a final 65% saturation. The precipitate is dissolved in TED and dialyzed against storage buffer (10 mM Tris–HCl, pH 7.9 at 4°, 10 mM MgCl$_2$, 0.1 M KCl, 0.1 mM EDTA, 1 mM DTT, 50% glycerol) for long-term storage. This sample containing holo RNA polymerase with various associated sigma factors is used later to purify various sigma factors by elution from the gel.

Superdex 200 Column

To further purify RNA polymerase, 5–10 mg of the ammonium sulfate precipitate is dissolved in 500 μl of TED buffer and centrifuged at 16,000 g for 10 min. The supernatant is loaded onto a 120-ml Hi-Load 16/60 Superdex 200 prep grade column (Pharmacia) and eluted with TG$_5$ED (TED containing 5% glycerol) buffer containing 0.3 M NaCl at a flow rate of 0.5 ml/min. Fractions (0.5 ml) are collected and monitored by an ultraviolet (UV) monitor. The RNA polymerase peak is eluted at about one-third column volume. The RNA polymerase fractions eluted from the Superdex 200 column are applied to a Mono Q column.

Mono Q Column

The core enzyme can be obtained from highly purified RNA polymerase preparations following the Mono Q column.[12] The diluted sample is loaded on to the 1-ml Mono Q HR 5/5 FPLC anion-exchange column (Pharmacia) preequilibrated with 5 ml of TG$_5$ED buffer containing 0.2 M NaCl. The column is washed with 5 ml of TG$_5$ED containing 0.2 M NaCl and is then eluted with a 10-ml shallow linear gradient of 0.3–0.6 M NaCl in TG$_5$ED at a flow rate of 0.5 ml/min. Fractions (0.5 ml) are collected and used in the *in vitro* transcription activity. The holoenzyme that can transcribe *hrdD* or *rrnD* promoters is eluted at 0.4 M NaCl. The trailing

---

[12] M. J. Buttner, A. M. Smith, and M. J. Bibb, *Cell* **52,** 599 (1988).

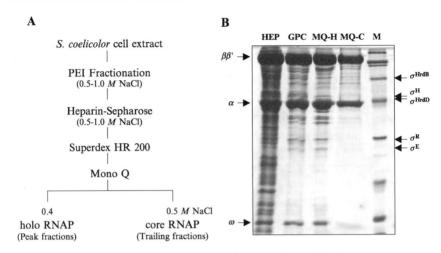

FIG. 2. Purification steps of RNA polymerase from *S. coelicolor*. (A) Schematic procedure for the purification of *S. coelicolor* RNA polymerase. (B) SDS–PAGE profile of eluates from Heparin-Sepharose 6B (lane HEP), Superdex 200 (lane GPC), and Mono Q columns (MQ-H and MQ-C). Peak holoenzyme activity is eluted from the Mono Q column at 0.4 *M* NaCl (lane MQ-H) and is followed by core RNAP fractions (lane MQ-C). Eluates are analyzed on a 0.1% SDS–13% polyacrylamide gel and stained with Coomassie brilliant blue R-250. Positions of RNA polymerase subunits and several associated sigma factors are indicated along with molecular mass markers (lane M; 14.4, 21.5, 31.0, 45.0, 66.2, 97.4, 116.3, and 200 kDa). (See color insert.)

fractions eluted at approximately 0.5 *M* NaCl have mainly nonspecific RNA-synthesizing activity with little promoter-specific activity. These fractions are devoid of most of the associated sigma as well as omega proteins and are used as the core enzyme in the *in vitro* transcription assay.

The purification procedure and the protein band patterns at several selected steps are summarized and shown in Fig. 2.

Identification of Associated Sigma Factors and ω Subunit from Holoenzyme

Reconstitution Assay

About 1 mg of RNA polymerase partially purified through a heparin–Sepharose column is subjected to preparative SDS–PAGE. The gel is stained with 0.25 *M* KCl and 1 m*M* DTT and proteins are eluted by the procedure of Hager and Burgess.[13] Gel slices (2–8 mm) are cut out and

---

[13] D. A. Hager and R. R. Burgess, *Anal Biochem.* **109,** 76 (1980).

placed in dialysis bags containing 300 μl of 2:5 diluted 0.1% (w/v) SDS-running buffer (25 mM Tris–HCl, pH 7.8, 250 mM glycine) and 3 μl of 10 mg/ml bovine serum albumin (BSA). Electroelution from the gel is carried out at 30 mA for 3 h at 4°. The eluates are recovered and precipitated by adding 1.2 ml of cold acetone. Renaturation of the eluted proteins is performed in the presence of *E. coli* GroEL as described.[14] The acetone precipitate is dissolved thoroughly in 20 μl of 6 M guanidine–HCl buffer (20 mM Tris–HCl, pH 7.8, 150 mM NaCl, 5 mM DTT, 0.1 mM EDTA, 6 M guanidine–HCl) and incubated at room temperature for 20 min. The resuspended pellet is then diluted into 1 ml of GroEL incubation buffer (50 mM Tris–HCl, pH 7.8, 12 mM MgCl$_2$, 9 μg/ml GroEL) and incubated at 22–25° for 2 h. It is then dialyzed against renaturation buffer [20 mM Tris–HCl, pH 7.8 at 4°, 10 mM MgCl$_2$, 10 mM KCl, 0.1 mM EDTA, 50% (v/v) glycerol] for 12–16 h with one change of buffer. Five microliters of the renatured protein is added to approximately 1 pmol of core RNA polymerase and the mixture is incubated on ice for 10 min. After the addition of 0.15 pmol of DNA template, the mixture is incubated at 30° for 30 min and subjected to the *in vitro* transcription assay as described later.[15]

N-Terminal Sequencing

Mono Q fractions exhibiting high transcription activity are pooled, electrophoresed on a SDS gel, and electroblotted onto a polyvinylidene difluoride (PVDF) membrane (pore size 0.1 μm, Millipore) in CAPS buffer [10 mM 3-(cyclohexylamino)-1-propanesulfonic acid, 10% methanol, pH 11.0]. The region in the PVDF membrane corresponding to associated protein bands such as σ^{31} (σ^R), σ^{28} (σ^E), and the 13-kDa band (ω) are cut out and the N-terminal amino acid sequences are analyzed by Edman degradation using systems such as the Procise protein sequencing system (Applied Biosystems).

Functional Analysis of RNA Polymerase Activity and Sigma/Antisigma Interaction

In Vitro *Transcription Assay*

The *in vitro* transcription assay can be used to monitor the activity of RNA polymerase, uncharacterized putative sigma, or antisigma factors. To assay sigma activity, holoenzyme reconstitution is performed as described earlier. The gel-eluted putative sigma factor (1.5–3 pmol) is

[14] K. L. Brown, S. Wood, and M. J. Buttner, *Mol. Microbiol.* **6**, 1133 (1992).

[15] J. G. Kang, M. Y. Hahn, A. Ishihama, and J. H. Roe, *Nucleic Acids Res.* **25**, 1566 (1997).

incubated with the RNA polymerase core enzyme (1.5 pmol Mono Q eluate) at 30° for 15 min. Then the mixture is subjected to an *in vitro* transcription assay containing various test promoter DNAs.[14] An example of assaying the antisigma activity of RsrA by *in vitro* transcription is given below.

σ^R and RsrA proteins are overproduced in *E. coli* BL 21 (DE3) pLysS using the pET system (Novagen) and purified through several columns.[6,16] The *in vitro* runoff transcription assay is performed using the combined conditions of Fujita *et al.*[10] and Buttner *et al.*[12] The core RNA polymerase (1.5 pmol) and σ^R (1.5–3 pmol) are incubated at 30° for 5 min in 16 μl of transcription buffer [40 mM Tris–HCl, pH 7.9, 10 mM MgCl$_2$, 0.6 mM EDTA, 0.4 mM potassium phosphate, 0.25 mg/ml BSA, 20% (v/v) glycerol] with 0.15 pmol of template DNA containing the *sigRp2* promoter. To examine the effect of RsrA, various amounts of RsrA are added in the transcription mixture. In a control experiment, we replace σ^R with σ^{HrdB}, the principal sigma factor of *S. coelicolor*, and substitute *sigRp2* with a DNA template containing the σ^{HrdB}-dependent promoter *rrnDp2*. In order to examine the effect of the redox condition, DTT is added at different concentrations from 0.1 to 5 mM. RNA synthesis is initiated by the addition of 3.5 μl substrate mixture containing 2 μCi of [α-^{32}P]CTP (400 Ci/mmol) and 0.4 mM each of UTP, ATP, and GTP. Heparin (3 μl) is then added to a 0.1-mg/ml final concentration after 2 min to prevent reinitiation and the incubation is continued for 5 min before adding 2.5 μl of cold CTP (1.2 mM final concentration). After 10 min of incubation, the reaction is terminated by adding 50 μl of stop solution (375 mM sodium acetate, pH 5.2, 15 mM EDTA, 0.15% SDS, 0.1 mg/ml calf thymus DNA). Transcripts are precipitated with ethanol, resuspended in formamide sample buffer [80% (v/v) formamide, 8% glycerol, 0.1% SDS, 8 mM EDTA, 0.01% bromphenol blue, 0.01% xylene cyanol] and electrophoresed on a 5% polyacrylamide gel containing 7 M urea. An example of the result is presented in Fig. 3.

Sigma and Antisigma Interaction Analyzed by Native PAGE

The interaction of sigma and antisigma factors can be monitored on a native polyacrylamide gel electrophoresis. The complex formed between sigma and antisigma factors can be detected as a new band. An example with the σ^R–RsrA interaction is presented (Fig. 4). Purified σ^R (4 μM) and RsrA (2–8 μM) are incubated in 25 μl of N$_2$-purged binding buffer [40 mM Tris–HCl, pH 8.0, 10 mM MgCl$_2$, 0.01 mM EDTA, 20% (v/v)

[16] M. S. Paget, J. K. Kang, J. H. Roe, and M. J. Buttner, *EMBO J.* **17,** 5776 (1998).

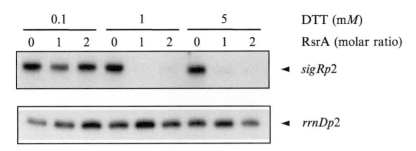

FIG. 3. Redox-dependent changes in antitranscriptional activity of RsrA. Inhibition of σ^R-directed transcription by purified recombinant RsrA under reducing conditions. Purified RsrA (100 or 200 nM final concentration) was added to the transcription buffer containing σ^R or σ^{HrdB} (100 nM) and DNA templates (5 nM) containing sigRp2 (for σ^R) or rrnDp2 (for σ^{HrdB}). The buffer contained 0.1, 1, or 5 mM DTT. Data are taken from Kang et al.[6]

glycerol] in the presence of DTT or H_2O_2 at 30° for 30 min. Reduced RsrA can be prepared by DTT treatment and subsequent dialysis in N_2-purged binding buffer to remove DTT. It is then incubated with σ^R in the binding buffer without DTT or with diamide (1 mM), a thiol oxidant. Samples are separated by electrophoresis on a native 10% polyacrylamide gel at 15 V for 16 h. The acrylamide gel solution is degassed extensively before gel casting and the electrophoresis buffer is saturated with N_2 gas. Proteins are visualized by Coomassie blue staining.

Sigma and Antisigma Interaction Analyzed by Surface Plasmon Resonance

The direct interaction of σ^R and RsrA can be analyzed by surface plasmon resonance (SPR) using a BIAcore optical biosensor (Pharmacia Biosensor AB) as described by Seok et al.[17] RsrA is immobilized to the carboxymethylated dextran surface of a CM5 Biacore sensor chip by amine coupling according to the manufacturer's instructions. The RsrA solution (10 µg/ml) in 40 µl coupling buffer (10 mM sodium acetate, pH 4.0) is allowed to flow over a sensor chip at 5 µl/min to couple the protein to the matrix by a NHS/EDC (a mixture of 0.1 M N-hydroxysuccinimide and 0.1 M 1-ethyl-3[(3-dimethylamino)propyl]carbodiimide) reaction. Unreacted NHS is inactivated by injecting 40 µl of 1 M ethanolamine–HCl, pH 8.0. A blank surface is prepared to examine nonspecific protein interactions with the carboxymethylated dextran surface, if any, by activation and inactivation of the sensor chip without any protein immobilization.

[17] Y. J. Seok, M. Sondej, P. Badawi, M. S. Lewis, M. C. Briggs, H. Jaffe, and A. Peterkofsky, *J. Biol. Chem.* **272,** 26511 (1997).

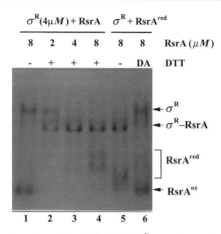

FIG. 4. Redox-dependent formation of the RsrA–σ^R complex monitored by native PAGE. σ^R (4 μM) and RsrA (2–8 μM, lanes 2–4) were incubated in 25 μl of N_2-saturated binding buffer. In a control reaction, DTT was omitted from the binding mixture (lane 1). Reduced RsrA (RsrA^{red}, 8 μM) was incubated with σ^R (4 μM) in the binding buffer without DTT (lane 5) or with added diamide (DA; 1 mM) (lane 6). Samples were separated by electrophoresis on a native 10% polyacrylamide gel and visualized by Coomassie blue staining. Positions of σ^R, the reduced and oxidized forms of RsrA (RsrA^{red}, RsrA^{ox}), and the σ^R–RsrA complex are indicated. Data are taken from Kang et al.[6] (See color insert.)

Assuming that 1000 resonance units (RU) correspond to a surface concentration of 1 ng/mm^2, RsrA is immobilized to a surface concentration of 2.5 ng/mm^2. HBS (10 mM HEPES, pH 7.2, with 150 mM NaCl) is used as a standard running buffer and is introduced at a flow rate of 10 μl/min. The sensor surface is regenerated between assays by sequential injections of 5 μl of 5 mM EDTA and 10 μl of 0.01% SDS to remove bound analyte. Different concentrations of purified σ^R solutions with or without DTT (1 mM) are injected to flow over immobilized RsrA on the sensor chip. No change in RU is observed in the absence of DTT, indicating no interaction in oxidized condition. Kinetic parameters for the interaction of σ^R with immobilized RsrA can be determined using the BIA-evaluation 2.1 software.

Acknowledgments

The authors are grateful to M. Buttner and J. Chun for helpful discussions and Y. Cho for help in drawing Fig. 1. This work was supported by a grant (2000-2-20200-001-1) from Korea Science and Engineering Foundation to J. H. Roe, M. Y. Hahn and J. B. Bae were recipients of BK21 fellowship from the Ministry of Education and Human Resources.

[8] Bacteriophage N4-Coded, Virion-Encapsulated DNA-Dependent RNA Polymerase

By ELENA K. DAVYDOVA, KRYSTYNA M. KAZMIERCZAK, and LUCIA B. ROTHMAN-DENES

Coliphage N4 utilizes the sequential activity of three different DNA-dependent RNA polymerases during its life cycle.[1] Unlike other DNA phages, which rely on the RNA polymerase of the host cell to transcribe their early genes,[2] N4 employs a phage-encoded enzyme designated virion RNA polymerase (vRNAP).[3] vRNAP is packaged into virion particles during morphogenesis and is injected into the host cell with a copy of the phage's 70.6-kb double-stranded linear DNA genome at the onset of infection. Products of early gene transcription include a heterodimeric DNA-dependent RNA polymerase (N4 RNAPII) and at least two accessory factors that together comprise the middle transcription machinery.[4–6] Middle genes encode, among others, proteins required for N4 DNA replication.[7] One of these proteins, the N4 single-stranded DNA-binding protein (N4 SSB), is also required for the transcription of N4 late genes, which is carried out by the *Escherichea coli* σ^{70}-RNA polymerase.[8–10]

N4 vRNAP purified from virions has an apparent molecular weight of 320,000.[11] In contrast to all other DNA-dependent RNA polymerases, vRNAP transcribes single-stranded promoter-containing templates with *in vivo* specificity; all determinants of promoter recognition are present on the template strand.[12,13] vRNAP promoters are composed of a 5-bp

[1] R. Zivin, W. Zehring, and L. B. Rothman-Denes, *J. Mol. Biol.* **152,** 335 (1981).

[2] D. Rabussay and E. P. Geiduschek, *in* "Comprehensive Virology" (H. Fraenkel-Conrat and R. R. Wagner, eds.), p. 1. Plenum, New York, 1977.

[3] S. C. Falco, K. VanderLaan, and L. B. Rothman-Denes, *Proc. Nat. Acad. Sci. USA* **74,** 520 (1977).

[4] S. C. Falco and L. B. Rothman-Denes, *Virology* **95,** 454 (1979).

[5] W. A. Zehring and L. B. Rothman-Denes, *J. Biol. Chem.* **258,** 8074 (1983).

[6] W. A. Zehring, S. C. Falco, C. Malone, and L. B. Rothman-Denes, *Virology* **126,** 678 (1983).

[7] D. Guinta, J. Stambouly, S. C. Falco, J. K. Rist, and L. B. Rothman-Denes, *Virology* **150,** 33 (1986).

[8] N.-Y. Cho, M. Choi, and L. B. Rothman-Denes, *J. Mol. Biol.* **246,** 461 (1995).

[9] M. Choi, A. Miller, N.-Y. Cho, and L. B. Rothman-Denes, *J. Biol. Chem.* **270,** 22541 (1995).

[10] A. A. Miller, D. Wood, R. E. Ebright, and L. B. Rothman-Denes, *Science* **275,** 1655 (1997).

[11] S. C. Falco, W. Zehring, and L. B. Rothman-Denes, *J. Biol. Chem.* **255,** 4339 (1980).

[12] L. L. Haynes and L. B. Rothman-Denes, *Cell* **41,** 597 (1985).

[13] M. A. Glucksmann, P. Markiewicz, C. Malone, and L. B. Rothman-Denes, *Cell* **70,** 491 (1992).

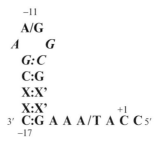

$$
\begin{array}{c}
-11 \\
A/G \\
A \quad\quad G \\
G{:}C \\
C{:}G \\
X{:}X' \\
X{:}X' \quad\quad +1 \\
3'\ \ C{:}G\ A\ A\ A/T\ A\ C\ C\ 5' \\
-17
\end{array}
$$

FIG. 1. N4 vRNAP promoters consensus sequence. +1 indicates site of transcription initiation. Bases in italics determine the unusual stability of the DNA hairpin (29,30). X:X' can be any base pair.

stem, three base loop hairpin structure. A number of positions within the promoter are also conserved at the sequence level (Fig. 1). *In vivo*, the activities of *E. coli* DNA gyrase and *E. coli* single-stranded DNA-binding protein (*Eco* SSB) are required for early transcription.[14] *In vitro*, vRNAP cannot utilize supercoiled, promoter-containing templates unless *Eco* SSB is present.[15,16]

Sequencing of the vRNAP gene revealed an open reading frame encoding a 3500 amino acid protein that contains no cysteine residues.[17] The sequence does not possess statistically significant similarity to members of either of the two families of DNA-dependent RNA polymerases.[17] However, sequence analysis identified matches to four motifs important for polymerase function in T7 RNAP-like enzymes.[18] These include motif B, which forms part of the nucleotide-binding pocket, and motifs A and C, which contain aspartate residues that chelate catalytically essential Mg^{2+} ions.[19] The fourth motif, T/DxxGR, is found in both DNA-dependent DNA polymerases and T7-like RNA polymerases.[20] In RNA polymerases, this motif plays a role in stabilization of the RNA–DNA hybrid.[21]

[14] S. C. Falco, R. Zivin, and L. B. Rothman-Denes, *Proc. Natl. Acad. Sci. USA* **75,** 3220 (1978).
[15] M. A. Glucksmann-Kuis, X. Dai, P. Markiewicz, and L. B. Rothman-Denes, *Cell* **84,** 147 (1996).
[16] P. Markiewicz, C. Malone, J. W. Chase, and L. B. Rothman-Denes, *Genes Dev.* **6,** 2010 (1992).
[17] K. M. Kazmierczak, E. K. Davydova, A. A. Mustaev, and L. B. Rothman-Denes, *EMBO. J.* **21,** 5815 (2002).
[18] G. M. T. Cheetham and T. A. Steitz, *Curr. Opin. Struct. Biol.* **10,** 117 (2000).
[19] P. A. Osumi-Davis, M. C. de Aguilera, R. W. Woody, and A. Y. Woody, *J. Mol. Biol.* **226,** 37 (1992).
[20] C. M. Joyce and T. A. Steitz, *Annu. Rev. Biochem.* **63,** 777 (1994).
[21] D. Imburgio, M. Anikin, and W. T. McAllister, *J. Mol. Biol.* **24,** 37 (2002).

Surprisingly, vRNAP possesses the TxxGR sequence characteristic of DNA polymerases. These four polymerase motifs cluster in a central region of the vRNAP polypeptide.[17] This and the fact that vRNAP is over three times larger than the T7 RNA polymerase suggested that a transcriptionally active domain may exist within vRNAP. To explore this possibility, we performed controlled trypsin proteolysis of vRNAP followed by an autolabeling assay[22,23] to identify peptides retaining catalytic activity.[17] This procedure defined a 1106 amino acid long domain (mini-vRNAP, amino acids 998–2103 of vRNAP) containing the previously mentioned motifs.[17] Mini-vRNAP displays the same initiation, elongation, and termination properties as vRNAP.[17] Amino- and carboxy-terminal deletion analysis indicates that mini-vRNAP is the shortest polypeptide possessing transcriptional activity. Biochemical and genetic analyses indicate that the catalytic core of vRNAP must resemble that of T7 RNAP[17]; however, the amino- and carboxy-terminal regions appear to play different roles than the corresponding regions of T7 RNAP and are presumably involved in promoter recognition.[24] Phylogenetic analysis indicates that N4 mini-vRNAP is a highly evolutionarily diverged member of the single subunit RNAP family.[17]

The vRNAP and mini-vRNAP sequences have been cloned under the control of an arabinose-inducible, glucose-repressible promoter with either amino- or carboxy-terminal hexahistidine tags.[17] Mini-vRNAP was also cloned with a single vector-encoded initiating methionine residue and is referred to as "untagged" to differentiate it from the tagged recombinant polymerases that contain 37–38 amino acids of the vector-encoded sequence. Comparison of these enzymes with N4 vRNAP and mini-vRNAP lacking vector-encoded sequences indicates that the exogenous sequences do not affect enzyme activity.[24]

Two procedures are described for the purification of full-length N4 vRNAP: one dealing with purification of the native enzyme from N4 virions and the other dealing with purification of the recombinant hexahistidine-tagged vRNAP under overreproducing conditions. Two procedures are also described for the purification of recombinant mini-vRNAP: one for the hexaxistidine-tagged enzyme and the second for the "untagged" enzyme.

[22] M. Grachev, T. Kolocheva, E. Lukhtanov, and A. Mustaev, *Eur. J. Biochem.* **163,** 113 (1987).

[23] G. R. Hartmann, C. Biebricher, S. J. Glaser, F. Grosse, M. J. Katzameyer, A. J. Lindner, H. Mosig, H.-P. Nasheuer, L. B. Rothman-Denes, A. R. Schaffner, G. J. Schneider, K.-O. Stetter, and M. Thomm, *Biol. Chem. Hoppe-Seyler* **369,** 775 (1988).

[24] E. K. Davydova, unpublished results.

Polymerase Activity Assay

Transcription is measured in a 5-μl reaction volume containing 10 mM Tris–HCl, pH 7.9, 10 mM MgCl$_2$, 50 mM NaCl, 1 mM dithiothreitol (DTT), 1 mM each ATP, CTP, and GTP, 0.1 mM UTP, 0.5–1 μCi [α-^{32}P]UTP (3000 Ci/mmol, Amersham-Pharmacia), 0.05–2 μM DNA template, and 10–100 nM enzyme. The standard template is single-stranded, promoter-containing DNA. Double-stranded, promoter-containing templates must be denatured before use. This is achieved by heating the template at 96° for 3 min, followed by chilling on ice. The denatured template should be used promptly. Reaction mixtures are incubated at 37° for the desired time (typically 5–30 min), terminated by the addition of 7 μl of stop solution (95% formamide, 20 mM EDTA, 0.05% bromophenol blue, 0.05% xylene cyanol FF), and heated for 5 min at 96° before analysis by electrophoresis on 6–10% polyacrylamide/8 M urea gels. Reaction products are quantitated by phosphorimaging using ImageQuant software (Molecular Dynamics).

The product of both vRNAP and mini-vRNAP transcription on single-stranded templates is not displaced, resulting in formation of an extended RNA–DNA hybrid.[14] The addition of *Eco* SSB under these conditions activates vRNAP transcription by recycling of the template. Consequently, activation by *Eco* SSB is only observed at limiting template concentrations (1–10 nM DNA).[24a] When used, *Eco* SSB (Pharmacia) is added to a concentration of 1–10 μM.

Purification of N4 vRNAP from Virions by
  Conventional Chromatography

All procedures are performed at 4° unless otherwise stated. DTT is omitted from buffers due to the absence of cysteine residues in the vRNAP polypeptide.

Buffers

> Phage resuspension buffer: 10 mM Tris–HCl, pH 8; 10 mM MgSO$_4$; 50 mM NaCl
>
> Phage disruption buffer: 4 M guanidine–HCl; 10 mM Tris–HCl, pH 8; 10 mM DTT
>
> Hydroxyapatite equilibration buffer: 4 M guanidine–HCl; 10 mM potassium phosphate, pH 7
>
> Dialysis buffer: 10 mM Tris–HCl, pH 8; 1 mM EDTA, pH 8; 50 mM NaCl; 5% glycerol

[24a] E. K. Davydova and L. B. Rothman-Denes, *Proc. Natl. Acad. Sci. USA* **100**, 9250 (2003).

Dilution buffer: 10 mM Tris–HCl, pH 8; 1 mM EDTA, pH 8; 5% glycerol

Step 1. Growth of N4 Phage

Escherichia coli W3350 cells are grown at 37° in 500 ml brain–heart infusion medium to $OD_{600} = 0.5$ and infected with N4 phage at a multiplicity of infection of 10. Cultures are incubated at 37° with vigorous shaking (350–400 rpm) for 3.5 h, at which time 5 ml of chloroform ($CHCl_3$) and 29.5 g of NaCl are added. Chloroform addition is required because N4 does not actively lyse the infected cell.[25] Cultures are shaken for 10 min and inspected for lysis; more $CHCl_3$ is added as needed. After cooling on ice for 30 min, bacterial debris is removed by centrifugation for 15 min at 11,000 g in a Sorvall GS-3 rotor. The supernatant is decanted carefully to clean centrifuge bottles. Phage are precipitated by the addition of 50 g powdered PEG8000 and overnight incubation at 4° and are pelleted by centrifugation at 3200 g for 30 min. Pelleted phage are resuspended in 5 ml of phage resuspension buffer and extracted two to three times with an equal volume of $CHCl_3$ to remove residual PEG8000.

Step 2. Purification of Phage

Phage used as starting material for vRNAP purification are purified by cesium chloride (CsCl) buoyant density centrifugation. Solid CsCl is added to the phage solution to a final concentration of 1 M. This solution is centrifuged for 10 min at 12,000 g in a Sorvall SS34 rotor to remove residual bacterial debris. The supernatant is added gently to 12.5-ml polyallomer tubes containing a 2.5-ml bottom layer of CsCl, $\delta = 1.664$ g/ml, in 15 mM Tris–HCl, pH 8, and a 4-ml top layer of CsCl, $\delta = 1.377$ g/ml, in 15 mM Tris–HCl, pH 8. The tubes are placed in a Beckman SW41Ti swinging bucket rotor and centrifuged for 1 h at 40,000 rpm and 20°. The major opalescent white band containing phage ($\delta = 1.5$ g/ml) is withdrawn from the side of the tube using an 18G gauge needle attached to a syringe. Concentrated phage is dialyzed either in Spectra-Por 2 membrane tubing (Spectrum) or in a Slide-a-lyzer dialysis cassette (Pierce) against two changes of 4 liters of phage buffer. Phage stocks are titered and stored at 4°.

Step 3. Virion Disruption

Four volumes of disruption buffer are added to 5–10 ml of CsCl-purified phage (5×10^{13} plaque-forming units/ml, $OD_{260} \leq 200$). Phage particles are disrupted by three to five cycles of freezing in a dry ice bath and thawing in 20° water, followed by sonication in 20-s pulses to reduce viscosity.

[25] G. C. Schito, A. M. Molina, and A. Pesce, *G. Microbiol.* **15,** 229 (1967).

Step 4. Hydroxyapatite Chromatography

The disrupted phage solution is loaded onto a 50-ml Macro-Prep ceramic hydroxyapatite, type I resin (Bio-Rad) column preequilibrated with buffer, which is then washed with 150 ml of the same buffer. N4 vRNAP does not bind to the column under these conditions and is recovered in the flow-through fraction and the first 50 ml of wash, which are pooled and dialyzed against four changes of 4 liters of dialysis buffer. The dialysis buffer is changed every 4–6 h for best results.

Step 5. Heparin Chromatography

The dialyzed protein solution is loaded onto a 5-ml HiTrap heparin column (Amersham Pharmacia). After washing with 15 ml of dialysis buffer, the protein is eluted with a 50-ml linear gradient of 0.05–0.5 M NaCl in dialysis buffer. Fractions containing vRNAP are identified by SDS–PAGE and the activity assay and are pooled.

Step 6. DEAE Chromatography

The heparin chromatography pool is diluted slowly with dilution buffer to a NaCl concentration below 0.1 M and applied to a 30-ml POROS-50 diethylaminoethyl (PerSeptive Biosystems) column. After washing with 150 ml of dialysis buffer, the protein is eluted with a 300-ml linear gradient of 0.05–0.5 M NaCl in dilution buffer. Fractions containing vRNAP are identified by SDS–PAGE and are pooled.

Step 7. SP Chromatography and Mono Q Concentration

The DEAE chromatography pool is diluted with 2 volumes of dilution buffer and is loaded onto a 5-ml HiTrap SP column (Amersham Pharmacia). The column is washed with 15 ml of dialysis buffer, and protein is eluted with a 50-ml linear gradient of 0.05–0.5 M NaCl in dilution buffer. Fractions containing vRNAP are pooled, diluted again with 2 volumes of dilution buffer, and applied onto a 1-ml HiTrap Q HP column (Amersham Pharmacia). The column is washed with 3 ml of dilution buffer; protein is eluted with 3 ml of 500 mM NaCl in dilution buffer and collected in 0.5-ml fractions. Fractions containing polymerase (typically fraction 2) are identified by SDS–PAGE, frozen in 50-μl aliquots, and stored at $-80°$. Typically, 0.1–0.3 mg of 95% pure vRNAP is obtained from 1000–2000 OD_{260} of cesium chloride-purified phage.

Purification of Recombinant Hexahistidine-Tagged vRNAP by Metal
  Affinity Chromatography

All procedures are performed at 4° unless otherwise stated.

Buffers

  Sonication buffer: 20 mM Tris–HCl, pH 8.0; 20 mM NaCl
  Wash buffer: 20 mM Tris–HCl, pH 8.0; 1 M NaCl
  Elution buffer: 20 mM Tris–HCl, pH 8.0; 20 mM NaCl; 60 mM
  imidazole.

Step 1. Protein Overproduction and Cell Lysis

Expression of the recombinant full-length vRNAP is deleterious to the
cell.[26] Therefore, 1 liter of M9 minimal salts medium [27] supplemented
with 0.2% casaminoacids (Difco), 1 mM MgSO$_4$, 0.1 mM CaCl$_2$, 10 μg/ml
thiamine, 4% Lenox L broth (LB), 100 μg/ml ampicillin, and 0.4% glucose
is inoculated with 10 ml of an overnight culture of *E. coli* BL21[28] cells bearing pKMK7, which encodes amino-terminally tagged vRNAP.[17] The culture is grown at 37° to OD$_{600}$ = 0.3. Cells are pelleted by centrifugation
for 15 min at 4200 g and 4° in a Sorvall SLA-3000 rotor and resuspended
in 1 liter M9 medium supplemented with 0.2% arabinose as the carbon
source to induce the production of recombinant vRNAP. After 30 min of
growth at 37°, the cultures are transferred to chilled centrifuge tubes and
cells are pelleted as described earlier, vRNAP production is limited to
30 min because the recombinant polypeptide is unstable and is degraded
during longer growth periods.

The cell pellet is resuspended in 10 ml of sonication buffer containing
1× complete protease inhibitor, EDTA-free (Roche), and sonicated in 20-s
pulses on ice until lysis is observed. Debris is removed by centrifugation for
15 min at 25,000 g in a Sorvall HB-4 rotor.

Step 2. Co^{2+} Metal Affinity Chromatography

The cleared sonicate is added to 4 ml of a 50% slurry of TALON metal
affinity resin (BD Biosciences) in sonication buffer and incubated for
30 min at 4° with gentle mixing on a LabQuake rotator. The slurry is transferred to a 5-ml Qiagen polypropylene column and allowed to settle by

[26] K. M. Kazmierczak and I. Kaganman, unpublished results.
[27] J. H. Miller, "Experiments in Molecular Genetics." Cold Spring Harbor Laboratory, Cold
  Spring Harbor, NY, 1972.
[28] F. W. Studier and B. A. Moffatt, *J. Mol. Biol.* **189,** 113 (1986).

gravity flow. The column is washed sequentially with 6 ml of sonication buffer, 6 ml of 1 M NaCl in sonication buffer to remove contaminant proteins, and again with 6 ml of sonication buffer to reduce the salt concentration before elution. Protein is eluted with 5 ml of elution buffer by gravity flow and collected in 1-ml fractions. Fractions containing recombinant polymerase (typically fractions 2–4) are identified by SDS–PAGE and pooled.

Step 3. Mono Q Concentration

The pooled fractions are applied onto a 1-ml HiTrap Q HP column (Amersham Pharmacia), which is then washed with 3 ml of sonication buffer. Protein is eluted with 3 ml of 500 mM NaCl and 5% glycerol in sonication buffer and collected in 0.5-ml fractions. Fractions containing polymerase (typically fraction 2) are identified by SDS–PAGE, frozen in 50-μl aliquots, and stored at $-80°$.

Although purification of recombinant hexahistidine-tagged vRNAP by affinity chromatography might seem preferable to protein purification from virions, recombinant vRNAP is poorly expressed and is not stable after induction. Several proteolytic products copurify with the full-length recombinant enzyme. Therefore, this procedure yields only 0.1–0.2 mg of 60% pure recombinant vRNAP from 1 liter of induced cells.

Purification of Recombinant Hexahistidine-Tagged Mini-vRNAP

All procedures are performed at $4°$ unless otherwise stated.

Buffers

> Sonication buffer: 20 mM Tris–HCl, pH 8.0; 20 mM NaCl
> Wash buffer: 20 mM Tris–HCl, pH 8.0; 1 M NaCl
> Elution buffer: 20 mM Tris–HCl, pH 8.0; 20 mM NaCl; 60 mM imidazole

Step 1. Protein Overproduction and Cell Lysis

Cells bearing recombinant mini-vRNAP plasmids can be grown in LB medium without adverse effects. *E. coli* BL21 cells bearing pKMK22, encoding amino-terminally tagged mini-vRNAP,[17] or pKMK25, encoding carboxy-terminally tagged mini-vRNAP, are grown in 100 ml of LB media with ampicillin (100 μg/ml) at $37°$ to $OD_{600} = 0.5$, at which time 0.2% arabinose is added to induce expression of the recombinant polypeptides. After 1 h of growth at $37°$, the culture is transferred to chilled tubes and centrifuged for 10 min at 11,000 g in a Sorvall SLA-3000 rotor. Pelleted cells are resuspended in 1.5 ml of sonication buffer and sonicated on ice in three 20-s pulses or until lysis is observed. The lysed sample is centrifuged for 20 min at 30,000 g in a Sorvall SM-24 rotor to pellet cellular debris.

Step 2. Co^{2+} Metal Affinity Chromatography

The cleared lysate is applied to a 2-ml capacity TALON spin column (BD Biosciences) prepacked with 0.5 ml TALON-NX metal affinity resin, which has been equilibrated with sonication buffer. The column is resealed and incubated for 30 min at 4° with gentle mixing on a LabQuake rotator. The column is allowed to settle and is washed sequentially with 1.5 ml of sonication buffer, 1.5 ml of wash buffer, and again with 1.5 ml of sonication buffer to reduce the salt concentration before elution. Protein is eluted with 5 ml of elution buffer by gravity flow and collected in 0.5-ml fractions. Fractions that contain recombinant polymerase (typically fractions 2–4) are identified by SDS–PAGE and pooled. Mini-vRNAP is concentrated on a 1-ml HiTrap Q HP column and stored as described for recombinant vRNAP or diluted 1:1 (vol:vol) with sterile glycerol and stored at −20°.

Up to 1 mg of 96% pure mini-vRNAP is obtained from 100 ml of induced cell culture. Polymerase retains its initial activity for at least 6 months when stored at −20° in 50% glycerol. The concentration of mini-vRNAP can be quantitated by UV spectrometry of a 1/100 dilution of the Mono Q eluate (1.0 A_{280} = 1.48 mg/ml = 12 μM of amino-terminally tagged mini-vRNAP; 1.0 A_{280} = 1.39 mg/ml = 11 μM of carboxy-terminally tagged mini-vRNAP, as calculated from sequence composition using the Schepartz Laboratory biopolymer calculator (available at http://paris. chem.yale.edu/).

Purification of Untagged Mini-vRNAP by DNA Affinity Chromatography

Mini-vRNAP displays high binding affinity for promoter-containing deoxyoligonucleotides (K_d 1 nM).[24] This property is exploited for the affinity purification of untagged mini-vRNAP. Untagged mini-vRNAP is retained specifically and completely on a promoter DNA affinity column (Fig. 2). Surprisingly, neither high salt nor high or low pH buffers elute the polymerase from the column, revealing the unusual stability of the mini-vRNAP–promoter complex. However, ddH_2O prewarmed to 25° elutes 90% of the bound polymerase, presumably due to melting of the promoter hairpin in the absence of metal ions and at an elevated temperature. Elution using ddH_2O prewarmed to 40° results in dissociation of the remaining protein, which is slightly degraded.

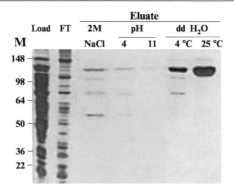

FIG. 2. DNA-Affinity Chromatography purification of untagged mini-vRNAP (see text).

Buffers

Sonication buffer: 20 mM Tris–HCl, pH 8.0; 20 mM NaCl; 2 mM EDTA
Wash buffer: 20 mM Tris–HCl, pH 8.0; 1 M NaCl; 2 mM EDTA

Step 1. Preparation of DNA Affinity Column

The DNA affinity column is prepared by adsorption of 5′ biotinylated promoter-containing deoxyoligonucleotides onto the matrix of a 1-ml Hi-Trap streptavidin HP column (Amersham Pharmacia) according to the manufacturer's instructions. The following promoter-containing biotinylated oligonucleotides have been tested (underlined sequences comprise the promoter hairpin inverted repeats):

A 5′-/5Bio/GGCATTACTTCATCCAAAAGAAGCGGAGCTTC-3′
B 5′-/5Bio/GGCATTACTTCATCCAAAAGAAGCTGAGCTTC-3′
C 5′-/5Bio/GGCATTACTTCATCCAAAAGAAGCGGAGC-3′

The best purification yield is achieved using oligonucleotide **A**. However, oligonucleotides **B** and **C** require lower temperatures than **A** for complete elution of the protein, in agreement with the decreased stability of the respective promoter hairpins.[29,30]

Step 2. Protein Overproduction and Cell Lysis

Escherichia coli BL21 cells bearing pKMK55, encoding "untagged" mini-vRNAP, are grown to OD$_{600}$ = 0.5 in 100 ml of LB medium with

[29] S. Yoshizawa, G. Kawai, K. Watanabe, K. Miura, and I. Hirao, *Biochemistry* **36,** 4761 (1997).
[30] X. Dai, M. Greizerstein, K. Nadas-Chinni, and L. B. Rothman-Denes, *Proc. Natl. Acad. Sci. USA* **94,** 2174 (1997).

ampicillin (100 μg/ml) at 37°. Induction of polymerase production and the cell sonicate preparation are as described for the purification of hexahistidine-tagged mini-vRNAP.

Step 3. DNA Affinity Chromatography

The cleared sonicate is applied to a 1-ml DNA affinity column preequilibrated with sonication buffer. The column is washed sequentially with 3 ml of sonication buffer, 3 ml of wash buffer, and again with 3 ml of sonication buffer. Mini-vRNAP is eluted with 3 ml of ddH$_2$O prewarmed to 25°. Fractions (0.5 ml) are collected, and 10 μl of 1 M Tris–HCl (pH 8.0) and 10 μl of 5 M NaCl are added to each fraction. Fractions containing mini-vRNAP are identified by SDS–PAGE and stored as described for hexahistidine-tagged mini-vRNAP.

A 100-ml culture of polymerase-expressing cells yields up to 0.5 mg of 90% pure mini-vRNAP in a single purification step using a 1-ml DNA affinity column (Fig. 2).

Concluding Remarks

Bacteriophage N4 vRNAP is unique in its ability to transcribe promoter-containing single-stranded DNAs with specificity, providing a tool for the synthesis of RNAs from synthetic oligonucleotide templates or from templates cloned into M13 DNA. It is an ideal enzyme for the synthesis of short RNAs in large amounts at limiting template concentrations in the presence of *Eco* SSB, for several reasons. N4 vRNAP displays high fidelity[17] and, in contrast to T7 RNAP, does not appear to produce abortive products.[24] Moreover, vRNAP can initiate transcription with any template-directed ribonucleoside triphosphate, although it is most efficient on templates containing C at positions +1 and +2.[24] The length of the transcripts that can be synthesized by mini-vRNAP is limited only by the length of the available single-stranded DNA template. Finally, vRNAP is useful for the synthesis of specific RNA–DNA hybrids when transcription is carried out in the absence of *Eco* SSB.

Both wild-type vRNAP and mini-vRNAP discriminate against the incorporation of deoxynucleoside triphosphates into transcripts. A mini-vRNAP Tyr→Phe mutation (Y678F), analogous to T7 RNAP Y639F,[31] allows the enzyme to incorporate dNTPs.[17]

The discovery of mini-vRNAP has provided us with a useful tool to study polymerase–promoter, polymerase–substrate, and polymerase–product

[31] L. G. Brieba and R. Sousa, *Biochemistry* **39**, 919 (2000).

interactions. We have pinpointed sequences within early promoters required for polymerase recognition. Conversely, we have identified regions within mini-vRNAP that interact with the promoter hairpin loop, hairpin stem, the initiating nucleotide, and the template and RNA product in an elongation complex.[24] We expect that mini-vRNAP will allow us to further characterize the steps of transcription initiation and elongation at N4 early promoters.

As mentioned earlier, biochemical and genetic analysis suggests that the catalytic center of vRNAP must resemble that of T7 RNAP. However, several lines of evidence indicate that vRNAP differs in its mechanism of promoter recognition.[24] Ongoing structural analysis will reveal both similarities and differences with T7 RNAP. Finally, identification of the vRNAP transcriptionally active domain raises questions about the possible roles of the amino- and carboxy-terminal domains of the polypeptide. Experiments in progress have indicated roles for these domains in virion morphogenesis and the initial steps of phage infection.[32]

Acknowledgment

This work was supported by NIH Grant AI 12575.

[32] I. Kaganman and E. K. Davydova, unpublished results.

[9] Preparation and Characterization of Recombinant *Thermus aquaticus* RNA Polymerase

By Konstantin Kuznedelov, Leonid Minakhin, and Konstantin Severinov

Most bacterial RNA polymerase (RNAP) core enzymes consist of five core subunits (β', β, a dimer of identical α subunits, and the ω subunit). Binding of one of the several σ factors converts the catalytically proficient core into the holoenzyme, which is able to initiate transcription from promoters. High-resolution structures of RNAP core and medium-resolution structures of RNAP holoenzyme from thermophilic eubacterium *Thermus aquaticus* (*Taq*) have became available.[1-3] A low-resolution model of *Taq* RNAP holoenzyme complexed with a model promoter substrate was also obtained.[4] In addition, a structure of highly related *T. thermophilus* RNAP holoenzyme was solved to high resolution.[5] Parallel advances in structural analysis of yeast RNA polymerase II had revealed an outstanding degree of structural similarity between bacterial and eukaryotic RNAP, which could not have been predicted based on primary sequence comparisons alone.[6,7]

Superposition of structural data with biochemical and genetic data obtained with *Escherichia coli* RNAP, which is functionally the best-studied enzyme of its class, allowed the building of structure–functional models of transcription initiation and transcription elongation complexes and the assigning of specific biochemical functions to RNAP functional domains.[8,9] The availability of high-resolution structures qualitatively raises the importance of rational structure-based mutational analysis of RNAP, which should verify functional inferences experimentally.

Mutational analysis of yeast RNAP II is complicated by the fact that this enzyme, which is indispensable for cell viability, cannot be obtained by *in vitro* reconstitution from isolated subunits. Therefore, only mutations that do not interfere with vital enzyme functions (and therefore leave most partial biochemical activities of RNAP intact) can be analyzed.

The *in vitro* activity of several bacterial RNAPs, including *Taq* RNAP, can be recovered after the separation of individual subunits in the presence of denaturing agents and subsequent mixing of the subunits and dialysis at controlled conditions.[10,11] *In vitro* reconstitution of *E. coli* RNAP from cloned and individually overexpressed and purified subunits provided means of obtaining RNAP harboring lethal mutations in quantities sufficient for biochemical analyses.[12] Studies of reconstituted recombinant

[1] G. Zhang, L. Campbell, L. Minakhin, C. Richter, K. Severinov, and S. A. Darst, *Cell* **98**, 811 (1999).

[2] E. A. Campbell, N. Korzheva, A. Mustaev, K. Murakami, S. Nair, A. Goldfarb, and S. A. Darst, *Cell* **104**, 901 (2001).

[3] K. S. Murakami, S. Masuda, and S. A. Darst, *Science* **296**, 1280 (2002).

[4] K. S. Murakami, S. Masuda, E. A. Campbell, O. Muzzin, and S. A. Darst, *Science* **296**, 1285 (2002).

[5] D. G. Vassylyev, S. Sekine, O. Laptenko, J. Lee, M. N. Vassylyeva, S. Borukhov, and S. Yokoyama, *Nature* **417**, 712 (2002).

[6] P. Cramer, D. A. Bushnell, J. Fu, A. L. Gnatt, B. Maier-Davis, N. E. Thompson, R. R. Burgess, A. M. Edwards, P. R. David, and R. D. Kornberg, *Science* **288**, 640 (2000).

[7] R. H. Ebright, *J. Mol. Biol.* **304**, 687 (2000).

[8] N. Korzheva, A. Mustaev, M. Kozlov, A. Malhotra, V. Nikiforov, A. Goldfarb, and S. A. Darst, *Science* **289**, 619 (2000).

[9] V. Mekler, E. Kortkhonjia, J. Mukhopadhyay, J. Knight, A. Revyakin, A. N. Kapanidis, W. Niu, Y. W. Ebright, R. Levy, and R. H. Ebright, *Cell* **108**, 599 (2002).

[10] W. Zillig, P. Palm, and A. Heil, *in* "RNA Polymerase" (R. Losick and M. Chamberlin, eds.), p. 101. Cold Spring Harbor Laboratory, Cold Spring Harbor, NY, 1976.

[11] L. Minakhin, S. Nechaev, E. A. Campbell, and K. Severinov, *J. Bacteriol.* **183**, 71 (2001).

[12] K. Zalenskaya, J. Lee, C. N. Gujuluva, Y. K. Shin, M. Slutsky, and Goldfarb, *Gene* **89**, 7 (1990).

wild-type *E. coli* RNAP and mutants harboring lethal mutations provided crucial insights into transcription mechanism and regulation and are largely responsible for our increased understanding of this enzyme (see, e.g., refs. 13–15). Despite obvious homology between RNAPs from *Taq* and *E. coli*, it would be highly desirable to perform mutational and structural studies using *Taq* RNAP rather than the *E. coli* counterpart, for which no structure is available.

Important advantages of studying *Taq* RNAP include the ability to (i) reduce assembly defects of mutations to a minimum by designing structure-based mutations, (ii) perform structural analysis of mutants, and (iii) reduce contamination with highly active, wild-type chromosomally encoded RNAP, which becomes a persistent problem when mutant enzymes with low specific activity are analyzed. Disadvantages of the *Taq* system are the lack of functional assays, the general absence of data on gene transcription in this organism, and, more importantly, the inability to manipulate *Taq* RNAP subunit (*rpo*) genes. To overcome these limitations and to make full use of the available structural information, we cloned *Taq rpo* genes on *E. coli* expression plasmids and developed a recombinant *Taq* RNAP transcription system.[11]

Methods

Plasmids Expressing Taq rpo *Genes*

Taq rpo genes are cloned in T7 RNAP-driven expression plasmids of the pET series (Novagene) between *Nde*I and *Eco*RI sites of the pET polylinker. Plasmids, pET28*Taq*A, pET28*Taq*B, pET28*Taq*C, and pET28*Taq*Z, express *Taq* RNAP α, β, β', and ω, respectively, with N-terminal hexahistidine tags. Corresponding plasmids based on the pET21 vector express untagged subunits. Plasmids containing *Taq rpo* genes cloned between *Nde*I and *Eco*RI sites of the pT7Blue cloning plasmid (Navagene) were also created. These plasmids can be used to introduce site-specific mutations in *Taq rpo* genes (see later).

Plasmid pET28*Taq*ABCZ, coexpressing *Taq rpoA*, *rpoB*, *rpoC*, and *rpoZ* genes, is constructed from the pET28*Taq*C plasmid, which is treated with the *Hind*III and DNA polymerase Klenow fragment, followed by

[13] M. Kashlev, J. Lee, K. Zalenskaya, V. Nikiforov, and A. Goldfarb, *Science* **248,** 1006 (1990).

[14] K. Igarashi and A. Ishihama, *Cell* **65,** 1015 (1991).

[15] M. Orlova, J. Newlands, A. Das, A. Goldfarb, and S. Borukhov, *Proc. Natl. Acad. Sci. USA* **92,** 4596 (1995).

*Eco*RI treatment, and ligated to the *Taq rpoB* containing fragment that is prepared from pET21*Ta*B by treating with the *Bgl*II and Klenow fragment, followed by *Eco*RI. The resultant plasmid, pET28*Ta*BC, has two genes, *rpoB* and *rpoC*, in an opposite orientation; each gene is preceded by the T7 RNAP promoter. pET28*Ta*BC is treated with *Eco*RI and ligated with the *Eco*RI-treated polymerase chain reaction (PCR) fragment of pET21*Ta*A. Oligonucleotides used for PCR are designed to anneal upstream of the T7 RNAP promoter of pET21*Ta*A and downstream of the *Taq rpoA* termination codon, and each contains an engineered *Eco*RI site. The resultant plasmid, pET28*Ta*ABC, has *Taq rpoA* under T7 RNAP promoter control inserted between *rpoB* and *rpoC* genes in the same orientation as *rpoC*. Plasmid pET28*Ta*ABCZ is constructed by ligating the *Sgr*AI-treated pET28*Ta*ABC with the *Sgr*AI-treated PCR fragment of pET21*Ta*Z. PCR primers are designed to anneal upstream of the T7 RNAP promoter of pET21*Ta*Z and downstream of the *Taq rpoZ* termination codon and each contains an engineered *Sgr*AI site. The resultant plasmid, pET28*Ta*ABCZ, has three genes, *rpoZ*, *rpoC*, and *rpoA*, in the same orientation and *rpoB* in the opposite orientation. The *rpoC* gene is fused to the N-terminal hexahistidine tag, and other genes are untagged. A derivative of pET28-*Ta*ABCZ that allows the preparation of untagged *Taq* RNAP core is also constructed.

Although the pET28*Ta*ABCZ plasmid provides adequate yields of wild-type *Taq* RNAP, expression levels of the two largest subunits are poorly coordinated, and some of the RNAP mutants constructed in this plasmid could only be prepared in a very low yield. A new, improved *Taq rpo* cooverexpression plasmid has been developed (Fig. 1). Plasmids pET21*Ta*A, pET21*Ta*B, and pET28*Ta*C are used as starting material. As the first step, a plasmid cooverexpressing *Taq rpoA* and *rpoB* is constructed. To this end, an *Sph*I–*Sal*I fragment of pET21*Ta*A containing *Taq rpoA* is prepared, and the *Sal*I sticky end is filled in partially by the Klenow fragment enzyme in the presence of dCTP and dTTP. Plasmid pET21*Ta*B is treated with *Sph*I and *Bgl*II, and the sticky end generated by *Bgl*II is filled in partially in the presence of dATP and dGTP. Cohesive-end ligation of the insert containing *Taq rpoA* and the pET21-*Ta*B vector results in the pET21*Ta*AB plasmid, with the *rpoA* gene followed by *rpoB* and each gene having an upstream T7 RNAP promoter and the *rpoB* gene followed by a downstream transcription terminator. As the second step, the *rpoAB* cassette from pET21*Ta*AB is inserted into pET28*Ta*C, between the *lacI* and the *rpoC* genes. The *rpoAB* cassette is excised from pET21*Ta*AB by treating with *Not*I and partially filling in the sticky end with the Klenow fragment enzyme in the presence of

FIG. 1. Construction of *E. coli* plasmid coexpressing *Taq rpo* genes. See text for details.

dGTP, followed by digestion with *Mlu*I (cleaves once in the *lac*I gene). The pET28*Ta*C plasmid is treated with *Sgr*A I, and the sticky end is filled in partially in the presence of dCTP, followed by *Mlu*I digestion. Ligation of the *Taq rpoAB* cassette and pET28*Ta*C vector yields plasmid pET28ABC$_{Taq}$, which contains three *Taq rpo* genes whose products are sufficient for active RNAP core assembly in the order *rpoA*, *rpoB*, and *rpoC*. Every gene is preceded by the T7 RNAP promoter, but only the *rpoC* gene is followed by the transcription terminator. Only the β' subunit expressed from this plasmid is tagged with an N-terminal hexahistidine tag.

Overexpression of Recombinant Taq *RNA Polymerase Subunits and RNA Polymerase*

Escherichia coli BL21(DE3) cells transformed with *Taq rpo* expression plasmids overproduce individual *Taq* core RNAP subunits at a high level on induction with 1 mM Isopropyl-β-D-thiogalactoside (IPTG), and the overexpressed proteins, including the α subunit, segregate into inclusion bodies. The following induction procedure works well. *E. coli* BL21(DE3)-competent cells are transformed with *Taq rpo* expression plasmids and plated on appropriate selective medium. When using ampicillin-resistant pET21-based plasmids, high (200 μg/ml) concentrations of ampicillin are recommended to prevent growth of satellite colonies. After overnight growth at 37°, colonies are scraped off the plate surface, resuspended in a small volume of LB broth, and used to inoculate prewarmed LB broth with appropriate antibiotics. A plate containing several hundred ~1-mm colonies is sufficient to inoculate 1 liter of liquid medium. The use of fresh transformants is strongly recommended for optimal induction. Cells are allowed to grow on a rotary shaker at 37° until OD_{600} reaches 0.8–1.0 and are induced with 1 mM IPTG for 5–6 h. The induction of *Taq* RNAP core subunit overexpression is confirmed by SDS–PAGE, and cells are collected and stored at −80° until further use.

To purify the *Taq* RNAP core from *E. coli* cells coexpressing *Taq rpo* genes, BL21(DE3) cells harboring pET28*Ta*ABCZ or pET28ABC$_{Taq}$ are inoculated in 1 liter of LB containing appropriate antibiotics. Cells harboring pET28*Ta*ABCZ are induced as described earlier. The induction level is checked by SDS–PAGE. *E. coli* BL21(DE3) cells transformed with pET28*Ta*ABCZ express *Taq* RNAP subunits at low levels. The amount of overproduced *Taq* β and β' is only slightly higher than the amount of endogenous *E. coli* RNAP largest subunits, which form a characteristic low-mobility double band on SDS gels of whole cell lysates.

The addition of IPTG to cells harboring pET28ABC$_{Taq}$ decreases cell growth rate and also increases unwanted segregation of proteins into inclusion bodies. Growing of cells at 37° for 18–20 h with vigorous shaking and without IPTG induction results in optimal yields of *Taq* RNAP from cells harboring this cooverexpression plasmid.

Taq *RNAP* σ^A *Subunit*

The specificity σ subunits are often interchangeable between RNAP core enzymes prepared from evolutionarily distant bacteria. When the *Taq* RNAP core is combined with *E. coli* σ^{70} at 45°, a condition when both *Taq* and *E. coli* enzymes are active, no promoter-specific initiation from the strong *E. coli* σ^{70}-dependent T7 A1 promoter is observed.[11]

Protein–protein band shift experiments show that *E. coli* σ^{70} binds to the *Taq* core (Fig. 2A, lane 9), but the resultant complex is evidently inactive. To overcome the problem of sigma incompatibility, we cloned the homologue of the *E. coli rpoD* (σ^{70}) gene from *Taq* using degenerate oligonucleotide probes based on *T. thermophilus sigA* gene[16] sequences in the vicinity of highly conserved regions 1.2 and 4.2. A single positive clone is

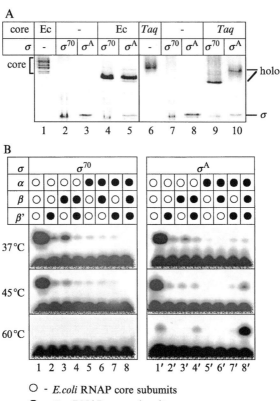

FIG. 2. Formation and activity of recombinant *Taq*, *E. coli*, and *Taq–E. coli* chimeric holoenzymes. (A) The indicated proteins were combined together in transcription buffer (40 m*M* Tris–HCl, pH 7.9, 40 m*M* KCl, and 10 m*M* MgCl₂) and were incubated for 10 min at 42°, and reaction products were separated by native gel-electrophoresis on precast Phast 4–15% gels (Pharmacia). (B) RNAP core enzymes were reconstituted from the indicated subunits as described in the text. Reconstitution mixtures were supplemented with recombinant *E. coli* σ^{70} or *Taq* σ^A as indicated and were used in an abortive transcription initiation assay using the T7 A1 promoter-containing DNA fragment as a template, CpA as a primer, and [α-³²P]UTP as a substrate. Reactions proceeded at the indicated temperatures, and products were resolved by denaturing PAGE and revealed by autoradiography.

obtained. Sequence alignments reveal that the product of the *Taq sigA* (σ^A) gene is highly similar to the published sequence of σ^A from *T. thermophilus*,[16] as expected. *Taq* σ^A is also highly similar to σ^{70} in conserved regions 2 and 4, which are responsible for promoter recognition. Therefore, well-characterized *E. coli* RNAP σ^{70} holoenzyme promoters can be used for promoter-dependent transcription initiation by *Taq* RNAP σ^A holoenzyme (see later). In striking contrast, *Taq* σ^A completely lacks the N-terminal conserved region 1.1. Instead, it contains a $\sim$100 amino acid long segment without homology to any of the published sequences.

The entire *Taq sigA* gene is subcloned into pET expression vectors. The pET28*Ta*σ plasmid, which expresses N-terminally hexahistidine-tagged *Taq* σ^A, is transformed in *E. coli* BL21(DE3). Fresh transformants are inoculated in 1 liter of LB containing 25 μg/ml kanamycin and are grown at 30°. IPTG (1 m*M*) is added when the culture OD$_{600}$ reaches 0.8–1.0. After 4 h of vigorous agitation at 30°, cells are harvested, and the cell pellet ($\sim$2 g) is resuspended in 15 ml of buffer **A** [10 m*M* Tris–HCl, pH 7.9, 5% glycerol, 500 m*M* NaCl, and 0.2 m*M* phenyl methyl sulfonyl fluoride (PMSF)]. Cells are lysed by sonication as described.[17] Cell debris is removed by low-speed centrifugation (twice for 30 min at 15,000 rpm), and the lysate is diluted twofold with 8 *M* urea, loaded onto a 1-ml chelating Sepharose Hi-Trap column (Pharmacia) loaded with Ni^{2+} according to the manufacturer's instructions, and equilibrated in buffer **A** containing 4 *M* urea. The column is washed with buffer **A** containing 4 *M* urea and 20 m*M* imidazole, and *Taq* σ^A is eluted with 100 m*M* imidazole in the same buffer. Fractions containing *Taq* σ^A are pooled and dialyzed against TGE buffer (20 m*M* Tris–HCl, pH 7.9, 1 m*M* EDTA, 5% glycerol, and 2 m*M* 2-mercaptoethanol) and loaded onto a 1-ml Resource Q column (Amersham) attached to Fast Protein Liquid Chromatography (FPLC) and equilibrated in TGE buffer. Pure *Taq* σ^A is eluted using a linear 40-ml gradient of NaCl (from 0.2 to 0.4 *M*) in TGE buffer. The protein is concentrated to the final concentration of $\sim$1.5–2 mg/ml and stored at $-20°$ in the presence of 50% glycerol.

The following procedure is used to purify untagged *Taq* σ^A overproduced by *E. coli* BL21(DE3) cells harboring the pET21*TaqRpo*D plasmid. Fresh transformants are inoculated in 50 ml of LB containing 200 μg/ml ampicillin and grown for 2–3 h at 37° with vigorous shaking until OD$_{600}$ reaches 1.5. Cells are collected by a 10-min room temperature centrifugation at 2500 rpm, resuspended in a small volume of LB, and the suspension

[16] M. Nishiyama, N. Kobashi, K. Tanaka, H. Takahashi, and M. Tanokura, *FEMS Microbiol. Lett.* **172,** 179 (1999).
[17] S. Borukhov and A. Goldfarb, *Prot. Exp. Purif.* **4,** 503 (1993).

is used to inoculate 1 liter of LB containing 200 μg/ml ampicillin. IPTG (1 mM) is added when the OD_{600} of the culture reaches 0.5–0.6. After 3 h of vigorous agitation at 37° or when OD_{600} reaches ~2.5, cells are harvested by centrifugation. The cell pellet (3.5 g) is resuspended in 20 ml of TGE buffer containing 200 mM NaCl and 0.2 mM PMSF. Cells are lysed by sonication as described.[17] Cell debris is removed by low-speed centrifugation, and the volume of the lysate is adjusted to 40 ml with TGE buffer. The lysate is transferred into a 50-ml screw-cap polypropylene tube and is incubated for 30 min at 70° with occasional mixing. The massive pellet formed during 70° is removed by low-speed centrifugation and discarded. The supernatant is diluted twofold with TGE buffer, loaded onto a 20-ml DE-52 cellulose column equilibrated in TGE buffer, and attached to an FPLC. The column is developed stepwise with TGE buffer containing 0.1, 0.2, 0.3, 0.4, and 1 M NaCl. Most of Taq σ^{A} is eluted at 0.2 M NaCl and is precipitated with ammonium sulfate (40 g/100 ml). The ammonium sulfate pellet is dissolved in 0.35 ml of TGE and is loaded onto a Superose-6 column (Pharmacia) equilibrated in TGE containing 0.2 M NaCl. The column is developed isocratically with the same buffer, and σ^{A}-containing fractions are pooled and dialyzed against storage buffer containing 40 mM Tris–HCl, pH 7.7, 0.2 M KCl, 50% glycerol, 1 mM dithiothieitol (DTT), and 1 mM EDTA.

When recombinant Taq σ^{A} (with or without a hexahistidine tag) is combined with the Taq RNAP core enzyme, specific interaction with and transcription from the T7A1 promoter-containing DNA fragment are observed.[11] Taq σ^{A} also forms complexes with the $E.$ $coli$ RNAP core enzyme (Fig. 2A, lane 5), and the resultant holoenzyme is active on the T7 A1 promoter. The chimeric holoenzyme is highly active at 37°, when the Taq σ^{A} holoenzyme demonstrates only low levels of activity; the chimeric holoenzyme demonstrates low levels of activity at 60°, when the $E.$ $coli$ σ^{70} holoenzyme is completely inactivated (Fig. 2B, compare lanes 1 and 1′). Thus it appears that although the main contribution to high-temperature transcription is coming from the Taq core, σ^{A} also contributes somewhat.

In Vitro Taq RNAP Reconstitution

The RNAP reconstitution procedure developed for $E.$ $coli$ RNAP and described in detail elsewhere[17] can be used successfully for Taq RNAP reconstitution, especially wild-type Taq RNAP reconstitution. The most important modification is that the 30-min thermoactivation step, which is carried out at 30° for $E.$ $coli$ RNAP, is performed at 42° for Taq RNAP. In addition, and following the thermoactivation step, a half-hour incubation at 65° to destroy contaminating $E.$ $coli$ RNAP is used. After reconstitution, RNAP preparations can be used directly in transcription assays or stored under 30% ammonium sulfate until further use. Alternatively, RNAP can be purified further by gel filtration on Superose-6 and Resource Q columns (Pharmacia) as described.[17] The Superose-6 step separates the assembled enzyme from assembly intermediates and unassembled subunits and determines whether the assembly of recombinant *Taq* RNAP has taken place as judged by the appearance of characteristic chromatographic peaks in the course of chromatography. Chromatographic fractions containing RNAP are concentrated by filtration through a C-100 concentrator (Amicon) to ∼1 mg/ml and are stored in the presence of 50% glycerol at −20°.

We performed intergeneric (*Taq–E. coli*) RNAP reconstitutions to determine whether active hybrid RNAP can be prepared. One needs to have an answer to this question to determine whether the active chimeric enzyme can contaminate recombinant *Taq* RNAP preparation, as in our experience, even the best preparations of *Taq* RNAP subunit inclusion bodies contain small amounts of *E. coli* RNAP subunits and/or *E. coli* RNAP. The results of one such intergeneric RNAP reconstitution experiment performed at standard conditions used to reconstitute *E. coli* RNAP is presented in Fig. 2B. After reconstitution, RNAP core enzymes were supplemented with recombinant σ^{70} or σ^A, and an abortive transcription initiation reaction was performed on the T7 A1 promoter at 37, 45 and 60°. As can be seen, RNAP containing *Taq* β and *E. coli* α, β', and σ^{70} assembled with low efficiency and was active at 37° but not at 60° (Fig. 2B, lane 3). In addition, σ^A, *Taq* β and β', and *E. coli* α assembled with low efficiency into an enzyme that was equally active at 37 and 60° (Fig. 2B, lane 4'). Results thus indicate that certain chimeric enzymes are indeed active and may present a problem when *Taq* RNAP mutants with low specific activity are analyzed. However, the availability of active chimeric enzymes opens the way for analyzing species-specific regulators interacting with *Taq* or *E. coli* RNAPs.

Preparation of Taq *RNAP from* E. coli *Cells Cooverexpressing* Taq rpo *Genes*

Wild-type *Taq* RNAP prepared by *in vitro* reconstitution is adequate for most biochemical assays, and its specific activity is fairly similar to the activity of RNAP purified from *Taq* cells. However, we found that *Taq* RNAP subunits harboring mutations, particularly those removing substantial portions of RNAP structural domains, assemble inefficiently *in vitro*. The use of plasmids cooverexpressing *Taq rpo* genes allows one to overcome these limitations and to prepare large amounts of assembled wild-type or mutant *Taq* RNAP.

A 4 to 5 g pellet of cells cooverexpressing *Taq* RNAP is resuspended in 40 ml of grinding buffer containing 40 mM Tris–HCl, pH 7.9, 100 mM NaCl, 0.2 mM PMSF, 10 mM EDTA, 15 mM 2-mercaptoethanol, and 5% glycerol. Cells are lysed by sonication, and cell debris is removed by low-speed centrifugation. The cleared lysate is transferred into a 50-ml polypropylene tube, diluted twofold with grinding buffer, and incubated at 75° for 30 min with occasional mixing. The massive pellet that is formed during this stage is removed by centrifugation (30 min at 15,000 rpm at 4°) and discarded. The supernatant is loaded, at 1 ml/min, onto a 5-ml heparin Hi-Trap column equilibrated in TGE buffer containing 100 mM NaCl, attached to an FPLC, and kept at room temperature. After loading, the column is washed with TGE buffer containing 300 mM NaCl, and *Taq* RNAP is step eluted in ~7–10 ml of TGE buffer containing 600 mM NaCl. The column is washed with TGE containing 1 M NaCl. The same column can be used for multiple purifications. Proteins in the 0.6 M NaCl heparin column fraction are precipitated overnight with dry ammonium sulfate (0.3 g/ml) in a cold room.

The ammonium sulfate pellet is collected by centrifugation in a 15-ml Corex glass tube, drained thoroughly, dissolved in 250 μl of TGE buffer, loaded, at room temperature, on a Superose-6 (Pharmacia) column attached to an FPLC, and equilibrated in TGE containing 200 mM NaCl. Chromatography is conducted at 0.4 ml/min. RNAP is eluted in two peaks (Fig. 3). The first peak contains material absorbing at 260 nm and probably represents RNAP complexes with nucleic acids and is discarded. The second peak, which contains the free RNAP core, is made 50% with glycerol and RNAP is stored at −20°. Alternatively, RNAP can be concentrated approximately four-fold by dialyzing against the storage buffer (earlier discussion). Typical yields are 1–3 mg of *Taq* RNAP core from 4 to 5 g of wet *E. coli* cells cooverexpressing *Taq rpo* genes.

In Vitro *Transcription*

Most *in vitro* transcription assays developed for *E. coli* RNAP can be used with *Taq* RNAP with minor modifications. The following are the most significant differences. The most obvious one is temperature—at our standard reaction conditions, the *Taq* σ^A RNAP holoenzyme has an optimal transcription initiation activity at 75° and shows ~7, 50, and 85% activity at 25, 45, and 65°, respectively. The second difference is that at optimal conditions, the *Taq* RNAP core appears to terminate transcription at intriasic *E. coli* terminators much less efficiently than *E. coli* RNAP. Third, *Taq* RNAP is much more resistant to rifampicin than *E. coli* RNAP, and one should therefore use heparin rather than rifampicin to prevent reinitiation of transcription. Finally, elongation complexes formed by the *Taq* RNAP core enzyme appear to be highly active in intrinsic transcript cleavage reaction, even at pH below 8.0, which complicates "walking" along the template using immobilized *Taq* RNAP. Several simple protocols

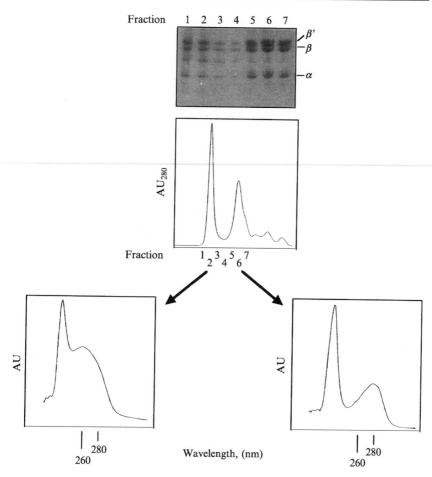

Fig. 3. Purification of recombinant *Taq* RNAP core enzyme from *E. coli* cells cooverproducing *Taq* RNAP subunits. Results of the final (gel filtration on Superose-6 column) purification step of the recombinant *Taq* RNAP core enzyme are presented. The protein content of chromatographic peaks shown in the middle was analyzed by SDS–PAGE and is shown at the top. PDA spectra of two main chromatographic peaks containing RNAP are shown at the bottom.

are presented that should provide a starting point for more complex *in vitro* transcription experiments involving *Taq* RNAP.

Standard abortive initiation reactions (10 μl) contain 40 mM Tris–HCl, pH 8.4, 40 mM KCl, 5 mM MgSO$_4$, 50 nM *Taq* RNAP core enzyme, and 100 nM of recombinant *Taq* σ^A. Reactions are preincubated for 10 min at 65°, followed by the addition of the 100 nM T7 A1 promoter-containing

DNA fragment and an additional 10-min incubation at 65°. Abortive transcription is initiated by the addition of 100 μM CpA and 5 μCi [α-^{32}P]UTP (3000 Ci/mmol) and is allowed to proceed for 10 min at 65°. Reactions are terminated by the addition of an equal volume of urea-containing loading buffer and are analyzed by denaturing gel electrophoresis (8 M urea, 20% polyacrylamide) and autoradiography.

For experiments involving transcription elongation, elongation complexes stalled at position +20 (EC^{20}) of the T7 A1 promoter-driven templates derived from the pAA141 plasmid originally developed in the Chamberlain's laboratory[18] are used. Reactions contain, in 15 μl, 40 mM Tris–HCl, pH 8.4, 40 mM KCl, 5 mM MgSO$_4$, 100 nM of the DNA fragment containing the T7 A1 promoter, 50 nM of RNAP, 100 nM σ^A, 10 μM CpApUpC, 50 μM ATP and GTP, and 5 μCi[α-^{32}P]CTP (3000 Ci/mmol). Reactions are incubated for no more than 10 min at 65° to allow EC^{20} formation (note that this complex contain 21 nucleotide long RNA, as the 5' end of the CpApUpC primer corresponds to position −1 of the template). Longer incubations lead to an accumulation of transcript cleavage products. EC^{20} can then be used to study transcription elongation, transcription termination, or transcription pausing by synchronously restarting EC^{20} by the addition of various concentrations of NTPs. After the addition of NTPs, reactions are allowed to proceed with reaction times and temperatures chosen by the experimenter. Reactions are terminated by addition of the formamide-containing loading buffer. Products are analyzed by urea–PAGE electrophoresis (8 M urea, 10% polyacrylamide), followed by autoradiography and PhosphorImager analysis.

Site-Specific Mutagenesis

Because of the very large size of plasmids cooverexpressing *Taq rpo* genes, site-directed mutagenesis becomes problematic due to the lack of unique restriction endonuclease recognition sites. Consequently, when constructing plasmids expressing mutant RNAP, a stepwise procedure needs to be adopted to construct cooverexpression plasmids bottom up. As an example, we present a scheme used to construct mutations in the evolutionarily conserved rudder element of the β' subunit (Fig. 4). Analysis of *Taq* RNAP harboring deletions in the β' rudder showed that this structural element plays no role in determining the length of the RNA–DNA hybrid, as was proposed originally based on structural analysis,[1] but instead plays a role in the elongation complex stability.[19] Our attempts to obtain rudder deletions in *E. coli* RNAP were unsuccessful

[18] C. K. Surratt, S. C. Milan and M. J. Chamberlin, *Proc. Natl. Acad. Sci. USA* **88**, 7983 (1991).
[19] K. Kuznedelov, N. Korzheva, A. Mustaev, and K. Severinov, *EMBO J.* **21**, 1369 (2002).

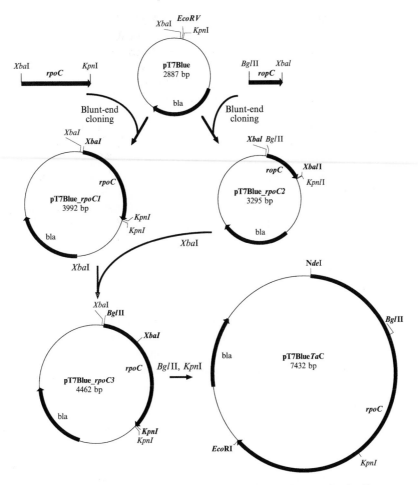

FIG. 4. Site-specific mutagenesis of the *Taq rpoC* gene. See text for details.

due to the high toxicity of the mutant *rpoC* gene (K. Kuznedelov and K. Severinov, unpublished observations).

To create plasmid pET28ABC$_{Taq}\Delta$R, cooverproducing *Taq* α, β, and His-tagged β' lacking the rudder, two PCR reactions are performed to amplify *Taq rpoC* regions at the left- and right-hand sides of the deletion site. The left-hand side nonmutagenic PCR primer corresponds to *Taq rpoC* positions 1280–1309 and contains a *Bgl*II site corresponding to a unique *rpoC Bgl*II site at position 1289. The right-hand side nonmutagenic primer is complementary to *rpoC* positions 2872–2899 and contains a *Kpn*I site corresponding to a unique *rpoC Kpn*I site at position 2885. The mutagenic primers correspond to *rpoC* positions 1729–1751 and 1804–1825 for the left-hand side PCR and the right-hand side PCR, correspondingly. Each mutagenic primer also contains an in-frame *Xba*I site. The products of each PCR reaction are purified and cloned into the pT7Blue blunt-end cloning vector. Recombinant plasmids containing inserts in appropriate directions (see Fig. 4) are selected by colony PCR. The *Bgl*II–*Kpn*I *rpoC* fragment containing the mutation is assembled by cloning the *Xba*I fragment containing the left-hand side *Taq rpoC* fragment into the pT7blue plasmid containing the *Xba*I–*Kpn*I *rpoC* fragment. The entire mutant *rpoC* gene is next assembled by cloning the *Bgl*II–*Kpu*I fragment harboring the mutation instead of the corresponding fragment of pT7blue*Ta*C. The entire *Bgl*II–*Kpu*I fragment harboring the mutation is next sequenced to confirm the presence of the desired mutation and the absence of undesirable mutations generated by PCR. The resultant plasmid contains the mutant *rpoC* gene bounded by unique *Nde*I and *Eco*RI sites, and these sites are used to reclone the mutant gene into the pET28 expression plasmid. Subsequent stages used to construct the cooverexpression plasmid are identical to the ones shown in Fig. 1.

The pT7blue*Ta*C plasmid has four unique sites close to or inside the *rpoC* gene, *Nde*I, *Bgl*II, *Kpn*I, and *Eco*RI, which divide the *rpoC* into approximately equal fragments suitable for site-directed mutagenesis. Therefore, the procedure outlined here can be used for the incorporation of site-specific mutants in any part of the *rpoC* gene.

Concluding Remarks

Overproduced *Taq* RNAP subunits assembled into functional RNAP *in vitro* and *in vivo* when coexpressed in *E. coli*. Although *Taq* RNAP differs from the prototypical *E. coli* enzyme in several important ways, transcription initiation, transcription termination, and transcription cleavage assays developed for *E. coli* RNAP can be adapted to study *Taq* RNAP transcription with very small modifications. The availability of recombinant *Taq* RNAP and discriminative transcription assays for this enzyme should make it possible to test, by means of genetic engineering and biochemical analysis, many of the predictions of current and future functional inferences derived from RNAP structures. The ability to purify mutant *Taq* RNAP directly from *E. coli* cells should also make the structural analysis of mutant RNAP possible.

Acknowledgments

This work was supported by the National Institutes of Health (GM64530).

[10] Purification and Assay of Upstream Activation Factor, Core Factor, Rrn3p, and Yeast RNA Polymerase I

By PRASAD TONGAONKAR, JONATHAN A. DODD, and MASAYASU NOMURA

In yeast *Saccharomyces cerevisiae*, 35S ribosomal RNA (rRNA) genes are arranged in tandem repeats of 100–150 copies on chromosome XII. These genes are transcribed by RNA polymerase I (Pol I), whose sole function is transcription of the 35S rRNA genes. Both *in vitro* and *in vivo* studies indicate that the 35S rRNA gene promoter extends about 155 bp upstream from the transcription initiation site. The core promoter, which supports a low level of transcription *in vitro*, extends from +5 to about −38 with respect to the transcription initiation site. A region upstream of the core promoter, the upstream element (UE), which extends to ∼−155, is required for higher levels of transcription *in vitro*.[1]

A genetic screen was used to identify mutants (*rrn* mutants) that are defective in Pol I transcription. These mutant yeast cells can grow in galactose media by synthesizing 35S rRNA by RNA polymerase II (Pol II) from the 35S rRNA gene fused to the *GAL7* promoter, but fail to grow in glucose media that repress transcription from the *GAL7* promoter. In addition to genes encoding subunits of Pol I, this genetic approach led to the identification of seven new genes (*RRN* genes for ribosomal RNA synthesis) that encode Pol I transcription factors or their subunits.[2]

The products of *RRN* genes have been expressed with epitope tags to aid their purification and characterization. From both *in vivo* and *in vitro* studies, it is now clear that in addition to Pol I, other transcription factors, namely core factor (CF), upstream activation factor (UAF), Rrn3p, and TBP are also required for the transcription of 35S rRNA genes. CF is composed of Rrn6p, Rrn7p, and Rrn11p, whereas UAF is composed of Rrn5p, Rrn9p, Rrn10p, UAF30p, and histones H3 and H4. Pol I, CF, and Rrn3p are sufficient for low levels of transcription from the core promoter *in vitro*. For higher levels of transcription, however, a template containing UE in addition to the core promoter is required, and transcription factors, TBP and UAF, are also required in addition to Pol I, Rrn3p, and CF. Rrn3p forms a complex with Pol I, and this complex formation is required to form an active preinitiation complex. The Pol I–Rrn3p complex and CF

[1] D. A. Keys, B.-S. Lee, J. A. Dodd, T. T. Nguyen, L. Vu, E. Fantino, L. M. Burson, Y. Nogi, and M. Nomura, *Genes Dev.* **10,** 887 (1996).

[2] Y. Nogi, L. Vu, and M. Nomura, *Proc. Natl. Acad. Sci. USA* **88,** 7026 (1991).

appear to bind the core promoter weakly. UAF is a sequence-specific DNA-binding factor and binds to the upstream regions of the 35S rRNA gene promoter in the absence of any other components. TBP interacts with the Rrn9 subunit of UAF and Rrn7p subunit of CF and appears to help, together with UAF, to recruit the Pol I–Rrn3p complex to the promoter, forming a stable preinitiation complex (for reviews, see ref. 3 and 4).

After identification and purification of these transcription factors and Pol I were achieved, specific *in vitro* Pol I transcription from 35S rRNA promoter was demonstrated using purified components.[5] The system should be useful to identify additional factors that may modulate Pol I transcription, as exemplified by the identification of Net1p as a factor stimulating Pol I activity *in vivo* and *in vitro*.[6] This article describes methods for the purification of Pol I and Pol I transcription factors. In addition, a protocol for *in vitro* transcription assay is also described.

Purification of Proteins for Reconstitution of Pol I Transcription *In Vitro*

Strains for Purification of Pol I and Pol I Transcription Factors

To facilitate the purification of Pol I and transcription factors, strains carrying epitope-tagged versions of genes encoding components of transcription factors or Pol I have been constructed. These strains are shown in Table I.

General Methods

Growth of Yeast and Escherichia coli *Cells for Protein Purification.* Yeast cells are grown in fermenters (VirTis, CF-CO Medical Industries, Inc.) containing YEPD medium (1% yeast extract, 2% peptone, and 2% dextrose) and 0.01% antifoam A (Sigma Chemical Co.). Yeast cells are grown in minimal medium to midlog phase and are added to fermenters so that A_{600} is ~0.006. Cells are then grown to a cell density of $A_{600} \sim 1$, harvested by centrifugation at 5000 rpm in a Sorvall H6000A swinging bucket rotor, washed, and stored at $-80°$ until required. The yield of cells is about 3.5–4 g/liter.

[3] M. Nomura, *in* "Transcription of Ribosomal RNA Genes by Eukaryotic RNA Polymerase I" (M. R. Paule, ed.), p. 155. Landes Bioscience, Austin, TX, 1998.

[4] M. Nomura, *Cold Spring Harb. Symp. Quant. Biol.* **66,** 555 (2001).

[5] J. Keener, C. A. Josaitis, J. A. Dodd, and M. Nomura, *J. Biol. Chem.* **273,** 33795 (1998).

[6] W. Shou, K. M. Sakamoto, J. Keener, K. M. Morimoto, E. E. Traverso, R. Azzam, G. J. Hoppe, R. M. R. Feldman, J. DeModena, D. Moazed, H. Charbonneau, M. Nomura, and R. J. Deshaies, *Mol. Cell* **8,** 45 (2001).

TABLE I
YEAST AND *E. COLI* STRAINS FOR PURIFICATION OF POL I, CF, UAF, RRN3P, AND TBP

Protein	Strain	Comments	Reference
Pol I	NOY760	*rpa135Δ::LEU2* has plasmid pNOY422, which expresses (His)$_6$-(HA1)$_3$-Rpa135p	5
CF	NOY797	*rrn7Δ::LEU2* has plasmid pNOY403, which expresses (HA1)$_3$-Rrn7p-(His)$_6$	5
UAF	NOY798	*rrn5Δ::TRP1* has plasmid pNOY402, which expresses Rrn5p-(HA1)$_3$-(His)$_6$	7
Rrn3p	*E. coli* BL21(DE3) carrying plasmid pNOY3162	The plasmid carries a gene for Rrn3p with a 27 amino acid residue N-terminal tag (MGS SHHHHHHSSGLVPRGSHMLEDPDK) cloned in the pET15b vector	5
TBP	*E. coli* BL21(DE3) carrying plasmid yIID/6HisT-pET11	The plasmid carries a gene for yeast TBP with an N-terminal MGSSHHHHHHSSGLVPRGSHM tag	8

For purification of recombinant proteins (Rrn3p and TBP) from *E. coli*, cells are grown in fermenters containing LB medium (1% tryptone, 1% NaCl, and 0.5% yeast extract) containing 50 μg/ml ampicillin at 37° to an $A_{600} \sim 0.8$. Cells are induced by the addition of 0.5 mM IPTG for 1.5 h, harvested by centrifugation and stored at $-80°$ until required. Typically, the yield of *E. coli* cells is about 2 g/liter.

Lysis of Cells. About 60 g of yeast cells is resuspended in $\sim$200 ml of the buffer as per the purification protocol given later. Cells are then lysed by passing through a French press (Aminco, Silver Spring, MD) at 20,000 psi twice. Lysis of *E. coli* cells is also carried out using a French press as described earlier but at 8000 psi pressure. The extracts are clarified by ultracentrifugation as given later.

Charging of Chelating Sepharose with Ni. Chelating Sepharose (Pharmacia, NJ) is charged with NiSO$_4$ by presoaking with 100 mM NiSO$_4$ containing 1% Tween 20 overnight. The resin is then centrifuged in a swinging bucket rotor, and the supernatant is removed. The resin is then packed in a column, washed with water, and then equilibrated with the required buffer. Buffers used with Ni-chelating Sepharose should not contain EDTA and reducing agents.

HA1 Antibody Resin. The HA1 antibody is linked to protein G Sepharose (Pharmacia) using dimethyl pimelimidate.[9] The resin may be

[7] J. Keener, J. A. Dodd, D. Lalo, and M. Nomura, *Proc. Natl. Acad. Sci. USA* **94,** 13458 (1997).
[8] A. Hoffmann and R. G. Roeder, *Nucleic Acids Res.* **19,** 6337 (1991).

regenerated after use by washing with a solution containing 100 mM glycine, pH 2.8, and 1.5 M KCl and then with 1 M Tris–HCl, pH 8, and then reused.

Protease Inhibitors. Phenylmethylsulfonyl fluoride (PMSF) is generally used while making extracts at a concentration of 0.2 mM. When required, protease inhibitors, chymostatin, antipain, leupeptin, aprotinin, bestatin, and pepstatin A, are used at a final concentration of 25 μg/ml each.

Pol I

Solutions

Wash buffer: 50 mM Tris–HCl, pH 7.8, 20% glycerol, and 300 mM (NH$_4$)$_2$SO$_4$

Ni breakage buffer: 50 mM Tris–HCl, pH 7.6, 20% glycerol, 500 mM KCl, 5 mM MgCl$_2$, 5 mM imidazole, 0.2 mM PMSF, and 0.1% Tween 20

Ni elution buffer: 20 mM Tris–HCl, pH.7.6, 20% glycerol, 100 mM KCl, 250 mM imidazole, 0.2 mM PMSF, and 0.1% Tween 20

Gradient buffer (GB): 20 mM Tris–HCl, pH 7.8, 20% glycerol, 0.1 mM EDTA, 0.5 mM dithiothreitol (DTT), and 0.2 mM PMSF. GB100, GB160, GB420, and GB750 are gradient buffers containing 100, 160, 420, and 750 mM KCl, respectively.

Purification of Pol I

Pol I is purified from ~85-g cells of strain NOY760, which expresses the (His)$_6$-(HA1)$_3$-tagged Rpa135p subunit. Cells are washed with wash buffer and are then resuspended in Ni breakage buffer and lysed using a French press as described earlier. The extracts are clarified by ultracentrifugation at 100,000 g for 1 h at 4°. The supernatent is mixed with 25 ml Ni-chelating Sepharose preequilibrated with Ni breakage buffer for 1 h at 4°. The resin is then packed in a column containing 5 ml Ni-chelating Sepharose. The column is washed with Ni breakage buffer and eluted by Ni elution buffer.

The eluate from the Ni-chelating Sepharose column is applied to a 1-ml heparin–Sepharose column (Hi-Trap, Pharmacia) preequilibrated with GB100. The column is washed with GB100, and the bound proteins are eluted with a gradient from 100 to 750 mM KCl in GB buffer. Pol I elutes at 375 mM KCl. Fractions containing Pol I are identified by sodium dodecyl sulfate-polyacrylamide gel electrophoresis (SDS-PAGE) followed by

---

[9] E. Harlow and D. Lane, "Antibodies: A Laboratory Manual." Cold Spring Harbor Laboratory, Cold Spring Harbor, NY, 1988.

Coomassie staining, pooled, and diluted with an equal volume of GB buffer and are applied to a 1-ml Mono Q column equilibrated in GB160 buffer. The column is washed with GB160, and Pol I is eluted with a 50-ml linear gradient from 160 to 420 mM KCl in GB buffer. Pol I elutes at around 330 mM KCl, and fractions containing Pol I are identified by SDS-PAGE followed by Coomassie staining. The protein concentration of fractions is determined by measuring A_{280}. The concentration of Pol I purified by this method in peak fractions is around 0.8 mg/ml. Peak fractions are assayed for activity in a transcription assay and stored at $-80°$.

Core Factor

Solutions

Wash buffer, Ni breakage buffer, and Ni elution buffer are the same as for the purification of Pol I.

Heparin wash buffer: Same as gradient buffer (GB) for Pol I, but in addition contains 200 mM KCl, 0.05% Tween 20, and 10 mM MgCl₂

Heparin elution buffer: Same as heparin wash buffer but contains 550 mM KCl instead of 200 mM

DB buffer: Composition of the DB buffer is the same as for gradient buffer (GB) Pol I, but also contains 0.05% Tween 20. DB buffers, DB240, DB250, DB550, and DB600 containing 240, 250, 550, and 600 mM KCl, respectively, are used for CF purification

Peptide elution buffer: Same as DB buffer but contains 240 mM KCl and 4 mg/ml HA1 peptide (peptide sequence: YPYDVPDYA)

Purification of CF

CF is purified from about 110 g of NOY797 cells. The initial steps in purification of CF until Ni elution buffer are the same as for Pol I. Eluates from two preparations are applied to a 5-ml heparin–Sepharose column (Hi-Trap, Pharmacia), which is preequilibrated in heparin wash buffer. The column is then washed with heparin wash buffer and step eluted using heparin elution buffer.

The eluate from the heparin column is mixed with 4 ml of HA1 anti-body resin for 1 h at 4°. The resin is then packed into a column and washed with DB550 and then with DB250. A Q–Sepharose column (1 ml Hi-Trap, Pharmacia) is linked to the HA1 antibody resin, and peptide elution buffer is circulated through the two columns for 1.5 h at 1 ml/min. The Q–Sepharose column is then transferred to an FPLC system and washed with DB240. CF is eluted from the column by applying a 15-ml DB240 to

DB600 linear gradient at 0.25 ml/min. Fractions are analyzed by sodium dodecyl sulfate–polyacrylamide gel electrophoresis (SDS-PAGE) followed by silver staining and stored at −80°. CF elutes at around 350 mM KCl. The purification of CF can be a challenging task because of its instability. An alternate protocol may be used in which extracts are first applied to the HA1 antibody column. The bound protein is eluted using HA1 peptide, and the eluate is applied to a heparin–Sepharose column. This is followed by another HA1 antibody column step.[10]

Upstream Activation Factor

Solutions

UAF extraction buffer: 200 mM Tris–HCl, pH 8.0, 400 mM (NH₄)₂SO4, 20 mM imidazole, 0.1% Tween 20, and 10% glycerol
TA buffer: 20 mM Tris–HCl, pH 8.0, 20% glycerol, and 0.1% Tween 20. KCl is added to 450 mM for TA450 and to 1000 mM for TA1000
TA450 imidazole: 20 mM Tris–HCl, pH 8.0, 450 mM KCl, 250 mM imidazole, 20% glycerol, and 0.1% Tween 20

Purification of UAF

For small-scale purification, NOY798 cells are grown in 30 liter YEPD medium, for large-scale UAF preparation, an ~150- to 200-liter culture may be used. Cells may be grown in several batches of fermenters, washed with UAF extraction buffer, and stored at −80°.

About 60 g of cells is suspended in 200 ml UAF extraction buffer and lysed using a French press as described earlier, PMSF is then added to 0.2 mM. The cell extracts are centrifuged at 11,000 g for 10 min and then at 100,000 g for 1 h. The supernatant is mixed with 20 ml of Ni-chelating Sepharose equilibrated in UAF extraction buffer. After 1 h at 4°, Ni-chelating Sepharose is packed into a column containing ~2 ml Ni-chelating Sepharose and washed with 30 ml TA450. A 5-ml heparin–Sepharose column (Hi Trap, Pharmacia), equilibrated previously with TA450, is attached to the outlet of the Ni-chelating Sepharose column. Proteins bound to Ni-chelating Sepharose are then eluted with 50 ml TA450 imidazole buffer. The heparin column, to which UAF eluted from Ni-chelating Sepharose column is bound, is transferred to the FPLC system. The column is washed with TA450 containing 0.5 mM DTT

---

[10] D. A. Keys, L. Vu, J. S. Steffan, J. A. Dodd, R. T. Yamamoto, Y. Nogi, and M. Nomura, *Genes Dev.* **8**, 2349 (1994).

(TA450–0.5 mM DTT), and bound proteins are eluted using a 22.5-ml linear gradient of TA450–0.5 mM DTT to TA1000–0.5 mM DTT at a flow rate of 0.5 ml/min. Fractions containing UAF are identified by dot blotting an aliquot of each fraction and probing the membrane with the α-HA1 antibody.

At this stage, fractions containing UAF from several batches may be pooled and protease inhibitors are added as mentioned earlier. For large-scale preparations of UAF, fractions containing UAF are mixed with 5 ml HA1 antibody beads for 3 h at 4°. The HA1 antibody beads are centrifuged at 1000 rpm in a Sorvall H6000A swinging bucket rotor and the supernatant is removed. Beads are then washed five times with TA450–0.5 mM DTT buffer containing protease inhibitors by centrifuging in a swinging bucket rotor and removing the supernatant. Beads are then packed in a column, and the outlet of the column is connected to a 5-ml heparin column (Hi Trap, Pharmacia). The outlet of the heparin column flows in a reservoir containing 10 ml TA450–0.5 mM DTT and HA1 peptide and protease inhibitors. The concentration of the HA1 peptide is adjusted so that the final concentration is about 1.0 mg/ml after taking into consideration the buffer present in the tubing, columns, and so on. The reservoir is connected to the HA1 antibody column through a pump, which circulates the reservoir buffer through the two columns overnight in a cold room. The heparin column is then removed and transferred to the FPLC system. The column is first washed with TA450–0.5 mM DTT, and bound proteins are eluted using a 18.5-ml linear gradient of TA450–0.5 mM DTT to TA1000–0.5 mM DTT at a flow rate of 0.5 ml/min. Fractions are analyzed by SDS–PAGE followed by silver staining. UAF elutes in a single peak at around 650 mM KCl.

Rrn3p

Solutions

Rrn3p Ni breakage buffer: This buffer is the same as the Ni breakage buffer for Pol I, but contains 200 mM KCl instead of 500 mM KCl
Ni wash buffer: 20 mM Tris–HCl, pH 7.6, 20% glycerol, 500 mM KCl, 5 mM imidazole, 0.2 mM PMSF, and 0.1% Tween 20
The Ni elution buffer is the same as for purification of Pol I
Q–Sepharose gradient buffer (QB): 20 mM Tris–HCl, pH 7.8, 20% glycerol, 0.1 mM EDTA, 0.5 mM DTT, and 0.1% Tween 20. QB100, QB190, and QB550 are Q–Sepharose gradient buffers containing 100, 190, and 550 mM KCl, respectively.

Purification of Rrn3p

Esherichia coli BL21(DE3) cells carrying the pNOY3162 plasmid for the expression of (His)$_6$-tagged Rrn3p are grown and induced as described earlier. Cells are then harvested by centrifugation. The cell pellet, $\sim$30 g, is resuspended in 120 ml Rrn3p Ni breakage buffer and lysed using a French press as described previously. The extracts are centrifuged at 30,000 g for 40 min, and the supernatant is mixed with 20 ml Ni-chelating Sepharose for 1 h at 4°. The resin is then centrifuged at 500 g, the supernatant is removed, and the resin is packed in a column containing 2 ml Ni-chelating Sepharose. The column is washed with 50 ml Ni wash buffer, and the bound protein is eluted with Ni elution buffer.

A 5-ml Q–Sepharose (Hi-Trap, Pharmacia) column is equilibrated with QB100, and the eluate from the Ni column is applied to the column. The Q–Sepharose column is washed with QB100, and the bound protein is eluted with a 60-ml QB190 to QB550 linear gradient at a flow rate of 1 ml/min. Rrn3p-containing fractions are identified by Coomassie staining. Rrn3p elutes at around 400 mM KCl. Fractions containing Rrn3p are reapplied to Ni-chelating Sepharose and eluted as described previously. The eluate from the Ni-chelating column is then applied to a 1-ml Mono Q (Pharmacia) column equilibrated previously with QB100. Rrn3p is eluted from the column using a 60-ml linear gradient of QB190 to QB550 as described earlier. The purity of fractions is estimated by Coomassie staining, and the protein concentration is determined by the Lowry method.[11] The concentration in the peak fraction is typically around 0.35 mg/ml, and fractions are assayed for activity in the transcription assay.

TBP

Solutions

Ni breakage buffer, Ni wash buffer, and Ni elution buffer are same as for Rrn3p purification. Q–Sepharose buffer is also the same as for Rrn3p, but the KCl concentrations are 100 mM (QB100), 240 mM (QB240), 360 mM (QB360), and 600 mM (QB600).

---

[11] O. H. Lowry, N. J. Rosenbrough, A. L. Farr, and R. J. Randolph, *J. Biol. Chem.* **193,** 265 (1951).

Purification of TBP

Purification of $(His)_6$-TBP from *E. coli* BL21(DE3) has been described previously.[8] The initial steps in the purification of TBP until elution from the Ni-chelating Sepharose column are the same as for Rrn3p. The eluate from the Ni-chelating Sepharose column is applied to a 5-ml heparin–Sepharose column (Hi-Trap, Pharmacia) equilibrated previously with QB240. The column is washed with QB240, and a 180-ml gradient from QB240 to QB600 is then applied to elute the bound proteins. TBP is eluted at around 430 mM KCl from this column. Fractions containing TBP are dialyzed to 100 mM KCl and are then applied to a Mono S column, equilibrated previously in QB100. The column is washed with QB100, and an 80-ml gradient from 100 to 360 mM KCl in QB is then applied at a flow rate of 0.5 ml/min to elute the bound proteins. Fractions containing TBP are identified by SDS-PAGE followed by Coomassie staining, pooled, and stored at $-80°$. TBP is eluted at around 200 mM KCl. The protein concentration is estimated by the Lowry assay,[11] and a typical yield from 30 g of cells is about 3 mg of TBP by the aforementioned method.

Reconstitution of Pol I Transcription *in Vitro*

Solutions

All solutions are made in sterile DEPC-treated water using RNase-free reagents. The solutions are then stored in small aliquots at $-80°$.

2.5X reaction buffer: The recipe to make 1 ml 2.5X reaction buffer is given (the final concentrations of reagents in this 2.5X reaction buffer are in parentheses): 50 μl 1 M Tris–acetate, pH 7.9 (50 mM), 125 μl 2 M potassium glutamate, pH 7.9 (250 mM), 67 μl 0.3 M Mg–acetate (20 mM), 2 μl 1 M DTT (5 mM), 2.5 μl 40 U/μl rRNAsin (Promega) (0.1 U/μl), 50 μl 10 mg/ml acetylated bovine serum albumin (BSA) (0.5 mg/ml) from New England Biolabs, and 703.5 μl water

Dilution buffer: 20 mM Tris–acetate, pH 7.9, 20% glycerol, 0.2 mg/ml acetylated BSA, and 1 mM DTT

20X NTP mix: 4 mM each of ATP, CTP, and UTP and 300 μM GTP

Glycogen precipitation mix: 2 mg/ml glycogen (Sigma type VII from mussel) in 50 mM EDTA, pH 7.5

Heparin solution: Heparin sodium salt (grade I, Sigma) is dissolved in water to make 1 mg/ml solution.

NH_4-acetate in ethanol: Water is added to 100% ethanol to make 95%, which is used to make 1 M NH_4-acetate solution

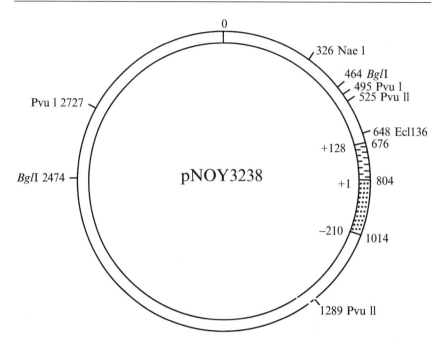

FIG. 1. The 3278-bp plasmid pNOY3238 was constructed by cloning the SmaI/XbaI fragment of the 35S rRNA gene into SmaI/XbaI sites of plasmid pBluescript KS(+). The plasmid was then digested with BglII/XbaI and religated after a fill-in reaction to give rise to the 35S rRNA gene spanning from −210 to +128. The transcription initiation site (+1), the −210 position at the 5′ end of the promoter, and the +128 position in the 35S rRNA gene are located at positions 804, 1014, and 676, respectively, of the plasmid as shown. Some restriction sites useful for linearizing the plasmid for use in *in vitro* transcription assays are also shown. The sequence of the pNOY3238 plasmid is available upon request.

    Gel-loading dye: 90% deionized formamide, 1X TBE (89 mM Tris base, 89 mM boric acid, 2 mM EDTA), 0.05% of xylene cyanol, and bromphenol blue

DNA template: A linear DNA template is used for runoff transcription reactions, which eliminates the effect of transcription termination. Plasmid pNOY3238, in which the 35S rRNA gene spanning from −210 to +128 is cloned, is digested with BglI (or other suitable restriction enzyme, see Fig. 1) and electrophoresed on a preparative agarose gel (other suitable linearized DNA templates may also be used, as in published papers[5,12]). The appropriate fragment

[12] D. L. Riggs and M. Nomura, *J. Biol. Chem.* **265**, 7596 (1990).

containing the 35S rRNA promoter is purified using Gene Clean as per the manufacturer's protocol. The concentration of the DNA is determined by measuring A_{260} of this solution.

PolI/Rrn3p complex: Molar concentrations of the Rrn3p and Pol I preparations are calculated from their protein concentrations. A slight molar excess of Rrn3p is mixed with Pol I and the mixture is incubated for 1–4 h at room temperature. Small aliquots are then made and stored at $-80°$. Excess Pol I in the reaction may lead to the formation of nonspecific transcripts.

Pol I Transcription Assay

Because very small amounts of factors are required for *in vitro* transcription experiments, it is best to make a master mix by adding reaction buffer, DNA template, and proteins and only varying the protein of interest, which minimizes experimental variations. In a typical transcription assay with reaction volume 20 μl, 8 μl of 2.5X reaction buffer, 6 ng DNA template, 0.1–0.2 μg Pol I (as Pol I/Rrn3 p complex), 20–50 ng TBP are used as a starting point. Peak fractions from the purification of CF and UAF are assayed for activity in *in vitro* transcription assays after diluting 5–30–fold with dilution buffer. For setting up transcription assays, it is advisable to first establish the transcription reaction using only the Pol I/Rrn3p complex and CF and later test for stimulation in the presence of TBP and UAF.

The volume of the master mix is adjusted by the addition of water so that 17 μl is used per reaction. Serial dilutions of the protein of interest are prepared using dilution buffer, and 1 μl of the diluted protein is added to the reaction tube containing the master mix and preincubated at room temperature for 10 min. During the preincubation period, the NTP mixture is prepared by mixing 25 μl 20X NTP mix, 1 μl 0.2 M ATP, and 20 μl $[\alpha\text{-}^{32}P]$ GTP (10 μC/μl; 3000 Ci/mmol). After preincubation, 2 μl of NTP mixture is added to the tubes (this makes up the total volume of the reaction 20 μl) to initiate transcription reaction, and incubation is continued at room temperature for 20 min. At the end of 20 min, 1 μl of heparin solution is added, and incubation is continued for an additional 5 min. Heparin binds to free Pol I and does not permit reinitiation of transcription, thus avoiding the formation of incomplete transcripts. The reaction is then stopped by adding 50 μl (1:1) phenol–chloroform and 60 μl glycogen precipitation mix. The sample is then vortexed, and the aqueous phase is withdrawn and added to tubes containing 330 μl NH_4-acetate in ethanol. The tubes are then placed in dry ice for 15–20 min, followed by centrifugation for 15 min at room temperature. At this stage, a small white glycogen pellet should be visible. Remove as much supernatant as possible without disturbing the pellet. The pellet is resuspended in 8 μl gel-loading dye, heated at 95° for 10 min, and analyzed on a 5% acrylamide TBE-urea denaturing gel.

5% Acrylamide TBE–Urea Gel Electrophoresis

For a 50-ml gel solution, 21 g urea is dissolved in a mixture of 8.3 ml 30:0.8 acrylamide:bisacrylamide solution and 5 ml 10X TBE, and water is added to make up the final volume to 50 ml. The solution is degassed, and 250 μl 10% ammonium persulfate solution and 50 μl TEMED are added to polymerize the acrylamide. The gels, with a thickness of 0.75 cm and measuring 15X26 cm, are allowed to polymerize for at least 2 h, and a prerun is done at 700 V for 1 h in 1X TBE running buffer. After loading the samples, electrophoresis is carried out at 700 V until bromphenol blue runs out of the gel. The gel is dried and subjected to autoradiography or phosphorimager analysis.

Concluding Remarks

As mentioned earlier, the *in vitro* Pol I transcription system described in this article should be useful in identifying additional factors modulating Pol I activity *in vivo*. In addition, some of the components in the system, such as Pol I and Rrn3p, are known to be phosphorylated, but the functional significance of such modification is just beginning to be studied.[13] UAF contains histones H3 and H4.[7] In view of a variety of known histone modifications, such as phosphorylation, acetylation, and methylation, and their physiological importance, possible modifications of histones in UAF might also influence Pol I activity. The methods described in this article should be useful in studies on these subjects.

[13] S. Fath, P. Milkereit, G. Peyroche, M. Riva, C. Carles, and H. Tschochner, *Proc. Natl. Acad. Sci. USA* **98**, 14334 (2001).

[11] Purification and Transcriptional Analysis of RNA Polymerase I Holoenzymes from Broccoli (*Brassica oleracea*) and Frog (*Xenopus laevis*)

By Julio Saez-Vasquez, Annie-Claude Albert, Keith Earley, and Craig S. Pikaard

Transcription of ribosomal RNA by RNA polymerase I (pol) is a rate-limiting process for ribosome production and hence for protein synthesis, cell growth, and cell proliferation. RNA polymerase I can be purified in a "holoenzyme" form capable of promoter recognition and accurate transcription initiation *in vitro*. Initial evidence for a pol I holoenzyme came from biochemical studies in plants using broccoli as an inexpensive and abundant source of proliferating cells. Pol I holoenzymes can also be purified from vertebrates. This article presents methods for the purification and transcriptional analysis of RNA polymerase I holoenzymes from broccoli and cultured *Xenopus* cells.

In eukaryotes, nuclear genes are transcribed by three RNA polymerases. RNA polymerase I (pol I) transcribes the repeated ribosomal RNA genes (rRNA genes) that are located at the nucleolus organizer regions.[1-5] Each rRNA gene encodes a primary transcript that is processed into the three largest structural RNAs (18S, 5.8S, 25–28S) of cytoplasmic ribosomes. RNA polymerase II transcribes mRNAs of protein-coding genes and the majority of small nuclear RNAs.[6-9] RNA polymerase III transcribes tRNAs, 5S ribosomal RNA, and other small RNAs.[4,10,11]

For all three polymerase systems, one can purify transcription factors separately from the RNA polymerase core enzyme (the form of the enzyme capable of RNA synthesis but not promoter recognition). However, all three nuclear polymerases can also be purified stably associated with most (or all) of the transcription factors required for promoter-dependent

[1] R. H. Reeder, *Prog. Nucleic Acid Res. Mol. Biol.* **62**, 293 (1999).
[2] K. M. Hannan, R. D. Hannan, and L. I. Rothblum, *Front. Biosci.* **3**, 376 (1998).
[3] I. Grummt, *Prog. Nucleic Acid Res. Mol. Biol.* **62**, 109 (1999).
[4] M. R. Paule and R. J. White, *Nucleic Acids Res.* **28**, 1283 (2000).
[5] T. Moss and V. Y. Stefanovsky, *Cell* **109**, 545 (2002).
[6] G. Orphanides, T. Lagrange, and D. Reinberg, *Genes Dev.* **10**, 2657 (1996).
[7] R. D. Kornberg, *Trends Cell Biol.* **9**, 46 (1999).
[8] T. I. Lee and R. A. Young, *Annu. Rev. Genet.* **34**, 77 (2000).
[9] A. Dvir, J. W. Conaway, and R. C. Conaway, *Curr. Opin. Genet. Dev.* **11**, 209 (2001).
[10] E. P. Geiduschek and G. A. Kassavetis, *J. Mol. Biol.* **310**, 1 (2001).
[11] Y. Huang and R. J. Maraia, *Nucleic Acids Res.* **29**, 2675 (2001).

transcription.[12–20] The latter protein complexes, termed RNA polymerase holoenzymes, may be responsible for promoter recognition and transcription initiation in the cell.

Purification of a cell-free rRNA gene transcription activity from broccoli (*Brassica oleracea*) provided initial evidence for a pol I holoenzyme.[16] Independent studies revealed pol I holoenzymes in *Xenopus laevis*,[17] mouse,[19] and rat.[18] In all cases, holoenzyme activity involves a small proportion (less than 15%) of the total extractable pol I activity. Pol I holoenzyme complexes have been estimated to be at least 2 MDa in mass, which is approximately three- to fourfold larger than the predicted mass of the pol I core enzyme.

In our laboratory, *Brassica* and *Xenopus* pol I holoenzymes have been purified to near homogeneity using, in sequence, preparative anion exchange, cation exchange, gel filtration, analytical anion exchange, and DNA affinity chromatography.[16,17,21] Individual fractions from each column program transcription initiation *in vitro* from a cloned rRNA gene promoter, starting at the same site, and requiring the same promoter sequences, as used *in vivo*. Pol I holoenzyme transcription is also insensitive to α-amanitin, a fungal toxin that inhibits RNA polymerases II and III.[22] Thus the purified holoenzymes retain the characteristics and promoter specificity of pol I transcription in intact cells. Although the identification of holoenzyme subunits is incomplete, biochemical and immunological assays have revealed protein kinase, histone acetyltransferase, topoisomerase, and DNA replication/repair activities.[17,18,23]

This article describes methods for the preparation of cell extracts and the purification and assay of holoenzyme activity. *Brassica* pol I holoenzyme purification is described in the greatest detail. Alterations in this protocol relevant to the purification of pol I holoenzyme activity from cultured *Xenopus* cells is provided in a later section.

[12] A. J. Koleske and R. A. Young, *Nature* **368**, 466 (1994).
[13] A. J. Koleske and R. A. Young, *Trends Biochem. Sci.* **20**, 113 (1995).
[14] V. Ossipow, J. P. Tassan, E. A. Nigg, and U. Schibler, *Cell* **83**, 137 (1995).
[15] Z. Wang, T. Luo, and R. G. Roeder, *Genes Dev.* **11**, 2371 (1997).
[16] J. Saez-Vasquez and C. S. Pikaard, *Proc. Natl. Acad. Sci. USA* **94**, 11869 (1997).
[17] A. C. Albert, M. Denton, M. Kermekchiev, and C. S. Pikaard, *Mol. Cell. Biol.* **19**, 796 (1999).
[18] R. D. Hannan, A. Cavanaugh, W. M. Hempel, T. Moss, and L. Rothblum, *Nucleic Acids Res.* **27**, 3720 (1999).
[19] P. Seither, S. Iben, and I. Grummt, *J. Mol. Biol.* **275**, 43 (1998).
[20] G. Gill, *Essay Biochem.* **37**, 33 (2001).
[21] J. Saez-Vasquez and C. S. Pikaard, *J. Biol. Chem* **275**, 37173 (2000).
[22] D. Michelot and R. Labia, *Drug Metab. Drug Interact.* **6**, 265 (1988).
[23] J. Saez-Vasquez, M. Meissner, and C. S. Pikaard, *Plant Mol. Biol.* **47**, 449 (2001).

Purification of *Brassica* Pol I Holoenzyme Activity

Solutions and Buffers Needed
Extraction buffer: 0.44 M sucrose, 1.25% Ficoll, 2.5% dextran T40, 20 mM HEPES–KOH, pH 7.4, 10 mM MgCl$_2$, 0.5% Triton X-100. Just before use, dithiothreitol (DTT) and protease inhibitors are added [final concentrations: 0.5 mM DTT; 1 mM phenylmethylsulfonyl fluoride (PMSF), 2.5 mg/ml antipain, 0.35 mg/ml bestatin, 0.5 mg/ml leupeptin, 4.0 mg/ml pepstatin A).

The following protease inhibitor stock solutions are needed.

PMSF (purchased from Sigma): 100 mM in isopropanol; store at room temperature. PMSF is highly toxic and should be handled with extreme care, as should all protease inhibitors.
Antipain (Sigma): 1 mg/ml in water; store at $-20°$
Bestatin (Sigma): 1 mg/ml in water; store at $-20°$
Leupeptin (Sigma): 5 mg/ml in water; store at $-20°$
Pepstatin A (Sigma): 1 mg/ml in methanol; store at $-20°$
Nuclei resuspension buffer (RB0): 50 mM HEPES–KOH pH7.9, 20% glycerol, 10 mM EGTA, 10 mM Mg(SO$_4$)$_2$

The following RB buffers are made by mixing RB0 and RB1000 (containing 0 or 1000 mM KCl, respectively) buffers in appropriate ratios.

RB100 buffer: Nuclei resuspension buffer containing 100 mM KCl
RB175 buffer: Nuclei resuspension buffer containing 175 mM KCl
RB400 buffer: Nuclei resuspension buffer containing 400 mM KCl
RB800 buffer: Nuclei resuspension buffer containing 800 mM KCl
CB100 (column buffer, 100 mM KCl): 20 mM HEPES–KOH, pH 7.9, 20% glycerol, 0.1 mM EDTA, 0.1 M KCl
CB1000: 20 mM HEPES–KOH, pH 7.9, 20% glycerol, 0.1 mM EDTA, 1.0 M KCl
Coomassie blue dye-binding protein assay[24] mix: 2 ml of Bio-Rad protein assay concentrate mixed with 7 ml H$_2$O. Then disperse 90 μl in wells of a microtiter dish.

Crude Nuclear Extract Preparation

The extract preparation is adapted from the method of Kieber and Signer.[25] Extracts are prepared in the cold room at 4° using prechilled equipment and glassware. Whenever possible, samples are kept on ice. Centrifuges and rotors are precooled to 4°.

[24] M. M. Bradford, *Anal. Biochem.* **72**, 248 (1976).

1. Choose broccoli at the local grocery store. If the proliferating floral buds are dark green and tightly closed, the broccoli is suitable for an extract (or a meal). Use a razor blade or scalpel to slice off the floral bud clusters that consist of the most rapidly proliferating cells. Avoid stem tissue.

2. Pour 100 g of tissue into a stainless steel Waring blender in the cold room. Add 200 ml of extraction buffer. Using the highest speed setting, homogenize with six to seven pulses of 5 s each, waiting 5 s between pulses.

3. Filter homogenate through two layers of Miracloth (Calbiochem) into a 500-ml glass beaker (baked, sterile). Wearing clean gloves, twist the Miracloth to compact the cell debris and squeeze out as much liquid as possible.

4. Centrifuge homogenate at 10,000 g (7500 rpm in Beckman rotor JA10) for 30 min at 4° to obtain a crude nuclear pellet that will also contain starch (white) and cell debris. Discard the supernatant.

5. Resuspend pellet in 20 ml of RB0 buffer, amended with DTT (0.5 mM) and PMSF (1 mM), using a 10-ml glass pipette to achieve a homogeneous suspension. Transfer to a sterile (baked) 150-ml glass beaker.

6. While swirling the beaker by hand, slowly add 1.16 g of solid NaCl (final concentration ~1.0 M). The solution will become viscous as nuclei lyse and release chromatin. Incubate for 10 min on ice, swirling the beaker occasionally.

7. Add a sterile (autoclaved) stir bar to the nuclear lysate and stir at low speed on a magnetic stirrer in the cold room. Add 0.4 ml of 50% (w/v) polyethylene glycol (PEG 8000, in water), one drop at a time. Stir gently for 20 min.

8. Pour lysate into a sterile oak ridge tube. Centrifuge at 19,000 g (12,500 rpm in a Beckman JA20 rotor) for 30 min at 4°.

9. Recover supernatant into a fresh oak ridge tube and repeat centrifugation.

10. Recover supernatant (20 ml) into a sterile graduated cylinder and dilute with RB0 to a final volume of 100 ml. Cover with parafilm and invert repeatedly to mix. Long strands of chromatin will be apparent.

11. Filter supernatant through two layers of Miracloth into a sterile glass beaker. Chromatin will adhere to the Miracloth. Wearing gloves, squeeze out as much liquid as is practical without squeezing chromatin through the Miracloth.

12. Measure the volume of the filtrate. Transfer to a chilled beaker on a stirring motor and stir at low speed. Over a 20- to 30-min period, add 0.33 g of solid $NH_4(SO_4)_2$ (ammonium sulfate) for every milliliter of solution, letting small amounts dissolve before adding more. The final

---

[25] J. J. Kieber, M. F. Lopez, A. F. Tissier, and E. Signer. *Plant Mol. Biol.* **18**, 865 (1992).

solution will be ~55% saturated for ammonium sulfate. The liquid will become cloudy as proteins precipitate. Stir for an additional 45 min after all ammonium sulfate has dissolved.

13. Pour into four oak ridge tubes. Centrifuge at 18,000 g for 30 min at $4°$.

14. After centrifugation, a pellet (green in color) forms along the walls of the centrifugation tube and a green pellicle sometimes floats at the surface. Save both the pellet and the pellicle by removing the supernatant using a pipette.

15. Resuspend each pellet in 3 ml of RB100 amended with 0.5 mM DTT and protease inhibitors (1 mM PMSF, 5.0 μg/ ml antipain, 3.5 μg/ ml bestatin, 5.0 μg/ml leupeptin, 4.0 μg/ ml pepstatin A). Pool the four resuspended pellets into one tube, giving a total volume of $\approx$12.5 ml. Using a sterile 10-ml glass pipette, pipette up and down to make the suspension homogeneous. If necessary, use a chilled 15-ml Dounce homogenizer.

16. Centrifuge the protein suspension at 3000 g for 25 min at $4°$ (5000 rpm in the JA20 rotor) to pellet insoluble material. Save the supernatant, avoiding the slimy pellet.

17. Dialyze the supernatant at $4°$ against 500 ml of RB100 amended with 0.5 mM DTT and 1 mM PMSF. After 2–3 h, replace with fresh dialysis buffer and dialyze overnight.

18. Centrifuge the dialyzed extract at 19,000 g for 30 min at $4°$ to pellet insoluble material. Save the pale green supernatant.

19. Begin chromatography immediately (for best results) or freeze the supernatant in liquid nitrogen and store at $-80°$. Thawed extract should be centrifuged at 19,000 g for 30 min at $4°$ to pellet any insoluble material.

Pol I Holoenzyme Purification

The following procedure results in highly purified pol I holoenzyme.[16,21,23] The full purification scheme involves DEAE-Sepharose (anion exchange), Biorex 70 (cation exchange), Sephacryl S300 (gel filtration), Mono Q (anion exchange), and DNA-cellulose (DNA affinity) chromatography (see Fig. 1). Purification through Mono Q is sufficient for most assays and provides multiple fractions for correlating holoenzyme activity with the presence of specific polypeptides (see Figs. 2A and 2B).

DEAE-Sepharose CL-6B Chromatography. For ~12.5 ml of protein extract (from 100 g broccoli; ~20 mg protein), use a column containing ~10 ml of DEAE-Sepharose CL-6B (Pharmacia). For larger volumes, scale up proportionately. We reuse DEAE columns a maximum of three times. Columns can be run using fast protein liquid chromatography

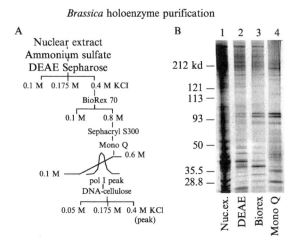

FIG. 1. Purification of RNA polymerase I holoenzyme activity from broccoli (*Brassica oleracea*). (A) Summary scheme for the purification procedure involving sequential chromatography using DEAE-Sepharose CL-6B, Biorex 70, Sephacryl S300, Mono Q, and double-stranded calf thymus DNA-cellulose. (B) Polypeptide composition of peak fractions at several steps in the purification procedure. Aliquots of the crude nuclear extract, DEAE 400 mM KCl fraction, Biorex 800 mM KCl fraction, and Mono Q peak were subjected to 10% SDS–polyacrylamide gel electrophoresis. Proteins were visualized by silver staining (Bio-Rad). Positions of molecular weight markers are indicated. The protein compositions of peak DNA-cellulose and Mono Q fractions are almost identical (data not shown).

(FPLC; Pharmacia in our case) or using a column clamped to a ring stand with tubing at the outlet linked to a fraction collector and with buffer flow controlled by gravity.

Wash the column with 5–10 column volumes of high salt (RB1000) buffer and then equilibrate with 5–10 volumes of low salt (RB 100) buffer. Load the nuclear extract and collect 2- to 4-ml fractions. Wash with RB100 until flow-through protein levels fall to baseline. If using FPLC, this is determined by monitoring UV_{280} absorbance. If using a column on a ring stand, 10-μl aliquots of each fraction are added to 90 μl of Bio-Rad protein assay mix in a well of a microtiter dish. Blue color intensity is proportional to protein content. Visual inspection is sufficient to determine when a return to baseline has been achieved.

Elute the column with RB175. Pool peak fractions for later analyses.

Elute pol I holoenzyme activity with RB400. Save peak fractions (~10–12 ml total for a 10-ml column) and dialyze against RB100 or use an FPLC desalting column (e.g., Pharmacia HiPrep 26/10) equilibrated and run in RB100 to accomplish buffer exchange.

A Testing mono Q fractions for promoter-dependent transcription (holoenzyme activity)

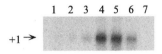

1 2 3 4 5 6 7

+1 →

Mono Q fraction: 11 12 13 14 15 16 17

B Immunoblotting of mono Q fractions to detect pol I subunits

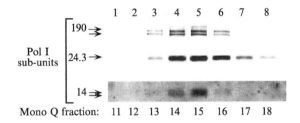

1   2   3   4   5   6   7   8

Pol I sub-units

190 →

24.3 →

14 →

Mono Q fraction: 11 12 13 14 15 16 17 18

C Testing DNA-cellulose fractions for holoenzyme activity

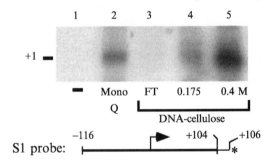

1   2   3   4   5

+1 ▬

▬ Mono   FT   0.175   0.4 M
   Q   └——————————————————┘
              DNA-cellulose

          −116   +104   +106
S1 probe: ├————————————————————┘   └─┐*

FIG. 2. RNA polymerase I subunits copurify with holoenzyme activity. (A) Mono Q fractions were tested for their ability to program accurate transcription initiation from a cloned *B. oleracea* rRNA gene promoter (pBor2). (B) Mono Q fractions (50 μl) were acetone precipitated, subjected to electrophoresis on a 12.5% SDS–polyacrylamide gel, and blotted to nitrocellulose membranes. Western blots were probed with antiserum raised against the last exon of the largest *A. thaliana* RNA polymerase I subunit (190-kDa subunit) or against the 24.3- and 14-kDa RNA polymerase subunits. (C) Comparison of the abilities of Mono Q and DNA-cellulose fractions to program accurate transcription initiation from a cloned *B. oleracea* rRNA gene promoter (lanes 2–5). RNAs transcribed *in vitro* were detected using the S1 nuclease protection assay. The probe was 5′ end labeled at a plasmid polylinker site located at +116. Transcripts initiated at the correct start site are labeled "+1." Lane 1 is a control reaction with no protein added. Data are reproduced, with permission from the publisher, from Saez-Vasquez and Pikaard.[21]

Biorex 70 Chromatography. Biorex 70 (Bio-Rad) is prepared according to the manufacturer's instructions. To maximize flow rates, fine particles should be removed by repeated sedimentation. Twelve milliliters of DEAE-400 fraction (in RB100) can be loaded onto ~2 ml Biorex 70 in an FPLC column (e.g., Pharmacia HR5/5) or a disposable column. Wash with 5 column volumes (10 ml) of RB100. Elute holoenzyme activity with RB800, collecting 0.5- to 1.0-ml fractions. Peak protein fractions are determined using UV absorbance or the Bio-Rad protein assay.

Sephacryl S300 Chromatography. To purify the holoenzyme according to mass while desalting the Biorex 800 fraction, we use gel filtration (size-exclusion) chromatography on a 190-ml Sephacryl S300 FPLC column (Pharmacia) equilibrated with CB100. The Biorex 800 fraction is loaded, and 1-ml fractions are collected and tested for polymerase I activity (see assays later). Pol I-containing fractions corresponding to masses larger than ~600 kDa (generally fractions 70–80 for a 190-ml Sephacryl S300 column) are pooled for subsequent Mono Q chromatography. Estimated masses are based on calibration of the column with Blue Dextran (average molecular mass, 2 MDa; to determine the void volume); thyroglobulin (670 kDa); apoferritin (440 kDa); catalase (250 kDa); and bovine serum albumin (66 kDa). Standards are purchased from Sigma.

Mono Q Chromatography. The approximately 10-ml Sephacryl S300 fraction representing masses from ~660 kDa to the void volume (in CB100) is loaded onto a 2 ml Mono Q FPLC column (Pharmacia) equilibrated in CB100 and run at a flow rate of 1 ml/min. The column is washed with CB100 and is then eluted using a 20-ml (10 column volumes) linear gradient from 100 to 600 mM KCl, collecting 1-ml fractions. Fractions are then dialyzed against RB100 (amended with 0.5 mM DTT) and tested using the promoter-independent polymerase assay as well as the promoter-dependent transcription activity (see Fig. 2A). Immunoblotting can also be used to verify the presence of polymerase subunits (Fig. 2B) or casein kinase subunits.[23] Peak pol I holoenzyme activity elutes at ~400 mM KCl.

DNA-Cellulose Affinity Chromatography. Peak Mono Q fractions dialyzed against RB100 are pooled, diluted with an equal volume of RB0 (to make KCl 50 mM), and loaded onto 0.3 ml calf thymus DNA-cellulose (Sigma) equilibrated in RB50 in a small disposable column (Bio-Rad). The flow through is reapplied several times to maximize DNA binding. The column is washed with 3 ml of RB50 and is then eluted sequentially with 0.5 ml RB175 and 0.5 ml RB400. Fractions are then dialyzed against RB100 and tested for polymerase activity. Some holoenzyme activity elutes in the 175 mM KCl fraction (~15–20%) but the majority elutes at 400 mM (Fig. 2C).

Transcription Assays

Promoter-Independent (Nonspecific) RNA Polymerase I Activity Assay

RNA polymerases in cell extracts or chromatography fractions are detected using a simple enzyme assay[26–29] based on promoter-independent transcription initiation at nicks or broken ends of DNA. Incorporation of one or more radioactive ribonucleotide triphosphates into RNA is monitored using anion-exchange paper (Whatman DE81) washed with 0.5 M sodium phosphate to elute unincorporated nucleotide triphosphates. Substituting manganese for magnesium reduces the specificity of the polymerase and increases nucleotide incorporation.

Inclusion of α-amanitin in the reactions at low concentration (2 μg/ml) will block RNA polymerase II transcription in both animals and plants.[22] At higher concentrations of α-amanitin (140 μg/ml), both RNA polymerases II and III are blocked in animals, allowing one to monitor only pol I transcription.[27] In plants, however, pol III is resistant to α-amanitin,[29] requiring such high concentrations of the toxin (>1 mg/ml) that its use is impractical to discriminate pol I from pol III. Note that α-amanitin is highly toxic and can cause death if ingested, thus it must be handled with extreme care.

Solutions Needed

Transcription reaction mix (2X concentrate; α-labeled ^{32}P-GTP used as the radioactive label): 100 mM HEPES–KOH, pH 7.9; 4 mM MnCl$_2$; 1 mM UTP; 1 mM rATP; 1 mM rCTP; 0.08 mM unlabeled (not radioactive) rGTP; 2 mg/ml acetylated bovine serum albumin (Boehringer); 10 μg/ml α-amanitin (Sigma); 50 μg/ml calf thymus DNA (Sigma); 100 mM KCl.

The transcription reaction mix is made just prior to use. Stock solutions of nucleotide triphosphates (100 mM in water), DNA (10 mg/ml), and α-amanitin (1 mg/ml in water) are stored frozen at $-20°$. Bovine serum albumin is stored in solution at $4°$. HEPES and MnCl$_2$ stocks (1 M) are stored at room temperature. DNA is dissolved in water at 10 mg/ml and is sheared by repeatedly forcing the solution through a hypodermic needle (25 gauge).

0.5 M sodium phosphate, dibasic (Na$_2$HPO$_4$); add 268 g of dibasic sodium phosphate very slowly to 2 liters of Milli Q-purified water

[26] L. B. Schwartz and R. G. Roeder, *J. Biol. Chem.* **249,** 5898 (1974).
[27] R. G. Roeder, *J. Biol. Chem.* **258,** 1932 (1983).
[28] D. R. Engelke, B. S. Shastry, and R. G. Roeder, *J. Biol. Chem.* **10,** 1921 (1983).
[29] T. J. Guilfoyle, *Plant Physiol.* **58,** 453 (1976).

that is stirring vigorously in a 2- to 4-liter beaker. If too much is added at one time, the crystals aggregate and do not dissolve.

Procedure

1. Dialyze all fractions into 100 mM KCl buffer (RB100 or CB100).
2. In a microcentrifuge tube, make a reaction master mix that contains 20 μl of 2X transcription reaction mix and 0.25 μl α-labeled ^{32}P-GTP (1 μCi/μl) for every reaction to be performed (e.g., for 10 reactions, mix 200 μl of 2X reaction mix and 2.5 μl of ^{32}P-GTP). Use appropriate precautions for the safe handling of radioactive compounds, including the use of protective clothing, eye protection, and shielding.
3. Pipette 20 μl of dialyzed fractions to be tested into numbered microcentrifuge tubes. Be sure to include a buffer (no added protein) control.
4. To each numbered microcentrifuge tube add 20 μl of reaction master mix. Vortex at low speed.
5. Incubate 20–30 min at room temperature.
6. Cut Whatman DE-81 (anion-exchange) chromatography paper into squares that are 1–2 cm on a side. Uniformity is needed because background radioactivity levels are proportional to filter size. Lay the filter squares on parafilm or aluminum foil. Number each square with a soft pencil. Pipette each 40-μl reaction onto a separate filter. Immerse the filter immediately in 50 ml 0.5 M Na$_2$HPO$_4$. All the filters can be pooled in the same beaker.
7. Wash the filters for 5 min, using an orbital shaker to achieve moderate agitation. Pour off the first wash solution into a radioactive waste container. Repeat the wash at least three more times to remove unincorporated ^{32}P-GTP.
8. Wash filters 5 min in 95% ethanol. Pour off the wash solution. Repeat ethanol wash.
9. Wash filters with a small volume as (usually ~20 ml) of diethyl ether to remove the ethanol and promote rapid drying of the filters. This step is optional, but saves time.
10. Lay filters on a sheet of aluminium foil and allow them to air dry for several minutes.
11. Place filters in scintillation vials and add scintillation cocktail.
12. Determine the number of radioactive decays per minute emitted from each filter using the ^{32}P channel of a scintillation counter.
13. If testing successive fractions eluted from a column, plot incorporated cpm versus fraction number to visualize elution of polymerase peaks (e.g., see Fig. 4A).

A *Xenopus* pol I holoenzyme purification scheme

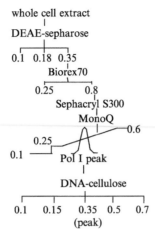

whole cell extract
|
DEAE-sepharose
|
0.1   0.18 0.35
|
Biorex70
|
0.25   0.8
|
Sephacryl S300
|
MonoQ
0.6
0.25
0.1 ———— Pol I peak
|
DNA-cellulose
|
0.1   0.15   0.35   0.5   0.7
                   (peak)

B Peak *Xenopus* pol I holoenzyme fractions

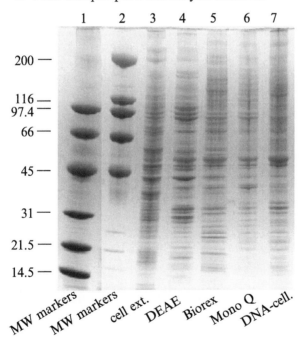

FIG. 3. *Xenopus* pol I holoenzyme purification. (A) Purification scheme involving DEAE-Sepharose CL-6B, Biorex 70, Sephacryl S300, Mono Q, and double-stranded calf thymus DNA-cellulose chromatography. (B) Polypeptide composition of peak fractions throughout

Testing mono Q column fractions for *Xenopus* pol I activity

A Promoter-independent polymerase activity

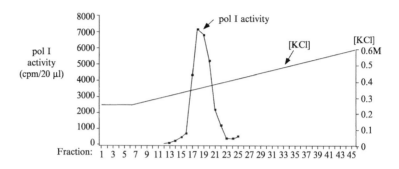

B Promoter-dependent transcription

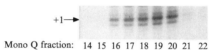

Mono Q fraction: 14 15 16 17 18 19 20 21 22

FIG. 4. Transcription assays to detect Mono Q-fractionated *Xenopus* pol I. (A) Detection of the pol I peak using the promoter-independent transcription assay. Fractions were tested for their ability to catalyze the incorporation of labeled nucleotide triphosphates into RNA. Transcription reactions were conducted in the presence of 200 μg/ml α-amanitin, which would block pol II or pol III (if they were present) and used sheared calf thymus DNA as the template. Counts per minute incorporated into RNA (in 20 μl of reaction mix) are plotted versus fraction number. (B) Detection of Mono Q fractions capable of accurate, promoter-dependent pol I transcription. Mono Q fractions were tested for their ability to program transcription using a supercoiled *X. laevis* minigene (Ψ40) that includes a complete promoter fused to plasmid sequences. Transcripts were detected using the S1 nuclease protection assay. Data are reproduced, with permission from the publisher, from Albert *et al.*

Promoter-Dependent Pol I Transcription and Detection of Transcripts

The promoter-independent polymerase assay described earlier allows peaks of polymerase activity to be identified but does not distinguish core polymerase activity from holoenzyme activity. The latter requires assaying fractions for the ability to catalyze transcription initiated from the correct

the purification procedure. Aliquots of whole cell extract, DEAE 350 mM KCl fraction, Biorex 800 mM KCl fraction, Mono Q peak fraction, and DNA-cellulose 350 mM KCl fraction were subjected to SDS–PAGE. Proteins were visualized by staining with Coomassie brilliant blue R-250. Masses of molecular weight markers are indicated to the left of the gel. Data are reproduced, with permission from the publisher, from Albert *et al.*

start site of a cloned rRNA gene promoter. We typically assay *Brassica* holoenzyme activity using the minigene construct pBor2.[21] This plasmid contains *B. oleracea* rRNA gene sequences from −517 to +104 (relative to the transcription start site, +1) cloned as an *Eco*RI–*Xba*I restriction fragment into pBluescript II KS⁻ (Stratagene). Our preferred assay for detecting transcripts is S1 nuclease protection using a DNA probe that spans the transcription start site and is 5′ end labeled on the antisense strand within plasmid polylinker sequences downstream of +104. The S1 nuclease protection assay allows accurate initiation and transcript abundance to be determined simultaneously.

Solutions Needed

2X transcription reaction mix: 30 mM HEPES, pH 7.9, 80 mM potassium acetate, 12 mM magnesium acetate, 1 mM DTT, 20 μg/ml α-amanitin (Sigma), 1 mM of each ribonucleotide triphosphate (ATP, UTP, CTP, GTP).

Transcription stop solution: 150 mM NaCl, 50 mM Tris–HCl, pH 8.0, 250 mM sodium acetate, pH 5.3, 3 mg/ml yeast tRNA, 6 mM EDTA, pH 8.0

Formamide hybridization buffer: 40 mM PIPES, pH 6.4, 400 mM NaCl, 1 mM EDTA, 80% deionized formamide

S1 digestion buffer: 5% glycerol, 1 mM zinc sulfate, 30 mM sodium acetate (pH 5.2), 50 mM NaCl, 125–250 units/ml S1 nuclease (Sigma)

Formamide-dye gel loading buffer: 90% deionized formamide, 10 mM NaOH, 1 mM EDTA, 0.1% (1 mg/ml) bromphenol blue, 0.1% (1 mg/ml) xylene cyanol

Procedure

1. Mix 20 μl of each purified fraction with ∼200 fmol (∼0.5 μg) of supercoiled pBor2 plasmid in a 1.5-ml microcentrifuge tube. Incubate 5 min at 25°.

2. Add 20 μl of 2X transcription mix. Mix by tapping the tube or by vortexing at low speed.

3. Incubate transcription reactions for 1–2 h at 25°.

4. Add 360 μl of transcription stop solution and mix well.

5. Extract reactions with an equal volume of phenol:chloroform:isoamyl alcohol (25:24:1) and transfer aqueous phase to a clean tube.

6. Add ∼50,000 cpm of probe DNA, which is the *Sph*I–*Xba* I (sequences −116 to +116) fragment of pBor2, 5′ end labeled at +116 using T4 polynucleotide kinase and γ-labeled ^{32}P-ATP. Sequences +104 to +116 of this fragment are derived from the pBluescript plasmid polylinker.

7. Precipitate nucleic acids with 2.5 volumes of ethanol.

8. Resuspend the RNA/probe pellet in 30 μl formamide hybridization buffer.

9. Cover hybridization reaction with 50 μl of mineral (paraffin) oil to prevent evaporation.

10. Incubate the hybridization reaction at 90–95° in a dry bath incubator for 15 min to denature the probe and RNA.

11. Quickly transfer tubes to a 37° water bath. Allow RNA and probe to hybridize 2 h to overnight.

12. Add to each tube 270 μl of S1 digestion buffer. Centrifuge tubes for 5 s in a microcentrifuge to get the S1 digestion buffer through the paraffin oil layer. Vortex briefly. Repeat the centrifugation step to obtain the layer of oil at the top again.

13. Incubate S1 nuclease digestion reactions for 30 min at 37°.

14. Stop reactions by removing 280 μl from the bottom of the tube (to avoid paraffin oil) to a fresh tube containing 10 μl of 10% SDS and 5 μl of 0.5 M EDTA. The EDTA chelates the zinc, stopping enzymatic activity. Vortex briefly to mix.

15. Add 30 μl of 7.5 M ammonium acetate and vortex to mix.

16. Add 1 ml cold ($-20°$) absolute ethanol to precipitate DNA/RNA hybrids. Vortex and store at $-80°$ for 30 min.

17. Collect pellets by centrifugation at 14,000 g for 15 min at room temperature. Wash pellets with 70% ethanol and then allow pellets to air dry.

18. Resuspend pellets in 6–8 μl of formamide-dye sample buffer (contains NaOH to degrade RNA such that only probe DNA remains). Heat samples 3 min at 95°.

19. Load samples onto a 6–8% urea-PAGE sequencing gel alongside molecular weight markers (a sequencing ladder using a primer 5' end labeled at the same nucleotide labeled for the S1 probe is ideal). Dry the gel and expose it to X-ray film or a phosphorimager screen.

Purification of *Xenopus laevis* Pol I Holoenzyme Activity

Cell extract preparation is the major difference between the protocols for isolating *Brassica* and *Xenopus* pol I holoenzyme activity. The column chromatography procedures are very similar, with only minor changes in salt concentrations needed. Promoter-independent and promoter-dependent transcription assays are also similar, although reaction buffers and S1 nuclease assay probes differ.

Xenopus laevis transcription extracts are prepared according to McStay and Reeder,[30] with only minor modifications. *X. laevis* cell line Xlk2 (established by Judy Roan and Ron Reeder; Fred Hutchinson Cancer

Research Center, Seattle) is grown as an adherent monolayer culture in glass or plastic roller bottles at room temperature in 50% Leibovitz's L-15 medium (+ glutamine; GIBCO), 5% fetal bovine serum (GIBCO), 5% NuSerum (Collaborative Research), and 1X penicillin (5 units)/streptomycin (5 μg/ml) (GIBCO; purchased as a 1000X stock). Cells are split approximately 1:8 every 10–14 days when the cultures become confluent. Alternatively, cells are harvested at confluence for preparation of transcription extracts.

Solutions Needed for Xenopus Extracts

Phosphate-buffered saline (PBS): 150 mM NaCl, 2 mM Na$_2$HPO$_4$. Adjust to pH 8.0 with 0.1 M NaH$_2$PO$_4$

Hypotonic buffer: 10 mM HEPES, pH 7.9, 10 mM KCl, 1.5 mM MgCl$_2$, 0.2 mM EDTA, 0.2 mM EGTA. Just before use, amend with 1 mM DTT, 0.1 mM PMSF, and 2 μg/ml aprotinin (final concentrations)

High salt buffer: 300 mM HEPES, pH 7.9, 1.4 M KCl, 30 mM MgCl$_2$. Just before use, add DTT to 1 mM

Dialysis buffer: 25 mM HEPES, pH 7.9, 20% glycerol, 100 mM KCl, 0.2 mM EDTA, 0.2 mM EGTA, 1 mM DTT, 0.1 mM PMSF

Procedure

1. Harvest cells using 1 mM EDTA in phosphate-buffered saline. The EDTA causes the cells to release cells from the walls of the culture vessels.

2. Collect cells in 250-ml conical centrifuge tubes by centrifugation at 350 g for 10 min at 4° (we use a tabletop centrifuge).

3. Resuspend cell pellets in PBS (without EDTA) using 100 ml of PBS for a packed cell volume of 10–15 ml. Divide into two 50-ml conical tubes and pellet the cells again by centrifugation.

4. Resuspend the pellet in each 50-ml conical tube using hypotonic buffer. The total cell plus buffer volume should be 50 ml. Shake the tubes vigorously for 15 s. Let the tubes incubate on ice for 15 min. The cells will swell in size two to three times their original packed cell volume.

5. Centrifuge the tubes as before to pellet the cells. Remove as much supernatant as possible using a pipette. Note the packed cell volume.

6. Add 2 packed cell volumes of hypotonic buffer. Shake tubes vigorously for 15 s and then incubate on ice for 15 min. Note the total volume.

7. Pour the swollen cells into a chilled 40-ml Dounce homogenizer on ice and homogenize with 15–20 strokes of an "A" (tight) pestle. Check cell

---

[30] B. McStay and R. H. Reeder, *Cell* **47**, 913 (1986).

lysis by examining aliquots under a microscope at 200X magnification (phase-contrast optics is helpful). Homogenize until at least 90% of the cells appear to be broken. With experience, one can see free nuclei as a result of homogenization.

8. Pour the homogenate into a 50-ml plastic conical tube. Add 0.16 volumes of high salt buffer and shake the tube vigorously for 5 s.

9. Transfer the homogenate to an oak ridge centrifuge tube and incubate on ice for 10 min. Invert tubes every 1–2 min to mix.

10. Centrifuge for 15 min at 25,000 g in a fixed angle rotor at $4°$ (14,200 rpm in a Beckman JA20 rotor).

11. Remove the lipid layer that forms at the top using a Pasteur pipette. Pour the supernatant into ultracentrifuge tubes and balance them. We use tubes appropriate for a swinging bucket rotor (Beckman SW 28).

12. Centrifuge the tubes at 100,000 g for 1.5 h at $2–4°$ (24,000 rpm in the Beckman SW 28 rotor).

13. Carefully aspirate the lipid layer off the top of the supernatant and then pour the supernatant into a 50-ml conical tube, avoiding the milky membranous layer at the bottom of the tube (just above a yellowish cell debris pellet), which will flow slowly when the supernatant is decanted.

14. Dialyze the extract against 1 liter of dialysis buffer for at least 4 h. Overnight dialysis is acceptable. A viscous precipitate will form during dialysis and the volume of the extract will decrease to approximately one-half the original volume.

15. Centrifuge the dialyzed extract for 10 min at 10,000 g at $4°$. The supernatant will be milky white.

16. The protein concentration in the extract is generally in the range of 15–25 mg/ml.

17. Begin chromatography procedures immediately or flash-freeze the extract in a polypropylene conical centrifuge tube placed in liquid nitrogen. Store frozen extracts at $-80°$.

Chromatography of Xenopus Pol I Holoenzyme Activity. Purification of the *Xenopus* pol I holoenzyme, like the *Brassica* holoenzyme, is performed using DEAE-Sepharose, Biorex 70, Sephacryl S300, Mono Q, and DNA-cellulose chromatography. The purification scheme is summarized in Fig. 3A. Cell extracts are first fractionated on DEAE. Proteins that elute in CB350 are diluted to 250 mM KCl and are loaded onto a Biorex column equilibrated in CB250. The Biorex column is then eluted with CB800 and the eluate is loaded onto a 190-ml Sephacryl S300 gel filtration column equilibrated in CB100. The holoenzyme peak that elutes from the Sephacryl column in fractions corresponding to masses greater than 1 MDa is then fractionated on Mono Q, using a 10 column volume gradient from 250 to 600 mM KCl. Holoenzyme activity peaks in fractions eluted at 0.35–0.38 M KCl. Peak Mono Q fractions are then pooled, dialyzed to 100 mM KCl (CB100), and subjected to chromatography on double-stranded DNA-cellulose (Sigma). The DNA-cellulose column is washed with CB100 and is then step eluted with CB150, CB350, CB500, and CB700. Holoenzyme activity elutes at 350 mM KCl.

Promoter-Dependent Transcription. Xenopus pol I holoenzyme transcription reaction conditions are adapted from McStay and Reeder.[30] Reactions are typically performed using 40-μl reactions containing 10% glycerol, 25 mM HEPES, pH 7.9, 90 mM KCl, 6 mM MgCl$_2$, 1 mM DTT, 150 μg/ml α-amanitin, 0.5 mM each of ATP, UTP, CTP, GTP, and 200–400 ng of *X. laevis* rRNA gene template DNA (typically, the minigene Ψ40[31]). Reactions are incubated 2 h at 25° and transcripts are detected using the S1 protection assay with a 5'end-labeled, single-stranded 65 nucleotide probe complementary to RNA strand sequences from −15 to +50. The RNA and probe are hybridized overnight at 65° in 0.3 M NaCl, 10 mM Tris–HCl, pH 7.5, 1 mM EDTA. S1 digestion is for 1 h at 37° in 5% glycerol, 50 mM NaCl, 30 mM sodium acetate pH 4.5, 1 mM zinc sulfate, and 100–150 units/ml S1 nuclease. Protected fragments are then subjected to denaturing polyacrylamide gel electrophoresis and are visualized by autoradiography or phosphorimaging (see Fig. 4B).

Activities Associated with Brassica *and* Xenopus *Pol I Holoenzymes*

Biochemical and immunological tests have detected several holoenzyme-associated activities that suggest that pol I holoenzymes are similar in animals and plants. For instance, casein kinase 2 copurifies with both *Xenopus* and *Brassica* pol I holoenzymes.[17,23] Holoenzyme fractions also contain at least one histone acetyltransferase activity in both the *Xenopus*[17] and *Brassica* (unpublished) systems. In *Xenopus*, TATA-binding protein has been shown to copurify with pol I holoenzyme activity, whereas the more abundant vertebrate transcription factor UBF does not.[17] It is not yet clear if TATA-binding protein is also part of the *Brassica* pol 1 holoenzyme, although this seems likely.

A priority for future research is to identify the polypeptides present in RNA polymerase I holoenzyme fractions, as this is likely to reveal proteins that integrate pol I transcription with other cellular processes, such as growth signaling, cell cycle control, and DNA repair mechanisms. The similarities between *Brassica* and *Xenopus* holoenzymes, including their similar chromatographic properties and their analogous associations with

[31] P. Labhart and R. H. Reeder, *Cell* **37**, 285 (1984).

casein kinase 2 and histone acetyltransferase activities, suggest that the identification of proteins in either system is likely to be relevant to the other. The fact that broccoli is readily available in large quantities favors its use for such studies. Furthermore, *Brassica* species are very closely related to *Arabidopsis thaliana*, for which complete genome sequence data exist, such that limited amino acid sequence data can be combined with database searching to aid in the identification of holoenzyme subunits.

[12] Assays and Affinity Purification of Biotinylated and Nonbiotinylated Forms of Double-Tagged Core RNA Polymerase II from *Saccharomyces cerevisiae*

By Maria L. Kireeva, Lucyna Lubkowska,
Natalia Komissarova, and Mikhail Kashlev

The molecular mechanisms of transcription by RNA polymerase II (Pol II) have been studied in various *in vitro* systems, which utilize a highly purified core enzyme of Pol II (with or without additional protein factors). The Pol II core enzyme is isolated using a combination of ion-exchange chromatography and immunoaffinity chromatography with an antibody against the Rpb1 subunit.[1,2] The alternative to the immunoaffinity approach is introduction of an affinity tag. This strategy was applied successfully to mammalian Pol II, where a FLAG epitope was inserted into the Rpb9 subunit, which allowed achieving a 1000-fold purification in one step.[3] We have introduced hexahistidine and biotin acceptor peptide tags into the amino terminus of the Rpb3 subunit of yeast Pol II.[4] According to the Pol II core enzyme[5] and elongation complex[6] crystal structures, tags are located on a surface of the enzyme opposite from the active center and sites of interaction with the DNA and RNA. Indeed, the tagged enzyme is indistinguishable from the wild-type Pol II in the *in vitro* activity test, and tagged Rpb3 rescues the *rpb3* temperature-sensitive phenotype as

[1] A. M. Edwards, S. A. Darst, W. J. Feaver, N. E. Thompson, R. R. Burgess, and R. D. Kornberg, *Proc. Natl. Acad. Sci. USA* **87**, 2122 (1991).
[2] N. E. Thompson and R. R. Burgess, *Methods Enzymol.* **274**, 513 (1996).
[3] E. Kershnar, S.-Y. Wu, and C.-M. Chiang, *J. Biol. Chem.* **273**, 34444 (1998).
[4] M. L. Kireeva, N. Komissarova, D. S. Waugh, and M. Kashlev, *J. Biol. Chem.* **275**, 6530 (2000).
[5] J. Fu, A. L. Gnatt, D. A. Bushnell, G. J. Jensen, N. E. Thompson, R. R. Burgess, P. R. David, and R. D. Kornberg, *Cell* **98**, 799 (1999).
[6] A. L. Gnatt, P. Cramer, J. Fu, D. A. Bushnell, and R. D. Kornberg, *Science* **292**, 1876 (2001).

efficiently as wild-type Rpb3.[4] The double tag insertion allowed us to develop and optimize a rapid purification protocol for Pol II: the biotinylated core enzyme is isolated in two steps by chromatography on a metal affinity column and monomeric avidin resin. An alternative protocol, which includes three chromatographic steps, was developed for the purification of nonbiotinylated histidine-tagged Pol II.

Along with the affinity purification opportunuties, the introduction of a tag provides a valuable tool for immobilization of the RNA polymerase in the solid phase and separation of the transcription complexes from the free nucleic acids, thus promoting analyses of RNA polymerase interactions with RNA and DNA. This approach has been used successfully for analyses of various aspects of transcription elongation and termination by *Escherichia coli* RNA polymerase[7–14] and yeast Pol II.[4,15] Another promising application of the tagged enzyme is a single molecule assay addressing the movement of the polymerase along the template.[16,17] This method requires the presence of a biotin residue on the surface of the enzyme for attachment of the polymerase molecule to the streptavidin-coated bead. Thus, introduction of the double (hexahistidine and biotin-containing) tag is exteremely useful for both purification of the enzyme and functional assays. This article is dedicated to purification of Pol II using hexahistidine and biotin tags and to the utilization of immobilized Pol II in various experimental systems.

Materials and Reagents

This section describes materials, reagents, and buffers required for purification of the double-tagged Pol II and characterization of the purified enzyme. It includes the protocol for expression and purification of biotin

[7] M. Kashlev, E. Martin, A. Polyakov, K. Severinov, V. Nikiforov, and A. Goldfarb, *Gene* **130**, 9 (1993).

[8] N. Komissarova and M. Kashlev, *Proc. Natl. Acad. Sci. USA* **94**, 1755 (1997).

[9] N. Komissarova and M. Kashlev, *J. Biol. Chem.* **272**, 15329 (1997).

[10] D. Wang, T. Meier, C. Chan, G. Feng, D. Lee, and R. Landick, *Cell* **81**, 341 (1995).

[11] I. Artsimovich and R. Landick, *Genes Dev.* **12**, 3110 (1998).

[12] I. Gusarov and E. Nudler, *Mol. Cell* **3**, 495 (1999).

[13] M. Kashlev and N. Komissarova, *J. Biol. Chem.* **277**, 14501 (2002).

[14] N. Komissariova, S. Solter, J. Becker, M. L. Kireeva, and M. Kashlev, *Mol. Cell* **10**, 1151 (2002).

[15] M. L. Kireeva, W. Walter, V. Tchenajenko, V. Bondarenko, M. Kashlev, and V. M. Studitsky, *Mol. Cell* **9**, 541 (2002).

[16] M. D. Wang, M. J. Schnitzer, H. Yin, R. Landick, J. Gelles, and S. M. Block, *Science* **282**, 902 (1998).

[17] R. J. Davenport, G. J. White, R. Landick, and C. Bustamante, *Science* **287**, 2497 (2000).

protein ligase BirA, which was used for Pol II biotinylation *in vitro*, the description of the yeast strain producing the double-tagged enzyme, and preparation of the cell lysate used as a source of Pol II.

All chemicals are from Sigma (St. Louis, MO) unless indicated otherwise. Ni^{2+}-NTA agarose and TALON resin are from Qiagen (Chatsworth, CA). SoftLink Soft Release avidin resin is from Promega (Madison, WI). GST-Sepharose, factor Xa protease, and the Mono Q HR5/5 column are from Amersham Pharmacia Biotech (Piscataway, NJ). The protein concentration is determined with the protein dye reagent (Bio-Rad, Hercules, CA) using bovine serum albumin (BSA) as the standard. Proteins are resolved under denaturing reducing conditions in the 4–12% gradient Nu-Page bis–Tris polyacrylamide gel with MES running buffer (all from Invitrogen, Carlsbad, CA) and stained with QuickBlue stain (Boston Biologicals, Wellesley, MA). RNA is resolved in a 20% acrylamide gel under denaturing conditions and quantified on a phosphorimager (Typhoon, Molecular Dynamics, Sunnyville, CA).

Buffers

All buffers used for protein purification are supplemented with the following protease inhibitors:[18] 1 mM phenylmethylsulfonyl fluoride (PMSF), 2 mM benzamidine, 2 μM pepstatin A, 0.6 μM leupeptin, 2 μg/ml chymostatin, and 5 μg/ml antipain.

> Lysis buffer: 150 mM Tris–acetate, pH 7.9, 50 mM potassium acetate, 5 mM $MgCl_2$, 10 mM $ZnCl_2$, 2 mM 2-mercaptoethanol, 0.5 mM EDTA
>
> BL buffer: 1×PBS (137 mM NaCl, 2.7 mM KCl, 4.3 mM Na_2-$HPO_4 \bullet 7H_2O$, 1.4 mM KH_2PO_4), 10 mM $MgCl_2$, 2 mM 2-mercaptoethanol
>
> BE buffer: 50 mM Tris–HCl, pH 8.0, 10 mM $MgCl_2$, 2 mM 2-mercaptoethanol, 10 mM glutathione
>
> TB buffer: 20 mM Tris–HCl, pH 7.9, 5 mM $MgCl_2$, 10 mM $ZnCl_2$, 2 mM 2-mercaptoethanol
>
> HB buffer: 20 mM Tris–HCl pH 8.0, 10 mM $ZnCl_2$, 0.5 mM EDTA, 1 mM 2-mercaptoethanol, 10% glycerol
>
> QB buffer: 20 mM Tris–acetate, pH 7.9, 10 mM $ZnCl_2$, 2 mM 2-mercaptoethanol, 0.5 mM EDTA 10% glycerol

The potassium chloride, ammonium sulfate, and potassium acetate concentrations in mM for buffers TB, HB, and QB, respectively, are indicated in parentheses in the text (e.g. TB(40) is TB containing 40 mM KCl).

---

[18] A. J. Koleske, D. M. Chao, and R. A. Young, *Method Enzymol.* **273,** 176 (1996).

GST-BirA Expression and Purification

The biotinylation of Pol II carrying the tagged Rpb3 subunit is performed *in vitro* by a recombinant GST-BirA fusion protein,[4,19] which is purified in one affinity chromatography step from *E. coli* BL21-based strain C14 [malE52::Tn10/dcm ompT lon λDE3] carrying pDW364 and pLysS (Novagen, Madison, WI) plasmids. pDW364 is a plasmid derived from the pGEX-3X vector (Amersham Pharmacia Biotech) by insertion of the BirA-coding sequence via *Bam*HI and *Eco*RI restriction sites. Here we provide a step-by-step protocol for GST-BirA expression and purification from this strain.

1. Inoculate 10 ml of LB containing 50 μg/ml ampicillin and 25 μg/ml chloramphenicol with a single colony of the C14-pDW364 strain, grow overnight at 37°, and start a 2-liter culture of LB with the same antibiotics.

2. Grow to OD_{600} of 0.6–0.7 at 30° (it usually takes 3–4 h) and induce with 1 mM IPTG for 2–3 h.

3. Harvest bacteria by centrifugation (10 min at 3000 rpm in a Sorvall SLA 3000 rotor) and resuspend in 30 ml of BL buffer.

4. Lyse bacteria on ice by sonication using a 500-W ultrasonic processor with a 13-mm probe (15–20 9-s pulses with 40% amplitude separated by a 10-s pause between the pulses is sufficient to achieve complete lysis).

5. Remove cell debris by centrifugation (15 min at 10,000 rpm in a Sorvall SLA 1500 rotor) and filter the supernatant through a PES Nalgene filter (Nalge Nunc International, Rochester, NY).

6. Incubate the lysate with 1 ml of GST Sepharose beads in a 50-ml conical tube on a rocking platform at 4°. Collect the beads by centrifugation (Sorvall R6000B table top centrifuge, 1000 rpm, 5 min), wash with 50 ml of lysis buffer, resuspend in 2 ml of lysis buffer, and pack into a disposable column.

7. Wash the column by gravity flow with 10 ml of BL buffer at room temperature.

8. Elute GST-BirA by gravity flow with BE buffer at room temperature. Collect 0.5-ml fractions. Fractions containing more than 1 mg/ml of protein should be combined. The typical yield is 15–20 mg of GST-BirA from 2 liters of culture. The protein in BE buffer can be frozen in liquid nitrogen and stored at −70° for more than a year without a detectable loss of activity.

[19] K. L. Tsao, B. DeBarbieri, H. Michel, and D. S. Waugh, *Gene* **169,** 59 (1996).

9. To obtain BirA free of the GST domain, treat the fusion protein with factor Xa protease, which cleaves between GST and BirA domains, according to the manufacturer's instructions. Isolate BirA by chromatography on a Mono-Q HR5/5 column exactly as described by O'Callaghan et al.[20] Note. We found that removal of the GST tag increases the activity of BirA only slightly and, therefore, is not necessary for the efficient biotinylation of Pol II.

BJ5464 Rpb3 His-Bio Strain

The BJ5464 Rpb3 His-Bio strain was derived from a protease-deficient BJ5464 strain[21] in which the wild-type Rpb3 open reading frome (ORF) was substituted by the tagged Rpb3 ORF. The tag contains six histidine residues followed by a 26 amino acid substrate for a bacterial biotin protein ligase BirA.[4,19] The complete translated sequence of the tag is MGSHHHHHHSNSGLNDIFEAQ**K**IEWHEDTG. The biotinylated lysine residue is shown in bold. Gene replacement was done according to the standard protocol,[22] and substitution of the wild-type copy by a tagged copy was verified by Southern hybridization.

Yeast Whole Cell Lysate Preparation

1. To prepare lysate from 100 to 120 g of wet yeast cells, start 500 ml of YPD culture with the Rpb3 His-Bio strain, grow overnight, inoculate a 10-liter fermentor (Biostat B, B. Braun Biotech, Melsungen, Germany), and grow to stationary phase (OD_{600} of 5). Concentrate the culture to 1 liter using a Sartocon slice filter cassette with a Hydrosart membrane (Sartorius, Goettingen, Germany), wash cells with water until the permeate runs clear, and collect by centrifugation (5–10 min, 3000 rpm in a Sorvall SLA 1500 rotor). At this point, cell paste can be frozen at $-70°$ for several months. However, because thawing and resuspension of the frozen yeast cells take at least 2 h, it may be more convenient to proceed with the lysis step immediately after cell harvesting.

2. Rinse 200 ml (dry volume) of acid-washed glass beads (400–600 μm, Sigma) with the lysis buffer. Resuspend the cells in 3× lysis buffer using 50 ml of the 3× lysis buffer per 100 g of cells. Pour the cell suspension into the stainless steel chamber of the BeadBeater apparatus (Biospec Products, Bartlesville, OK) and fill the chamber with wet glass beads.

[20] C. A. O'Callaghan, M. F. Byford, J. R. Wyer, B. E. Willcox, B. K. Jacobsen, A. J. McMichael, and J. I. Bell, *Anal. Biochem.* **266,** 9 (1999).
[21] E. W. Jones, *Methods Enzymol.* **194,** 428 (1991).
[22] S. Sherer and R. W. Davis, *Proc. Natl. Acad. Sci. USA* **76,** 4951 (1979).

Disrupt yeast by twenty 30-s pulses with 60-s pauses between runs using a digital controller (VWR International, West Chester, PA). The chamber should be chilled in an ice saltwater bath during the run. *Note.* The 400-ml stainless steel chamber is optimal for the disruption of cells obtained from a 10-liter culture. Lysis of larger amounts of yeast cells requires multiple runs. Processing of smaller amounts of yeast cells in the same chamber is undesirable because a significant fraction of protein will be lost on the beads or the lysate will be diluted excessively. A procedure described by Krogan *et al.*[23] may be used if processing of 10–20 g of yeast cells is preferred.

3. When cell disruption is completed, disassemble the chamber and let the beads settle. Collect the lysate using a 10-ml pipette. The beads can be rinsed once with 30–50 ml of the lysis buffer.

4. Remove insoluble material from the collected lysate by centrifugation (20 min, 10,000 rpm, Sorvall SLA-3000 rotor). We obtain approximately 200 ml of the lysate with 30–35 mg/ml total protein from 100 g of cell paste. The lysate can be frozen in liquid nitrogen in 40-ml aliquotes and stored at $-70°$.

Activity Assays and Optimization of Purification Conditions

This section includes the basic activity assay, which was used to detect transcriptionally active Pol II in the whole cell lysate and in fractions obtained throughout purification. We also included experimental protocols used for optimization of Pol II purification by affinity chromatography on Ni^{2+}-NTA agarose and on SoftLink Soft Release avidin resin. These protocols illustrate how insertion of the double affinity tag provides flexibility in manipulations with Pol II from the crude cell lysate and may serve as a basis for the design of various solid-phase systems with Pol II tailored to specific experimental goals. In addition, they can be adapted for isolation and activity testing of Pol II from several milliliters of culture and, therefore, may be especially valuable for simultaneous biochemical analyses of many Pol II mutants.

Activity Assay for Histidine-Tagged Pol II

The activity of the Pol II core enzyme can be determined by a factor-independent reconstitution of the elongation complex (EC) from the enzyme, RNA primer, and two DNA oligonucleotides.[4] However, the basic EC assembly protocol, described in detail elsewhere,[24] is not suitable for

[23] N. J. Krogan, M. Kim, S. H. Ahn, G. Zhong, M. S. Kobor, G. Cagney, A. Emili, A. Shilatifard, S. Buratowski, and J. F. Greenblatt, *Mol. Cell. Biol.* **22,** 6979 (2002).

activity testing during purification, as the crude yeast lysate contains multiple ribonucleases, which degrade the RNA primer instantly. This section outlines a modification of the basic assembly protocol (Fig. 1). The main differences between the basic[24] and the modified protocol are that the polymerase is immobilized on Ni^{2+}-NTA agarose and that the bulk of the lysate proteins is removed from the beads by a high-salt washing before RNA and DNA addition.

1. Incubate 1 μl of lysate with 5 μl of Ni^{2+}-NTA agarose [10 μl of 50% suspension of beads in TB(40)] for 30 min at room temperature with constant shaking in an Eppendorf thermomixer (Brinkmann, Westbury, NY) at 1400 rpm, incubate for 10 min in TB(1000), and wash with 1 ml of TB(40). All washes are done by resuspending the agarose in 1 ml of wash buffer, spinning down the pellet for about 10 s in a six-tube tabletop centrifuge (VWR, West Chester, PA) and removing all but 50–100 μl of supernatant.

2. All assembly steps are done at 25°. Combine the immobilized Pol II [10 μl of Ni^{2+}-NTA agarose suspension in TB(40)] with 1 pmol of the RNA:DNA hybrid[24] and incubate for 10 min. We routinely use 41 nucleotide template and nontemplate DNA strands and a 9 nucleotide RNA primer (Fig. 1). For quantitative analyses, it is convenient to use a RNA primer labeled by 5′ end phosphorylation with γ-[^{32}P]ATP (7000 Ci/mmol, ICN Biomedicals, Irvine, CA).[24]

3. Add 10 pmol of the nontemplate DNA oligonucleotide for 10 min.

4. Wash the EC with 1 ml of TB(40), incubate with 1 ml of TB(1000) for 10 min, and wash again with TB(40).

5. Incubate with 10 μM of each ATP, CTP, and GTP for 5 min so that the 9 nucleotide primer in catalytically active ECs is extended to 20 nucleotides. The known amount (3–100 fmol) of the free labeled RNA should be loaded on the same gel and used as a standard for EC quantification. *Note*: The efficiency of the EC formation by Pol II immobilized from the lysate is the same as that of the purified Pol II. The possible limitations of this assay for the quantitation of Pol II in the crude lysate are discussed in more detail later.

Binding Capacity of Ni^{2+}-NTA Agarose for Pol II

The first step in affinity purification of double-tagged Pol II is metal affinity chromatography. We used the EC reconstitution assay to determine the saturation point of Ni^{2+}-NTA agarose by Pol II from the whole

---

[24] N. Komissarova, J. Becker, M. L. Kireeva, and M. Kashlev, *Methods Enzymol.* **370**, 233 (2003).

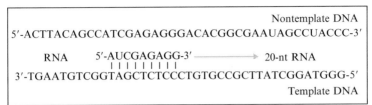

Nontemplate DNA
5′-ACTTACAGCCATCGAGAGGGACACGGCGAAUAGCCUACCC-3′

RNA   5′-AUCGAGAGG-3′ ——————▶ 20-nt RNA
        | | | | | | | | |
3′-TGAATGTCGGTAGCTCTCCCTGTGCCGCTTATCGGATGGG-5′
Template DNA

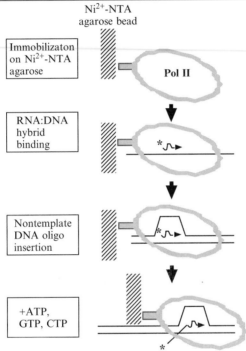

Ni²⁺-NTA
agarose bead

Immobilizaton on Ni²⁺-NTA agarose

Pol II

RNA:DNA hybrid binding

Nontemplate DNA oligo insertion

+ATP, GTP, CTP

FIG. 1. Assembly of EC with immobilized Pol II. (Bottom) A tagged enzyme (oval) immobilized on a surface of a Ni²⁺-NTA agarose bead, formation of the EC on a single-strand template (solid line) with annealed RNA primer (wavy line with arrow), and incorporation of a fully complementary nontemplate DNA strand to complete the transcription bubble formation. The position of the radioactive label on the 5′ end of the RNA is indicated by an asterisk. (Top) Sequences of RNA and DNA oligonucleotides, and the 20 nucleotide transcript synthesized by the active EC upon incubation with ATP, CTP, and GTP.

yeast cell lysate and optimized the binding conditions to increase the selectivity of the Pol II interaction with Ni²⁺-NTA agarose. For this purpose, we incubated 10 μl of the lysate with a decreasing amount of the Ni²⁺-NTA agarose (Fig. 2). Immobilization of histidine-tagged Pol II was done by a 30-min incubation at 25° with constant shaking with Ni²⁺-NTA agarose suspension in TB(40) in all experiments described later.

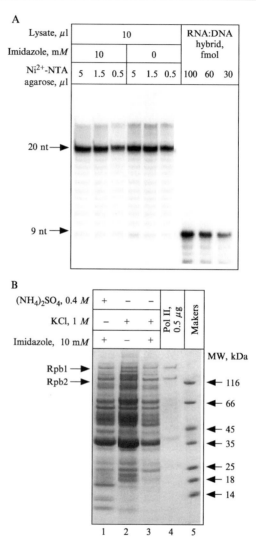

FIG. 2. Binding of hexahistidine-tagged Pol II from the yeast whole cell lysate to Ni^{2+}-NTA agarose. (A) Binding capacity of Ni^{2+}-NTA agarose. Ten microliters of lysate was mixed with 5 μl of 3 M KCl and 1.5 μl of 1 M imidazole (where indicated) and was incubated with 5, 1.5, or 0.5 μl of Ni^{2+}-NTA agarose for 30 min with constant shaking. The agarose was washed as described in the text, the volume of the suspension was adjusted to 10 μl, and the immobilized enzyme was used for EC formation with 2 pmol of the RNA:DNA hybrid and 20 pmol of the nontemplate DNA strand. The complex was treated exactly as described in the activity assay protocol (steps 2–5). RNA products are indicated by arrows. (B) Proteins bound to Ni^{2+}-NTA agarose from the whole cell lysate. Forty microliters of lysate was mixed with 20 μl of 3 M KCl, 6 μl of 100 mM imidazole, or 12.5 μl of 2 M ammonium sulfate where indicated,

Shorter incubation times may result in incomplete binding, especially if the amount of protein is close to saturation of the affinity resin.

The binding capacity of Ni^{2+}-NTA agarose for Pol II from the crude cell lysate was 100 pmol (50 μg) of active Pol II per 1 ml of Ni^{2+}-NTA agarose (note that here and below the volume of agarose pellet, not the 50% suspension, is indicated; the pellet is always resuspended in TB, and the total volume of the suspension, which may vary, will be indicated in the protocols) (Fig. 2A). It is significantly lower than the capacity reported by the manufacturer (5–10 mg of protein per 1 ml of Ni^{2+}-NTA agarose). Two batches of Ni^{2+}-NTA agarose tested in parallel were indistinguishable; the Co^{2+} – based TALON resin has an even lower binding capacity (data not shown).

The low binding capacity of the affinity resin can, at least in part, be caused by a competition of the lysate proteins with the histidine-tagged Pol II for binding to the Ni^{2+}-NTA groups. We performed the binding in the presence of 1 M KCl to prevent hydrophilic interactions of the lysate proteins with the agarose. To determine optimal conditions for Pol II purification by metal affinity chromatography, we examined the protein composition of the fraction bound to Ni^{2+}-NTA agarose in the presence of $(NH_4)_2SO_4$, imidazole, and glycine, which are known to inhibit interactions of histidine-containing proteins with the Ni^{2+} ion. Indeed, the addition of 10 mM imidazole to the lysate slightly reduced the binding of contaminating proteins (Fig. 2B, lanes 2 and 3). However, the addition of 500 mM $(NH_4)_2SO_4$ to the lysate increased the amount of some low molecular weight contaminating proteins, but, again, did not affect Pol II binding (lanes 1 and 2). Glycine at a concentration above 10 mM inhibited both Pol II and contaminating protein binding with equal efficiency (data not shown). In summary, we could not achieve any noticeable increase in the resin-binding capacity for Pol II from the whole cell lysate under any conditions tested. The size of the Ni^{2+}-NTA agarose column used for Pol II purification was determined on the basis of experimental results shown in Fig. 2A.

Biotinylation of Double-Tagged Pol II

The second step of double-tagged Pol II purification is affinity chromatography on a SoftLink Soft Release avidin resin, which is based on reversible interaction of the monomeric avidin with a biotin acceptor peptide,

---

incubated with 10 μl of Ni^{2+}-NTA agarose, and washed with TB(40). Proteins were eluted from the beads with 30 μl of TB(40) containing 100 mM imidazole. Lane 4 contains 0.5 μg of purified Pol II. The molecular weight of the markers (lane 6) is indicated at the right.

biotinylated *in vitro* with the BirA enzyme. A number of examples of the *in vitro* biotinylation of proteins carrying the biotin acceptor domain[25] or biotin acceptor peptides[26,27] have been described to date (reviewed by Cull and Schatz[28]). Notably, biotinylation of proteins in the crude cell lysate is not widely used, mostly due to the high activity of proteases and a low concentration of the substrate protein in the lysate, which may lead to a low efficiency of biotinylation.[28] However, in the case of tagged Pol II, efficient biotinylation in the crude cell lysate was critical for the development of a fast and simple affinity purification procedure. Our goal was to use the first purification step, metal affinity chromatography, not only to remove the bulk of the whole cell lysate proteins from Pol II, but also to separate biotinylated Pol II from free biotin and use affinity chromatography on the avidin column as the next (and the last) step. To achieve the highest possible yield of the biotinylated protein, we have tested multiple conditions of the biotinylation reaction.

The extent of biotinylation in various reaction conditions was addressed by comparing the amount of Pol II immobilized via the hexahistidine tag to Ni^{2+}-NTA agarose to the amount of Pol II immobilized via the biotin to avidin resin. The assay reproduces the first steps of the Pol II purification protocol: the biotinylation reaction is done in the crude cell lysate and is followed by binding of Pol II to Ni^{2+}-NTA agarose and subsequent elution of the bound proteins with imidazole performed in a batch format. Then, partially purified Pol II is reimmobilized on either avidin or Ni^{2+}-NTA resin, and activity of the immobilized Pol II is measured by the EC assembly. We applied this approach to determine the effect of the BirA concentration, biotin concentration, time, and temperature on biotinylation efficiency. A similar assay was also used to determine the size of the Soft-Link Soft Release avidin resin column for the preparative purification. Importantly, because BirA activity is inhibited by an increased ionic strength and glycerol,[26] the yeast lysis buffer has a low salt concentration and does not contain glycerol. The detailed protocol for the titration of GST-BirA in the biotinylation reaction is given next as a sample assay, and the result of the titration is shown in Fig. 3.

1. To 100 μl of the yeast whole cell lysate, add ATP to 1 mM from a 100 mM stock solution, D-biotin to 100 μM from a 200 mM stock solution

[25] J. E. Cronan, Jr., *J. Biol. Chem.* **265,** 10327 (1990).
[26] P. J. Schatz, *Bio/Technology* **11,** 1138 (1993).
[27] P. Saviranta, T. Haavisto, P. Rappu, M. Karp, and T. Lovgren, *Bioconj. Chem.* **9,** 725 (1998).
[28] M. G. Cull and P. J. Schatz, *Methods Enzymol.* **326,** 430 (2000).

in dimethyl sulfoxide (DMSO), and purified GST-BirA. Incubate for 90 min at $4°$.

2. Add 50 μl of 3 M KCl and combine with 50 μl of Ni^{2+}-NTA agarose. Incubate for 30 min with shaking at $25°$.

3. Wash five times with 1 ml of TB(300) to remove unbound proteins and free biotin.

4. Adjust the volume of the suspension to 90 μl, add 10 μl of 1 M imidazole (pH 7.9), and incubate for 10 min at $25°$.

5. Withdraw 40 μl of supernatant. Dilute to 400 μl with TB(40). Incubate 100 μl of the diluted imidazole eluate with 10 μl of Ni^{2+}-NTA agarose or with 10 μl of SoftLink Soft Release avidin resin for 30 min at $25°$ with constant shaking.

6. Add 1 ml of TB(1000), incubate for 10 min, and wash twice with 1 ml of TB(40). Adjust the volume of the Ni^{2+}-NTA agarose or SoftLink Soft Release avidin resin suspension to 100 μl.

7. Test the activity by reconstitution of the EC as described in steps 2–5 of the activity assay protocol. Use 5 μl of the agarose per 1 pmol of the RNA:DNA hybrid.

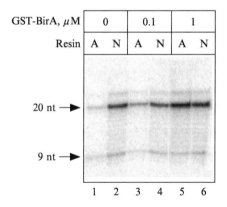

FIG. 3. Optimization of biotinylation conditions of the double-tagged Pol II: Titration of GST-BirA. One hundred-microliter aliquots of the yeast cell lysate were treated as described in the text (optimization of biotinylation protocol, step 1) without the addition of GST-BirA or using 0.1 and 1 μM of the enzyme. Free biotin was removed by Pol II immobilization on Ni^{2+}-NTA agarose and washing the beads (steps 2 and 3), and Pol II was eluted with imidazole (step 4) and reimmobilized on Ni^{2+}-NTA (lanes marked N) and SoftLink Soft release avidin (lanes marked A) resins (steps 5 and 6). EC assembly was done using one-tenth of the reimmobilized Pol II with 1 pmol of the RNA:DNA hybrid and 10 pmol of nontemplate DNA (step 7). The equal amount of ECs formed on the Ni^{2+}-NTA and avidin resins when the lysate was incubated with 1 μM GST-BirA (lanes 5 and 6) indicates that all molecules of Pol II carry both hexahistidine and biotin tags in these conditions.

Purification of Double-Tagged Pol II

The protocols described here were developed and tested for the medium-scale purification (from 20 g of wet yeast cell paste) of both biotinylated and nonbiotinylated forms of the double-tagged Pol II. One 40-ml aliquot of the frozen lysate described earlier was used for each run. All purification steps should be done at 4–8°.

Affinity Purification of Biotinylated Pol II

1. Equilibrate the 10-ml Ni^{2+}-NTA agarose column (2.6 × 4 cm) with lysis buffer containing 1 M KCl. Prepare the 5-ml column (1 × 7 cm) with SoftLink Soft Release avidin resin according to the manufacturer's instructions and equilibrate it with TB(300).

2. To the yeast cell lysate obtained from 20 g of cells, add purified GST-BirA to 1 μM (75 μg/ml) concentration, ATP to 1 mM from a 100 mM stock, and biotin to 100 μM from a 200 mM stock solution in DMSO. Incubate the lysate for 90 min with stirring to complete biotinylation. Slowly add 3 M stock of KCl to a final concentration of 1 M and 1 M stock of imidazole (pH 7.9) to final concentration of 10 mM.

3. Load the lysate to the Ni^{2+}-NTA column at 0.5 ml/min.

4. Wash the column with 100–200 ml of TB(1000) at 0.2–0.5 ml/min (the wash step can be done overnight).

5. Elute the protein from the Ni^{2+}-NTA agarose column with TB(300) containing 200 mM imidazole at 0.5 ml/min. The elution progress can be monitored by measuring UV absorbtion.

6. When the imidazole eluate starts to come out from the Ni^{2+}-NTA agarose column, it is convenient to connect it to the SoftLink Soft Release avidin resin column and load the imidazole eluate directly to the avidin column.

7. When the imidazole elution from the Ni^{2+}-NTA agarose column is complete (15–20 ml of the elution buffer), separate the columns and wash the avidin resin with 100 ml of TB(300) at 1 ml/min.

8. Elute Pol II with TB(300) containing 5 mM biotin at 0.5 ml/min. The peak can be monitored by measuring UV absorption or 5-ml fractions should be collected and tested for activity.

9. Concentrate the eluted protein by ultrafiltration in a Centiprep (Millipore, Bedford, MA) with a 50- or 100-kDa molecular weight cut-off membrane to approximately 0.5–1 mg/ml, add glycerol to 20% final concentration, and freeze in aliquots in liquid nitrogen. Store the protein at −70°. The typical yield is 0.2–0.3 mg of Pol II from 40 ml of lysate (20 g of wet cells).

Purification of Nonbiotinylated Pol II Core Enzyme

Parts of the following protocol for the purification of Pol II are adapted from the protocols for the Pol I, II, and III purification reported by Huet *et al.*[29] and from the Pol II holoenzyme purification procedure by Koleske *et al.*[18]

1. Add 2 *M* stock of ammonium sulfate to the yeast whole cell lysate obtained from 20 g of cells to achieve 0.4 *M* final concentration. Clarify the lysate by centrifugation in a Beckman 45Ti rotor for 2 h at 35,000 rpm.

2. Prepare a 40-ml heparin HyperD (BioSepra, Cergy-Saint-Christophe, France) column (2.5 × 8 cm) and equilibrate it with buffer HB(100).

3. Dilute the lysate fourfold with HB(0) buffer and load it to the heparin HyperD column at 1 ml/min.

4. Wash the column with 100 ml of HB(100) buffer and develop with a 400-ml gradient from 100 to 800 m*M* ammonium sulfate in buffer HB at 1 ml/min. The Pol II activity peak is eluted at 200 m*M* ammonium sulfate.

5. Pool fractions containing Pol II activity determined by the EC assembly from the preimmobilized Pol II and load to a 5-ml Ni^{2+}-NTA agarose column equilibrated with HB(200) at 0.5 ml/min.

6. Wash the column with 30 ml TB(300) and elute Pol II with 10 ml TB(300) containing 200 m*M* imidazole at 0.5 ml/min.

7. Dilute the eluate twofold with buffer QB(0) and load at 0.5 ml/min to a Mono Q HR5/5 column equilibrated with buffer QB(100).

8. Apply a 15-ml gradient from 100 to 2000 m*M* potassium acetate in buffer QB at 0.5 ml/min and collect 0.3-ml fractions. The Pol II peak is eluted at 0.8–0.9 *M* potassium acetate.

9. Pool three to five of the most active fractions, dilute threefold with buffer QB(0), and concentrate on Centricone (Millipore, Bedford, MA) with a MWCO 100 kDa. Add glycerol to 20% final concentration, and freeze in aliquots in liquid nitrogen. Store the protein at −70°. The yield is 0.2–0.3 mg of Pol II from 20 g of yeast cell paste.

The Pol II preparation obtained from the BJ5464 Rpb3His-Bio tag strain without BirA treatment contains approximately 15% of biotinylated enzyme (Fig. 3, lane 1), most likely due to endogenous biotinylation by the yeast biotin protein ligase.[30] If necessary, the biotinylated fraction can be removed by incubating the purified Pol II with 0.5 ml of avidin resin and collecting the unbound fraction.

[29] J. Huet, N. Manaud, G. Dieci, G. Peyroche, C. Conesa, O. Lefebvre, A. Ruet, M. Riva, and A. Sentenac, *Methods Enzymol.* **273**, 249 (1996).
[30] J. E. Cronan, Jr. and J. C. Wallace, *FEMS Microbiol. Lett.* **130**, 221 (1995).

Yield of Pol II Core Enzyme

The yield of the purified Pol II core enzyme that we report is about twofold higher than the yield of the immunoaffinity purification (compare 200–300 μg from 20 g of cells and 720 μg from 165 g of cells[1]). Do we isolate all Pol II from the yeast cells in the form of core enzyme or do we purify a fraction of Pol II? On the one hand, 30 pmol (at least 15 μg) of Pol II capable of active EC formation, as judged by the assembly assay (Fig. 2A), or 20–30 μg of Pol II protein, as estimated by the Coomassie blue-stained SDS gel (Fig. 2B and data not shown), is immobilized on Ni^{2+}-NTA agarose from 2 ml of lysate (1 g of cells). The longer incubation time or the addition of fresh resin does not increase the yield of Pol II. Pol II activity measured by the nonspecific initiation/chain elongation assay suggests that the cell lysate obtained from 165 g of cells contains 1.2 mg of Pol II (7 μg per 1 g of cells).[1] Thus, our results are sufficiently close to those reported by others and may indicate a complete recovery of Pol II from the lysate by metal affinity chromatography. On the other hand, if the total amount of Pol II is 30,000 copies per cell, as was determined by Borggrefe et al.,[31] 1 g of yeast cells should contain about 500 μg of core Pol II enzyme. They used denaturing conditions to extract total protein from the yeast cells grown in minimal media to the midlog phase. The difference between our estimate of Pol II amount in the crude cell lysate and data reported by Borggrefe et al.[31] may have different explanations. It is possible that yeast harvested in the stationary phase contains a lower amount of Pol II than cells from the culture grown to the midlog phase. However, it is most likely that the missing Pol II pool is associated with the insoluble fraction of the lysate. In addition, we cannot rule out that the cell lysate contains a pool of soluble Pol II molecules that are bound to transcription factors or nucleic acids, and therefore are inactive in the transcription assay and are inaccessible for immobilization via the affinity tag.

Properties of Affinity-Purified Double-Tagged Pol II

Binding Capacity of Ni^{2+}-NTA Agarose for Purified Pol II

Purified histidine-tagged Pol II was used in a solid-phase transcription system with the enzyme immobilized on Ni^{2+}-NTA agarose. Therefore, we analyzed the binding capacity of Ni^{2+}-NTA agarose for purified Pol II

[31] T. Borgreffe, R. Davis, A. Bareket-Samish, and R. D. Kornberg, J. Biol. Chem. 276, 47150 (2001).

in an experiment similar to the one described for the crude cell lysate. We found that it is 0.2–0.3 mg of Pol II per 1 ml of Ni^{2+}-NTA agarose, as compared to 5–10 mg reported by the manufacturer. Notably, the capacity of the same resin for the histidine-tagged *E. coli* RNA polymerase is about 0.1 mg of protein per 1 ml of agarose (M. Kashlev, unpublished observation). The possible (but unproven experimentally) explanation of such a low capacity of Ni^{2+}-NTA agarose to hexahistidine-tagged RNA polymerases is that these enzymes do not easily penetrate the pores on the surface of agarose beads due to their large size and/or formation of multimers.

The capacity of Ni^{2+}-NTA agarose for the EC, preassembled in the solution, is the same as the capacity for free RNA polymerase for both yeast and bacterial enzymes. The immobilization of Pol II or the EC on a limiting amount of resin may lead to the appearance of catalytically inactive complexes (M. L. Kireeva, unpublished observation). Therefore, we routinely immobilize 1 μg (2 pmol) of purified Pol II (or EC formed with 1 μg of the enzyme) on 10 μl of Ni^{2+}-NTA agarose beads. The increase of the Ni^{2+}-NTA agarose amount up to 100 μl of agarose beads per 1 μg of polymerase does not affect the complex formation or enzyme activity. Such an excess, for instance, is used in the activity assay performed with the Pol II from the crude extract to ensure the complete immobilization of the protein.

Subunit Composition of Affinity-Purified Pol II

We have compared biotinylated and nonbiotinylated Pol II preparations purified by the two protocols described earlier from two separate batches of yeast cells (Fig. 4). One batch was kept in stationary phase overnight and another was cultured without media change for 3 days. All preparations contained Pol II core enzymes of approximately 90% purity (Fig. 4) with 10–12 subunits, which could be identified by their electrophoretic mobility.[1,32] We have not addressed the phosphorylation state of the CTD subunit of Rpb1 directly; however, we believe that it is not phosphorylated in our Pol II preparations because no phosphatase inhibitors were used throughout the purification. Reconstitution of the ECs showed similar specific activity in all four of these Pol II preparations (data not shown).

Analyses of the subunit composition revealed that Pol II obtained from the overnight culture contains smaller amounts of Rpb4 and Rpb7 subunits as compared to the preparation purified in the same way from the cells that were cultured for 3 days (Fig. 4, compare lanes 1 and 5). This result could be expected, as it is known that the amount of Rpb4 and Rpb7 in the yeast

[32] M. H. Sayre, H. Tschochner, and R. D. Kornberg, *J. Biol. Chem.* **267,** 23367 (1992).

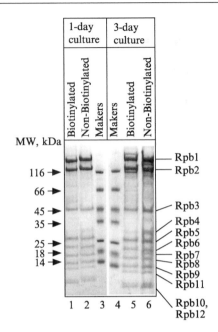

FIG. 4. Affinity-purified double-tagged Pol II. Fifteen micrograms each of biotinylated (lanes 1 and 5) and nonbiotinylated (lanes 2 and 6) Pol II purified from overnight (lanes 1 and 2) and 3-day cultures (lanes 5 and 6) of Rpb3 His-Bio strain were resolved in the 4–12% SDS polyacrylamide gel and stained as described in the text. The molecular weight of the markers (lanes 3 and 4) is indicated at the left. Pol II subunits are identified at the right.

cell is significantly affected by growth conditions, and Pol II purified from logarithmically growing cells contains substoichiometric amounts of these subunits.[33] We also observed that biotinylated Pol II purified by affinity chromatography on the SoftLink Soft Release avidin resin contained less Rpb4 and Rpb7 than nonbiotinylated Pol II purified from the same lysate by heparin and metal affinity chromatography (compare lanes 1 and 2 or 5 and 6). Such a depletion might be caused by the effect of the methacrylate matrix, which is used for avidin immobilization in the SoftLink Soft Release avidin resin, on the interaction of the Rpb4/Rpb7 dimer with Pol II. The Rpb4/Rpb7 dimer is usually dispensable for *in vitro* transcription in most experimental systems.[5] However, if proper stoichiometry of the dimer is essential in a particular experimental setup, we recommend purifying Pol II from the culture, which has been kept in a stationary phase for 3 days, and using the purification protocol suggested for the nonbiotinylated enzyme.

[33] M. Choder and R. A. Young, *Mol. Cell. Biol.* **13**, 6984 (1993).

The highly purified histidine- or histidine- and biotin-tagged Pol II core enzyme can be used in a variety of *in vitro* transcription assays. Assembly of the EC provides a convenient tool for the promoter- and factor-independent initiation of transcription. The variety of the templates utilized in this system can be expanded significantly using EC ligation to a linear DNA fragment.

Acknowledgments

We thank Mikhail Bubunenko for critical reading of the manuscript. The contents of this chapter do not necessarily reflect the views or policies of the Department of Health and Human Services, nor does mention of trade names, commercial products, or organizations imply endorsement by the U. S. Government.

[13] Dephosphorylation of the Carboxyl-Terminal Domain of RNA Polymerase II

By Patrick S. Lin and Michael E. Dahmus

The carboxyl-terminal domain (CTD) of the largest RNA polymerase (RNAP) II subunit is composed of tandem repeats of the consensus sequence $Tyr_1Ser_2Pro_3Thr_4Ser_5Pro_6Ser_7$. The consensus repeat has been well conserved, although the number of repeats varies from 52 in mammals to 26–27 in yeast. Reversible phosphorylation of the CTD, especially at Ser positions 2 and 5, plays an important role in the regulation of gene expression. RNAP IIA, which contains an unmodified CTD, and RNAP IIO, which contains a hyperphosphorylated CTD, have distinct functions in the transcription cycle. RNAP IIA is recruited actively to the promoter as part of the preinitiation complex, whereas RNAP IIO is responsible for transcript elongation. The phosphorylated CTD mediates promoter escape and the recruitment of factors necessary for processing of pre-mRNA. CTD kinases and phosphatase(s) modulate the state of CTD phosphorylation and accordingly play a critical role in regulating the synthesis and processing of the primary transcript. Previously, procedures were described for the *in vitro* phosphorylation of RNAP IIA with various CTD kinases.[1] This article describes procedures for various assays involving CTD dephosphorylation. These procedures provide the tools for

[1] M. E. Dahmus, *Methods Enzymol.* **273,** 185 (1996).

the identification and characterization of enzymes that dephosphorylate the CTD of RNAP II.

Basic Assay Procedures: General Comments

A variety of approaches have been developed for measuring CTD phosphatase activity. These approaches differ primarily in the nature of the substrate used and the method of following phosphate release. The choice of substrate depends on the experimental objective and, to a certain extent, the nature of the substrate available to the investigator. The nature of the substrate generally dictates the method to be utilized for measuring phosphate release.

Consideration of Substrate

The *in vivo* substrate for CTD phosphatase is RNAP II. Accordingly, this would seem to provide the most reliable substrate for the identification, purification, and characterization of the CTD phosphatase(s). RNAP II is a complex multisubunit enzyme with a molecular weight in excess of 550,000. It is not available in recombinant form, and the phosphorylated form of the enzyme, RNAP IIO, is difficult to purify from most sources. However, RNAP IIO can be prepared by the *in vitro* phosphorylation of RNAP IIA with a variety of CTD kinases.[2] An additional complication is that RNAP IIO can exist either free or within macromolecular complexes *in vivo*. The ability of CTD phosphatase to dephosphorylate the CTD is influenced by the nature of the complex in which RNAP IIO is contained.[3–5]

Early observations establish that mammalian CTD phosphatase (FCP1) is not active in the dephosphorylation of either the isolated largest RNAP IIO subunit or recombinant CTDo (rCTDo) under the reaction conditions employed for the dephosphorylation of native RNAP IIO.[6] An additional complication is the fact that rCTD is not a good substrate for several CTD kinases that actively phosphorylate RNAP II *in vitro*.[2] Accordingly, it is difficult to obtain fully phosphorylated rCTD that might be utilized as a substrate. These studies suggest that determinants outside the CTD influence the activity of both FCP1 and a variety of CTD kinases.

[2] P. S. Lin, M. F. Dubois, and M. E. Dahmus, *J. Biol. Chem.* **277,** 45949 (2002).
[3] E.-J. Cho, M. S. Kobor, M. Kim, J. Greenblatt, and S. Buratowski, *Genes Dev.* **15,** 3319 (2001).
[4] H. Cho, T. K. Kim, H. Mancebo, W. S. Lane, O. Flores, and D. Reinberg, *Genes Dev.* **13,** 1540 (1999).
[5] N. F. Marshall and M. E. Dahmus, *J. Biol. Chem.* **275,** 32430 (2000).
[6] R. S. Chambers, B. Q. Wang, Z. F. Burton, and M. E. Dahmus, *J. Biol. Chem.* **270,** 14962 (1995).

Synthetic peptides containing multiple copies of the consensus repeat and phosphorylated on Ser 2, Ser 5, or at both positions have been used to examine the specificity of CTD phosphatase.[7] The conditions required to observe activity with synthetic peptide substrates, specifically pH and CTD phosphatase concentration, differ significantly from reactions in which native RNAP IIO serves as a substrate.

Measurement of Phosphate Release

The two basic procedures for measuring CTD phosphatase activity include a direct measure of phosphate release and a change in the electrophoretic mobility in SDS–PAGE that accompanies dephosphorylation of either the largest RNAP IIO subunit or rCTDo. Most CTD kinases appear to phosphorylate the CTD in a processive manner at either Ser 2 or Ser 5. In mammalian RNAP II, the incorporation of 50+ phosphates within the CTD results in a marked reduction in the electrophoretic mobility of the largest subunit.[8] Accordingly, the reversal of this reaction by CTD phosphatase results in a marked increase in electrophoretic mobility.[9] It is important to note that in assays that utilize synthetic peptides, activity can be followed by the release of phosphate from a fraction of substrate, whereas in assays that utilize RNAP IIO, activity is followed by a change in electrophoretic mobility that results from the complete conversion of RNAP subunit IIo or rCTDo to the unphosphorylated form.

Dephosphorylation of Synthetic CTD Phosphopeptides

CTD phosphopeptides composed of four tandem YSPTSPS repeats, phosphorylated at Ser 2, Ser 5, or both Ser 2 and Ser 5, or even Tyr 1, are made using an automated peptide synthesizer.[10] CTD phosphopeptides are purified by reversed phase high-performance liquid chromatography (HPLC) on a Vydac C18 column developed with a linear acetonitrile gradient. The peak UV-absorbing fractions are lyophilized, dissolved in TE buffer (10 mM Tris–HCl pH 8.0, 1.0 mM EDTA), and stored at 4°. The CTD phosphopeptides can be biotinylated prior to purification for use in affinity chromatography or immunoprecipitation.

The standard reaction (25 μl) contains 25 μM CTD phosphopeptide and 1–5 μg FCP1 in 50 mM Tris–acetate, pH 5.5, 10 mM MgCl$_2$.[7] The

[7] S. Hausmann and S. Shuman, J. Biol. Chem. 277, 21213 (2002).
[8] J. M. Payne and M. E. Dahmus, J. Biol. Chem. 268, 80 (1993).
[9] R. S. Chambers and M. E. Dahmus, J. Biol. Chem. 269, 26243 (1994).
[10] C. K. Ho and S. Shuman, Mol. Cell 3, 405 (1999).

reaction is incubated at 37° for 60 min and is quenched by the addition of 0.5 or 1.0 ml of malachite green reagent (BIOMOL green reagent, BIOMOL Research Laboratories). Phosphate release is determined by measuring A_{620} and is quantified by comparison to a phosphate standard curve.

Preparation of rCTDo and RNAP IIO Substrates

General Considerations

The CTD of mammalian RNAP II contains at its very C terminus a serine flanked by acidic residues. This serine lies within a consensus casein kinase II (CKII) phosphorylation site and is C-terminal to the last consensus CTD repeat. The *in vitro* phosphorylation of RNAP II with CKII in the presence of $[\gamma^{-32}P]ATP$ has been used extensively to label RNAP II for a variety of studies.[8,9,11] The subsequent incubation of RNAP II with CTD kinase in the presence of excess unlabeled ATP results in the formation of RNAP IIO, which can serve as a substrate for CTD phosphatase.[6,9] Because only the most carboxyl-terminal serine (CKII site) is labeled with ^{32}P and lies outside the consensus repeat, complete dephosphorylation by a protein phosphatase specific for the consensus CTD repeat results in an electrophoretic mobility shift of subunit IIo back to the position of subunit IIa without the loss of label.

rCTD

Mammalian GST-CTD (90 kDa) is expressed in *Escherichia coli* and partially purified over a glutathione–agarose column with a step elution of 15 mM glutathione and by ammonium sulfate fractionation. ^{32}P-labeled GST-CTDo (apparent M_r of 116,000) is prepared by phosphorylation at the most carboxyl-terminal serine with recombinant CKII in the presence of $[\gamma^{-32}P]ATP$ (6000 Ci/mmol), followed by phosphorylation in the presence of 2 mM unlabeled ATP with recombinant MAPK2/ERK2 (Upstate Biotechnology)[2] (Fig. 1A, lanes 11–15). GST-CTDo is purified over a glutathione–agarose column. Conditions have not been defined that result in the efficient phosphorylation of rCTD with either mammalian TFIIH or P-TEFb (Fig. 1A, lanes 1–5 and 6–10, respectively).

---

[11] M. E. Kang and M. E. Dahmus, *J. Biol. Chem.* **268,** 25033 (1993).

A

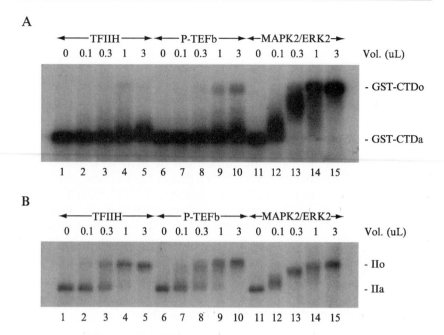

FIG. 1. Preparation of GST-CTDo and RNAP IIO. Recombinant GST-CTD and purified calf thymus RNAP IIA were labeled with ^{32}P by phosphorylation with CKII and subsequently incubated with 2 mM unlabeled ATP in the presence of increasing amounts of TFIIH (lanes 1–5), P-TEFb (lanes 6–10), and MAPK2/ERK2 (lanes 11–15). Reactions contain equimolar amounts of GST-CTD (A) and RNAP II (B). The difference in band intensity is a result of differences in the efficiency of CKII labeling. Reaction products were resolved by a 5% SDS–PAGE gel. The positions of GST-CTDa and GST-CTDo and subunits IIa and IIo are indicated. From Lin et al.[2]

RNAP IIO

Calf thymus RNAP IIA is purified by the method of Hodo and Blatti[12] with modifications as described by Kang and Dahmus.[11] Specific isozymes of ^{32}P-labeled RNAP IIO can be prepared by the phosphorylation of purified RNAP IIA with CKII and [γ-^{32}P]ATP, followed by CTD phosphorylation in the presence of 2 mM ATP with different CTD kinases.[2] The apparent M_r of subunit IIo is 240,000 relative to subunit IIa, which is 214,000. RNAP IIO is purified by step elution from DE53 as described by Chesnut et al.[13] Preparation of RNAP IIO by the phosphorylation of ^{32}P-labeled RNAP IIA with TFIIH, P- TEFb, and MAPK2/ERK2 is shown in Fig. 1B. Note the marked shift in electrophoretic mobility of subunit IIa to the position of subunit IIo that accompanies CTD phosphorylation.

[12] H. D. Hodo and S. P. Blatti, *Biochemistry* **16,** 2334 (1977).

[13] J. D. Chesnut, J. H. Stephens, and M. E. Dahmus, *J. Biol. Chem.* **267,** 10500 (1992).

Dephosphorylation of rCTDo

Buffer

The standard CTD phosphatase buffer contains 50 mM Tris-HCl, pH 7.9, 10 mM MgCl$_2$, 20% glycerol, 0.025% Tween 80, 0.1 mM EDTA, and 5 mM dithiothreitol (DTT).

Assay Procedure

CTD phosphatase assays are carried out in 20 μl CTD phosphatase buffer.[9] Reactions contain 75 fmol ^{32}P-labeled GST-CTDo and 7 pmol RAP74. Reactions are initiated by the addition of CTD phosphatase and are incubated at 30° for 30 min. Assays are terminated by the addition of 5× concentrated Laemmli buffer and GST-CTDa/o resolved on an 8% SDS–PAGE gel. Gel images are scanned with a phosphor imager. Although mammalian FCP1 does not dephosphorylate rCTDo, it may well serve as a substrate for yet to be discovered CTD phosphatases.

Dephosphorylation of Purified Calf Thymus RNAP IIO

Reactions (20 μl) contain 18 fmol ^{32}P-labeled RNAP IIO and 7 pmol RAP74 in the standard CTD phosphatase buffer described earlier. RAP74 stimulates FCP1 activity with RNAP IIO as a substrate by 20-fold.[6] Reactions are initiated by the addition of CTD phosphatase and are incubated at 30° for 30 min. Assays are terminated by the addition of 5× concentrated Laemmli buffer, and RNAP II subunits are resolved on a 5% SDS–PAGE gel. Gel images are scanned on a phosphor imager. The dephosphorylation of ^{32}P-labeled calf thymus RNAP IIO prepared by TFIIH, P-TEFb, and MAPK2/ERK2 is shown in Fig. 2.

An equimolar mixture of RNAP IIO and GST-CTDo can be used in assays to examine the dependence of CTD phosphatase(s) activity on native RNAP II. For assays involving both substrates in the same reaction, a 5–6% SDS–PAGE gel permits the best resolution of the phosphorylated and unphosphorylated species.

Whereas the pH optimum for the dephosphorylation of CTD phosphopeptides is 5.5, the pH optimum for the dephosphorylation of RNAP IIO is 7.9. Mammalian FCP1 does not dephosphorylate RNAP IIO at pH 5.5 nor does it dephosphorylate CTD phosphopeptides or GST-CTDo at pH 7.9. Therefore, RNAP IIO is the preferred substrate of mammalian FCP1 at physiological pH. Results utilizing nonnative substrates for the identification and characterization of novel CTD phosphatase(s) must be interpreted with some caution.

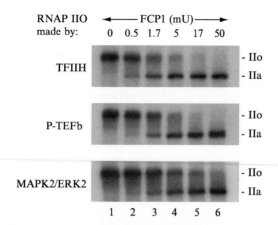

FIG. 2. Dephosphorylation of RNAP IIO prepared by phosphorylation with TFIIH, P-TEFb, and MAPK2/ERK2. Different isozymes of RNAP IIO, prepared as shown in Fig. 1, were incubated with increasing amounts of FCP1 and analyzed by a 5% SDS–PAGE gel. The positions of subunits IIa and IIo are indicated. From Lin et al.[2]

Dephosphorylation of RNAP IIO Contained in Nuclear Extracts

A variety of studies have shown that changes in growth conditions can alter the level and pattern of CTD phosphorylation. Therefore, HeLa nuclear extracts from differentially treated cells provide an excellent source of heterogeneous, endogenous RNAP IIO. Although these assays utilize "native" RNAP IIO, extracts can contain endogenous CTD phosphatase activity as well as factors that influence the reaction in unknown ways. In experiments using HeLa nuclear extracts, CKII labeling results in the incorporation of radiolabeled phosphates onto a myriad of proteins. Therefore, the mobility shift that results from dephosphorylation can be followed most easily by Western blotting.[2] Antibodies directed against an epitope on the largest RNAP II subunit that lies outside of the CTD, such as POL3/3,[14] are most useful in that their reactivity is not influenced by the state of CTD phosphorylation. A shift in the mobility of subunit IIo to that of subunit IIa is a direct measure of CTD phosphatase activity. Monoclonal antibodies specific for Ser 2 and Ser 5, H5, and H14, respectively, can also be utilized to examine the relative activity of a given phosphatase for Ser 2 and Ser 5 phosphates.[2,15] The disappearance of

[14] R. E. Kontermann, Z. Liu, R. A. Schulze, K. A. Sommer, L. Queitsch, S. Duebel, S. M. Kipriyanov, F. Breitling, and E. K. F. Bautz, Biol. Chem. Hoppe-Seyler 376, 473 (1995).
[15] M. Patturajan, R. J. Schulte, B. M. Sefton, R. Berezney, M. Vincent, O. Bensaude, S. L. Warren, and J. L. Corden, J. Biol. Chem. 273, 4689 (1998).

immunoreactivity corresponding to subunit IIo is a direct measures of CTD phosphatase activity. Control reactions in which the largest subunit is followed by POL3/3 immunoreactivity are necessary to ensure that the loss of H5 and H14 reactivity is not the consequence of limited proteolysis removing the CTD.

The dephosphorylation of RNAP IIO contained in nuclear extracts is carried out as described earlier with minor modifications. Reactions are performed in 20 μl of modified CTD phosphatase buffer without Tween 80 and in the presence of 20 mM KCl. Tween 80 can decrease the efficiency of membrane transfer of the largest RNAP II subunit, whereas KCl prevents the precipitation of proteins in HeLa nuclear extracts. The removal of Tween 80 and the addition of KCl do not appear to affect CTD phosphatase activity.

Reactions involving endogenous RNAP II in HeLa nuclear extracts contain approximately 200–250 fmol of RNAP IIO. Reactions are carried out in the presence of 7 pmol RAP74 and are initiated by the addition of CTD phosphatase. Following an incubation at $30°$ for 30 min, reactions are terminated by the addition of 5× concentrated Laemmli buffer. The large RNAP II subunits are resolved on a 6% SDS–PAGE gel prior to blotting. The use of Western blots to follow the dephosphorylation reaction is not limited to RNAP IIO contained in cell extracts. It can also be applied in assays involving purified RNAP IIO.[4]

The dephosphorylation of RNAP IIO contained in a HeLa nuclear extract upon incubation with FCP1 is shown in Fig. 3. The top image shown in Fig. 3 is an immunoblot probed with monoclonal antibody POL3/3 and shows the mobility shift of subunit IIo to the position of subunit IIa. The bottom two images in Fig. 3 are identical samples probed with monoclonal antibodies H5 and H14. The disappearance of immunoreactivity with H5 and H14 demonstrates the removal of phosphates from Ser 2 and Ser 5 by FCP1.

Dephosphorylation of RNAP IIO in Elongation Complexes

The assembly of RNAP II into macromolecular complexes at specific points in the transcription cycle can influence the activity of CTD phosphatase. Of special interest is the turnover of phosphate during the elongation phase of transcription. Procedures have been developed for the preparation of elongation complexes and the examination of their sensitivity to dephosphorylation.[4,5,16] To establish the effect of CTD phosphatase on RNAP II in early elongation complexes (EECs), RNAP II is first

---

[16] A. L. Lehman and M. E. Dahmus, *J. Biol. Chem.* **275,** 14923 (2002).

100mU FCP1   –   +

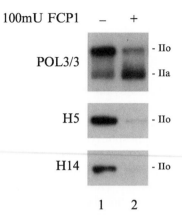

POL3/3   - IIo / - IIa

H5   - IIo

H14   - IIo

1   2

FIG. 3. Dephosphorylation of RNAP IIO contained in a nuclear extract. A HeLa nuclear extract containing about 200–250 fmol of RNAP II was incubated with FCP1. The reaction products were resolved by a 6% SDS–PAGE gel, transferred, and probed with monoclonal antibodies POL3/3, H5, and H14. From Lin et al.[2]

assembled onto the promoter. ^{32}P-labeled RNAP IIA is prepared as described previously and is incubated with the HeLa nuclear extract on ice for 10 min. To form preinitiation complexes on immobilized templates, the ^{32}P-labeled RNAP IIA/HeLa nuclear extract mixture is incubated with 100 ng of immobilized template on Dynabeads (Dynal Inc.) for 30 min at 30° in a 12 μl reaction containing 20 mM HEPES, pH 7.9, 7 mM MgCl$_2$, 55–60 mM KCl, and 7 mM DTT. For transcription of the adenovirus 2 major late (Ad2-ML) template, the pulse labeling of transcripts is initiated in the presence of 600 μM ATP, GTP, and UTP and 0.6 μM CTP to generate elongation complexes paused at G11. Pulse labeling is carried out for 30 s at 30° and is terminated by the addition of 0.5 μl of 0.5 M EDTA. Beads containing the paused EECs are then concentrated magnetically and washed once with 12 μl HMKT (20 mM HEPES, pH 7.9, 7 mM MgCl$_2$, 55 mM KCl, 0.1% Triton X-100). The beads are washed twice successively with 12 μl HMKS (20 mM HEPES, pH 7.9, 7 mM MgCl$_2$, 55 mM KCl, 1% Sarkosyl) and then twice with 12 μl HMKT. EECs are then resuspended in 12 μl HMKT. Elongation complexes can be walked down the template in the presence of required nucleotides to give a final concentration of 20 μM for each nucleotide present.[5] After 5 min at 30°, the complexes are concentrated magnetically, washed twice with 12 μl HMKT, and resuspended in 12 μl HMKT. The cycle of chase–wash–resuspension is repeated as many times as needed to advance complexes to the desired position.

For CTD phosphatase assays involving EECs, 24 μl complex (twice the standard transcription reaction) in HMKT is washed twice with 20 μl aliquots of CTD phosphatase buffer and resuspended in 18 μl phosphatase buffer. CTD phosphatase buffer containing RAP74 and CTD phosphatase (2 μl) is added to initiate the reaction. Reactions are incubated for 30 min at 30°. Assays are terminated by the addition of 5× concentrated Laemmli buffer, and RNAP II subunits are resolved on a 5% SDS–PAGE gel. The gel images are scanned on a phosphor imager.

Figure 4 shows the dephosphorylation of free RNAP IIO and RNAP IIO contained in a G11 EEC in the presence of increasing amounts of FCP1. The bottom part of Fig. 4 is a quantitation of subunits IIa and IIo and demonstrates that RNAP IIO in an elongation complex is about 100-fold more resistant to dephosphorylation than free RNAP IIO.

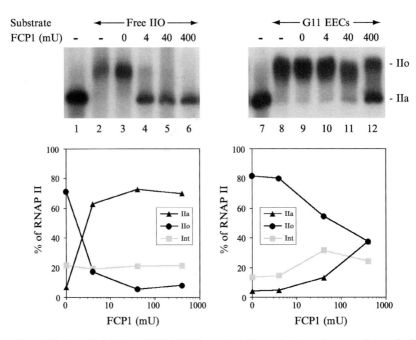

Fig. 4. Dephosphorylation of RNAP IIO contained in early elongation complexes. Early elongation complexes, paused at G11, were prepared on an immobilized Ad2-ML template and incubated with increasing amounts of FCP1. Transcription was initiated by the addition of ^{32}P-labeled RNAP IIA. Input RNAP II is shown in lane 7. Lanes 2 and 8 show free RNAP IIO and RNAP IIO contained in G11 complexes, respectively. Lanes 3–6 and 9–12 show the reaction products after incubation with increasing amounts of FCP1. (Left) Free RNAP II and (right) RNAP II contained in EECs. The lower panels are a quantitation of the results. From Marshall and Dahmus.[5]

Acknowledgments

We gratefully acknowledge our colleagues Alexandre Tremeau-Bravard for his careful review of the manuscript and Grace Dahmus for her continuing technical support. Research in our laboratory was supported by the National Institutes of Health (GM33300).

[14] RNA Polymerase III from the Fission Yeast, Schizosaccharomyces pombe

By Ying Huang, Mitsuhiro Hamada, and Richard J. Maraia

RNA polymerase III (Pol III) is a nuclear enzyme that has been specialized to produce small nontranslated RNAs in great abundance. Analyses reveal that the Pol III system of the fission yeast *Schizosaccharomyces pombe* differs from the well-characterized *S. cerevisiae* and human Pol III systems[1,2] in that it requires precisely positioned (i.e., at −30) upstream TATA elements for function.[3,4] These observations, together with observed differences among different model eukaryotes, in Pol III termination[5] indicate that *S. pombe* will provide novel insights into transcription mechanisms. This article describes the immunoaffinity purification of epitope-tagged Pol III from *S. pombe*, the use of *Tsp*RI-generated templates that contain a 9 nucleotide 3′ overhang for use in promoter-independent transcription, and other methods for functional analyses *in vitro*.

Materials and Solutions

Strains and Media

The wild-type allele of *rpc53*, which encodes a Pol III-specific subunit of *S. pombe* Pol III, was replaced by a double epitope-tagged allele of *spRPC53*, by homologous recombination, in the fission yeast strain, yYH3282 [*h⁺ his3-D1 leu1-32 ura4-D18 ade6-M216 rpc53::(FH-rpc53, ura4⁺)*]. In this strain, the nmt1 promoter controls RPC53 expression. The resulting RPC53 protein contains a FLAG epitope and six histidines at its N terminus. Cells are grown in YES (5 g/liter yeast extract, 30 g/liter

[1] E. P. Geiduschek and G. A. Kassavetis, *J. Mol. Biol.* **310,** 1 (2001).
[2] Y. Huang and R. J. Maraia, *Nucleic Acids Res.* **29,** 2675 (2001).
[3] M. Hamada, Y. Huang, T. M. Lowe, and R. J. Maraia, *Mol. Cell. Biol.* **21,** 6870 (2001).
[4] Y. Huang, E. McGillicuddy, M. Weindel, and R. J. Maraia, *Nucleic Acids Res.* **31,** 2108 (2003).
[5] M. Hamada, A. L. Sakulich, S. B. Koduru, and R. Maraia, *J. Biol. Chem.* **275,** 29076 (2000).

glucose, supplemented with 225 mg/liter each of adenine, histidine, leucine, uracil, and lysine).

Resins

Anti-FLAG M2 agarose is from Sigma (St. Louis, MO). Ni-NTA resin is from Qiagen (Valencia, CA). Streptavidin resin is from Pierce (Rockford, IL).

Other Reagents

FLAG peptide (DYKDDDDK) and imidazole are from Sigma. Protease inhibitors are from Roche. Ribonucleotides are from Pharmacia (Piscataway, NJ). Radiolabeled $[\alpha$-^{32}P]GTP (10 mCi/ml, 3000 Ci/mmol) is from ICN (Costa Mesa, CA). Biotin-14-dATP is from InVitrogen (Carlsbad, CA). *Escherichia coli* tRNA is from Roche, (Indianapolis) and RNasin is from Promega (Madison). *Tsp*RI, *Eco*RI, *Bss*HII, *Xho*I, and the Klenow fragment are from New England Biolabs (Beverly, MA).

Solutions

Buffer A: 200 mM Tris–HCl, pH 7.9, 10% glycerol (v/v), 10 mM MgCl$_2$, 2 mM dithiothreitol (DTT), 1 mM phenylmethylsulfonyl fluoride (PMSF), 2.5 μg/ml antipain-dihydrochloride, 1 μg/ml aprotinin, 0.35 μg/ml Bestatin, 2 μg/ml chymostatin, 0.5 μg/ml leupeptin, 0.5 μg/ml pepstatin A, and 0.1 mg/ml Pefabloc SC

Buffer B: 20 mM HEPES, pH 7.9, 20% glycerol (v/v), 2 mM DTT, 0.1% NP-40, and 0.1 mM PMSF

Buffer C: 20 mM HEPES, pH 7.9, 60 mM KCl, 10 mM MgCl$_2$, 0.2 mM EDTA, 0.5 μl of RNasin, 0.6 mM ATP, CTP, and UTP, 0.05 mM GTP, and 5 μCi $[\alpha$-^{32}P]GTP TB: 20 mM HEPES, pH 7.9, 60 mM KCl, 10 mM MgCl$_2$, 0.2 mM EDTA, and 0.5 μl of RNasin

Buffer D: 20 mM Tris–HCl, pH 7.4, 7.5 M guanidine–HCl, 0.5% sarkosyl, and 5 μg/ml of *E. coli* tRNA

Buffer E (99% deionized formamide, 0.02% SDS, bromphenol blue, and xylene cyanole to faint color intensity)

TBE: 90 mM Tris–borate and 2 mM EDTA, pH 8.0

Methods

Method I. Preparation of Fission Yeast Extracts

1. Grow yYH3282 cells in YES at 32° to an OD$_{600}$ of 0.8–1.0. Examine under a microscope to ensure a healthy culture of rod-shaped fission yeast cells.

2. Collect cells by centrifugation at 3000 rpm for 15 min. Wash the pellet (~20 g) once in distilled water and suspend cells in buffer A.
3. Break yeast cells in a French pressure cell (Thermo Spectronic, Rochester, NY). Pass through a French press twice to increase cell breakage.
4. Slowly add cold 5 M NaCl to a final concentration of 400 mM and incubate the lysate for 20 min at $4°$, with rotation.
5. Clear the lysate by centrifugation at 36,000 rpm for 1 h at $4°$ in a 70Ti Beckman rotor (Beckman, Fullerton, CA).
6. Collect the "S-100" supernatant, excluding the top lipid layer, and pass it through a Miracloth filter (Calbiochem, Darmstadt, Germany) to remove any cloudy material.
7. Slowly add cold and finely ground ammonium sulfate at a ratio of 0.35 g/ml of the supernatant.
8. Incubate the precipitate at $4°$ overnight for complete solubilization of the salt.
9. Collect proteins by centrifugation at 10,000 rpm for 30 min at $4°$ in a Sorvall SS-34 rotor (Sorvall, Charlotte, NC).
10. Discard the supernatant. Remove residual supernatant by a second centrifugation at 10,000 rpm for 10 min at $4°$ in a Sorvall SS-34 rotor.
11. Solubilize the precipitated protein pellet in 20 ml of buffer B and dialyze the solubilized extract into buffer B containing 500 mM NaCl. Note that the amount of buffer added at this step will determine the final concentration of the extract.
12. Remove insoluble material by centrifugation at 10,000 rpm for 10 min at $4°$ in a Sorvall SS-34 rotor.
13. Aliquot portions of the extract in small tubes (e.g., 1.5 ml) and store at $-80°$. The concentration is typically 20 mg/ml.

Method II. Construction of TspRI-Generated 9 Nucleotide 3′ Overhang (i.e., 3′-Tailed Templates)

A series of tRNA^{Ser}UGAM genes that differ only in the Pol III terminator that follows the tRNA^{Ser}UGAM sequence contain (dT)2–7 residues at the test terminator.[5] These genes also contain an efficient Pol III terminator, (dT)8, located approximately 100 bp downstream of the test terminator. The 3′-tailed tRNA^{Ser}UGAM templates are constructed as follows. A primer containing a *Tsp*RI recognition sequence is used to amplify the tRNA^{Ser}UGAM-2T, -3T, -4T, -5T, -6T, and -7T genes by the polymerase chain reaction (PCR) using the primers (Mserup2, AC-CTGCAGTGAAATCAACCATTT GAGCATTGGAAACT, the *Tsp*RI

site is underlined, and Mserdow2, TGG<u>GAATTC</u>GAGACCTCCTTTCC
GAGCCCGG CCTTTTTG GCCGAG, *Eco*RI site underlined), and the
products were cloned into the TA-cloning vector (In Vitrogen) and named
pTA-tRNA^{Ser}UGAM-2T, -3T, etc. PCR parameters are one cycle at 94° for
1 min, 25 cycles at 94° for 15 sec, 55° for 30 s, 72° for 1 min, and one cycle
at 72° for 10 min. To generate the 3′-tailed tRNA^{Ser}UGAM templates, PCR
is performed on the pTA-tRNA^{Ser}UGAM templates using the M13 forward
and reverse primers. The 3′-tailed templates are generated by digestion of
these PCR products with *Eco*RI and *Tsp*RI, followed by gel purification.
*Tsp*RI leaves a 9 nucleotide 3′ overhang upstream of the tRNA sequence,
whereas *Eco*RI leaves a 5′ overhang just beyond the efficient (dT)8 terminator. Because Pol III will initiate from the 3′ overhang but not the 5′ overhang (R. Maraia, unpublished result), the tailed template will efficiently
load Pol III for unidirectional transcription (see Fig. 1).

A

Purification scheme for tandem epitope-tagged *S. pombe* RNA Polymerase III

Ammonium sulfate (40%) precipitation of S100

Anti-FLAG M2 agarose chromatography

Nickel-NTA agarose chromatography

Glycerol gradient (10-40%) sedimentation

B   *Tsp*RI-generated 3'-tailed template constructs

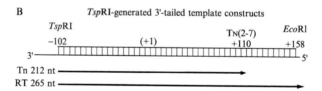

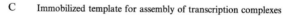

C   Immobilized template for assembly of transcription complexes

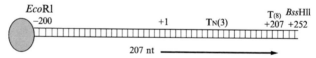

FIG. 1. (A) Purification scheme for double epitope-tagged *S. pombe* RNA polymerase III.
(B) Schematic representation of the *Tsp*RI-generated 9 nucleotide, 3′ overhang-containing
template used to examine promoter-independent transcription of naked DNA constructs by
SpPol III. (C) Schematic representation of the tRNA gene construct used to examine
promoter-dependent transcription of preassembled transcription complexes by SpPol III.

Method III. Promoter-Dependent Transcription by S. pombe RNA
Polymerase III of Preassembled, Immobilized tRNA Transcription
Complexes (Pol III Recycling Assay)

Accurate initiation of the *S. pombe* tRNA^{Ser}UGAM-3T gene produces a ~210 nucleotide transcript that terminates at the downstream (dT)8 terminator.[5] The tRNA gene is immobilized as follows.

1. Digest 4 μg of plasmid pMser3T carrying the tRNA^{Ser}UGAM-3T gene with *Eco*RI.
2. Biotinylate the DNA with the Klenow fragment and biotin-14-dATP. Unincorporated nucleotides are removed by a G-50 quick spin high-capacity column (Roche).
3. Digest with *Bss*HII and *Xho*I.
4. Immobilize the 469-bp DNA fragment containing the tRNA^{Ser} UGAM-3T gene by overnight incubation with the streptavidin-agarose resin as described.[6]
5. Quantitate the concentration of immobilized DNA per volume of agarose by ethidium bromide staining and comparison to known amounts of standard DNA.
6. Store at 4° as a 50% slurry of agarose-DNA in TE containing 0.1% sodium azide.

Assembly of the Preinitiation Complex

1. Incubate 10 μl of extract with 20 μl of streptavidin resin containing 0.5 μg immobilized template in 200 μl of TB containing 0.5 mM ATP, CTP, and GTP for 30 min at 25° (to compare different ratios of immobilized DNA to your extract).
2. Remove the supernatant after centrifugation at 1000 rpm at 4° for 5 min. Wash the resin twice with 0.5 ml of TB containing 400 mM KCl and twice with TB.

Promoter-Dependent Transcription Reaction and RNA Purification

1. Make a mixture of 25 μl of buffer C containing 200 ng of the immobilized template and 1–5 μl of affinity-purified *S. pombe* RNA pol III.
2. Incubate at 25° for 1 h.
3. Stop the reaction by adding 87.5 μl of buffer D, followed by 87.5 μl of H$_2$O and 200 μl of ultrapure phenol:chloroform:isoamyl alcohol (25:24:1, v/v). Vortex and/or nutate for at least 5 min. Centrifuge at 14,000 rpm for 10 min at 25°.

[6] R. J. Maraia, D. J. Kenan, and J. D. Keene, *Mol. Cell. Biol.* **14**, 2147 (1994).

4. Extract the aqueous phase, add 0.1 volume of 2 M NaOAc, pH 5.2, and 0.5 ml of ethanol, vortex, and incubate the mixture at $-20°$ for 2 h.
5. Collect the pellet by centrifugation at 10,000 rpm for 30 min. Discard the supernatant, and dry the pellet in a SpeedVac Plus centrifuge (Savant, Hicksville, NY) for 15 min.
6. Resuspend the pellet in 20 μl of buffer E.
7. Heat at $85°$ for 10 min.
8. Analyze transcription products by electrophoresis on prerun, prewarmed 6% polyacrylamide gels containing 8 M urea in TBE.
9. Fix the gel in 30% methanol and 5% acetic acid for 30 min, dry the gel with a gel dryer apparatus (Bio-Rad, Hercules, CA), and expose the gel to a FUJIFILM screen (Fuji, Stamford, CT) for autoradiography.
10. Analyze data with Image Reader software (Fuji).

Method IV. Purification of Double Epitope-Tagged S. pombe RNA Polymerase III Affinity Purification

1. Incubate 100 mg of extract containing the epitope-tagged subunit of Pol III with 100 μl of anti-FLAG M2 agarose at $4°$ for 4 h with constant rotation.
2. Wash the resin five times with 15 volumes of buffer B containing 500 mM NaCl.
3. Elute bound material twice for 30 min at $4°$ in 10 μl of buffer B containing 500 mM NaCl and 200 μg/ml FLAG peptide that has been neutralized to pH 7.4.
4. Combine both eluates and add buffer B containing 500 mM NaCl to a final volume of 300 μl and imidazole to a final concentration of 5 mM.
5. Add 30 μl of Ni-NTA resin and incubate for 4 h at $4°$.
6. Wash the resin three times with 15 volumes of buffer B containing 500 mM NaCl and 5 mM imidazole, followed by 15 volumes of buffer B containing 500 mM NaCl and 20 mM imidazole.
7. Elute bound material twice for 30 min at $4°$ in 20 μl of buffer B containing 100 mM NaCl and 300 mM imidazole.
8. To remove imidazole, dilute the preparation 10-fold in buffer B containing 100 mM NaCl and then concentrate with a Microcon YM-3 (Millipore, Bedford, MA). Repeat this procedure three times.
9. Aliquot and store the Pol III preparation at $-80°$.

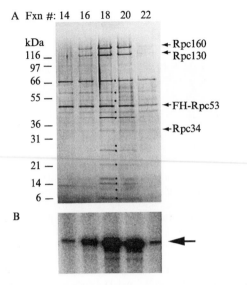

FIG. 2. SDS–PAGE, silver staining analysis of tandem epitope-tagged, affinity-purified *S. pombe* RNA Pol III. (A) SpPol III was isolated from yYH3282 cells by anti-FLAG M2 chromatography, followed by purification on Ni-NTA agarose (lane 5). The affinity-purified SpPol III was then subjected to centrifugation through a 10 to 40% glycerol gradient. Gradient fractions are collected from top to bottom and analyzed by 4–20% SDS–PAGE followed by silver staining. Lanes 1–5 are even numbered fractions 14–22. Positions of four *S. pombe* RNA pol III subunits, spRpc160, spRpc130, FH-spRpc53, and spRpc34, detected by Western blotting (data not shown) are indicated by arrows, and positions of associated proteins are indicated by dots at the right of lane Fxn #18. Molecular mass markers are indicated to the left. (B) Products of *in vitro* transcription reactions with 10 μl of the glycerol gradient fractions 14 through 22 and the *Tsp*RI-generated 3′-tailed template (corresponding to the −3T construct, see Fig. 3) shown in Fig. 1B. The arrow indicates the position of the major transcription product of SpPol III transcription.

10. Analyze the preparation by 4–20% SDS–PAGE (In Vitrogen) followed by silver staining with the SilverXpress kit (In Vitrogen). A typical sample of the Pol III prepared this way (1 μl) is shown in Fig. 2A.

Glycerol Gradient Sedimentation

1. Prepare a 10–40% glycerol (v/v) gradient in buffer B containing 100 mM NaCl in a 3-ml Beckman centrifuge tube using the Biologic program on a Biologic workstation (Bio-Rad).

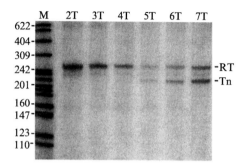

FIG. 3. Examination of termination by SpPol III using *Tsp*RI-generated 3′-tailed templates. Digestion of DNA with *Tsp*RI yields a 3′ overhang of 9 nucleotides and serves to load Pol III for initiation at the junction of the 3′ overhang (the tail) and the dsDNA. Immunoaffinity-purified SpPol III was used to transcribe 3′-tailed templates that differed only in the number of dTs at the test terminator, $T_{N(2-7)}$ (from 2T to 7T, as labeled above the lanes). Polymerases that terminate at T_N produce a band of ~210 nucleotides, whereas polymerases that continue beyond T_N produce read-through (RT) transcripts of ~260 nucleotides (i.e., that extend to the *Eco*RI end of the template). Note the appearance of and increasing amount of T1 band with increasing dT tract length of T_N. Both RT and T_N bands were sensitive to the Pol III-specific inhibitor tagetitoxin (not shown).

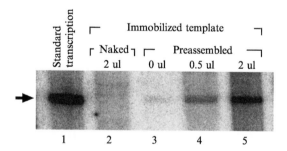

FIG. 4. Promoter-dependent transcription by purified, tandem epitope-tagged SpPol III. *In vitro* transcription was performed with supercoiled plasmid pMser-3T and 5 μl of *S. pombe* extract to serve as a positive control (lane 1) or using the immobilized pMser-3T template shown in Fig. 1C preassembled with transcription complexes (lanes 3–5) or without preassembly (i.e., naked immobilized DNA, lane 2). One-half microliter (lane 4) or 2 μl (lane 5) of affinity-purified SpPol III was used in the reactions shown in lanes 4 and 5 as indicated. Two microliters of bovine serum albumin (250 ng/ml) served as control for lane 3. Arrow points to the 207 nucleotide RNA.

2. Gently layer 200 μl of the purified Pol III onto the top of the 10–40% glycerol gradient and centrifuge at 55,000 rpm for 5 h at 4° in a SW60Ti rotor using a L8 ultracentrifuge (Beckman).

3. Collect 100-μl fractions from the top of the centrifugation tube using the density gradient fractionator (Brandel, Gaithersburg, MD) and the continuous fraction collector (Brandel). Fractions (5 μl) when analyzed by SDS–PAGE followed by silver staining typically show the pattern as in Fig. 2B.

Discussion

Our analyses reveal that *SpP*ol III is a multisubunit complex as expected (Fig. 2). Characterization of the individual polypeptide bands by sequencing should allow the identification of each subunit of *S. pombe* RNA polymerase III and any associated factors. Comparison of our preparation to Pol III from human and *S. cerevisiae* is consistent with a high degree of conservation.

The affinity-purified SpPol III (Fig. 2) is active in (i) promoter-independent transcription of naked DNA containing a 3′ overhang, (ii) recognition of the Pol III-specific termination signal (Fig. 3), and (iii) promoter-dependent tRNA gene transcription (Fig. 4). However, the relatively low level of termination efficiency of isolated Pol III as compared to that observed from intact *S. pombe* transcription complexes assembled via a promoter-dependent mechanism on the same templates[5] suggests a role for promoter-mediated initiation and/or involvement of TFs III in increasing termination efficiency. Additional characterization of the affinity-purified, epitope-tagged RPC53 Pol III also exhibits sensitivity to the Pol III-specific inhibitor tagetitoxin (data not shown).

Isolation of SpPol III represents a necessary step toward the reconstitution of an *in vitro* Pol III system from *S. pombe*. The method described here should also be applicable to other multisubunit transcription factors, e.g., the Pol III-specific factor TFIIIC, and to identify factors associated with the TFIIB-related factor (Brf) variant that is involved in the TATA-dependent transcription of Pol III-dependent genes in *S. pombe*.

The *Tsp*RI-mediated generation of 3′-overhang templates described here, for the first time, provides a homogeneous preparation of a template containing an extended, i.e., 9 nucleotide 3′ overhang that should be generally applicable to other templates (i.e., the *Tsp*RI recognition sequence can accommodate a large number of unique sequence variations) for the study of other RNA polymerases.

[15] Purification of Highly-Active and Soluble *Escherichia coli* σ^{70} Polypeptide Overproduced at Low Temperature

By HUIJUN ZHI and DING JUN JIN

The sigma subunit of bacterial RNA polymerase (RNAP) is required for promoter-specific transcription initiation.[1] Among the multiple sigma factors in *Escherichia coli*, σ^{70} is the primary sigma factor that is responsible for the transcription of house keeping genes.[2,3] In traditional RNAP purification procedures, σ^{70} is tightly associated with core RNAP forming holoenzyme and it can be separated from core RNAP by multiple chromatography involving Bio-Rex 70/DEAE-cellulose and Ultragel AcA44.[4] For rapid purification of large amounts of the σ^{70} polypeptide, several expression systems have been developed to overproduce native or recombinant σ^{70} polypeptide.[5–10] In general, expression of the σ^{70} polypeptide was induced at 37° or higher. Under these conditions, however, the overproduced σ^{70} formed insoluble inclusion bodies inside cells[5–10] (see Fig. 2). Thus, to purify σ^{70} from inclusion bodies, one must solubilize the σ^{70} in denaturant solutions such as 8 M urea or 6 M guanidinium hydrochloride and then allow the solubilized σ^{70} polypeptide to refold by gradually removing the denaturants through dialysis or multiple dilutions. However, it has been found that only a small fraction (5–10%) of the refolded σ^{70} polypeptide is monomeric and fully active.[5,11]

It is a very common phenomenon that high-level expression of desired proteins is accompanied by a deposition of overproduced products into insoluble inclusion bodies in *E. coli*-based expression systems. The reasons for misfolding and subsequent aggregation of overproduced proteins in

[1] R. R. Burgess, A. A. Traver, J. J. Dunn, and E. K. F. Bautz, *Nature (Lond.)* **221,** 43 (1969).

[2] J. D. Helmann and M. J. Chamberlin, *Annu. Rev. Biochem.* **57,** 839 (1988).

[3] M. Lonetto, M. Gribskov, and C. A. Gross, *J. Bacteriol.* **174,** 3843 (1992).

[4] P. A. Lowe, D. A. Hager, and R. R. Burgess, *Biochemistry* **18,** 1344 (1979).

[5] M. Gribskov and R. R. Burgess, *Gene* **26,** 109 (1983).

[6] A. Kumar, B. Grimes, N. Fujita, K. Makino, R. A. Malloch, R. S. Hayward, and A. Ishihama, *J. Mol. Biol.* **235,** 405 (1994).

[7] A. J. Dombroski, *Methods Enzymol.* **273,** 134 (1996).

[8] C. Wilson and A. J. Dombroski, *J. Mol. Biol.* **267,** 60 (1997).

[9] L. Rao, D. P. Jones, L. H. Nguyen, S. A. McMahan, and R. R. Burgess, *Anal. Biochem.* **241,** 3 (1996).

[10] T. M. Gruber, D. Markov, M. M. Sharp, B. A. Young, C. Z. Lu, H. J. Zhong, I. Artsimovitch, K. M. Geszvain, T. M. Arthur, R. R. Burgess, R. Landick, K. Severinov, and C. A. Gross, *Mol. Cell* **8,** 21 (2001).

[11] R. R. Burgess, *Methods Enzymol.* **273,** 145 (1996).

Overnight culture in LB with 2% glucose, ampicillin (100 μg/ml) and tetracycline
(10 μg/ml) at 30°C or 37°C

↓

Cell culture growing at 30°C or 37°C is induced at mid-log phase with 1 m*M* IPTG,
followed by growing cells at 30°C. Cells are harvested 3 hrs after the induction

↓

Cells are disrupted by pressure disruption
Collect soluble fractions after centrifugation

↓

Metal affinity resin
Bound σ^{70} is eluted with 50 m*M* imidazol

↓

Resource Q column
0.1–0.5 *M* NaCl gradient

↓

σ^{70} fractions are concentrated and stored in protein storage buffer

FIG. 1. Schematic for overproduction and purification of soluble His-tagged σ^{70} from *E. coli*
XL-1 blue/pQE30-*rpoD*.

E. coli are not well understood thus far; however, it has been reported that
the solubility and activity of overproduced proteins can be increased by
induction at low temperature,[12–14] fusion with thioredoxin,[13,14] or coexpression with some molecular chaperones, such as GroEL/GroES, ClpB/
DnaK, and DnaJ.[14–16] This article describes a simple method for the purification of large amounts of soluble σ^{70} polypeptide with high activity. The
key element in obtaining overproduced soluble σ^{70} polypeptide inside cells
is to induce the expression of the protein at a low temperature (30°) instead
of at 37° (Fig. 2). By this method, we can isolate highly pure and active His-
tag-σ^{70} recombinant proteins in larger quantities using only two simple
steps. A schematic of the overproduction and preparation procedure is
shown in Fig. 1

Cell Growth and Induction

The *E. coli* strain (AD19275) that overproduces the histidine-tagged σ^{70}
protein was kindly provided by Dr. Alicia Dombroski. The *rpoD* gene
is cloned into *Kpn*I and *Hind*III sites in the polylinker of the pQE-30

[12] R. T. Aplin, J. E. Baldwin, S. C. Cole, J. D. Sutherland, and M. B. Tobin, *FEBS Lett.* **319**, 166 (1993).

[13] K. X. Huang, Q. L. Huang, M. R. Wildung, R. Croteau, and A. I. Scott, *Protein Expr. Purif.* **13**, 90 (1998).

[14] Y. P. Chao, C. J. Chiang, T. E. Lo, and H. Fu, *Appl. Microbiol. Biotechnol.* **54**, 348 (2000).

[15] K. Yokoyama, Y. Kikuchi, and H. Yasueda, *Biosci. Biotechnol. Biochem.* **62**, 1205 (1998).

[16] C. Schlieker, B. Bukau, and A. Mogk, *J. Biotechnol.* **96**, 13 (2002).

expression vector (Qiagen, Chatsworth, CA), resulting in pQE30-*rpoD* and the recombinant σ^{70} protein containing a six-histidine tag at its N terminus.[7,8] The host for the pQE30-*rpoD* is XL1-Blue [*recA1*(F', *proAB*, *lacI*^{q}*Z* ΔM15, Tn*10*)] (Stratagene, La Jolla, CA). The overnight culture of AD19275 (XL1-Blue/pQE30-*rpoD*, also renamed as DJ420 in our strain collection) is grown overnight in 25 ml LB medium plus 2% glucose, 100 μg/ml ampicillin, and 10 μg/ml tetracycline at 37°. The next day, the entire 25-ml overnight culture is diluted 1:10 in fresh LB medium (250 ml) plus 2% glucose and 100 μg/ml ampicillin and is grown at 30° or at 37° until midlog phase ($OD_{600} \sim 0.5$–0.7). Isopropyl-β-D-thiogalactoside (IPTG) is added to a final concentration of 1 mM, and the cultures are grown at 30° for an additional 3–4 h. Cells are harvested by centrifugation at 3500 rpm for 15 min at 4° in a cold Sorvall H-6000 rotor. The cell pellets are stored at −70° or used immediately. *Note.* To monitor the expression of *rpoD*, 1 ml of cells from cultures immediately before the addition of IPTG and 1 ml of cells from cultures after induction for 3–4 h are spun down and resuspended in 100 μl of 1× NuPAGE LDS sample buffer (Invitrogen, Carlsbad, CA). After heating at 90° for 5 min, samples (10 μl) from equivalent cells (OD_{600} units) are resolved on a 10% Bis–Tris SDS–PAGE gel and stained with Simply Blue SafeStain (Invitrogen).

Comments. Glucose is included in the medium to minimize leaky expression of σ^{70} under noninducing conditions. Because an excessive amount of soluble σ^{70} from leaky expression is detrimental to cell growth (at 30°), it is better to grow preinduction cultures at 37° to expedite the growth of cells followed by induction at 30°.

Purification Procedure

The following are carried out at 4° or on ice with precooled buffers.

1. The cell pellet from the 250-ml culture is resuspended in 20 ml ice-cold binding buffer (BB): 20 mM Tris–HCl, pH 8.0, 5 mM imidazole, 500 mM NaCl, 1 mM 4-(2-aminoethyl)benezenesulfonyl fluoride hydrochloride (AEBSF; ICN Biomedicals Inc.). Cells are lysed by two passes of the cell suspension through a French press at 1000 psi. The cell lysate is centrifuged for 20 min at 12,000 rpm in a Sorvall SS-34 rotor, and the supernatant is saved, as it contained most of overproduced σ^{70} (Fig. 2). *Note.* When needed, the pellet is resuspended in 20 ml binding buffer containing 8 M urea. To monitor the distribution of the σ^{70} in different fractions, the samples of total cell lysate, supernatant, and 8 M urea extract are mixed with equal volume of 2× NuPAGE LDS sample buffer, resolved on 10% Bis–Tris SDS–PAGE gels, and stained (Fig. 2).

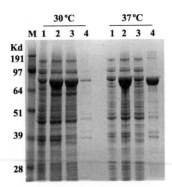

FIG. 2. SDS–PAGE analysis for solubility of His-tagged σ^{70} overproduced at 30 and 37°. Lane M: SeeBlue Plus2 prestained protein markers (Invitrogen); their sizes are indicated on the left. Lane 1: cell lysate before induction. Lane 2: cell lysate after induction for 4 h with 1 mM IPTG. Lane 3: the supernatant of cell lysate. Lane 4: 8 M urea extract of the pellet. Each lane contains the equivalent of 100 μl of cells at $OD_{600} = 0.5$.

2. Six milliliters of 50% slurry of Talon Superflow metal affinity resin (Clontech, Palo Alto, CA) is placed in a gravity flow column (1 × 20 cm, Econo column, Bio-Rad, Hercules, CA). The column is washed twice with 10 ml of distilled water, followed by equilibrium with BB. The supernatant containing σ^{70} is passed two times by gravity flow through the metal affinity column. The column is washed four times with 5 ml of BB. Bound proteins are eluted in 20 ml of elution buffer (20 mM Tris–HCl, pH 8.0, 50 mM imidazole, 500 mM NaCl, and 1 mM AEBSF). Most of the His-tagged σ^{70} polypeptide elutes under this condition (Fig. 3).

3. The His-tagged σ^{70} fraction from Talon metal affinity column is purified further by the application of a Resource Q column (1 ml, Amersham, Piscataway, NJ) using the FPLC system. The eluate from the metal affinity column is dialyzed overnight against 1 liter of TGED buffer [50 mM Tris–HCl, pH 8.0, 10% glycerol, 0.1 mM EDTA, 0.1 mM dithiothreitol (DTT)] plus 0.01% Triton X-100 and is then loaded on a Resource Q column, which is equilibrated with TGED. A total of 40 ml of the linear gradient from 0.1 to 0.5 M NaCl in TGED buffer is used to separate the His-tagged σ^{70} at 1 ml/min flow rate. The fractions (2 ml each) are collected and subjected to SDS–PAGE analysis. More than 50% of input σ^{70} is eluted at 0.35 M NaCl from the Resource Q column with high purity (>98%) (Fig. 4). Fractions containing σ^{70} are concentrated to about 1–2 mg/ml by application of Centricon YM-30 concentrators (Amicon, Bedford, MA) mixed with an equal volume of pure glycerol and stored at −70° for long-term storage. The typical yield is about 5–10 mg of σ^{70} from a 250-ml culture.

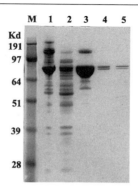

FIG. 3. Purification of soluble His-tagged σ^{70} using Talon Superflow metal affinity resin. Lane M: SeeBlue Plus2 prestained protein markers. Lane 1: the soluble fraction of cell lysate. Lane 2: flow through from the metal affinity column. Lanes 3–5: the eluate of 50, 100, and 1000 mM imidazole, respectively. Each lane contains the equivalent of 100 μl of cells at OD$_{600}=0.5$.

In Vitro Transcription Assay for Sigma Activity

The activity of the purified σ^{70} is evaluated by its ability to stimulate promoter-specific transcription. First, the holoenzyme for the in vitro transcription assay is reconstituted by mixing purified core enzyme at a concentration of 100 nM with various concentrations of the purified σ^{70}. For example, to reconstitute the holoenzyme at a 1:1 molar ratio of core RNA polymerase to σ^{70}, about 4 μg core RNA polymerase and 0.7 μg σ^{70} are mixed in 100 μl of protein storage buffer (50 mM Tris–HCl, pH 8.0, 50% glycerol, 0.1 mM EDTA, 0.1 mM DTT, and 50 mM NaCl) at room temperature (25°) for 10 min and then stored at −20°. The soluble σ^{70} is purified as described in this article, whereas the σ^{70} from inclusion bodies (they are induced at 37°) is prepared as described previously.[7,8]

For in vitro transcription, a reaction mixture is prepared as follows: 2 μl of 5× transcription buffer (200 mM Tris–HCl at pH 8.0, 5 mM DTT, 0.5 mg/ml purified bovine serum albumin, 50 mM magnesium chloride, and 750 mM NaCl), 1 μl of plasmid DNA template containing the Tac promoter (pDJ631, 0.1 μg/μl), and 4 μl of H$_2$O. The 7-μl aliquots of the reaction mixture are mixed with 1-μl aliquots of the reconstituted RNAP holoenzyme ($\sim$0.04 μg/μl) to give the final concentration of $\sim$10 nM (calculated for a final reaction volume of 10 μl). Following a 15-min preincubation at 37°, the transcription reactions are initiated by the addition of 2 μl of 5× NTP mixture (1 mM for ATP, CTP, and GTP, 0.1 mM for UTP) containing 1 μCi of [α-^{32}P]UTP. After incubation at room temperature (25°) for 15 min, the transcription reactions are terminated by the addition

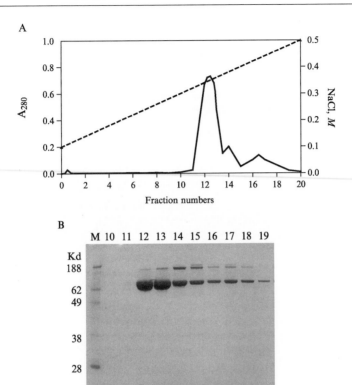

FIG. 4. Purification of soluble His-tagged σ^{70} using the Resource Q column. (A) The protein elution profile. (B) SDS–PAGE analysis of fractions from the Resource Q column. Ten-microliter samples of each fraction were mixed with 10 μl of 2× sample buffer, and 10 μl of the mixture was loaded on a 10% NuPAGE Bis–Tris gel (Invitrogen). The fraction numbers are labeled on the top. Lane M: SeeBlue prestained protein markers (Invitrogen) and their sizes are indicated on the left.

of 5 μl of stop solution (250 mM EDTA at pH 8.0, 50% glycerol, 0.05% xylene cyanol, and bromphenol blue). Aliquots of the terminated reactions (3 μl) are analyzed on a 6% sequencing gel. The RNA transcripts are visualized and quantified by exposure of the gel to a PhosphorImager screen (Amersham, Piscataway, NJ).

As shown in Fig. 5, the reconstituted RNAP from the soluble σ^{70} polypeptide was highly active. The RNAP reconstituted at a 1:1 molar ratio of core RNA polymerase to σ^{70} was nearly as active as the holoenzyme purified by the conventional method,[17] and the RNAP reconstituted at a

---

[17] M. V. Sukhodolets and D. J. Jin *J. Biol. Chem.* **273,** 7018 (1998).

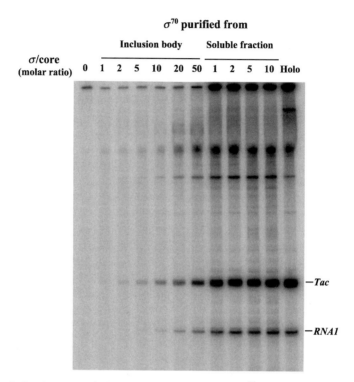

FIG. 5. *In vitro* transcription assays using His-tagged σ^{70} polypeptides purified from soluble or insoluble fractions. The soluble σ^{70} was purified as described in the text, whereas the σ^{70} from inclusion bodies was prepared as described previously.[7,8] The *in vitro* transcription assay was carried out in the presence of about 10 nM core RNAP and indicated amounts of purified σ^{70}. The same amount of purified holo RNAP (labeled as Holo, 10 nM) was used as control. Transcripts from *Tac* and *RNA1* promoters are indicated.

1:2 molar ratio of core RNA polymerase to σ^{70} was a active as the holoenzyme. However, the recombinant σ^{70} polypeptide purified from insoluble inclusion bodies had very low activity (Fig. 5), in agreement with the notion that the majority of the refolded σ^{70} polypeptides purified from inclusion bodies were multimeric and inactive.[5,11]

Acknowledgments

We thank Dr. Alicia Dombroski for the gift of the strains that overproduce the recombinant σ^{70} polypeptide and for many helpful discussions.

[16]   Expression, Purification of, and Monoclonal Antibodies to σ Factors from Escherichia coli

By Larry C. Anthony, Katherine M. Foley, Nancy E. Thompson, and Richard R. Burgess

Escherichia coli DNA-dependent RNA polymerase is the sole enzyme responsible for the synthesis of messenger, transfer, and ribosomal RNA. This multisubunit enzyme consists of four different polypeptide chains with subunit architecture $\alpha_2\beta\beta'\omega$. While the RNA polymerase core enzyme is capable of elongation and termination of transcription, an additional subunit, σ, is required to initiate transcription selectively at specific promoter sequences.[1,2] The *E. coli* housekeeping σ factor, σ^{70}, was the first prokaryotic σ factor to be purified and characterized.[1] Since then, six additional σ factors have been found in *E. coli* K12. All seven sigma factors have been categorized into two families by means of sequence similarity.[3] The σ^{70} family contains σ^{70}, σ^S, σ^{32}, σ^F, σ^E, and σ^{FecI}. However, σ^{54}, because of differences in sequence, promoter recognition, and function, is relegated to its own separate family.[4] Proteins from the σ^{70} family have been shown to interact with two conserved DNA hexamers centered approximately 10 and 35 bp upstream of the transcription start site.[5,6] In addition to unique spacing requirements within promoters, each sigma factor recognizes specific promoter sequences, allowing *E. coli* the means to regulate gene expression.[7,8] For example, σ^{32} initiates transcription of genes expressed during heat shock,[9] whereas σ^S directs expression of genes required for stationary phase survival (Table I).[10]

The major sigma factors of *E. coli* (σ^{70}, σ^{54}, σ^S, σ^{32}, and σ^F) can be purified using the basic principles of overexpression, isolation of inclusion bodies, denaturation and protein refolding, and purification over an

---

[1] R. R. Burgess, A. A. Travers, J. J. Dunn, and E. K. Bautz, *Nature* **221,** 43 (1969).

[2] J. D. Helmann and M. J. Chamberlin, *Annu. Rev. Biochem.* **57,** 839 (1988).

[3] M. Lonetto, M. Gribskov, and C. A. Gross, *J. Bacteriol.* **174,** 3843 (1992).

[4] W. Cannon, M. F. Claverie, S. Austin, and M. Buck, *Mol. Microbiol.* **8,** 287 (1993).

[5] T. Gardella, H. Moyle, and M. M. Susskind, *J. Mol. Biol.* **206,** 579 (1989).

[6] D. A. Siegele, J. C. Hu, W. A. Walter, and C. A. Gross, *J. Mol. Biol.* **206,** 591 (1989).

[7] C. A. Gross, C. Chan, A. Dombrowski, T. Gruber, M. Sharp, J. Tupy, and B. Young, *Cold Spring Harb. Symp. Quant. Biol.* **63,** 141 (1998).

[8] A. A. Travers and R. R. Burgess, *Nature* **222,** 537 (1969).

[9] A. D. Grossman, D. B. Straus, W. A. Walter, and C. A. Gross, *Genes Dev.* **1,** 179 (1987).

[10] R. Lange and R. Hengge-Aronis, *Mol. Microbiol.* **5,** 49 (1991).

TABLE I

EXPRESSION STRAINS AND PROPERTIES OF *E. coli* σ FACTORS

σ factor	Overexpression strain (plasmid)[a]	Function	Calculated MW	No. of amino acids	1 $A_{280\ nm}$[b] (mg/ml)	Isoelectric point[b]	Charge (pH 7)[b]	Refs.
σ70 (σD)	LCA57 (pLA4)	Housekeeping	70,261	613	1.77	4.59	−42.5	14
σ54 (σN)	LCA7 (pLA5)	Nitrogen regulation	54,011	477	1.26	4.54	−30.6	18
σ38 (σS)	LN14 (pLHN30)	Stationary phase	37,956	330	1.46	4.79	−17.3	16
σ32 (σH)	LN9 (pLHN16)	Heat shock	32,468	284	0.75	5.72	−5.0	12, 17
σ28 (σF)	LCA17 (pLN17)	Flagellar synthesis	27,520	239	1.01	5.16	−8.2	11, 15
σE	LCA30 (pLA17)	Periplasmic stress	21,695	191	1.47	5.31	−2.9	This work; 19
σFecI	LCA35 (pJP2)	Iron transport	19,479	173	1.08	6.01	−2.5	This work; 20

[a] BL21(DE3)pLysS was used as the expression strain for σ70, σ54, σ38, σ32, and σF. BL21(DE3) was used for σE and BL21(DE3)Tuner was used for σFecI overexpression. All BL21 derivatives were obtained from Novagen (Madison WI).

[b] Absorption coefficients, isoelectric points, and protein charge were calculated for the gene sequence using the DNASTAR Protean module (DNASTAR, Madison, WI).

ion-exchange column. Their significant negative charge at physiological pH (Table I) allows for purification on anion-exchange resin such as PorosHQ or Mono Q resins (Amersham). Protocols describing the purification of these sigma factors have been published previously.[11–19] Therefore, the purification protocols described in this article merely summarize and refine the established protocols using the most current chromatography techniques.

Because of lower expression levels and their specific properties, the extracytoplasmic sigma factors (σ^E and σ^{FecI}) cannot be purified using the basic technique described earlier for the purification of the other *E. coli* σ factors. Neither σ^E nor σ^{FecI} has a significant charge at physiological pH (Table I) and will not bind tightly to anion or cation-exchange resin. Additionally, upon overexpression of σ^E from a T7 promoter, it was found that 90% of the σ^E protein remained with the cellular membrane fraction following cell disruption and was not suitable for purification.[19] Due to its extreme insolubility and aggregation during refolding from denatured inclusion bodies, σ^{FecI} has required untraditional means of purification, such as size exclusion chromatography in the presence of mild denaturant.[20] Clearly, alternative methods for purification are required for these σ factors. Our laboratory has found that purification of σ^E can be accomplished easily through use of a His_6-thrombin cleavage tag overexpression vector (pET15b, Novagen). Through the use of the Fold-It kit (Hampton Research), we found the optimal refolding conditions for σ^{FecI} and developed a method of purification based on one single chromatographic step.

We also have generated a library of monoclonal antibodies to each of the seven *E. coli* σ factors (Table II). This antibody library provides useful tools for performing Western blot assays, immunoprecipitation assays, and analysis of transcription in *E. coli*.

[11] D. N. Arnosti and M. J. Chamberlin, *Proc. Natl. Acad. Sci. USA* **86,** 830 (1989).
[12] R. R. Burgess, *Methods Enzymol.* **273,** 145 (1996).
[13] R. R. Burgess and M. W. Knuth, in "Strategies for Protein Purification and Characterization: A Laboratory Course Manual," p. 205. Cold Spring Harbor Press, Cold Spring Harbor, NY, 1996.
[14] M. Gribskov and R. R. Burgess, *Gene* **26,** 109 (1983).
[15] T. K. Kundu, S. Kusano, and A. Ishihama, *J. Bacteriol.* **179,** 4264 (1997).
[16] L. H. Nguyen and R. R. Burgess, *Protein Expr. Purif.* **8,** 17 (1996).
[17] L. H. Nguyen, D. B. Jensen, and R. R. Burgess, *Protein Expr. Purif.* **4,** 425 (1993).
[18] D. Popham, J. Keener, and S. Kustu, *J. Biol. Chem.* **266,** 19510 (1991).
[19] S. Raina, D. Missiakas, and C. Georgopoulos, *EMBO J.* **14,** 1043 (1995).
[20] A. Angerer, S. Enz, M. Ochs, and V. Braun, *Mol. Microbiol.* **18,** 163 (1995).

TABLE II

σ-SPECIFIC MONOCLONAL ANTIBODIES AND THEIR PROPERTIES

Antibody	σ	Isotype	Epitope (amino acids)	Western blot?	Immunoprecipitates?	Polyol responsive for IAC?[a]	Refs.[b]
2G10	σ^{70}	IgG3	470–486	Yes	Holoenzyme	No	1, 2
6RN3	σ^{54}	IgG2a	88–137	Yes	No	No	3
1RS1	σ^{S}	IgG1	33–256	Yes	No	No	4
3RH3	σ^{32}	IgG1	1–25	Yes	Holoenzyme	ND	5
1RF18	σ^{F}	IgG1	ND[c]	Yes	ND	ND	This work
1RE53	σ^{E}	IgG1	ND	Yes	σ^{E}, holoenzyme	Yes	This work
1FE16	σ^{FecI}	IgG1	ND	Yes	σ^{FecI}, holoenzyme	Yes	This work

[a] Antibodies that are "polyol" responsive may be used for immunoaffinity chromatography (IAC) by attaching the antibodies to a chromatography resin (i.e., CNBr-activated Sepharose).[6]

[b] 1. M. S. Strickland, N. E. Thompson, and R. R. Burgess, *Biochemistry* **27**, 5755 (1988); 2. L. Rao, D. P. Jones, L. H. Nguyen. S. A. McMahan, and R. R. Burgess, *Anal. Biochem.* **241**, 173 (1996); 3. S. A. Lesley, Ph.D. Thesis, University of Wisconsin-Madison; 4. L. H. Nguyen, D. B. Jensen, N. E. Thompson, D. R. Gentry, and R. R. Burgess, *Biochemistry* **32**, 11112 (1993); 5. S. A. Lesley, N. E. Thompson, and R. R. Burgess, *J. Biol. Chem.* **262**, 5404 (1987); 6. N. E. Thompson, D. A. Hager, and R. R. Burgess, *Biochemistry* **31**, 7003 (1992).

[c] Not determined.

Purification of σ^{70}, σ^{54}, σ^{S}, σ^{32}, and σ^{F}

Buffer A: 50 mM Tris–HCl, pH 7.9; 100 mM NaCl; 0.1 mM EDTA; 0.1 mM dithiothreitol (DTT); 5% glycerol
Buffer B: 50 mM Tris–HCl, pH 7.9; 6 M guanidine–HCl; 100 mM NaCl; 0.1 mM DTT
Buffer C: 50 mM Tris–HCl, pH 7.9; 50 mM NaCl; 0.1 mM EDTA; 0.1 mM DTT; 5% glycerol
Buffer D: 50 mM Tris–HCl, pH 7.9; 500 mM NaCl; 0.1 mM EDTA; 0.1 mM DTT; 5% glycerol
Storage buffer: 50 mM Tris–HCl, pH 7.9; 100 mM NaCl; 0.1 mM EDTA; 0.1 mM DTT; 50% glycerol

1. *Escherichia coli* BL21(DE3)pLysS cells containing the appropriate overexpression vector are subcultured into 500 ml LB medium supplemented with antibiotics at 37° to $A_{600\ nm}$ 0.5–0.6 and induced by the addition of 1 mM isopropylthio-β-D-galactoside (IPTG). Cells are allowed to grow for an additional 3 h before harvesting by centrifugation at 8000 g for 25 min at 4°. Cells can be stored at −20° until use.

2. Cells (~0.7 g wet weight) are resuspended in 10 ml of buffer A. Cells are sonicated for 90 s and centrifuged at 20,000 g for 15 min at 4°.

3. The inclusion body pellet is washed with 10 ml of buffer A + 1% Triton X-100 to remove cell membranes and membrane proteins and is then centrifuged as described earlier. The inclusion body pellet is then washed twice with 10 ml of buffer A to remove residual Triton X-100.

4. The wash inclusion body pellet is solubilized in 5 ml of buffer B and incubated at room temperature for 30 min. Insoluble material is removed by centrifugation at 20,000 g for 15 min at 4°.

5. The solubilized inclusion bodies are slowly drip diluted over 5 min into 300 ml of buffer C at 4° with constant stirring and allowed to refold for 1 h at 4°. The solution is then centrifuged at 8000 g for 25 min to remove any insoluble material.

6. The clarified solution is applied to a 10-ml PorosHQ50 anion-exchange column preequilibrated with buffer C at a flowrate of 2–3 ml/min. The column is then washed with buffer C for 10 minutes and eluted over 30 min with a linear salt gradient of 0–100% buffer D.

7. Peak fractions are pooled and dialyzed against storage buffer and stored at −80°.

Purification Notes

- A SDS–PAGE gel showing the purified σ factors after dialysis into storage buffer is presented in Fig. 1.

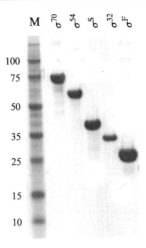

FIG. 1. SDS–PAGE of the purified major σ factors from *E. coli*. All proteins were purified from inclusion bodies, solubilized and refolded from 6 M guanidine hydrochloride, and purified by anion-exchange chromatography. Lane M is broad range protein molecular weight markers (Novagen). Sizes in kilodaltons are indicated on the left.

- We have found that while PorosHQ50 resin works well for the purification of σ^{70}, it does not have the resolution necessary to separate σ^{70} monomer from dimer. For this reason, we recommend using Mono Q resin, which is easily able to separate the two σ^{70} species.

- For overexpression of σ^{32}, rifampicin was added (150 μg/ml final concentration) to the LB culture media 30 min after the addition of IPTG. This step is necessary to inhibit the production of heat shock proteins, which would otherwise degrade the overproduced σ^{32}.[12,13]

- Because of both positive and negative protein surfaces, σ^{32} can be purified over either anion- or cation-exchange columns. A detailed protocol on refolding σ^{32} from Sarkosyl and use of cation exchange has been published previously.[12,13]

Purification of σ^{E}

Buffer E: 50 mM Tris–HCl, pH 7.9; 500 mM NaCl; 10 mM imidazole; 0.1% Tween 20; 5% glycerol
Storage buffer: 50 mM Tris–HCl, pH 7.9; 200 mM NaCl; 0.1 mM EDTA; 0.1 mM DTT; 50% glycerol

1. *Escherichia coli* cells LCA30 [BL21(DE3), pLA17] are subcultured into 200 ml LB medium supplemented with 100 μg/ml ampicillin at 37° to $A_{600\ nm}$ 0.5–0.6 and induced by the addition of 1 mM IPTG. Cells are allowed to grow for an additional 3 h before harvesting by centrifugation at 8000 g for 25 min at 4°. Cells can be stored at −20° until use.

2. Cells (0.4 g wet weight) are resuspended in 20 ml of buffer E and lysozyme is added to facilitate cell breakage. Cells are sonicated for 90 s, and the sample is centrifuged at 20,000 g for 15 min 4°.

3. The supernatant containing soluble His$_6$-thrombin-σ^E is transferred to a 50-ml polypropylene tube and diluted with 20 ml of buffer E to reduce sample viscosity.

4. The solution is applied to a 0.5-ml Ni-NTA agarose column (Qiagen) with a flow rate of approximately 0.5–1.0 ml/min. The Ni-NTA column is washed with 20 ml of buffer E to remove material that may bind nonspecifically to the Ni-NTA resin. His$_6$-thrombin-σ^E is eluted with the addition of buffer E + 300 mM imidazole. Typical yields from this step are around 8–10 mg of His$_6$-thrombin-σ^E.

5. If desired, the His$_6$-tag can be removed easily with the Thrombin Capture Cleavage kit (Novagen). Purified σ^E (~3 mg) is added to a reaction consisting of 300 μl thrombin digest buffer, 2.6 ml of H$_2$O, and 6 U of biotinylated thrombin. The reaction is incubated for 2.5 h at 25°.

6. To remove the biotinylated thrombin, 300 μl of streptavidin agarose (50% slurry) is added to the reaction and allowed to incubate for 30 min at 25° on a rotary shaker.

7. After incubation, the reaction is transferred to spin filters (included in the kit) and centrifuged at 500 g for 5 min. The filtrates should contain a mixture of digested His$_6$-tag fragment and σ^E and be free of thrombin.

8. The filtrates (~3 ml) are pooled, diluted with 12 ml of buffer E, and poured over a 0.5-ml Ni-NTA column to remove His$_6$-tag fragments. The flow through, which should contain pure σ^E, is collected.

9. Fractions are analyzed for protein concentration by the Bradford assay and analyzed for purity by SDS–PAGE.

Purified σ^E	Starting amount of His$_6$-σ^E	Yield σ^E
2.65 mg	3.00 mg	88%

10. The purified σ^E is dialyzed against storage buffer and stored at −80°.

Purification Notes

- A SDS–PAGE gel showing the protein yields at each purification step is presented in Fig. 2.
- The second Ni-NTA column (step 8) removes undigested His$_6$-thrombin-σ^E as well as any protein contaminants that bind nonspecifically to the Ni-NTA resin.
- This procedure should be scalable for purifying larger amounts of σ^E. However, we would recommend optimization of the thrombin digest conditions to achieve the maximum amount of thrombin digestion and final yield.
- Because the purified σ^E is relatively dilute ($\sim$200 μg/ml), the flow through can be concentrated in a 4-ml 10-kDa Ultrafree-4 concentrator (Millipore) by spinning at 7500 g for 90 min at 4°. We recommend keeping protein concentrations below 0.75 mg/ml to avoid potential aggregation of the sample during storage.

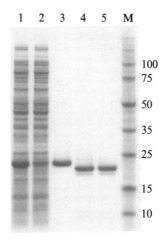

FIG. 2. SDS–PAGE of the purification of σ^E. After overexpression, His$_6$-thrombin-σ^E was purified from the soluble fraction by Ni-NTA chromatography. The N-terminal His$_6$-thrombin tag was removed from σ^E by thrombin digest. Thrombin was removed by the use of streptavidin agarose, and undigested protein was removed by a second Ni-NTA column. Lane 1, cell supernatant; lane 2, flow through from the first Ni-NTA column; lane 3, Ni-NTA eluate from the first Ni-NTA column; lane 4, eluate from the streptavidin agarose column; lane 5, pure σ^E from the second Ni-NTA column flow through. Lane M is broad range protein molecular weight markers (Novagen). Sizes in kilodaltons are indicated on the right.

Purification of σ^{FecI}

Buffer F:	50 mM Tris–HCl, pH 7.9; 500 mM NaCl; 0.1 mM EDTA; 0.1 mM DTT; 5% glycerol
Buffer G:	50 mM Tris–HCl, pH 7.9; 6 M guanidine–HCl; 100 mM NaCl; 0.1 mM DTT
Buffer H:	55 mM Tris, pH 8.2; 264 mM NaCl; 11 mM KCl; 0.05% PEG3350; 550 mM GuHCl; 2.2 mM MgCl$_2$; 2.2 mM CaCl$_2$; 440 mM sucrose; 550 mM arginine; 30 mM lauryl maltoside; 0.1 mM DTT
Buffer I:	50 mM Tris–HCl, pH 7.9; 200 mM NaCl; 0.1 mM EDTA; 0.1 mM DTT; 5% glycerol
Storage buffer:	50 mM Tris–HCl, pH 7.9; 200 mM NaCl; 0.1 mM EDTA; 0.1 mM DTT; 50% glycerol

1. *Escherichia coli* cells LCA35 [BL21(DE3) Tuner, pJP2] are subcultured into 500 ml LB medium supplemented with 50 μg/ml kanamycin at 37° to $A_{600\ nm}$ 0.5–0.6 and induced by the addition of 1 mM IPTG. Cells are allowed to grow for an additional 3 h before harvesting by centrifugation at 8000 g for 25 min at 4°. Cells can be stored at -20° until use.

2. Cells (1 g wet weight) are resuspended in 10 ml of buffer F and lysozyme is added to facilitate cell breakage. Cells are sonicated for 90 s and Triton X-100 is added to a final concentration of 1%. The solution is mixed well and centrifuged at 20,000 g for 15 min at 4°.

3. The inclusion body pellet is washed three times with 10 ml of buffer F and centrifuged as described earlier.

4. The washed inclusion body pellet (80 mg wet weight) is solubilized in 1 ml of buffer G and incubated at room temperature for 30 min. Insoluble material is removed by centrifugation at 20,000 g for 15 min at 4°.

5. The solubilized inclusion bodies ($\sim$3.35 mg protein) are drip diluted into 7 ml of buffer H at 4° and allowed to refold for 1 h at 4°.

6. Half of the sample (4 ml = 1.6 mg) is injected onto a Superdex 200 HR 10/60 gel filtration column (Amersham) preequilibrated with buffer I at a flow rate of 0.75 ml/min. One-milliliter fractions are collected while monitoring at $A_{280\ nm}$ and $A_{260\ nm}$. The column profile is shown in Fig. 3A.

7. The remaining sample (4 ml) is injected in a second run on the Superdex 200 column as described previously.

8. Fractions are analyzed for protein concentration by the Bradford assay and analyzed for purity by SDS–PAGE.

	Purified σ^{FecI}	Protein loaded	Yield σ^{FecI}
Run 1	0.71 mg	1.58 mg	44%
Run 2	0.86 mg	1.58 mg	55%

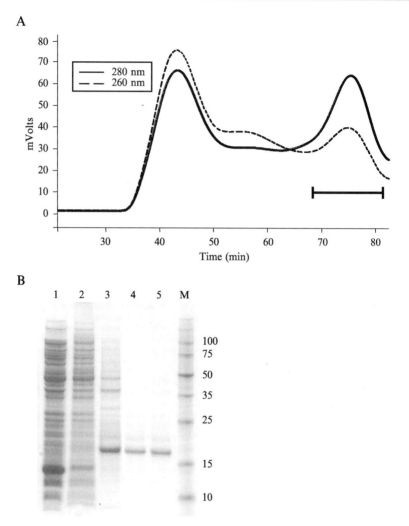

FIG. 3. Chromatography and purification of σ^{FecI}. σ^{FecI} was purified from inclusion bodies, solubilized in 6 M guanidine hydrochloride, refolded, and purified by gel filtration on a Superdex 200 HR 10/60 column. (A) Chromatograph from a Superdex 200 column. Absorption was monitored at $A_{280\ nm}$ (solid line) and $A_{260\ nm}$ (dashed line). Horizontal bar indicates fractions pooled for dialysis into storage buffer. (B) SDS–PAGE gel of each purification step. Lane 1, cell extract after overexpression; lane 2, cell supernatant; lane 3, inclusion bodies; lane 4, pooled fractions from first Superdex 200 column; lane 5, pooled fractions from second Superdex 200 column. Lane M is broad range protein molecular weight markers (Novagen). Sizes in kilodalton are indicated on the right.

9. Pooled σ^{FecI} fractions are dialyzed against storage buffer and stored at $-80°$. The final yield of σ^{FecI} is 1.57 mg.

Purification Notes

- Expression of σ^{FecI} was performed in the BL21(DE3)Tuner strain (Novagen), which contains a mutation in the lactose permease gene *lacY*. We were unable to obtain any overexpression of σ^{FecI} in the normal BL21(DE3)pLysS strain. The reason for this is unknown.
- A SDS–PAGE gel showing the sample at each purification step is presented in Fig. 3B.
- σ^{FecI} appears to be unstable at concentrations over 0.5 mg/ml and precipitates readily during refolding. The optimal refolding conditions (buffer H) were identified by use of the Fold-It kit from Hampton Research.
- The difference between the weights of steps 4 and 5 is primarily due to water contamination. The wet weight of the inclusion body pellet (step 4) was done on a Mettler balance, whereas the protein concentration of the solubilized inclusion bodies (step 5) was determined by the Bradford assay, which measures only protein content.
- The purified inclusion body pellet is contaminated with nucleic acids. However, the nucleic acids are removed by the Superdex 200 column and do not coelute with the purified σ^{FecI}. Typical $A_{280\ nm/260\ nm}$ ratios for the purified σ^{FecI} were from 1.6–1.8, indicating that nucleic acid contamination of the purified sample is minimal.

Production of σ-Specific Monoclonal Antibodies

To prepare monoclonal antibodies against the seven σ factors of *E. coli*, each recombinant σ factor (holoenzyme was used for σ^{32}) was used to immunize Balb/c ByJ mice (Jackson Laboratories). The mouse serum was tested by the ELISA assay and Western blot assay to ensure a high titer and specific immune response. Spleen cells from the antigen-stimulated mouse were fused with either NS-1 or SP2/0 plasmacytoma cells using a standard hybridoma technique.[21] The fusions were screened by the ELISA assay and Western blot assay. Hybridomas were subisotyped using an ELISA-based kit (American Qualex). Hybridoma cells were injected into Pristane-primed Balb/c ByJ mice for the production of ascites fluid.

[21] E. Harlow and D. Lane, "Antibodies: A Laboratory Manual." Cold Spring Harbor Press, Cold Spring Harbor, NY, 1988.

A description of the monoclonal antibodies and their usefulness in common biochemical assays is presented in Table II.

Acknowledgments

Research was supported by grants from NIGMS and NCI. In compliance with University of Wisconsin-Madison's Conflict of Interest Committee oversight policies, RRB and NET acknowledge financial interest in the company NeoClone, LLC., Madison, WI, which sells the antibodies listed in Table II.

[17] Studying Sigma–Core Interactions in *Escherichia coli* RNA Polymerase by Electrophoretic Shift Assays and Luminescence Resonance Energy Transfer

By VEIT BERGENDAHL and RICHARD R. BURGESS

Introduction to Protein-Binding Studies

There is great interest in identifying and characterizing protein–protein interactions. Many biological processes, such as cell signaling, metabolism, enzymatic activity, and specificity, are controlled by the modulation of protein structure and function upon binding to another protein. Transcriptional regulation, for example, is mediated significantly by protein–protein interaction because binding of transcription factors to RNA polymerase confers its specificity for promoters of transcribed genes. We use the transcription machinery from *Escherichia coli* in particular to illustrate two very useful strategies to investigate protein binding. Our work focuses on the interaction of the main transcription factor $\sigma 70$ with core RNA polymerase (RNAP). Their major interaction interface is part of region 2.1 to 2.2 (residues 360–421) in $\sigma 70$ and a coiled–coil at residues 260–309 in the β' subunit of RNAP.[1–3]

There are many approaches to studying protein–protein interactions. This article describes electrophoretic mobility shift (EMS) assays and fluorescence resonance energy transfer (FRET) for several reasons. EMS assays

[1] M. M. Sharp, C. L. Chan, C. Z. Lu, M. T. Marr, S. Nechaev, E. W. Merritt, K. Severinov, J. W. Roberts, and C. A. Gross, *Genes Dev.* **13**, 3015 (1999).
[2] T. M. Gruber, D. Markov, M. M. Sharp, B. A. Young, C. Z. Lu, H. J. Zhong, I. Artsimovitch, K. M. Geszvain, T. M. Arthur, R. R. Burgess, R. Landick, K. Severinov, and C. A. Gross, *Mol. Cell* **8**, 21 (2001).
[3] T. M. Arthur and R. R. Burgess, *J. Biol. Chem.* **273**, 31381 (1998).

are fairly simple and quick to perform with the equipment present in most biologically oriented laboratories. At the same time, they can give useful initial information about a protein binding to another protein, DNA, or RNA. EMS assays are based on the change of mobility of a protein during polyacrylamide gel electrophoresis (PAGE) upon binding to DNA, RNA, or another protein. Because binding of a protein to its binding partner predominantly requires that its higher-order structure is intact, nondenaturing or native gels are used. Crucial to EMS assays is the fact that the procedure involves separating the complex from the unbound binding partner by size and charge differences. This changes the equilibrium conditions at which initial binding occurs and thus weak interactions, such as in complexes that have a half-life shorter than the time scale of the separation step, are underrepresented or cannot be detected, as the interaction does not persist throughout the procedure. This is true for all nonhomogeneous techniques, such as surface plasmon resonance (SPR, used in BIAcore instruments), coimmunoprecipitation or "pull-down" assays, ELISA assays, Far-western blotting, and size exclusion and affinity chromatography, as well as sucrose gradient centrifugation.

In a homogeneous binding assay, all components of the assay are free in solution and no separation steps are used. A key benefit of such an assay is that the binding measurement is done without perturbing the binding equilibrium. This often allows more accurate assessment of the binding forces and mechanisms, thus resulting in most cases in the superior although technically more demanding way of investigating binding. The most common techniques involve spectroscopic methods because the use of light as a probe does not change the conditions in most systems. These methods use nuclear magnetic resonance (NMR), light scattering, and very often fluorescent probes, such as in fluorescence polarization (FP) and FRET assays. Recent developments of FRET also allow analysis of protein–protein interactions *in vivo*.[4] Together with two-hybrid screens and functional assays, these are currently the only established methods to obtain reliable *in vivo* data.

In a FRET-based assay, the binding partners have to be labeled with two different fluorophores. In order to allow resonance energy transfer between the two, the emission spectrum of the fluorophore serving as a donor has to overlap sufficiently with the excitation spectrum of the other fluorophore serving as the acceptor. Resonance (i.e., acceptor frequency equals donor frequency) facilitates a radiationless energy transfer through dipole–dipole interactions. As described by the Förster[5] theory, the amount

[4] M. Elangovan, R. N. Day, and A. Periasamy, *J Microsc.* **205**, 3 (2002).
[5] T. Förster, *in* "Modern Quantum Chemistry," Vol. 3, Academic Press, New York, 1965.

of transferred energy decays with the inverse sixth power of the distance between the dyes. The intensities of the donor or the acceptor emission can therefore give information about the distance of the two dyes and ultimately tell whether the labeled proteins are bound to each other.

We decided to focus on luminescence resonance energy transfer (LRET)-based assays for a homogeneous assay to measure formation of the $\sigma70-\beta'$ complex. LRET is a recent modification of FRET[6-9] that uses a lanthanide-based donor fluorophore. The more general term luminescence instead of fluorescence (as in FRET) indicates that lanthanide emission is technically not fluorescence (i.e., arising from a singlet to singlet transition). Its most characteristic property is a very long fluorescent lifetime of up to several milliseconds, compared to the short lifetime of most organic-based fluorophores in FRET (picoseconds up to a microsecond). The conventional dipole–dipole theory of Förster is still applicable for LRET, so that energy transfer from a lanthanide donor to a generally short-lived (nanoseconds) acceptor fluorophore results in its prolonged emission (milliseconds).

We labeled $\sigma70$ with a Europium–DTPA–AMCA complex as a donor and the β' fragment with IC5–PE–maleimide. Because fluorescence detection is usually very sensitive, the labeled proteins can also be used very efficiently for their detection in gels resulting from EMS assays, avoiding the need for other labeling methods, such as ^{32}P labeling.

Introduction to Fluorescence Labeling of Proteins

For uniform labeling of proteins with only one fluorophore, we exploited the specific reaction of maleimide-linked dyes with cysteine residues in proteins with only one cysteine residue. The maleimide moiety of the dye reacts readily in sulfhydryl-free buffers and is highly specific to cysteine residues at pH values between 6.5 and 8. If uniform labeling is not critical, succinimide derivatives of the dyes can be used to randomly label lysine residues in amine-free buffers at pH 7.5 to 9.0. Our studies addressed a few challenges in fluorescence labeling of proteins. For example, because oligomerization and oxidation of the sample can lead to a significant loss of labeling, a reduction step before derivatization can increase yields. Unfortunately, reducing agents such as 2-mercaptoethanol (2-ME)

[6] T. Heyduk, *Methods* **25**, 44 (2001).
[7] T. Heyduk, *Curr. Opin. Biotechnol.* **13**, 292 (2002).
[8] P. R. Selvin, *Annu. Rev. Biophys. Biomol. Struct.* **31**, 275 (2002).
[9] P. R. Selvin, *in* "Applied Fluorescence in Chemistry, Biology and Medicine" (W. Rettig, B. Strehmel, and S. Schrader, eds.), p. 457. Springer Verlag, New York, 1999.

and Tris(2-carboxyethyl)phosphine (TCEP) have to be removed before the labeling reaction, as they inhibit maleimide labeling.[10] Another challenge when labeling proteins is the separation of labeled product from the excess dye. This is especially important for samples used in FRET and LRET assays, as too much free dye can cause an additional signal by diffusion-controlled resonance energy transfer when working at free dye concentrations above 200 n*M*. For cyanine dyes such as IC5, it is very difficult to remove excess dye by dialysis and it usually requires an extra gel filtration step.

This article describes a procedure for purification and labeling where reduction, labeling, and washing away excess dye are carried out while the protein is bound to the column. Additional information about the labeling procedures presented here can be found in Bergendahl *et al.*[11] In case of the hexahistidine-tagged β prime (RpoC) fragment (residues 100–309), which has a single naturally occurring cysteine at residue 198, we worked on a Ni-NTA column. We purified and labeled a single cysteine σ70 mutant (C132S, C291S, C295S, S442C)[12] on a DE-52 anion-exchange column. To circumvent reducing the Ni resin by 2-ME, we used TCEP instead. After reduction, TCEP was washed away and on-column derivitization was carried out immediately to avoid oxidation. Excess dye could be washed from the column while the labeled product remained bound to the resin.

The following protocols are presented.

A. Ni-NTA purification and labeling of HMK-His$_6$-β'(100–309)
B. DE52 purification and DTPA–AMCA–Europium derivatization of σ70(C132S, C291S, C295S, S442C)
C. Electrophoretic mobility shift (EMS) assay for complex formation of labeled σ70 and β' fragment
D. LRET assay for protein–protein interaction of labeled σ70 and β' fragment

All chromatographic procedures are carried out by gravity flow at room temperature. The assays described in C and D can be performed with native labeled proteins or with denatured proteins added directly into the reaction mixture, which by dilution of GuHCl in the final assay allows instant refolding. This can be done to prevent precipitation of the refolded labeled protein upon storage before the assay. In order to ensure that refolding occurs and to confirm the results from the experiments with

---

[10] E. B. Getz, M. Xiao, T. Chakrabarty, R. Cooke, and P. R. Selvin, *Anal. Biochem.*, **273**, 73 (1999).

[11] V. Bergendahl, L. C. Anthony, T. Heyduk, and R. R. Burgess, *Anal. Biochem.* **307**, 368 (2002).

[12] E. Heyduk and T. Heyduk, *J. Biol. Chem.* **274**, 3315 (1999).

denatured protein being added, controls with native or refolded proteins should be carried out.

Materials and Chemicals

IC5–PE–maleimide is from Dojindo Molecular Technologies, Inc. (Gaithersburg, MD). DTPA–AMCA–maleimide is produced as described by Heyduk[12] (CAS numbers of the used starting materials: DTPA–anhydride: 23911-26-4, Maleimidopropionic acid, NHS ester: 55750-62-4, AMCA: 106562-32-7). These so-called Selvin chelates for Europium are available commercially from PanVera (Madison, WI). Other Europium labels are sold by Packard. Any of these labels should work in these protocols as long as they contain the maleimide group for derivatization. Disposable columns (10 ml PolyPrep, 0.8×4 cm) are from Bio-Rad. Ni-NTA resin is from Qiagen (Hilden, Germany). TCEP and Triton X-100 are from Pierce. $EuCl_3$ and all other chemicals are from Sigma unless indicated otherwise in the text. Precast PAGE gels are from Invitrogen. A multiplate reader (Wallac, VictorV^2 1420) is used to perform the LRET assays.

Buffers and Solutions

Native PAGE buffer: 200 mM Tris–HCl, pH 7.5, 20% glycerol, 0.005% bromphenol blue

NTG buffer: 50 mM NaCl, 50 mM Tris, 5% glycerol; pH 7.9

TGE buffer: 50 mM Tris–HCl, pH 7.5, 5% glycerol, 0.1 mM EDTA

NTTw buffer: 500 mM NaCl, 50 mM Tris–HCl, pH 7.9, 0.1% (v/v) Tween 20

NTGED buffer: 50 mM NaCl, 50 mM Tris–HCl, pH 7.5, 5% glycerol, 0.1 mM EDTA, 0.1 mM dithiothreitol (DTT)

Storage buffer: 0.2 M NaCl, 50 mM Tris–HCl, pH 7.5, 50% glycerol, 0.1 mM EDTA, 0.1 mM DTT

A. Ni-NTA Purification and Labeling of HMK-His$_6$-β'(100–309)

Overproduction of HMK-His$_6$-β' (100–309) and Purification of Inclusion Bodies

HMK-His$_6$-β'(100–309) is overproduced in *Escherichia coli* BL21 (DE3) from pTA133[10] and is a chimera (25 kDa) with the amino acid sequence MA<u>RRASV</u>HHHHHHM fused N-terminally to β'(100–309). The underlined sequence represents a heart muscle kinase (HMK) recognition site and was used in the original construct to allow ^{32}P labeling of the protein, but we did not make use of it in this article. Cells are grown in Luria-Bertani broth (LB) at 37°, induced for 2 h with 0.5 mM IPTG, harvested, and lysed. Inclusion bodies are isolated and washed in 10 ml NTGED buffer + 0.1% (v/v) Triton X-100. Washed inclusion bodies are stored in aliquots at −20° until use.

Procedure

1. One aliquot of about 20 mg (wet weight, from 100 ml of culture) of β' inclusion bodies is solubilized in 1.5 ml NTTw buffer + 6 M GuHCl + 5 mM imidazole.

2. The precipitate from the solubilized inclusion bodies is spun down in a microfuge at 18,000 g (14,000 rpm) for 5 min.

3. The supernatant is loaded on a Bio-Rad column (10 ml PolyPrep, 0.8 × 4 cm) containing 0.5 ml Ni-NTA matrix (Qiagen) that has been equilibrated with 5 ml NTTw buffer + 6 M GuHCl + 5 mM imidazole. That way the column is saturated with approximately 5 mg of the target protein.

4. The column is washed with 5 ml NTTw buffer + 6 M GuHCl + 20 mM imidazole to remove unbound protein.

5. To reduce any disulfide bonds, the column is washed with 5 ml of freshly prepared NTTw buffer + 6 M GuHCl + 20 mM imidazole and 2 mM TCEP.

6. Excess TCEP is removed by washing with 3 ml NTTw buffer + 6 M GuHCl + 20 mM imidazole.

7. The bound protein is derivatized with IC5–maleimide by loading 2 ml of freshly prepared NTTw buffer + 6 M GuHCl + 20 mM imidazole and 0.1 mM IC5–PE–maleimide.

8. The flow through is reloaded onto the column twice before excess dye is removed by washing with 3 ml NTTw buffer + 6 M GuHCl + 20 mM imidazole.

9. The derivatized protein is refolded while column bound by washing with 3 ml NTTw buffer + 20 mM imidazole to wash away the denaturant GuHCl.

10. Derivatized protein is eluted with NTTw buffer + 200 mM imidazole, dialyzed into storage buffer, and stored at −20° at concentrations of around 0.5 mg/ml.

Comments

The procedure can be done under denaturing conditions throughout, eluting the labeled protein denatured. Another option is to load the protein onto the Ni-NTA resin after refolding the β' fragment into a 65-fold excess of TGE buffer. A third variation is to refold the β' fragment on the column just before labeling with the dye. TCEP is used as a reducing agent because it does not interfere with the Ni-NTA resin like 2-ME. TCEP is an odorless white powder that can be added as a solid just before use. It dissolves readily in aqueous solutions and is very stable unless chelators like EDTA are present. IC5–PE–maleimide is dissolved in chloroform and divided into aliquots of 0.1 μmol. The aliquots are then dried down in a SpeedVac (Speed-Vac SVC100H, Savant Instruments, Farmingdale, NY). The use of water is not recommended as it may cause hydrolysis of the maleimide moiety over time, especially upon storage.

B. DE52 Purification and DTPA–AMCA–Europium Derivatization of $\sigma70$(C132S, C291S, C295S, S442C)

Purification and derivatization of $\sigma70$ are done after refolding and in a similar fashion as the β' procedure, this time using Whatman DE52 resin instead of the Ni-NTA agarose and using TGE buffer instead of NTTw buffer.

Overproduction of $\sigma70$ (C132S, C291S, C295S, S442C) and Purification of Inclusion Bodies

Sigma70(442C) is overproduced from a plasmid derived from the $\sigma70$ expression system pGEMD[13,14] that had a *Hind*III fragment containing the *rpoD* gene from *E. coli* cloned into a pGEMX-1 (Promega) vector. The cells are grown, induced, harvested, and lysed. Inclusion bodies are isolated and washed in 10 ml NTGED buffer + 0.1% (v/v) Triton X-100. Washed inclusion bodies are frozen in aliquots at $-20°$ until use.

Procedure

1. One aliquot of about 20 mg (wet weight, from 100 ml of culture) of $\sigma70$ inclusion bodies is solubilized in 5 ml TGE buffer + 6 M GuHCl.
2. To refold proteins, the denaturant is diluted 100-fold by dripping into 500 ml of chilled TGE buffer + 0.01% Triton X-100 slowly stirring on ice. If precipitation occurred, the precipitate is removed by centrifugation at 25,000 g (15,000 rpm, SS-34 rotor) for 15 min at $4°$.

---

[13] K. Igarashi and A. Ishihama, *Cell* **65**, 1015 (1991).
[14] Y. Nakamura, *Mol. Gen. Genet.* **178**, 487 (1980).

3. The soluble refolded protein is then bound to an anion-exchange resin by adding 1 g DE52 (Whatman) dry resin as a suspension in 5 ml TGE buffer + 0.01% Triton X-100 directly into the 500 ml of diluted, renatured protein.

4. After stirring slowly for 15 min, the suspension is poured into an empty 25-ml Econo-Pack column (Bio-Rad) and washed with 5 ml NTG buffer + 0.01% Triton X-100.

5. Five milliliters of NTG buffer + 2 mM TCEP + 0.01% Triton X-100 is loaded onto the column to reduce any disulfides formed by the dimerization of sigma.

6. The column is washed with 5 ml NTG buffer + 0.01% Triton X-100 to remove TCEP.

7. To label σ70, 1 ml NTG buffer + 0.01% Triton X-100 + 1 mM DTPA–AMCA is loaded onto the resin and flow through is reloaded onto the column twice to ensure maximum labeling.

8. The Europium complex is formed by loading 5 ml NTG buffer + 0.01% Triton X-100 + 1 mM EuCl$_3$ onto the resin containing the derivatized protein. (Phosphate buffers should be avoided when working with lanthanides because insoluble complexes can lead to undesired precipitation.)

9. The column is washed with 5 ml NTG buffer + 0.01% Triton X-100.

10. Labeled protein is eluted with TGE buffer + 500 mM NaCl. The eluted fractions are analyzed by SDS–PAGE and pooled according to their purity and protein content.

11. Fractions of labeled σ70 are pooled and brought up to 50% glycerol or dialyzed against storage buffer (final protein concentration is about 2 mg/ml). The labeled protein is stored at -20° until use.

Comments

Labeling can also be done after purification by adding the purified protein to an aliquot of the dye (same buffers and conditions). That way samples of 10–100 μl can be labeled. Excess label and Eu ions are then removed either by dialysis or by using a Pharmacia G50 spin column. A SDS gel of a typical purification is shown in Fig. 1.

C. Electrophoretic Mobility Shift Assay for Complex Formation of Labeled σ70 and β' Fragment

For the performance of an EMS assay, it is important to use the appropriate buffer conditions. The salt concentration has to be chosen wisely, as some protein–protein interactions are sensitive to high salt. The pH of the

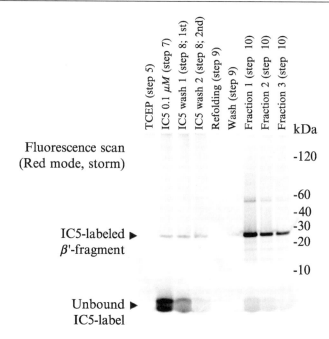

FIG. 1. SDS–PAGE gel of β' purification and derivatization steps. The fluorescence scan of a gel with samples taken after each step throughout the procedure illustrates purification of the β' fragment (refolding after fluorescence labeling). Labeling above lanes represents step numbers in the procedure at which the samples were taken. Purification from minor impurities can be observed, as well as very low elution of the β' fragment at any time except for fractions 1 to 3 eluted with 200 mM imidazole. The IC5 scan was performed with a Molecular Dynamics Storm system in the red fluorescence mode and thus shows IC5-labeled proteins. The gel was scanned before staining with Coomassie blue. Note the low level of free label in the eluted fractions.

buffer has to be chosen according to the pI of the proteins. Because the sample buffer and the native gel contain no SDS, which confers the proteins with negative charges, the proteins themselves have to have a negative charge in order to migrate into the gel during electrophoresis. Most sequence analysis software allows an estimate of the pI of a protein and even the net charge at a given pH. EMS assays are performed in a total volume of 20 μl. The buffer conditions are 5% glycerol, 50 mM Tris–HCl, pH 8.8, 50 mM NaCl, and 0.005% (w/v) bromphenol blue. Standard final protein concentrations are 250 nM of σ70 (Eu chelate-labeled protein) and 0.1 to 2 μM β' fragment (IC5-labeled protein). Concentrations can be lowered to 10 nM. While the detection limit of the Coomassie blue stain is reached at these concentrations, the fluorescence scan still gives good readouts.

Procedure

1. Labeled σ70 is added to the appropriate amount of water and 5 μl of native PAGE buffer (to result in a final volume of 20 μl after all components are added).
2. For competition experiments, unlabeled β' fragment is added.
3. The labeled β' fragment is added last.
4. The mixture is incubated for 5 min at room temperature.
5. Fifteen microliters is loaded on a precast native PAGE gel (12 well, 12%, Tris/glycine).
6. The gel is run at constant 120 V in the cold room at 4° for 4 h.

Comments

The concentration of the added components should be adjusted by dilution so that the added volume does not exceed 2 μl. That way mixing and final concentrations are not varied much by the addition of each component. The solutions have to be mixed well after the addition of each component by pipetting. This is especially important to allow good refolding when denatured proteins are added. The electrophoresis should be run with prechilled buffers, gels, and apparatus in the cold room (4°) at a constant voltage of 120 V (5–20 mA, variable). In some cases, running the gel at 4° might not be crucial, but in our case, it increased observable complex significantly. Running times may vary according to the charge and therefore the mobility of the proteins. In our case, running at 2–4 h gave best results in terms of separation and sensitivity. The IC5 emission was scanned on a Storm system (Molecular Dynamics) in the red fluorescence mode. The Europium emission can be scanned by using an Fotodyne UV light box ($\lambda_{excit.} = 312$ nm) with 6 s of acquisition time and the common filter for scanning ethidium bromide-stained gels. Total protein was stained with Coomassie blue stain using the Gel Code staining solution (Pierce) according to the procedure of the manufacturer and scanned with a Hewlett Packard flat bed scanner (ScanJet 6200C). For a typical gel from a EMS assay, see Fig. 2. We have also used EMS assays successfully with core RNA polymerase to investigate sigma binding. Because native gel multiple bands arise from core due to oligomerization, it is recommended to use only the σ70 in the labeled form and core RNA polymerase unlabeled.

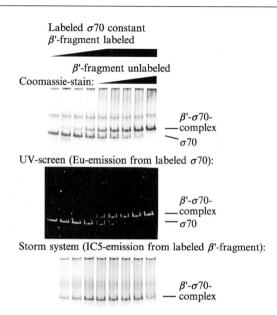

FIG. 2. Result of an EMS assay. The three pictures of the same gel taken by different staining and scanning techniques show increasing amounts of the labeled β' fragment (0, 50, 100, 250 nM) that can shift the labeled $\sigma70$ (250 nM) into the upper band. In the five lanes to the right, the unlabeled β' fragment (100, 250, 500, 750, 1000 nM) was added to compete with the IC5-labeled β' fragment for Eu-labeled $\sigma70$, which were kept at a constant concentration in a ratio of 1:1 (250 nM). The upper frame shows the Coomassie-stained gel. Labeled $\sigma70$ shows up in the lower band and the complex in the top band. The extra band above the complex arises from $\sigma70$ dimers linked by a disulfide bond, which prevented their fluorescence labeling. The middle frame represents the same gel, but was acquired before Coomassie staining with a UV box and an orange filter on the camera that can only visualize Eu emission due to the excitation wavelength (312 nm). It confirms that both bands contain Eu-labeled $\sigma70$. The bottom frame of the same gel was taken with a Storm imager (Molecular Dynamics) that can only visualize the IC5 label. It confirms that only the upper band contains labeled β' fragment. The free β' fragment runs as a diffuse band barely migrating into the gel. The unlabeled β' fragment can clearly compete for the labeled $\sigma70$, as can be seen by a loss of the band five lanes to the right.

D. LRET Assay for Protein–Protein Interaction of Labeled $\sigma70$ and β' Fragment

The format of the assay presented here was first designed to be suitable for high-throughput screening for inhibitors of the protein–protein inter-action. Thus it is robust, quick to perform, and can give useful initial data about binding, although the use of a screening device, such as the multiwell

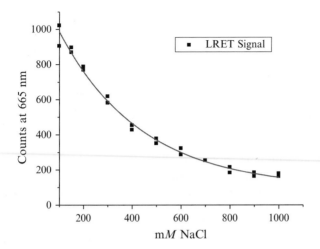

FIG. 3. Dependence of binding on NaCl concentration in the LRET assay. With increasing amounts of salt, the LRET signal decreases significantly due to the decreased amount of the $\sigma70/\beta'$ complex formed. The assay was performed in the fashion described in the text with the following protein concentrations: 20 nM Eu-labeled $\sigma70$ and 40 nM IC5-labeled β' fragment.

plate reader, does not fully exploit the potential of LRET. Much higher accuracy and sensitivity can be achieved by the use of a suitably equipped fluorimeter like the one described in Heyduk and Heyduk[15] Furthermore, such a setup allows accurate measurement of the distance between the dyes, which, in the case of allosteric binding, can give valuable information about mechanistic details. Nevertheless, the cost, operation, and access to such instruments are much more demanding compared to plate readers.

The assay was performed in a total volume of 200 μl in NTG buffer [plus 2.5% dimethyl sulfoxide (DMSO) when library samples were used] with typical concentrations of 10 to 100 nM labeled protein. It is very important to keep salt concentrations constant, as many protein–protein interactions are salt dependent. Therefore, amounts of salt added with the proteins and samples should be corrected for when adding NaCl to give the final concentration of 50 mM. Stock solutions (200 nM) of $\sigma70$ are prepared prior to the assay by dilution with NTG of labeled protein (40 μM) that is stored at $-20°$ in storage buffer. Because the labeled β' fragment shows a tendency to aggregate upon storage, we prefer to store it denatured in 6 M GuHCl and refold it by dilution into the assay. A stock

[15] E. Heyduk and T. Heyduk, *Anal. Biochem.* **248,** 216 (1997).

solution (1.25 μM) of the β' fragment is prepared by a 10-fold dilution of labeled denatured protein (stored at 75 μM in NTTw buffer containing 6 M GuHCl) with 6 M GuHCl and 6-fold dilution to 1 M GuHCl with NTG. First, 10 μl of the $\sigma 70$ stock solution is mixed with NTG buffer (amount adjusted to give a final volume of 200 μl), then the potential inhibitor (5 μl in DMSO), salt, or solvent is added, and finally the 5 μl of denatured labeled β' stock solution is added. Mixing (pipetting up and down three times) after the addition of each component is very important for reproducible results. The mixture is incubated for 30 min at room temperature and measured in a 96-well plate (Costar 3650) with a multiplate reader (Wallac, VictorV^2 1420). For this time-resolved fluorescence measurement, the manufacturer's protocol (LANCE high count 615/665) is used (excitation was with 1000 flashes at 325 nm; measurement is delayed by 50 μs and data acquired for 100 μs at 615 and 665 nm). The time for one measurement cycle is set for 1000 μs. Typical results from a LRET measurement showing salt dependence of binding are shown in Fig. 3.

Procedure

1. The appropriate amounts of 500 mM NaCl, 500 mM Tris–HCl, pH 7.9, 50% glycerol, and H$_2$O are mixed to result in the final concentrations 50 mM NaCl, 50 mM Tris–HCl, pH 7.9, 5% glycerol, and 200 μl total volume after samples and proteins have been added.
2. One to 10 pmol of labeled $\sigma 70$ (10–100 nM final concentration) is added in a 10-μl volume.
3. For competition or inhibition experiments, unlabeled proteins or samples are added.
4. Five to 20 pmol labeled β' fragment is added.
5. The mixture is incubated for 30 min at room temperature.
6. Samples are read in the multiplate reader (Wallac, VictorV^2 1420) with a TR-LRET protocol (here: LANCE high count 615/665).

Comments and Variations

Without going into detail, we would like to point out that for controlled labeling of the target protein, alternative methods can be used. Site-specific protein lableing using biarsenical compounds is described in Gaietta et al.[16] The Terbium donor and its compatible biarsenic compound FlAsH-EDT2 are available from PanVera. Site-specific intein-mediated

---

[16] G. Gaietta, T. J. Deerinck, S. R. Adams, J. Bouwer, O. Tour, D. W. Laird, G. E. Sosinsky, R. Y. Tsien, and M. H. Ellisman, *Science* **296,** 503 (2002).

C-terminal 5-(L-cysteinylamido-acetamido)-fluorescein (Cys-F) labeling, which can be used in combination with Terbium chelates as donors, is described in Mekler *et al.*[17] and Tolbert and Wong,[18] respectively. For most of the experiments, it is actually not necessary to label the proteins at defined positions. This is only necessary for measurements determining the distance between fluorescent probes. For general binding studies, random labeling of lysine residues using succinimidyl ester or isothiocyanides of the acceptor is absolutely sufficient. We have successfully labeled core RNA polymerase with IC5–OSu while it was bound to a DE52 column and used it in LRET assays with this strategy. Because lysine residues are very abundant on most protein surfaces, random labeling with a three- to five-fold excess of label over protein resulted in protein mostly active in binding. Molecules inactivated through labeling can be neglected because they cannot create a signal in the LRET assay. Thus a loss of binding activity has to be corrected for when looking at quantitative results. This approach has the advantage that the target protein does not have to be mutated to a single cysteine variant, which would still have to have its activity confirmed. Generally, any protein could be studied that way by labeling it in its purified form. Therefore, the calculated amount of a three- to five-fold molar excess dye over 100 μl protein (1 mg) is added dissolved in either DMSO or water. The reaction time is usually not more than 15 min at room temperature. Unbound label can be removed afterward by dialysis. In the case of very "sticky" dyes, such as cyanine dyes, a chromatography step to remove free dye is highly recommended.

A specific antibody can be labeled in the same way. It can then be used for indirect labeling of its antigen and measured with the binding partner of the fluorescence-labeled antigen in a time-resolved LRET experiment. Our experiments show comparable results with this approach, although it is critical to keep the concentration of labeled antibodies below the antigen concentration, as the addition of "unlabeled" antigen in competition experiments will increase the signal until the antigen is in excess. It is obvious that in screens for inhibitors of a protein–protein interaction of interest, compounds that interfere with the antibody–antigen interaction would show up as false positives.

[17] V. Mekler, E. Kortkhonjia, J. Mukhopadhyay, J. Knight, A. Revyakin, A. N. Kapanidis, W. Niu, Y. W. Ebright, R. Levy, and R. H. Ebright, *Cell* **108,** 599 (2002).
[18] T. J. Tolbert and C.-H. Wong *J. Am. Chem. Soc.* **122,** 5423 (2000).

[18] Assay of *Escherichia coli* RNA Polymerase: Sigma–Core Interactions

By TANJA M. GRUBER and CAROL A. GROSS

The sigma subunit (σ) of prokaryotic RNA polymerase confers promoter specificity to the enzyme. It not only recognizes promoter DNA, but also triggers a series of structural transitions required for initiation. Members of the σ^{70} family of proteins are modular, consisting of four conserved regions and their subregions (reviewed in Gross *et al.*[1]). Two DNA-binding determinants, located in regions 2.4 and 4.2, recognize the conserved −10 and −35 regions of the prokaryotic promoter. Region 1.1, conserved only among primary σ's, is an autoinhibitory domain that masks DNA-binding determinants in free σ^{70} but also promotes open complex formation.[2,3]

Genetic and biochemical methods have shown that the σ subunit can form contacts with core RNA polymerase (core) along the length of the protein.[4] Crystal structures of bacterial holo RNA polymerase show the extensive contacts between σ and core in the initiation complex.[5–7] Three stably folded domains, which correspond to the respective conserved regions of σ factors, are connected by flexible linkers; the self-inhibitory region 1.1 was not resolved in the crystal structure. The three resolved domains of σ are spread out across one face of core. Evidence shows that significant conformational changes occur in both σ and core upon binding, as well as from the transition from initiation to elongation phase. Based on the comparison of crystal structures of core versus holoenzyme, each σ domain interacts with and alters the position of a mobile domain within core (reviewed in Young *et al.*[8]).

[1] C. A. Gross, C. Chan, A. Dombroski, T. Gruber, M. Sharp, J. Tupy, and B. Young, *Cold Spring Harb. Symp. Quant. Biol.* **63,** 141 (1998).

[2] A. J. Dombroski, W. A. Walter, M. T. J. Record, D. A. Siegele, and C. A. Gross, *Cell* **70,** 501 (1992).

[3] A. J. Dombroski, W. Walter, and C. A. Gross, *Genes Dev.* **7,** 2446 (1993).

[4] M. M. Sharp, C. L. Chan, C. Z. Lu, M. T. Marr, S. Nechaev, E. W. Merritt, K. Severinov, J. W. Roberts, and C. A. Gross, *Genes Dev.* **13,** 3015 (1999).

[5] K. S. Murakami, S. Masuda, E. A. Campbell, O. Muzzin, and S. A. Darst, *Science* **296,** 1285 (2002).

[6] K. S. Murakami, S. Masuda, and S. A. Darst, *Science* **296,** 1280 (2002).

[7] D. G. Vassylyev, S. Sekine, O. Laptenko, J. Lee, M. N. Vassylyeva, S. Borukhov, and S. Yokoyama, *Nature* **417,** 712 (2002).

[8] B. A. Young, T. M. Gruber, and C. A. Gross, *Cell* **109,** 417 (2002).

Various other approaches have also been used to look at conformational changes following σ–core binding. For example, fluorescence resonance energy transfer (FRET) analyses showed that interaction with core reorients regions 1.1 and 3.1–4.2 of σ^{70} relative to the central 1.2–2.4 domain in a dramatic fashion.[9,10] In one report,[11] we used protein interaction studies with wild-type and mutationally altered proteins, previously known to impair σ^{70} function *in vivo* and binding to core *in vitro*,[4] to identify a number of interaction sites between σ^{70} and RNA polymerase. We also provided evidence that σ^{70} and RNA polymerase interact in a multistep process.

This article focuses on the methods employed in that study. The approach is useful to study interactions of regions not resolved in the crystal structure, as well as the determination of interactions that occur at an initial stage versus interactions that occur subsequent to conformational changes of the two binding partners. Domains masking interaction sites within the protein can also be identified. Furthermore, interactions of core with σ factors other than housekeeping σ factors can be investigated and compared. This approach can, in principle, be used to examine any protein–protein interaction.

Experimental Approach

The interactions of fragments of core with σ and its fragments are compared by GST pull-down assays. Mutations that are known to decrease interactions in the context of the intact holoenzyme[4] are introduced into the various fragments. Only interactions that are affected by these mutations are considered for further analyses. This prevents the consideration of potentially nonspecific interactions. Also, previous mutational analysis validates the biological relevance of the interactions observed.

Plasmids and DNA Manipulations

A large number of commercial vector systems are available to attach affinity tags to proteins, which allows rapid purification of large quantities of proteins. An exhaustive description of different peptide and protein tags is described by Hearn and Acosta.[12] In our applications, we used GST-tagged σ factor constructs and His_6-tagged core fragments.

[9] S. Callaci, E. Heyduk, and T. Heyduk, *J. Biol. Chem.* **273**, 32995 (1998).

[10] S. Callaci, E. Heyduk, and T. Heyduk, *Mol. Cell* **3**, 229 (1999).

[11] T. M. Gruber, D. Markov, M. M. Sharp, B. A. Young, C. Z. Lu, H. J. Zhong, I. Artsimovitch, K. M. Geszvain, T. M. Arthur, R. R. Burgess, R. Landick, K. Severinov, and C. A. Gross, *Mol. Cell* **8**, 21 (2001).

[12] M. T. Hearn and D. Acosta, *J. Mol. Recogn.* **14**, 323 (2001).

Boundaries of the core fragments represented either naturally occurring split sites in some organisms or functional splits in proteolytically sensitive regions. They are able to assemble into functional RNA polymerase.[13,14] σ^{70} fragments were also believed to be independently folding domains, as they showed by previous studies that they have sufficient structural integrity to be functionally active (see Gruber et al.[11]). With the crystal structures of core RNA polymerase and σ factors available now,[5–7,15,16] fragments can be designed to even greater accuracy to represent functional domains.

The GST tag has been used successfully in overexpressing and purifying GST–σ fusion constructs.[17] Some advantages of using the GST tag with σ constructs are the increased solubility and stability of the fusion proteins compared to σ constructs alone. A disadvantage can be the large size (26 kDa) of the GST moiety, which can interfere with certain analyses. This moiety, however, can be removed by protein cleavage through a protease site between GST and σ. We generated expression vectors for GST–σ fusion proteins by inserting the DNA fragments downstream and in-frame of the GST gene of pGEX-2T (Pharmacia, Piscataway, NJ[18]). With pGEX vectors, many different Escherichia coli strains can be used for overexpression because the plasmid contains a copy of the lacI gene, which keeps the promoter repressed until the inducer Isopropyl-β-D-thiogalactoside (IPTG) is added.

The core fragments were fused to the N-terminal His$_6$ tag derived from the pET15b vector (Novagen, Madison, WI). The His$_6$ tag can be removed subsequently by a protease cleavage site situated between the tag and the core construct. In pET vectors the target genes are cloned under the control of bacteriophage T7 transcription and translation signals. The promoter is not recognized by E. coli RNA polymerase and virtually no expression occurs until T7 RNA polymerase is provided to the BL21 expression host (Novagen) containing a chromosomal copy of the T7 RNA polymerase gene under lacUV5 control. Thus, genes cloned into the pET system do not cause plasmid instability due to the production of proteins potentially toxic to the cell. Furthermore, the BL21 family of strains is

[13] K. Severinov, A. Mustaev, M. Kashlev, S. Borukhov, V. Nikiforov, and A. Goldfarb, J. Biol. Chem. 267, 12813 (1992).

[14] K. Severinov, A. Mustaev, A. Kukarin, O. Muzzin, I. Bass, S. A. Darst, and A. Goldfarb, J. Biol. Chem. 271, 27969 (1996).

[15] E. A. Campbell, O. Muzzin, M. Chlenov, J. L. Sun, C. A. Olson, O. Weinman, M. L. Trester-Zedlitz, and S. A. Darst, Mol. Cell 9, 527 (2002).

[16] G. Zhang, E. A. Campbell, L. Minakhin, C. Richter, K. Severinov, and S. A. Darst, Cell 98, 811 (1999).

[17] A. J. Dombroski, Methods Enzymol. 273, 134 (1996).

[18] D. B. Smith and K. S. Johnson, Gene 67, 31 (1988).

deficient in the two proteases lon and ompT, which can further stabilize the produced proteins.

Mutant σ and core constructs were created using the QuickChange site-directed mutagenesis kit (Stratagene, La Jolla, CA). This system requires no specialized vectors, can be performed on miniprep DNA, and can be completed in as little as 1 day. It can be used to make point mutants or to delete or insert single or multiple amino acids. A polymerase chain reaction is carried out with oligonucleotide primers containing the desired mutation on double-stranded plasmid DNA. Subsequently, the product is treated with the restriction enzyme *Dpn*I, which specifically digests the methylated parental DNA template. The remaining plasmid DNA is transformed into a standard *E. coli* cloning strain and sequenced to confirm the mutation. The mutant proteins should be checked for major misfolding by standard methods such as a CD scan.[11,19]

Overproduction and Purification of Fusion Proteins

Overexpression of σ constructs often results in the formation of inclusion bodies that localize to the insoluble fraction. Growing the cells at 32° lowered the fraction of inclusion bodies and we were able to get at least 50% of the overexpressed protein yield in the soluble fraction. GST–σ fusion constructs are purified in batch using glutathione-agarose affinity beads (sulfur linkage, Sigma, St. Louis, MO).

A 5-ml culture of DH5α cells (or any common *E. coli* cloning strain) containing the σ–GST fusion construct is grown overnight at 37° in LB medium plus ampicillin (100 μg/ml) and transferred to 1 liter of LB plus ampicillin (100 μg/ml). The culture is grown at 32° to midexponential phase (OD$_{600}$ ~ 0.5) and induced with 1 mM IPTG. After growing for 2 h at 32°, the cells are harvested by centrifugation and resuspended in ~10 ml lysis buffer [50 mM Tris, pH 8.0, 150 mM NaCl, 1 mM EDTA, 1 mM dithiothieitol (DTT), 0.1 mM p-Aminoethyl benzene sulfonyl Fluoride (AEBSF), 1 mg/ml lysozyme]. After sonication, Triton X-100 is added to 1% (v/v) and the lysate is mixed on a rocker for 1 h at 4°. The mixture is centrifuged and 1 ml of a 50% slurry of glutathione-agarose beads is added to the soluble fraction and rocked overnight at 4°. The beads are prepared by resuspending 400 mg in 10 ml of wash solution (150 mM NaCl, 16 mM Na$_2$HPO$_4$, 4 mM NaH$_2$PO$_4$, pH 7.4, 0.5% Triton X-100, 1 mM EDTA). The beads are washed four times with wash solution for 5 min at 4° and finally resuspended in a total volume of 10 ml wash solution. Note that

[19] B. A. Young, L. C. Anthony, T. M. Gruber, T. M. Arthur, E. Heyduk, C. Z. Lu, M. M. Sharp, T. Heyduk, R. R. Burgess, and C. A. Gross, *Cell* **105**, 935 (2001).

the beads are spun down by centrifugation at a maximum of 500 g. The lysate–bead mixture is washed four times for 5 min each at 4° with wash solution and one time with GST pull-down buffer [500 mM NaCl, 50 mM Tris, pH 7.4, 20% (v/v) glycerol, 1 mM DTT, 1 mM EDTA, 0.5% Triton X-100, 0.1 mM AEBSF]. As purified GST–σ constructs were used directly in GST pull-down assays, the constructs were not eluted off the beads and were stored in a 50% slurry at 0°.

Fragments of core were found to be mainly in insoluble inclusion bodies upon overexpression. The best results have been obtained by denaturation of the protein with urea. The denatured proteins are batch bound to metal affinity resin, eluted, and refolded by dialysis.

A 5-ml culture of BL21(DE3)-pLysS $E.$ $coli$ cells (Novagen) containing the His$_6$ tag core constructs is grown overnight at 37° in LB medium plus ampicillin (100 μg/ml) and chloramphenicol (20 μg/ml) and transferred to 1 liter of LB plus ampicillin (100 μg/ml) and chloramphenicol (20 μg/ml). The culture is grown at 37° to midexponential phase (OD$_{600}$ $\sim$ 0.5) and induced with 1 mM IPTG. After 2 h at 37°, the cells are harvested by centrifugation, and the resulting cell pellet is stored at $-20°$. The thawed cells are resuspended at 5 ml per gram of wet cells in denaturing buffer (8 M urea, 0.1 M NaH$_2$PO$_4$, 0.01 M Tris–Cl, pH 8.0) and sonicated. The cellular debris is pelleted by centrifugation. One milliliter of a 50% slurry of Ni-NTA resin (Qiagen, Santa Clarita, CA) in denaturing buffer is added to each 10 ml of lysate and mixed for 1 h at room temperature. The lysate–resin mixture is loaded onto an empty column and washed extensively with His wash buffer (8 M urea, 0.1 M NaH$_2$PO$_4$, 0.01 M Tris–Cl, pH 6.3). The core constructs are eluted six times with 1 ml elution buffer (8 M urea, 0.1 M NaH$_2$PO$_4$, 0.01 M Tris–Cl, pH 4.5). Fractions containing protein are dialyzed into GST buffer [500 mM NaCl, 50 mM Tris, pH 7.4, 20% (v/v) glycerol, 1 mM DTT, 1 mM EDTA, 0.5% Triton X-100, 0.1 mM AEBSF] and stored as aliquots at $-80°$. A number of fragments aggregated upon dialysis; it is thus advisable to refold the core fragments at low protein concentrations (0.1 mg/ml or less).

GST Pull-Down Experiments

In our experiments, the σ–GST constructs remained on beads and the His$_6$-tagged core fragments were added in solution. After incubation and washing, the eluted His$_6$-tagged core fragments were visualized by an anti-His antibody, thus ensuring that signal intensity is proportional to the moles of core fragment retained. In principle, the pull-down reactions could have been carried out by affixing the core fragments to the resin and detecting the σ fragments with an anti-GST antibody.

Our assay conditions include high salt concentrations and detergent to preclude weak ionic and hydrophobic interactions, making it likely that we are observing bona fide interactions between the two partners. In general, 250 nM immobilized GST–σ fusions are incubated with 750 nM His$_6$-tagged core fragments for at least 2 h at 4° in a total volume of 200 μl GST buffer. Because the concentration of each fragment varies from one protein preparation to the next, both σ and core fragments are stored in GST buffer. This allows varying quantities to be added to the pull-down reaction while maintaining identical buffer conditions in each reaction. As a control, GST affixed to beads is incubated with the various core constructs to rule out interactions of the core fragments with the GST moiety. The resin beads are washed three times with 600 μl GST buffer and resuspended and boiled in SDS–PAGE sample buffer. Note that the beads should not be spun faster than 500 g. The proteins are fractionated on 10 or 12% SDS–PAGE gels and transferred to a nitrocellulose membrane. Western blots are incubated with antipenta-His primary antibody (Qiagen) and a secondary antimouse antibody (Promega, Madison, WI). His$_6$-tagged core fragments are visualized using the ECL system (Amersham Life Science, Piscataway, NJ). For graphical representation of data, bands are scanned and analyzed by Alpha Innotech densitometry software (Alpha Innotech, San Leandro, CA). Because the efficiency of transfer and exposure time vary between blots, signals can be compared only within a particular blot and not among blots.

For quantitation of the signal, a chemiluminescent substrate can be used, such as SuperSignal West Dura Extended Duration Substrate (Pierce, Rockford, IL). The light-emitted signal in the linear range is read with a CCD camera or a phosphor-imaging device.

Data Analysis

Our experiments focused on the comparisons of interactions between full-length σ and its fragments with fragments of core.[11] This allowed us to determine which interactions of σ are capable of being carried out in the context of free σ and which interactions can occur only after new determinants have been exposed due to a conformational change within σ following initial binding to core.

Initial Interface

The initial interface consists of core-binding determinants exposed in free σ. These show a positive signal in the pull-down experiments when full-length σ is used as the binding substrate. Furthermore, interactions are only considered as biologically relevant if a mutation known to disrupt the σ–core interface in the context of holoenzyme also disrupts the interaction in question.

Subsequent Interactions

The holoenzyme interface consists of binding determinants initially masked in free σ but exposed as a consequence of interactions in the initial interface. The following criteria are used to place binding determinants in this category: (1) Interactions of the σ fragment construct with the core fragment must be stronger than the interaction of full-length σ with that same fragment. (2) Mutations must alter the interaction with core fragments in the GST pull-down assay only in the context of the σ fragment, and not in full-length σ. (3) Mutations altering a binding determinant must significantly decrease the ability of the mutated σ to compete with wild-type σ for core in the context of holoenzyme, indicating that the site is used during the normal interaction process. Combining these data allows the establishment of a temporal model of interactions.

Acknowledgments

Work in this laboratory was supported in part by the NIH (GM30477) to CAG and Fellowship 2-19-99 of the American Cancer Society–California Division to TMG. We thank Brian Young and David Steger for critical reading of the manuscript.

[19] Purification and Characterization of Bacteriophage-Encoded Inhibitors of Host RNA Polymerase: T-Odd Phage gp2-like Proteins

By Sergei Nechaev, Diane Imburgio, and Konstantin Severinov

Infection of *Escherichia coli* with bacteriophage T7 leads to rapid and selective inactivation of host RNA polymerase (RNAP) by the product of viral gene 2, a 64 amino acid residue-long protein (gp2). Experiments *in vitro* with gp2 purified from T7-infected *E. coli*[1–3] and with the recombinant T7 gp2[4] showed that gp2 inhibited transcription by binding to the

[1] B. Hesselbach and D. Nakada, *Nature* **258**, 354 (1975).
[2] B. Hesselbach and D. Nakada, *J. Virol.* **24**, 746 (1977).
[3] M. A. DeWyngaert and D. C. Hinkle, *J. Biol. Chem.* **254**, 11247 (1979).
[4] S. Nechaev and K. Severinov, *J. Mol. Biol.* **289**, 815 (1999).

RNAP σ^{70} holoenzyme and rendering the resulting gp2–holoenzyme complex unable to recognize promoters. However, gp2 had no effect on transcription once an open promoter complex had formed.

Point mutations and deletions in the *rpoC* gene, which codes for the largest RNAP subunit, β', abolish the binding of gp2 to RNAP and make the mutant enzymes resistant to the inhibitor.[4] The region of β' defined by the gp2 resistance mutations is evolutionarily variable and functionally dispensable and likely forms the gp2-binding site. In the *Thermus aquaticus* RNAP core enzyme structure, the corresponding region of β' forms a poorly structured domain, which protrudes into the DNA-binding channel downstream of the catalytic center of the enzyme.[5]

Evidently, the domain contacts downstream DNA, either specifically or nonspecifically, during promoter complex formation and transcript elongation. The contact prevents binding of gp2 to the transcription complexes. Conversely, interaction of gp2 with this domain in free RNAP prevents contacts between RNAP and downstream DNA, which are necessary for stable transcription complex formation.

This article presents methods of overexpression and assays of recombinant T7 gp2, which is the best-studied protein of its class. It also presents a general protocol for the purification of viral-encoded inhibitory proteins that interact with host RNAP using the case of *X. oryzae* infection with bacteriophage Xp10 as an example. The protocol allowed us to purify Xp10-encoded protein p7, which is unrelated to gp2 proteins of T7-like phages (Fig. 1) but likely serves a similar function during the bacteriophage development.[6]

Overexpression and Purification of Recombinant T7 gp2

T7 gene 2 is polymerase chain reaction-amplified from T7 DNA and cloned between the *Nde*I and the *Xho*I sites of *E. coli* pET15b expression plasmid (Novagene). The recombinant protein expressed from the plasmid pET15_T7gp2 has an N-terminal hexahistidine tag. The tag does not interfere with the ability of gp2 to bind to and inhibit transcription by *E. coli* RNAP.[4] The recombinant protein is overexpressed and purified to homogeneity by a combination of Ni^{2+}-ion affinity and anion-exchange chromatography.

[5] K. Severinov, *Curr. Opin. Microbiol.* **3**, 118 (2000).
[6] M. I. Pajunen, S. J. Kiljunen, M. E. Soderholm, and M. Skurnik, *J. Bacteriol.* **183**, 1928 (2001).

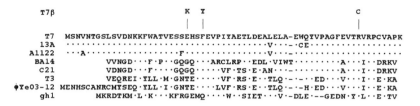

```
T7β                          K   Y                              C
                            |   |                              |
T7       MSNVNTGSLSVDNKKFWATVESSEHSFEVPIYAETLDEALELA-EWQYVPAGFEVTRVRPCVAPK
13A      ·······················································V···-·CE·················
A1122    ···A·····················F····················V···-·················
BA14     VVNGD···F·P··GQGQ···ARCLRP··EDL·VIWT·········A···I··DRKV
C21      VDNGD···F····GQGQ·····VF·TS·E·AN···-·········A···I··DRKV
T3       VEQREI·YLL·M·GNTE·····VF·RS·E··TLQ·--·ED···V···I·E·KA
◆Ye03-12 MENHSCANRCMYSEQ·YLL·I·GNTE····LVF·RS·E··TLQ·--H·ED···V···I·E·KA
 gh1      MKRDTKM·L·K···KFRGEMQ····W··SIET···V·-DLE·--GEDN·Y·L··E·TV


10  p7   MNEFTQISGYVNAFGSQRGSVLTVKVENDEGWTLVEEDFDRADYGSDPEFVAEVSSYLKRNGGIKDLTKVLTR
```

FIG. 1. Sequence similarities between gp2 proteins. Deduced amino acid sequences of several gp2-like proteins are presented in a single amino acid code and aligned to the T7 gp2 sequence (top). The Xp10 p7 sequence, which is unrelated to gp2 sequences, is shown at the bottom. Dots represent identical amino acids, hyphens represent gaps. The triple substitution found in T7β, a mutant T7 that is able to grow on *E. coli* harboring an *rpoC* (RNAP β') mutation that abolishes *E. coli* RNAP-binding and transcription inhibition by wild-type gp2,[4] is shown above the T7 gp2 sequence. Several *E. coli*, *S. flexneri*, *S. sonnei*, and *Y. enterocolitica* phages from the Republic of Georgia Eliava Bacteriophage Institute collection encode gp2 proteins whose sequences are identical to either T7 or T3 gp2 sequences.

Protocol

Inoculate 1 liter of LB medium containing 200 μg/ml ampicillin with *E. coli* SRB9 (DE3) cells[4] transformed with pET15_T7gp2 plasmid by transferring ~100 individual colonies from a fresh (plated the night before) plate into the medium. Grow cells with aeration at 37° until OD_{600} reaches ca. 0.4. Add Isopropyl-β-D-thiogalactoside (IPTG) to a final concentration of 1 mM and allow the culture to grow for another 3 h. Collect cells by centrifugation (Sorvall GS-3 rotor or similar, 5000 rpm for 10 min at 4°). Resuspend the cell pellet in 20 ml of buffer A (10 mM Tris–HCl, pH 8.0, 5% glycerol, 500 mM NaCl) containing 2 mM imidazole, chill on ice for 10 min, and transfer into a prechilled 45-ml centrifuge tube. Disrupt cells by sonication (five to six 30-s pulses at maximum power output, with 60-s rest intervals between the pulses, keeping the tube in an ice bath all the time). Remove cell debris by centrifugation at 15,000 rpm for 15 min (Sorvall SS-34 rotor, 4°). Transfer the supernatant into a fresh 45-ml centrifuge tube and repeat the centrifugation. Load the supernatant onto a 1-ml Hi-Trap chelating column (Amersham Biosciences) attached to an FPLC system, charge with Ni^{2+} according to manufacturer's recommendations, and equilibrate in buffer A. Wash the column with buffer A containing 20 mM imidazole until the baseline is stabilized (usually 5 to 10 column volumes is sufficient). Elute his-tagged gp2 with buffer A containing 200 mM imidazole. Dilute the 200 mM imidazole fraction threefold with buffer B (10 mM Tris–HCl, pH 8.0, 5% glycerol, 5 mM 2-mercaptoethanol) and load onto a 1-ml Resource Q column (Amersham Biosciences) equilibrated with buffer B and attached to an FPLC system. Elute proteins, at 0.5 ml/min, with a 40-ml linear gradient of NaCl (from 200 to 500 mM) in buffer B, collecting 1.5-ml fractions. Analyze 5-μl aliquots of each fraction by 15% SDS–PAGE (gp2 elutes at about 300 mM NaCl). Pool fractions containing gp2 and concentrate on a Centricon-3 centrifugal filter device (Amicon) to approximately 1 mg/ml, add glycerol to a final concentration of 50%, and store at $-70°$. The yield is $\sim$1 mg of gp2 (>95% pure as judged by visual inspection of an SDS gel).

Notes

T7 and T3 gp2 inactivate *E. coli* RNAP very efficiently and so gp2 expression plasmids are highly toxic. pET-based gp2 expression plasmids can be maintained without problems in *E. coli* cells that lack T7 RNAP, such as JM101, but cannot be transformed into BL21(DE3) *E. coli* cells that harbor the T7 RNAP gene on the λ (DE3) prophage. To overcome this limitation and allow high-level overproduction of gp2, the *E. coli* SRB9 strain[7] harboring an *rpoC* point mutation that renders RNAP resistant to gp2[4] was lysogenized with the DE3 phage using the DE3 lysogenization kit (Novagen) by following the manufacturer's instructions. The resulting strain was named SRB9(DE3).[4] To overexpress T7 gp2β,[7] a triple mutant[8] (Fig. 1) that inhibits RNAP from SRB9 cells but is less effective against the wild-type *E. coli* RNAP (unpublished observations), *rpoC*$^+$ *E. coli* BL21(DE3) strain harboring the pLysE plasmid may be used. However, with this system, which offers tight control of T7 RNAP expression, the level of gp2β overexpression and the yield of pure protein are $\sim$20-fold lower that those achieved for wild-type gp2 in SRB9(DE3). Viral proteins that either do not bind to *E. coli* RNAP (e.g., Xp10 p7) or bind to *E. coli* RNAP with relatively low efficiency (e.g., gh1 gp2, Fig. 2) are not toxic to *E. coli*, and the BL21(DE3) strain may be used as an expression host.

Overexpression and Purification of RNAP Subunits or Their Fragments

Plasmids used for overexpression of *E. coli* and *X. oryzae* RNAP subunits and subunit fragments have been described previously.[4,9,10] Most RNAP subunits and their fragments are insoluble and therefore can be

[7] M. Chamberlin, *J. Virol.* **14,** 509 (1974).

[8] M. P. Schmitt, P. J. Beck, C. A. Kearney, J. L. Spence, D. DiGiovanni, J. P. Condreay, and I. J. Molineux, *J. Mol. Biol.* **193,** 479 (1987).

[9] H. Tang, K. Severinov, A. Goldfarb, and R. Ebright, *Proc. Natl. Acad. Sci. USA* **92,** 4902 (1995).

[10] K. Severinov, A. Mustaev, A. Kukarin, O. Muzzin, I. Bass, S. A. Darst, and A. Goldfarb, *J. Biol. Chem.* **271,** 27969 (1996).

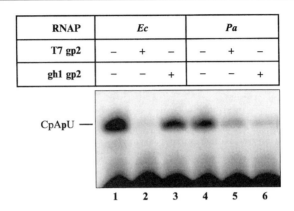

FIG. 2. Effect of T7 and gh1 gp2 proteins on abortive transcription initiation by *E coli* and *P. aeruginosa* RNAPs from the T7 A1 promoter. *E. coli* (Ec) or *P. aeruginosa* (*Pa*) σ^{70} RNAP holoenzymes alone (lanes 1 and 2) or RNAP holoenzymes with the indicated gp2 proteins were combined with the T7 A1 promoter-containing DNA fragment, followed by the addition of transcription substrates CpA and [α-^{32}P]UTP to initiate the abortive synthesis of CpA**p**U (bold lettering indicates the position of the radioactive phosphate). Reactions were performed and reaction products were analyzed as described in the text. An autoradiograph of a 20% denaturing polyacrylamide gel is presented.

purified easily as inclusion bodies. In contrast, overexpressed RNAP α subunits are usually soluble. For convenience, they are overproduced as hexahistidine tag fusions and purified using the protocol described above for hexahistidine-tagged gp2 purification.

Protocol

Grow 1 liter of BL21(DE3) cells carrying appropriate RNAP subunit or subunit fragments expression plasmids, induce protein overexpression by IPTG, and collect cells as described above. Resuspend cell pellets in 20 ml of buffer B containing 200 mM NaCl and 0.2% sodium deoxycholate. Transfer the suspension into a 45-ml polypropylene centrifuge tube, place the tube in an ice bath, and disrupt cells by sonication (four cycles of 1 min sonication at 50% duty cycle and maximal output power followed by 1-min rest periods). Centrifuge cell suspension at 15,000 rpm for 15 min in a Sorvall SS34 rotor at 4° and discard supernatant. Add 20 ml of the same buffer to the pellet and sonicate as described above to wash inclusion bodies. Collect the pellet, which contains inclusion bodies, by centrifugation (Sorvall SS34 rotor, 15,000 rpm for 5 min) and wash two more times by sonication and centrifugation. After the final centrifugation, resuspend the pellet containing washed inclusion bodies in 5 ml of the same buffer by brief sonication. Aliquot the suspension into 1.5-ml microcentrifuge tubes, transferring ~0.5 ml in each tube. Pellet inclusion bodies by 1-min centrifugation in a microcentrifuge and discard the supernatant. Store inclusion bodies at −70°. A typical yield is 50 mg of 90% pure *E. coli* RNAP β or β' subunit from 1 liter of culture.

Purification of RNA Polymerase

The following procedure, essentially a slightly modified protocol originally developed for the purification of *E. coli* RNAP,[11] has been used successfully for preparation of RNAP from several gram-negative bacteria, including *P. aeruginosa* and *X. oryzae*. All procedures, unless indicated otherwise, are performed on ice or in a cold room.

Protocol

Grow bacteria in 4 liters of appropriate medium and collect as described above. Resuspend cells in 30 ml of buffer C [50 mM Tris–HCl, pH 8.0; 5% glycerol, 5 mM 2-mercaptoethanol, 1 mM EDTA, 1 mM phenylmethylsulfonyl fluoride (PMSF)] containing 200 mM NaCl. Disrupt cells by sonication (five to six 30-s pulses at maximum power output, with 60-s rest intervals between the pulses, keeping the tube in an ice bath all the time) or by passage through a French press. Transfer the cell lysate into a 45-ml centrifuge tube and remove cell debris by 15-min centrifugation (15,000 rpm, Sorvall SS34 rotor). Measure the volume of the supernatant and transfer the supernatant into a fresh 45-ml centrifuge tube. Add 10% polyethylenimine P pH 8.0 (Polimin P, Sigma) solution to the supernatant to a final concentration of 0.8%. Add Polimin P dropwise, mixing the extract continually. Upon Polymin P addition, the transparent cell extract should turn cloudy and have an appearance of egg yolk soup. After a 10 min incubation on ice, collect the pellet by 10 min centrifugation (5000 rpm, Sorvall SS34 rotor). Discard the supernatant. Wash the pellet with 20 ml of buffer C containing 500 mM NaCl. For best results, it is essential to thoroughly resuspend the Polimin P pellet during the wash steps. We usually first mash the pellet with a glass rod in a minimal volume of buffer (~1 ml), add the rest of the buffer, and sonicate briefly to obtain a homogeneous suspension. After resuspension, collect the pellet by centrifugation (as described above) and repeat the wash step. After the second wash, elute RNAP by resuspending the pellet in 20 ml of buffer C

[11] D. A. Hager, D. J. Jin, and R. R. Burgess, *Biochemistry* **29,** 7890 (1990).

containing 1 M NaCl. Centrifuge as described previously and transfer the supernatant into a fresh 45-ml centrifuge tube; precipitate the RNAP by the addition of 0.45 g ammonium sulfate powder per 1 ml of the supernatant. Once the ammonium sulfate is dissolved, the clear supernatant should turn opaque. The precipitate is allowed to form by incubating the tube on ice for 1 h or overnight in a cold room.

Collect the ammonium sulfate pellet by 30-min centrifugation (Sorvall SS34 rotor, 15,000 rpm) (if precipitate floats at the top of the tube after centrifugation, add several drops of buffer C, mix, and repeat the centrifugation step). Discard the supernatant and allow the pellet to drain thoroughly. Dissolve the pellet in 50 ml of buffer B. Although this amount of buffer is usually sufficient to ensure complete binding of RNAP to a heparin column, it is advisable to check the conductivity of the protein solution to make sure that the salt concentration is below 200 mM. Centrifuge the sample twice for 20 min (Sorvall S334 rotor, 15,000 rpm) to remove undissolved residue. There should be no visible pellet after the last centrifugation.

Load the sample, at 1–1.5 ml/min, onto a 5-ml Hi-Trap heparin–Sepharose column (Amersham Biosciences) equilibrated in buffer B and attached to an FPLC system. After loading, wash the column with at least 3 column volumes of buffer B containing 300 mM NaCl. Elute RNAP with buffer B containing 600 mM NaCl. (The column can be regenerated by washing with buffer B containing 1000 mM NaCl.) Analyze fractions by 8% SDS–PAGE and precipitate the RNAP-containing fraction with ammonium sulfate as described above.

Collect the ammonium sulfate pellet by centrifugation, remove the supernatant, and dissolve the pellet in 400 μl of buffer B containing 200 mM NaCl. Clarify the sample by several centrifugation steps in a microcentrifuge at maximum speed until no visible pellet is formed and load, at 0.4 ml/min, onto a Superose 6 column (Amersham Biosciences) equilibrated with the same buffer and attached to FPLC. Develop the column isocratically with the same buffer at 0.4 ml/min, collecting 1.2-ml fractions. Analyze fractions by SDS–PAGE, pool fractions containing RNAP, and load, at room temperature at 0.2 ml/min, onto a 1-mL Protein-Pak Q 8HR column (Waters) equilibrated in buffer B containing 200 mM NaCl. Develop the column with a 10-ml linear gradient of NaCl (200–600 mM) in buffer B, collecting 0.5-ml fractions. Analyze 10-μl aliquots of each fraction by SDS–PAGE. Pool fractions containing RNAP core and holoenzymes separately, concentrate on a Centricon-100 centrifugal filter device (Amicon) to ~1 mg/ml, adjust glycerol concentration to 50%, and store at $-20°$.

In Vitro Transcription

To study the effect of gp2 on RNAP function *in vitro*, the abortive transcription initiation assay is performed in the presence of purified recombinant gp2. Because many bacteriophage inhibitors are specific for RNAP prepared from bacteria that host the virus (Fig. 2), it is essential to use RNAP from the corresponding host bacterium rather than *E. coli* RNAP. The choice of promoters is less essential, as most bacterial σ^{70}-type RNAP holoenzymes are active on the strong T7 A1 promoter.

Protocol

In a 1.5-ml Eppendorf tube, combine 50 ng of RNAP holoenzyme and 10 ng of recombinant gp2 or 20 ng of the DNA fragment containing the T7 A1 promoter[4] in 9 μl of transcription buffer (10 mM Tris–HCl, pH 8.0, 40 mM NaCl, 10 mM MgCl$_2$, 5 mM 2-mercaptoethanol). Incubate reactions for 5 min at 30°. Add the remaining substrates (DNA template or gp2, as appropriate) and incubate for an additional 5 min at 30°. Initiate transcription by the addition of 5 μl of substrate mix containing 300 μM CpA dinucleotide primer and 10 μCi [α-^{32}P]UTP (3000 Ci/mmol) substrate and continue incubation for another 10 min. Terminate the reactions by the addition of 15 μl of sample buffer (1xTBE, 7 M urea, 0.02% bromphenol blue), heat for 3 min at 90°, and resolve reaction products on a 20% polyacrylamide (19:1 acrylamide/bisacrylamide ratio) 6 M urea gel. We use the Bio-Rad Mini-Protean electrophoresis apparatus, which can complete electrophoretic separation of abortive initiation products in 20 min (at 400 V). Reaction products are revealed by brief (5–10 min) autoradiography. The transcription product, an RNA trinucleotide CpApU, migrates just above the bromphenol blue tracking dye, whereas unicorporated substrates run below bromphenol blue.

Coimmobilization of RNA Polymerase Subunits with gp2

Recombinant gp2 is immobilized on Ni^{2+}-NTA agarose to obtain gp2-affinity resin. The resin is incubated with proteins being tested for gp2 binding. Retention of a protein on gp2-affinity resin but not on Ni^{2+}-NTA agarose without gp2 indicates that gp2 and the protein under study interact (Fig. 3).

Protocol

Dissolve an aliquot of inclusion bodies containing RNAP subunit in 500 μl of denaturing buffer (10 mM Tris–HCl, pH 8.0, 6 M urea, 10% glycerol, 1 mM EDTA). Spin in a microcentrifuge at maximum speed for

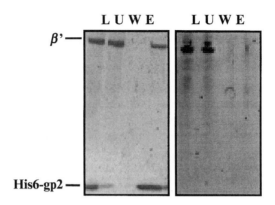

FIG. 3. Specific interaction of T7 gp2 with *E. coli* RNAP β' subunit. Wild-type β' was renatured from inclusion bodies and mixed with Ni^{2+}-NTA agarose beads in the presence (left) or absence (right) of His_6-gp2. Reactions were treated as described in the text, and proteins were analyzed on a 4–20% gradient SDS gel. L, load, protein composition of reactions prior to fractionation; U, unbound fraction; W, wash fraction; E, elution.

5 min at 4°. Transfer the supernatant into a fresh tube. Measure protein concentration by the Bradford assay, dilute with denaturing buffer to a final protein concentration of 0.2–0.5 mg/ml, and dialyze, at 4°, against two changes of 100 volumes of reconstitution buffer (50 mM Tris–HCl, pH 8.0, 200 mM KCl, 10 mM $MgCl_2$, 5 mM 2-mercaptoethanol, 20% glycerol). Remove the pellet that sometimes forms during dialysis by centrifugation, aliquot samples, flash-freeze in liquid nitrogen, and store at −70°.

To prepare the gp2-affinity resin, combine 10 μg of recombinant hexa-histidine-tagged gp2 and 70 μl of binding buffer (10 mM Tris–HCl, pH 8.0, 100 mM NaCl, 5% glycerol) and 20 μl of 50% Ni^{2+}-NTA agarose suspension in a siliconized 1.7-ml microcentrifuge tube. Allow gp2 to bind to the resin by incubating the test tube on a rotating platform for 30 min at 4°. As a control, incubate 20 μl of Ni^{2+}-NTA agarose with the buffer but without gp2. Spin for 10 s in a microcentrifuge to pellet Ni^{2+}-NTA agarose beads. Remove the supernatant. Wash the beads by the addition of 1 ml of binding buffer and brief centrifugation. Remove the supernatant, leaving the beads in a small volume (∼20 μl). Add 80 μl of binding buffer and ∼10 μg of protein of interest (i.e., RNAP or renatured RNAP subunit). Incubate reactions on a rotating platform for 30 min at 4°. Centrifuge briefly, remove, and save the supernatant (the unbound fraction). Wash the beads once with 1 ml of binding buffer and then twice with 1 ml of binding buffer containing 10 mM imidazole. After the final wash, resuspend the beads in

50 μl of binding buffer containing 10 mM imidazole, spin briefly, and save the supernatant as a wash fraction. At this step, make sure that the supernatant is removed completely from Ni^{2+}-NTA agarose beads. Elute bound proteins by the addition of 30 μl of the buffer containing 200 mM imidazole and incubating for 5 min at 4° on a rotating platform. Spin briefly and save the supernatant (Elution). Analyze all fractions by 8–25% SDS–PAGE.

Screening Bacteriophages for the Presence of RNAP-Binding Inhibitors

It has been estimated that bacteriophages constitute the most abundant life form on the planet.[12] The variety of host shut-off mechanisms that bacteriophages have evolved must be staggering. Gaining insights into these mechanisms will increase our understanding of bacteriophage evolution and the mechanism and regulation of bacterial macromolecular synthesis. Because of the rapid rate of bacteriophage evolution, molecular biology tools alone are often not sufficient for the identification of genes coding for inhibitors of bacterial transcription. For example, oligonucleotide primers specific for T7 gene 2 will not reveal the presence of gene 2 in bacteriophage T3 (unpublished observations). Therefore, it appears that the biochemical approach is the most reliable way of identifying and characterizing phage-encoded RNAP-binding factors. The approach consists of three steps. First, one has to establish that the bacteriophage under study encodes an activity that inhibits bacterial host transcription. Second, the phage-encoded inhibitor is purified from phage-infected cell extracts using an RNAP affinity purification step. Finally, the protein is identified by N-terminal amino acid sequence analysis, the corresponding gene is cloned, the recombinant protein is overexpressed, and properties of the recombinant inhibitor are studied *in vitro*. This section describes the application of this approach for the purification of bacteriophage Xp10 protein p7, a novel inhibitor of bacterial RNAP.[13]

Establishing the Presence of Host RNAP Inhibitor in Phage-Infected Cells

To establish that the phage under study encodes a host RNAP-binding transcription inhibitor, cell extracts from phage-infected and control, mock-infected cells are prepared, followed by the addition of small

---

[12] H. Brussow and R. W. Hendrix, *Cell* **108**, 13 (2002).

[13] S. Nechaev, Y. Yuzenkova, A. Niedziela-Majka, T. Heyduk, and K. Severinov, *J. Mol. Biol.* **320**, 11 (2002).

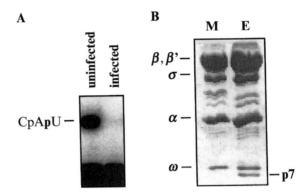

FIG. 4. Purification of bacteriophage Xp10-encoded *X. oryzae* RNAP-binding transcription inhibitor p7. (A) An RNAP-binding inhibitor of *X. oryzae* RNAP is present in Xp10 phage-infected *X. oryzae*. A whole cell lysate of Xp10-infected and control uninfected *X. oryzae* cells was combined with heparin–agarose beads containing bound *X. oryzae* RNAP, the beads were washed, and abortive transcription initiation reactions were performed using the T7 A1 promoter-containing DNA fragment as a template and CpA and [α-^{32}P]UTP as substrates. Reaction products were separated by denaturing PAGE and visualized by autoradiography. (B) A 7-kDa protein p7 is present in inactivated *X. oryzae* RNAP preparations. Heparin–agarose flow-through fractions prepared from a XP10-infected *X. oryzae* extract were passed through a heparin–agarose column containing a bound *X. oryzae* RNAP holoenzyme. The column was washed, and bound material was eluted and analyzed by SDS–PAGE (lane E). M is a control lane, which shows the subunit composition of RNAP loaded onto a heparin–agarose column to prepare a RNAP affinity column. Transcription assays established that material in lane M is active, whereas material in lane E is not.[13]

amounts of cell lysates to *in vitro* transcription reactions carried out with batch-purified host RNAP. In the case of lysates prepared from T7 and Xp10-infected cells, an obvious inhibitory effect on an *in vitro* abortive transcription initiation is seen, provided cell extracts are added prior to promoter complex formation (Fig. 4A).

Protocol

Grow *X. oryzae* on a rotary shaker in 50 ml of TGSC medium [1% (v/w) Bacto-tryptone, 0.5% Bacto-soytone, 0.5% NaCl, 0.2% dextrose, 0.05% Ca(NO$_3$)$_2$] at 28° until OD$_{600}$ reaches ~0.5. Transfer 40 ml of culture into a fresh flask and add Xp10 phage lysate to achieve the multiplicity of infection (MOI) of ~10 and resume the growth of both infected and uninfected cells. Every 5 min, withdraw a 5-ml aliquot, transfer into a glass tube, and place the tube on ice to chill cells rapidly. Xp10 lyses *X. oryzae* in approximately 50 min at 28° so up to eight aliquots are withdrawn. Testing cell extracts for the presence of host RNAP inhibitory activity at various stages of infection allows one to estimate whether the viral inhibitor is an early, middle, or late protein. Pellet cells in each aliquot by centrifugation, and resuspend cell pellets in 200 μl of buffer B containing 100 mM NaCl. Transfer cell suspensions into Eppendorf tubes and disrupt cells by sonication. Remove cell debris by 5-min centrifugation at 4° and transfer supernatants to fresh Eppendorf tubes. Add 60 μl of ~50% heparin–agarose (Amersham Biosciences) suspension in the same buffer to the extract prepared from uninfected cells. To allow RNAP to bind to heparin–agarose, incubate the test tube on a rotating platform for 30 min at 4°. Pellet heparin–agarose by brief centrifugation and discard the supernatant. Wash heparin–agarose twice with 1 ml of the same buffer containing 100 mM NaCl. Distribute heparin–agarose beads containing bound RNAP into several Eppendorf tubes and add 200 μl of infected cell extracts. As a control, add 200 μl of uninfected cells extract to one aliquot of heparin–agarose. Incubate reactions for 5 min at 4°, wash twice with 1 ml of the buffer, and elute bound proteins in 30 μl of buffer B with 500 mM NaCl. Use 3 μl of the eluate in an abortive transcription initiation assay (see earlier discussion). Figure 4A demonstrates results of the abortive transcription assay by *X. oryzae* RNAP in the presence of cell extracts prepared from uninfected cells (left lane) or from cells collected 15 min postinfection (right lane). The absence of abortive transcription in this sample indicates the presence of a bacteriophage-encoded host RNAP-binding inhibitory activity.

Affinity Purification of Host RNAP Inhibitor

Once it is established that host transcription shut off does occur during viral infection, phage-encoded factors that interact with host RNAP can be purified using host RNAP affinity chromatography. First, the Xp10-infected *X. oryzae* cell extract depleted of host RNAP is prepared by repeatedly passing the cell extract through a heparin–Sepharose column at conditions when RNAP binds to the column, while most proteins flow through the column. The RNAP-depleted extract is then passed through a column containing immobilized *X. oryzae* RNAP to purify RNAP-binding proteins.

Protocol

Grow *X. oryzae* in 4 liters of TGSC medium, infect with Xp10 at MOI of 5, and allow the infection to proceed for 10 min. Collect cells and disrupt by sonication in 50 ml of buffer B containing 100 mM NaCl and 1 mM PMSF. After low-speed centrifugation to remove cell debris, centrifuge the cell extract for 30 min in an ultracentrifuge (Beckman Ti50 rotor, 50,000 rpm). Load the clarified cell extract onto a 5-ml heparin–Sepharose Hi-Trap column (Amersham Biosciences) equilibrated in buffer B containing 100 mM NaCl. Collect the flow-through fraction. Wash the column with buffer B containing 1000 mM NaCl and then reequilibrate it in buffer B containing 100 mM NaCl. Reload the flow-through fractions on the same column, collect the material that passes through, and regenerate the column by washing with 1000 mM NaCl buffer followed by equilibration in 100 mM NaCl buffer. Repeat two or three times until little or no protein (as monitored by OD$_{280}$) is present in the 1000 mM NaCl wash. The column flow-through is now depleted of host RNAP but contains the Xp10-encoded host RNAP inhibitor.

To prepare an RNAP affinity column, dilute 200 μg of purified *X. oryzae* RNAP (a 1:1 mixture of core and holoenzymes) with buffer B containing 100 mM NaCl to a final volume of 2 ml and load onto a new 1–ml heparin–Sepharose Hi-Trap column equilibrated in buffer B containing 100 mM NaCl and attached to an FPLC system. Wash the column with the same buffer until a steady baseline is reached (there should be little or no unbound protein because pure RNAP is used). Pass the RNAP-depleted extract prepared from phage-infected cells through the RNAP affinity column. Wash the column with at least 20 volumes of buffer B containing 100 mM NaCl. Elute RNAP and other bound proteins with buffer B containing 1000 mM NaCl. Analyze the protein composition of the eluate by SDS–PAGE alongside an aliquot of RNAP used to prepare the RNAP affinity column (Fig. 4B). The gel should be overloaded with (at least) 10 μg of RNAP to allow visualization of minor bands and/or small proteins. Coomassie staining is generally preferable, as many small phage-encoded proteins do not stain well with silver stain.

Notes

The procedure presented above allowed us to purify Xp10 p7, determine the N-terminal sequence of the protein, and clone its gene.[13] Biochemical studies of recombinant p7 showed that transcription inhibition by this novel protein proceeds through a unique mechanism, thus demonstrating that studies of bacteriophages other than the few relatively well-studied *E. coli* phages can serve as a source of new paradigms of transcription regulation. The procedure works well for several bacteriophage inhibitors that bind host RNAP tightly and that are produced in excess of host RNAP during infection. When purifying an unknown RNAP-binding protein, at each step of purification, RNAP-containing fractions must be assayed for transcription activity and add-back experiments must be performed to locate the inhibitory activity if it had dissociated from RNAP during purification. In the case of Xp10 p7, a preliminary experiment established that phage-encoded *X. oryzae* RNAP inhibitory activity flowed through heparin–agarose column, which formed the basis for purification strategy. However, because most proteins do not bind to heparin–agarose, the procedure should be fairly general and could be applied, with minor modifications, for the purification of other RNAP-binding proteins.

Concluding Remarks

Inhibitors of bacterial host RNAP appear to be widespread among bacteriophages that encode single subunit RNAPs (Fig. 1). Gp2 homologs were found in *E. coli* bacteriophage T3,[14] 13A, and BA14 (unpublished observations), *Yersinia enterocolitica* phage ϕYe03-12,[6] *Y. pestis* phages ϕA1122 (I. Molineux, unpublished results), and C21 (unpublished observations), as well as *Pseudomonas aeruginosa* bacteriophage gh1 (A. Kropinski, submitted) and several T7-like phages from the Eliava Bacteriophage Institute (Republic of Georgia) specific for *E. coli*, *Shigella flexneri*, *S. sonnei*, and *Y. enterocolitica* (unpublished results). In addition, *Xanthomonas oryzae* bacteriophage Xp10, which relies on a single-subunit RNAP for its development as well, also encodes a small protein that inhibits *X. oryzae* RNAP.[13] To the extent these viral proteins have been studied, they all inhibit host RNAP transcription by interacting with the RNAP β' subunit.[4,13] Moreover, at least some of these viral proteins appear to be specific to RNAP prepared from cognate host bacteria and do not bind to and/or inhibit RNAPs from nonhost bacteria. This interesting finding suggests that deciphering molecular details of RNAP–inhibitor interactions would make it possible to design species-specific inhibitors of bacterial RNAPs.

Acknowledgments

Work in our laboratory has been supported by the Burroughs Wellcome Career Award and NIH Grant GM 59295, DHHS USDH Biotechnology Engagement Grant, and NRC fellowship (to KS). DD was partially supported by Charles and Johanna Busch Biomedical Fund postdoctoral fellowship. We are grateful to Drs. Ian Molineux and A. Kropinski for allowing us to cite their unpublished data.

[14] S. P. Mahadik, B. Dharmgrongartama, and P. R. Srinivasan, *Proc. Natl. Acad. Sci. USA.* **69**, 162 (1972).

Section II

Analyses of Promoter and Transcription Patterns

[20] A Procedure for Identifying Loosely Conserved Protein-Binding DNA Sequences

By MICHAEL C. O'NEILL

There are different types of searches for protein-binding DNA sequences. In some cases, one knows only a region of sequence that contains somewhere within it the desired binding site of an unknown character, e.g., most eukaryotic promoters. Given several such regions, it may be possible to determine the sequence using a search method that looks for improbable levels of local complexity in the DNA. The more common type of search is typical of a more advanced stage of knowledge in which case one has several examples of known binding sites for a particular protein and wishes to find other members of that sequence family. This latter case is discussed here.

Searching by Homology

The starting point for this search procedure is, therefore, an aligned set of six or more examples of a particular sequence motif. What information can be drawn from this collection? With the sequences aligned, it is simple to determine the consensus base at each position, allowing an immediate homology search with the consensus sequence. If the binding sequences are tightly conserved, say like the lambda operator set, this may be sufficient. (Indeed, if there is only one example of the desired sequence, one is limited to some type of homology search.) However, given that much of the regulatory range of a given binding protein is supplied by deviations in the binding sequence away from the optimum sequence, homology searches often bring inadequate information to the search task; the result is that criteria allowing the capture of the known sites result in very high false-positive levels.

Motif Redundancy Measured by the Reduction in Shannon Entropy

What other information does our sequence collection provide? It gives us the distribution of bases at each position within the motif. A consensus sequence search cannot be an optimal search because it has discarded this information. Staden[1] and Mulligan et al.[2] were among the first to employ

---

[1] R. Staden, *Nucleic Acids Res.* **12,** 505 (1984).
[2] M. E. Mulligan, D. K. Hawley, R. Entriken, and W. R. McClure, *Nucleic Acids Res.* **12,** 789 (1984).

0076-6879/03 $35.00

this information in a distribution-weighted search using base frequencies at each position. A decade earlier, Gatlin[3] had suggested that some of the work that Shannon had done on string transmission in noisy communication channels might well be applied to the strings of the genetic code. This suggestion, made before any DNA regulatory sequence was available, lay dormant until the work of Schneider et al.[4] in which the authors applied Shannon's entropy function to the characterization of binding sequence families. The function that they proposed was

$$RI = S_{bg} - S_{p-\text{motif}} \qquad (1)$$

which is the difference between the Shannon entropy of the background and the Shannon entropy of a given position within the motif, to be calculated at each position within the motif. Here the observed frequency of a particular base at a given position of the current sequence collection is taken as the probability, p_b, that the base will occur at that position in members of this family. For a running example, a family of six three base sequences will be employed:

$$AGT$$
$$AAT$$
$$GGT$$
$$AGC$$
$$CAT$$
$$AGA$$

for equiprobable bases, A=C=G=T in the genome

$$S_{bg} - S_p = -(4 * 0.25\ln_2 0.25)$$
$$-\left[-\left(p_a \ln_2 p_a + p_c \ln_2 p_c + p_g \ln_2 p_g + p_t \ln_2 p_t\right)\right] \qquad (2)$$

$$= 2 + \sum_b p_b \ln_2 p_b \qquad (3)$$

where p_b is the frequency of the base in question drawn from the sequence collection at the position being calculated.

$$\text{position } 1 = 2 + 0.67(-0.578) + 0.167(-2.59) + 0.167(-2.59) + 0(0) \qquad (4)$$

$$= 0.732 \text{ bits} \qquad (5)$$

[3] L. L. Gatlin, "Information Theory and the Living System," Columbia Univ. Press, New York, 1972.
[4] T. D. Schneider, G. D. Stormo, L. Gold, and A. Ehrenfeucht, J. Mol. Biol. 188, 415 (1986).

$$\text{position } 2 = 2 + 0.33(-1.6) + 0.67(-0.578) + 0(0) + 0(0) \tag{6}$$

$$= 1.068 \text{bits} \tag{7}$$

$$\text{position } 3 = 2 + 0.167(-2.59) + 0.167(-2.59) + 0(0) + 0.67(-0.578) \tag{8}$$

$$= 0.732 \text{bits} \tag{9}$$

for skewed base ratios

$$S_{bg} - S_p = \left[\sum_b p_b \ln_2(p_b/f_b) \right] \tag{10}$$

where f_b is the genomic base ratio for the base in question.

This function, Eqs. (1) and (10), represents the loss in Shannon entropy moving from background DNA to the fixation of the motif as represented in the current sequence collection. This function is determined for each position and assumes positional independence (linearity) for these regulatory regions. It provides a particular metric for averaging the base information at each position of the motif. Schneider et al.[4] called it information content, but as it is the difference of two information contents, it is better named an index of redundancy (RI). Because sequence conservation requires the constant application of energy for maintenance, redundancy can reasonably be equated with importance. Thus the redundancy index gives us the relative importance of each position across the motif. Schneider et al.[4] also suggested that the sum of the RI across the whole motif provided an estimate of the frequency of occurrence of motif members within a target sequence, taken as random:

$$\text{Freq} = (\text{target size})/(2^{\Sigma RI}) \tag{11}$$

$$\text{for our example} = (\text{target size})/2^{2.53} = (\text{target size})/5.8 \tag{12}$$

What can be learned from this? Figure 1 shows the RI profile of a collection of 26 CRP-binding sequences from *Escherichia coli*. Because *E. coli* has equiprobable base ratios, the maximum RI value is 2 (this maximum is somewhat reduced for small sample size) representing total conservation at a position, and the lowest value is 0, representing background base frequencies at that position. We see that (a) total conservation is not seen at any position, (b) there is substantial variation in the relative importance of positions within the motif, (c) some internal positions are random, and (d) despite being an inverted repeat sequence, the symmetry within the collection is, like many other inverted repeats, quite unbalanced with a strong

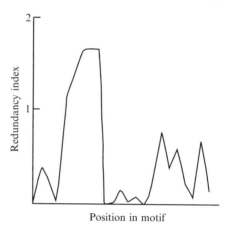

FIG. 1. Redundancy index profile. The redundancy index is plotted for the 22 positions of the CRP motif based on the base frequencies of 26 CRP sites in *E. coli*.

half and a weak half. Does this bring us any closer to the goal of being able to search for motif members? It does not appear to since it is only the sequence group which is characterized; there is no link here between the ideal motif profile and how well individual sequences of the family fit that ideal.

Functional Ranking of Individual Sequences

Various functions have been proposed to fill this gap and thus provide a rank ordering of the family members. If one can do a rank ordering, the only additional requirement to do a search is a cutoff value for acceptance. Berg and von Hippel[5] first proposed the function

$$\text{Index} = \log \left[\left(p_{\text{opt}} + 1/N \right) / \left(p_{\text{obs}} + 1/N \right) \right] \tag{13}$$

where p_{opt} is the consensus base in the current collection at the position being calculated, p_{obs} is the frequency at which the base found in the query sequence is found in the equivalent position of the current collection, and N is the number of examples in the current collection. This value would be determined independently at each position and summed over the whole sequence, the lower the value the better. This index discards information about at least two base frequencies at each position and so cannot be optimal. It is a measure of how well the position is filled without reference to how important the position is within the motif, a shortcoming duly noted

[5] O. G. Berg and P. H. von Hippel, *J. Mol. Biol.* **193**, 723 (1987).

by the authors. Consider two frequency distributions for a position: A=0.5 C=0.45 G=0.05 T=0 and A=0.3 C=0.27 G=0.21 T=0.22 with N very large in both cases. The sequence being evaluated starts with C. In the first case, we have log[0.5/0.45] or log[1.11]; in the second case we have log[0.3/0.27] or log[1.11]. In both cases the Berg von Hippel function gives the same answer for a C in the query sequence. However, in the first collection, C is one of a highly conserved pair of base choices for this position and in the second collection C is a member of an almost random distribution at this position. Clearly C should not carry the same weight in the two instances. If there were a way to factor in the positional importance, this situation might be remedied.

Another function that has been proposed by various investigators at various times,[1,6–8] but each time without any proof of its efficacy, is

$$\text{Index} = \ln_2 (p_b/f_b) \qquad (14)$$

where p_b and f_b are as defined earlier. As before, this function would be determined for each position of the query sequence and summed over the whole sequence. This index discards base frequency information for the other three bases at each position; it cannot, therefore, be optimal. In *E. coli*, a background choice would be 0; a totally conserved base at a position would score a 2. However, the function blows up for any base choice that is unrepresented in the current collection; while this tends to keep the false-positive level low, it has the effect that no "new" variants can ever be found. Again consider two different frequency distributions corresponding to different sequence collections: A=0.60 C=0.40 and A=0.20 C=0.40 G=0.20 T=0.20. The base being evaluated is C. It gets the same ranking value in each case, but in the first case it is one of two bases in a conserved position that has an RI of 1.03 bits and in the second case it is one of four bases in a nearly randomized position having an RI of 0.08 bits. If tested on the same data sets used by Berg and von Hippel,[5] this index underperforms the Berg von Hippel index (unpublished data[9]).

None of this is encouraging in our effort to develop an accurate search procedure. However, the hint of the way out was given earlier. To achieve an accurate ranking, one must provide two pieces of information for each position of the sequence. The first is how well is the position filled with respect to the motif distribution and the second is how important is that position within the motif? One might think that these are one and the same,

[6] P. Bucher, *J. Mol. Biol.* **212,** 563 (1987).
[7] G. D. Stormo, *Methods Enzymol.* **208,** 458 (1991).
[8] T. D. Schneider, *J. Theor. Biol.* **201,** 87 (1999)
[9] M. C. O'Neill, unpublished data.

but the frequency examples given earlier show that this is not true. The consensus base at a highly conserved position should receive a high weighting, whereas the consensus base at a randomized (RI near 0) position should not. The Berg von Hippel function, Eq. (13), lacked a sense of positional importance; however, that is precisely what RI purports to measure. Because each function is evaluated independently at each position, it is feasible to factor them. As was shown in earlier work,[10] the product (RI^*BvH) summed across the sequence is a very effective ranking function that, when combined with a cutoff value, can also be used as a free-standing search function. Let us return to the six-sequence example, supposing our query sequence to be CAT. RI^*BvH for

$$\text{Position } 1 = 0.732 * \log[(0.67 + 0.166)/(0.166 + 0.166)] = 0.29 \tag{15}$$

$$\text{Position } 2 = 1.07 * \log[(0.67 + 0.166)/(0.33 + 0.166)] = 0.24 \tag{16}$$

$$\text{Position } 3 = 0.732 * \log[(0.67 + 0.166)/(0.67 + 0.166)] = 0.0 \tag{17}$$

$$\text{total} = 0.29 + 0.24 + 0 = 0.53 \tag{18}$$

Similar considerations led to a second ranking function that performs about as well as the RI^*BvH function. If RI is determined for a given position of the sequence collection, it can be designated RI_-. If the sequence to be evaluated is then added to the sequence collection and RI is redetermined, an additional A,C,G, or T has been added at that position. The new RI can be designated RI_+. The difference RI_+ -RI_- is a measure of how well the position has been filled in the query sequence. If the new base is better than the average base at that position in the sequence collection, the difference will be positive; if it is worse than average, the difference will be negative. Again the measure of positional importance is just RI_-, the value for the current collection. This suggests

$$(RI\delta RI) = RI_-(RI_+ - RI_-) \tag{19}$$

as a ranking function to be determined at each position of the query sequence and summed across the whole sequence. Its effectiveness has earlier been shown to be very similar to the RI^*BvH function.[11] Again, if we work through our example with the query sequence CAT, $RI\delta RI$ for position

$$\begin{aligned} 1 &= 0.732 * (2 + 0.57 * \ln_2 0.57 + 0.29 * \ln_2 0.29 \\ &\quad +0.14 * \ln_2 0.14 + 0.0 - 0.732) \end{aligned} \tag{20}$$

$$= -0.08 \tag{21}$$

[10] M. C. O'Neill, *J. Mol. Biol.* **207**, 301 (1989).
[11] M. C. O'Neill, *Proc. Natl. Acad. Sci. USA* **95**, 10710 (1998).

$$2 = 1.07*(2 + 0.43 * \ln_2 0.43 + 0.0 + 0.57 * \ln_2 0.57 + 0.0 - 1.07) \quad (22)$$

$$= 0.05 \quad (23)$$

$$3 = 0.732 * (2 + 0.14 * \ln_2 0.14 + 0.14 * \ln_2 0.14$$
$$+ 0.0 + 0.71 * \ln_2 0.71 - 0.732) \quad (24)$$

$$= 0.09 \quad (25)$$

$$\text{total} = 0.04 \quad (26)$$

Both of the aforementioned functions, Eqs. (14) and (19), leave room for improvement when used as search functions in that the false-positive level can be 0.1% (i.e., percentage of total bases searched).

Sequence Analysis with Neural Networks

A very different approach to sequence identification is provided by the use of backpropagation neural networks[12,13] to find additional members of an existing sequence collection. In the simplest form, one has merely to code the sequence into binary form, e.g., A=0001 C=0010 G=0100 T=1000. The network is then trained by supplying coded examples of the sequence type sought along with the desired answer, 1, for yes this is a family member or 0, for this is not a member. Random sequence and double-down mutant sequences can be used for negative inputs. During training the network will learn to distinquish the two classes of input sequences. If the training sample has been sufficiently representative, the network can then be used to evaluate a sequence it has not seen before to find other members of the family. If one wished to search the *E. coli* genome, the coding program would set a window equal to the length of the query sequence to code that length at the beginning of the genome sequence as the first test example; the program would then shift forward one base to code the second test example and so on until the genome sequence had been exhausted (note that when coding genomes the file size limit of 32-bit processors can be exceeded easily). It has been shown earlier that this can be a highly effective search procedure.[14,15] In the case of a very difficult sequence family, such as the σ 70 promoters of *E. coli*, it was estimated to find about 85% of true promoters with a false-positive level on the order of 0.1%.

[12] P. J. Werbos, *in* "The Roots of Backpropagation." Wiley, New York, 1994.
[13] D. E. Rumelhart, G. E. Hinton, and R. J. Williams, *in* "Parallel Distributed Processing: Explorations in the Microstructures of Cognition," p. 318. MIT Press, Cambridge, MA, 1986.
[14] M. C. O'Neill, *Nucleic Acids Res.* **19**, 313 (1991).
[15] M. C. O'Neill, *Nucleic Acids Res.* **20**, 3471 (1992).

Ranking Function Preprocessing for Neural Networks

Inasmuch as the use of a ranking function and the use of neural networks are complementary, it seemed reasonable to preprocess the neural net input by providing the ranking function information at each position of the input sequence rather than the raw base together with an evaluation of the whole sequence. Caution is advised when preprocessing data for a neural net because such processing may lead to the loss of information. In this case, additional information is supplied on the fitness of the base relative to the motif ideal together with an overall evaluation of the sequence.

Returning to our example, let us see how our query sequence, CAT, might be coded using the RI^*BvH function and the six sequence collection. The worst value possible for a position given our sequence collection would be a C or T at

$$\text{Position } 2 = 1.07 * \log[(5/6)/(1/6)] = 1.07 * 0.699 = 0.75. \quad (27)$$

The worst possible sequence total would be 1.77, the best being 0.0. If 6 bits (input positions) are devoted to a qualitative scoring of the position value and 4 bits to the sequence total, each three base sequence could be coded in $3 * 6+4$ or 22 bits. Our code could be $< = 0.05 = 000001, >0.05< = 0.1 = 000010, > 0.1< = 0.2 = 000100, >0.2< = 0.3 = 001000, >0.3< = 0.5 = 010000, >0.5 = 100000$ for positions and $<0.5 = 0001, >0.5< = 1.0 = 0010, >1.0< = 1.5 = 0100, >1.5 = 1000$ for sequence total. The input for the query sequence would then be 0010000010000000010010, corresponding to 0.29, 0.24, 0.0, and 0.53. All other sequences would be coded accordingly. The $RI\delta RI$ function could be used in the same fashion.

This preprocessing step had the effect of reducing the false-positive level by fourfold or more, dropping it to 0.025% or less.[11] This level can be reduced even further if one has some other limiting information about the query sequence. For example, if one is looking for a transcriptional factor-binding site in *E. coli*, one might be willing to add the condition that it is expected to lie with 300 bases of the start point of translation. Assuming 4300 promoters, $300 * 4300/2 * 4600000$ suggests that only 1/7 of the genome satisfies this condition, for a false-positive level approaching $1/7*0.025\%$ or 0.003%. If one were doing a genomic search for promoters, this would still yield 1 false positive for each 15 true promoters (see files p16, p17, p18 at http://research.umbc.edu/~moneill for the results of such a search).

This procedure provides a general search method that is powerful enough to be of real utility in the case of prokaryotic genomes or smaller searches. Unfortunately, protein-binding sequences in eukaryotes do not appear to be distinctly larger or more diverse than those of prokaryotes, yet the genome (=total search space) may be more than a thousand times larger. If the cell were considered as a bag with binding proteins diffusing freely within it, there would not appear to be sufficient information for these proteins to find their specific targets. This paradox would be avoided if the proteins always bound as part of a larger complex or if the proteins were concentrated in specific subdomains, making the cellular concentrations relatively irrelevant.

Ending on a practical note, the software for every type of search described here—homology, up to 6 million bases under Windows and alloted memory under Unix with helix search formatting, specific limit on base errors, zero-weighting specific positions, and indeterminate homology search size among other features, RI characterization of sequence families, RIδRI or RI*BvH rank ordering of sequence families or sequence searches with a user-set cutoff, neural network searches using base sequences, and neural network searches using ranking function preprocessing—are available, with advice, from the author (moneill@umbc.edu).

[21] Computational Detection of Vertebrate RNA Polymerase II Promoters

By Vladimir B. Bajic and Vladimir Brusic

The sequence of the human genome[1,2] and sequences of other vertebrate genomes (including mouse, rat, fugu fish, zebra fish) represent raw material for the identification and annotation of complete lists of their genes. The basic, and largely incomplete, annotations are available at the major repositories, such as NCBI[3] or Ensembl.[4] Gene annotations should provide information on coding, noncoding, and regulatory regions of genes, as well as their regulatory elements. A promoter region is an integral part of a gene that is responsible for the initiation and partial regulation of the gene transcription process.[5] Each eukaryotic gene must have at least one

[1] E. S. Lander *et al.*, *Nature* **409**, 860 (2001).
[2] J. C. Venter *et al.*, *Science* **291**, 1304 (2001).
[3] M. Feolo, W. Helmberg, S. Sherry, and D. R. Maglott, *Rev. Immunogenet.* **2**, 461 (2000).
[4] T. Hubbard *et al.*, *Nucleic Acids Res.* **30**, 38 (2002).
[5] R. O. J. Weinzierl, "Mechanism of Gene Expression," Imperial College Press, London, 1999.

promoter. A promoter contains the transcription start site (TSS)[6,7] and the region around it, mainly upstream of the TSS. The complexes comprising RNA polymerase II and general transcription factors bind DNA in the promoter region, enabling transcription of mRNA-coding genes.[5,8,9] Promoters of most human and other vertebrate genes have not been yet well characterized,[10,11] and detailed understanding of gene expression regulation remains a main challenge in genomic research.

The eukaryotic promoter database[12] (EPD) is a major resource containing entries of eukaryotic promoters. Each EPD entry is ideally a fragment of DNA comprising 500 nucleotides upstream of the TSS and 100 nucleotides downstream. EPD data have been used in the development of a range of computational methods for the prediction of eukaryotic promoters. Promoter prediction methods attempt to identify various signals within the DNA sequence to locate potential TSSs or to estimate promoter regions. The earliest approaches for promoter prediction were based on the identification of a TATA box motif,[13] composition of oligomers,[14] or recognition of other characteristic motifs (such as transcription factor-binding sites) in promoters.[15–18]

Other methods utilize artificial neural networks,[19–21] Markov models,[22] or a combination of interpolated Markov models and artificial neural networks.[23] Some specialized methods focus on subsets of promoters, such as those that recognize CpG island-related promoters.[24–26] The

[6] P. F. Johnson and S. L. McKnight, *Annu. Rev. Biochem.* **58**, 799 (1989).
[7] J. W. Fickett and A. G. Hatzigeorgiou, *Genome Res.* **7**, 861 (1997).
[8] A. G. Pedersen, P. Baldi, Y. Chauvin, and S. Brunak, *Comp. Chem.* **23**, 191 (1999).
[9] J. Greenblatt, *Curr. Opin. Cell Biol.* **9**, 310 (1997).
[10] R. Liu and D. J. States, *Genome Res.* **12**, 462 (2002).
[11] T. Werner, *In Silico Biol.* **2**, 0023 (2002).
[12] V. Praz, R. Périer, C. Bonnard, and P. Bucher, *Nucleic Acids Res.* **30**, 322 (2002).
[13] P. Bucher, *J. Mol. Biol.* **212**, 563 (1990).
[14] G. B. Hutchinson, *CABIOS* **12**, 391 (1996).
[15] Y. V. Kondrakhin, A. E. Kel, N. A. Kolchanov, A. G. Romaschenko, and L. Milanesi, *CABIOS* **11**, 477 (1995).
[16] D. S. Prestridge, *J. Mol. Biol.* **249**, 923 (1995).
[17] Q. K. Chen, G. Z. Hertz, and G. D. Stormo, *CABIOS* **13**, 29 (1997).
[18] V. Solovyev and A. Salamov, *ISMB* **5**, 294 (1997).
[19] S. Knudsen, *Bioinformatics* **15**, 356 (1999).
[20] D. Kulp, D. Haussler, M. G. Reese, and F. H. Eeckman, *Proc. Int. Conf. Intell. Syst. Mol. Biol.* **4**, 134 (1996).
[21] N. Mache, M. Reczko, and A. Hatzigeorgiou, unpublished.
[22] S. Audic and J. M. Claverie, *Comput. Chem.* **21**, 223 (1997).
[23] U. Ohler, H. Niemman, G. C. Liao, and G. M. Rubin, *Bioinformatics* **17**, S199 (2001).
[24] I. P. Ioshikhes and M. Q. Zhang, *Nature Genet.* **26**, 61 (2000).
[25] S. Hannenhalli and S. Levy, *Bioinformatics* **17**, S90 (2001).
[26] L. Ponger and D. Mouchiroud, *Bioinformatics* **18**, 631 (2002).

PromoterInspector[27] program considerably reduced the level of false-positive predictions relative to other promoter recognition systems. This method predicts, with a relatively high accuracy, broad promoter regions by identifying a set of overrepresented sequence motifs in promoters, but it does not predict the TSS locations. Comparative reviews and a comparative analysis of promoter prediction methods have been published.[7,10] A main challenge for the prediction of eukaryotic TSSs from genomic sequence is a low specificity of predictions resulting from a high number of false-positive (FP) predictions for any useful level of true positive (TP) predictions.[7,10] The promoter prediction methods were shown to suffer from low sensitivity (SE), high false-positive rate–low specificity (SP), and low positional accuracy of TSS predictions.[28] These problems have been addressed in a new generation of programs, such as Eponine,[28] CONPRO,[10] FirstEF,[29] and Dragon Promoter Finder.[30–32] Eponine utilizes a relevance vector machine method of machine learning aiming to predict the exact position of TSSs. It predicts TATA box-containing promoters in G + C-rich regions well, but cannot achieve sensitivities over 45% even on G + C-rich human chromosome 22. The CONPRO combines five prediction methods and predicts TSS if three or more predictions (the majority vote) fall within a region of 300 bp. The CONPRO aims at producing high specificity and low false-positive rate predictions that can be used with genomic sequence, but because it is a consensus-based predictor, it is expected to have a relatively low positional accuracy of TSS predictions. Moreover, the CONPRO requires initial Genscan[33] predictions, plus the support of EST/cDNA/mRNA sequence data, which limits its generality considerably. Even with the support of Genscan and EST/cDNA/mRNA sequences, it cannot achieve sensitivity higher than 37% if only 3′ ESTs are known. The FirstEF, a gene-finding program, uses a broad context of gene start to infer the location of promoters and actually belongs to the category of gene-finding programs. Its performance is good on G + C-rich chromosomes (sensitivity 77% on chromosome 22 and ppv = 40%).

This article describes the system and method termed Dragon Promoter Finder 1.3 (hereafter referred to as DRAGON) program. The DRAGON

[27] M. Scherf, A. Klingenhoff, and T. Werner, *J. Mol. Biol.* **297,** 599 (2000).
[28] T. A. Down and T. J. P. Hubbard, *Genome Res.* **12,** 458 (2002).
[29] R. V. Davulari, I. Grosse, and M. Q. Zhang, *Nature Genet.* **29,** 412 (2001).
[30] V. B. Bajic, S. H. Seah, A. Chong, G. Zhang, J. L. Y. Koh, and V. Brusic, *Bioinformatics* **18,** 198 (2002).
[31] V. B. Bajic, A. Chong, S. H. Seah, and V. Brusic, *IEEE Intel. Syst.* **17,** 64 (2002).
[32] V. B. Bajic, S. H. Seah, A. Chong, S. P. T. Krishnan, J. L. Y. Koh, and V. Brusic, *J. Mol. Graph. Model.* **21**(5), 323–332 (2003).
[33] C. Burge and S. Karlin, *J. Mol. Biol.* **268,** 78 (1997).

aims at producing both high accuracy predictions of vertebrate promoters and high positional accuracy of TSS predictions. The DRAGON allows users to select among five specificity levels of predictions, ranging from a very high specificity to a very high sensitivity. Very high sensitivity predictions are suitable for screening relatively short fragments of a genomic sequence, whereas at the other extreme, very high specificity predictions are suitable for genome-scale predictions because of the significantly reduced number of false-positive predictions.

Dragon Promoter Finder

The DRAGON is an improved version of the Dragon Promoter Finder program.[30,31] It was designed to search for promoters and to locate TSSs within anonymous DNA sequences. DRAGON predictions are strand specific, making it different from several other recently released promoter prediction programs.[24-27] The DRAGON algorithm has been tuned to enable the user selection of sensitivity of predictions, ranging from 25 to 80%. The DRAGON algorithm[32] combines respective modules specialized for promoter prediction in the G + C-rich (G + C content ≥ 0.5) and in A + T-rich (G + C content < 0.5) sequences. The DRAGON requires user input for the DNA sequence and for selection of the sensitivity level (Table I). The DNA sequence can be pasted in the sequence window or loaded from the local file. The DRAGON accepts the following data formats of the input DNA sequence: flat text, Fasta, GenBank, EMBL, IG, and GCG. Users are required to select one of the five predefined sensitivity levels as shown in Table I, which will be described in more detail later. In addition to TSS predictions, the DRAGON provides a unique feature for the analysis of the region around the predicted TSS for potential transcription factor-binding sites using the Match tool from the TRANSFAC package.[34] The DRAGON is accessible through web browser access at sdmc.lit.org.sg/

TABLE I
SENSITIVITY/SPECIFICITY LEVELS OF INDIVIDUAL BASIC
MODELS IN THE DRAGON

% True positives	Sensitivity	Specificity
80	Very high	Very low
65	High	Low
50	Moderate	Moderate
35	Low	High
25	Very low	Very high

promoter. The restrictions that apply for Web access include maximal length of the query DNA sequence of 35000 and a daily maximum of 500 requests. Users who require greater resources should contact the authors.

Dragon Promoter Finder 1.3 Model

The overall structure of the DRAGON system is shown in Fig. 1. The system comprises five independent models, optimally tuned for particular sensitivity levels. Activation of the relevant model is user driven through the selection of the desired "expected sensitivity" level. The system takes the query DNA and analyzes the content of the sliding window of length L along the query DNA. The step length for window sliding is a single nucleotide, whereas the window length has been optimized for each available sensitivity level. Each independent model comprises two submodels and the logic for selection of the submodel based on the composition of the DNA window (Fig. 2). The two submodels are derived for G + C-rich and A + T-rich DNA sequences. The system analyzes the content of the data window and automatically selects the appropriate submodel. Each submodel comprises three parallel sensors (Fig. 3). Each of the three sensors was developed to recognize distinct functional regions of a gene:

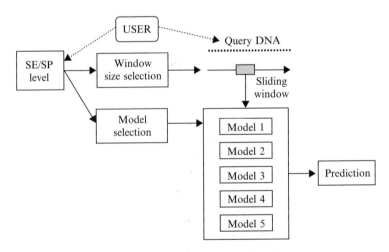

FIG. 1. Schematic representation of the overall structure of Dragon Promoter Finder 1.3. Query DNA is supplied by users, and the sensitivity/specificity (SE/SP) level is selected by users before predictions can be done.

---

[34] E. Wingender *et al.*, *Nucleic Acids Res.* **29,** 281 (2001).

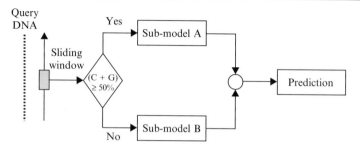

FIG. 2. Schematic representation of the structure of the individual models in Dragon Promoter Finder 1.3. Submodel is automatically selected for a C + G-rich or A + T-rich data window.

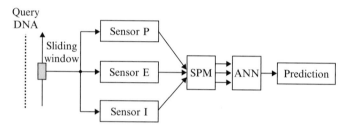

FIG. 3. Schematic representation of the submodel structure in Dragon Promoter Finder 1.3. Sensors P, E, and I are trained to recognize signals characteristic for promoters, coding exons, and introns. The signal processing module (SPM) transforms sensor outputs and feeds them into an artificial neural network (ANN), which performs integration of sensor signals, signal smoothing, and prediction of promoter sequences.

promoter, coding exon, or intron. Sensors for these three regions were derived from positional distributions of overlapping nucleotide pentamers (sequences of five consecutive nucleotides) from training data. Here we describe how the DRAGON system was built.

The positional distributions of pentamers were represented as positional weight matrices (PWM). Each PWM was constructed from training data for its corresponding functional region. For a given data window of length L, a PWM of overlapping pentamers had dimensions $1024 \times (L\text{-}4)$. PWMs were used for the calculation of scores defining the values of sensor signals. A data window of length L contained sequence $W = n_1 n_2 \ldots n_{L-1} n_L$ of nucleotides where n_j represented any of the four nucleotides. The corresponding sequence of overlapping pentamers $P = p_1 p_2 \ldots p_{L-5} p_{L-4}$ inside the data window W had length $K = L\text{-}4$. The score for the data window W was calculated by

$$S = \frac{\left(\sum_{i=1}^{L-4} p_j^i \otimes f_{j,i}\right)}{\left(\sum_{i=1}^{L-4} \max_j f_{j,i}\right)}, \quad p_j^i \otimes f_{j,i} = \begin{cases} f_{j,i}, & \text{if } p_i = p_j^i \\ 0, & \text{if } p_i \neq p_j^i \end{cases}, \tag{1}$$

where p_j^i was the jth pentamer at position i and $f_{i,j}$ was the frequency of the jth pentamer at position i. The sensor signals could take values between 0 and 1—the higher the score, the higher the likelihood that the data window represented the corresponding functional region. The signal values of the promoter, exon, and intron sensors—σ_p, σ_e, and σ_i, respectively—represent input for the signal processing module (SPM). The output of the SPM are three signals S_E, S_I, and S_{EI}:

$$\begin{aligned} s_E &= sat(\sigma_p - \sigma_e, a_e, b_e) \\ s_I &= sat(\sigma_p - \sigma_i, a_i, b_i) \\ s_{EI} &= sat(\sigma_e - \sigma_i, a_{ei}, b_{ei}) \end{aligned} \tag{2}$$

$$sat(x, a, b) = \begin{cases} a, & \text{if } x > a \\ x, & \text{if } b \leq x \leq a \\ b, & \text{if } b > x \end{cases} \tag{3}$$

The a and b were part of the tuning parameters of the system that were determined for each submodel. The three SPM signals were transformed using principal component analysis and used as inputs to an artificial neural network (ANN), which served as a multisensor integrator. The signals for functional regions, particularly those for introns and promoters, were quite noisy. To address this problem, we combined a nonlinear signal SPM for signal preprocessing and a simple ANN for classification. A feed-forward ANN was trained for each submodel using the Bayesian regularization method for best separation of classes of input signals. Each ANN had three layers, three input nodes, and a single output node. The activation of hidden- and output-layer nodes was defined by tansig function:

$$\tanh(x) = \frac{e^x - e^{-x}}{e^x + e^{-x}} \tag{4}$$

ANNs were designed to be simple and of low complexity to perform signal smoothing efficiently and predict promoter regions accurately. Each independent model (Fig. 1) has two SPM/ANN blocks—one block for each submodel. The ANN for each submodel was trained to separate promoter from nonpromoter regions in input data. The output range for each ANN was between -1 and 1, and the best separation threshold was selected for the working system. ANN output above the separation threshold represented positive prediction, indicating that the data window was predicted to belong to the promoter region.

System Training and Tuning

Positive training data were derived from the EPD[12] release 67. We extracted 793 vertebrate promoter sequences that had a clearly indicated TSS. We used the region of 250 nucleotides consisting of 200 nucleotides upstream of the TSS, the TSS, and 49 nucleotides downstream of the TSS. Nonredundant (less than 50% mutual identity for any two sequences) negative training data were selected randomly from the GenBank database[35] release 121 entries. The final negative training data set comprised 250 nucleotide long nonoverlapping 800 sequences of coding exons and 4000 introns. Promoter and nonpromoter data were first combined and then split into C + G-rich and A + T-rich training sets for training respective submodels A and B (Fig. 2).

The tuning data set was used for determining system parameters, including submodel tuning parameters [Eq. (2)], sensor signal thresholds, and ANN classification thresholds. The tuning set comprised the complete training set and the additional sequences were each 250 nucleotides long. The additional nonpromoter sequences were assembled from 1600 3′UTR UTR regions extracted from the UTRdb database,[36] and 500 human intron and 500 human exon sequences were extracted from the GenBank release 121. These nonpromoter sequences were also nonredundant, nonoverlapping, and mutually exclusive vs the training set data. The tuning set also contained 20 promoter sequences representing known full-length vertebrate genes with known TSS. Positive tuning data and the EPD set were mutually exclusive.

All 10 submodels were trained using the relevant training data set (C + G rich or A + T rich). Each submodel was tuned for best performance at its relevant SE/SP level (Table I). We used the sensitivity (SE) measure and positive predictive value (ppv), which is used commonly as a proxy for true specificity (SP) and is often called specificity in bioinformatics[37]:

$$SE = \frac{TP}{TP + FN},$$
$$ppv = \frac{TP}{TP + FP} \qquad (5)$$

[35] D. A. Benson, I. Karsch-Mizrachi, D. J. Lipman, J. Ostell, B. A. Rapp, and D. L. Wheeler, *Nucleic Acids Res.* **28**, 15 (2000).

[36] G. Pesole, S. Liuni, G. Grillo, F. Licculli, A. Larizza, W. Makalowski, C. Saccone, *Nucleic Acids Res.* **28**, 193 (2000).

[37] V. B. Bajic, *Brief. Bioinform.* **1**, 214 (2000).

TP stands for true positives (number of real promoters predicted as promoters), FP for false positives (number of nonpromoters predicted as promoters), and FN for false negatives (number of real promoters predicted as nonpromoters). The goal of tuning was the selection of model parameters that provide the best performance for the selected sensitivity level. Based on simulation results (data not shown) the optimized data window length for sensitivity models with SE > 25% was 200 nucleatides and 250 for very high specificity models (SE ≤ 25%).

Model Testing

The test set was compiled from published data sets used in the earlier development of gene finding, gene analysis, and promoter prediction systems. These include sequences used in training gene finding and analysis programs such as Genie,[38] Genscan,[33] NetGene,[39] GeneId,[40] and some others.[41,42] We also included sequences used in testing a promoter recognition program[21] and those used in the analysis of TATA box motifs and TSSs.[43] Sequences used in training the Genie program made up the majority of the test set. The test set was compiled to (a) represent gene diversity as much as possible, (b) not overlap with the training and tuning set, (c) have sufficiently detailed annotation so that the location of TSS could be inferred with confidence, and (d) not contain redundant sequences. The test set was additionally checked against relevant literature for the accuracy of annotations. The final test set comprised 147 human and human–virus sequences containing 159 TSSs. The cumulative length of the sequences in the test set was in excess of 1.15 million base-pairs.

Predictions of the DRAGON were compared to those of three promoter recognition systems, Promoter2.0,[19] NNPP2.1,[20] and Promoter-Inspector.[27] The PromoterInspector has been shown to outperform significantly five other promoter prediction systems.[27] The comparative analysis of promoter predictions of DRAGON, the earlier version Dragon Promoter Finder 1.2, and three other promoter prediction systems is shown in Fig. 4. For PromoterInspector, the prediction was considered correct if the TSS was within the predicted region. For other prediction systems, including DRAGON, the prediction was considered correct, as defined

[38] M. Reese, D. Kulp, A. Gentles, and U. Ohler, (1999) www.fruitfly.org/seq_tools/datasets/ Human.

[39] S. Brunak, J. Engelbrecht, and S. Knudsen, *J. Mol. Biol.* **220**, 49 (1991).

[40] R. Guigo, S. Knudsen, N. Drake, and T. Smith, *J. Mol. Biol.* **226**, 141 (1992).

[41] M. Gelfand, A. A. Mironov, and P. A. Pevzner, *Proc. Natl. Acad. Sci. USA*, **93**, 9061 (1996).

[42] R. Farber and A. Lapedes, *J. Mol. Biol.* **226**, 471 (1992).

[43] F. E. Penoti, *J. Mol. Biol.* **213**, 37 (1990).

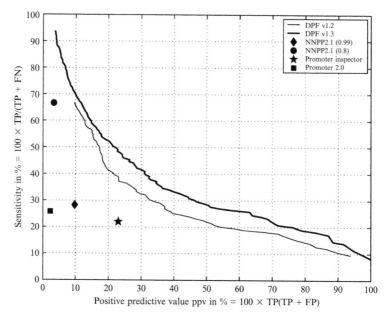

FIG. 4. Accuracy of DRAGON predictions estimated using a test set and comparison to the performance of other promoter prediction systems, including an earlier version of DRAGON.

elsewhere,[7] if predicted TSS was within the region encompassing 200 nucleotides upstream and 100 nucleotides downstream of the real TSS. One of our initial aims was minimization of the number of false-positive predictions. The improvement in FP predictions of the DRAGON is shown in Table II. We demonstrated that DRAGON showed significantly improved prediction accuracy relative to the selected comparison programs, ranging from 3.8 to 57 times improvement in the number of FP predictions.

To test the suitability of the DRAGON for predicting promoters at the genomic scale, we tested the performance on known genes of the chromosomes 21[44] and 22.[45] The total length of chromosomes 21 and 22 used in the analysis was 33,981,048 and 34,553,830 bp, respectively.

Prediction results are shown in Table III. We considered only known genes (i.e., those whose functional product was known and confirmed experimentally). Human chromosome 21 had 153 known genes (58% CpG island related), whereas chromosome 22 had 339 known genes (65% CpG

[44] www.ncbi.nlm.nih. gov/PMGifs/Genomes/euk_g.html (*Homo sapiens* Chromosome 21, Aug. 1st 2002).

[45] http://www.sanger. ac.uk/HGP/Chr22/ (Human Chromosome 22, August 1st 2002).

TABLE II
RELATIVE IMPROVEMENT IN FP PREDICTIONS OF THE
DRAGON RELATIVE TO THE PERFORMANCE OF
COMPARISON SYSTEMS[a]

Comparison system	SE (%)	FP/FP$_{DRAGON}$
PromoterInspector	22	6.88
NNPP2.1 (threshold 0.8)	28	3.8
NNPP2.1 (threshold 0.99)	66	8.82
Promoter2.0	25	56.9

[a] The measure is the ratio of FP predictions of the comparison program vs DRAGON (times improvement). The sensitivity of predictions was matched for comparison.

TABLE III
DRAGON PREDICTIONS ON ANNOTATED GENES OF HUMAN
CHROMOSOMES 21 AND 22

Chromosome 21		Chromosome 22	
SE (%)	ppv (%)	SE (%)	ppv (%)
50	41	49	48
58	29	58	42
61	25	64	33
73	18	74	30
79	12	80	23
91	9	95	11

island related). Chromosome 22 is rich in G + C content: it has the G + C content of 48%, while the average for the human genome is 42%. Chromosome 21 has a G + C content of 41%. Results of promoter predictions on chromosomes 21 and 22 indicate that DRAGON enables high sensitivity predictions that can exceed 90%, which is higher than the sensitivity of predictions by any other currently available TSS predictor. The price for such high sensitivity is an increased number of FP predictions. Results on chromosome 22 indicate that at achieved sensitivity of 80% for each TP prediction there are 3.34 FP predictions, and at achieved sensitivity of 49% for each TP prediction there are 1.1 FP predictions. Chromosome 22 has a higher proportion of CpG-related genes than other genes. CpG

TABLE IV
PROMOTER PREDICTIONS (DRAGON) IN THE
MOUSE GENOME[a]

SE/SP level (%)	Prediction frequency	Total number of predictions
80	4,973	1,114,806
65	16,433	337,399
50	40,957	135,377

[a] The Frequency of Predictions was measured in one prediction per number of nucleotides.

islands are associated with promoters in virtually all housekeeping genes and in approximately 50% of all known mammalian genes.[46–48] We have thus demonstrated that DRAGON predicts promoters in genomic sequences with a good consistency for C + G-rich, A + T-rich, and CpG island related and CpG island nonrelated promoters and can be considered as a general promoter predictor.

We applied DRAGON on the nonmasked mouse genome assembly[4] (v. 3. May 1, 2002) and produced promoter predictions (Table IV). Predictions at a very high sensitivity setting produced more than a million guesses, whereas a moderate setting produced approximately 135,000 guesses. Taking into account the expected one FP prediction for each TP prediction, the latter number is broadly in agreement with the range of predicted number of genes (ranging from 30,000 to 100,000). We can therefore conclude that the current technology can be applied to the large-scale genomic sequence analysis, in which case the sensitivity range of 0.4–0.65 should be selected. Genscan makes a slightly lower number of TSS predictions than DRAGON at a sensitivity of 50%, whereas Fgenesh, Grail, and GeneId make more predictions. However, predictions using DRAGON consider only 200 nucleotides in the analysis process and thus represent a very localized assessment of the promoter presence. Thus, the DRAGON can be used easily in combination with some of the gene-finding programs to support predictions of TSS, promoters, and genes.

To assess the positional robustness of the DRAGON predictions of TSS locations, we analyzed them in simulated noisy promoter sequences. We used 945 vertebrate promoter sequences from the EPD (rel. 71),

[46] M. Gardiner-Garden and M. Frommer, *J. Mol. Biol.* **196,** 261 (1987).
[47] F. Larsen, G. Gundersen, R. Lopez, and H. Prydz, *Genomics* **13,** 1095 (1992).
[48] S. H. Cross and A. P. Bird, *Curr. Opin. Genet. Dev.* **5,** 309 (1995).

TABLE V
TSS Predictions (DRAGON with SE = 0.8) of Noisy
Promoter Sequences

Noise level (%)	SE with correct TSS position prediction	Other predictions within [−100,50] relative to TSS	Total number of predictions
5	0.86	533	1403
10	0.82	554	1393
15	0.77	595	1397
20	0.70	649	1380
25	0.52	802	1388
30	0.46	828	1351

covering nt range [−300, +100] relative to the TSS and altered the DNA content by randomly changing nucleotides in the promoter sequences. The sensitivity of DRAGON predictions (the SE setting of 80%) in altered promoter sequences is shown in Table V. Even when 30% of the promoter sequence content was changed randomly, the DRAGON achieved SE = 46% with the positional accuracy of 100% (no positional error in TSS predictions), whereas approximately 90% of other predictions fell within [−100, 50] relative to the TSS. At the 5% noise level, the DRAGON achieved SE = 86%, with the positional accuracy of 100%.

Concluding Remarks

The original aim behind the DRAGON was the development of a promoter recognition system that provides a high accuracy of predictions, namely high sensitivity with a relatively low number of false-positive predictions (high specificity), as well as a high positional accuracy of predictions. The testing results indicate that DRAGON represents a solution that provides high-accuracy predictions of TSS (Fig. 4 and Tables II and III) relative to other solutions. The DRAGON predicted with consistent accuracy TSSs in the test set and in the annotated genes in human chromosomes 21 and 22. The DRAGON can be considered a universal promoter prediction program, rather than a specialized solution, such as CpG-related predictors[24–26] or those that predict TATA box-related promoters in C + G-rich regions.[28] The positional accuracy of the predictions of DRAGON (Table V) is, to our knowledge, the best of all current promoter prediction programs. In addition, DRAGON performs well in noisy promoter sequences.

The DRAGON allows a reasonable trade-off between the sensitivity and the specificity of predictions. When the promoter region is localized and the main question is the position of the TSS, users can use higher sensitivity settings, whereas if a genomic search is required, a moderate sensitivity setting is recommended. In addition, DRAGON provides means for the analysis of promoter content of predicted TSSs that no other system currently allows. The DRAGON is therefore a comprehensive tool for promoter prediction and promoter content analysis.

A major problem associated with the design of promoter prediction systems is a relatively low number of well-annotated promoters. Currently, less than 2000 promoter sequences of all species are deposited in the EPD database and an additional several hundred promoters are scattered across the literature and Web resources. We expect that improvements in computational promoter analysis will follow two paths. First, the increased number of experimentally verified promoters and their TSSs will help further refine promoter prediction methods and increase the overall accuracy. Second, the improved analysis of promoter structure will help understand the diversity of promoters and identification of specific patterns that regulate gene expression.

The analysis of promoters will remain a critical issue in the analysis of genes and genomes, particularly for the study of regulation of gene expression and gene networks. Because of the large size of vertebrate genomes, computational analysis will remain the key technology in the analysis of functional sites in DNA and we can expect the emergence of even more specialized computational tools.

[22] Detection of DNA-Binding Helix-Turn-Helix Motifs in Proteins Using the Pattern Dictionary Method

By KALAI MATHEE and GIRI NARASIMHAN

Motifs are small conserved regions in related proteins that exhibit similar three-dimensional folds and similar functional properties. The helix-turn-helix (HTH) motif was the first protein motif to be discovered for site-specific DNA recognition. Other examples of DNA-binding motifs include (but are not limited to) zinc finger, leucine zipper, steroid receptor, and homeodomain. Existing methods for HTH motif detection are not entirely satisfactory. This article discusses a recent Web-based tool called GYM that uses pattern discovery techniques. The resulting program can be accessed from a website, which also provides additional information on HTH motifs.

The expression of most prokaryotic genes, ensuring opportune expression of an apposite amount of proteins, is modulated by the interaction of regulatory proteins with promoter elements. This interaction of regulatory proteins and DNA appears to use a highly specialized structure called the HTH motif. Based on sequence analysis, a large number of regulator families containing HTH motifs such as AraC, ArsR, AsnC, CRP, DeoR, GntR, IclR, LacI, LuxR, LysR, MarR, MerR, TetR, and sigma factors have been described. Thus, prediction of the presence of a HTH motif in a newly discovered protein suggests that it is a DNA-binding regulator protein. Because motifs are often associated with functional features, motif detection in proteins is an important aspect of protein classification.

In the past, many of the HTH motifs in protein sequences were inferred from genetic, mutational, and sequence analysis. Lately, they have also been confirmed by protein structure determination methods. The characteristic features of the HTH motif have been reviewed in detail.[1-5] Figure 1 shows a helix followed by a turn followed by a second helix. The

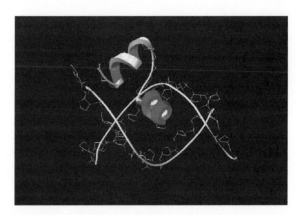

FIG. 1. A helix-turn-helix motif in lambda Cro repressor with the second helix interacting with elements of the major groove of the DNA molecule. Visualization was done using Swiss-PDB Viewer.[6]

[1] R. G. Brennan and B. W. Matthews, *J. Biol. Chem.* **264,** 1903 (1989).
[2] S. C. Harrison and A. K. Aggarwal, *Annu. Rev. Biochem.* **59,** 933 (1990).
[3] H. C. Nelson, *Curr. Opin. Genet. Dev.* **5,** 180 (1995).
[4] C. O. Pabo and R. T. Sauer, *Annu. Rev. Biochem.* **53,** 293 (1984).
[5] C. O. Pabo and R. T. Sauer, *Annu. Rev. Biochem.* **61,** 1053 (1992).
[6] N. Guez and M. C. Peitsch, *Electrophoresis* **18,** 2714 (1997).

second helix is nestled in the major groove of the DNA double helix, which is shown traversing horizontally across the figure.

Figure 2 shows examples of amino acid residues in the HTH motif in several proteins. The motif is about 22 residues in length.[1] The residues are numbered 0 through 21. The turn consists of about four amino acids, and the two helices make an angle of approximately $120°$.[3] In most proteins, the second of the two helices is used for binding to DNA in a sequence-specific manner and is referred to as the *recognition helix*. Residues of the recognition helix interact directly with bases in the major groove of the DNA. Residues in both the helices are believed to be responsible for maintaining the appropriate angle between the two helices. Proteins with HTH motifs share only limited sequence similarity in the motif region; dissimilarity could be attributed to the different sequence-specific interactions with the bases in the DNA exhibited by the various motifs. The AraC and sigma family of proteins are examples of proteins that have more than one HTH motif, although many regulator proteins are known to have at most one HTH motif. Motif recognition is further complicated by the fact that variations in the length of the motif and in the turn angle exist; these variations extend the range of interactions made by these proteins.[3]

Protein Name	Helix 2								Turn					Helix 3								
	0	1	2	3	4	5	6	7	8	9	10	11	12	13	14	15	16	17	18	19	20	21
Cro	F	G	Q	E	K	T	A	K	D	L	G	V	Y	Q	S	A	I	N	K	A	I	H
434 Cro	M	T	Q	T	E	L	A	T	K	A	G	V	K	Q	Q	S	I	Q	L	I	E	A
P22 Cro	G	T	Q	R	A	V	A	K	A	L	G	I	S	D	A	A	V	S	Q	W	K	E
Rep	L	S	Q	E	S	V	A	D	K	M	G	M	G	Q	S	G	V	G	A	L	F	N
434 Rep	L	N	Q	A	E	L	A	Q	K	V	G	T	T	Q	Q	S	I	E	Q	L	E	N
P22 Rep	I	R	Q	A	A	L	G	K	M	V	G	V	S	N	V	A	I	S	Q	W	E	R
CII	L	G	T	E	K	T	A	E	A	V	G	V	D	K	S	Q	I	S	R	W	K	R
LacR	V	T	L	Y	D	V	A	E	Y	A	G	V	S	Y	Q	T	V	S	R	V	V	N
CAP	I	T	R	Q	E	I	G	Q	I	V	G	C	S	R	E	T	V	G	R	I	L	K
TrpR	M	S	Q	R	E	L	K	N	E	L	G	A	G	I	A	T	I	T	R	G	S	N
BlaA Pv	L	N	F	T	K	A	A	L	E	L	Y	V	T	Q	G	A	V	S	Q	Q	V	R
TrpI Ps	N	S	V	S	Q	A	A	E	Q	L	H	V	T	H	G	A	V	S	R	Q	L	K

FIG. 2. Examples of HTH motifs.

The properties mentioned earlier make automatic recognition of HTH motifs a nontrivial algorithmic problem.

Motif Detection

Motif detection methods have been reviewed extensively and can be classified broadly as consensus-based methods, profile-based methods, and pattern discovery-based methods.[7,8] We briefly summarize some of the motif detection methods from the literature. Bork and Gibson also discuss in detail techniques to make the most of existing motif detection programs and to interpret appropriately the evaluation of the outputs.[7]

Consensus Methods

In this classical method, known motifs are aligned and residues conserved in the alignment are identified. By checking which of these conserved residues appear in a new protein sequence, it is possible to devise simple detection schemes.[9,10] For HTH motifs, the positions numbered 5, 6, 10, and 16 are particularly well conserved. Position 10 is usually occupied by a glycine, position 6 consists of an alanine or glycine, and finally positions 5 and 16 are usually occupied by an uncharged hydrophobic residue (such as leucine, valine, or isoleucine). Other moderately conserved residues include a charged residue (aspartic or glutamic acid) in position 4 and a hydrophobic residue (valine, leucine, or isoleucine) in positions 9 and 11.

Brennan and Matthews[1] used the consensus information mentioned previously and devised a simple scoring scheme for HTH motifs. A score called the AAC (average amino acid change per codon) score was computed for each 22 amino acid subsequence of a given protein. The AAC score is determined by summing the total number of amino acid residues in a carefully designed master set that differ from the given subsequence. This value is then normalized to a value between 0 and 1. A value of less than 0.79 was considered a strong score for a subsequence to be a HTH motif.

Generalized consensus sequences or signatures for HTH and other motifs (as well as for protein families) are cataloged by PROSITE.[11,12] Such generalized consensus sequences consist of a sequence of sets of

[7] P. Bork and T. J. Gibson, *Methods Enzymol.* **266,** 162 (1996).
[8] W. R. Taylor, *Protein Eng.* **2,** 77 (1988).
[9] L. Patthy, *J. Mol. Biol.* **198,** 567 (1989).
[10] L. Patthy, *Methods Enzymol.* **266,** 184 (1996).
[11] A. Bairoch, *Nucleic Acids Res.* **20** (Suppl.), 2013 (1992).
[12] A. Bairoch, P. Bucher, and K. Hofmann, *Nucleic Acids Res.* **24,** 189 (1996).

amino acids where amino acids within the same set could substitute each other in that position. In particular, many of the protein families with HTH motifs have PROSITE signature sequences that overlap the HTH motif. For example, the signature pattern for theDeoR family of proteins is listed as R-x(3)-[LIVM]-x(3)-[LIVM]-x(16,17)-[STA]-x(2)-T-[LIVMA]-[RH]-[KRNA]-D-[LIVMF]. The part of the pattern that overlaps the HTH motif is emphasized with a bold font. In order to interpret these signatures, note that amino acids enclosed in square brackets such as [LIVM] indicate that the location is occupied by one of the amino acids listed within the brackets. The character x indicates that the position is occupied by any amino acid. The number in parentheses such as (3) indicates that it is repeated three times. When the number appears as a pair, such as (16, 17), it indicates that the number of repeats is from the range 16 through 17.

Earlier, Wu and Brutlag[13] had showed a way of constructing substitution sets in a statistically significant manner. Nevill-Manning et al.[14] devised a method to automatically construct generalized consensus sequences such that any one of the sequences in the set could describe the motif. Additionally, the software based on their method (**EMOTIF**) provided the user with parameters to tradeoff sensitivity for specificity.

Profile Methods

A number of other sophisticated detection schemes that are statistically motivated have been reported. These are typified by the *Profile* method described by Gribskov et al.[15–17] The first step, once again, involved making a multiple alignment of known motifs. The next step typically involved computing a probability matrix or a position-specific score matrix, which assigns a different score to each possible residue at each position in the motif. Intuitively, the entries of this matrix represent a measure of the probability that a certain residue occurs in that location normalized by the background frequencies for that residue. Minor variants exist in the methods employed to compute the scoring matrix, as well as in scoring a match (see, e.g., Gusfield[18]). Given a score matrix, the detector, when given

[13] T. D. Wu and D. L. Brutlag, *Proc. Int. Conf. Intell. Syst. Mol. Biol.* **4**, 230 (1996).
[14] C. G. Nevill-Manning, T. D. Wu, and D. L. Brutlag, *Proc. Nat. Acad. Sci USA* **95**, 5865 (1998).
[15] M. Gribskov, A. D. McLachlan, and D. Eisenberg, *Proc. Nat. Acad. Sci. USA* **84**, 4355 (1987).
[16] M. Gribskov, R. Luthy, and D. Eisenberg, *Methods Enzymol.* **183**, 146 (1990).
[17] M. Gribskov and S. Veretnik, *Methods Enzymol.* **266**, 198 (1996).
[18] D. Gusfield, "Algorithms on Strings, Trees, and Sequences." Cambridge Univ. Press, Cambridge, 1997.

an input protein sequence, computes a weighted score for every sub-sequence of the input sequence and reports the subsequence with the highest score as the detected motif, as long as this score is above a certain threshold.

Dodd and Egan[19,20] used a similar scheme for the HTH motif. Their contribution lay in carefully picking a set of 91 HTH motifs from known proteins so that the position-specific scoring matrix could be computed with little bias and in a way that resulted in an accurate detection method for HTH motifs. Enhancements on the profile method and its variants have been incorporated into several programs.[21–24] Related methods were also used to detect coiled–coil motifs.[25–30]

Other statistically based methods for detecting motifs include that of using *neural networks*,[31,32] *hidden Markov models*,[33] and *Gibbs sampling*.[34]

Pattern Discovery Methods

The basic idea behind these techniques is that of finding frequently occurring patterns in known motifs and then using the presence or absence of these patterns as a basis for motif detection. We have described a new approach to the problem of automatic motif detection. We implemented the approach successfully in a program called GYM and tested it on both prokaryotic HTH and eukaryotic homeodomain motifs.[35,36]

[19] I. B. Dodd and J. B. Egan, *Protein Eng.* **2,** 174 (1988).
[20] I. B. Dodd and J. B. Egan, *Nucleic Acids Res.* **18,** 5019 (1990).
[21] J. D. Thompson, D. G. Higgins, and T. J. Gibson, *Comput. Appl. Biosci.* **10,** 19 (1994).
[22] R. Luthy, I. Xenarios, and P. Bucher, *Protein Sci.* **3,** 139 (1994).
[23] E. V. Koonin, R. L. Tatusov, and K. E. Rudd, *Methods Enzymol.* **266,** 295 (1996).
[24] R. L. Tatusov and E. V. Koonin, *Comput. Appl. Biosci.* **10,** 457 (1994).
[25] E. Wolf, P. S. Kim, and B. Berger, *Protein Sci.* **6,** 1179 (1997).
[26] B. Berger, D. B. Wilson, E. Wolf, T. Tonchev, M. Milla, and P. S. Kim, *Proc. Nat. Acad. Sci. USA* **92,** 8259 (1995).
[27] B. Berger, *J. Comput. Biol.* **2,** 125 (1995).
[28] B. Berger and M. Singh, *J. Comput. Biol.* **4,** 261 (1997).
[29] M. Singh, B. Berger, P. S. Kim, J. M. Berger, and A. G. Cochran, *Proc. Natl. Acad. Sci. USA* **95,** 2738 (1998).
[30] M. Singh, B. Berger, and P. S. Kim, *J. Mol. Biol.* **290,** 1031 (1999).
[31] D. Frishman and P. Argos, *J. Mol. Biol.* **228,** 951 (1992).
[32] J. Hanke, G. Beckmann, P. Bork, and J. G. Reich, *Protein Sci.* **5,** 72 (1996).
[33] W. N. Grundy, T. L. Bailey, C. P. Elkan, and M. E. Baker, *Biochem. Biophys. Res. Commun.* **231,** 760 (1997).
[34] C. E. Lawrence, S. F. Altschul, M. S. Boguski, J. S. Liu, A. F. Neuwald, and J. C. Wootton, *Science* **262,** 208 (1993).
[35] Y. Gao, K. Mathee, G. Narasimhan, and X. Wang, *in* "String Processing and Information Retrieval (SPIRE)," p. 63, Cancun, Mexico, 1999.
[36] G. Narasimhan, C. Bu, Y. Gao, X. Wang, N. Xu, and K. Mathee, *J. Compu. Biol.* **9,** 707 (2002).

Similar techniques can also be found in the work of Rigoutsos and Floratos,[37] who devised a method to discover unknown motifs without doing alignment, i.e., the training set for their program is a set of unaligned protein sequences. Their method is based on similar ideas of generating patterns, which in turn could be used to perform detection. Other related methods are reviewed by Brazma et al.[38] The GYM algorithm differs from others in that it detects known motifs after being trained on a set of aligned sequences for the same motif, thus making use of all available knowledge about the motif. Our methods also share some overlap with that of Nevill-Manning et al.[14] The fundamental difference lies in the way the threshold is used; they require that their motifs *cover* some percentage of the sequences in the training set. The idea of correlations between residues in specific locations was also explored by Berger et al.,[26] however, only pair-wise correlations were considered.

Pattern Dictionary Method

Our approach to the problem of automatic motif detection is referred to as the *pattern dictionary* method.[36] The resulting GYM algorithm was based on techniques from the fields of *data mining* and *knowledge discovery*. Compared to existing algorithms, GYM displays increased sensitivity while maintaining high accuracy and also providing additional information about a given protein sequence. Our algorithm is not based on statistical methods but requires a *training set* of aligned sample motifs. Some key features and assumptions are listed.

1. An appropriate length of the motif needs to be selected beforehand. This is true for many known motifs. In the case of HTH motifs, ample evidence shows that the motif lies within a window of size about 22.[1]
2. A reasonably large number of verified examples of the motif are needed for this method. The training set can then be chosen from these known motifs. Again this is true for HTH motifs.
3. It is assumed that a combination of key residues is sufficient to constitute the necessary physical structure and to confer upon it the appropriate function; the rest of the parts of the motif may serve other purposes. In other words, it is assumed that different functional and structural constraints lead to the appearance of different

---

[37] I. Rigoutsos and A. Floratos, *in* "Proceedings of the 4th Annual International Conference on Computational Molecular Biology (RECOMB)," p. 221, ACM, New York, 1998.
[38] A. Brazma, I. Jonassen, I. Eidhammer, and D. Gilbert, *J. Comput. Biol.* **5,** 279 (1998).

combinations of residues to within the motif. This is where the GYM algorithm differs from assumptions made by other methods. While many of the methods attach separate significance to the occurrence of specific residues in specific locations in the motif, they do not necessarily account for the *reinforcing effect of a combination of specific residues*. For example, one of the most frequently occurring patterns in the training set that was used was **A6, L9, G10**, i.e., an alanine in position 6, a glycine in position 10, and a leucine in position 9. It is known that position 11 (in the turn) is often occupied by a hydrophobic residue (such as valine, leucine, or isoleucine). However, our patterns indicate that a valine or isoleucine in that position favors a glutamic acid residue in position 4 over aspartic acid. Leucine in position 11, however, does not favor glutamic acid over aspartic acid in position 4 (or vice versa). A favorable combination in many motifs consists of **L5, A6, Y17**, i.e., a leucine in position 5, an alanine in position 6, and a tyrosine in position 17. It is likely that the patterns in the pattern dictionary discovered by our method represent such reinforcing combinations, helping in the detection of new motifs.

4. A *good* combination of residues must occur *frequently enough* to be called a valid pattern for the motif. To account for relatively rare reinforcing combinations, the GYM algorithm sets absolute threshold values to decide whether a combination occurs frequently enough, as opposed to a requirement that a combination occurs in a specified percentage of the sequences in the training set.

Using the aforementioned criteria, the GYM algorithm searches for patterns from the sample training set that are present in a new protein sequence. GYM discovers patterns in known motifs to compute a pattern dictionary. Detection of a motif in a new protein sequence is then a function of which patterns from the dictionary are present in the new protein sequence. The algorithm consists of the following two phases (Fig. 3).

1. *Pattern mining phase.* This preprocessing step involves the input of a set of known and aligned motifs, or the master set. The preprocessing step needs to be performed only once. The output is a pattern dictionary consisting of frequently occurring patterns within the master set.

2. *Detection phase.* The input to this phase consists of the pattern dictionary output from the preprocessing step and the input protein for which the actual detection of motif takes place. The output of the detection algorithm indicates whether the protein sequence contains a motif, states the location of this motif, provides a score indicating the confidence of the prediction, and also gives a list of proteins from the master set that share

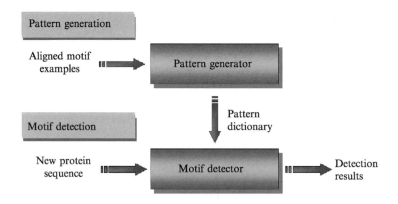

Fig. 3. Schematic representation of the GYM algorithm. The algorithm has two phases: pattern generation phase and motif detection phase.

high sequence homology with the detected motif as inferred from matching patterns from the dictionary.

Pattern Mining Phase

The input to this phase is a master set of aligned motifs without gaps. The success or failure of the motif detection depends on a careful selection of the training set or master set. In GYM 2.0, we used only 88 of the 91 proteins listed in Dodd and Egan's master set because there was no consensus between various programs on the correct location of the motif in the three remaining proteins, and experimental evidence defining the precise locations was not available in the literature.[20,36] For the current version of GYM, we also added five more motif sequences to the master set, as several groups of proteins were not represented in the master set chosen by Dodd and Egan.[20] Details about our master set can be found at the GYM website (see section on web page).

Detection Phase

It takes as input the dictionary of significant patterns output by the pattern-mining phase, and the given protein sequence to be examined for the motif. We slide a window of length 22 across the input sequence **P**. The subsequence of **P** that lies in the window is then matched against every significant pattern in the dictionary. The match is scored and the two best matches are reported as possible motif locations, if their quality exceeds a prespecified *threshold*. The match procedure returns several parameters that indicate the quality of the match: (1) **LPM**, the length of the longest pattern matched; (2) **NPM**, the number of distinct maximal patterns matched; and (3) **WSP**, a weighted score for all the distinct positions from the window that matched some pattern. It quantifies how well the window matched against the patterns in the dictionary. The residues were weighted by values from the BLOSUM62 substitution matrix for substituting each amino acid in the pattern by itself.[39]

The choice of threshold values is critical because they represent a trade-off between selectivity and sensitivity. When a threshold is set too high, the results are likely to be more selective and conservative, resulting in a higher number of false negatives. However, if it is set too low, the sensitivity is greater but with less accuracy and more false positives. The threshold values used for our experiments were either computed statistically or were set to the maximum values that optimized the detection of motifs from the training set itself, i.e., if the threshold were any higher, then the detection algorithm would fail in some instances from the master set itself.

The match procedure also compares two matches and makes decisions about which are the two best matches in the protein sequence. When comparing matches, the length of the longest pattern (LPM) was considered the most important criterion followed by the weighted score (WSP). A significant pattern from the dictionary represents a combination of amino acids in specific locations that (potentially) positively reinforce the motif structure. Thus, the longer the pattern, the greater the number of positive reinforcements to the structure and, consequently, the better the quality of the match. More precisely, a match is said to exceed the threshold if both of the following are true: (a) LPM is at least 4 and (b) WSP is at least 29.

Output of Algorithm

The Web-based GYM program is at http://www.cs.fiu.edu/~giri/bioinf/GYM/welcome.html. It implements the pattern discovery algorithm mentioned earlier.[35,36] The profile-based algorithm, as proposed by Dodd and Egan,[20] is also implemented and available from the same URL; this program is referred to as the DE program. For a given protein, the algorithm is designed to locate all possible HTH motifs, as long as the scores are above a prespecified threshold. The algorithm will output the location of the motif as well as the residues in the motif. It also prints the match parameters, i.e., LPM, NPM, and WSP, for the two best matches. Output includes the set of patterns that are matched at the predicted motif

[39] S. Henikoff and J. G. Henikoff, *Adv. Protein Chem.* **54,** 73 (2000).

location. It also indicates which particular residues in the motif are present in the patterns matched. Finally, the output gives a list of proteins from the training set that exhibit the same patterns found in the motif of the input protein sequence. For new protein sequences, this could provide clues to the family to which this protein may belong in terms of its function. This information would become more valuable when it is combined with similar information from other motifs found in the same protein sequence.

When the amino acid sequence for the RNA polymerase sigma factor from *Escherichia coli* (Rpoh or sigma 32—Heat shock regulatory protein; 284 amino acids; gi:133287) is submitted to GYM, two HTH motifs are predicted: one at location 251 and the other at location 152 (Fig. 4). Members of the sigma family are known to have two HTH motifs, occurring in regions labeled as 3.1 and 4.2[40,41]; these regions are separated by about 90–100 residues and are involved in recognition of the −10 and −35 promoter regions, respectively. The GYM program clearly identified the weaker HTH motif in region 3.1, which is often missed by other prediction programs.

For the motif predicted in location 251, the GYM method had scores of 5, 72, and 80 for the match parameters LPM, NPM, and WSP, respectively. Because the threshold for the WSP score is 29, a score of 80 indicates a strong motif at this location. The DE method also had a strong score of 2163 for this location (threshold for the DE method was 1050). For the second best motif predicted in location 152, the GYM method had scores of 4, 8, and 41 for the match parameters LPM, NPM, and WSP, respectively. The DE method was not able to predict this HTH motif, as the score was only 513. Results indicate that the two predicted locations exhibit motifs of differing strengths.

Results also show the patterns that were matched from the dictionary. For the example used earlier, 72 patterns were matched at the HTH motif located at 251. These patterns covered a total of 17 out of 22 positions; these are marked by a "∧" under the corresponding positions in the motif. For each of the 72 patterns, proteins from the master set that also matched that pattern are provided as part of the output (data not shown).

Analysis of Families of DNA-Binding Proteins

The GYM program was tested on 1068 sequences downloaded from GenBank and SWISSProt databases. The programs GYM and DE disagreed on 80 (approximately 8%) of the 1068 sequences. The percentage

[40] T. M. Gruber and D. A. Bryant, *J. Bacteriol.* **179,** 1734 (1997).
[41] M. Lonetto, M. Gribskov, and C. A. Gross, *J. Bacteriol.* **174,** 3843 (1992).

GYM Version 3.0, February 16, 2003

Results for Sequence (gi: 133287)

Pattern Discovery (GYM 3.0)

Pick	Location	Score	Detected	Sequence
I	251	80	+	STLQELADRYGVSAERVRQLEK
II	152	41	+	DEVEMVARELGVTSKDVREMES

---- GYM's BEST MATCH ----

Start Location: 251
Length of Longest Pattern Matched (LPM): 5
Number of Patterns Matched (NPM): 72
Number of Locations Matched: 17
Weighted score (WSP): 80
Best Motif Sequence:

```
0123456789 0123456789 01
STLQELADRY GVSAERVRQL EK
_^^_^^^^^_ ^^^_^_^^^^
```

---- GYM's 2nd Best Match ----

Start Location: 152
Length of Longest Pattern Matched (LPM): 4
Number of Patterns Matched (NPM): 8
Number of Locations Matched: 9
Weighted score (WSP): 41
Possible Motif Sequence:

```
0123456789 0123456789 01
DEVEMVAREL GVTSKDVREM ES
_____^^__^ ^^^___^^__ ^_
```

---- Profile (Dodd/Egan) Method ----

Pick	Location	Score	Detected	Sequence
I	251	2163	+	STLQELADRYGVSAERVRQLEK
II	152	513	-	

FIG. 4. Output from the GYM program for *E. coli* RpoH.

of false positives was about 7%. The sequences represent several families of proteins such as sigma factors,[40,41] arabinose operon regulatory proteins and related subfamilies (AraC),[42,43] two-component response regulators,[44] cAMP receptors and related subfamilies (CRP),[45] deoxyribose operon repressors and related subfamilies (DeoR),[46] transcriptional repressor of the gluconate operon and related subfamilies (GntR),[47,48] lactose operon repressor and related subfamilies (LacI),[49] diaminopimelate decarboxylase regulator proteins and related subfamilies (LysR),[50,51] mercuric resistance operon regulatory proteins and related subfamilies (MerR),[52,53] and regulators from enterobacterial transposons (TetR).[54] Among the proteins selected, 93 are proteins involved in metabolic pathways and other enzymatic reactions. These are assumed to not have a HTH motif and are unlikely to bind to DNA. We refer to this family as the *Negates* family. Results are summarized on the Web page set up for this program (see section on Web page).

In about 9% of the sequences with annotations, GYM predictions disagreed with the annotations on the location of the HTH motif. Of the input sequences, the genomic databases had the motif location annotated in only about 56% of the cases. However, the database had no information on how these locations were determined, e.g., what program was used to determine the motif location and were the claims verified experimentally? In many cases, the annotations stated simply that the motif was determined by sequence analysis. Thus, we consider these annotations to be unconfirmed, as are the predictions of our program.

[42] M. T. Gallegos, R. Schleif, A. Bairoch, K. Hofmann, and J. L. Ramos, *Microbiol. Mol. Biol. Rev.* **61**, 393 (1997).
[43] M. T. Gallegos, C. Michan, and J. L. Ramos, *Nucleic Acids Res.* **21**, 807 (1993).
[44] J. A. Hoch and T. J. Silhavy, "Two-Component Signal Transduction," ASM Press, 1995.
[45] A. Kolb, S. Busby, H. Buc, S. Garges, and S. Adhya, *Annu. Rev. Biochem.* **62**, 749 (1993).
[46] K. Hammer, L. Bech, P. Hobolth, and G. Dandanell, *Mol. Gen. Genet.* **237**, 129 (1993).
[47] A. Reizer, J. Deutscher, M. H. Saier, Jr., and J. Reizer, *Mol. Microbiol.* **5**, 1081 (1991).
[48] D. J. Haydon and J. R. Guest, *FEMS Microbiol. Lett.* **63**, 291 (1991).
[49] M. J. Weickert and S. Adhya, *J. Biol. Chem.* **267**, 15869 (1992).
[50] S. Henikoff, G. W. Haughn, J. M. Calvo, and J. C. Wallace, *Proc. Natl. Acad. Sci. USA* **85**, 6602 (1988).
[51] M. A. Schell, *Annu. Rev. Microbiol.* **47**, 597 (1993).
[52] J. D. Helmann, B. T. Ballard, and C. T. Walsh, *Science* **247**, 946 (1990).
[53] J. D. Helmann, L. M. Shewchuk, and C. T. Walsh, *Adv. Inorg. Biochem.* **8**, 33 (1990).
[54] P. Orth, D. Schnappinger, W. Hillen, W. Saenger, W. Hinrichs, *Nat. Struct. Biol.* **7**, 184 (2000).

Concluding Remarks

Most prokaryotic transcriptional regulator proteins interact with DNA using a HTH motif. An algorithm based on data mining and knowledge discovery to detect HTH motifs in a given protein was designed. This algorithm assumes that a motif is constituted by the presence of a "good" combination of residues in appropriate locations of the motif. The algorithm attempts to compile such good combinations into a "pattern dictionary" by processing an aligned training set of protein sequences. The dictionary is subsequently used to detect motifs in new protein sequences. Statistical significance of the detection results is ensured by statistically determining the various parameters of the algorithm.

The GYM program has an excellent ability to predict HTH motifs. It appears to have an increased sensitivity relative to the DE program and can detect motifs with greater divergence from the training set. The number of false positives was lowered by using a score weighted by the use of BLOSUM62 substitution matrix and by careful choices of threshold values resulting from sound statistical experiments. The number of false negatives is improved constantly by updating the master set carefully with a subset of the false negatives and by recomputing the pattern dictionary. The GYM program also provides additional useful information about the HTH motifs in a given protein sequence. This includes a list of patterns present in the motif, all the residues from the motif that were part of some pattern in the dictionary, and for each pattern a list of proteins from the master set that shared the pattern.

A number of proteins involved in plant and animal development contain a canonical motif called the homeodomain motif, which is almost three times as long as the HTH motif.[55] Modification of the GYM program to detect the 60 residue homeodomain motifs was also successful and resulted in very high agreement with the DE program and the database sequence annotations.[36]

Web Page

The URL for the GYM program is http://www.cs.fiu.edu/~giri/bioinf/GYM/welcome.html. It can therefore be used conveniently over the internet. The Web page also contains supplementary information on our experiments with GYM and the sequences used for training and testing, as well as other useful links.

[55] M. P. Scott, J. W. Tamkun, and G. W. Hartzell, 3rd, *Biochim. Biophys. Acta* **989**, 25 (1989).

Acknowledgments

   The authors gratefully acknowledge the contributions of Yuan Gao, Changsong Bu, Ning Xu, Xuning Wang, Gaolin Zheng, Tom Milledge, Yanli Sun, Zhengyue Deng, Chengjun Zhan, and Na Zhao to the implementation of old and new versions of the GYM program. We also thank Elaine Newman and S. Padmanabhan for careful reading of this manuscript.

[23]   DNA Microarrays and Bacterial Gene Expression

By Robert A. Britton

Since the first full bacterial genome sequence was released in 1995,[1] the amount of sequence information for bacterial species has exploded. Currently there are 209 publicly available completed and partially completed bacterial genome sequences, with more being added each month. To best utilize the increasing amount of sequence information, numerous computational and molecular biological methods for studying organisms on the genome-wide level have been developed. The most popular of these methods are DNA microarrays, used primarily in the analysis of gene expression. Since this technology emerged in the late 1990s it has rapidly become an important tool in studies from bacteria to humans. In addition to expression profiling, DNA microarrays have been used in many applications, including genome organization,[2,3] small RNA identification,[4] and identification of DNA-binding sites for proteins.[5,6] In systems for which no genetic tools are available, DNA microarrays offer researchers a valuable resource to address questions about important processes that otherwise could not be approached.

   One straightforward and widely used application of DNA microarrays is determination of the roles of known and putative transcription factors in gene expression. Entire regulons involved in processes that have been studied for decades are now being examined on the genome-wide level (i.e., SOS,[7] development of genetic competence,[8] sporulation,[9] environmental

---

[1] R. D. Fleischmann *et al., Science* **269,** 496 (1995).

[2] N. Salama *et al., Proc. Natl. Acad. Sci. USA* **97,** 14668 (2000).

[3] J. C. Smoot *et al., Proc. Natl. Acad. Sci. USA* **99,** 4668 (2002).

[4] K. M. Wassarman, F. Repoila, C. Rosenow, G. Storz, and S. Gottesman, *Genes Dev.* **15,** 1637 (2001).

[5] V. R. Iyer *et al., Nature* **409,** 533 (2001).

[6] B. Ren *et al., Science* **290,** 2306 (2000).

[7] J. Courcelle, A. Khodursky, B. Peter, P. O. Brown, and P. C. Hanawalt, *Genetics* **158,** 41 (2001).

stresses[10,11]). Because many of these regulons are well characterized, the array approach in these systems was validated easily.

Bacteria are particularly amenable to microarray analysis because of their small genome size and ease of identification of genes by sequence analysis. In genetically tractable organisms, researchers can mutate and over-express transcription factors to investigate their role in gene expression. An approach combining gene expression data with whole genome sequence analysis of a transcription factor (TF)-binding site has been successful in characterizing a number of *Bacillus subtilis* regulons, including SpoOA[9] (master regulator of sporulation), Sigma-B[10] (general stress response), and Sigma-H[12] (stationary phase). Two types of experimental designs were utilized in each case. First, the transcriptome of a wild-type strain was compared to a TF mutant under conditions where the TF is normally active. This experiment gives an overall view of the regulon, revealing both directly and indirectly regulated genes. Second, the transcriptome of a wild-type strain was compared to a strain that overexpresses the TF under conditions where the TF is not normally active. If overexpression of the TF is sufficient to activate gene expression, then candidates for direct regulation by the TF can be identified. The upstream regions of differentially expressed genes identified in the two approaches were scanned for potential TF-binding sites. Those genes that were expressed differentially and had a putative TF-binding site were considered to be candidate genes for direct regulation by the TF.

This article focuses on the construction and utilization of two-color-spotted DNA microarrays in the analysis of bacterial gene expression. These microarrays consist of either polymerase chain reaction (PCR) products or long oligonucleotides (50–70 nucleotides in length) spotted onto an amine-coated glass surface. Samples to be compared on the array are labeled differentially with fluorescent dyes, usually Cy3 or Cy5, and hybridized competitively to a single array. The following methods detail the construction, hybridization, and analysis of DNA microarrays. Most of the protocols are versions of methods developed originally by Patrick Brown and colleagues (http://cmgm.stanford.edu/pbrown/protocols/index.html), Quakenbush and Colleagues at TIGR (http://www.tigr.org/tdb/microarray/protocolsTIGR.shtml), and J. L. DeRisi and Colleagues at UCSF (http://

[8] R. M. Berka *et al.*, *Mol. Microbiol.* **43**, 1331 (2002).
[9] P. Fawcett, P. Eichenberger, R. Losick, and P. Youngman, *Proc. Natl. Acad. Sci. USA* **97**, 8063 (2000).
[10] C. W. Price *et al.*, *Mol. Microbiol.* **41**, 757 (2001).
[11] J. D. Helmann *et al.*, *J. Bacteriol.* **185**, 243 (2003).
[12] R. A. Britton *et al.*, *J. Bacteriol.* **184**, 4881 (2002).

www.microarrays.org) that were modified for use with *B. subtilis* DNA microarrays. The design of microarray experiments and, particularly, the analysis of microarray data are still emerging fields for which no single best approach exists. Throughout the article, the reader is referred to various tools used in the production and analysis of DNA microarrays. A more complete listing of tools available for all steps of microarray analysis was recently reviewed.[13]

Microarray Construction

PCR versus Long Oligonucleotide Arrays

DNA printed on glass slides can be either double-stranded DNA (such as a PCR product) or a single-stranded oligonucleotide. Initial studies using microarrays utilized PCR-amplified DNA for each gene; in many cases the entire coding sequence of a gene was used. While studies using these arrays have been very successful, it was clear early on that significant cross-hybridization could occur between similar genes. Cross-hybridization can lead to the false identification of genes as being expressed differentially, costing researchers valuable time when following up leads from a microarray experiment. An additional problem is that double-stranded PCR products will hybridize to transcription occurring on both the coding or the noncoding strand. For example, consider a spot (PCR product) on a microarray for gene X. The transcription of gene X is monitored by hybridization of the labeled cDNA to the PCR product on the array. Labeled cDNA from transcription of an unknown gene or read-through from an operon occurring on the *opposite strand* of the gene of interest will also hybridize and be identified incorrectly as expression of gene X. This can often lead to incorrect conclusions about the expression of gene X.

An alternative approach gaining popularity is the use of long single-stranded oligonucleotides, generally 50–70 nucleotides in length, as probes on microarrays. There are substantial time and cost savings when using oligonucleotides as probes when compared to PCR-based arrays. Oligonucleotides can be printed directly from the plates in which they were synthesized. However, PCR-based arrays require a significant amount of time and money in generating the PCR products, documenting the success of each reaction and in the purification of each product. An additional advantage of oligonucleotide arrays is that only the coding sequence is

[13] A. J. Holloway, R. K. van Laar, R. W. Tothill, and D. D. Bowtell, *Nature Genet.* **32**(Suppl.) 481 (2002).

printed as a probe so that hybridization does not occur from transcription on the opposite strand.

Protocol for Printing Slides (PCR Based or Long Oligonucleotides)

Laboratories that choose to print their own slides will require access to an arrayer. Many universities have a genomics core facility that provides a microarray robot for this purpose. Because printing of DNA on slides can be a significant source of variation, it is important to control the humidity and temperature carefully during the arraying process. For example, maintaining the humidity between 40 and 50% and the temperature between 22 and 26° is desirable when printing in dimethyl sulfoxide (DMSO) (see later).

DNA used for microarrays may be prepared in either a DMSO or SSC buffer. DMSO may be preferred as it is hygroscopic and limits evaporation of the sample during the printing process. DMSO offers an added benefit over SSC buffer in printing cDNA arrays because it denatures the DNA before attaching to the slide. To prepare DNA in DMSO, the DNA is resuspended in 50% DMSO to a concentration of 200 ng/μl for cDNA arrays and between 10 and 50 μM for oligonucleotide arrays. Alternatively, DNA may be resuspended in 3× SSC buffer supplemented with betaine (1.5 M for cDNA arrays and 0.2 M for oligonucleotide arrays). Betaine limits the evaporation of samples during printing.

Amine-coated slides are the most widely used for DNA microarray applications and can be purchased from several manufacturers. Protocols for coating slides exist; however, the quality from batch to batch is difficult to control and therefore commercially prepared slides are recommended. Many companies that produce oligonucleotide libraries for microarrays add a 5′ amino linker on the end that can be attached covalently to specially treated slide surfaces. At a cost of up to $5 per oligonucleotide, this modification can add significant cost to the array. Because unmodified oligonucleotides printed on amine-coated slides bind well and yield excellent results, the addition of a modification to attach the oligonucleotide to the slide covalently may not be necessary.

Slides should be numbered in the order they were printed using a diamond pen. If problems arise during the use of the slides it is often useful to know the order in which they were printed. Etching is also useful in noting which side of the slide the DNA is printed on and the orientation of the array. Care should be taken to not get any glass shards on the arrayed DNA. Slides are stored in a dust-free container in the dark. The DNA should be cross-linked to the slide using a 90-mJ dose with a UV transilluminator. Slides can be cross-linked immediately after printing and stored or can be cross-linked later, just prior to prehybridization.

Verification of Array Quality

The quality of DNA arrays should be determined prior to experimental use. To verify that DNA is spotted on the array and is available for hybridization, a mixture of random nonamers end labeled with Cy3 is hybridized to two of the slides of the print run. A signal for each DNA spot placed on the array should be visible. The random nonamers will not bind to each spot equally and therefore the spots will not appear of equal intensity. It is recommended that the first and last slides of a print run be hybridized to assess any changes in spot presence or morphology during the print. Alternatively, Cy3- or Cy5-labeled genomic DNA can be hybridized to the slide.

Random Nonamer Protocol

1. Prehybridize slides as described in the *Prehybridization* section of Labeling and Hybridization.
2. Prepare hybridization probe of random nonamers [5× SSC, 0.1 μg/μl poly(A), 50 mM Tris–HCl, pH 8, 0.2% Sodium Dodecyl Sulfate (SDS), 150 pmol of Cy3 labeled nonamers].
3. Place at 100° for 5 min. Spin down briefly.
4. Apply probe to array and cover with coverslip.
5. Hybridize 10 min at room temperature.
6. Perform first wash at room temperature for 5 min in 2× SSC, 0.2% SDS.
7. Perform second wash in at room temperature for 5 min in 0.05× SSC.
8. Dry slides in 50-ml conical tube as described in Labeling and Hybridization.

Genomic DNA Protocol

1. Digest 10 μg of genomic DNA with a restriction enzyme that recognizes a 4-bp sequence (i.e., *Hpa*II).
2. Purify digested DNA using the MinElute PCR purification kit (Qiagen).
3. Add H_2O to a final volume of 21 μl.
4. Add 15 μl of 1 μg/μl random hexamers.
5. Add 5 μl of 10X reaction buffer (500 mM Tris–HCl, pH 7.0, 50 mM MgCl$_2$, 100 mM 2-mercaptoethanol).
6. Incubate at 100° for 5 min and place on ice.
7. Add 5 μl of 10X dNTP mix (1.2 mM dATP, dGTP, dCTP, 0.6 mM dTTP).
8. Add 3 μl of Cy3-dUTP.

9. Add 1 μl of high-concentration Klenow (50 U/μl). Incubate at 37° for 2 h.
10. Add 5 μl of 0.5 M EDTA, pH 8, to stop the reaction.
11. Remove unincorporated nucleotides using the MinElute PCR purification kit (Qiagen).
12. Hybridize to array as described in Labeling and Hybridization.

Control Spots

Control spots on DNA microarrays are generally sequences that have no significant similarity to the genome sequence of the organism being studied. They can serve as either positive and/or negative controls on the array. To assess labeling efficiency, *in vitro*-transcribed RNA corresponding to a control spot can be added to the labeling reaction. Printing a single control several times in a random pattern throughout the array can detect spatial bias in the hybridization. If the expression ratio of the control spot changes greatly depending on its location on the slide, then there is likely a spatial bias in the expression ratios. Using multiple controls can help determine the limit of detection on the array by doping in decreasing amounts of control in the samples. Control spots can also serve as a negative control in that they should not hybridize to any sequence present in the experimental sample. Hybridization to a negative control indicates that a significant nonspecific hybridization is occurring and the stringency of the washing conditions should be increased.

Experimental Design

A carefully crafted experimental design is critical to achieving meaningful results from a microarray experiment. Consideration must be given to the question being asked as well as to how the comparisons on the microarrays will be performed. Is the question being asked most easily answered using DNA microarrays? Which samples should be compared on the arrays? How should the hybridizations be performed? How will the data be analyzed? This section gives a brief discussion of different technical designs for microarray experiments. A more thorough discussion can be found elsewhere.[14,15]

Three types of experimental design are used in microarray experiments: direct comparison, reference comparison, and loop design. The first method, direct comparison, simply compares two different samples directly

[14] Y. H. Yang and T. Speed, *Nature Rev. Genet.* **3,** 579 (2002).
[15] G. A. Churchill, *Nature Genet.* **32**(Suppl.), 490 (2002).

on a single array (i.e., wild type vs mutant). This method is the easiest to understand in that it produces meaningful differential expression ratios from a single array. A potential drawback of this method is cost; when the number of samples to be analyzed is large, it becomes too costly to do a direct comparison of all samples. In the second method, reference comparison, all experimental samples are hybridized to a common reference RNA or DNA. The benefit of using a common reference is that many experimental samples can be compared to one another simply by subtracting the expression ratio of one array from another (after log transformation). In using this method it is important to consider whether to use a reference RNA or DNA and how plentiful is the supply of the reference. Genomic DNA has been shown to work well as a reference for bacteria.[16] A significant trade-off in using a reference system is that measurement errors are magnified (two arrays are needed for generation of an expression ratio), causing data to be less statistically robust. Finally, loop designs are an alternative to the designs just discussed. For more information about loop designs, see Yang and Speed.[14] Regardless of the experimental method chosen for performing the hybridizations, the most important factor in experimental design is generating biological replicates.[15]

Labeling and Hybridization

Growth of Cells and RNA Preparation

The single most important factor in achieving reproducible results is to ensure that all of the biological replicates (independently grown bacterial cultures) are grown as close to identically as possible. Factors such as the temperature of the water bath, the speed at which the cultures were shaken, size of the flask and the volume of medium in the flask, and the cell density at which the cultures were taken for RNA isolation should be monitored and recorded carefully. The amount of bacteria needed for a sufficient amount of RNA for array analysis should be determined empirically for each experimental system.

Bacteria respond rapidly to their environment and therefore it is important to arrest cellular activity when taking samples for microarrays, as results may reflect variations in the response to liquid handling and centrifugation rather than the biological question being addressed. To overcome this problem for *B. subtilis*, cell samples are mixed immediately with an equal volume of $-20°$ methanol to cease growth and metabolic activity.

[16] A. M. Talaat, *et al.*, *Nucleic Acids Res.* **30**, e104 (2002).

Other methods such as freezing cultures directly in liquid nitrogen or direct submersion in hot phenol can also be used.

Growth and Preparation of Cells

1. Grow cells to the desired point.
2. Mix an equal volume of the sample with $-20°$ methanol. Let sit for 1–2 min.
3. Centrifuge to pellet cells for 5 min.
4. Decant supernatant and freeze pellet at $-80°$.
5. Isolate RNA using a preferred method or kit. Be sure to remove any excess methanol prior to RNA isolation.

Total RNA works well for most applications, and although stable RNA can make up 80–90% of the total RNA in the cell, it is not necessary to remove stable RNA before labeling. Kits designed for rapid isolation of DNA-free bacterial total RNA are available. The RNeasy kit from Qiagen yields high-quality RNA from *B. subtilis* that performs well in microarray applications. Often the concentration of the RNA isolated is too low to be used in subsequent labeling. Use the following steps to precipitate and concentrate RNA.

Concentration of RNA

1. Mix the RNA sample with an equal volume of 4 M LiCl$_2$.
2. Add 2–3 volumes of 100% ethanol. Mix well.
3. Incubate at $-20°$ for at least 1 h.
4. Pellet RNA by centrifugation for at least 15 min at $4°$.
5. Wash pellet with cold 70% ethanol.
6. Pellet RNA by centrifugation for 5 min at $4°$.
7. Remove supernatant and air dry.
8. Resuspend pellet in RNase-free water to a concentration of 2–5 μg/μl.

Labeling: Indirect Incorporation Using Aminoallyl dUTP

The two methods available for generating labeled cDNA from RNA are direct labeling incorporating Cy3 or Cy5 dUTP during reverse transcription and indirect labeling using aminoallyl dUTP followed by chemical coupling of Cy3 or Cy5 dyes via the aminoallyl group. Indirect labeling offers several advantages over direct incorporation, including (1) Cy3 and Cy5 are not incorporated equally into cDNA during direct incorporation, (2) dye swapping experiments do not require a new reverse transcription step, and (3) alternative dyes to Cy3 and Cy5, such as Alexa546 and Alexa647 (Molecular Probes), can be used without major changes to the protocol.

Materials

1. 25X dNTP stock solution with aminoallyl-dUTP: 12.5 mM dATP, dCTP, dGTP, 2.5 mM dTTP, 10 mM aminoallyl-dUTP. (Optimal ratio of dTTP:aminoallyl-dUTP should be determined empirically for each system).
2. Superscript II reverse transcriptase (200 U/μl), 5X RT buffer, 0.1 mM dithiothreitol (DTT) (GIBCO).
3. RNase Out (40 U/μl) (GIBCO)
4. 0.1 M NaOH
5. 0.1 M HCl
6. Qiagen QIAQuick PCR purification kit and Minelute PCR cleanup kit. Substitute 75% ethanol for PE buffer. Do not use PE buffer. The free amines from the Tris in the buffer could interfere with the coupling reaction.
7. 1 M NaHCO$_3$, pH 9.0
8. Monofunctional NHS-ester Cy3 and Cy5 (Amersham). Dyes are resuspended in 20 μl of DMSO; 1 μl of freshly dissolved dye is used for coupling. Unused dye is aliquoted into 2-μl aliquots in separate tubes and dried down in a Speed-Vac and stored descicated, light protected, at $-20°$.
9. 4 M hydroxylamine

Reverse Transcription Reaction

1. Combine in PCR tubes: 10 μg of total RNA, 5 μl of random hexamers (0.5 μg/μl), 1 μl control RNA (optional), and x μl of H$_2$O to a final volume of 17.8 μl.
2. Incubate at 70° for 10 min. Cool on ice for 5 min.
3. Make nucleotide mix for reactions and store on ice. Components of NT mix for one reaction are 6 μl RT buffer, 1.2 μl 25X dNTP stock solution, 3 μl 0.1 M DTT, 1.5 μl Superscript II RT (200 U/μl), and 0.5 μl RNase Out (optional).
4. Add 12.2 μl of nucleotide mix to each reaction and mix.
5. Incubate at 25° for 10 min at 42° for 70 min, and at 70° for 15 min.

Hydrolysis of RNA

1. Degrade RNA by adding 15 μl of 0.1 M NaOH. Incubate at 70° for 10 min.
2. Neutralize by adding 15 μl of 0.1 M HCl.
3. Add 40 μl H$_2$O to each reaction to achieve a final volume of 100 μl.

Cleanup: Modified Qiagen Minelute QIAQuick Protocol

1. Add 500 μl of buffer PB to the reaction and mix.
2. Apply to spin column and centrifuge for 1 min. Discard flow through.
3. Wash with 750 μl 75% ethanol. Centrifuge for 1 min. Discard flow through and repeat. Do not use PE buffer as it may contain free amines that will interfere with the coupling reaction.
4. Centrifuge column an additional 1 min. Transfer column to a 1.5-ml microfuge tube.
5. Add 10 μl of H_2O. Wait 1–2 min. Centrifuge for 1 min.

At this point the reaction may be stored at $-20°$ or may be used immediately.

Coupling. All subsequent steps should be carried out in the dark or protected from direct fluorescent light or sunlight to limit photobleaching of the dyes.

1. Add 0.5 μl of 1 M NaHCO$_3$ to the 10-μl reaction.
2. Add 1 μl of freshly dissolved Cy3 or Cy5 dye. If using previously dried Cy dye, then transfer reaction to a dried aliquot of Cy dye and mix.
3. Incubate at room temperature for 1 h. Mix every 15 min. Reactions should be stored in the dark.

Quenching and Cleanup

1. Add 4.5 μl 4 M hydroxylamine. Mix. Incubate reaction in the dark at room temperature for 15 min.
2. Use the QIAQuick MinElute PCR purification kit (Qiagen) to remove unincorporated Cy dyes.

   A. Add 70 μl of H_2O to Cy3 or Cy5 reaction and mix. Combine with the other Cy5 or Cy3 reaction.

   B. Add 500 μl of buffer PB.

   C. Apply to the QIAQuick column and centrifuge for 1 min. Discard flow through.

   D. Wash with 750 μl of buffer PE. Centrifuge for 1 min. Discard flow through.

   E. Centrifuge for 1 min to dry column. Transfer column to a 1.5-ml microfuge tube.

   F. Add 12 μl of H_2O to membrane and leave at room temperature for 3 min.

   G. Centrifuge for 1 min.

   H. Reuse flow through from step F and repeat elution.

Hybridization of Probe to Microarrays Slides

Materials

20× SSC
10% SDS
Bovine serum albumin (BSA) (Sigma)
Nitrogen gas for drying slides
Salmon testes DNA (9.9 mg/ml, Sigma)
Yeast tRNA (10 mg/ml, Invitrogen)
2X hybridization buffer (0.2% SDS, 10× SSC, 50% formamide)
Hybridization chambers (Corning)
Staining jars (Sigma)
Coverslips

Prehybridization

1. Prehybridize slide in 5X SSC, 0.1% SDS, 1% BSA. Incubate at 42° for at least 45 min.
2. Wash slide by dunking several times in ddH₂O.
3. Stuff a Kim Wipe in the bottom of a 50-ml conical tube. Place slide in the tube and centrifuge at low speed in a swinging bucket rotor for 3–5 min. If slides are still wet, blow nitrogen gas over the surface of the slide until completely dry.
4. Place slide in hybridization chamber. Use within 1 h.

Hybridization. The following protocol works well for an array that can be covered with a 22×40-mm coverslip (32 pin print layout). The amounts listed here may be scaled up or down for different sized arrays.

1. Use labeled probe eluted from the MinElute column. Adjust to 11 μl with H₂O if necessary.
2. Place 10 μl in a screw cap tube. Save 1 μl to determine amount of cDNA and Cy dye incorporation.
3. Add 1 μl each of salmon sperm DNA and yeast tRNA.
4. Denature by placing at 100° for 5 min. Spin down briefly.
5. Add 12 μl of 2x hybridization buffer (stored at 52°). Mix by pipetting.
6. Place probe in the center of the array. Having a template slide with the position of the printed area marked with a pen to identify the position of the array on the slide is useful.
7. Place coverslip on top of the array. The probe will spread across the entire coverslip. Adjust the coverslip to fit over the array properly. Most bubbles that develop will float to the edge of the coverslip and air will escape. Watch bubbles carefully; if they do not move, tap

the top of the coverslip lightly until they work to the edge of the coverslip.

8. Add 10 μl of water to the wells at each end of the hybridization chamber. Close chamber carefully.

9. Incubate overnight in a 42° water bath or incubator.

Washing

1. Prepare four solutions in staining jars.
   Solution 1: 1X SSC, 0.2% SDS; prewarm to 42°
   Solution 2: 0.1X SSC, 0.2% SDS
   Solutions 3 and 4: 0.1X SSC

2. Place slide into a slide rack and submerge up and down into solution 1 until coverslip falls off.

3. Leave slide submerged in solution 1 at 42° for 4 min with light shaking.

4. Place in solution 2, dip slide up and down five times, and then place at room temperature for 4 min with light shaking.

5. Wash in solution 3 by dunking up and down at least 10 times. Place in solution 4 and incubate 4 min with light shaking at room temperature. The reason for having both solutions is to be sure that no SDS will carry over to the final wash.

6. Place slide in a 50-ml conical tube that has a KimWipe in the bottom of the tube.

7. Spin at low speed in a swinging bucket rotor to remove liquid.

8. If necessary, blow nitrogen gas over slide to dry completely.

9. Slide is now ready for scanning.

Data Analysis and Storage

Microarray experiments yield tremendous amounts of information and the analysis and storage of the data can be overwhelming. Although there is no agreed upon "gold standard" for the analysis of microarray data, many public and commercial tools are now available. Some brief observations about the analysis of microarray data are discussed.

Normalization

Microarray data must be normalized after scanning to account for a number of variables, including the amounts of starting RNA and the labeling efficiency of samples. Normalization of microarray data is relatively straightforward for complete genome bacterial DNA microarrays. Because all genes (excluding stable RNA genes) are being probed on the array, the readout of a single channel is representative of the mRNA pool from that sample. Therefore, the channels can be normalized to the total signal of each channel on the array. Consider an example where two samples are being compared on a microarray, sample A and sample B. Sample A is labeled with Cy3 and sample B is labeled with Cy5. After scanning and image analysis, the sum intensity from all of the genes from sample A is 1,000,000 and from sample B is 800,000. By multiplying the signal for each spot in sample B by 1.25 (sample A/sample B), the total signal of sample B will be 1,000,000 and now a meaningful comparison of the expression ratios can be made. This type of normalization does not account for intensity and spatial biases in data. Methods for dealing with these problems (such as Lowess normalization) are now being developed and utilized.[17]

Because virtually every gene from the genome is present on the array, the total signal from a sample can be considered the entire mRNA pool from that sample. Thus each gene is being represented as a fraction of the total mRNA pool. Direct comparisons regarding differential gene expression can be made by normalizing each sample to the same total signal. This should not be confused with the total mRNA per cell. Because the fraction of stable RNA:mRNA can vary greatly depending on growth conditions, it is not advisable to extrapolate array data to transcripts per cell without additional experimentation. For further information about data normalization please see a review by Quackenbush.[18]

Identification of Differentially Expressed Genes

A major challenge of DNA microarray expression analysis is determining which genes are significantly differentially expressed when comparing one sample to another (often referred to as *outliers*). Many projects utilize arrays to identify leads for further study and therefore the need to limit the amount of false positives is important to prevent focusing on irrelevant genes. Large changes in gene expression are generally easy to detect and confirm. However, smaller changes (two-fold) in gene expression that can have profound effects on phenotype are often difficult to distinguish from the inherent noise of microarrays. Thus the ability to detect these small changes must be balanced with the number of false positives identified in the data set.

Currently there is no single best approach for the identification of differentially expressed genes. Initial studies arbitrarily assigned fold changes of anywhere from two- to fivefold fold as being significant. This approach

[17] Y. H. Yang et al., Nucleic Acids Res. 30, e15 (2002).
[18] J. Quackenbush, Nature Genet. 32(Suppl.), 496 (2002).

should no longer be used because it provides no information about the reproducibility of the measurements or the quality of data. More recent studies have implemented a wide variety of statistical methods for determining outliers. One program for determining outliers that is gaining popularity and merits mention is significance analysis for microarrays (SAM). SAM acts as a plug-in to the program EXCEL, is relatively easy to implement, and can be downloaded from http://www-stat.stanford.edu/~tibs/SAM/ free of charge for academic users. For more information on detecting differential gene expression and clustering (not discussed in this article), see a review by Slonim.[19]

Many other approaches are being developed, far more than can be discussed here. New methods are being reported at a rapid pace and the number of choices can be overwhelming. In choosing analysis methods, it is recommended that biologists with limited statistical backgrounds consult with a statistician in making this decision. From a practical standpoint, it is important that the user be comfortable with the statistical methods being used to analyze data.

Data Storage

Effective data storage is a major problem in the microarray field. Many laboratories rely on the individual user to store and manage the data, causing at least three considerable data management problems. First, other members of the laboratory cannot access their colleagues, experimental data without having to find the primary user and have them find the appropriate files. Second, consistent and accurate record keeping to relate microarray data with the experimental methods used to generate the biological samples is difficult. Third, when individuals leave the laboratory, it is almost impossible to keep track of data they generated if not stored in a database. Thus it is advisable to set up a database for microarray data management. There are several commercially available products for storing microarray data that also include analysis tools. For a complete list of commercial and academic database products, refer to Holloway et al.[13]

In an effort to establish standards in the microarray field, the Microarray Gene Expression Data society was formed (http://www.mged.org). The society is developing the MIAME standard for microarray data (Minimal Information About a Microarray Experiment). This standard is being adopted by many journals for greater ease of interpretation and comparison of microarray experiments. Any database program designed for storing microarray data and information should strongly consider incorporating these standards in their database design.

[19] D. K. Slonim, *Nature Genet.* **32**(Suppl.), 502 (2002).

Concluding Remarks

DNA microarrays hold much promise as a tool to approach biological problems on a genome-wide level. The ability to investigate regulatory networks on a whole-genome level is already providing key insights into transcription factor function. The microarray field is a fast-changing area in which many new technical and statistical tools are being developed at a rapid pace. Because of the diverse technical and analytical challenges encountered when using microarrays, a team effort with expertise in biology, bioinformatics, and statistics is required to successfully utilize this technology and incorporate new advances.

Acknowledgments

I thank Elke Kuster-Shuck and Laura Schaefer for comments and editing of this chapter.

[24] Analysis of Microarray Data for the *marA*, *soxS*, and *rob* Regulons of *Escherichia coli*

By ROBERT G. MARTIN and JUDAH L. ROSNER

An important problem in gene regulation is how to determine which genes are regulated transcriptionally by a given signal. For this, microarray technology has great promise because it has the potential to identify, in just a few experiments, a large number of the genes/promoters affected by the environmental stimulus or the regulatory protein or RNA.[1,2] However, it is often difficult to sift through the noise generated by technical factors (e.g., the isolation of the mRNA, the hybridization to the DNA) and focus on the significantly affected promoters. New experimental, statistical, and computational methods are expected to improve the technique but may not resolve all issues. Ultimately, each candidate gene or promoter has to be confirmed directly.

Study of the *marA*, *soxS*, and *rob* regulons of *Escherichia coli* illustrates one approach for using microarray results.[3] These regulons are highly overlapping (and therefore referred to here as the *marA/soxS/rob* regulon) because the transcriptional activators that directly control them, MarA, SoxS,

[1] M. Schena, D. Shalon, R. W. Davis, and P. O. Brown, *Science* **270**, 467 (1995).
[2] C. S. Richmond, J. D. Glasner, R. Mau, H. Jin, and F. R. Blattner, *Nucleic Acids Res.* **27**, 3821 (1999).
[3] R. G. Martin and J. L. Rosner, *Mol. Microbiol.* **44**, 1611 (2002).

and Rob, are highly homologous and bind with roughly similar affinities to a 20-bp asymmetric degenerate sequence.[4–7a] However, each activator is regulated differently. *marA* expression is repressed by MarR and is derepressed by the interaction of MarR with compounds such as salicylate.[3,8] *soxS* expression is activated by an oxidized form of SoxR, which can be achieved by treating cells with the superoxide-generating agent paraquat.[9] Rob activity is enhanced posttranslationally by treatment of cells with 2,2'- or 4,4'-dipyridyl, an effect mediated by the C-terminal domain of Rob.[10] Thus, each treatment turns on one of the activators, which can be expected to transcriptionally activate one set of genes that will be common to all treatments and separate sets of genes peculiar to each treatment.

Data were pooled from three microarray conditions, which used arrays from the same manufacturer (Sigma-Genosys) and followed similar protocols. The microarray study of Barbosa and Levy[11] compared transcripts from a *mar*-deleted strain and a strain expressing MarA constitutively and found that 47 genes had elevated transcription. The microarray study of Pomposiello *et al.*[12] examined two conditions: the effects of salicylate and of paraquat treatments, which increase the expression of *marA* and *soxS*, respectively. Elevated transcription of 84 and 66 genes was found for salicylate- and paraquat-treated cells, respectively. Together, the mRNA levels of 153 different genes (~136 promoters) were elevated significantly under at least one of three conditions in the two studies.

Only 28 promoters were identified by at least two of the microarray conditions.[3] Of these, 8 were among the 16 promoters identified previously by standard techniques as being activated directly by MarA, SoxS, or Rob. (The transcript for a 17th regulon member, the RNA gene, *micF*, could not have been found, as the microarrays did not include the complementary single-stranded DNA.) In fact, only 7 of the 20 promoters proved to be activated directly when tested by standard molecular biology methods. However, the remaining 13 cloned promoters did not respond to salicylate,

[4] B. Dangi, P. Pelupessey, R. G. Martin, J. L. Rosner, J. M. Louis, and A. M. Gronenborn, *Mol. Biol.* **314,** 113 (2001).
[5] K. L. Griffith and R. E. Wolf, Jr., *Mol. Microbiol.* **40,** 1141 (2001).
[6] R. G. Martin, W. K. Gillette, S. Rhee, and J. L. Rosner, *Mol. Microbiol.* **34,** 431 (1999).
[7] R. G. Martin, W. K. Gillette, and J. L. Rosner, *Mol. Microbiol.* **35,** 623 (2000).
[7a] R. G. Martin and J. L. Rosner, *Proc. Natl. Acad. Sci. USA* **92,** 5456 (1995).
[8] M. N. Alekshun and S. B. Levy, *Antimicrob. Agents Chemother.* **41,** 2067 (1997).
[9] B. Demple, *Gene* **179,** 53 (1996).
[10] J. L. Rosner, B. Dangi, A. M. Gronenborn, and R. G. Martin, *J. Bacteriol.* **184,** 1407 (2002).
[11] T. M. Barbosa and S. B. Levy, *J. Bacteriol.* **182,** 3467 (2000).
[12] P. J. Pomposiello, M. H. J. Bennik, and B. Demple, *J. Bacteriol.* **183,** 3890–3902.

paraquat, or dipyridyl, which suggests that they are not activated even indirectly by these treatments.

A bioinformatic approach can also be useful for analyzing microarrays. Because the MarA, SoxS, or Rob activator-binding site must be configured appropriately relative to the RNA polymerase-binding site,[6,13] we could devise an algorithm for finding such activator sites relative to the first nucleotide of the open reading frame. This proved fairly effective. Of the 9 genes judged positive by the algorithm, 6 were found to be activated transcriptionally by MarA, SoxS, or Rob, and of the 11 judged negative, 10 were found not to be activated.[3]

Application of the bioinformatic algorithm to 118 genes identified in only one of the three microarray conditions now suggests that about 30 may prove to be authentic members of the regulon. This would increase the regulon size to about 55 rather than our previous estimate of less than 40.[3] Thus, while the majority of genes identified by the microarray studies do not appear to be true regulon members, bioinformatic and standard molecular biological approaches can be used to help distinguish signal from noise. However, until such analysis is undertaken, caution should be exercised in interpreting microarray data.

[13] T. I. Wood, K. L. Griffith, W. P. Fawcett, K.-W. Jair, T. D. Schneider, and R. E. Wolf, Jr., Mol. Microbiol. **34,** 414 (1999).

Section III

Polymerase Associated Factors

[25] Purification and Activity Assays of RapA, the RNA Polymerase-Associated Homolog of the SWI/SNF Protein Superfamily

By Maxim V. Sukhodolets, Susan Garges, and Ding Jun Jin

The RNA polymerase (RNAP)-associated protein, ATPase RapA[1] is a bacterial homolog of the SWI/SNF protein family.[2,3] Eukaryotic representatives of this family have been implicated in chromatin remodeling and regulation of gene expression.[4-6] Studies indicate that eukaryotic SWI/SNF proteins can alter the configuration of naked DNA[7,8]; however, the specific role of helicase-like motifs—a characteristic feature of SWI/SNF proteins—has yet to be clarified. Under appropriate conditions, RapA stimulates RNAP activity dramatically by enabling RNAP recycling[9]; models for RapA catalysis suggest that the enzyme promotes dissociation of one or more components of the transcript RNA–RNAP–DNA complex.[9]

The protocol for purification of native RapA from the RNAP holoenzyme–RapA complex described in this article incorporates a number of classic purification steps that have been gradually introduced into laboratory routine over a period of decades. In 1975, Burgess and Jendrisak used a combination of polymin P precipitation and chromatography on immobilized single-stranded DNA for the purification of *Escherichia coli* RNA polymerase[10]; further improvement of the purification procedure was made in 1990 by Hager *et al.*[11] with the introduction of Mono Q chromatography that allowed separation of various forms of RNAP, such as core RNAP, RNAP holoenzyme, and the RNAP holoenzyme–RapA complex.[1] The purification procedure described here has also revealed a number of accessory proteins that copurify with RNAP; some of these proteins (such as

[1] M. V. Sukhodolets and D. J. Jin, *J. Biol. Chem.* **273,** 7018 (1998).
[2] P. Bork and E. V. Koonin, *Nucleic Acids Res.* **21,** 751 (1993).
[3] A. Kolsto, P. Bork, K. Kvaloy, T. Lindback, A. Gronstadt, T. Kristensen, and C. Sander, *J. Mol. Biol.* **230,** 684 (1993).
[4] C. L. Peterson, *Curr. Opin. Genet. Dev.* **6,** 171 (1996).
[5] C. Muchardt and M. Yaniv, *J. Mol. Biol.* **293,** 187 (1999).
[6] M. J. Pazin and J. T. Kadonaga, *Cell* **88,** 737 (1997).
[7] K. Havas, A. Flaus, M. Phelan, R. Kingston, P. A. Wade, D. M. Lilley, and T. Owen-Huges, *Cell* **103,** 1133 (2000).
[8] I. Gavin, P. J. Horn, and C. L. Peterson, *Mol. Cell* **7,** 97 (2001).
[9] M. V. Sukhodolets, J. E. Cabrera, H. Zhi, and D. J. Jin, *Genes Dev.* **15,** 3300 (2001).
[10] R. R. Burgess and J. J. Jendrisak, *Biochemistry* **21,** 4634 (1975).
[11] D. A. Hager, D. J. Jin, and R. R. Burgess, *Biochemistry* **29,** 7890 (1990).

NusA and RapA) have been shown to form complexes with RNAP and participate in various stages of the transcription cycle; a number of other protein contaminants identified by N-terminal protein sequencing (whose copurification with RNAP may be the result of interaction with polymerase) can be considered a subject for more detailed future studies. A complete list of proteins that copurify with RNA polymerase in a similar purification protocol is provided elsewhere.[12]

Purification of RapA from RapA–RNAP Holoenzyme Complex

Approximately 100 g of frozen cell pellets, harvested from *E. coli* MG 1655 cell cultures grown in superbroth to an OD_{600} 4–6, is mixed with 300 ml of grinding buffer [50 mM Tris–HCl, pH 8, 5% glycerol, 2 mM EDTA, 230 mM NaCl, 1 mM 2-mercaptoethanol, 0.023 mg/ml phenylmethylsulfonyl fluoride, 0.015 mg/ml dithiothreithol (DTT), 0.26 mg/ml lysozyme] in a Waring blender and homogenized at low speed for 2–3 min until the cells are completely resuspended and the temperature has increased to 2–5°. After 20 min, 8 ml of 10% deoxycholate (sodium salt) is gradually added to lyse the cells, and the mixture is blended again for 30 s at low speed. After 20 min at 8–12°, the mixture is blended at high speed for 1 min to shear the DNA; then 400 ml of TGED buffer (10 mM Tris–HCl, pH 8, 5% glycerol, 0.1 mM EDTA, 0.015 mg/ml DTT) containing 0.2 M NaCl is added, and the mixture is blended again at high speed for 1 min. The cell extract is centrifuged at 4° for 1 h at 10,000 g, and the supernatant is collected, typically yielding 730–750 ml. The extract is again placed in a Waring blender, and 29 ml of 10% polymin P (polyethyleneimine; Sigma P3143), pH 7.8, is added gradually with a constant slow stirring. After 5 min of slow stirring, the mixture is centrifuged for 15 min at 6000 g, and the supernatant fraction is poured off and discarded. The drained pellet is resuspended in 600 ml of TGED containing 0.5 M NaCl. After 8 min of slow stirring, the suspension is centrifuged again for 15 min at 6000 g, and the supernatant (the 0.5 M NaCl wash) is discarded. The washed pellet is then again scraped into the blender and resuspended in 500 ml of TGED containing 1 M NaCl with gentle stirring for 8 min. The mixture is centrifuged for 30 min at 10,000 g, and the supernatant is collected, yielding 460–480 ml of a 1 M NaCl eluate. Solid ammonium sulfate (36–38 g/100 ml) is then added to the 1 M NaCl eluate; at this stage the precipitated protein can be left overnight. Approximately 30 ml of DNA agarose (single-stranded DNA agarose, Pharmacia) is then packed in an XK-26 column (Pharmacia) and preequilibrated with TGED buffer

---

[12] D. J. Jin, H. Zhi, and W. Yang, *Methods Enzymol.* **370**, [26], this volume (2003).

containing 0.2 M NaCl. The ammonium sulfate pellets from the previous step are dissolved in TGED buffer to a conductivity corresponding to that of TGED containing 0.2 M NaCl, and the protein solution is loaded on the DNA agarose column at a flow rate of 1.5–2 ml/min. The DNA agarose column is washed with 300–400 ml of TGED containing 0.2 M NaCl, and the bound protein (predominantly RNAP, plus some DNA-binding proteins) is eluted with TGED containing 1 M NaCl; the protein peak (80–100 ml) is subsequently collected, and the protein is precipitated with ammonium sulfate (36–38 g/100 ml). A Mono Q HR 10/10 column (Pharmacia) is then preequilibrated with TGED buffer containing 0.1 M NaCl. The ammonium sulfate pellets from the previous step, obtained by a brief centrifugation for 10 min at 6000 g, are dissolved in TGED containing 0.1 M NaCl (the undissolved protein can be removed by centrifugation for 15 min at 10,000 g) and loaded on the Mono Q column at a flow rate of 1 ml/min. Once the protein fractions have been loaded, the column is washed with ~100 ml of TGED containing 0.1 M NaCl, and the Mono Q-bound proteins are eluted with a shallow linear NaCl gradient, typically 0.1 to 0.5 M NaCl in 500 ml of TGED; 10-ml fractions are collected. Fractions containing the protein peaks (Fig. 1B) identified from the A_{280} profile (Fig. 1A) are then pooled and concentrated (typically about 20-fold) using Centriprep YM-10 concentrators (Amicon). The concentrated RNAP–RapA fractions (fractions 37–39 in Fig. 1C) are then dialyzed against two changes (1 L/8 h each) of 50 mM MOPS, 0.2 mM EDTA, 0.1 mM DTT (buffer B) containing 1 M ammonium sulfate. The protein is then loaded on a phenyl-Superose 5/5 column (Pharmacia) preequilibrated with buffer B supplemented with 1 M ammonium sulfate. Bound proteins are then eluted with a reverse linear gradient of ammonium sulfate in buffer B (1 to 0 M ammonium sulfate in 50 ml of buffer B), and 2.5-ml fractions are collected (Fig. 1D, top). The RapA peak is typically eluted at 0.32–0.25 M ammonium sulfate. Fractions representing the RapA peak (fractions 13–15 in Fig. 1D, top) are concentrated to 100–150 μl using Centricon 10 concentrators (Amicon), and 50-μl aliquots are subjected to gel filtration on a Superose 6 HR 10/30 column in TGED containing 50 mM NaCl to remove traces of RNAP from the RapA preparation. The column is typically run at a flow rate of 0.25 ml/min, and 0.5-ml fractions are collected. The purified protein (fraction 30 in Fig. 1D, bottom)—nearly homogeneous as judged from silver-stained SDS gels—is concentrated to 1–2 mg/ml using Microcon YM-10 concentrators (Millipore), mixed with an equal volume of 100% glycerol, and stored at $-20°$. Typical yields are 200–400 μg of RapA per 100 g of cells.

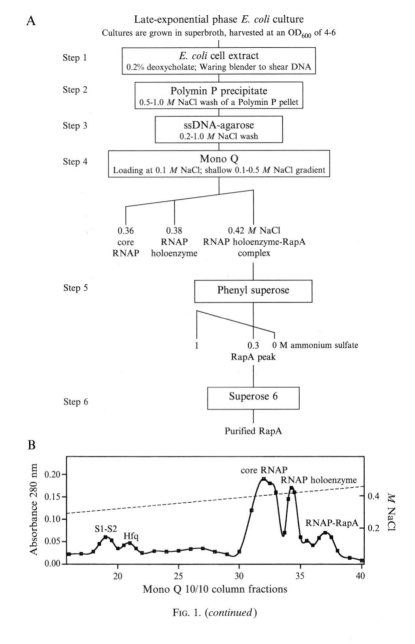

A

Late-exponential phase *E. coli* culture

Cultures are grown in superbroth, harvested at an OD_{600} of 4-6

Step 1
E. coli cell extract
0.2% deoxycholate; Waring blender to shear DNA

Step 2
Polymin P precipitate
0.5-1.0 *M* NaCl wash of a Polymin P pellet

Step 3
ssDNA-agarose
0.2-1.0 *M* NaCl wash

Step 4
Mono Q
Loading at 0.1 *M* NaCl; shallow 0.1-0.5 *M* NaCl gradient

0.36 core RNAP

0.38 RNAP holoenzyme

0.42 *M* NaCl RNAP holoenzyme-RapA complex

Step 5
Phenyl superose

1   0.3   0 M ammonium sulfate
RapA peak

Step 6
Superose 6

Purified RapA

B

FIG. 1. (*continued*)

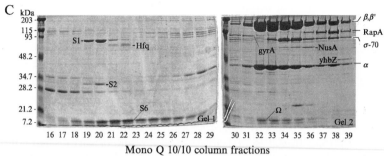

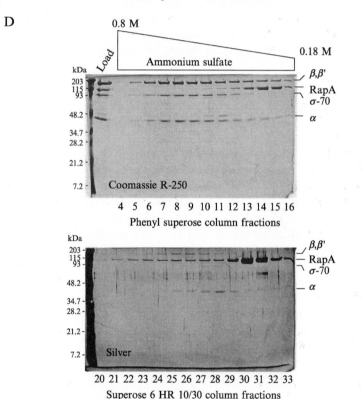

Fig. 1. (A) Schematic for the purification procedure of native RapA from the RNAP holoenzyme–RapA complex. (B) Mono Q column protein profile. (C) Mono Q column fractions stained with Coomassie brilliant blue R-250; 10 μl from each fraction was mixed with an equal volume of 2× Laemmli sample buffer and loaded per lane of a 10% SDS–polyacrylamide gel. Prestained protein markers (Bio-Rad, broad range) were loaded in the first lane of each gel. The S1, S2, S6, Hfq, GyrA, NusA, RNAP Ω subunit, and yhbZ proteins were identified based on the following N-terminal protein sequences: MTESFAQLFE (S1),

Catalytic Activities of RapA: RNAP-Stimulatory Effect during *in Vitro* Transcription and RNAP-Stimulated ATP Hydrolysis

RapA is capable of dramatic stimulation of RNAP transcriptional activity *in vitro*.[9] The stimulatory effect of RapA is not promoter specific; however, several other conditions have to be met in order to assay this enzymatic activity of RapA, including (i) a relatively high concentration of salt in the transcription reaction (various salts show a distinct, somewhat narrow range of concentrations for optimal RapA-mediated stimulation of transcription,[9] thus it is advisable to test a range of various salt concentrations in order to find the optimal conditions) and (ii) conditions favoring multiple-round transcription reactions, such as low molar ratios (<2) of RNAP:DNA template, the omission of heparin—a DNA competitor—or protein factors that may contribute to RNAP recycling.

In Vitro *Transcription*

The assay described here is performed with the supercoiled DNA template pCPGλtr2,[13] which contains the T7A1 promoter and the λtr2 terminator approximately 300 bp downstream. Mixing purified RNAP holoenzyme (a by-product of the RapA purification procedure), supercoiled DNA template, unlabeled rNTPs, and $[\alpha^{32}\text{-P}]$ATP or any other α-labeled rNTP of choice, in the absence or presence of RapA, results in labeled RNA synthesis *in vitro*. The transcription reaction products are then separated on a denaturing polyacrylamide gel in order to estimate the stimulatory effect of RapA on the levels of the promoter-specific transcript (Fig. 2A).

The 15-μl master reaction contains 2 μl of 20X NBT (NTPase-binding transcription) buffer (1X NBT buffer contains 50 mM Tris–HCl, pH 7.5, 2 mM MgCl$_2$, 0.1 mM EDTA, and 0.1 mM DTT), 2.2 μl of supercoiled plasmid DNA pCPGλtr2 (0.64 mg/ml), 2 μl of the RNAP holoenzyme at 0.1 mg/ml, 2 μl of purified RapA at ~0.5 mg/ml (or an equivalent amount of the storage buffer—1X TGED containing 100 mM NaCl and 40%

[13] R. Reynolds, R. M. Bermudez-Cruz, and M. J. Chamberlin, *J. Mol. Biol.* **224,** 31 (1992).

XTVSMRDMLK (S2), MRHYEIVFM (S6), AKGQSLQDPFL (Hfq), XDLAREITPVNI (GyrA), MNKEILAVVEAVSNE (NusA), ARVTVQDAVEKIGNR (Ω), and MKFVDEA-SILVVA (yhbZ). (D) Phenyl-Superose 5/5 column fractions stained with Coomassie brilliant blue R-250 (top) and Superose 6 HR 10/30 column fractions stained with silver (bottom). Protein samples for denaturing electrophoresis were made as described (C), and the Superose 6 fraction containing the purified RapA protein used in the enzymatic assays is indicated. (See color insert.)

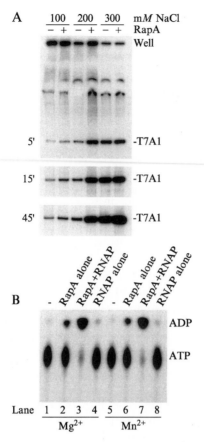

FIG. 2. The catalytic activities of RapA: RNAP-stimulatory effect during *in vitro* transcription and RNAP-stimulated ATP hydrolysis. (A) *In vitro* transcription reactions show a salt-specific stimulatory effect of RapA on the levels of promoter-specific transcript synthesized. The position of the RNA transcript initiated from the T7A1 promoter is indicated. (B) *In vitro* ATPase reactions performed in the presence of 1 mM Mg^{2+} (lanes 1–4) or 1 mM Mn^{2+} (lanes 5–8) demonstrate ATP hydrolysis by RapA (lanes 2 and 6) and the RNAP–RapA complex (lanes 3 and 7). The specific activity of the RapA ATPase calculated from these data was $\sim$ 24 pmol of ATP hydrolyzed μg protein/min.

glycerol—in the control reaction), and 6.8 μl of purified water. The master reaction may also include 0.5 μl of RNAsin. Four-microliter aliquots of the master reaction (with or without RapA) are then mixed with 4-μl aliquots of 0.25 M NaCl, 0.5 M NaCl, and 0.75 M NaCl (for a total of six independent transcription reactions), and the reactions are preincubated for 15 min at 37°. (Preincubation is typically performed to achieve "open" complex formation; omitting this step will not have any significant impact on the transcript yields.) The *in vitro* transcription reaction is then initiated by the addition of 2 μl of 5X rNTP mix containing 1 mM each of ATP, GTP, CTP, and UTP and 0.2–1 μCi of [α^{32}-P]ATP. After 5, 15, or 45 min at 37°, a 2-μl aliquot is taken from each of the reactions and mixed with 4 μl of stop solution (250 mM EDTA, pH 8.0, 50% glycerol, 0.1% bromphenol blue) and 4 μl of formamide; 3-μl aliquots of the terminated reactions are then analyzed on 8% polyacrylamide–6 M urea gels. The labeled RNA transcripts are visualized by exposing imaging film to the gel (covered with plastic wrap) at −70° (Fig. 2A); the levels of the transcripts can be estimated using a phosphorimager.

ATPse Assay

The nature of the biochemical reaction coupled to RapA-mediated ATP hydrolysis remains unclear; however, formation of a RapA–RNAP complex moderately stimulates the ATPase activity of RapA.[1] In order to assay this effect, the products of *in vitro* ATPase reactions containing the purified enzyme(s) and [α^{32}-P]ATP are separated by thin-layer chromatography on polyethyleneimine (PEI)-cellulose plates, and the plates are developed in 1 M formic acid/1 M LiCl. This technique allows efficient separation of ADP (the reaction product) from ATP (Fig. 2B); quantitation of the amount of ATP hydrolyzed gives the specific ATPase activity of RapA, which is typically 20–30 pmol ATP hydrolyzed per μg protein/min.[1,9]

The 10-μl ATPase reactions contain 1 μl of 10X NBT buffer (reactions 1–4) or 1 μl of 10X NBT/Mn^{2+} buffer (reactions 5–8; NBT/Mn^{2+} buffer is similar to NBT buffer, but contains MnCl$_2$ instead of MgCl$_2$), 1 μl of 0.5 M NaCl, 1 μl of RapA (at ~0.1 mg/ml; reactions 2, 3, 6, and 7), 1 μl of the RNAP holoenzyme (at 0.5 mg/ml; reactions 3, 4, 7, and 8), 1 μl of 1 mM ATP (containing 0.05–0.1 μCi of [α^{32}-P]ATP), and purified water to a final volume of 10 μl. Reactions are incubated for 60 min at 37°, and a 2-μl aliquot from each reaction is spotted on a PEI-cellulose plate. The plate is air dried for 15 min, placed in a tray and rinsed briefly (using a tabletop shaker) once with water and once with methanol, dried for 30 min, and then developed in 1 M formic acid/1 M LiCl. The developed plate is air dried briefly (10–15 min) and autoradiographed at −70° (Fig. 2B; the typical exposure times are 5–30 min). The specific ATPase activity of RapA can be calculated after quantitation of the amount of ATP hydrolyzed in the reaction using a phosphorimager.

[26] *Escherichia coli* Proteins Eluted from Mono Q Chromatography, a Final Step During RNA Polymerase Purification Procedure

By HUIJUN ZHI, WENXUE YANG, and DING JUN JIN

Preparation of highly pure and active RNA polymerase (RNAP) is an essential step for biochemical and biophysical analysis of transcription. In *Escherichia coli*, the basic transcription machinery is core RNAP, which consists of $\alpha_2\beta\beta'$. Although core RNAP is capable of elongation and termination at simple terminators, it requires the binding of a sigma factor to form a holoenzyme ($\alpha_2\beta\beta'\sigma$) in order to be able to initiate transcription at promoters on DNA templates.[1,2] Among many sigma factors in *E. coli*, σ^{70} is the major one and is responsible for the expression of housekeeping genes.[3] Thus, a typical RNAP preparation procedure usually results in the purification of core RNAP and holoenzyme containing the σ^{70} polypeptide.[4] The major principle for the purification of RNAP is to take advantage of its ability to bind to DNA [either double stranded or single stranded (ss)]. Because core RNAP is in excess of σ^{70} in the cell (about 2:1), in order to obtain a holoenzyme with a stoichimetric amount of σ^{70}, it is important to separate core RNAP from the holoenzyme during preparations. This can be accomplished by the use of Mono Q chromatography with a shallow linear gradient.[5]

We have been routinely purifying RNAP from wild-type cells and different RNAP mutants to study the mechanisms of transcription using a preparation procedure outlined in Fig. 1. This procedure is based on the established methods as described,[5] with some modifications: (1) a single-stranded DNA agarose column is used instead of the DNA-cellulose column, (2) the Sephacryl S-300 size column is omitted before the Mono Q column, and (3) the gradient condition for Mono Q chromatography is changed. The reason for the modifications is to speed up the purification process while maintaining the high quality of RNAP. This article describes the detailed procedure of purification of RNAP.

[1] R. R. Burgess, A. A. Traver, J. J. Dunn, and E. K. F. Bautz, *Nature (Lond.)* **221,** 43 (1969).
[2] J. D. Helmann and M. J. Chamberlin, *Annu. Rev. Biochem.* **57,** 839 (1988).
[3] C. A. Gross, C. Chan, A. Dombroski, T. Gruber, M. Sharp, J. Tupy, and B. Young, *Cold Spring Harb. Symp. Quant. Biol.* **63,** 141 (1998).
[4] R. R. Burgess and J. J. Jendrisak, *Biochemistry* **14,** 4634 (1975).
[5] D. A. Hager, D. J. Jin, and R. R. Burgess, *Biochemistry* **29,** 7890 (1990).

Cell extract
Cells disrupted with lysozyme and NaDOC
DNA sheared by high speed blending

↓

Polymin P precipitation
RNAP precipitated as RNAP-DNA-polymin P complexes
0.5 M NaCl wash; 1 M NaCl elution

↓

ssDNA-agarose affinity column
RNAP bound to the column at 0.2 M NaCl; eluted in 1 M NaCl

↓

Mono Q chromatography
Linear gradient: 0.2 M to 0.5 M NaCl
Core RNAP eluted at 0.36 M, holoenzyme at 0.38 M, holo-RapA at 0.42 M NaCl

FIG. 1. Schematic for the purification of E. coli RNA polymerase.

In the final step of the RNAP purification procedure, Mono Q chromatography, the peaks corresponding to core RNAP, holoenzyme, and holoenzyme + RapA, which is a bacterial homolog of SWI2/SNF2,[6,7] were eluted at ~0.36, 0.38, and 0.42 M NaCl, respectively (Fig. 2), as reported.[5,6] In addition, a relatively small amount of σ^{32} was found to be present in the leading shoulder of the core peak (Fig. 2B, fraction 43), consistent with the observation of Hager et al.[5] However, many proteins other than RNAP are being eluted from the Mono Q column. The profiles of these proteins eluted from the Mono Q column are very reproducible. A representative elution profile of Mono Q chromatography is shown in Fig. 2A, and SDS-PAGE gels of the proteins from different fractions are shown in Fig. 2B. A list of these proteins is summarized in Table I.

Several proteins eluted from the Mono Q column are known transcription factors, including NusA, GreA, and ribosomal protein S4. NusA binds preferentially to core RNAP[8,9] and can be copurified with RNAP using an affinity-based purification procedure.[10] Results showed that NusA was eluted mainly with a holoenzyme peak (see Fig. 2B, fractions 49–52). It is possible that the core RNAP–NusA complex was eluted at the same position as the holoenzyme during Mono Q chromatography. GreA interacts with

[6] M. V. Sukhodolets and D. J. Jin, J. Biol. Chem. 273, 7018 (1998).
[7] M. V. Sukhodolets, J. E. Cabrera, H. Zhi, and D. J. Jin, Genes Dev. 15, 3330 (2001).
[8] J. Greenblatt and J. Li, Cell 24, 421 (1981).
[9] S. C. Gill, S. E. Weitzel, and P. H. von Hippel, J. Mol. Biol. 220, 307 (1991).
[10] N. E. Thompson, D. A. Hager, and R. R. Burgess, Biochemistry 31, 7003 (1992).

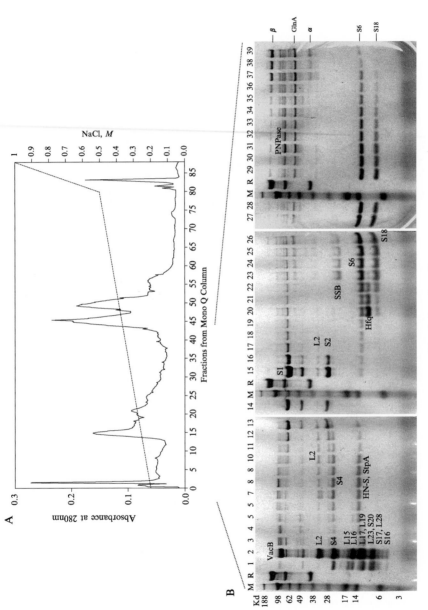

FIG. 2. (continued)

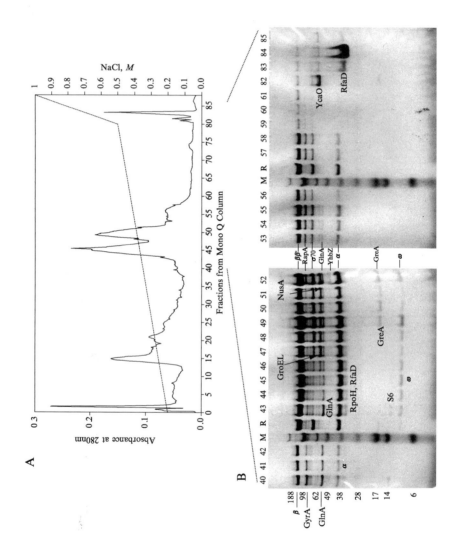

RNAP *in vitro*.[11,12] Like NusA, GreA was coeluted with the RNAP holoenzyme peak during Mono Q chromatography (see Fig. 2B, fractions 49–52). Ribosomal protein S4, which is reported to bind to RNAP and have antitermination function,[13] is eluted at the beginning of the gradient, far away from RNAP-containing peaks. Interestingly, some fractions of S4 are coeluted with HN-S and the HN-S homolog StpA (see Fig. 2B, fractions 4–11).

All other proteins eluted from the Mono Q column have no reported transcription function yet, and at present it is not clear if any of them would interact with RNAP with any biological significance. These proteins can be divided into two groups depending on their properties in nucleic acid (DNA, RNA) binding. One group contains proteins that bind to nucleic acid, a property shared with RNAP. These proteins include many ribosomal proteins, VacB (RNase R), HN-S (histone-like nucleoid structuring protein) and its homolog StpA (suppresser of Td^- phenotype), Hfq (host factor for phage Qβ replication), PNPase (polynucleotide phosphorylase), SSB (single-stranded DNA-binding protein), GyrA (DNA gyrase subunit A), and YhbZ (or ObgE, a GTP-binding protein). Hfq forms a hexamer in solution.[14,15] In our SDS–PAGE analysis, Hfq runs almost exclusively as a monomer with an apparent molecular mass of 12 kDa (fractions 20 to 22 in Fig. 2B), as reported.[16,17] However, it was observed that Hfq runs almost exclusively as a hexamer with an apparent molecular mass of 70 kDa by SDS–PAGE analysis.[18] At present, we do not know the reasons that caused the apparently different behaviors of Hfq in SDS–PAGE analyses. YhbZ (ObgE), a DNA-binding protein with GTPase activity,[19] was

[11] S. Borukhov, A. Polyakov, V. Nikiforov, and A. Goldfarb, *Proc. Natl. Acad. Sci. USA* **89**, 8899 (1992).

[12] S. L. Travlglia, S. A. Datwyler, D. Yan, A. Ishihama, and C. F. Meares, *Biochemistry* **38**, 15774 (1999).

[13] M. Torres, C. Condon, J. M. Balada, C. Squires, and C. L. Squires, *EMBO J* **20**, 3811 (2001).

[14] A. Zhang, K. M. Wassarman, J. Ortega, A. C. Steven, and G. Storz, *Mol. Cell* **9**, 11 (2002).

[15] M. A. Schumacher, R. F. Pearson, T. Moller, P. Valentin-Hansen, and R. G. Brennan, *EMBO J* **21**, 3546 (2002).

[16] A. Takada, M. Wachi, A. Kaidow, M. Takamura, and K. Nagai, *Biochem. Biophys. Res. Commun.* **236**, 576 (1997).

[17] H. Ueno and T. Yonesaki, *Genes Genet. Syst.* **77**, 301 (2002).

[18] M. V. Sukhodolets, S. Garges, and D. J. Jin, *Methods Enzymol.* **370**[25], this volume (2003).

[19] G. Kobayashi, S. Moriya, and C. Wada, *Mol. Microbiol.* **41**, 1037 (2001).

FIG. 2. (A) Protein profile of Mono Q chromatography during the *E. coli* RNA polymerase purification procedure. (B) SDS–PAGE analysis of proteins from different fractions. Aliquots (7.5 μl) of samples were resolved on a 10% bis–Tris SDS–PAGE gel and stained with Simply Blue SafeStain (Invitrogen, Carlsbad, CA). Lane M: Protein markers with their sizes indicated on the left. Lane R: 2 μg RNAP holoenzyme.

TABLE I

Gene	Protein	Function	N-terminal sequence[a]
rnr	VacB (RNase R)	3'-5' exoribonuclease	sqdpfqereaekyanpi
rplB	50S ribosomal subunit protein L2	Protein synthesis	avvckckptspgr
rpsD	30S ribosomal subunit protein S4	Protein synthesis, Transcription regulation	arylgpklklsr
rplO	50S ribosomal subunit protein L15	Protein synthesis	mrlntlspaegs
rplP	50S ribosomal subunit protein L16	Protein synthesis	mlqpkrtkfrkm
hns	DNA-binding protein, H-NS	Regulator for gene expression	sealkilnni
rplQ	50S ribosomal subunit protein L17	Protein synthesis	mrhrksgrql
rplS	50S ribosomal subunit protein L19	Protein synthesis	sniikqleqe
rplW	50S ribosomal subunit protein L23	Protein synthesis	mireerllkvlr
rpsT	30S ribosomal subunit protein S20	Protein synthesis	aniksakkraiq
rpsQ	30S ribosomal subunit protein S17	Protein synthesis	tdkirtlqgrvv
rpmB	50S ribosomal subunit protein L28	Protein synthesis	srvcqvtgkrp
rpsP	30S ribosomal subunit protein S16	Protein synthesis	mvtirlarhga
stpA	H-NS-like protein (StpA)	Regulator for gene expression, RNA chaperone	svmlqslnni
rpsA	30S ribosomal subunit protein S1	Protein synthesis	mtesfaqlfe
rpsB	30S ribosomal subunit protein S2	Protein synthesis	atvsmrdmlk
ssb	ssDNA-binding protein (SSB)	DNA replication, repair and restriction/modification	asrgvnkvil

Gene	Protein	Function	Sequence
rpsF	30S ribosomal subunit protein S6	Protein synthesis	mrhyeivfmv
hfq	Host factor I for bacteriophage Qβ replication (Hfq)	RNA binding	akgqslqdpfln
rpsR	30S ribosomal subunit protein S18	Protein synthesis	yfrrkfcrfta
pnp	Polynucleotide phosphorylase (PNPase)	RNA modification	mlnpivrkfq
glnA	Glutamine synthetase (GlnA)	Glutamine biosynthesis	saehvltmln
rpoB	RNA polymerase, β subunit	Transcription	mvysytekkr
gyrA	DNA gyrase subunit A (GyrA)	DNA modification	msdlareitpvn
rpoA	RNA polymerase, α subunit	Transcription	mqgsveflk
mopA	GroEL, chaperone Hsp60	Chaperones, ATPase	aakdvkfgndarvkm
greA	Transcript cleavage factor (GreA)	RNA cleavage, transcription	mqaipmtlrg
ycaO	Hypothetical protein (YcaO)	Unknown	tqtfipgkda
rfaD	ADP-L-glycero-D-mannoheptose-6-epimerase (RfaD)	Enzyme; surface polysaccharides and antigens	miivtggagf
yhbZ	YhbZ (ObgE)	GTPase, chromosome partition	mkfvdeasil
nusA	NusA (L factor)	Transcription elongation/antitermination	mnkeilavve
hepA	RNA polymerase-associated protein (RapA)	Homolog of SWI2/SNF2 ATPase, RNAP recycling	pftlgqrwiv
rpoH	RNA polymerase, σ³² factor	Regulation of heat shock genes	tdkmqslala
rpoZ	RNA polymerase, ω subunit	Transcription?	arvtvqdave

[a] The sequences of N-terminal residues were determinated by Edman degradation.

eluted mainly with the holoenzyme + RapA peak. Although gyrase consists of four subunits, two subunits A and two subunits B, we found that only GyrA (subunit A) was eluted from the Mono Q column. GyrA was eluted immediately before the core RNAP peak (probably at the same positions as the $\alpha_2\beta$ subassembly of RNAP), whereas the other proteins in this group were eluted at the beginnings of the gradient, which were separated from the RNAP-containing peaks.

The remaining proteins have no known nucleic acid-binding properties and we do not know how they "copurify" with RNAP during the preparations. GlnA (glutamine synthetase) was eluted broadly from the Mono Q column at 0.3 to 0.4 M NaCl, whereas GroEL eluted mainly with holoenzyme. The other two proteins, YcaO and RfaD (ADP-L-glycero-D-manno-heptose-6-epimerase), were eluted mainly at about 0.55 and 0.7 M NaCl, respectively. It is possible that some of these proteins may bind to DNA.

Protocol for RNAP Purification from *Escherichia coli*

Unless otherwise mentioned, all steps are carried out at 4° or on ice with precooled buffers.

Cell Growth

Escherichia coli cells (MG1655) are grown at 37° with superbroth in a 10-liter fermentor (BioFlo IV, New Brunswick Scientific, Edison, NJ), which is inoculated with a 500-ml overnight culture grown in LB. Cells are harvested at late-log growth phase ($OD_{600} \sim 8$), washed once with M63 salt, and stored at $-70°$. About 150 g wet cell pastes are obtained.

Cell Disruption

The frozen cell pastes (100 g) are broken into pieces 2 cm or less in diameter and placed in a 1-liter Waring blender with 400 ml of grinding buffer (50 mM Tris–HCl, pH 8.0, 2 mM EDTA, 230 mM NaCl, 1 mM dithiothreithol, 0.2 mM phenylmethylsulfonyl fluoride, and 5% glycerol). Cells are stirred at low speed for 5–10 min until the cells are resuspended completely. Lysozyme (100 mg) is added to give a final concentration of 0.25 mg/ml, followed by stirring at low speed for 20 min at room temperature. About 10 ml of 10% (w/v) sodium deoxycholate (NaDOC) is added gradually while stirring to give a final concentration of 0.25%; stirring is maintained at low speed for 10 min. *Note:* At this stage, the lysate becomes very viscous. Then the lysate is blended at high speed for 30 s (twice) to shear the DNA. An additional 400 ml of TGED buffer containing 0.2 M NaCl is added, and the mixture is blended at low speed for 2 min. The mixture is divided into two portions with about 400 ml in each portion. For each portion, the mixture is blended at high speed for 30 s (twice). *Note:* At this stage, the lysate should be nonviscous. The cell extract is centrifuged at 8000 rpm for 45 min using a SLA-3000 rotor in a Sorvall RC5B Plus centrifuge. The supernatant (850 ml) is collected for the next step.

Polymin P Precipitation

The supernatant is placed into the blender, and about 30 ml of 10% (w/v) polyethyleneimine (polymin P, pH 7.9) is added slowly while stirring to a final concentration of ~0.35% to precipitate DNA (and DNA-binding proteins, including RNAP). After continuing to stir for another 5 min, the mixture is centrifuged for 15 min at 6000 rpm using a SLA-3000 rotor. The pellet is collected, which contains RNAP–DNA–polymin P complexes.

0.5 M NaCl Wash

The pellet is scraped into the blender and resuspended in 400 ml of TGED + 0.5 M NaCl with gentle stirring at low speed for 10–15 min. *Note:* The proteins bound to DNA less tightly than RNAP will be washed off at this stage. The pellet containing RNAP is collected by centrifugation at 6000 rpm for 15 min using a SLA-3000 rotor.

1 M NaCl Elution

The resultant pellet is again placed in the blender and is resuspended in 400 ml of TGED + 1 M NaCl by stirring gently for 20 to 30 min to elute RNAP and other DNA-binding proteins. The mixture is centrifuged at 8000 rpm for 30 min using a SLA-3000 rotor. The supernatant is collected, yielding ~400 ml of 1 M NaCl eluate. To precipitate RNAP, solid ammonium sulfate is added gradually to the 1 M NaCl eluate in a clean container while stirring to give 50% saturation (35 g/100 ml). After continuous stirring for 30 min, the mixture is left overnight at 4°. The next day, the mixture is centrifuged at 8000 rpm for 45 min using a SLA-3000 rotor. The drained pellet is dissolved in TGED (~250 to 300 ml) with a final conductivity less than that of TGED + 0.2 M NaCl, followed by centrifugation at 10,000 rpm for 20 min using a SLA-3000 rotor in a Sorvall RC5B Plus centrifuge to get rid of undissolvable particles.

Single-Stranded DNA-Agarose Chromatography

Approximately 30 ml of ssDNA agarose (Pharmacia) is packed in an XK-26 column (Pharmacia) and preequilibrated with TGED + 0.2 M NaCl. The clear protein solution from the 1 M NaCl elution step is loaded onto the column at a flow rate of 1.5 ml/min, and the column is then washed with $\sim$100 ml of TGED + 0.2 M NaCl. RNAP and other proteins that bound to single-stranded DNA are eluted with TGED + 1 M NaCl at a flow rate of 1 ml/min. The protein peak is collected, usually yielding 100 to 120 ml of eluate. RNAP in the eluate is precipitated by ammonium sulfate (50% saturation, or 35 g/100 ml) and then dissolved in TGED ($\sim$40–50 ml) to give a final conductivity less than that of TGED + 0.2 M NaCl, followed by centrifugation to obtain a clear protein solution as described previously.

Mono Q Chromatography

The protein sample is loaded onto a Mono Q HR column (10/10, Pharmacia), which is preequilibrated with TGED + 0.2 M NaCl, at a flow rate of 1 ml/min. The column is then washed with 30 ml TGED + 0.2 M NaCl. The proteins are eluted with a shallow linear gradient, 400 ml from 0.2 to 0.5 M NaCl in TGED, at a flow rate of 1 ml/min, and 5-ml fractions are collected. After the shallow gradient, a steep linear gradient, 40 ml from 0.5 to 1 M NaCl in TGED, is applied to the column to elute tightly bound proteins at a flow rate of 1 ml/min in 5-ml fractions.

Peaks corresponding to core RNAP, holoenzyme, and holoenzyme + RapA are eluted at $\sim$ 0.36, 0.38, and 0.42 M NaCl, respectively (Fig. 2). Fractions containing RNAP are concentrated $\sim$ 8 to 10-fold using Centricon YM-30 concentrators (Amicon, Bedford, MA). The concentrated RNAP fractions are mixed gently with an equal volume of highly pure glycerol (to give a final concentration $\sim$50%) and kept at $-70°$ for long-term storage.

[27] Techniques for Studying the Oxygen-Sensitive Transcription Factor FNR from *Escherichia coli*

By VICTORIA R. SUTTON and PATRICIA J. KILEY

FNR is an Fe–S cluster-containing global transcriptional regulator that controls the expression of genes necessary for the growth of *Escherichia coli* under most anaerobic respiratory conditions. FNR regulates the transcription of over 100 genes, including the activation of genes whose products are involved in anaerobic respiration and metabolism, and the repression of genes whose products are involved specifically in aerobic respiration.[1-3] This article focuses on the unique conditions required for the isolation of this protein in its active form.

[1] S. Spiro and J. R. Guest, *FEMS Microbiol. Rev.* **6**, 399 (1990).

In *E. coli*, FNR is inactivated by O_2 and therefore must be isolated under anaerobic conditions.[4] The active form of FNR (abbreviated 4Fe-FNR) is a homodimer that contains one $[4Fe-4S]^{2+}$ cluster per subunit[4,5] and binds to an approximately 22-bp DNA site with the consensus sequence TTGAT-N_4-ATCAA.[1] The location and/or number of binding sites within a promoter region is likely to dictate whether FNR serves as an activator or a repressor at a given promoter. FNR typically binds to FNR-activated promoters at sites centered at about -41.5 relative to the transcription start site, while the location of binding sites in FNR-repressed promoters is more variable.[2] Much of the interest in FNR stems from the O_2-dependent conversion of the $[4Fe-4S]^{2+}$ cluster to a $[2Fe-2S]^{2+}$ cluster, which transforms FNR to an inactive monomeric species and thereby allows FNR to function as a direct O_2 sensor in *E. coli*.[4–6] In addition, the profound effect that changes in FNR activity have on cellular metabolism has made its transcriptional control properties of widespread interest.

FNR belongs to a family of transcription factors that include CRP,[7] the cyclic AMP receptor protein, whose three-dimensional structure has been solved.[8] All of the secondary structural elements of the CRP structure are predicted to be conserved in FNR, although this has not yet been confirmed directly by structural studies of FNR. FNR is predicted to contain a helix-turn-helix DNA-binding motif within the C-terminal domain, as well as an α helix that forms a coiled–coil structure between the two subunits in the active homodimer.[1] Numerous genetic and biochemical studies support the proposed structural and functional assignment of these regions.[2,9,10] In addition, FNR has an N-terminal regulatory domain analogous to CRP. However, in the case of FNR, this domain does not bind cAMP as is well known for CRP, but instead ligates the $[4Fe-4S]^{2+}$ cluster through four cysteines (residues C20, C23, C29, and C122).[5,11,12]

[2] J. R. Guest, J. Green, A. S. Irvine, and S. Spiro, in "Regulation of Gene Expression in *Escherichia coli*" (E. C. C. Lin and A. S. Lynch, eds.), p. 317. R. G. Landes Company, Austin, TX, 1996.

[3] T. Patschkowski, D. M. Bates, and P. J. Kiley, in "Bacterial Stress Responses" (G. Storz and R. Hengge-Aronis, eds.), p. 61. ASM Press, Washington, DC, 2000.

[4] B. A. Lazazzera, H. Beinert, N. Khoroshilova, M. C. Kennedy, and P. J. Kiley, *J. Biol. Chem.* **271**, 2762 (1996).

[5] N. Khoroshilova, C. Popescu, E. Münck, H. Beinert, and P. J. Kiley, *Proc. Natl. Acad. Sci. USA* **94**, 6087 (1997).

[6] P. A. Jordan, A. J. Thomson, E. T. Ralph, J. R. Guest, and J. Green, *FEBS Lett.* **416**, 349 (1997).

[7] D. J. Shaw, D. W. Rice, and J. R. Guest, *J. Mol. Biol.* **166**, 241 (1983).

[8] I. T. Weber and T. A. Steitz, *J. Mol. Biol.* **198**, 311 (1987).

[9] P. J. Kiley and H. Beinert, *FEMS Microbiol. Rev.* **22**, 341 (1999).

[10] L. J. Moore and P. J. Kiley, *J. Biol. Chem.* **276**, 45744 (2001).

The predicted conservation of structure between FNR and CRP has been important in the identification of three activating regions (AR) of FNR that are hypothesized to stimulate transcription activation by contacting RNA polymerase. For all three ARs of FNR, amino acid substitutions have been identified that affect transcriptional regulation and consequently support the hypothesized roles of these regions.[13] AR1 of FNR maps primarily to proposed surface-exposed loops (residues 181–191, 71–75, and 116–121) and is believed to function in transcription activation by contacting the C-terminal domain of the α subunit of RNA polymerase.[14,15] AR2 (residues 49 and 50) is hypothesized to interact with the N-terminal domain of the α subunit of RNA polymerase.[16] AR3 maps primarily to residues 80–89, which are predicted to form a surface-exposed loop, and to residue F112; AR3 is proposed to form contacts with the σ subunit of RNA polymerase.[13,14,17]

The mechanism by which FNR activity is regulated has been an important area of investigation. The requirement for the $[4Fe\text{-}4S]^{2+}$ cluster in active FNR can be explained by its role in dimerization.[4] Apparently, the $[4Fe\text{-}4S]^{2+}$ cluster promotes homodimerization along a coiled–coil formed from an α helix to produce a dimer analogous to that of CRP.[10] Because only dimeric FNR is competent to bind DNA site-specifically through the helix-turn-helix DNA-binding domain and regulate transcription, FNR activity is dependent on the $[4Fe\text{-}4S]^{2+}$ cluster.[4] The reaction of O_2 with the $[4Fe\text{-}4S]^{2+}$ cluster inactivates FNR rapidly by converting each $[4Fe\text{-}4S]^{2+}$ cluster to a $[2Fe\text{-}2S]^{2+}$ cluster, which facilitates the dissociation of dimeric FNR into monomers.[4–6] A further step (as yet poorly understood) in aerobic cells results in the destruction of the $[2Fe\text{-}2S]^{2+}$ cluster to produce a monomeric clusterless form. Neither of the monomeric forms of FNR can bind DNA site-specifically;[4,18] therefore, both are inactive for transcriptional regulation. The redox chemistry of the $[4Fe\text{-}4S]^{2+}$ cluster of FNR is unusual when compared to other proteins containing $[4Fe\text{-}4S]^{2+}$ clusters in that it is not reduced easily and that oxidation by oxygen induces conversion to a $[2Fe\text{-}2S]^{2+}$ cluster.[9]

[11] S. B. Melville and R. P. Gunsalus, *J. Biol. Chem.* **265,** 18733 (1990).

[12] A. D. Sharrocks, J. Green, and J. R. Guest, *FEBS Lett.* **270,** 119 (1990).

[13] D. Browning, D. Lee, J. Green, and S. Busby, in "Signals, Switches, Regulons and Cascades: Control of Bacterial Gene Expression" (D. A. Hodgson and C. M. Thomas, eds.). Proceedings of the 150th ordinary meeting, Society for General Microbiology, Reading, UK, 2002.

[14] B. Li, H. Wing, D. Lee, H. C. Wu, and S. Busby, *Nucleic Acids Res.* **26,** 2075 (1998).

[15] S. M. Williams, N. J. Savery, S. J. W. Busby, and H. J. Wing, *Nucleic Acids Res.* **25,** 4028 (1997).

[16] T. Blake, A. Barnard, S. J. Busby, and J. Green, *J. Bacteriol.* **184,** 5855 (2002).

[17] K. E. Lamberg, C. Luther, K. D. Weber, and P. J. Kiley, *J. Mol. Biol.* **315,** 275 (2002).

[18] B. A. Lazazzera, D. M. Bates, and P. J. Kiley, *Genes Dev.* **7,** 1993 (1993).

FNR homologs, classified by their participation in oxygen-dependent processes and through similarity in amino acid sequence (particularly in the helix-turn-helix DNA-binding motif), have been identified in many species of eubacteria, including both gram-negative and gram-positive bacteria, and both pathogenic and nonpathogenic bacteria. The FNR superfamily contains transcription factors that regulate a variety of processes, including virulence, photosynthesis, nitrogen metabolism, catabolite repression, and anaerobic respiration.[3] Furthermore, several bacteria have multiple FNR homologs,[3,19] which raises questions as to how the specificity for their target genes is achieved and whether they all respond to oxygen in a similar way. Our objective is to provide information that can promote further study of this interesting family of proteins. This article is intended to broadly describe the special problems involved in purifying and characterizing oxygen-sensitive proteins and to describe explicitly the procedures used to study FNR from *E. coli*.

Anaerobic Isolation of FNR

Achieving an Oxygen-Free Environment

The need for oxygen-free or anaerobic conditions is most often met through the use of an anaerobic glove-box chamber (Coy Laboratory Products, Model A) that is filled with a gas mixture composed of 80% N_2, 10% CO_2, and 10% H_2. In the presence of hydrogen, a palladium catalyst (Coy Laboratory Products) promotes the reduction of any oxygen that enters the chamber to produce H_2O, which is then adsorbed by a desiccant. To simplify the execution of multifaceted experiments in an anaerobic environment, the chambers are fitted with equipment that includes an incubator, a small microcentrifuge, HPLC and FPLC purification systems, and a power source for electrophoresis, as well as other standard molecular biology tools.

Growth of Cells

Several methods for overexpressing 4Fe-FNR in *E. coli* have been compared, and the method found to work most reproducibly and with the highest yield of Fe-S cluster/protein (defined as occupancy) is reported here.[20] All of the strains used for protein isolation are OmpT⁻, as SDS–PAGE analysis suggests that the cleavage of the N terminus of FNR (which

[19] S. Spiro, *Antonie van Leeuwenhoek* **66,** 23 (1994).
[20] C. V. Popescu, D. M. Bates, H. Beinert, E. Münck, and P. J. Kiley, *Proc. Natl. Acad. Sci. USA* **95,** 13431 (1998).

TABLE I

BUFFERS AND GROWTH MEDIA

Glucose minimal medium: Minimal M9 medium supplemented with 0.2% casamino
  acids, 0.2% glucose, 1 mM MgCl$_2$, 100 μM CaCl$_2$, 2 μg/ml thiamine, 20 μM
  ferric ammonium citrate
Buffer A: 50 mM KPO$_4$, pH 6.8, 10% glycerol, 0.1 M KCl
Buffer B: 50 mM KPO$_4$, pH 6.8, 10% glycerol, 0.1 M KCl, 1 mM DTT, 1.7 mM DTH,
  0.1 mM phenylmethylsulfonyl fluoride
Buffer C: 50 mM KPO$_4$, pH 6.8, 10% glycerol, 0.8 M KCl

is only observed following cell breakage) is reduced in these strains. Typically, strain PK22 [derived from BL21(DE3), Novagen],[18] containing
pET-11a (Novagen) with *fnr* cloned into the *NdeI–Bam*HI sites, is grown
aerobically in four 2-liter flasks, each of which contains 1 liter of glucose
minimal medium (see Table I) with 50 μg/ml ampicillin, with shaking at
250 rpm at 37°. When the cell culture reaches OD$_{600}$ $\cong$ 0.3 (Spectronic
spectrophotometer) or about 5 × 10^8 cells/ml, the expression of FNR is induced with 0.4 mM Isopropyl-β-D-thiogalactoside (IPTG) for 1 h. The
cultures are then combined in a 9-liter glass carboy that is fitted with a stopper pierced by two pieces of glass tubing: the first piece connects a line from
an argon gas tank to a gas dispersion tube in the sample and the second permits gas release from the sparging vessel. The cultures are sparged with
argon overnight at 4°, which increases the yield of 4Fe-FNR by allowing
the insertion of [4Fe-4S]$^{2+}$ clusters into the accumulated FNR apoprotein
and by stabilizing the 4Fe-FNR that forms. After the overnight sparging
step, cells are poured into a container of suitable size and transferred immediately into the anaerobic chamber, where the cells are dispensed into
500-ml centrifuge bottles that seal tightly with lids containing O rings
(Beckman). The bottles are removed from the chamber and centrifuged
to pellet the cells (Beckman Avanti J-25 centrifuge, JLA10.500 rotor, 4°,
15 min at 8000 rpm). The centrifuge bottles are then returned to the chamber, where the supernatant is poured off and the process is repeated until
all of the cells have been collected. At this point, the anoxic cell pellets
can either be stored at −80°C or used immediately.

Preparation of Cell Lysate

    Within the anaerobic chamber, the pelleted cells are resuspended (to
0.5% of their original volume) in anaerobic buffer B (see Table 1) and then
pipetted immediately into a cold French press cell that has been transferred
into the chamber and rinsed with anaerobic buffer to remove any traces of

oxygen. The French press cell containing the cell suspension is sealed and the sample outlet tubing is fitted with a capped 18-gauge needle before the cell is removed from the anaerobic chamber. A butyl rubber-stoppered glass vial is used as the anaerobic receptacle for the French press. Butyl rubber fittings are used because butyl rubber is relatively impermeable to oxygen.[21] After a single pass through a cold French press at 20,000 psi, the anaerobic lysate is brought back into the anaerobic chamber, transferred to ultracentrifuge tubes fitted with a gas-tight lid, removed from the chamber, and clarified in an ultracentrifuge (Beckman Optima LE-80K, 70.1Ti rotor, 4°, 45 min at 50,000 rpm). The lysate samples are then returned to the chamber, where the supernatant is used as a source of 4Fe-FNR.

Protein Purification

Initially, we purified FNR using a Pharmacia FPLC equipped with a 5-ml BioRex-70 cation-exchange column (BioRad Laboratories).[5] The Biorex-70 column was equilibrated in buffer A (see Table I), and FNR was eluted with a 50-min gradient from 0.1 to 0.55 M KCl at a flow rate of 10 ml/h. We have since found that newer batches of Biorex-70 do not work as efficiently. Therefore, we have developed an equally effective method that uses a PolyCAT A cation-exchange column (Nest Group) connected to a Beckman HPLC. This method uses buffer A (see Table I) with a 15-ml gradient from 0.1 to 1.0 M KCl at 0.5 ml/min. FNR elutes at about 0.4 M KCl from both resins, as measured with a conductivity meter. Both the HPLC and the FPLC methods utilize the anaerobic chamber, as described under Protein Purification Equipment.

Because the $[4Fe-4S]^{2+}$ cluster absorbs visible light maximally at a wavelength of about 420 nm, purified 4Fe-FNR is dark green, which makes fractions containing concentrated 4Fe-FNR simple to identify. In addition, 4Fe-FNR can be detected electronically (e.g., with an in-line photo diode array detector) by its absorbance peak at 420 nm and can therefore be distinguished from the majority of cell proteins, which do not absorb light of this wavelength.

Many of our applications require concentrated protein solutions (>400 μM FNR). Therefore, we concentrate the fractions containing 4Fe-FNR by diluting them to 0.1 M KCl and loading the mixture onto a 1-ml gravity-flow Biorex-70 column in an anaerobic chamber. The 4Fe-FNR is eluted from the column as a concentrated eluate using 1–2 column volumes of 0.8 M KCl (buffer C, Table I). As an alternative, 4Fe-FNR

---

[21] R. E. Hungate, W. Smith, and R. T. Clarke, *J. Bacteriol.* **91**, 908 (1965).

can be concentrated using a 1-ml Hi-Trap heparin column (Amersham Biosciences) by equilibrating with buffer A and eluting with buffer C.

The isolated FNR fractions are more than 95% pure, as estimated visually by SDS–PAGE. The fraction of FNR containing $[4Fe\text{-}4S]^{2+}$ clusters is determined by measuring the moles of sulfide per mole of FNR protein in each sample.[22] The occupancy (defined as the percentage of FNR molecules containing a $[4Fe\text{-}4S]^{2+}$ cluster) of purified 4Fe-FNR typically ranges from 50 to 80%. As a control for these measurements, the iron concentration of the samples is also determined, usually resulting in values slightly higher than the sulfide concentrations (due to the presence of a small amount of iron that apparently is not bound in clusters). 4Fe-FNR preparations are stored in glass vials topped with butyl rubber stoppers (see Preparation of the Cell Lysate) and crimp sealed, which can be removed from the anaerobic chamber for storage and/or manipulation.

Special Procedures/Equipment Required for Working with Oxygen-Sensitive Proteins

Protein Purification Equipment

The HPLC (Beckman System Gold Nouveau) is contained entirely within the anaerobic chamber and is fitted with PEEK tubing (Upchurch Scientific) throughout the system. The solvents are kept on ice, and the cold cell lysate is loaded through injection into a 10-ml superloop. The PolyCAT A cation-exchange column (Nest Group), is kept in a column chiller (Jones Chromatography) at $4°$ throughout the purification.

The FPLC used with the chamber (Pharmacia LCC-501 Plus system) has been assembled with most of the system (including the column) outside of the chamber but with the solvents and the fraction collector inside the chamber. The entire system is fitted with PEEK tubing, which is impermeable to oxygen so that the contents of the system remain anaerobic. The path of solvent through the FPLC is from the solvent bottles inside the chamber, out to the pumps and onto the column, and back into the chamber to the fraction collector. The cell lysate is loaded onto the column from a 50-ml Superloop (Pharmacia) that is filled with anaerobic lysate and sealed inside the chamber and then removed from the chamber and connected to the system. Collected fractions are transferred to glass vials with butyl rubber septa (see Preparation of the Cell Lysate) for storage.

---

[22] H. Beinert, *Anal. Biochem.* **131**, 373 (1983).

Apparatus for Gas Exchange

Liquid samples must be sparged with oxygen-free gas before being brought into the anaerobic chamber (see Achieving an Oxygen-Free Environment). For this purpose, we use an apparatus composed of copper tubing that delivers copper-scrubbed argon gas (Sargent-Welch furnace) via a three-position valve to a custom-built manifold, composed of multiple lines that are capped by either gas dispersion tubes (Upchurch Scientific) or needles. A vacuum pump also connects to the three-position valve, which can be turned to permit the flow of gas or vacuum or neither throughout the manifold. Solutions to be brought into the chamber are sparged (via gas dispersion tubes) with argon gas for 20 min before being immediately capped and cycled through the vacuum air lock and are then left uncapped in the chamber overnight before being used; this method effectively removes oxygen from solutions. Alternatively, the needle attachments on the manifold may be used to remove oxygen from butyl rubber-sealed glass vials containing either solid reagents or liquids with volumes too small for sparging. After inserting the needle into the vial, the manifold valve is slowly turned to allow a vacuum to build in the vial (while at the same time the sample is constantly agitated), and then the valve is slowly turned to fill the vial with argon gas; this cycle is repeated three to five times or as necessary to remove the oxygen in the sample.

Anaerobic Procedures

The extent to which methods must be modified for the analysis of oxygen-sensitive samples depends on the technique involved. Some methods, including simple sample handling (such as pipetting and microcentrifugation), PAGE analysis, and culture growth, can be performed as for aerobic samples but within the anoxic environment of the chamber. Other methods require specialized equipment. For example, a spectrophotometric analysis of anaerobic samples is performed using quartz cuvettes (Starna) manufactured to seal with O ring-adapted screw caps; the samples are simply prepared in the chamber and then sealed and removed to a spectrophotometer for analysis. To monitor the effect of oxygen introduction to an anaerobic mixture spectrophotometrically, we use a specially made glass cuvette with a long neck ending in a stopcock that is turned to seal the cuvette tightly. The sample is added to the cuvette in the anaerobic chamber, a vacuum is generated in the space above the sample, and the sample is monitored anaerobically in the spectrophotometer. To introduce oxygen, the cuvette is removed from the spectrophotometer and the stopcock is opened slowly. The sample is then shaken to equilibrate the solution with air, and the cuvette is returned to the spectrophotometer for further analysis. Experiments designed to investigate the effects of oxygen in solution are constrained by the limited solubility of oxygen at room temperature ($\sim$220 μM); we choose detection methods that are sensitive to low concentrations of FNR, which thereby allow us to maintain oxygen in 10-fold excess over FNR.

While handling cells and purified protein samples in the anaerobic chamber largely protects them from oxygen, the introduction of small amounts of oxygen is unavoidable during normal use of the chamber. To inhibit oxygen-induced destruction of samples, dithionite (DTH) can be added to solutions to produce reducing conditions that readily destroy any oxygen molecules that enter a sample. Dithiothreitol (DTT) can also be added to keep sulfhydryls reduced. The storage of DTT requires simple desiccation, but because DTH is highly reactive in solution, the powder is stored under an argon atmosphere (see Apparatus for Gas Exchange) to maintain a dry and oxygen-free environment; both DTT and DTH solutions are made fresh for each use.

Analysis of FNR Function *In Vitro*

In Vitro *Assays of FNR Function*

The anaerobic analysis of FNR activity *in vitro* has been performed with gel shift assays and *in vitro* transcription assays. The *in vitro* transcription reactions are executed within the anaerobic chamber, after which the samples are removed and terminated immediately by the addition of stop solution. The samples are then analyzed by PAGE.[22a] Anaerobic gel shift assays, however, must be performed entirely within the anoxic environment of the anaerobic chamber, as the interaction of 4Fe-FNR with the DNA depends on the continuous absence of oxygen. To perform these assays, acrylamide gels are brought into the chamber after polymerization (which is an oxygen-dependent reaction) and prerun with anaerobic buffers containing DTH before samples are loaded onto the gel.[23] When FNR preparations are used for DNA-binding or transcription assays, we purify FNR from a strain (PK4529)[17] with reduced exonuclease and endonuclease activity. In addition, sterile tubes, buffers, and so on are used to both collect and store purified FNR.

FNR* Mutants

Some experiments can be performed under aerobic conditions by using FNR variants that are active in the presence of oxygen (FNR* mutants), provided that data suggest that the variants do not exhibit properties

[22a] R. K. Karls and P. J. Keiley, unpublished.

[23] N. Khoroshilova, H. Beinert, and P. J. Kiley, *Proc. Natl. Acad. Sci. USA* **92,** 2499 (1995).

different than the wild-type protein for the assay(s) in question. The two most commonly used variants are FNR-L28H and FNR-D154A. FNR-L28H contains a stabilized $[4Fe-4S]^{2+}$ cluster with greatly reduced oxygen sensitivity; this variant retains the cluster as well as DNA-binding, dimerization, and transcriptional activity even after 1 h of oxygen exposure, whereas the wild-type protein loses 50% activity within a few minutes of oxygen exposure. L28H-FNR is normally purified under anaerobic conditions to increase the yield of 4Fe-FNR.[24] FNR-D154A contains an amino acid substitution in the putative dimerization helix that induces dimerization and transcriptional regulation even in the absence of $[4Fe-4S]^{2+}$ clusters.[18,23,25] We normally perform gel shift assays, in vitro transcription assays, and DNA footprinting assays using the FNR-D154A variant under aerobic conditions using standard methods.[25,26]

Analysis of FNR Function In Vivo

FNR activity is usually measured by using β-galactosidase assays in strains containing single copy operon fusions of lacZ to FNR-dependent promoters.[17] The promoters used routinely to examine in vivo function are P_{dmsA} and P_{narG}, which are activated by FNR, and P_{ndh}, which is repressed by FNR.

While in vivo FNR function can be analyzed aerobically using FNR* variants (see FNR* mutants) the analysis of wild-type FNR in vivo requires the maintenance of anaerobic conditions, which can be met in a variety of ways. Cells can be grown anaerobically on solid media or with several different methods in liquid culture.

Solid Media

The anaerobic growth of E. coli on agar plates can be accomplished either by growing plates in a 37° incubator contained within the anaerobic chamber or by using the GasPak system (BD Biosciences). The GasPak system consists of a jar into which the plates are placed for incubation along with commercially available GasPaks (BD Biosciences) that catalytically remove oxygen from the atmosphere of the containers; the entire jar is placed in an incubator to promote cell growth.

[24] D. M. Bates, C. V. Popescu, N. Khoroshilova, K. Vogt, H. Beinert, E. Münck, and P. J. Kiley, J. Biol. Chem. **275**, 6234 (2000).
[25] E. C. Ziegelhoffer and P. J. Kiley, J. Mol. Biol. **245**, 351 (1995).
[26] K. E. Lamberg and P. J. Kiley, Mol. Microbiol. **38**, 817 (2000).

Screening cells for a lactose utilization phenotype under anaerobic conditions is complicated by the fact that some screens depend on oxygen-dependent reactions. For example, X-gal cannot be used anaerobically, as formation of the blue color is oxygen dependent; however, S-gal (Sigma) is a modified chromogenic substrate whose color development is oxygen independent and can therefore be used to screen for alterations of the lactose phenotype of colonies incubated under either aerobic or anaerobic conditions.[27] Alternatively, because the center of colonies becomes oxygen limited even in the presence of air, lactose utilization can be monitored using X-gal and MacConkey agar media. The X-gal and MacConkey agar. The color production by cells under these conditions produces a "fish-eye" colony phenotype.[28]

Liquid Media

The anaerobic growth of *E. coli* in liquid media is usually accomplished without the use of the anaerobic chamber by subculturing aerobically grown cells into screw-cap tubes filled to the brim with appropriate medium, which are then capped and incubated. The growing culture quickly exhausts the oxygen available in the medium and in the headspace under the cap, and consequently becomes anaerobic. There is no significant difference between the levels of FNR activity in cells grown by these means and those grown with the Hungate method.[29]

Sparging Station

For experiments that require sampling of cultures or the introduction of additives during cell growth, use of the aforementioned growth regimen is not adequate. Instead, cultures are grown in a medium while being sparged with either aerobic or anaerobic gas mixtures using the sparging station described later (see Fig. 1). In addition, the sparging station can be used for experiments that monitor the effects of the introduction or loss of oxygen from a growing culture.

Our sparging station uses tanks of N_2, O_2, and CO_2 gases that are attached to a custom-made gas mixer designed to create a combined gas flow of the desired composition and flow rate. Aerobic conditions are achieved by sparging cultures with a gas mixture of 70% N_2, 25% O_2, and 5% CO_2; for anaerobic conditions, the mixture is 95% N_2 and 5% CO_2. The flow rate of each gas is adjusted using a flow controller (VICI Condyne, Model 100

[27] E. L. Mettert and P. J. Kiley, unpublished
[28] M. J. Lombardo, D. Bagga, and C. G. Miller, *J. Bacteriol.* **173,** 7511 (1991).
[29] R. E. Hungate, *Bacteriol. Rev.* **14,** 1 (1950).

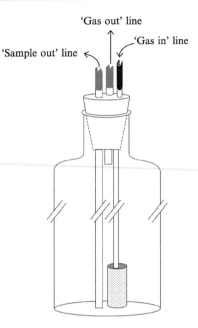

'Gas out' line

'Gas in' line

'Sample out' line

FIG. 1. Diagram of the culture vessel used in the sparging station for cell growth under aerobic and aerobic conditions. Mixed gases are bubbled into the growing culture through the "gas in" line, and gas is released from the container by the "gas out" line. Samples are removed by clamping the "gas out" line, which builds positive pressure in the bottle and forces liquid to flow out through the unclamped "sample out" line.

and Vishay Spectrol Model 15), and then the gases are combined and flow into a manifold that dispenses the gas mixture via butyl rubber tubing to up to four sparging vessels. A sparging vessel (see Fig. 1) is composed of a bottle fitted with a butyl rubber stopper, through which appropriately sized holes permit the passage of three pieces of glass tubing. The first piece of tubing connects to the butyl rubber hose that carries the gas mixture and, below the stopper, leads to a gas dispersion tube through which gases are bubbled into the culture. The second piece of glass tubing extends to the bottom of the bottle in order to draw up cell culture, as will be explained, and is capped above the stopper by flexible tubing whose end can be placed in tubes for collecting the samples; this forms the "sample line." The final piece of glass tubing forms the "gas out" line, connecting to a piece of flexible tubing above the stopper and ending just below the stopper, to allow the release of gases from the tube as sparging occurs. The sparger apparatus allows gas mixtures to bubble through the gas dispersion tube into the culture, and gas flows out of the "gas out" line; the flexible tubing of the sample line is kept clamped except when samples are being drawn. To draw a sample from the culture, the "gas out" line is clamped and then the sample line is unclamped, at which point the positive pressure built up from the unreleased gas forces the sample to flow out through the sample line. When enough sample has been collected, the sample line is reclamped and then the "gas out" line is unclamped to allow gas release. Throughout the experiment, the sparging vessel stands in a water bath, secured by attachment to a ring stand, for incubation at the appropriate temperature during growth.

Summary

A large variety of techniques can be adapted for use with oxygen-sensitive samples. The growth of cells and *in vivo* analyses, as well as protein purification and *in vitro* assays, can be executed either by performing necessary steps in anaerobic environments (ranging from simple closed containers to the anaerobic chamber) or by circumventing the need for anaerobiosis with the use of oxygen-resistant protein variants.

Acknowledgments

The authors thank past and present members of the Kiley laboratory and Dr. Helmut Beinert for critical reading of the manuscript and for their contributions to the work described herein. We particularly acknowledge the work of Kevin Vogt, whose efforts produced major improvements in our methods. We also thank Dr. Tim Donohue for critical reading of the manuscript. This work was supported by NIH Molecular Biosciences Training Grant GM07215 (to VRS) and by NIH Grant GM45844 (to PJK).

[28] Assay of Transcription Modulation by Spo0A of *Bacillus subtilis*

By STEVE D. SEREDICK, BARBARA M. TURNER, and GEORGE B. SPIEGELMAN

Spo0A is a two-domain, response regulator transcription factor whose activity is regulated by phosphorylation of an aspartate residue in its conserved N-terminal receiver domain. Activation of Spo0A enhances its affinity for DNA sites, 0A boxes. 0A boxes occur primarily as tandem pairs and in both orientations (forward or reverse) at different positions relative to the start site of transcription.[1,2]

---

[1] J. M. Baldus, B. D. Green, P. Youngman, and C. P. Moran, Jr., *J. Bacteriol.* **176**, 296 (1994).

This variety in 0A box positioning is indicative of the functional complexity of Spo0A. At the promoter for the *spoIIG* operon, Spo0A~P binds to a pair of reverse 0A boxes overlapping the -35 element, thereby contacting σ^A of RNA polymerase holoenzyme to activate transcription.[3-6] At the σ^H-dependent *spoIIA* operon promoter, multiple forward and reverse 0A boxes are found 50 to 100 bp upstream of the initiation site, a region known to be important for activating transcription.[7] Spo0A~P also acts as a repressor; at the promoter for the *abrB* gene, Spo0A~P binds to a pair of forward 0A boxes downstream of the transcription start site and prevents open complex formation by the polymerase.[8,9]

This article describes methods used to purify Spo0A and some of the techniques used to study the mechanism by which Spo0A modulates transcription. We use methods developed studying the stimulation of *spoIIG* transcription as an example.

Purification of Spo0A

Escherichia coli BL21 (pLysS) freshly transformed with pKK0A[10] is grown in LB containing 100 μg/ml of ampicillin at 30°. At an OD_{600} of ~0.9, the culture is induced with Isopropyl-β-D-thiogaloctoside (IPTG) (final concentration 0.5 mM) and allowed to grow for a further 5 h. The cells are harvested by centrifugation and stored at $-20°$ until required.

Pellets from 1 liter of culture are thawed and resuspended in 20 ml of buffer B [20 mM Tris–HCl pH 8.3, 1 mM EDTA, 150 mM KCl, 10 mM MgCl$_2$, 1 mM dithiothreitol (DTT)] and 1 mM phenylmethylsulfonyl floride (PMSF) and are sonicated to lyse the cells. The broken cell extract is centrifuged for 30 min at 15,000 g. The supernatant is diluted with buffer B so the absorbance at 280 nm is approximately 6. Saturated $(NH_4)_2SO_4$ (4.1 M) is added slowly to bring the solution to 20% saturation and then

[2] G. B. Spiegelman, T. H. Bird, and V. Voon, *in* "Two-Component Signal Transduction" (J. A. Hoch and T. J. Silhavy, eds.), p. 159. American Society for Microbiology, Washington, DC, 1995.

[3] S. Satola, P. A. Kirchman, and C. P. Moran, Jr., *Proc. Natl. Acad. Sci. USA* **88**, 4533 (1991).

[4] C. M. Buckner, G. Schyns, and C. P. Moran, Jr., *J. Bacteriol.* **180**, 3578 (1998).

[5] G. Schyns, C. M. Buckner, and C. P. Moran, Jr., *J. Bacteriol.* **179**, 5605 (1997).

[6] S. W. Satola, J. M. Baldus, and C. P. Moran, Jr., *J. Bacteriol.* **174**, 1448 (1992).

[7] J. J. Wu, P. J. Piggot, K. M. Tatti, and C. P. Moran, Jr., *Gene* **101**, 113 (1991).

[8] M. Strauch, V. Webb, G. Spiegelman, and J. A. Hoch, *Proc. Natl. Acad. Sci. USA* **87**, 1801 (1990).

[9] E. A. Greene and G. B. Spiegelman, *J. Biol. Chem.* **271**, 11455 (1996).

[10] H. Zhao, T. Msadek, J. Zapf, Madhusudan, J. Hoch, and K. Varughese, *Structure (Camb).* **10**, 1041 (2002).

it is stirred on ice for 15 min, and the precipitate is separated by centrifugation at 15,000 g for 15 min. The supernatant is brought to 50% $(NH_4)_2SO_4$ with solid $(NH_4)_2SO_4$ and stirred on ice for 15 min. The precipitate is collected by centrifugation at 15,000 g for 15 min, and the Spo0A-containing pellet is resuspended in 10 ml of buffer C (20 mM NaPO$_4$, pH 8.0, 1 mM EDTA, 0.1 mM DTT, and 1 mM PMSF) + 150 mM NaCl and dialyzed overnight against 2 liters of buffer C + 150 mM NaCl. The dialysate is centrifuged for 30 min at 15,000 g to remove any insoluble material and is loaded directly onto a 10-ml heparin–agarose column (Sigma-Aldrich Corp., St. Louis, MO) equilibrated with buffer C + 150 mM NaCl. The column is washed with the same buffer until the OD$_{280}$ of the wash is less than 0.1. The column is then washed with buffer C + 250 mM NaCl until the wash OD$_{280}$ is approximately 0.1, and Spo0A is eluted with 60 ml of a 250–850 mM NaCl linear gradient in buffer C collecting 1-ml fractions. Fractions are stored at $-20°$ while the protein content is analyzed by SDS–PAGE. Fractions containing Spo0A are pooled and concentrated by placing a dialysis bag containing the fractions in a dish and surrounding it with PEG$_{20,000}$ (Sigma-Aldrich Corp.). The concentrated fractions are dialyzed against 2 liters of buffer C + 150 mM NaCl and 0.1 mM DTT and glycerol is added to 30%. Samples are aliquoted and stored at $-20°$. The final product is approximately 85–90% pure and may be used directly for *in vitro* studies.

Purification of Spo0ABD

This protocol is based on a procedure developed by Zhao *et al.*[10] Cell growth, induction, and harvesting are identical to that for the purification of Spo0A except that plasmid pTM0A15 is used. pTM0A15 encodes the *B. subtilis* spo0A ribosome-binding site and the first two codons of the *spo0A* gene fused to codon 143, followed by the entire coding sequence for the C-terminal domain, as well as the *spo0A* stop codon and transcription terminator.

Cell pellets from 6 liters of culture are resuspended in 100 ml of buffer D (20 mM NaPO$_4$, pH 6.8, 1 mM EDTA) + 150 mM NaCl, and PMSF is added to a final concentration of 1 mM. Cells are disrupted by sonication on ice and centrifuged for 20 min at 15,000 g. The supernatant is loaded directly onto a Macroprep High Capacity S cation-exchange resin column (BioRad, Hercules, CA) equilibrated in buffer D + 150 mM NaCl and washed with the same buffer until the OD$_{280}$ of the effluent begins to drop. Washing is continued with buffer D + 250 mM NaCl until the OD$_{280}$ drops below 0.1, and Spo0ABD is eluted with a 250–800 mM NaCl gradient in buffer D (500 ml). Relevant fractions based on SDS–PAGE analysis are pooled, dialyzed overnight against buffer D + 150 mM NaCl, and loaded onto a heparin–Sepharose column. The column is washed and eluted exactly as for the cation-exchange column. Fractions containing Spo0ABD are ~90% pure at this step and may be used for *in vitro* studies. Spo0ABD may be purified further using a FPLC HiPrep 16/60 Sephacryl S200 gel filtration column (Amersham Biosciences, Piscataway, NJ) in buffer D + 150 mM NaCl.

RNA Polymerase Holoenzyme Purification

RNA polymerase is purified essentially as described[11] with several minor modifications. Sodium acetate (NaAc) has been substituted for NaCl in the glycerol gradients to reduce the nonspecific inhibition of transcription due to chloride anions[12-15] and the final heparin–Sepharose column chromatography step has been eliminated. Peak fractions from the glycerol gradient step of the purification are tested for DNase activity by incubating with an end-labeled DNA fragment. Fractions with high specific activity are adjusted to 30% glycerol and stored at $-70°$. RNA polymerase purified from 25 to 30 g of early logarithmic cells yields sufficient active enzyme for 3 to 6 months of use. The specific activity of the RNA polymerase in 30% glycerol held at $-70°$ falls to half its original value in about 8 months. Other methods exist for the purification of highly active *B. subtilis* RNA polymerase suitable for sensitive *in vitro* assays.[16]

Activation of Spo0A

Activation of DNA binding by the full-length protein requires phosphorylation. In our laboratory, activation is achieved by reconstituting the phosphorelay that contains three additional proteins, KinA, Spo0F, and Spo0B. All three proteins can be purified from *E. coli* containing expression vectors for these proteins using published methods.[17-19]

[11] K. F. Dobinson and G. B. Spiegelman, *Biochemistry* **26**, 8206 (1987).
[12] D. A. Rowe-Magnus, Ph.D. thesis, University of British Columbia, Vancouver, 1998.
[13] T. Arakawa and S. N. Timasheff, *Biochemistry* **23**, 5912 (1984).
[14] T. Arakawa and S. N. Timasheff, *J. Biol. Chem.* **259**, 4979 (1984).
[15] S. Leirmo, C. Harrison, D. S. Cayley, R. R. Burgess, and M. T. Record, Jr., *Biochemistry* **26**, 2095 (1987).
[16] J. D. Helmann, F. R. Masiarz, and M. J. Chamberlin, *J. Bacteriol.* **170**, 1560 (1988).
[17] C. E. Grimshaw, S. Huang, C. G. Hanstein, M. A. Strauch, D. Burbulys, L. Wang, J. A. Hoch, and J. M. Whiteley, *Biochemistry* **37**, 1365 (1998).
[18] J. W. Zapf, J. A. Hoch, and J. M. Whiteley, *Biochemistry* **35**, 2926 (1996).
[19] X. Z. Zhou, Madhusudan, J. M. Whiteley, J. A. Hoch, and K. I. Varughese, *Proteins* **27**, 597 (1997).

To activate Spo0A *in vitro*, a 20-μl reaction containing 4 μM Spo0A, 1 μM of the histidine kinase KinA, 2 μM of the single domain response regulator Spo0F, 0.2 μM of the phosphotransferase Spo0B, and 10 mM ATP in 10 mM HEPES, pH 8.0, 80 mM KAc is composed and allowed to sit for at least 1 h at room temperature. This system has proven relatively robust, with typical phosphorylation efficiencies estimated at greater than 50%.[20] Typically, 400 nM Spo0A$\sim$P promotes maximum stimulation of transcription from the *spoIIG* promoter, whereas 200 nM is required for the inhibition of transcription from the *abrB* promoter. Phosphorelay reactions may be stored at $-20°$, but lose the ability to stimulate upon repeated freeze/thawing. The direct stability of Spo0A$\sim$P in solution has not been reported.

Alternately, Spo0ABD, the constitutively active DNA-binding domain of Spo0A, may be used for assays. The stimulatory properties of Spo0ABD at *spoIIG* are similar to Spo0A$\sim$P but not identical.[21,22] In general, more Spo0ABD is required for maximum DNA binding or transcription stimulation, with the caveat that there is no method for determining the specific activity of the proteins independent of interaction with RNA polymerase. Some experiments have also suggested a direct interaction of the N-terminal receiver domain of Spo0A$\sim$P with the polymerase.[23] The major advantage of Spo0ABD is in experiments testing the effects of the addition of ATP on initiation, as Spo0ABD would not contain ATP. Typically, 800 nM Spo0ABD yields maximum stimulation of transcription from the *spoIIG* promoter.

Response regulators, including Spo0A, may be phosphorylated by small molecule phosphodonors, such as acetyl phosphate or phosphoamidate. Phosphoramidate stimulation has been reported,[24] although the product has not been tested in transcription assays. Unphosphorylated full-length Spo0A preparations sometimes stimulate transcription because they contain Spo0ABD. Because the linker region between the N terminus and the C terminus appears to be unusually labile, storage even at $-20°$ results in the accumulation of transcriptionally active Spo0ABD. Preparations of Spo0A that stimulate transcription in the absence of phosphorylation should be checked by high percentage acrylamide SDS–PAGE to allow the detection of 14.1-kDa Spo0ABD.

[20] T. H. Bird, Ph.D. thesis, University of British Columbia, Vancouver, 1994.
[21] D. A. Rowe-Magnus and G. B. Spiegelman, *J. Biol. Chem.* **273**, 25818 (1998).
[22] J. K. Grimsley, R. B. Tjalkens, M. A. Strauch, T. H. Bird, G. B. Spiegelman, Z. Hostomsky, J. M. Whiteley, and J. A. Hoch, *J. Biol. Chem.* **269**, 16977 (1994).
[23] M. A. Cervin and G. B. Spiegelman, *Mol. Microbiol.* **31**, 597 (1999).
[24] R. J. Lewis, D. J. Scott, J. A. Brannigan, J. C. Ladds, M. A. Cervin, G. B. Spiegelman, J. G. Hoggett, I. Barak, and A. J. Wilkinson, *J. Mol. Biol.* **316**, 235 (2002).

In Vitro Transcription Assay

Rationale

The most sensitive functional assay is a single-round *in vitro* transcription system developed by Bird *et al.*,[25] which is based on the observation that initiated RNA polymerase is resistant to the inhibitor heparin. We have used this assay with minor modifications to study the effects of protein and promoter mutations, and reaction conditions on Spo0A-stimulated transcription from *spoIIG* and *spoIIA* promoters, Spo0A-mediated repression of the *abrB* gene promoter, and Spo0A-independent transcription from the *B. subtilis* phage ϕ29 promoters A2 and G2.[9,21,23–31] Transcription from most promoters can be studied using isolated linear DNA fragments; however, the *spoIIA* promoter requires supercoiling to serve as an RNA synthesis initiation site.[32]

Protocol

The *spoIIG* template is isolated from plasmid PUCIIG*trpA*[3] on a 600-bp DNA fragment generated by digestion with *Pvu*II. This fragment contains the *spoIIG* promoter, with 100 bp of the upstream sequence containing two tandem sets of reverse 0A boxes and the *trpA* terminator approximately 160 bp downstream from the transcription start site. The fragment is isolated by agarose gel electrophoresis, recovered by either gel extraction or electroelution. The residual ethidium is extracted by repeated butanol extraction, precipitated twice with ethanol to remove butanol, and dissolved and stored in 10 mM HEPES pH 8.0, 30 mM NaAc, 0.1 mM EDTA at 4°. Large quantities of the fragments are prepared so that the concentration may be determined using absorbance at 260 nm.

The *B. subtilis* RNA polymerase does not make classical stable complexes at a number of promoters, including the Spo0A~P-dependent *spoIIG* and *spoIIA* promoters and the Spo0A~P-independent ϕ29 A2

[25] T. H. Bird, J. K. Grimsley, J. A. Hoch, and G. B. Spiegelman, *Mol. Microbiol.* **9,** 741 (1993).

[26] T. H. Bird, J. K. Grimsley, J. A. Hoch, and G. B. Spiegelman, *J. Mol. Biol.* **256,** 436 (1996).

[27] M. A. Cervin, R. J. Lewis, J. A. Brannigan, and G. B. Spiegelman, *Nucleic Acids Res.* **26,** 3806 (1998).

[28] M. A. Cervin and G. B. Spiegelman, *J. Biol. Chem.* **275,** 22025 (2000).

[29] M. A. Cervin, G. B. Spiegelman, B. Raether, K. Ohlsen, M. Perego, and J. A. Hoch, *Mol. Microbiol.* **29,** 85 (1998).

[30] D. A. Rowe-Magnus, M. Mencia, F. Rojo, M. Salas, and G. B. Spiegelman, *J. Bacteriol.* **180,** 4760 (1998).

[31] D. A. Rowe-Magnus and G. B. Spiegelman, *Proc. Natl. Acad. Sci. USA* **95,** 5305 (1998).

[32] T. Bird, D. Burbulys, J. J. Wu, M. A. Strauch, J. A. Hoch, and G. B. Spiegelman, *Biochimie* **74,** 627 (1992).

promoter. At these promoters, RNA polymerase resistance to heparin occurs only after synthesis of an RNA trimer[12] so that heparin is added with nucleotides that allow elongation of RNA chains from complexes that have initiated transcription. Transcription reactions are carried out in $1\times$ transcription buffer [10 mM HEPES, pH 8.0, 10 mM MgAc, 1 mM DTT, 0.1 mg/ml bovine serum albumin (BSA) and KAc below 120 mM] in a final reaction volume of 10 μl and with a DNA template concentration of between 4 and 10 nM. When using the *spoIIG* promoter as a template, transcription assays are carried out by composing 8 μl of an initiation mix that contains 1 μl of $10\times$ transcription buffer, 1 μl of 40 nM template DNA, the desired form of Spo0A, 1 μl of 4 mM ATP, 50 μM GTP, and approximately 3 μCi of $[\alpha^{32}$-P]GTP (800 Ci/mmol; Perkin-Elmer, Boston, MA) in 0.65-ml microfuge tubes on ice. Reaction tubes are warmed at 37° for 90 s before the addition of 1 μl of RNA polymerase (freshly diluted in a dilution buffer of $1\times$ transcription buffer with 30% glycerol). Reactions are vortexed briefly upon addition of the polymerase and returned to 37°. After a further 1 min to allow formation of an initiated polymerase–promoter complex, RNA elongation is allowed with the addition of 1 μl of a mixture containing 100 μg/ml heparin, 4 mM UTP, and 4 mM CTP. The reaction is vortexed briefly and elongation is allowed for 5 min. min. The reaction is stopped with 5 μl of loading buffer (7.0 M urea, 90 mM Trizma base, 90 mM boric acid, 2 mM EDTA, 0.1% bromphenol blue, and 0.1% xylene cyanol) and loaded into wells formed in a 8% polyacrylamide slab gel (40% acrylamide:1.38% bisacrylamide; 0.75 mm $\times$ 10 cm $\times$ 12 cm) containing 7 M urea (prerun for at least 5 min). Transcripts are separated from unincorporated radionucleotides by electrophoresis at 450 V until the unincorporated nucleotides have been run off the gel. Transcripts are detected by autoradiography using Kodak XAR film (Perkin-Elmer, Boston, MA) overnight at -70°, and promoter activity is quantified on a PhosphorImager SI (Molecular Dymamics; Amersham Biosciences, Piscataway, NJ) using ImageQuant software.

Notes

1. The protocol listed uses a 10-μl reaction volume, but the assay can be scaled up to 20 μl if reproducibility is problematic.

2. Binding of the polymerase to the *spoIIG* promoter is very rapid. Extended incubation times prior addition to elongation nucleotides will lead to loss of products. Elongation times greater than 5 min have no additional effect on transcription in a single-round assay.

3. Potassium acetate concentrations greater than 120 mM in the transcription reactions reduce transcription, and the reaction is sensitive

to chloride ions at low concentration. Low ionic strength in the reaction appears to reduce the specificity of initiation. The useful range of ionic strength is 10 to 80 mMKAc.

4. 10× transcription buffer, 1 mg/ml heparin, and NTP stocks in Tris, pH 7.4 are stored at −20°. Heparin solutions lose effectiveness over a period of 12 months. Nucleotides are purchased as 100 mM stocks (Amersham Biosciences, Piscataway, NJ) and are diluted to 40 nM working stocks.

5. In testing order of addition experiments, the RNA polymerase should not be incubated at 37° in the absence of DNA as it loses activity rapidly.

Electrophoretic Mobility Shift Assays

Rationale

The electrophoretic mobility shift assay (EMSA) was developed by Garner and Revzin[33] and Fried and Crothers.[34] Excellent treatments of the theory underlying this assay system may be found elsewhere.[35,36] We note that we have yet to find conditions under which activated Spo0A shifts the mobility of DNA fragments reproducibly. Furthermore, RNA polymerase binds relatively poorly to promoters such as *spoIIG*, presumably because of the nonconsensus nature of the promoter elements so that care must be used in interpreting results from EMSA. EMSA has been useful in examining the effects of other regulators on Spo0A~P–RNA polymerase complexes.[27,29]

Protocol

Template DNA is prepared by end labeling PUCIIG*trpA*. The plasmid (25 μg) is digested with *Bam*HI (which cuts 135 bp downstream of the transcription start site) in a final volume of 50 μl. The template is then dephosphorylated and labeled using the forward reaction as described by Sambrook *et al.*[37] The DNA fragment containing the *spoIIG* promoter is then released from the vector by digestion with *Hind*III, and the labeled fragments are separated on a 5% nondenaturing polyacrylamide slab gel

[33] M. M. Garner and A. Revzin, *Nucleic Acids Res.* **9**, 3047 (1981).
[34] M. Fried and D. M. Crothers, *Nucleic Acids Res.* **9**, 6505 (1981).
[35] D. Lane, P. Prentki, and M. Chandler, *Microbiol. Rev.* **56**, 509 (1992).
[36] L. D. Kerr, *Methods Enzymol.* **254**, 619 (1995).
[37] J. Sambrook, E. F. Fritsch, and T. Maniatis, "Molecular Cloning: A Laboratory Manual." Cold Spring Harbor Laboratory Press, Cold Spring Harbor, NY, 1989.

(1.5 mm × 16.0 cm × 10.0 cm, sample loaded in two wells) in 0.5 × TBE buffer[37] by electrophoresis at 15 W until the xylene cyanol has migrated three-fourths the length of the gel. The 235-bp fragment containing the end-labeled promoter is localized by exposure to an X-ray film for less than 1 min and excised, and the DNA is either electroeluted into a dialysis bag[37] or recovered by chopping the gel slice into small pieces with a razor, transferring the pieces to a 1.7-ml microfuge tube and incubating in 800 μl of elution buffer (500 mM NH$_4$Ac, 10 mM MgAc, 10 mM Tris, pH 7.9 and 1 mM EDTA) for 16 to 18 h at 37°. After elution, the DNA is ethanol precipitated and dissolved in 10 mM HEPES, pH 8.0, 30 mM NaAc, and 1 mM EDTA. The efficiency of end labeling is determined by measuring the Cerenkov radiation in a diluted sample.

Complexes of RNA polymerase, Spo0A, and DNA are formed in 1× transcription buffer in a final reaction volume of 10 μl. Reactions (9 μl) containing end-labeled template DNA (2 × 10^4 cpm), different forms of Spo0A, and initiating nucleotides (final concentrations of 0.4 mM when present) are prepared in 0.65-ml microfuge tubes and held on ice. The reactions are incubated at 37° for 2 min before the addition of 1 μl of RNA polymerase. After 2 min to allow for the equilibration of ternary complex formation, a one-third volume of loading buffer containing either heparin [10 μg/ml heparin, 20% glycerol (v/v) in 1× transcription buffer] or sonicated calf thymus DNA [300 μg/ml, 20% glycerol (v/v) in 1× transcription buffer] is added. The times of addition of RNA polymerase are staggered so that each sample can be loaded immediately after the addition of loading buffer onto a running (12 V/cm) 5% polyacrylamide gel (40% polyacrylamide:1.38% bisacrylamide) containing 40 mM Trizma base, 40 mM acetic acid, 1 mM EDTA, and 2% glycerol. Electrophoresis is continued until the xylene cyanol dye front is about 1 cm from the bottom edge. Gels are dried and exposed to X-ray film or a PhosphorImager screen.

Notes

1. Sonicated calf thymus (or salmon sperm) DNA acts as a weak competitor and dissociates nonspecific protein–DNA interactions. Adding calf thymus DNA in the binding reaction versus the loading buffer yields no appreciable differences.

2. Because bromphenol blue and xylene cyanol appear to inhibit the polymerase in binding assays, they are not added to EMSA-loading buffers. To mark electrophoretic mobility, 3 μl of 30% glycerol, 1× TBE, and 0.1% xylene cyanol are loaded onto one lane on the slab gel immediately before beginning the incubation of the binding reactions so

that the dye will migrate into the gel before a reaction is applied to that lane.

3. The inclusion of EDTA in the reaction, gel, and electrophoresis buffer can be detrimental to complex formation of certain proteins and may be omitted.[27]

4. The samples should be layered very gently at the bottom of the wells using protein gel-loading pipette tips (Diamed, Mississauga, ON). This is facilitated by marking and labeling the bottom of the wells for each lane on the glass plates. Numbering the wells is highly useful, as the samples do not have tracking dye and are therefore invisible after loading.

DNase I Footprinting

Rationale

This technique was developed by Galas and Schmitz.[38] Our current protocol is based on minor modifications of a procedure first published by Bird *et al*,[26] to investigate structural transitions in the Spo0A∼P–RNA polymerase–*spoIIG* complex during transcription initiation. Comprehensive discussion of the quantitative treatment of data gained by this method may be found elsewhere.[39–41]

Protocol

The end-labeled template DNA is prepared as described for EMSA reactions except that the *Pvu*II–*Bam*HI fragment is used instead of the *Hind*III–*Bam*HI fragment. DNase assays are performed in 1× transcription transcription buffer with end-labeled template (2×10^5 cpm) in a final reaction volume of 20 μl. When appropriate, different forms of Spo0A and the initiating nucleotides (final concentrations of 0.4 mM) are included and the reaction mixture is held on ice. Samples are incubated at 37° for 90 s before the addition of 2 μl of RNA polymerase. After allowing 2 min min for equilibration of complex formation, 4 μl of a dilution of DNase I (Sigma-Aldrich Corp.) is added, the sample is vortexed briefly, returned to 37° quickly, and digestion is allowed to proceed for 10 s. The reaction

[38] D. J. Galas and A. Schmitz, *Nucleic Acids Res.* **5**, 3157 (1978).

[39] M. Brenowitz, D. F. Senear, M. A. Shea, and G. K. Ackers, *Methods Enzymol.* **130**, 132 (1986).

[40] M. L. Craig, O. V. Tsodikov, K. L. McQuade, P. E. Schlax, Jr., M. W. Capp, R. M. Saecker, and M. T. Record, Jr., *J. Mol. Biol.* **283**, 741 (1998).

[41] O. V. Tsodikov, M. L. Craig, R. M. Saecker, and M. T. Record, Jr., *J. Mol. Biol.* **283**, 757 (1998).

is terminated by the addition and rapid vortexing of 75 μl stop buffer (0.1% SDS, 4 mM EDTA, 400 nM NaCl, and 40 μg/ml sonicated salmon sperm DNA) from a "preloaded" micropipette. The samples can be returned to 37° while others are being processed or placed on ice. The DNA is ethanol precipitated, vacuum dried, and dissolved in 3.0 μl of formamide loading buffer (98% formamide, 1× TBE, 0.1% bromphenol blue, and 0.1% xylene cyanol). The dissolved DNA is transferred to a clean 0.65-ml microfuge tube, and the level of Cerenkov radiation in the sample is measured (LS 6000IC; Beckman, Fullerton, CA). Samples are then heated at 90° for 2 min and placed immediately in an ice bath, centrifuged briefly, and vortexed. An equal number of cpm for each reaction is loaded by adjusting the volume of sample applied onto a prerun 6% polyacrylamide sequencing gel containing 7 M urea in 0.5 × TBE.[37] DNA fragments are separated by electrophoresis at 2000 V and the gel is dried and exposed to Kodak XAR film or a PhosphorImager SI screen.

Notes

1. The kinetics of RNA polymerase binding and isomerization has been followed by scaling up the initial reaction mixture and volume of RNA polymerase added and by then withdrawing a 20-μl volume at appropriate time points to add to microfuge tubes containing 4 μl of DNase I. Reactions are terminated with 75 μl of stop buffer and processed as described.

2. The specific activity of DNase I preparations varies. The enzyme is titrated by performing trial inputs with serial dilutions over 2–3 orders of magnitude to determine a working concentration such that less than 70% of the DNA is digested. DNase I is diluted in a buffer containing 1× transcription buffer and 30% glycerol. Powdered DNase I is dissolved at 2 mg/ml in dilution buffer and is stored at −20°.

Permanganate Footprinting

Rationale

Permanganate footprinting was first developed by Sasse-Dwight and Gralla[42,43] to investigate DNA strand separation mediated by RNAP and takes advantage of the finding that permanganate ions react principally with unstacked thymidine bases. Our procedure was developed by Rowe-Magnus and Spiegelman[31] and relies on the modification of end-labeled

[42] S. Sasse-Dwight and J. D. Gralla, *Proc. Natl. Acad. Sci. USA* **85**, 8934 (1988).
[43] S. Sasse-Dwight and J. D. Gralla, *J. Biol. Chem.* **264**, 8074 (1989).

fragments rather than modification of unlabeled DNA to be used as a template for Klenow extension of a labeled primer.

Protocol

$KMnO_4$ modification reactions are performed in $1\times$ transcription buffer containing end-labeled template DNA (1.5×10^5 cpm), different forms of Spo0A, and various nucleotide combinations (final concentrations of 0.4 mM when present) in a final volume of 20 μl. Eighteen-microliter reactions containing all components except RNA polymerase are prepared in 1.7-ml microfuge tubes and placed on ice. Samples are incubated at 37° for 2 min before the addition of 2 μl of RNA polymerase. After allowing 2 min for the formation of complexes, permanganate modification is carried out by the addition of 1 μl of freshly prepared 200 mM $KMnO_4$ solution ("AnalaR"; VWR, Missassauga, ON). After 3 min, the reactions are terminated by the addition of 50 μl of stop buffer (1.5 M 2-mercaptoethanol, 0.1 mM EDTA, 420 mM NaAc pH 7.0, 120 μg/ml sonicated salmon sperm DNA). The DNA is ethanol precipitated, dissolved in 100 μl of 1 M piperidine, incubated at 90° for 30 min, and placed on ice. The DNA is butanol precipitated,[37] dissolved in 100 μl of 1% SDS, butanol precipitated a second time, and residual piperidine removed by vacuum centrifugation for 30 min. Following resuspension in 3.0 μl of formamide loading buffer (88% formamide, 1X TBE, 0.1% bromphenol blue, 0.1% xylene cyanol), the samples are heated, separated by electrophoresis, and analyzed as described in the DNase assay reaction.

Notes

1. The *spoIIG* promoter is relatively weak even when fully stimulated by Spo0A~P. Neither increasing the concentration of potassium permanganate in the reaction nor decreasing the concentration of DTT in the transcription buffer increases the extent of modification of exposed thymidine residues.

2. We have found that the permanganate sensitivity assay is more sensitive to the concentration of potassium acetate that other assays. Thus it is prudent to examine the reactivity at various ionic strengths.

Acknowledgments

Work in the GBS laboratory is supported by grants from Natural Sciences and Engineering Research Council of Canada and from the Canadian Institutes of Health Research to GBS.

[29] Assay of Prokaryotic Enhancer Activity over a Distance *In Vitro*

By Vladimir Bondarenko, Ye V. Liu, Alexander J. Ninfa, and
Vasily M. Studitsky

Regulation of expression of eukaryotic genes depends almost entirely on enhancers—30- to 200-bp DNA sequences usually composed of several binding sites for an activator protein(s). The landmark of enhancers is their ability to activate target genes over a large distance (up to 60 kb).[1] Analysis of eukaryotic transcriptional enhancers is complicated by the fact that current RNA polymerase (RNAP) II-dependent experimental systems *in vitro* are very inefficient (less than 1% of templates are transcribed[2]). In contrast, transcriptional enhancers of *Escherichia coli* are extremely efficient and, under the appropriate conditions, can activate transcription of the majority of DNA templates *in vitro*.[3,4] Prokaryotic and eukaryotic enhancers share several key properties: activation over a large distance, tight coupling of DNA melting with ATP hydrolysis, high stability of the initiation complexes, and absolute dependence of transcription on the presence of an activator are crucial features of eukaryotic transcription machinery shared only by enhancer-dependent promoters in *E. coli* (see Buck *et al.*[5] for review). These properties of prokaryotic enhancers make them a valuable system for the investigation of enhancer function, providing data complementary to *in vivo* studies of their eukaryotic counterparts.

The mechanism of action of bacterial transcriptional enhancers has been studied intensely using the *glnAp2* promoter of *E. coli* as a model system. Activity of this promoter is entirely dependent on the activator, σ^{54} and, under the appropriate conditions *in vitro*, is dependent on the enhancer.[3,4,6] The enhancer is bound by the phosphorylated form of the activator protein NtrC, which is phosphorylated by the NtrB protein kinase, as well as by acetyl phosphate.[7,8] When phosphorylated, NtrC

---

[1] E. M. Blackwood and J. T. Kadonaga, *Science* **281**, 61 (1998).
[2] J. A. Knezetic, G. A. Jacob, and D. S. Luse, *Mol. Cell. Biol.* **8**, 3114 (1988).
[3] A. J. Ninfa, L. J. Reitzer, and B. Magasanik, *Cell* **50**, 1039 (1987).
[4] D. L. Popham, D. Szeto, J. Keener, and S. Kustu, *Science* **243**, 629 (1989).
[5] M. Buck, M. T. Gallegos, D. J. Studholme, Y. Guo, and J. D. Gralla, *J. Bacteriol.* **182**, 4129 (2000).
[6] S. Sasse-Dwight and J. D. Gralla, *Proc. Natl. Acad. Sci. USA* **85**, 8934 (1988).
[7] A. J. Ninfa and B. Magasanik, *Proc. Natl. Acad. Sci. USA* **83**, 5909 (1986).
[8] J. Keener and S. Kustu, *Proc. Natl. Acad. Sci. USA* **85**, 4976 (1988).

forms homooligomers, interacts with the $E\sigma^{54}$ holoenzyme, and stimulates conversion from the closed (RP_c) to the open (RP_o) complex.[4,6,9–12] During the enhancer–promoter interaction, intervening DNA is looped out transiently,[13,14] placing enhancer and promoter in close proximity to each other.[15]

The progress of studies of the mechanism of prokaryotic enhancer action was somewhat limited by the lack of an experimental system supporting multiple-round transcription. In particular, some models for the action of eukaryotic enhancers (such as the "hit-and-run" and "stable DNA loop" models[16]) propose existence of a "molecular memory" facilitating enhancer action during multiple-round transcription (see Blackwood and Kadonaga[1] for review). This class of the models can only be analyzed properly using a multiple-round transcription assay.

We have established an experimental system supporting the action of prokaryotic enhancers over a large distance (more than 2.5 kb) and allowing multiple rounds of transcription per template in vitro.[17–19] Using this system, we have demonstrated that DNA supercoiling greatly facilitates enhancer–promoter communication over a large distance and eliminated other previously proposed "memory" models for enhancer action over a distance.[17,18] We have also shown that low-affinity NtrC-binding sites between the enhancer and the promoter act as a "governor" to reduce the rate of transcription initiation at very high activator concentrations.[19] This article describes key experimental techniques used for the analysis of the mechanism of enhancer action.

Materials

MicroSpin G-50 column, Sephacryl S-400, and heparin–Sepharose (Pharmacia); hydroxyapatite, Bio-Rex 70, and silver stain kit (BioRad); plasmid purification mega kit (Qiagen)

[9] M. Buck and W. Cannon, Mol. Microbiol. **6**, 1625 (1992).
[10] S. C. Porter, A. K. North, A. B. Wedel, and S. Kustu, Genes Dev. **7**, 2258 (1993).
[11] A. Wedel and S. Kustu, Genes Dev. **9**, 2042 (1995).
[12] C. Wyman, I. Rombel, A. K. North, C. Bustamante, and S. Kustu, Science **275**, 1658 (1997).
[13] W. Su, S. Porter, S. Kustu, and H. Echols, Proc. Natl. Acad. Sci. USA **87**, 5504 (1990).
[14] K. Rippe, M. Guthold, P. H. von Hippel, and C. Bustamante, J. Mol. Biol. **270**, 125 (1997).
[15] A. Wedel, D. S. Weiss, D. Popham, P. Droge, and S. Kustu, Science **248**, 486 (1990).
[16] V. M. Studitsky, FEBS Lett. **280**, 5 (1991).
[17] Y. Liu, V. Bondarenko, A. Ninfa, and V. M. Studitsky, Proc. Natl. Acad. Sci. USA **98**, 14883 (2001).
[18] V. Bondarenko, Y. Liu, A. Ninfa, and V. M. Studitsky, Nucleic Acids Res. **30**, 636 (2002).
[19] M. R. Atkinson, N. Pattaramanon, and A. J. Ninfa, Mol. Microbiol. **46**, 1247 (2002).

Reagents

Ethidium bromide; butyl alcohol; ethanol; NaCl; 2-mercaptoethanol; urea; heparin; chloroform; $KMnO_4$ (Sigma); RNase inhibitor (Roche); ATP, UTP, CTP, GTP, dATP, dUTP, dCTP, dGTP (Pharmacia); $[\alpha\text{-}^{32}P]UTP$, 3000 Ci/mmol, and $[\gamma\text{-}^{32}P]ATP$, 6000 Ci/mmol (NEN); Tris-saturated phenol (Invitrogen); sheared DNA (Intergen)

Buffers

Transcription buffer (TB): 50 mM Tris–OAC pH 8.0, 100 mM KOAc, 8 mM Mg $(OAc)_2$, 27 mM NH_4OAc, 0.7% PEG (8000), and 0.2 mM dithiothreitol (DTT)
$2\times$ transcription stop buffer (2XTSB): 200 μg/ml sheared DNA, 40 mM EDTA
TMD buffer: 50 mM Tris–Cl, pH 7.2, 10 mM $MgCl_2$, 0.2 mM DTT
Quench buffer (QB): 4 M NH_4OAc, 20 mM EDTA
Protein purification buffer (PPB): 50 mM Tris, pH 8.0, 5% glycerol, 10 mM EDTA, 1 mM DTT

Enzymes

Escherichia coli enzymes: Core RNAP; σ 54; NtrC; NtrB (purified as described later); Klenow fragment (DNA polymerase I large fragment, Invitrogen); DNase I (Sigma); calf thymus DNA topoisomerase I (Invitrogen); T4 polynucleotide kinase (New England Biolabs)

Purification of DNA, Proteins, and Protein Complexes

Plasmid templates are purified using routine techniques.[17,18] NtrC and σ^{54} are purified using previously published protocols.[20,21] Methods for isolation of core RNAP and NtrB[20,22] are modified considerably and are described in detail later.

[20] T. P. Hunt and B. Magasanik, *Proc. Natl. Acad. Sci. USA* **82,** 8453 (1985).
[21] L. J. Reitzer and B. Magasanik, *Proc. Natl. Acad. Sci. USA* **82,** 1979 (1985).
[22] A. J. Ninfa, S. Ueno-Nishio, T. P. Hunt, B. Robustell, and B. Magasanik, *J. Bacteriol.* **168,** 1002 (1986).

Purification of E. coli Core RNAP

This is a modification of the protocol described by Burgess and Jendrisak.[23] This method involves French press lysis, polymin precipitation, elution of the enzyme from the polymin precipitate with high salt, ammonium sulfate precipitation, and chromatography on heparin–Sepharose and Bio-Rex 70. Core RNA polymerase is purified to electrophoretic homogeneity (more than 95% purity) with a yield of 32 mg of core enzyme from 100 g of E. coli cells. All procedures are conducted at 4°.

This method provides core polymerase that is suitable for transcription studies. For purer core polymerase, we used a BioGel 0.5 M gel filtration step between the heparin–Sepharose and Bio-Rex 70 steps, as described.[23] As an alternative to using polymerase purified as described later, core RNAP from Epicentre Technologies may be used. However, we have observed that this preparation of core polymerase is contaminated with σ^{54}, such that it is not possible to conduct control experiments lacking σ^{54}.

1. Thaw 100 g of frozen E. coli cells (JM105 strain), homogenize in 200 ml ice-cold PPB buffer supplemented with 0.2 mM phenylmethylsulfonyl fluoride (PMSF), and lyse the cells by passing twice through a French press at 12,000 psi.

2. Remove cell debris by centrifugation on a Sorvall rotor GS3 at 8000 rpm for 45 min. Carefully transfer the clear supernatant into a glass beaker and measure its volume.

3. Slowly add (with constant stirring) 10% polymin P, pH8.0, to a final concentration 0.4%. Continue stirring for 10 min.

4. Precipitate aggregated DNA–protein complexes by centrifugation at 8000 rpm for 15 min. Remove the supernatant, scrape the pellet into a prechilled low-speed Hamilton blender with 250 ml of PPB containing 0.5 M NaCl, and resuspend pellet with gentle stirring for 5–10 min. Remove supernatant and wash the pellet one more time with PPB buffer containing 0.5 M NaCl.

5. Remove supernatant and dissolve precipitate in 200 ml of PPB buffer containing 1 M NaCl using a blender as described earlier. Add ammonium sulfate powder to the 1 M NaCl eluate with stirring to 50% saturation (35 g 100 ml). When salt is dissolved completely, continue stirring for 20 min.

---

[23] R. R. Burgess and J. J. Jendrisak, *Biochemistry* **14**, 4634 (1975).

6. Centrifuge the mixture at 8000 rpm for 15 min. Discard supernatant and dissolve drained pellet in 25 ml of PPB buffer containing 10 mM MgCl$_2$ and 50 mM NaCl. Dialyze against 1 liter of the same buffer for 6 h.

7. Clear the dialyzed solution to remove formed precipitate at 15,000 rpm on a Sorvall SS-34 rotor for 10 min and load cleared supernatant on a 15-ml (8.5 × 1.5 cm) heparin–Sepharose column equilibrated with PPB buffer containing 10 mM MgCl$_2$ and 50 mM NaCl buffer. Wash column with 60 ml of the same buffer and elute RNA polymerase with a 100-ml linear salt gradient from 0.05 to 0.5 M NaCl at a flow rate of 0.5 ml/min collecting 2-ml fractions. A peak of RNA polymerase, identified by 10% SDS–PAGE (AA:Bis ratio 29:1), elutes at about 0.25 M NaCl. Pool fractions containing RNA polymerase and dialyze overnight against 2 liters of PPB buffer containing 0.15 M NaCl at 4°.

8. Load dialyzed fractions on a 125-ml (30 × 4 cm) Bio-Rex 70 column equilibrated with PPB buffer containing 0.15 M NaCl. Wash the column with 200 ml of the same buffer and elute with 400 ml of a linear salt gradient from 0.15 to 1.0 M NaCl at a flow rate of 1 ml/min collecting 4-ml fraction. Analyze eluate by 10% SDS–PAGE to identify fractions containing core RNA polymerase. The σ^{70} subunit comes out in a flow-through fraction, and the core RNA polymerase is eluted at about 0.4 M NaCl.

9. Combine fractions containing core enzyme and dialyze against 10 mM Tris, pH 8.0, 50% glycerol, 0.1 mM EDTA, 0.1 mM DTT, 100 mM NaCl overnight. Freeze fractions in liquid nitrogen and store at −70°.

Expression and Purification of Recombinant E.coli *NtrB*

The NtrB protein is isolated from *E.coli* strain RB9131R carrying plasmid pLOP, which contains the *glnL* (encoding the NtrB protein) gene under control of the λP_L promoter and temperature-sensitive λ repressor. Ammonium sulfate fractionation is followed by DE52 and hydroxyapatite-chromatography. The expression of NtrB is induced by raising the temperature during growth of the bacterial culture (in the middle of log phase) from 30° to 44° to release the inhibitory effect of the λ repressor on transcription of the *glnL* gene. The recombinant NtrB protein is isolated with >95% purity, resulting in a yield of about 12 mg of the enzyme from 8 g of induced cells. All procedures are conducted at 4°.

The NtrB purified as described later is free of detectable nuclease activity or other confounding activities and is suitable for transcription studies. The strongest step in the fractionation procedure is ammonium sulfate fractionation, after which NtrB is about 70–80% pure. As an alternative to the hydroxyapatite step, gel filtration on Sephadex G-100 may be used as the final step. For optimal purity of NtrB, chromatography on agarose-ethane is used as an initial chromatography step[22]; however, this is not required for transcription studies.

1. Grow RB9131R/pLOP in 1 liter of LB medium containing 100 μg/ml ampicillin at $30°$ to $OD_{600} \sim 1.0$, raise the temperature to $44°$, and continue incubation for additional 4 h to induce expression of NtrB.

2. Harvest cells by centrifugation on a Sorvall GS-3 rotor at 8000 rpm for 15 min.

3. Resuspend cells ($\sim$8 g total) in 25 ml of PPB buffer containing 50 mM Tris–HCl, pH 8.0, 5% glycerol, 1 mM EDTA, and 0.1 mM DTT. Lyse cells by passing twice through a French press at 12,000 psi and remove cell debris by centrifugation at 12,000 rpm (20,000 g) for 30 min on a Sorvall SS-34.

4. Slowly add ammonium sulfate powder to the cleared lysate, with constant stirring, until the first precipitate is formed (when the concentration of ammonium sulfate reaches $\sim$2 %) and continue stirring for an hour or overnight at $4°$. Spin down precipitate at 12,000 rpm for 30 min.

5. Discard the supernatant and resuspend drained pellet in 15 ml of PPB buffer. Dialyze mixture overnight against 0.5 liter of the same buffer and centrifuge dialyzes solution at 12,000 rpm for 20 min to remove precipitate formed during dialysis.

6. Load 15 ml of the cleared supernatant containing NtrB on a 35-ml (12 $\times$ 2 cm) DE52 column at a flow rate of 0.5 ml/min equilibrated with PPB buffer containing 25 mM NaCl, wash column with 70 ml of the same buffer, and elute with a 200-ml linear salt gradient from 0.025 to 1.0 M NaCl at a flow rate of 0.5 ml/min. NtrB is eluted at $\sim$0.25 M NaCl.

7. Combine fractions containing NtrB (identified by SDS–PAGE) and add an equal volume of 100% saturated ammonium sulfate with constant stirring. Incubate in ice for an hour and spin down precipitate at 20,000 g for 30 min. Discard supernatant and resuspend drained pellet in 12 ml of PPB buffer. Spin mixture to remove insoluble material at 12,000 rpm for 30 min and dialyze supernatant overnight against 1.0 liter of 10 mM Na-phosphate buffer, pH 7.0 at $4°$.

8. Load the dialyzed solution on a 12-ml (4 $\times$ 2 cm) hydroxyapatite column at a flow rate of 0.3 ml/min equilibrated with 10 mM Na-phosphate, pH 7.0, and elute with 100 ml of linear gradient of Na-phosphate, pH 7.0, from 10 to 500 mM collecting 2-ml fractions.

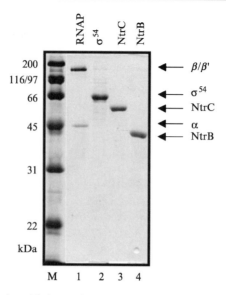

FIG. 1. Analysis of purified proteins used for reconstitution of enhancer-dependent transcription *in vitro*. Proteins were separated in 10% PAGE and stained with Coomassie. Mobilities of protein subunits are indicated on the right. M, protein molecular mass markers.

9. Pool fractions containing NtrB (identified by 10% SDS–PAGE) and dialyze overnight against buffer containing 50 mM Tris–HCl, pH 7.5, 50% glycerol, and 0.1 mM EDTA. Freeze in liquid nitrogen and store at $-70°$ (up to several years without significant loss of activity).

10. Analyze purified proteins in 10 % SDS–PAGE (Fig. 1).

Single- and Multiple-Round Assays for Enhancer-Dependent Transcription

After purification of the proteins required for specific enhancer-dependent transcription, the whole system has to be characterized extensively using a single-round transcriptional assay (Fig. 2A). Single- and multiple-round transcription assays are also useful for analysis of different aspects of enhancer action[3,4,17,18] (see later); an example is shown in Fig. 2B. The multiple-round transcription assay is particularly useful for analysis of kinetic aspects of enhancer action. Thus it has been shown that the first and subsequent rounds of enhancer-dependent transcription *in vitro* occur with similar rates, suggesting that no "memory" is established during the first round[18] (Fig. 3).

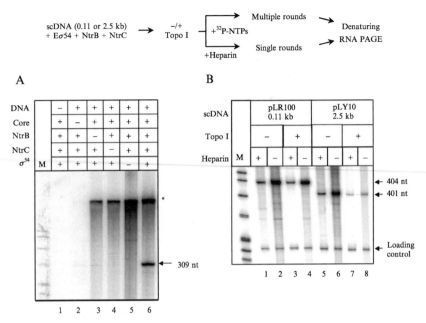

FIG. 2. Analysis of *glnAp2* promoter activation by NtrC-dependent enhancer using single- and multiple-round transcription assays. The experimental approach is outlined at the top. (A) Characterization of the enhancer-dependent transcription system *in vitro* using super-coiled plasmid pTH8 (0.11-kb enhancer–promoter spacing). Different combinations of the purified proteins were analyzed in a single-round transcription assay (lanes 1 to 6). The specific product (309 nucleotide transcript) was generated only in the presence of the full set of proteins (lane 6) and ATP (not shown). Asterisk indicates position of nonspecific transcript that is present when the core RNAP and a plasmid template are added to the reaction.[3] Note that specific transcription entirely depends on the presence of each component in the reaction. M, end-labeled pBR322–*Msp*I digest (it was used in all experiments, which include analysis of labeled RNA or DNA). (B) DNA supercoiling is required for activation of transcription over a 2.5-kb distance[17] (copyright 2001 National Academy of Sciences, USA). sc plasmid templates having 0.11- or 2.5-kb enhancer–promoter spacing (pLR100 and pLY10 plasmids, respectively[17]) were incubated in the presence or in the absence of topo I (+/− topo I); incubation with calf thymus topo I converts DNA into a completely relaxed state. Then transcription was conducted under single-round (+ heparin) or multiple-round (−heparin) conditions. The loading control (227-bp end-labeled DNA fragment) was added to the reaction mixtures immediately after terminating the reaction.

The single-round assay is based on the observation that pre-formed RP_o is stable in the presence of heparin (80 μg/ml); however, its formation *de novo* is strongly inhibited.[24] First, the components are incubated in the presence of ATP as the sole nucleotide. ATP is required at two points in

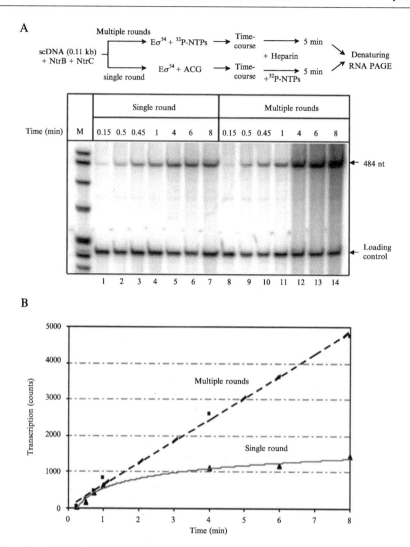

Fig. 3. Quantitative analysis of the rates of single-and multiple-round enhancer-dependent transcription of the glnAp2 promoter[18] (copyright 2002 Oxford University Press). The experimental strategy for comparison of the rates of single- and multiple-round transcription is outlined at the top. (A) Time courses of single-round and multiple-round transcription of supercoiled pLR100 plasmid having 110-bp enhancer–promoter spacing. Reaction mixtures were incubated in the presence of all nucleotides (multiple round) or with ACG mixture only (single round). Use of the ACG mixture instead of ATP prevents conversion of the RP_o back to RP_c[24] and thus allows comparison of single- and multiple-round transcription under similar conditions. Labeled transcripts were analyzed in a denaturing PAGE. The loading control (a 227-bp end-labeled DNA fragment) was added to the reaction mixtures immediately after

the transcription cycle to obtain the open complex. First, ATP is required for the phosphorylation of NtrC by NtrB to form the active form of the activator.[7,8] Second, the cleavage of ATP by oligomerized NtrC~P bound at the enhancer is required for the formation of the open complex by RNA polymerase.[25] In the absence of additional nucleotides, the open complex accumulates. Transcripts are then obtained by adding a mixture of the remaining nucleotides (CTP, GTP, UTP) and heparin, which prevents reinitiation. In some experiments, DNA (that is highly supercoiled after its isolation from bacterial cells) is completely relaxed with calf thymus topo I.[17,18]

To obtain early elongation complexes [RP_{el}], the components are incubated in the presence of all nucleotides except UTP. During this incubation, open complexes are formed, transcription initiation proceeds, and elongation complexes are stalled at positions requiring the incorporation of UTP (at nucleotide 18[18,26]). These ternary complexes of template, polymerase, and mRNA are extremely stable; thus, their accumulation serves as a useful reporter of open complex formation. To obtain transcripts, the ternary complexes are then given a mixture of UTP and heparin, allowing for the formation of full-length transcripts and preventing reinitiation. The protocol provided here is most useful for studies of enhancer action and better reflects the situation *in vivo*, where initiation occurs readily on open complex formation.

The multiple-round assay is similar to the single-cycle assays, except that all components are present from the start and heparin is not included.

1. Form initiation complexes in 50 μl TB at the following final concentration of the components: 2.8 nM supercoiled (sc) plasmid DNA, 500 nM core RNAP, 1000 nM σ^{54}, 120 nM NtrC, 400 nM NtrB. Incubate at 37° for 15 min to form RP_c. Under these conditions, NtrC binds quantitatively to the enhancer and RNAP forms the RP_c at the promoter, but enhancer–promoter communication does not occur.[3,4,6,9,17,18]

[24] J. Feng, T. J. Goss, R. A. Bender, and A. J. Ninfa, *J. Bacteriol.* **177**, 5523 (1995).
[25] D. S. Weiss, J. Batut, K. E. Klose, J. Keener, and S. Kustu, *Cell* **67**, 155 (1991).
[26] Y. Tintut, J. T. Wang, and J. D. Gralla, *Genes Dev.* **9**, 2305 (1995).

terminating the reaction. (B) Quantitative analysis of data shown in Fig. 4A. The intensities of the bands containing 484 nucleotide transcripts were quantified using a PhosphorImager. The half-time for transcription initiation in the first round was ~1 min and during multiple-round transcription one transcript was synthesized every 2 min. Thus transcription initiation during the first and subsequent rounds occurs with similar rates.

2. To completely relax plasmid DNA, add topo I to a final concentration of 0.1 units/μl and incubate at $37°$ for 30 min.

3. Add ATP to a final concentration of 4 mM and incubate at $37°$ for another 15 min to form RP$_o$.

4. To form RP$_{el}$, add GTP and CTP to final concentration of 50 μM each and incubate at $37°$ for 10 min.

5. For single-round transcription, add heparin to a final concentration of 80 μg/ml.

6. Add RNase inhibitor to a final concentration of 0.2 unit μl, GTP, UTP, and CTP to final concentrations of 80 μM together with 2.5 μCi [α-^{32}P]UTP. Incubate at $37°$ for various times.

7. Add equal volume of 2X TSB to terminate the reaction.

8. Add loading control (a labeled DNA fragment having a distinct mobility in a denaturing gel).

9. Extract the samples with 100 μl phenol:chloroform (1:1).

10. Precipitate with ethanol, wash with 70% ethanol, and dissolve in 5 μl 100% formamide.

11. Denature the sample at $90°$ for 1 min, cool on ice, and load 1–3 μl on 8% denaturing urea-containing PAGE.

Analysis of the Rate of Enhancer–Promoter Communication

The rates of enhancer–promoter communication on supercoiled and relaxed DNA templates can be compared quantitatively using a single-round transcription assay. In this case, communication is initiated by adding ATP after preformation of RP$_c$ and NtrC–DNA complex. Because both RNAP and the enhancer-binding protein (NtrC) are prebound to DNA, measurements of the rate of enhancer–promoter communication are not complicated by processes of establishing DNA–protein interactions[17] (Fig. 4). The addition of ATP results in NtrC phosphorylation that initiates NtrC–RNAP communication and eventually leads to RP$_o$ formation.

1. Form RP$_c$ in 50 μl of TB as described earlier. Add ATP to a final concentration 0.5 mM to initiate enhancer–promoter communication. Incubate at $37°$ for different times (usually from 1 to 30 min).

2. Terminate enhancer–promoter communication and start single-round transcription by adding heparin to a final concentration of 80 μg/ml and NTPs to a final concentration of 80 μM [together with 2.5 μCi of [α-^{32}P]UTP and RNase inhibitor (final concentration 0.2 units/μl)].

3. Incubate the samples at $37°$ for 10 min to complete transcription.

A

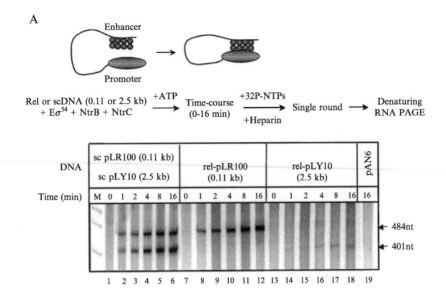

B

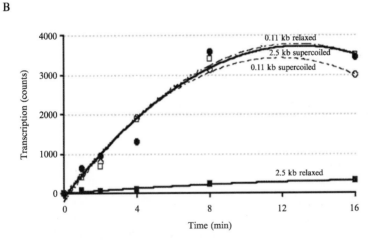

FIG. 4. Analysis of the rates of enhancer–promoter communication over a large distance on relaxed and supercoiled DNA[17] (copyright 2001 National Academy of Sciences, USA). (A) Time courses of enhancer–promoter communication on relaxed and sc plasmids having 0.11- or 2.5-kb enhancer–promoter spacing. Analysis of labeled transcripts in a denaturing PAGE. The experimental strategy for comparison of the rates of enhancer–promoter communication on relaxed (topo I+) and sc (topo I−) plasmids having 0.11- or 2.5-kb enhancer–promoter spacing is outlined at the top. The control pAN6 plasmid does not contain the enhancer. See text for detail. (B) Quantitative analysis of data shown in A. The intensities of the bands containing 484 and 401 nucleotide transcripts were analyzed using a PhosphorImager. Transcription is saturated after three to four rounds, probably because ATP pool is depleted by NtrC.

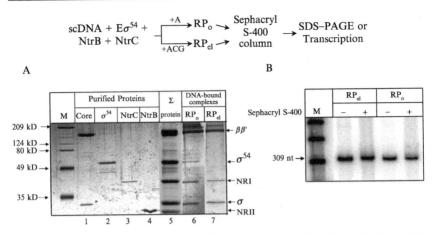

FIG. 5. Analysis of purified RP_o and RP_{el}[18] (copyright 2002 Oxford University Press). The experimental strategy for purification of RP_c and RP_o complexes is outlined at the top. (A) RNA polymerase subunit σ^{54} is depleted from the early elongation complex. Complexes were purified from DNA-free proteins on a Sephacryl S-400 column, separated in an SDS–PAGE, and silver stained (lanes 6 and 7). Purified proteins were loaded as additional markers (lanes 1 to 4). Total proteins present in the reaction mixture were stained with Coomassie (lane 5). M, protein molecular mass markers. (B) Functional RP_o and RP_{el} complexes survive the Sephacryl S-400 column. RP_o and RP_{el} complexes were analyzed using a single-round transcription assay before (−) or after (+) fractionation on a Sephacryl S-400 column. No DNA-free proteins were added to the reaction after the column. The majority of the complexes (about 75%) remain functionally active after purification on the column.

4. Add an equal volume of 2XTSB to terminate the reaction.

5. Prepare the samples and run denaturing PAGE as described earlier.

Purification of Initiation and Early Elongation Complexes by Gel Filtration

The RP_o and RP_{el} formed on plasmid DNA can be purified from proteins that are not bound or loosely bound to the DNA by gel filtration chromatography. Remarkably, both complexes remain functionally active after the chromatography[18,27] (Fig. 5B; RP_c does not survive the chromatography). This procedure is particularly useful for the analysis of protein composition of the complexes (Fig. 5A). Thus it has been shown that σ^{54} is dissociated from the template after the RNAP leaves the GlnAp2 promoter; this experiment provided structural evidence for the

[27] A. J. Ninfa, E. Brodsky, and B. Magasanik, in "DNA–Protein Interactions in Transcription." A. R. Liss, New York, 1989.

lack of the "memory" during enhancer action.[18] This technique is complementary to the analysis of the complexes by native PAGE and DNase I footprinting.[3,4,18,26]

1. For the experiments, including purification of RP_o and RP_{el} by gelfiltration chromatography, the complexes are prepared as described earlier but in larger amounts (in 200 μl) with the following concentrations of the components: 100 nM supercoiled plasmid DNA, 1 μM core RNA polymerase, 6 μM σ^{54}, 1.5 μM NtrC, and 1 μM NtrB.

2. Equilibrate a 4-ml (14 × 0.6 cm) Sephacryl S-400 column by washing with 8 ml TB at the rate of 0.2 ml/min.

3. Load 200 μl RP_o or RP_{el}. Wash the column with 1 ml TB.

4. Elute the complexes with 1 ml TB. Collect eluate in 100-μl aliquots.

5. Measure OD_{260} to identify the DNA peak. The DNA–protein complexes should coelute with plasmid DNA in the void volume while unbound proteins are retained on the column.

6. Pool peak fractions having a DNA concentration of 0.5 A_{260} or higher. At this point, functional activities of the complexes or their protein composition can be analyzed.

7. Analysis of functional activity: Add all NTPs to a final concentration of 0.2 mM to the pooled fractions to begin (RP_o) or resume (RP_{el}) a single-round transcription as described previously. Stop the reaction and analyze the transcripts as described earlier.

8. Analysis of protein composition: Add one-fourth volume of 5X SPLB and incubate the samples at 95° for 3 min. Separate the proteins in a 10% SDS–PAGE. Stain the gel with the silver stain kit to visualize the proteins (Fig. 5A).

Acknowledgment

The work was supported in part by NIH Grant GM58650 to V.M.S.

[30] DnaA as a Transcription Regulator

By Walter Messer and Christoph Weigel

Functions of DnaA

DnaA is the key protein for the initiation of chromosome replication in *Escherichia coli* and probably most other eubacteria.[1–4] Also, the replication of various bacteriophages and plasmids requires host DnaA as well as the cognate initiator protein. In addition, DnaA is a transcription factor involved in the regulation of several *E. coli* genes.[5] DnaA can act as a repressor [*dnaA*,[6–9] *mioC* (for review see Messer and Welgel[2]), *rpoH*,[10] *uvrB*,[11] *proS*[12]] or it can be a transcriptional activator (*nrd*,[13] *gua*,[14,15] *glpD*,[2] *fliC*,[16] *polA*,[17] $\lambda\ p_R$[18]). DnaA can also terminate transcription,[19,20] provided two suitably located DnaA boxes mediate loop formation of the template.[21] The transcription regulator function of DnaA is not essential

[1] K. Skarstad and E. Boye, *Biochim. Biophys. Acta* **1217**, 111 (1994).
[2] W. Messer and C. Weigel, in "*Escherichia coli* and *Salmonella*, Cellular and Molecular Biology" (F. C. Neidhardt, R. Curtiss III, J. Ingraham, E. C. C. Lin, K. B. Low, B. Magasanik, W. S. Reznikoff, M. Riley, M. Schaechter, and H. E. Umbarger, eds.), p. 1579. ASM, Washington, DC, 1996.
[3] J. M. Kaguni, *Mol. Cells* **7**, 145 (1997).
[4] W. Messer, *FEMS Microbiol. Rev.* **26**, 355 (2002).
[5] W. Messer and C. Weigel, *Mol. Microbiol.* **24**, 1 (1997).
[6] T. Atlung, E. Clausen, and F. G. Hansen, *Mol. Gen. Genet.* **200**, 442 (1985).
[7] R. E. Braun, K. O'Day, and A. Wright, *Cell* **40**, 159 (1985).
[8] C. Kücherer, H. Lother, R. Kölling, M. A. Schauzu, and W. Messer, *Mol. Gen. Genet.* **205**, 115 (1986).
[9] C. Speck, C. Weigel, and W. Messer, *EMBO J.* **18**, 6169 (1999).
[10] Q. Wang and J. M. Kaguni, *J. Biol. Chem.* **264**, 7338 (1989).
[11] E. A. van den Berg, R. H. Geerse, J. Memelink, R. A. L. Bovenberg, F. A. Magnée, and P. van de Putte, *Nucleic Acids Res.* **13**, 1829 (1985).
[12] Z. Zhou and M. Syvanen, *Mol. Gen. Genet.* **172**, 281 (1990).
[13] L. B. Augustin, B. A. Jacobson, and J. A. Fuchs, *J. Bacteriol.* **176**, 378 (1994).
[14] F. Tesfa-Selase and W. T. Drabble, *Mol. Gen. Genet.* **231**, 256 (1992).
[15] C. Schaefer, A. Holz, and W. Messer, in "DNA Replication: The Regulatory Mechanisms" (P. Hughes, E. Fanning, and M. Kohiyama, eds.), p. 161. Springer, Berlin, 1992.
[16] T. Mizushima, A. Tomura, T. Shinpuku, T. Miki, and K. Sekimizu, *J. Bacteriol.* **176**, 5544 (1994).
[17] A. Quinones, G. Wandt, S. Kleinstauber, and W. Messer, *Mol. Microbiol.* **23**, 1193 (1997).
[18] A. Szalewska-Palasz, A. Wegrzyn, A. Blaszczak, K. Taylor, and G. Wegrzyn, *Proc. Natl. Acad. Sci. USA* **95**, 4241 (1998).
[19] A. Gielow, C. Kücherer, R. Kölling, and W. Messer, *Mol. Gen. Genet.* **214**, 474 (1988).
[20] C. Schaefer and W. Messer, *EMBO J.* **8**, 1609 (1989).

for survival, as *dnaA*(Null) mutants in which the replication defect is suppressed are viable.

Structure of DnaA

Escherichia coli DnaA (SWISS accession number P03004) is a basic (*pl* 8.77) DNA-binding protein with a molecular mass of 52.5 kDa. The protein is composed of four structural and functional domains, which were identified by genetic and comparative biochemical studies of the *E. coli, B. subtilis, S. coelicolor,* and *T. thermophilus* DnaA proteins.[4,22] The four-domain structural model of DnaA could be confirmed by secondary structure analysis of the >100 *dnaA* sequences known at the end of 2001 (C. Weigel and W. Messer, http://www.molgen.mpg.de/~messer/). No candidate for a eukaryotic DnaA homologue could yet be identified, but all sequenced eubacterial genomes contain a *dnaA* gene(s) with the exception of the tsetse fly endosymbiont *Wigglesworthia glossinidia* genome.[23]

During the initiation of chromosome replication in *E. coli*, DnaA interacts with other replisomal proteins, including DnaA itself,[24–26] DnaB,[27,28] and the clamp/clamp loader complex of polymerase III in cooperation with the Hda protein.[29–31] It is presently not known whether DnaA—as a transcription factor—interacts physically with RNA polymerase, but the observed allele-specific suppression of temperature-sensitive *dnaA* mutants by *rpoB* mutants strongly supports this hypothesis.[32,33]

The N-terminal domain 1 (86 residues) promotes oligomerization of DnaA as well as interaction with DnaB and is linked to the core domains

[21] G. Konopa, A. Szalewska-Palasz, A. Schmidt, S. Srutkowska, W. Messer, and G. Wegrzyn, *FEMS Microbiol. Lett.* **174**, 25 (1999).

[22] W. Messer, F. Blaesing, J. Majka, J. Nardmann, S. Schaper, A. Schmidt, H. Seitz, C. Speck, D. Tüngler, G. Wegrzyn *et al.*, *Biochimie* **81**, 819 (1999).

[23] L. Akman, A. Yamashita, H. Watanabe, K. Oshima, T. Shiba, M. Hattori, and S. Aksoy, *Nature Genet.* **32**, 402 (2002).

[24] C. Weigel, A. Schmidt, H. Seitz, D. Tuengler, M. Welzeck, and W. Messer, *Mol. Microbiol.* **34**, 53 (1999).

[25] W. Messer, F. Blaesing, D. Jakimowicz, J. Majka, J. Nardmann, S. Schaper, H. Seitz, C. Speck, C. Weigel, G. Wegrzyn *et al.*, *Biochimie* **83**, 1 (2001).

[26] D. Jakimowicz, J. Majka, G. Konopa, G. Wegrzyn, W. Messer, H. Schrempf, and J. Zakrzewska-Czerwinska, *J. Mol. Biol.* **298**, 351 (2000).

[27] M. D. Sutton, K. M. Carr, M. Vicente, and J. M. Kaguni, *J. Biol. Chem.* **273**, 34255 (1998).

[28] H. Seitz, C. Weigel, and W. Messer, *Mol. Microbiol.* **37**, 1270 (2000).

[29] Z. Kelman and M. O'Donnell, *Annu. Rev. Biochem.* **64**, 171 (1995).

[30] T. Katayama and K. Sekimizu, *Biochimie* **81**, 835 (1999).

[31] J. Kato and T. Katayama, *EMBO J.* **20**, 4253 (2001).

[32] T. Atlung, *Mol. Gen. Genet.* **197**, 125 (1984).

[33] M. M. Bagdasarian, M. Izakowska, and M. Bagdasarian, *J. Bact.* **130**, 577 (1977).

3+4 by a flexible loop, domain 2 (48 residues). *E. coli* DnaA binds tightly to ATP and has a weak intrinsic ATPase activity.[34] Domain 3 (239 residues) shows a specific $[\alpha\beta]_5$ Rossman fold-type ATPase motif,[35] which makes DnaA a member of the so-called AAA+ family of NTPases.[36] Members of this family include replication proteins (e.g., DnaC, the γ, δ, and δ' subunits of DNA polymerase III), recombination proteins (e.g., RuvB, ClpX), and transcription factors (e.g., NtrC). A characteristic of this protein family, and also of DnaA, is the coupling of nucleotide hydrolysis with conformational changes, which in turn modify their oligomerization and DNA-binding properties (see later). The published crystal structure of domains 3 and 4 of one DnaA protein, from *Aquifex aeolicus*, shows the residues involved and their interactions.[37]

Different DNA-Binding Properties of DnaA

The C-terminal domain 4 of DnaA (94 residues) is responsible for DNA binding.[38] Monomeric DnaA binds with high affinity ($k_D \approx 1$ n*M*) to the nonpalindromic motif 5'-TT^A/$_T$TNCACA, the so-called DnaA box.[39] This motif is generally used when searching for putative DnaA-binding sites in genomic sequences. Degenerate DnaA boxes with one or two mismatches bind DnaA as well, but they require the cooperation of two monomers, one bound to a consensus DnaA box.[9] The efficiency of DnaA binding to the DnaA box is context dependent[39] and involves both DNA strands.[40] In contrast to ADP-DnaA, ATP-DnaA also binds to the degenerate 6-mer ATP-DnaA box motif 5'-Agatct ($k_D \approx 0.4$ μM).[9] Binding of ATP-DnaA to ATP-DnaA boxes strictly requires the presence of a DnaA box close by and DnaA oligomerization.[9] ATP-DnaA also binds to single-stranded ATP-DnaA boxes ($k_D \approx 40$ n*M*).[41] Although DnaA binding to single-stranded DNA does not require the presence of a strong dsDNA-binding site close by, it probably requires DnaA oligomerization for stabilization. Although DnaA binding to single-stranded DNA has been described so far only for the replication origin of *E. coli*, *oriC*, it should be noted that it could, in principle, occur temporarily also at DnaA-regulated promoters.

[34] K. Sekimizu, D. Bramhill, and A. Kornberg, *Cell* **50**, 259 (1987).
[35] M. G. Rossmann, D. Moras, and K. W. Olsen, *Nature* **250**, 194 (1974).
[36] A. F. Neuwald, L. Aravind, J. L. Spouge, and E. V. Koonin, *Genome Res.* **9**, 27 (1999).
[37] J. P. Erzberger, M. M. Pirruccello, and J. M. Berger, *EMBO J.* **21**, 4763 (2002).
[38] A. Roth and W. Messer, *EMBO J.* **14**, 2106 (1995).
[39] S. Schaper and W. Messer, *J. Biol. Chem.* **270**, 17622 (1995).
[40] C. Speck, C. Weigel, and W. Messer, *Nucleic Acids Res.* **25**, 3242 (1997).
[41] C. Speck and W. Messer, *EMBO J.* **20**, 1469 (2001).

Sequence-independent binding of nucleotide-free DnaA to DNA may be considered unphysiological.[42]

The DnaA proteins of *E. coli, S. coelicolor*, and *T. thermophilus* differ considerably in their requirements for the number and quality of DnaA boxes for efficient binding, probably due to differences in their capacity to oligomerize. Comparative studies allowed the derivation of a set of "rules for binding" for the individual DnaA proteins, which were described in detail by Messer *et al.*[25]

Regulation of Transcription by DnaA *in Vivo*

Simple and reliable genetic methods are available for a preliminary analysis whether a promoter is regulated by DnaA. The easiest way to obtain indications for a regulation by DnaA is to measure the change in expression level of a given gene in a *dnaA*(ts) mutant on a temperature shift. If the results from such preliminary experiments are inconclusive due to small variations in the expression level, they might be improved by overexpressing DnaA. Transcript levels can be measured using standard techniques, such as Northern blots, primer extension, or S1 mapping.[43] A genome-wide survey for DnaA-regulated promoters can, in principle, be done using the array technique. To our knowledge, this has not yet been done.

The use of monitor genes is a frequently used quantitative approach to determine the regulation of a promoter by DnaA. It requires the construction of a reporter gene fusion to the promoter in question. A *dnaAp::lacZ* reporter gene fusion was used to characterize the autoregulation of DnaA expression.[6,7] Although easier to construct than λ-based reporter gene fusions, plasmid-borne reporter gene fusions have the disadvantage that the copy number of the reporter plasmid may vary with growth conditions and thus demand tedious copy number measurements for normalization. A convenient method for the transfer of a reporter gene to the chromosome has been described by Diederich *et al.*[44]; the outline of this method is described in the following.

Experimental Protocol

1. Grow a pLDR8 carrying strain with kanamycin (50 $\mu g \cdot ml^{-1}$) at 30° and check for kanamycin resistance at 30° and sensitivity at 42° (important step!). The low copy-number plasmid pLDR8 carries a temperature-sensitive

[42] M. Makise, T. Tsuchiya, and T. Mizushima, *J. Biochem.* **131**, 419 (2002).
[43] J. Sambrook, E. F. Fritsch, and T. Maniatis, "Molecular Cloning." Cold Spring Harbor Laboratory Press, Cold Spring Harbor, NY, 1989.
[44] L. Diederich, L. J. Rasmussen, and W. Messer, *Plasmid* **28**, 14 (1992).

replicon derived from pSC101. Caution: the temperature-sensitive mutation has a high reversion rate. Prepare plasmid DNA by any miniscale method.[43]

2. Choose an appropriate lac^- host strain, e.g., CSH26 [ara, Δ(lac-pro), thi],[45] which would later carry the chromosomal lacZ reporter gene fusion with the promoter in question, and transform with pLDR8. Incubate plates for colony formation at 30° on L-medium plates[45] containing kanamycin. Check kan^R transformants for temperature sensitivity (42°).

3. Grow a culture of a positive clone to an OD_{600} of 0.2 and then for an additional 30 min at 37° to allow for the efficient expression of integrase. Prepare competent cells according to standard protocols.[43]

4. Construct a transcriptional–translational fusion of the promoter to be assayed for regulation by DnaA and the N terminus of this gene with an N-terminally truncated lacZ gene, e.g., from pTAC3734[46] or pCB267.[47]

5. Reclone an appropriate restriction fragment carrying the lacZ reporter gene fusion into the multiple cloning site of plasmid pLDR9 (or pLDR10, pLDR11, respectively). Note: if for a later step a DnaA-overexpressing plasmid requires selection by ampicillin, the bla gene of pLDR9 needs to be replaced by a less commonly used marker, e.g., the hygB gene from pUK4H conferring resistance to hygromycin (50 μg · ml^{-1}).[48]

6. Restrict the pLDR9 derivative with NotI in order to separate the fragment carrying the neo gene and the plasmid replicon from the fragment carrying the reporter gene fusion adjacent to attP and the selective marker. Isolate the desired fragment from agarose gel using QIAQuick kits (Qiagen, Hilden, D.). Determine DNA concentration.

7. Religate the purified fragment at a DNA concentration $\leq$ 30 ng · μl^{-1}. Note that higher DNA concentrations ($\geq$50 ng · μl^{-1}) result in the formation of linear concatemers, whereas low DNA concentrations favor the desired intramolecular recircularization.

8. Transform an aliquot of the competent cells of the pLDR8-containing host strain (step 3) with the entire ligation mix. Allow transformed cells to regenerate for > 1 h at 37°. Plate the regenerated cells on selective plates and incubate at 37°. Recheck single transformants for temperature resistance (growth at 42°) and kanamycin sensitivity in order to verify loss of pLDR8.

[45] J. H. Miller, "Experiments in Molecular Genetics." Cold Spring Harbor Laboratory, Cold Spring Harbor, NY, 1972.
[46] L. Brondsted and T. Atlung, J. Bacteriol. 176, 5423 (1994).
[47] K. Schneider and C. F. Beck, Gene 42, 37 (1986).
[48] F. Blaesing, C. Weigel, and W. Messer, Mol. Microbiol. 36, 557 (2000).

9. Verify the chromosomal integration of the reporter gene fusion by polymerase chain reaction (PCR). Using one primer derived from the attλ sequence of the *E. coli* genome and a pair of *lacZ*-specific primers allows determination of the orientation of the reporter gene fusion with respect to neighboring genes.

10. Transform the host strain carrying the *lacZ* reporter gene fusion with a DnaA-overexpressing plasmid, e.g., pdnaA116.[49]

11. Grow the reporter strain carrying the inducible DnaA-overexpressing plasmid under the desired conditions, induce, and perform LacZ assays according to Miller.[45]

A kit with strains and plasmids required (pLDR8–pLDR11) can be obtained from ATCC. Using this method, a *mioC::lacZ* translational fusion was constructed and used successfully as a reporter gene for the characterization of DNA-binding mutants of the *E. coli* DnaA protein.[48] The *mioC* promoter is repressed by DnaA (see earlier discussion).

Cloning and Expression of DnaA

Several DnaA-overexpressing plasmids from *E. coli* using different promoters have been described: pdnaA116,[49] pKC596,[50] and pKA231.[51]

Purification of *E. coli* DnaA

In vitro analysis of DnaA protein is usually performed with DnaA purified by either of two methods: (i) the method developed by Sekimizu in A. Komberg's laboratory,[52] which includes a denaturation step, followed by renaturation in the presence of nucleotide or (ii) the nondenaturing method, which is described in its optimized version elsewhere.[39,49] Methods for protein purification under nondenaturing conditions have also been developed for the DnaA proteins of *B. subtilis*,[49,53] *T. thermophilus*,[54] and *S. coelicolor*,[55] in the latter case making use of a C-terminal his tag.

[49] M. Krause, R. Lurz, B. Rückert, and W. Messer, *J. Mol. Biol.* **274**, 365 (1997).
[50] K. M. Carr and J. M. Kaguni, *Mol. Microbiol.* **20**, 1307 (1996).
[51] T. Mizushima, S. Nishida, K. Kurokawa, T. Katayama, T. Miki, and K. Sekimizu, *EMBO J.* **16**, 3724 (1997).
[52] K. Sekimizu and A. Kornberg, *J. Biol. Chem.* **263**, 7131 (1988).
[53] T. Fukuoka, S. Moriya, H. Yoshikawa, and N. Ogasawara, *J. Biochem.* (*Tokyo*) **107**, 732 (1990).
[54] S. Schaper, J. Nardmann, G. Lüder, R. Lurz, C. Speck, and W. Messer, *J. Mol. Biol.* **299**, 655 (2000).
[55] J. Majka, W. Messer, H. Schrempf, and J. Zakrzewska-Czerwinska, *J. Bact.* **179**, 2426 (1997).

Experimental Protocol

1. Transform a suitable *E. coli* strain with a DnaA-overproducing plasmid, e.g., pdnaA116.[49]

2. Inoculate 10 ml L medium[45] containing 50–75 $\mu g \cdot ml^{-1}$ ampicillin with a single transformed colony and incubate the preculture at $37°$ overnight.

3. Inoculate 2 liters of L medium containing 50–75 $\mu g \cdot ml^{-1}$ ampicillin 1:500 with the fresh overnight culture and incubate at $37°$ with strong aeration.

4. Induce the culture with 1 m*M* Isopropyl-β-D-thiogalactoside (IPTG) at an OD_{600} of 0.6–0.8.

5. Incubate the culture for an additional 2–3 h at $37°$ and harvest the cells by low-speed centrifugation.

6. Resuspend the cell pellet in sonication buffer [25 m*M* HEPES/KOH, pH 7.6, 100 m*M* K-glutamate, 1 m*M* dithiothreitol (DDT), 1 m*M* *p*-aminobenzamidine] at 1–2 ml per gram of wet weight.

7. Sonicate the cell suspension on ice (e.g., Branson sonifier 450; 5× : 10-s bursts, 20 s cooling, 200–300 W) and centrifuge for 30 min at $4°$ (SS34 rotor, 16,000 rpm). Collect the supernatant.

8. Precipitate the supernatant with 0.28 g solid $(NH_4)_2SO_4$ per milliliter supernatant and centrifuge for 1 h at $4°$ (SS34 rotor, 16,000 rpm). Discard the supernatant.

9. Resuspend the protein pellet in LG buffer [45 m*M* HEPES/KOH, pH 7.6, 100 m*M* K-glutamate, 10 m*M* Mg-acetate, 1 m*M* DDT, 0.5 m*M* EDTA, 20% (w/v) sucrose] to a final volume of 2.5 ml and desalt the protein solution by passing over a NAP25 column (Pharmacia); elute with LG buffer.

10. Load the protein solution onto a SP-Sepharose or Mono S column (Pharmacia; 1 × 10 cm) equilibrated previously with LG buffer. Wash the column with 40 ml LG buffer and elute DnaA with a K-glutamate gradient of 0.1–1.0 *M*. DnaA is usually the last protein eluted from the column and contains up to 1 *M* salt. If for some reason DnaA is not pure enough, desalt the solution and apply it to a Mono P, Mono Q, or DEAE-Sepharose column (Pharmacia) in LG buffer. DnaA does not bind to these columns and can be collected in the flow-through.

11. Freeze aliquots in liquid nitrogen and store at $-70°$. Storage buffer: LG buffer + 0.5–1 *M* K-glutamate. *Note*: Concentrated DnaA solutions $(0.5–1.0 \text{ mg} \cdot ml^{-1})$ remain fully active for up to a week when stored on ice in high salt buffer. In low salt buffer, however, DnaA tends to precipitate within a few hours.

This protocol results in the purification of DnaA in its ATP form; the ADP form is obtained by dialysis as described.[9]

A simple method for the purification of DnaA has been published by Sekimizu *et al.*[56] Dialysis of a cell extract against low salt buffer precipitates DnaA, presumably due to the presence of phospholipids. The precipitate is collected by low-speed centrifugation and solubilized with 4 M guanidine–HCl, which is subsequently removed by gel filtration or dialysis. DnaA protein purified with this technique is free of nucleotide, has a distorted secondary and tertiary structure,[57] and binds DNA unspecifically.[42,58,59] ATP (or ADP) must therefore be added in order to regenerate, at least partially, the native conformation.

Purified DnaA protein can be quantified using the Bradford technique.[60] In mixtures with other proteins, DnaA is quantified using Western blots calibrated with purified DnaA.

Specific Binding of DnaA to DNA

All activities of DnaA, including its role as a transcription regulator, require binding to its cognate-binding site. For a detection of DnaA binding, the same techniques are applied that are generally used for DNA-binding proteins. The oldest method is filter binding, such as binding of a radioactive DNA or oligonucleotide to a nitrocellulose filter via a DnaA protein.[61] However, relatively high concentrations of DnaA are required, and the method is sensitive to artifacts, for example, the presence of contaminating proteins or aggregation of DnaA. Therefore, other techniques, such as gel retardation, are preferable. If the DnaA boxes on the target DNA have a low binding affinity, the stability of the complexes may be insufficient, and they dissociate upon entering the gel. In this case, DNase I footprinting,[9] electron microscopy,[62] or surface plasmon resonance[9,48] should be employed. All these techniques measure binding of DnaA to its DNA-binding site. A more complete set of activities is analyzed with the unwinding of the AT-rich *oriC* region[49,63,64] or with the complete *in vitro* replication reaction.[65,66]

[56] K. Sekimizu, B. Y. Yung, and A. Kornberg, *J. Biol. Chem.* **263**, 7136 (1988).
[57] T. Kubota, T. Katayama, Y. Ito, T. Mizushima, and K. Sekimizu, *Biochem. Biophys. Res. Commun.* **232**, 130 (1997).
[58] D. S. Hwang and A. Kornberg, *J. Biol. Chem.* **267**, 23083 (1992).
[59] C. Margulies and J. M. Kaguni, *J. Biol. Chem.* **271**, 17035 (1996).
[60] M. M. Bradford, *Anal. Biochem.* **72**, 248 (1976).
[61] R. S. Fuller and A. Kornberg, *Proc. Natl. Acad. Sci. USA* **80**, 5817 (1983).
[62] C. Weigel, A. Schmidt, B. Rückert, R. Lurz, and W. Messer, *EMBO J.* **16**, 6574 (1997).
[63] D. Bramhill and A. Kornberg, *Cell* **52**, 743 (1988).
[64] H. Gille and W. Messer, *EMBO J.* **10**, 1579 (1991).
[65] R. S. Fuller, J. M. Kaguni, and A. Kornberg, *Proc. Natl. Acad. Sci. USA* **78**, 7370 (1981).
[66] B. E. Funnell, T. A. Baker, and A. Kornberg, *J. Biol. Chem.* **261**, 5616 (1986).

Gel Retardation

The DNA-binding activity of purified DnaA is analyzed most conveniently by electrophoretic mobility shift (gel retardation) assays (EMSA) on agarose gels.[62]

Experimental Protocol

1. Choose a ∼500-bp DNA fragment as a target that contains at least one well-characterized consensus DnaA box (5′-TT^A/$_T$ TNCACA),[39] e.g., a fragment from the *E. coli dnaA* promoter region.[9]

2. Prepare the fragment by PCR or isolate a restriction fragment from a plasmid.[62] The quality of the fragments obtained by one-step purification with QIAQuick kits is sufficient. Precipitate DNA, dissolve in H_2O, and adjust the concentration to 1 ng $\cdot \mu l^{-1}$.

3. Prepare a restriction analysis-type 1.2% agarose gel (Seakem LE; FMC, Rockland, ME) in 0.5x TBE buffer (22.5 mM Tris–borate, 0.5 mM EDTA, pH 8.0).[43]

4. Thaw an aliquot of the DnaA preparation on ice. Adjust to a final concentration of 100 μM ATP, incubate for 30 min on ice, and dilute into binding buffer [20 mM HEPES/KOH, pH 8.0, 5 mM Mg-acetate, 1 mM EDTA, 4 mM DTT, 0.2% Triton X-100, 5 mg $\cdot$ ml^{-1} bovine serum albumin (BSA), 5% glycerol, 100 μM ATP].[67] Prepare DnaA dilutions to cover a range of 0.2 to 2 with respect to the molar ratio of DnaA to DNA boxes. DnaA dilutions remain active for 1 day when kept on ice.

5. Assemble ≤20-μl reaction mixes by adding an equal volume of the desired protein dilution to 10 ng DNA in H_2O on ice. Mix carefully by pipetting up and down twice to avoid bubbles. Vortexing is not recommended.

6. Incubate reaction mixes for 10 min at 37°. Add 2 μl loading buffer (0.25% bromphenol blue, 30% glycerol) and place on ice.

7. Load samples onto the gel and carry out electrophoresis at room temperature for approximately 1.5 h in 0.5x TBE buffer at 4 V $\cdot$ cm^{-1}.

8. Stain the gel with SYBR-Green (Applied Biosystems, Foster City, CA) or an equivalent fluorescent dye for 10 min in electrophoresis buffer according to the manufacturer's instructions. Place staining tray on a slowly rotating shaker at room temperature for even staining.

9. Scan the stained gel with a Molecular Dynamics FluorImager 575 or an equivalent device and process the image with appropriate software, e.g., Molecular Dynamics ImageQuantNT software (Amersham Biosciences, Little Chalfont, Buckinghamshire HP7 9NA, GB).

---

[67] C. A. Parada and K. J. Marians, *J. Biol. Chem.* **266,** 18895 (1991).

Due to its basic character (pl 8.77), *E. coli* DnaA does not migrate into the gel. For the same reason, DNA–DnaA complexes migrate well behind the unbound DNA and give a sharp banding pattern, which allows their unambiguous detection. A DnaA preparation can safely considered to be active (for DNA binding) if the formation of DNA–DnaA complexes is already observed at molar ratios (DnaA/DNA boxes) of well below 1. A slightly blurred banding pattern indicates unspecific binding or formation of higher-order complexes and can usually be observed only at ratios $\geq$2. DnaA preparations should be discarded if they give rise to fuzzy band shifts at low protein concentrations repeatedly. BSA gives a diffuse band with SYBR-Green, which, however, can be distinguished easily from DNA–DnaA bands.

The high affinity of DnaA for the consensus DnaA box ($k_D \approx 1$ nM)[39] makes it possible to detect specific binding in the absence of competitor DNA. To increase the specificity of the assay, however, unrestricted or *Hae*III-digested ϕX174 DNA (depending on the size of the chosen target fragment) can be added to the reaction mixes as a competitor.[68]

If no device for the detection of fluorescent dyes is available, ethidium bromide staining of the gel is also possible, albeit with a loss of sensitivity. In this case, the amount of target DNA in the reaction mixes should be raised to 100 ng for a ~500-bp DNA fragment and the amounts of protein accordingly.

To speed up this assay, a heat-inactivated restriction digest of a plasmid can be used directly, i.e., without fragment isolation or purification, as a target for DnaA binding if at least one fragment contains a well-characterized DnaA box. Note, however, that the amount of plasmid DNA in the assay has to be adjusted such that the amount of the DnaA box-containing fragment(s) is ~10 ng.[62] In this setup of the assay, the restriction fragment(s) not containing DnaA boxes represents competitor DNA.

If the aim of the experiment is the determination of dissociation constants or if small DNA fragments are to be analyzed, radioactively labeled DNA and native polyacrylamide gels should be used instead of agarose gels.[39]

DNase 1 Footprinting

In DNase I footprinting, a protein-binding site is detected by limited cleavage of an end-labeled DNA fragment by DNase I, followed by gel electrophoresis in a sequencing gel and autoradiography or phosphorimaging.[9,69]

---

[68] U. Langer, S. Richter, A. Roth, C. Weigel, and W. Messer, *Mol. Microbiol.* **21,** 301 (1996).
[69] D. Galas and A. Schmitz, *Nucleic Acids Res.* **5,** 3157 (1978).

Bound protein, such as DnaA, protects the binding site(s) against cleavage and leaves a footprint. The conditions of limited DNase I digestion must be determined carefully. Ideally, the DNase I concentration and incubation time should be such that the intensity of bands in the gel is uniform over the whole length of the fragment and a small amount of undigested full-length fragment is left. It is possible to bind two proteins to a fragment, e.g., DnaA and RNA polymerase[70] or IHF and RNA polymerase,[71] and determine their mutual effects on binding using DNase I footprinting.

Experimental Protocol

1. The DNA fragment to be analyzed is prepared by PCR amplification with one primer $5'$-^{32}P-labeled using T4 polynucleotide kinase.

2. Binding reactions with DnaA protein are typically carried out for 10 min at $37°$ with 5 ng of an end-labeled PCR fragment in 20 μl binding buffer (25 mM HEPES, pH 7.6, 100 mM potassium acetate or 50 mM potassium glutamate, 5 mM magnesium acetate, 5 mM calcium acetate, 4 mM DTT, 0.2% Triton X-100, 0.5 mg · ml^{-1} BSA) and increasing DnaA concentrations (10 to 300 nM in DnaA storage buffer).

3. DNase I is added according to the pilot experiment (typically 1.5 pg), and incubation is continued at $37°$ for 2–4 min.

4. The reaction is quenched by an equal volume of stop solution (1% SDS, 200 mM NaCl, 20 mM EDTA, pH 8.0, 1 mg · ml^{-1} mussel glycogen), the DNA is extracted by phenol-chloroform, precipitated with ethanol, and resuspended in 1 μl TE buffer (10 mM Tris–HCl, pH 8.0, 0.1 mM EDTA).

5. Five microliters of gel-loading buffer (98% formamide, 0.025% bromphenol blue, 0.025% xylene cyanol) is added, and the samples are heated to $96°$ for 5 min and loaded onto 8% sequencing gels.

6. After electrophoresis, gels are dried and subjected to autoradiography or analyzed in a phosphorimager. The best size standard is a sequencing reaction using the same labeled primer.

In Vitro Transcription

A direct estimate of the effect of DnaA on transcription from a given promoter can be obtained from runoff transcripts synthesized *in vitro* in the presence of DnaA protein.[9,18]

---

[70] H. Giladi, S. Koby, M. E. Gottesman, and A. B. Oppenheim, *J. Mol. Biol.* **224,** 937 (1992).
[71] M. Glinkowska, J. Majna, W. Messer, and G. Wegrzyn, *J. Biol. Chem.* **278,** 2225 (2003).

Experimental Protocol

1. The template can be prepared from a plasmid or amplified by PCR.

2. Preincubate the DNA (10 ng) for 5 min at $37°$ with DnaA protein (30–300 ng) in 20 μl transcription buffer (25 mM HEPES, pH 7.6, 100 mM potassium acetate, 5 mM magnesium acetate, 4 mM DTT, 0.2% Triton X-100, 0.5 mg $\cdot$ ml^{-1} BSA).

3. Add RNA polymerase to a concentration of 66 nM, followed by another 5-min incubation.

4. Start transcription by adding nucleotides: [α-^{32}P]UTP, unlabeled UTP (60 μM), ATP (200 μM), and CTP and GTP (400 μM each). If single-round transcripts are measured, heparin (200 μg+ml^{-1}) is added before the addition of nucleotides.

5. Incubate mixtures for 5 to 15 min. The reactions are terminated by adding stop solution and processed as described earlier for DNase I footprints (steps 4–6).

Repression of the *dnaA* promoter by ADP- and ATP-complexed DnaA has been measured using this technique.[9]

[31] Analysis of Transcription Factor Interactions at Sedimentation Equilibrium

By Margaret A. Daugherty and Michael G. Fried

Large numbers of proteins participate in the assemblies that regulate and catalyze transcription.[1–5] Among methods available for characterizing their interactions, sedimentation equilibrium (SE) ultracentrifugation stands out as a direct and rigorous means of determining molecular masses, interaction stoichiometries, association constants (hence free energies of association), and the influences of low molecular weight effectors, ions, and crowding on the stabilities of protein complexes. The method is useful for characterizing molecules and complexes with masses between ~100 and ~50 × 10^6 Da in a wide variety of buffers and over a very large range of protein concentration. Relatively little material is required, and because

[1] A. Hochschild and S. L. Dove, *Cell* **92**, 597 (1998).
[2] T. I. Li and R. A. Young, *Annu. Rev. Genet.* **34**, 77 (2000).
[3] D. Beckett, *J. Mol. Biol.* **314**, 335 (2001).
[4] A. Dvir, J. W. Conaway, and R. C. Conaway, *Curr. Opin. Genet. Dev.* **11**, 209 (2001).
[5] G. Gill, *Essays Biochem.* **37**, 33 (2001).

the method is nondestructive, samples can often be recovered for further use. The availability of modern instrumentation and the development of improved analysis methods have resulted in an upsurge of interest in SE during the past decade. This article describes the application of SE techniques to the characterization of transcription factors and their interactions.

Overview of the Sedimentation Equilibrium Method

The SE technique has been in use since the mid-1920s,[6] and the theoretical basis of SE analysis has been under development continually since that time. As a result, the literature describing applications of the technique is vast. Many excellent reviews have been published during the past decade,[7–13] and earlier work is also a valuable resource.[14–16] This article focuses on three situations that are encountered most frequently in studies of transcription factors: self-association, heteroassociation, and the presence of inactive components. Analysis of such interactions provides crucial information about the role of protein–protein interactions in the assembly of transcription and transcription–regulatory complexes.[17]

At sedimentation equilibrium, the concentration of species i at a specified position (r) in the solution column is given[7,9,12] by

$$c_i(r) = c_{i,0} \exp\left[\sigma_i\left(r^2 - r_0^2\right)\right] \tag{1}$$

Here $c_{i,0}$ is the concentration of the ith species at the reference position r_0 (typically close to the meniscus), the reduced molecular weight, σ_i, is equal to $M_i(1 - \bar{v}_i\rho)\omega^2/2RT$ with M_i the molecular weight, $\bar{v}_i$ the partial specific volume (in ml/g), ρ the *solvent* density (g/ml), ω the rotor angular velocity

[6] T. Svedburg and K. O. Pedersen, "The Ultracentrifuge." Johnson Reprint Corp., New York, 1956.

[7] D. K. McRorie and P. J. Voelker, "Self-associating Systems in the Analytical Ultracentrifuge" Beckman Instruments, Inc., Palo Alto, CA, 1993.

[8] J. C. Hansen, J. Lebowitz, and B. Demeler, *Biochemistry* **33**, 13155 (1994).

[9] T. M. Laue, *Methods Enzymol.* **259**, 427 (1995).

[10] A. P. Minton, *Progr. Colloid Polym. Sci.* **107**, 11 (1997).

[11] G. Rivas, W. F. Stafford III, and A. P. Minton, *Methods* **19**, 194 (1999).

[12] T. M. Laue and W. F. Stafford III, *Annu. Rev. Biophys. Biomol. Struct.* **28**, 75 (1999).

[13] J. Lebowitz, M. Lewis, and P. Schuck, *Protein Sci.* **11**, 2067 (2002).

[14] H. K. Schachman, "Ultracentrifugation in Biochemistry" Academic Press, New York, 1959.

[15] J. W. Williams, "Ultracentrifugation of Macromolecules" Academic Press, New York, 1972.

[16] D. C. Teller, *Methods Enzymol.* **27**, 346 (1973).

[17] D. F. Senear, J. B. Ross, and T. M. Laue, *Methods* **16**, 3 (1998).

($= \text{rpm} \cdot \pi/30$), R the gas constant (8.314×10^7 erg mol^{-1} K^{-1}), and T the absolute temperature.[18]

For a solution of several components, the *total* concentration at position r is given by

$$c(r) = \sum_i c_{i,0} \exp\left[\sigma_i(r^2 - r_0^2)\right] \tag{2}$$

Here the summation is over all species. This expression embodies both the power and the challenge of the SE method. For solutions with a small number[19] of components, direct fitting of Eq. (2) to experimental data can yield relative concentrations and buoyant molecular weights of each component. Often the stoichiometry of an assembly can be deduced from its molecular weight whereas the reference concentrations of sedimenting species ($c_{i,0}$) can provide data allowing calculation of the equilibrium constant(s) governing the behavior of the system (see later). The challenge comes from the fact that values of the $c_{i,0}$ and σ_i terms must be extracted by curve fitting. Variants of Eq. (2) must be chosen that are appropriate to the molecular system, but the quality and quantity of data can limit the ability to discriminate between competing models. This problem rapidly becomes severe as the number of species included in the model (and thus the number of parameters to be determined by fitting) is increased. Examples presented here demonstrate these features and some methods for coping with these challenges.

Applications

Self-Association

Increasing molecular weight with increasing [protein] is indicative of a mass action association.[4] For a protein (A) undergoing *reversible* monomer-nmer self-association according to $nA \leftrightarrows A_n$, with apparent association constant $K_{obs} = [A_n]/[A]^n = c_{A_n,0}/(c_{A,0})^n$, Eq. (2) becomes[20]

$$c(r) = c_{A,0} \exp\left[\sigma_A(r^2 - r_0^2)\right] + K_{obs}(c_{A,0})^n \exp\left[n\sigma_A(r^2 - r_0^2)\right] \tag{3}$$

[18] Some authors prefer to define the reduced molecular weight as $\sigma_i \equiv M_i(1 - \bar{v}_i \rho)\omega^2/RT$. This is larger by a factor of two than the σ_i defined in the text and requires inclusion of an additional factor of 1/2 in Eq. (1). Often data will contain a concentration-independent component due to a constant absorbance or refractive index difference between reference and sample sectors. For this reason, expressions describing absorbance or refractive index gradients commonly contain an additional constant "baseline offset" term.

[19] Typically ≤ 3, although under favorable conditions, systems containing more species are amenable to analysis.

[20] Because the term $c_{A,0}$ appears elsewhere in Eq. (3), the substitution of $c_{A_n,0}$ by $K_{obs}(c_{A,0})^n$ does not change the number of adjustable parameters.

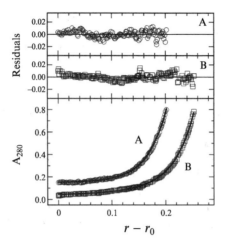

FIG. 1. Sedimentation equilibrium analyses of *E. coli* RNA polymerase holoenzyme. Data obtained at $4°$ in buffers consisting of 40 mM Tris, pH 7.9, 10 mM MgCl$_2$, 0.1 mM dithiothreitol, 5% glycerol, and 300 mM KCl (curve A) or 100 mM KCl (curve B). Sample A was centrifuged at 9000 rpm; data are offset by 0.1 absorbance unit for clarity. The smooth curve is the global fit of the ideal single-species model [Eq. (2) with a single term] to six data sets, of which only one is shown here. The value of M_r returned by this analysis was 454,000 ± 6000, in good agreement with that expected for the enzyme. Sample B was centrifuged at 7000 rpm. The smooth curve is the global fit of the monomer–dimer model [Eq. (3), with $n = 2$] to six data sets, with monomer molecular weight set at 454,000. In both cases the curve-fitting residuals (upper panels) are small and lack obvious systematic dependence on radial position, demonstrating that the corresponding models are consistent with the mass distributions present in these samples. Data from Dyckman and Fried,[30] with permission.

Often, data of reasonable quality will allow assignment of n and K_{obs} for a two-species model, even if the buoyant molecular weight of the monomer is unknown (Fig. 1). For more than two species, the situation can be more challenging. Shown in Fig. 2A are data for the self-association of the full-length yeast TATA-binding protein.[21,22] Here, the model that consistently fit the data, over a wide range of [salt], pH, temperature, and [protein], was one in which monomers, tetramers, and octamers were in equilibrium[23]:

$$c(r) = c_{TBP,0} \exp\left[\sigma_{TBP}(r^2 - r_0^2)\right] + K_{1-4}(c_{TBP,0})^4 \exp\left[4\sigma_{TBP}(r^2 - r_0^2)\right]$$
$$+ k_{1-8}(c_{TBP,0})^8 \exp\left[8\sigma_{TBP}(r^2 - r_0^2)\right] \tag{4}$$

[21] M. A. Daugherty, M. Brenowitz, and M. G. Fried, *J. Mol. Biol.* **285**, 1389 (1999).

[22] M. A. Daugherty, M. Brenowitz, and M. G. Fried, *Biochemistry* **39**, 4869 (2000).

[23] This model assumes that σ_{TBP} (and hence $\bar{v}_{TBP}$) is unchanged by self-association. If σ is allowed to float, there are four adjustable parameters. This can be reduced to three if the value of σ is fixed at the known monomer molecular weight.

A

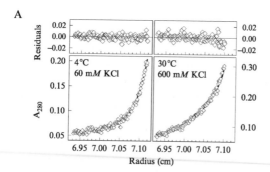

B

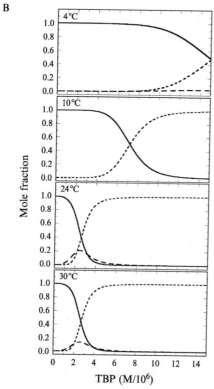

FIG. 2. (A) Yeast TATA-binding protein at sedimentation equilibrium. The buffer contained 20 mM HEPES/KOH, pH 7.9, 1 mM EDTA, 1 mM dithiothreitol (DTT) plus KCl as indicated. Rotor speed: 16,000 rpm. Solid lines represent global fits of Eq. (4) to data sets obtained with three starting concentrations of TBP and two rotor speeds (16,000 and 24,000 rpm). In both cases the curve-fitting residuals (upper panels) are small and appear to be distributed randomly, demonstrating that the monomer–tetramer–octamer model is consistent with the mass distributions present in these samples. (B) Dependence of the mole fractions of monomeric, tetrameric, and octameric TBP species on TBP concentration and

The tetrameric species was a minor component under all conditions examined; however, at low temperature, where [tetramer] was low (mole fraction ≤ 0.2; Fig. 2B), satisfactory fits were also obtained with a monomer–octamer association model [Eq. (4) without the middle term]. Models containing a term for dimeric TBP returned concentrations for this species that were indistinguishable from zero under all conditions tested.[21,22] This outcome is intriguing because some TBP and related TFIID preparations have been shown to form dimers.[24–26] These examples demonstrate the challenge of distinguishing between competing models and the importance of testing several sets of experimental conditions (T, [protein], [salt], etc.) before concluding that a particular association mechanism is correct.

The equilibrium constants given in Eqs. (3) and (4) are in the concentration units (absorbance, refractive index difference) appropriate to the detection system of the centrifuge. Equation (5) converts absorbance-scaled values of K_{obs} to the more familiar molar concentration scale:

$$K_{molar} = K_{obs}\left(\frac{\varepsilon^{n-1}l^{n-1}}{n}\right) \tag{5}$$

Here n is the association stoichiometry and ε and l are the molar extinction coefficient and the optical path length, respectively.[27] For refractive index-scaled values, the corresponding equation is[9]

$$K_{molar} = K_{obs}\left(\frac{M_A^{n-1}}{n}\right)\left(\frac{Y_T}{C_T}\right)^{n-1} \tag{6}$$

Here M_A is the monomer molecular weight, n is the assembly stoichiometry, and Y_T/C_T is the conversion factor relating signal-to-weight concentration.[28]

---

[24] R. A. Coleman, A. K. P. Taggart, L. R. Benjamin, and B. F. Pugh, *J. Biol. Chem.* **270**, 13842 (1995).

[25] A. K. P. Taggart and B. F. Pugh, *Science* **272**, 1331 (1996).

[26] K. M. Campbell, R. T. Ranallo, L. A. Stargell, and K. J. Lumb, *Biochemistry* **39**, 2633 (2000).

[27] Two optical path lengths are common: that of standard cells is 1.2 cm and that of short optical path cells is 0.3 cm.

[28] For proteins, the refractive increment dn/dc is equivalent to ~ 3.33 fringes ml/mg using light of 670 nm and a 1.2-cm optical path length[4].

---

temperature. The buffer contained 20 mM HEPES/KOH, pH 7.9, 1 mM EDTA, 1 mM DTT, and 120 mM KCl. Mole fractions of the monomer (solid line), tetramer (dashed line), and octamer (dotted line) were calculated using association constants determined at sedimentation equilibrium. Data from Daugherty *et al.*,[22] with permission.

Heteroassociation

For two *different* proteins A and B associating *reversibly* according to the mechanism $nA + mB \rightleftharpoons A_nB_m$, with apparent association constant $K_{obs} = [A_nB_m]/[A]^n[B]^m$, Eq. (2) becomes[11]

$$c(r) = c_{A,0} \exp[\sigma_A(r^2 - r_0^2)] + c_{B,0} \exp[\sigma_B(r^2 - r_0^2)] \\ + c_{AB,0} \exp[\sigma_{AB}(r^2 - r_0^2)] \tag{7}$$

Here there is one term for each molecular species, and the parameters characterizing the complex are $c_{AB,0} = K_{obs}(c_{A,0})^n(c_{B,0})^m$ and $\sigma_{AB} = (nM_A + mM_B)(1 - \overline{v_{AB}}\rho)\omega^2/2RT$. The partial specific volume of the complex is usually approximated by the weight average:

$$\overline{v_{AB}} = \frac{nM_A\overline{v_A} + mM_B\overline{v_B}}{nM_A + mM_B} \tag{8}$$

This is equivalent to the assumption that complex formation is not accompanied by a significant change in the partial specific volumes of the components. This is a robust assumption, as shown by the accuracy with which $\overline{v}$ may be estimated from amino acid composition for many proteins.[29] However, if highly accurate measurement of M_{AB} is required, experimental determination of $\overline{v}$ is worthwhile. A method for measuring $\overline{v}$ is discussed later.

Under favorable conditions, the stoichiometry of a complex can be inferred from its molecular weight. For example, Fig. 3 shows data for the association of the *Escherichia coli* CAP (M_r 47,238/dimer) and *lac* repressor (M_r 154,520/tetramer) proteins. The fit to Eq. (7) returned M_r (complex) 240,800 ± 14,100, a value indistinguishable from that predicted for a 2CAP dimer:1 repressor tetramer complex (M_r 248,996). The small residuals attest to the compatibility of this model to data, although their upward deviation at the bottom of the cell suggests that higher molecular weight species can form at very high [protein].

Inference of stoichiometry from molecular weight is more difficult when the interacting partners have similar M_r (in such a case, for example, the trimolecular complexes A_2B and AB_2 would have indistinguishable molecular weights). It is also difficult when one interacting partner is so much larger than the other that the difference in M_r(complex) expected for two possible stoichiometries is smaller than the uncertainty in the measured value of M_r. Thus, data[30] for the binding of *E. coli* CAP (M_r 47,238/

[29] H. Durschlag, *in* "Thermodynamic Data for Biochemistry and Biotechnology" (H.-J. Hinz, ed.), p. 45. Springer-Verlag, Berlin, 1986.
[30] D. Dyckman and M. G. Fried, *J. Biol. Chem.* **277**, 19064 (2002).

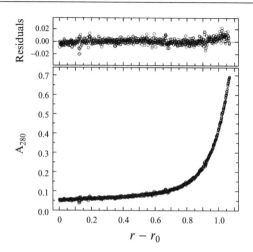

FIG. 3. Association of *E. coli* CAP and *lac* repressor proteins. A solution containing a 2.4 × 10^{-6} *M* CAP dimer and a 1.2 × 10^{-6} *M lac* repressor tetramer was centrifuged to equilibrium at 8000 rpm and 4°. The smooth curve is the least-squares fit of Eq. (7) to data, with M_r values of CAP and *lac* repressor fixed, and the $\bar{v}_{complex}$ given by the weight average of the constituent proteins. This analysis gave M_r (complex) 240,820 ± 14,140, consistent with a 2 CAP dimer:1 repressor tetramer complex. Data from Fried and Daugherty,[36] with permission.

dimer) with RNA polymerase holoenzyme ($M_r \sim 455,000$) gave M_r(complex) 1,020,000 ± 57,400. The uncertainty in this value prevents direct inference of the stoichiometry. In such cases, a continuous variation experiment can often specify the stoichiometry or narrow the range of alternatives.

Continuous variation (Job) analyses[31] allow determination of the ratio of reactants that maximize product formation at constant total [reactant]. For complexes in which one molecule of type A is bound by several of B, this "optimal combining ratio" is equivalent to stoichiometry. For complexes containing several molecules of each type (A_nB_m), the stoichiometry is a multiple of the optimal combining ratio. This experiment requires a series of samples in which the sum of protein concentrations is constant, but the mole fraction of each protein varies between 0 and 1. These are obtained conveniently from stock solutions of each protein adjusted to the same molar concentration. Samples are prepared by combining different volumes of each stock, maintaining a constant final volume. Analysis of SE data using Eq. (7) provides a measure of the concentration of complex in each sample; this value is maximized at the optimal combining ratio. For the CAP–RNA polymerase system described earlier, the yield of complex

[31] C. Y. Huang, *Methods Enzymol.* **87,** 509 (1982).

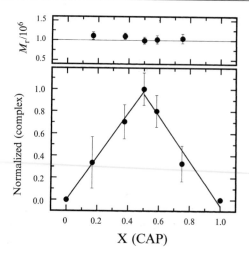

X (CAP)

FIG. 4. Continuous variation (Job) plot for the CAP–RNA polymerase interaction. (Bottom) Dependence of normalized [complex] ($\pm$ 95% confidence limits) on mole fraction of CAP. Normalized [complex] is the concentration of complex in a given sample divided by that in the sample with the greatest amount of association. The total protein concentration was fixed ([CAP] + [RNA polymerase] = $7.2 \times 10^{-7}M$), but the mole fraction was allowed to vary as indicated. (Top) Molecular weights ($\pm 95\%$ confidence limits) returned by these analyses. The horizontal line indicates the molecular weight expected for a 2:2 CAP:polymerase complex (M_r 1.004×10^6). Data from Dyckman and Fried,[30] with permission.

is maximized at a CAP mole fraction 0.5 [equivalent to a 1:1 combining ratio (Fig. 4, bottom) panel]. The constant molecular weight at all combining ratios is a characteristic of mechanisms in which only one kind of complex is formed (Fig. 4, top). Because the molecular weight of the complex is slightly more than twice that of the RNA polymerase monomer, the stoichiometry of the complex[32] must be 2:2.

For heteroassociation reactions, the conversion of apparent equilibrium constants from absorbance units to the molar concentration scale is shown as

$$K_{molar} = K_{obs} \left(\frac{\varepsilon_A^n \varepsilon_B^m}{\varepsilon_{A_n B_m}} \right) 1^{n+m-1} \qquad (9)$$

Here K_{obs} is the equilibrium constant in absorbance units, n and m are the stoichiometries of species A and B, respectively, and ε_A, ε_B, and $\varepsilon_{A_n B_m}$

---

[32] If the stoichiometry can be established from M_r and/or combining ratio, reanalysis of original data using a model in which monomer molecular weights and stoichiometries are fixed can allow the extraction of K_{obs} while minimizing the number of adjustable species in the fit.

are the molecular extinction coefficients of species A, B, and the complex. If extinction coefficients do not change with association, $\varepsilon_{A_nB_m} = n\varepsilon_A + m\varepsilon_B$. The conversion of apparent equilibrium constants from refractive index units (fringes) to the molar concentration scale is given by

$$K_{molar} = K_{obs} \left(\frac{(M_A)^n (M_B)^m}{M_{A_nB_m}} \right) \left(\frac{Y_T}{C_T} \right)^{n+m-1} \tag{10}$$

As before, Y_T/C_T is the conversion factor relating signal-to-weight concentration.

Noninteracting components

Equation (2) can describe independently sedimenting species as well as ones that associate. This is valuable, as it is unusual to find samples that are both highly pure and 100% active. An interacting system containing an additional independent species might be described as $nA + mB + C \leftrightarrows A_nB_m + C$; here component C can be an inactive fraction of one of the interacting species (A or B), or an unrelated contaminant. At sedimentation equilibrium the concentration distribution is given by

$$c(r) = c_{A,0} \exp\left[\sigma_A (r^2 - r_0^2)\right] + c_{B,0} \exp\left[\sigma_B(r^2 - r_0^2)\right] + \\ c_{C,0} \exp\left[\sigma_C(r^2 - r_0^2)\right] + c_{AB,0} \exp\left[\sigma_{AB}(r^2 - r_0^2)\right] \tag{11}$$

As described earlier, $c_{AB,0} = K_{obs}(c_{A,0})^n (c_{B,0})^m$, but $c_{C,0}$ does not depend[33] on $c_{A,0}$ or $c_{B,0}$. If species C is an inactive fraction of component A, for example, analysis of SE data according to a model in which all components equilibrate [e.g., Eq. (7)] will result in values of K_{obs} that decrease with increasing [A] and with increasing rotor speed.[9] Inclusion of a separate term for species C results in values of K_{obs} that are *independent* of reactant concentrations and rotor speed, as expected for equilibrium constants.[34] As mentioned previously, distinguishing between alternative models becomes more difficult as the number of species increases. When this becomes a problem, data from other techniques (such as sedimentation velocity) can provide valuable guidance in selection of the simplest realistic model of the system.

---

[33] Under dilute solution conditions. In concentrated solutions, steric interactions and other nonideal effects can prevent independent sedimentation.

[34] The presence of inactive species increases the number of adjustable terms in the sedimentation equation that must be evaluated by fitting. Because acquiring a meaningful fit becomes more difficult as the number of species increases, systems containing several inactive species should be purified further before a detailed SE analysis is attempted.

Thermodynamic Linkage

Often, low molecular weight solutes affect the stabilities of transcription factor assemblies. Such solutes include physiological signals such as cAMP and cGMP but also buffer ions and osmolytes. When this is the case, solute binding and macromolecular assembly are linked thermodynamically.[35] Sedimentation equilibrium analysis can provide quantitative information about the stoichiometries and consequences of solute participation in the protein assembly reaction. At the relatively slow rotor speeds appropriate for SE analyses of proteins, the concentrations of low molecular weight solutes are virtually unchanged from meniscus to the bottom of the solution column. This allows one to treat macromolecular equilibria as if they are taking place at a constant concentration of the low molecular weight solute. An example of this is analysis of the cAMP-dependent association of the *E. coli* CAP and *lac* repressor proteins, for which the mechanism

$$2 \text{ CAP} + n(\text{cAMP}) + lac \text{ repressor} \leftrightarrows \text{CAP}_2 \bullet \text{cAMP}_n \bullet lac \text{ repressor} \quad (12)$$

appears to operate.[36] Here the equilibrium constant $K = [(\text{CAP}_2 \bullet \text{cAMP}_n) \bullet lac \text{ repressor}]/[\text{cAMP}]^n[\text{CAP}]^2[lac \text{ repressor}]$, and the *observable* macromolecular equilibrium quotient $K_{obs} = [(\text{CAP}_2 \bullet \text{cAMP}_n) \bullet lac \text{ repressor}]/[\text{CAP}]^2[lac \text{ repressor}]$. cAMP stoichiometry[37] can be evaluated using the relation

$$\frac{\partial \log K_{obs}}{\partial \log [\text{cAMP}]} = n \quad (13)$$

A graph of log K_{obs} as a function of log [cAMP] is shown in Fig. 5. The slope (1.8 ± 0.3) indicates that $\sim$2 equivalents of cAMP are *bound* for each CAP$_2\bullet$repressor complex formed. While these data do not specify the distribution of cAMP within the complex, in the absence of repressor, native CAP dimers bind one equivalent of cAMP over the concentration range $0 < [\text{cAMP}] \leq 20 \ \mu M$ spanned in this experiment.[38] It is thus possible to speculate that each molecule of CAP in the (CAP$_2\bullet$cAMP$_2$)$\bullet$repressor complex binds one molecule of cAMP.

---

[35] J. Wyman and S. J. Gill, "Binding and Linkage," University Science Books, Mill Valley, CA, 1990.

[36] M. G. Fried and M. A. Daugherty, *J. Biol. Chem.* **276,** 11226 (2001).

[37] For the reaction shown in Eq. (12), net cAMP uptake will yield $n > 0$, net release $n < 0$, and a cAMP-independent reaction, $n = 0$.

[38] T. Takahashi, B. Blazy, and A. Baudras, *Biochemistry* **19,** 5124 (1980).

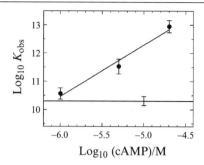

Log_{10} (cAMP)/M

FIG. 5. Dependence of log K_{obs} on log [cAMP] in formation of the CAP2•cAMP$_n$•*lac* repressor complex. Values of log K_{obs} measured in the presence of cAMP are indicated by filled symbols. The horizontal line gives the value of log K_{obs} at [cAMP] = 0. Error bars represent 95% confidence limits of plotted values. The slope (1.8 ± 0.3) is a measure of the number (*n*) of cAMP molecules bound per molecule of complex formed. Data from Fried and Daugherty,[36] with permission.

Temperature Dependence of Association

Because the temperature of samples can be controlled with reasonable precision (typically ±0.1°), it is sometimes possible to extract thermo-dynamic information from the temperature dependence of equilibrium constants measured at sedimentation equilibrium.[39] At any temperature, changes in standard free energy, enthalpy, and entropy that accompany a protein interaction are described by $-RT \ln K_{obs} = \Delta G° = \Delta H° - T\Delta S°$. Hydrophobic interactions, which remove large amounts of nonpolar surface from water, are characterized by positive values of $\Delta H°$ and $\Delta S°$ and negative values of $\Delta C_p°$, the heat capacity difference.[40] A nonzero standard heat capacity difference contributes to enthalpy and entropy differences as shown by

$$\Delta H° = \Delta H_\theta° - \Delta C_{p,\theta}° \cdot (T - \theta) \tag{14}$$

$$\Delta S° = \Delta S_\theta° - \Delta C_{p,\theta}° \cdot \ln\left(\frac{T}{\theta}\right) \tag{15}$$

Here, θ is a reference temperature and $\Delta S_\theta°$ and $\Delta H_\theta°$ are the enthalpy and entropy differences at that temperature. The temperature dependence of the association constant can be written[41]

[39] For accurate measurements, the operating temperature of the centrifuge should be calibrated as described by S. Liu and W. F. Stafford III, *Anal. Biochem.* **224,** 199 (1995).
[40] R. S. Spolar, J.-H. Ha, and M. T. Record, Jr., *Proc. Natl. Acad. Sci. USA* **86,** 8382 (1989).
[41] E. C. W. Clarke and D. N. Glew, *Trans. Faraday Soc.* **62,** 539 (1966).

$$R \ln K_{obs} = -\frac{\Delta G^o_\theta}{\theta} + \Delta H^o_\theta \left(\frac{1}{\theta} - \frac{1}{T}\right) + \Delta C^o_{p,\theta} \left(\frac{\theta}{T} - 1 + \ln\left(\frac{T}{\theta}\right)\right) \quad (16)$$

An example of a graph of $R\ln K_{obs}$ against temperature is shown in Fig. 6. Nonlinearity in such plots is evidence of nonzero values of $\Delta C^o_{p,\theta}$.

One caveat to the use of SE in investigating association thermodynamics is that analytical ultracentrifuges are typically limited to temperatures between 4 and 40°. This narrow range can result in inaccurate estimates of van't Hoff enthalpy and heat capacity changes. An offsetting advantage is that data for individual association reactions can often be obtained in solutions in which more than one interaction is taking place (e.g., the monomer–tetramer–octamer association of TATA-binding protein; Fig. 2). Thus, while the SE approach may not be appropriate for all systems, the thermodynamic information that can be obtained under favorable conditions is often unique and is generally complementary to that provided by calorimetry.

Sample Preparation

Volume

The most widely used cells are the six-channel short-column cell, containing three samples and three reference solutions, and the two-channel long-column cell, holding one sample and one reference solution. For most purposes, centerpieces with a thickness of 12 mm are used, although a two-channel centerpiece with 3 mm thickness is also available. Six-channel,

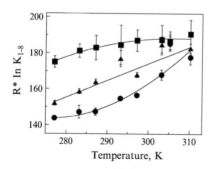

FIG. 6. Temperature dependence of yeast TATA-binding protein self-association. Data for the overall monomer–octamer reaction is shown. Smooth curves are fits of Eq. (16) to data using $\theta = 303$ K (30°) as reference temperature. Binding reactions were carried out in 20 mM HEPES/KOH, pH 7.9, 1 mM EDTA, 1 mM dithiotrietol supplemented with 120 (■), 300 (●), and 1 (▲) M KCl. Data from Daugherty et al.,[22] with permission.

12-mm cells accommodate samples $\leq$ 120 μl, whereas two-channel, 12-mm cells accept samples of $\leq$ 450 μl. Much smaller sample volumes have been used successfully in both cell types, and techniques for doing so have been described.[9] In addition, special very short-column centerpieces, requiring very small sample volumes, have been described.[9,42,43] Short-column experiments take significantly less time to reach sedimentation equilibrium than long-column experiments.[9,44] However, long-solution columns provide many more data points in a single scan and can allow a single scan to span a wider range of concentrations than the short-column format. These features are of value in data analysis, especially when samples are heterogeneous.

Sample Concentration

For experiments carried out with absorbance detection, the optimal signal/noise is obtained between 0 and 1 AU at the selected wavelength. To obtain a concentration gradient that spans that range, the starting absorbances are typically between 0.1 and 0.35. Studies that require higher concentrations can be performed without exceeding the optimal range of the detector by tuning the monochromator away from the λ_{max} of the sample. Studies at very low concentrations may require greater sensitivity than that available with intrinsic chromophores. Such work is often possible if the molecule(s) of interest is labeled with an extrinsic chromophore that has a high extinction coefficient. In addition, labeling with an extrinsic chromophore can allow an individual protein to be detected in complex mixtures where the number of species detectable at 215 or 280 nm is large. An elegant example of this approach takes advantage of the changes in absorbance spectra that result from the biosynthetic substitution of tryptophan with analogues such as 5-hydroxytryptophan.[17,45] Methods for the simultaneous analysis of multiple components with differing absorbance spectra have been described.[46] For experiments carried out with interference detection, protein concentrations in the range of 0.1–5 mg/ml are appropriate.

[42] J. J. Correia and D. A. Yphantis, *in* "Analytical Ultracentrifuation in Biochemistry and Polymer Science" (S. E. Harding, A. J. Rowe, and J. C. Horton, eds.), p. 237. Royal Society of Chemistry, Cambridge, 1992.

[43] T. M. Laue, Technical Information, Vol. DS-835, Beckman Instruments, Palo Alto, CA, 1992.

[44] G. Ralston, "Introduction to Analytical Ultracentrifugation." Beckman Instruments, Inc, Fullerton, 1993.

[45] T. M. Laue, D. F. Senear, S. Eaton, and J. B. Ross, *Biochemistry* **32,** 2469 (1993).

[46] M. S. Lewis, R. I. Shrager, and S. J. Kim, *in* "Modern Analytical Ultracentrifugation" (T. M. Schuster and T. M. Laue, eds.), p. 94, Birkhauser, Boston, 1994.

Sample Integrity

As described earlier, classical sedimentation equilibrium analysis is limited by the number of species that can be resolved. For this reason, the best samples are of high purity. Because centrifuge runs can be time-consuming, we recommend prerun verification that sample proteins are intact, in the correct state of posttranslational modification, and exhibit appropriate activities (e.g., DNA binding, enzyme activity). Many proteins have the potential to form more than one noncovalent complex (e.g., a specific assembly and a nonspecific aggregate). As these are not usually detectable by SDS–PAGE, but are detectable readily by sedimentation velocity analysis,[13,47–49] measurement of the samples' s value distribution can provide guidance in the choice of models for analysis of SE data. Postrun analyses (e.g., enzyme assays, SDS–PAGE) are also valuable, as they allow detection of any changes in sample integrity during sedimentation. After a run, samples are recovered easily through the filling holes using a micropipettor equipped with a gel-loading tip or a 1-ml syringe fitted with a capillary tube. For cells that lack filling holes (e.g., early vintage six-channel cells), samples can be recovered by careful removal of one window.

Absence of Nonprotein Contaminants

Nonprotein sample components (e.g., nucleic acids, starches, dextrans, and polyacrylamide) can complicate the sedimentation equilibrium analysis of a protein by contributing to the absorbance (or refractive index) gradient, by binding the proteins(s) of interest, or affecting activity coefficients through macromolecular crowding. If these components are not part of the system under study, it is worthwhile taking steps to ensure that they are not present in significant concentrations.

Buffer

Because sedimentation equilibrium experiments can be performed under a very wide range of solution conditions, it is often possible to tailor buffer compositions to the needs of a given system. However, some buffer properties can affect the outcome of a SE experiment. Important among these are transparency, salt concentration, and ability to suppress proteolysis. Data acquisition requires the transmission of light so buffer components that absorb strongly at the experimental wavelength(s) should be

[47] P. Schuck, M. A. Perugini, N. R. Gonzales, G. J. Howlett, and D. Schubert, *Biophys. J.* **82,** 1096 (2002).
[48] J. J. Correia, *Methods Enzymol.* **321,** 81 (2000).
[49] L. M. Carruthers, V. R. Schirf, B. Demeler, and J. C. Hansen, *Methods Enzymol.* **321,** 66.

avoided. Proteins far from their isoelectric pH values can have significant charge; electrostatic interactions can bias the equilibrium concentration gradient, resulting in incorrect apparent molecular weights. For moderately charged proteins, concentrations of 1:1 salts in the 100–300 mM range are often enough to suppress this effect. Attainment of sedimentation equilibrium can require many hours, giving proteolytic contaminants ample time to degrade samples. The treatment of samples with phehylmethylsulfonyl fluoride and/or inclusion of low molecular weight protease inhibitors (e.g., aprotinin, leupeptin) can minimize this problem.

A critical requirement for analysis of multicomponent systems is that all macromolecules must be at exchange (dialysis) equilibrium with the buffer.[50] A typical dialysis consists of two or more changes of solvent in a 1000:1 solvent- to-sample ratio. The dialysate is then used in the reference sectors of the centerpieces. Alternatively, samples can be equilibrated rapidly with buffer using centrifugal gel-filtration techniques.[51,52]

Rotor Speed and Duration of Run

If a molecular weight range can be predicted for the system, a useful estimate of rotor speeds can be obtained from data of Chervenka,[53] reprinted more recently in the manual by McRorie and Voelker.[7] These speeds are especially useful for experiments with long solution columns, although we often find that they are slower than optimal for short-column experiments. However, they are a valuable point for departure, as data obtained at several rotor speeds can allow discrimination of polydispersity and reversible association.[9] In addition, data sets obtained over a range of speeds can be usefully combined in a global analysis using programs such as NONLIN.[54] Because attainment of equilibrium following a speed reduction can be slow, multispeed experiments in which rotor speeds increase can take less time to complete than ones in which rotor speeds decrease.

The time needed to reach equilibrium is proportional to the square of the solution column height.[9,12] Thus short-column experiments reach equilibrium much faster than long-column runs. Attainment of equilibrium can be verified by subtracting successive scans.[55] In short-column experiments,

[50] E. F. Casassa and H. Eisenberg, *Adv. Prot. Chem.* **19**, 287 (1964).
[51] H. S. Penefsky, *Methods Enzymol.* **56**, 527 (1979).
[52] K. Struhl, *in* "Current Protocols in Molecular Biology" (F. M. Ausubel *et al.*, eds.), p. 3.4.8, Wiley, New York, 1989.
[53] C. H. Chervenka, "A Manual of Methods for the Analytical Ultracentrifuge," Spinco Division, Beckman Instruments, Palo Alto, CA 1969.
[54] M. L. Johnson, J. J. Correia, D. A. Yphantis, and H. R. Halvorson, *Biophys. J.* **36**, 575 (1981).
[55] The program MATCH by D. Yphantis and J. Lary (available on the web at ftp:// rasmb.bbri.org) facilitates this comparison.

scans separated by ≥ 1 h are an adequate test; for intermediate- and long-column experiments scans should be separated by at least 6 h. Very large molecules and ones undergoing slow association reactions can require many hours to reach equilibrium; for these systems, longer intervals are sometimes needed. Under favorable conditions, the time required to reach equilibrium can be reduced by "overspeeding" (a brief period at a higher speed followed by gentle deceleration to the final equilibrium speed).[56]

Determination of Buffer Density

The buffer density contributes to the buoyancy term, $1 - \bar{v}\rho$, for each sedimenting species. Because dilute aqueous buffers have densities close to 1 g/ml and typical, unmodified proteins have $\bar{v} \sim 0.73$ ml/g, a 1% error in solvent density results in $\sim 3\%$ error in M_r. Thus, buffer density must be known with accuracy. The density of a solution can be estimated by summing density increments for each component.[11] Extensive published data allow accurate calculation of densities for dilute aqueous buffers with common components.[7,29,57] This procedure is automated by the program SEDNTERP (Table I). However, oscillating densimeters allow solvent densities to be measured at high precision over the range of temperatures accessible to the centrifuge.[58] Because these values represent the actual densities of the buffer preparations, they should be used wherever possible.

Determination of Partial Specific Volume

An accurate value of the partial specific volume $(\bar{v})$ is also needed for evaluation of the buoyancy term. For proteins without posttranslation modification and/or nonprotein cofactor, $\bar{v}$ can be estimated from the amino acid composition according to the relation[59]

$$\bar{v}_{est} = \frac{\sum_i N_i M_i \bar{v}_i}{\sum_i N_i M_i} \tag{17}$$

where N_i is the number of residues of type i, M_i is the component molecular weight (amino acid molecular weight minus 18), and $\bar{v}_i$ is the tabulated

[56] D. E. Roark, *Biophys. Chem.* **5**, 185 (1976).
[57] T. M. Laue, B. D. Shah, T. M. Ridgeway, and S. L. Pelletier, *in* "Analytical Ultracentrifugation in Biochemistry and Polymer Science" (S. E. Harding, A. J. Rowe, and J. C. Harding, eds.), p. 90. The Royal Society of Chemistry, Cambridge, England, 1992.
[58] O. Kratky, H. Leopold, and H. Stabinger, *Methods Enzymol.* **27**, 98 (1973).
[59] E. J. Cohn and J. T. Edsall, "Proteins, Amino Acids and Peptides as Ions and Dipolar Ions," Reinhold, New York, 1943.

TABLE I
REPRESENTATIVE SOFTWARE FOR EDITING, DISPLAY, AND ANALYSIS OF SEDIMENTATION
EQUILIBRIUM DATA[a]

Application	Function	Platform (source)
NONLIN[b]	Analysis of reversibly associating homogeneous systems	PC,[c,d] Mac,[d] DEC[d]
Beckman software	Analysis of reversibly associating homogenous systems	PC[e]
SEDEQ[f]	Analysis of up to three independently sedimenting species	PC (DOS)[g]
TWOCOMP[f]	Analysis of two species that may reversibly self- or heteroassociate	PC (DOS)[h]
Omega Analysis[i]	Analysis of SE data	PC[d]
Ultrascan II	Nonlinear analysis of SE and sedimentation velocity data; estimation of solution density and protein $\bar{v}$	PC, UNIX, LINUX[j]
Ultraspin	Analysis of SE data; 20 models including homogeneous and some heterogeneous systems	PC[k]
WinMatch	Test for attainment of equilibrium	PC[c]
WinREEDIT	Editing raw data files from XL-I or XL-A	PC[c]
XLGRAPH	Data graphing and transformation of XL-A and XL-I data	PC[d,l]
SEDNTERP[m]	Estimation of solution density and protein $\bar{v}$	PC[d,l]

[a] The Reversible Associations in Structural and Molecular Biology (RASMB) group provides a web-based resource for researchers interested in the use and advancement of analytical ultracentrifugation (http://www.bbri.org/rasmb). A software archive is maintained at that address.

[b] M. L. Johnson, J. J. Correia, D. A. Y phantis, and H. R. Halvorson, *Biophys. J.* **36,** 575 (1981).

[c] http://vm.uconn.edu/~wwwbiotc.uaf.html.

[d] http://www.biochem.uthscsa.edu/auc/

[e] Available from Beckman Instruments, Inc.

[f] G. Rivas, W. F. Stafford, III, and A. P. Minton, *Methods* **19,** 194 (1999).

[g] ftp://bbri.harvard.edu/rasmb/spin/ms_dos/sedeq-minton/sedeq.doc; ftp://bbri.harvard. edu/rasmb/spin/ms_dos/sedeq-minton/sedeq.exe

[h] ftp://bbri.harvard.edu/rasmb/spin/ms_dos/twocomp-minton/sedeq.doc; ftp://bbri.harvard. edu/rasmb/spin/ms_dos/twocomp-minton/sedeq.exe

[i] G. B. Ralston and M. B. Morris, *in* "Analytical Ultracentrifugation in Biochemistry and Polymer Science" (S. E. Harding, A. J. Rowe, and J. C. Horton, eds.), p. 243, Royal Society of Chemistry, Cambridge, 1992.

[j] http://www.ultrascan.uthscsa.edu/.

[k] http://www.mrc-cpe.cam.ac.uk/ultraspin/.

[l] http://jphilo.mailway.com/default.htm.

[m] T. M. Laue, B. D. Shah, T. M. Ridgeway, and S. L. Pelletier, *in* "Analytical Ultracentrifugation in Biochemistry and Polymer Science" (S. E. Harding, A. J. Rowe, and J. C. Harding, eds.), p. 90, The Royal Society of Chemistry, Cambridge, England, 1992.

partial specific volume. Tabulated values of $\bar{v}_i$ have been published in several places.[7,29] Values of $\bar{v}_i$ are tabulated at a reference temperature (frequently 25°). If the experimental temperature differs from the reference, the value of $\bar{v}_{est}$ should be corrected as described.[57] Values of $\bar{v}_{est}$ obtained in this way are frequently accurate to 1–2%. Alternatively, if the sequence molecular weight $N_i \bullet M_i$ is known, the remaining unknown, $\bar{v}$, can be estimated from the experimental reduced molecular weight, σ. While both approaches assume that the component is not undergoing self-association (which can change $\bar{v}$), the second method has the advantage that the $\bar{v}$ estimate is obtained for the exact solution conditions used in the experiment.

For proteins of unknown amino acid composition, with substantial post-translational modification, or nonprotein prosthetic groups, partial specific volume can be measured experimentally by performing parallel experiments in H_2O- and D_2O-containing buffers. The value of $\bar{v}$ can be obtained by simultaneous solution of the equations[60]

$$M_{\text{bouyant, } H_2O} = M(1 - \bar{v}\rho_{H_2O}) \tag{18}$$

$$M_{\text{bouyant, } D_2O} = M(1 - \bar{v}\rho_{D_2O}) \tag{19}$$

in which M_{bouyant,H_2O} and M_{bouyant,D_2O} are the observed buoyant molecular weights and ρ_{H_2O} and ρ_{D_2O} are the measured densities of the H_2O- and D_2O-containing buffers.[61]

Data Analysis

Two general approaches have been used for the analysis of SE data. These are graphical methods[62–66] and nonlinear least-squares fitting.[67] Although each has advantages, graphical methods generally require transformation of data, which can complicate error analysis.[68] Direct fitting

[60] S. J. Edelstein and H. K. Schachman, *Methods Enzymol.* **27**, 82 (1973).
[61] This calculation ignores changes in M that result from H-D exchange, which will depend on the identity of amino acids and nonprotein components of the molecule(s) in question.
[62] D. E. Roark and D. A. Yphantis, *Ann. N.Y. Acad. Sci.* **164**, 245 (1969).
[63] R. H. Haschemeyer and W. F. Bowers, *Biochemistry* **9**, 435 (1970).
[64] G. J. Howlett, *Chemtracts-Biochem. Mol. Biol.* **11**, 950 (1998).
[65] D. J. Winzor and P. R. Wills, *in* "Modern Analytical Ultracentrifugation: Acquisition and Interpretation of Data for Biological and Synthetic Polymer Systems" (T. M. Schuster and T. M. Laue, eds.), p. 66. Birkhauser, Boston, 1994.
[66] D. J. Winzor, M. P. Jacobsen, and P. R. Wills, *Biochemistry* **37**, 2226 (1998).
[67] M. L. Johnson and M. Straume, *in* "Modern Analytical Ultracentrifugation: Acquisition and Interpretation of Data for Biological and Synthetic Polymer Systems" (T. M. Schuster and T. M. Laue, eds.), p. 37, Birkhauser, Boston, 1994.
[68] M. L. Johnson, *Anal. Biochem.* **206**, 215 (1992).

using functions that relate concentration to radial position avoids this drawback. Criteria for judging the adequacy and consistency of fitting results have been reviewed.[67,68] The past decade has seen an upsurge in the development of software for analysis of SE data, and more programs are under development at the time of this writing. A representative list of currently available software is given in Table I. The most powerful programs allow simultaneous fitting of several data sets, with some parameters (e.g., $c_{0,i}$) considered as local variables evaluated for each data set and some (e.g., M_i, K_{obs}) considered as global, evaluated for the entire data ensemble. Data collected at several rotor speeds, wavelengths, or starting concentrations can be combined, as can data obtained by absorbance and interference detectors. The analysis of large amounts of data obtained over a wide range of conditions makes it possible to discriminate between competing models describing the sedimentation behavior of moderately complex systems like those described earlier.

Prospects

The past decade has seen a renaissance in SE analysis, catalyzed by the availability of modern instrumentation and software for data analysis and the increasing availability of biologically important macromolecules. Several developments suggest that this trend will continue. Fluorescence detection[69] has lowered the concentration requirements of the SE technique, allowing study of hard-to-acquire macromolecules and equilibria with large association constants. Multiwavelength analysis[46] and methods for labeling proteins with chromophoric amino acid analogues (e.g., 5-hydroxytryptophan[45,70]) allow the simultaneous measurement of several proteins in complex mixtures, as well as the analysis of labeled proteins in the presence of several unlabeled species. Postcentrifugation fractionation methods[10,71] allow the detection of components by tracer methods, including direct counting of radioisotopes, radioimmunoassay, and real-time polymerase chain reaction. With these techniques, SE analyses can be carried out on target species at very low concentrations, in highly complex mixtures, and in the presence of high background concentrations of other macromolecules. These developments will make possible the analysis of systems of life-like complexity, spanning the concentration ranges that occur *in vivo*.

[69] T. M. Laue, A. L. Anderson, and B. J. Weber, *in* "Ultrasensitive Clinical Laboratory Diagnostics" (G. Cohn, ed.), Vol. 2985, p. 196. SPIE, Bellingham, WA, 1997.

[70] J. B. Ross, D. F. Senear, E. Waxman, B. B. Kombo, E. Rusinova, Y. T. Huang, W. R. Laws, and C. A. Hasselbacher, *Proc. Natl. Acad. Sci. USA* **89,** 12023 (1992).

[71] S. Darawashe and A. P. Minton, *Anal. Biochem.* **220,** 1 (1994).

Acknowledgments

We gratefully acknowledge valuable discussions with Drs. Jack Correia, Michael Johnson, Jacob Lebowitz, and Allen Minton and thank the many people who sent reprints and preprints.

[32] Single-Molecule Studies of DNA Architectural Changes Induced by Regulatory Proteins

By LAURA FINZI and DAVID DUNLAP

Single-molecule techniques are now established approaches in the investigation of the molecular mechanisms involved in nucleic acids/protein interactions. The advantages of single-molecule approaches are related to the fact that the observation is performed on just one molecule at a time. This allows the study of the behavior of a molecule as it goes through different conformational states in time, thus revealing states that are averaged out in bulk experiments where unsynchronized molecules are observed simultaneously. The tethered particle motion (TPM) method is a general method used to measure the contour length of a suitably bead-labeled and surface-immobilized DNA fragment. As such, it can potentially be used in a variety of single-molecule kinetics experiments and can yield mechanistic information. It can be used to reveal and monitor, dynamically, events that induce large conformational changes in single DNA molecules, such as those induced by regulatory proteins, or to study the mechanisms by which enzymes move in a directed fashion along DNA (polymerases, helicases, etc). For example, it is now clear that TPM assays can provide much information about the mechanisms by which transcriptional regulatory proteins act.[1–3]

Furthermore, the effect of supercoiling on DNA/protein interactions or the change in DNA-linking number induced by regulatory proteins can be characterized in a TPM experiment using magnetic tweezers.[4,5] This article concentrates on the protocol used to prepare most TPM experiments. This consists mainly of three stages: (i) preparation of DNA constructs suitably labeled with biotin at one end and digoxigenin at the

[1] Finzi and Gelles, *Science* **267**, 378, (1995).
[2] Lia *et al.*, *PNAS* **100**, 11373–11377 (2003).
[3] J. Gelles, personal communication.
[4] T. R. Strick, J.-F. Allemand, D. Bensimon, A. Bensimon, and V. Croquette, *Science* **271**, 1835 (1996).
[5] T. R. Strick, V. Croquette, and D. Bensimon, *Nature* **404**, 901 (2000).

other end, (ii) preparation of the microscope flow chamber, and (iii) data registration and analysis. Protocols differ for measurements with and without manipulation by magnetic tweezers.

Preparation of DNA Constructs

DNA Fragments with One Biotin and One Digoxigenin Molecule per End

These fragments are used in TPM experiments without torsion and are synthesized by a polymerase chain reaction (PCR) starting from a plasmid containing the sequence of interest and a pair of primers 5' labeled with a biotin and a digoxigenin, respectively.

DNA Fragments with Several Biotin and Digoxigenin Labels per End

These fragments are used in TPM experiments where torsional stress is applied (magnetic tweezers). These fragments are synthesized in steps. First, the sequence of interest is amplified by PCR from plasmidic DNA using nonlabeled primers in such a way that each end of the amplified, linear fragment contains a site for one of a pair of restriction enzymes. Second, a digestion is performed with the pair of different restriction enzymes and the clipped ends are eliminated by filtration. Third, two different DNA "tails" are synthesized by PCR using biotin-labeled nucleotides in one and digoxigenin-labeled nucleotides in the other. In so doing, the labels will be on both strands of the DNA tail. The biotinated tail and the digoxigeninated one, approximately 1000 bp long, will contain, respectively, a restriction site identical to one of the two already used to prepare the fragment of interest. In this way, after digestion of the tails, they can be ligated to opposite ends of the fragment of interest, which will then have one biotinated end while the opposite end is digoxigeninated. To test labeling, a gel mobility shift assay[6] may be used. Excess streptavidin is added to 150- to 200-bp-long digests of the DNA, incubated for 30 min at 37°, and loaded onto a nondenaturing 8% polyacrylamide gel (bisacrylamide 1:19) in 0.5 TBE buffer. Antidigoxigenin is used to check labeling with digoxigenin.

Preparation of Microscope Flow Chamber

Microscope flow chambers can be made out of (i) two glass coverslips or (ii) a glass capillary.

---

[6] M. G. Fried and D. M. Crothers, J. Mol. Biol. **172,** 263 (1984).

Assembly

Flow chambers from coverslips are made creating a channel between two coverslips of different size (Menzel-Gläser). One coverslip is approximately 40–50 × 24 mm, while the other is 22 × 22 mm. They are held at a slight separation by two strips (approximately 22 × 3 mm) of double-sided tape. Between the tape spacers, lines of high vacuum grease dispensed through a syringe ending with a 0.5- to 10-μl pipette tip are used to create a channel approximately 3 mm wide and 100 nm deep, resulting in a volume of 15–20 μl. Fluid placed at one end of the channel may be drawn into it by capillary force and withdrawn for washings and changes of buffer at the other end of the flow chamber using filter paper. A line of high vacuum grease is also used to create a barrier on the top coverslip, right above the inlet, to avoid overflow of the liquid being introduced into the flow chamber. Finally, with the same grease, a small reservoir is created in front of the inlet. The flow chamber described is best assembled on an aluminum support as described[7] previously.

Flow chambers from capillaries are used traditionally in measurements involving magnetic tweezers. Capillaries can be bought from "VitroCom, precision glass products" 1 mm × 1 mm × 3–7 cm depending on the microscope stage. Independently of the microscope used, the stage will have to be engraved or modified slightly to hold a capillary perfectly horizontally and stably. The ends of the capillaries are connected through plastic tubing (Tygon), about 1 mm in diameter, to inlet and outlet reservoirs. Gravity or a pump can regulate the flow.

Cleaning

Glass surfaces of the flow chamber need cleaning before sample incubation. Different laboratories use different cleaning procedures and standardization would be a welcome development. Coverslips may be cleaned simply by washing them with pure ethanol, washing with deionized, Milli Q H_2O, and finally by wiping them with Kodak optical paper. At this point the glass is ready to be coated with antidigoxigenin (see following section). Capillaries may be cleaned in any of the following different ways.

1. Soak them in NH_4:H_2O_2:H_2O (1:1:5) to render them hydrophilic and then soak them in "Sigmacote" to make them hydrophobic. At this point, biotinated bovine serum albumin (BSA) may be inserted followed by avidin.[8]

[7] B. J. Schnapp, *Methods Enzymol.* **134,** 561 (1986).
[8] B. M. Jaffar Ali, R. Amit, I. Braslavsky, A. B. Oppenheim, O. Gileadi, and J. Stavans, *Proc. Natl. Acad. Sci. USA* **98,** 10658 (2001).

2. Soak them in aminopropyl-trietoxy-silano (APTES), dry them in a clean atmosphere, and then coat them with antidigoxigenin.
3. Dip capillaries in toluene containing 1% polystyrene (MW 45,000 g/mol), dry them with inert gas, keep dry for at least 2 days, and insert 50 μl of antidigoxigenin[2] (see following section).

Preparation of the Sample

In order to produce beads tethered by single DNA molecules, the latter must bind specifically to the bead at one end and to the glass surface at the other end. The two kinds of attachments should be different to maximize the yield of tethers. This is accomplished by labeling linear DNA constructs at one end with biotin and at the other end with digoxigenin (see earlier discussion). These ligands are particularly suited for the applications described here. The streptavidin/biotin system, for example, has one of the largest free energies of association yet observed for noncovalent binding of a protein and small ligand in aqueous solution ($K_{assoc} = 10^{14}$). The complexes are also extremely stable over a wide range of temperature and pH. It is irrelevant which protein/ligand pair is used at which end of the DNA. Traditionally, the antidigoxigenin/digoxigenin pair is used to attach DNA to the glass surface, whereas the avidin/biotin pair is used to link DNA to the bead. This is due mostly to the larger variety of avidin rather than antidigoxigenin-coated beads available on the market.

Antidigoxigenin IgG (Roche Applied Science) should be resuspended in phosphate-buffered saline (PBS) at 200 μg/ml, aliquoted, and stored at $-20°$. The working solution of antibody is 20 μg/ml.

Flow Chambers

Flow chambers are coated with antidigoxigenin by filling them with a solution of antibody and incubating for 20 min. After washing the flow chamber with 800 μl of buffer supplemented with 0.1 mg/ml α-casein (to prevent nonspecific adhesion of DNA to the glass surface, see later), biotin/digoxigenin-labeled DNA in the appropriate reaction buffer is inserted and incubated for 1 h. After incubation, the unbound DNA is washed out with 800 μl of buffer and avidinated beads are inserted and incubated for 30 min. Unbound beads are washed away with buffer with or without protein. Some researchers label DNA with beads before introducing the solution of bead-labeled DNA molecules in the flow chamber. The preparation is conducted at ambient temperature. The sample is then observed in a microscope, and images of beads tethered by DNA are recorded for control measurements. Finally, protein-containing buffer is added and registration resumes. Image analysis of the recordings (see later) reveals protein-induced DNA architectural changes.

Capillaries

Capillaries are coated with antidigoxigenin by filling them with a solution of antibody and incubating for a day in a humid chamber to avoid evaporation. The inside is passivated with BSA in protein buffer for another day and washed with 1 ml of standard buffer (10 mM phosphate buffer, 0.1% Tween 20, 0.1 mg/ml BSA, 10 mM sodium azide). At this point, the capillaries can be stored at 37° until used. DNA, approximately 1 nM, is then incubated with approximately 10^{16} beads in a volume of approximately 10 μl for about 1 min. In order to stop the DNA bead-labeling reaction, the DNA/bead mixture is diluted 25 times in standard buffer or in the reaction buffer appropriate for the DNA/protein interaction. Twenty microliters is inserted in the capillary and incubated for 20 min or until a sufficient number of tethers are observed in the microscope.

Nonspecific Sticking of DNA Tethers to the Glass Surface

Nonspecific sticking can occur if the glass surface is not completely passivated. In order to check for nonspecific DNA sticking to the glass surface, the range of Brownian motion observed for the tethered bead is compared with previously calibrated values. If the motion is smaller, the DNA may be partially stuck to the glass or several tethering molecule may have become attached to a particular bead.

In order to avoid nonspecific sticking of DNA to the glass surface, carrier proteins such as α-casein at 0.1 mg/ml or BSA at 0.1–10 mg/ml are used. α-Casein may be prepared at 5 mg/ml and then centrifuged at 34,540 g for 10 min to remove insoluble material.[9] It is useful to determine the effective concentration of carrier protein in the stock solution by measuring its OD at 280 nm. D-Biotin (Roche Applied Science) is used for bead passivation after tethers are formed. BSA may also be used to reduce aggregation of the DNA-binding protein.

Surface Density

When more than one DNA molecule tethers the same bead to the glass surface, that bead displays reduced Brownian motion. TPM experiments require specimens in which each bead is attached to a single DNA molecule. However, because the avidin-coated beads may have multiple biotin-binding sites, they can potentially attach to multiple DNA molecules. To

[9] H. Yin, R. Landick, and J. Gelles, *Biophys. J.* **67,** 2468 (1994).

minimize this possibility, it is useful to use a low surface density (D) of DNA on the flow chamber walls as described previously.[1] Two DNA molecules cannot attach to the same bead unless $L \leq 2l + d$, where L is the distance between the two DNA/bead complexes, l is the initial tether length of the complexes, and d is the bead diameter. The fraction (P) of tethering DNA molecules that have the nearest neighbors such that $L \leq 2l + d$ is given by the Poisson distribution as $P = 1 - \exp[-D\pi(2l + d)^2]$. Usually the experiments are performed with $P < 0.10$. This can be measured in an experiment that used DNA radiolabeled by inclusion of $[\alpha\text{-}^{32}P]$deoxycytidine 5'-triphosphate (dCTP) in the PCR. For example, when chambers were incubated with DNA at $<2 \times 10^{-9}\ M$, the surface density was proportional to the applied DNA concentration.[1] The DNA concentration (approximately $1 \times 10^{-12}\ M$) used for the microscope specimens was chosen to give a surface density of 0.01 molecules/μm^2. At that density, $\leq 3\ \%$ of DNA molecules have the nearest neighbor sufficiently close (within 1.04 μm) so that two molecules could attach to the same bead.

Beads

Paramagnetic and nonparamagnetic beads functionalized differently may be bought from a variety of manufacturers, such as BangsLabs, Indicia Diagnostics, Dynal, Seradyn, Polyscience, Molecular Probes, and Interfacial Dynamics Corporation (IDC). Avidin-, streptavidin-, and digoxigenin-labeled beads are available, but for some applications it may be preferable to prepare them in the laboratory.

In the case of nonparamagnetic beads, coating with streptavidin may be achieved as follows. Carboxylated polystyrene beads (Polyscience Inc., Warrington, PA) may be biotinylated covalently, coated with avidin, and purified by gel filtration as described.[10] Streptavidin-coated beads can be prepared by first conjugating 0.01 mg/ml biotin x-cadaverin (Molecular Probes) to 0.25% (w/v) 100- to 600-nm carboxylated polystyrene beads with 0.5 mg/ml 1-ethyl-3-(3-dimethylaminopropyl)carbodiimide HCl, 0.625 mg/ml N-hydroxysulfosuccinimide in 10 mM sodium phosphate, pH 7.0, and 0.1% Tween 20. After 1.5 h, the reaction is quenched by the addition of 0.2 M glycine, pH 8.0. Excess biotin-x-cadaverine is removed by dialysis in 10 mM sodium phosphate, pH 7.0, 50 mM NaCl, and 0.1% Tween 20. Streptavidin is then added to 20 mg/g of beads and the beads are stored at 4$°$. Before use, unbound streptavidin may be removed by gel filtration (Sepharose 4B) in 50 mM imidazole–Cl$^-$, pH 6.7, 50 mM NaCl, 2 mM EGTA, 4 mM MgCl$_2$, and 0.1% Tween 20. The final bead

[10] E. Berliner, E. C. Young, K. Anderson, H. K. Mahtani, and J. Gelles, *Nature* **373**, 718 (1995).

concentration may be determined by comparing A_{260} to that of standard dilutions of unconjugated beads.

In the case of paramagnetic beads, coating with streptavidin may be achieved as follows. A washing procedure is necessary to equilibrate the beads in the appropriate buffer.

1. Resuspend the Tosyl-activated beads by pipetting and vortexing for 1–2 min. Avoid foaming.
2. Immediately pipette the volume to be used into the desired test tube.
3. Place the tube in a magnet (Dynal MPC) for 1 min.
4. Aspirate the supernatant carefully, leaving beads undisturbed.
5. Remove the test tube from the Dynal MPC and resuspend the beads carefully in an ample volume ($1000\times$ original volume) of buffer A or B (see later). Mix gently for 2 min.
6. Place the tube in the magnet for 1 min, pipette off the supernatant, and resuspend the washed beads in the same volume of buffer A or B as that removed in step 4 or to the desired concentration (see later for recommended coating concentration). The Tosyl-activated beads are now washed and ready for coating.

The following steps are used for the coating procedure.

1. Make a homogeneous suspension of beads Tosyl activated by using a pipette and by vortexing for 1–2 min. Pipette out the desired number of Dynabeads and wash as described previously.

2. Add 5 μg pure antibodies or other ligands/10^7 beads and continue to vortex for 1–2 min. A concentration of 4–10×10^8 Dynabeads/ml final coating solution is used (i.e., including antibody).

3. Incubate for 16–24 h at 37° with slow tilt rotation. Lower temperatures may be used for temperature-sensitive antibodies. *Note.* Do not let the beads settle during the incubation period.

4. After incubation, place the tube in the magnet for 2–3 min and remove the supernatant. Generally, 40–80% of the added antibody will bind.

5. Wash the coated beads four times:
Twice in buffer C for 5 min at 4°
Once in buffer D for 24 h at 20° or for 4 h at 37° (Tris will block free Tosyl groups), do not let the beads settle during this period either
Once in buffer C for 5 min at 4°

6. Tosyl-activated beads are now coated and ready to use.

7. Store the coated beads in buffer C at a known concentration. Coated beads can usually be stored for several months at 4°, depending on the

stability of the immobilized ligand. Sodium azide [0.02% (w/v) (final concentration)] may be added as a bacteriostatic agent.

Buffer A: 0.1 M phosphate buffer, pH 7

Buffer B: 0.1 M borate buffer, pH 9.5

Note. Buffers A and B are used for prewashing and coating of Tosyl-activated beads. Do not add any protein, sugar, and so on to these buffers.

Buffer C: PBS, pH 7.4, with 0.1% (w/v) BSA(HSA) (phosphate-buffered saline)

Buffer D: 0.2 M Tris, pH 8.5, with 0.1% (w/v) BSA(HSA). Make sure the HSA/BSA is sterile and free from endotoxins

Buffer E: 0.2 M Tris, pH 8.5, with 0.1% (w/v) BSA(HSA)

Tosyl groups on the beads tend to hydrolyze in time. Many bead manufacturers provide protocols for the functionalization of their beads, which can be found on their web sites.

Data Collection

Microscope

A well-maintained, vibrationally isolated microscope is essential for high-quality analysis. X–Y tracking has been implemented using either differential interference contrast or bright-field imaging. While low illumination suffices for bright-field imaging, DIC contrast improves with intensity. High-magnification, plan objectives with minimal color correction work well for either contrast method, but oil immersion slightly enlarges and adds contrast, especially to bright-field images due to the spherical aberration associated with differences between oil (immersion) and water (specimen) refractive indices. With either technique, a video rate camera with the low sensitivity necessary for transmitted light microscopy may be used, although higher temporal resolution is possible with double-speed, CCD cameras from JAI (Copenhagen, Denmark), Pulnix (Sunnyvale, CA), Sentech (Sensor Technologies America, Inc., Carrollton, TX), or Sony (Tokyo, Japan). CCD cameras with small pixel areas enhance the optical magnification. X–Y calibration requires standard specimens.

Data Recording

VHS video tape may still be the least expensive and simplest recording medium. However, super VHS equipment can enhance the resolution of the recording and is preferred. Tape recording is being supplanted rather quickly by digital video storage, which has become accessible technologically and economically with the advantage that it can be stored virtually indefinitely with no loss in reproduction quality. Because long digital video recordings may require several gigabytes of disk space, video compression algorithms are usually necessary, and professional-quality compression hardware for personal computers (Pinnacle Systems, Inc., Mountain View, CA) is available to optimize the quality of compressed data. In any case, a unique compression/decompression algorithm should be chosen and utilized for recording and reproduction to ensure high fidelity. To further reduce disk space requirements, images may be cropped to include only the area of interest.

Image Processing

Most optimization of the images should be done optically, but background subtraction may improve contrast. In particular, differential interference contrast images can be enhanced considerably by illuminating with alternating orthogonal polarizations of light and subtracting the resulting images.[11]

Particle Tracking

Fundamental to accurate analysis is a frame of reference because vibrations and thermally induced size fluctuations of microscope parts give rise to artifacts. This is established most easily by determining the positions of mobile, tethered beads in reference to a fixed object (e.g., stuck bead) in the image.[12]

To analyze the effective tether length associated with a particle, one can either determine the particle positions in each image of the video sequence and calculate the average displacement from a mean position (tether point) during a selected interval of time or create an average image from the images in a selected interval and determine the broadening of the object in the average image that correlates with the range of Brownian motion.[13] The former method requires accurate recognition of the particle in each image of the sequence, which may be blurry in the case of a small rapidly moving microsphere, but gives high temporal resolution and direct detection of discretely distributed tether lengths. Time-averaged tether lengths may be compared to theoretical worm-like chain values (unpublished data). Frame rate tracking can be accomplished using convolution (Meta-Morph, Universal Imaging, Downingtown, PA) or custom routines that exploit the symmetry of rings surrounding defocused images of microspheres to simplify the problem into two one-dimensional convolutions.[14] Instead,

[11] G. Holzwarth, S. C. Webb, D. J. Kubinski, and N. S. Allen, *J. Microsc.* **188**, 249 (1997).
[12] J. Gelles, B. J. Schnapp, and M. Sheets, *Nature* **331**, 450 (1988).
[13] D. A. Schafer, J. Gelles, M. P. Sheetz, and R. Landick, *Nature* **352**, 444 (1991).
[14] C. Gosse and V. Croquette, *Biophys. J.* **82**, 3314 (2002).

the broadening of the average image method has lower time resolution and does not reveal either the actual length or the distribution of tether lengths but is immune to blurring. Accuracy and precision of this TPM method have been published[9] and can be implemented using routines available under a general public license software agreement (www.bio.brandeis.edu/~gelles/software/index.html).

[33] Assay of an Intrinsic Acetyltransferase Activity of the Transcriptional Coactivator CIITA

By JOCELYN D. WEISSMAN, APARNA RAVAL, and DINAH S. SINGER

Posttranslational modification of proteins is increasingly being recognized as an important regulatory mechanism in transcription. The best characterized of the posttranslational modifications are the phosphorylations of proteins involved in regulating signal transduction pathways and glycosylation of membrane proteins. However, studies have demonstrated that acetylation and methylation play pivotal roles in transcriptional regulation.[1] Acetylation of nucleosomal histones is associated with transcriptionally active chromatin, whereas methylation correlates primarily with silent regions of chromatin. A number of nonhistone proteins have been demonstrated to undergo acetylation, although the function of this modification is not understood. Among the proteins shown to be acetylated are p53, ATF2, HIV Tat, and the coactivator CIITA.[2–4]

A number of proteins with acetyltransferase enzymatic (AT) activity have been identified. Most of these proteins, such as CBP, p300, and PCAF, were first identified as transcriptional coactivators and later shown to acetylate histones.[5] Other proteins such as TAFII250 and Tip60, which mediate transcription but do not function as coactivators, also have been shown to have AT activity.[6,7] Interestingly, the substrates of these acetyltransferases have not been well characterized. Although the coactivators appear to primarily acetylate histones, some are able to acetylate

[1] T. Jenuwein and C. D. Allis, *Science* **294**, 2477 (2001).
[2] B. Cullen, *FASEB J.* **5**, 2361 (1991).
[3] V. Ogrysko, *Cell. Mol. Life Sci.* **58**, 683 (2001).
[4] A. Raval, T. K. Howcroft, J. D. Weissman, S. Krishner, X. S. Zhu, K. Yokoyama, J. Ting, and D. S. Singer, *Mol. Cell.* **7**, 105 (2001).
[5] V. Ogryzko, L. Schiltz, V. Russanova, B. Howard, and Y. Nakatani, *Cell* **87**, 953 (1996).
[6] C. Mizzen, X. Yang, T. Kokubo, J. Brownell, A. Bannister et al. *Cell* **87**, 1261 (1996).
[7] Y. Yamamoto and M. Horikoshi, *J. Biol. Chem.* **272**, 30595 (1997).

       0076-6879/03 $35.00

nonhistone proteins. For example, CBP and PCAF have been shown to acetylate CIITA.[8,9] Similarly, the AT activity of TAFII250 is required for its function in *in vitro* transcription of naked DNA in the absence of nucleosomal histones.[10]

The transcriptional coactivator CIITA mediates the interferon-γ response by nucleating formation of the transcriptional enhanceosome that activates MHC class I and class II transcription.[11] We have demonstrated that CIITA, like other coactivators, contains intrinsic AT activity.[4] Unlike other coactivators, this AT activity is stimulated by GTP. The CIITA enzymatic activity appears to depend on protein conformation, as it is only detected in protein purified either from transfected mammalian cells or from insect Sf9 cells; bacterially derived CIITA does not contain detectable AT activity.

The methods given here describe the purification of both insect- and mammalian-derived CIITA, the acetyltransferase assay using histones as a substrate, and Western blotting to confirm purification of the protein. In all of these protocols, we have used a flag-tagged CIITA. It is also important to note that the AT activity of CIITA as measured with histone substrates is considerably less than that of CBP or p300. This may reflect either an intrinsic difference in enzymatic activity or that histones are not the native substrate of CIITA.

Preparation of Recombinant CIITA from Baculovirus

rCIITA purified from a baculovirus that has been harvested from insect Sf9 cells, as described later, reproducibly displays readily detectable histone acetyltransferase activity. To facilitate purification and monitoring of the CIITA, the protein is flag tagged at the 5' end of the molecule. However, purification of enzymatically active CIITA depends on the conditions in which Sf9 cells are grown and maintained and the conditions of transfection and viral infection. The following protocols describe the growth of the cells, their transfection with the CIITA gene cloned into a baculoviral vector to generate viral stocks, the isolation of viral stocks, the infection of Sf9 cells for the quantitative isolation of virus containing CIITA, and purification of the CIITA from the virus.

[8] J. D. Fontes, S. Kanazawa, D. Jean, and B. M. Peterlin, *Mol. Cell Biol.* **19,** 941 (1999).
[9] A. Kretsovali, T. Agalioti, C. Spilianakis, E. Tzortzakaki, M. Merika, and J. Papamatheakis, *Mol. Cell Biol.* **18,** 6777 (1998).
[10] J. D. Weissman, J. Brown, T. K. Howcroft, J. Hwang, A. Chawla, P. Roche, L. Schiltz, Y. Nakatani, and D. S. Singer, *Proc. Natl. Acad. Sci. USA* **95,** 11601 (1998).
[11] K. Masternak, A. Muhlethaler-Mottet, J. Villard, M. Zufferey, V. Steimle, and W. Reith, *Genes Dev.* **14,** 11566 (2000).

Reagents

SF9 culture medium: TMN-FH (Gemini Bio-Products)
Baculogold transfection kit: Pharmingen kit
Phosphate-buffered saline (PBS)
Lysis buffer: 50 mM Tris–Cl, pH 8.0, 5 mM MgCl$_2$, 150 mM KCl, 0.1% NP-40, 10% glycerol, freshly added 10 μg/ml pepstatin, 5 μg/ml leupeptin, 20 μg/ml aprotinin, 1 mM phenylethylsulforyl fluoride (PMSF)
Antiflag agarose beads: Sigma A1205
TBS: 50 mM Tris, pH 8.0, 150 mM NaCl
0.1 M glycine, pH 3.5
Flag peptide stock solution: 5 mg/ml in TBS
Buffer D: 20 mM HEPES, pH 7.9, 100 mM KCl, 0.2 mM EDTA, 0.5 mM dithiothreitol (DTT), 20% glycerol (for Hi-Trap columns)

Growth of Sf9 Cells

Sf9 insect cells can be grown as either a monolayer in a tissue culture flask or in suspension culture in spinner flasks. In either case, care should be taken while growing Sf9 cells, as they are extremely sensitive to fluctuations in temperature, cell density, and agitation.

Spinner SF9 cultures have to be maintained at 27° at a density of 5 × 10^5 to 2 × 10^7 cells/ml and under gentle shaking. Cells are harvested by centrifugation at 500 rpm or less for 10 min with no brake.

Monolayer cultures should be plated at a density of 10^7 cells in 30 ml medium in a T100 plate. Cells should be transferred to fresh medium every 2–3 days by harvesting with gentle agitation and dilution at about 1:3 into a clean plate to reestablish the original seeding density. (The cells are extremely fragile and may be broken by scraping with a rubber policeman.)

Transfection of Sf9 Cells to Generate Viral Seed Culture

The following protocol describes transfection into SF9 cells of the baculovirus expression vector, PVL 1393 with the inserted CIITA cDNA, recovery of the recombinant virus, and expansion of the virus. Then, expression of the CIITA recombinant protein is analyzed by Western blotting of infected SF9 cell extracts.

Day −2: Transfer Sf9 cells to fresh medium before transfection to ensure that cells are growing in an exponential phase.

Day 1: Harvest cells and plate at 3 million cells/10 ml in a T25 flask. Plate an additional flask for a nontransfected control. Close and set it aside for 1 h to let cells adhere. Check under a microscope for adherence before proceeding. Bring buffers A and B (from Pharmingen Baculovirus Gold kit) to room temperature. Combine 0.5 μg of Baculovirus Gold DNA from the kit with 4 μg of CIITA expression vector PVL 1323 (instead of the 2 μg suggested by the kit protocol) in an Eppendorf tube. Leave at room temperature for 5 min. While waiting, aspirate the medium from Sf9 cells in the T25 flask. Add 1.25 ml of buffer A. After the 5-min incubation, add 1.25 ml buffer B to the DNA mixture and mix well. Add this mix dropwise to the flask containing the cells. Leave the flask at 27° for 4 h. At this point, the supernatant should look cloudy. Remove the supernatant by aspiration and add 4 ml fresh SF9 medium. Incubate for 48 h at 27°.

Day 3: Examine the cells in the flask under a microscope. Transfected cells should appear larger than those in the nontransfected control flask. Replace 2 ml of the old culture medium with 2 ml fresh SF9 medium.

Day 5: At this point, most of the remaining cells should be floating, not adherent, although a few may remain attached to the flask. This is an indication of a successful transfection. Using sterile technique, remove the medium with the floating cells in a sterile 15-ml tube and centrifuge at 1200 rmp for 10 min to clear debris. Store the cleared supernatant at 4°. If large numbers of cells remain attached to the flask, add an additional 4 ml of fresh medium and continue to incubate the cells for 2 more days.

Day 7: Remove the medium, spin, and store the supernatant at 4°. Do not combine with first supernatant.

Viral Assay

To assay for the presence of viral particles in the culture supernatants, seed 3 million Sf9 cells in a small T25 flask, allowing them to adhere at 27° for 1 h. Remove medium. Sterilely mix 600 μl of supernatant from day 5 with 600 μl of fresh medium and add to flask. Incubate for 1 h. Every 10–15 min rotate the flask gently to make sure that the entire surface of the flask is covered. After 1 h add 3 ml medium and incubate at 27° for 48 h. After 48 h, harvest the cells and make lysate (see later). Run SDS–PAGE and Western to determine if protein has been made. Because there is considerable variation in the rate and the stability of the produced recombinant proteins, a time course should be run to determine the optimal time for harvesting of the supernatants after transfection.

Preparing a Large-Scale Viral Stock

If the transfection has been successful, as measured by the production of CIITA protein, a viral stock is generated.

Take 30 million cells in a large T flask and let them adhere for 1 h. Add 2 ml of the viral seed culture (supernatant from day 5 or 7) and 8 ml fresh medium. Incubate at 27° for 1 h, rotating every 10–15 min. After 1 h, add 30 ml fresh medium. Harvest cells on day 5 in a 50-ml sterile tube and spin at 1200 rpm for 10 min and store the supernatant at 4°. This is the viral stock.

Protein Production and Purification

Place 200 million cells (1–1.5 million cells/ml) in a spinner flask and add 10 ml of viral stock. (At some point, you will need to make Sf9 cell extract from a nontransfected culture to use as control.) Incubate at 27° for 48 h. Harvest cells and spin at room temperature, 500 rpm with no brake. Remove the supernatant and wash the pellet once with room temperature PBS. Immediately freeze the pellet on dry ice. In general, the pellet should be processed immediately, although some proteins may be left at −70° for a few days. Thaw the pellet at 37° for 5 min. Add 3–4 ml lysis buffer and incubate on ice for 20 min. Break open the cells with passage through a 23-gauge needle twice. Spin lysate at 40,000 rpm at 4° for 20 min to remove debris. Collect the supernatant, which can be stored at −70°. Repeated freezing and thawing may lead to protein degradation.

Antiflag Bead Activation

The CIITA protein used in our studies has an N-terminal flag tag that is used as the basis of the protein purification. During ultracentrifugation of the lysate described earlier activate the antiflag beads. Wash 500 μl of anti-flag agarose beads three times with TBS as follows: place beads in a 1.5-ml Eppendorf tube, add 1 ml of TBS, and invert several times. Spin at 2000 rpm for 2–3 min. Discard the supernatant and repeat the washing step. Then wash three times with 0.1 *M* glycine, pH 3.5, to activate the beads. Wash again with TBS. Finally wash once in lysis buffer with protease inhibitors. Resuspend in 1 ml lysis buffer (i.e., 2X original bead volume).

Notes. The beads should be activated fresh everytime, before the immunoprecipitation. The beads should not be left in the glycine buffer more than 20 min.

Protein Immunoprecipitation

Add 1 ml of the activated antiflag agarose bead slurry to the 3–4 ml of cleared viral lysate prepared in earlier. (A nontransfected control lysate should be processed in parallel.) Rotate the mixture at 4° for 4 h and then spin down the bead at 2000 rpm for 2–3 min. Wash the beads five times with 10 ml of lysis buffer (with protease inhibitors). Elute the bound protein with 500 μl of the flag peptide solution, freshly diluted to 200 μg/ml in TBS. Elute four times by adding 500 μl each time of this 200-μg/ml solution (before eluting, check pH of flag peptides, which should not be acidic). After each 500-μl addition, spin in a horizontal rotor at 2000 rpm for 15 min at 4°. Collect the supernatant in a fresh tube. The first two elutions have most of the proteins. Therefore, combine the first and second elutions separately from the third and fourth. This results in two aliquots of protein of 1 ml each (four elutions of 500 μl each). Dialyze each aliquot against three changes of buffer of interest for 4 h to remove flag peptides.

If flag peptides remain in the protein preparation after the dialysis, they can be removed by gel chromatography or Hi-Trap columns. For Hi-Trap column chromatography, the Hi-Trap column is washed with buffer D and elution is also done in buffer D. After elution, concentrate the eluate in a 30 or 50K cutoff Microcon or Centricon and then add fresh protease inhibitors, pepstatin, leupeptin, and aprotinin. Make aliquots and store at −70°.

To prevent loss of protein due to denaturation, the entire process should be completed in 1 day.

Preparation of CIITA from Transfected HeLa Cells

CIITA is prepared from HeLa cells, transfected by $CaPO_4$ precipitation, or any other method of choice. For the assays described here, 10 150-mm plates seeded with HeLa cells are each transfected with 10 μg of the CIITA mammalian expression vector DNA. Transfection is for 48 h. After transfection, the cell lysate is prepared and the flag-tagged CIITA is immunoprecipitated using antiflag agarose beads, as described later. Immunoprecipitation is done in two parts: an aliquot of the lysate is used for Western immunoblotting and the remainder is used in the acetyltransferase assay.

Immunoprecipitation of CIITA from Transfected Hela Cells

Reagents

Lysis buffer: 50 mM Tris–Cl, pH 8.0, 5 mM $MgCl_2$, 150 mM KCl, 0.1% NP-40, 10% glycerol, freshly added 10 μg/ml pepstatin, 5 μg/ml leupeptin, 20 μg/ml aprotinin, 1 mM PMSF
Agarose beads: Antiflag (Sigma, A1205)
Bead wash buffer: Cold TBS (50 mM Tris, pH 7.4, and 150 mM NaCl) 0.1 M glycine, pH 3.5
Antibody: Monoclonal Ab (M2, antiflag Sigma F3165), antimouse HRP conjugate (Santa Cruz).

After 48 h of transfection, adherent cells are rinsed once with cold PBS and then harvested in 2 ml cold PBS by scraping with a rubber policeman. Cells are pooled and centrifuged at 1200 rpm at 4° for 5 min. The cell pellet is resuspended at a volume equal to two times the packed cell volume in cold lysis buffer. Cells are swollen on ice for 20 min and then lysed by passage twice through a 23-gauge needle. The lysate is centrifuged at 40,000 rpm for 20 min at 4°. The supernatant is collected and stored at −70°.

For Western immunoblotting, 25 μl of the cleared cell lysate is combined with 10–15 μl of activated antiflag agarose beads (activated as described previously) for assay of acetyltransferase activity. Two hundred microliters of cleared cell lysate is combined with 60 μl of activated beads. In both cases, the slurries are incubated overnight at 4° on a vertical rotating wheel. After overnight incubation, the sample is centrifuged for 10 min at 2000 rpm. The supernatant can be checked for residual, unbound protein and then discarded. The beads should be washed three more times with 1 ml of cold lysis buffer. The bulk of the protein should be attached to the beads, which are processed further.

Western Immunoblot Analysis of CIITA

Reagents

4× stacking buffer: 60.6 g Tris, 40 ml 10% SDS, pH 6.8, per liter

4× sample buffer: 250 μl 4× stacking buffer, 10 μl bromphenol blue (saturated solution), 90 μl H$_2$O, 50 μl glycerol. Immediately before use, combine 80 μl of 4× sample buffer with 20 μl 2-mercaptoethanol

Transfer buffer: for 1 liter, combine 5.92 g Tris, 2.93 g glycine, 10% SDS, and 200 ml methanol

TBS: 10 mM Tris, pH 8.0, 150 mM NaCl

TBST: 10 mM Tris, pH 8.0, 150 mM NaCl, 0.05% Tween

Blocking buffer: 5.0% nonfat dried milk in TBST or TBS

Primary antibody: Monoclonal antibody antiflag M2 (antiflag Sigma F3165), 10 μg/ml secondary antibody: antimouse HRP conjugate (Santa Cruz), 1:2000 dilution

Substrate: Pierce Supersignal Chemiluminescent Substrate

Agarose beads as described earlier, with the captured CIITA protein, are resuspended in 30 μl of lysis buffer and 10 μl of 4× sample buffer and heated at 55° for 20 min. After centrifugation, the immunoprecipitated proteins are separated on a 10% SDS–PAGE gel.

Following electrophoretic separation, proteins are transferred to a nitrocellulose membrane (Schleicher and Schuell, BA-S85, 0.45 μm). The membrane should be wet once in water and then in transfer buffer. Transfer is done using Transblot SD, semidry transfer cells from Bio-Rad at 15 V for 60 min. After the transfer, the membrane must be blocked for nonspecific proteins. Blocking can be done at room temperature for 4 h in 5% milk/TBST. Alternatively, block overnight at 4° in 5% milk/TBS only, followed by a 1-h incubation at room temperature 5% milk/TBST in the morning.

Incubate the membrane with the antiflag M2 antibody (10 μg/ml) in 5% milk/TBST for 60 min at room temperature on a rocker. Drain away antibody and wash four times for 10 min each with TBST. The membrane is next incubated with the antimouse HRP conjugate at 1:2000 dilution in TBST for 30 min at room temperature, followed by four washes for 5 min min each in TBST and one wash in TBS.

The HRP tag of the second antibody is developed using the Pierce Supersignal Westpico Chemiluminescent Substrate. Mix equal volumes of buffers A and B (as directed by manufacturer) and add to the membrane. Use enough to cover entirely. Rock for 5 min on a platform. Drain. Place the membrane on 3MM Whatman paper, cover with Saran wrap, and expose to X-ray film. Various times of exposure should be tested to yield maximum detection.

Histone Acetyltransferase Assay

Reagents

10× HAT buffer: 500 mM Tris, pH 8.0, 10 mM DTT, 1 mM EDTA
10× butyric acid: 100 mM in water
^{3}H Acetyl-CoA (10 Ci/pmol) (Perkin Elmer #290, 50 μCi)
^{14}C Acetyl-Co A (60 mCi/pmol) (from Amersham, CFA729, 10 μCi)
Histones H3/H4: 1 mg/ml (Boehringer Mannheim)

To assay the acetyltransferase activity of recombinant CIITA, prepare the master mix, as follows, per reaction:

3.0 μl 10× HAT buffer
3.0 μl 10× butyric acid
3.0 μl glycerol
0.7 μl ^{3}H acetyl-CoA
0.3 μl ^{14}C acetyl-CoA
5.0 μl lysis buffer (freshly added protease inhibitor)
1.25 μl (50 ng) of recombinant protein CIITA
28 μl total volume with water

To this mix add 2.0 μl of histone mixture [mix 1 μl of H3 (1 mg/ml) + 1 μl H4 (1 mg/ml)]. Incubate at 30° for 30 min and stop the reaction by adding 10 μl of 4× SDS sample buffer (without glycerol). Heat at 55° for 20 min.

Load on to a SDS–PAGE minigel (Bio-Rad), 16% resolving gel, and 5% stacking gel. Run at 150 V. Stain the gel in 35% methanol, 10% acetic acid with 0.25 mg Coomassie R250/100 ml for 1 h at room temperature. Destain the gel in 35% methanol and 10% acetic acid O/N. Dry gel under slow temperature rise settings to prevent cracking. Expose to a tritium enhancing screen. Visualize and quantitate on a Molecular Dynamics Storm phosphorimager or equivalent.

To assay the AT activity of transfected CIITA isolated from HeLa cells, combine 60 μl of activated antiflag agarose beads with 200–300 μl of the HeLa cell lysate and immunoprecipitate as described earlier. After the last wash, add 28 μl of the HAT assay master mix to the beads in each tube. Add 2.0 μl of histones (H3 + H4). Continue as described for recombinant CIITA.

[34] Purification and Assay of Saccharomyces cerevisiae Phosphatase That Acts on the C-Terminal Domain of the Largest Subunit of RNA Polymerase II

By Susanne Hoheisel, Michael S. Kobor, Erik Pierstorff, Jack Greenblatt, and Caroline M. Kane

The largest subunit of RNA polymerase II contains a heptad repeat with the consensus sequence YSPTSPS at its C terminus, referred to as the CTD. The heptad is repeated multiple times, with 26 repeats in Saccharomyces cerevisiae and 52 repeats in rodents and humans. The phosphorylation along this repeated sequence varies during transcription, during the cell cycle, and with changes in cellular metabolism.[1,2] Several kinases have been shown to participate in the differential phosphorylation of this repeat, but only one phosphatase been isolated that is specific for the polymerase (Fcp1p). Another phosphatase, PP1 in mammals, has been shown capable of participating in dephosphorylation of RNA polymerase II as well.[3] However, it is the phosphatase encoded by FCP1 in S. cerevisiae that is the topic of this article. Several assays used to quantitate the phosphatase activity are presented, as are several methods for purification of the phosphatase itself, from the more traditional biochemical method to

[1] M. S. Kobor and J. Greenblatt, Biochim. Biophys. Acta. 1577, 261 (2002).
[2] P. S. Lin, N. F. Marshall, M. E. Dahmus, Prog. Nucleic Acid Res. Mol. Biol. 72, 333 (2002).
[3] K. Washington, T. Ammosova, M. Beullens, M. Jerebtsova, A. Kumar, M. Bollen, S. Nekhai, J. Biol. Chem. 277, 40442 (2000).

methods used with recombinant proteins overexpressed in bacterial and insect cells. Also included is a method using tandem affinity purification (TAP) to isolate recombinant protein from the native environment of the yeast cells.

Assays for CTD Phosphatase Activity

Several assays have been described for the purification of phosphatases that act on the C-terminal domain of the largest subunit of RNA polymerase II.[4–9] The substrates have been intact RNA polymerase II that has been phosphorylated, multimers of the CTD repeat that have been phosphorylated, or *p*-nitrophenylphosphate, a nonspecific small molecule substrate. For the last assay, large amounts of Fcp1p are needed, probably reflecting the substrate preference for RNA polymerase II. Assays with RNA polymerase II or the multimeric CTD often also contain the isolated general transcription factor TFIIF to stimulate the activity of Fcp1p.

RNA Polymerase II as Substrate

In both assays described here utilizing intact yeast RNA polymerase II, the purified, predominantly hypophosphorylated (IIa) form of the polymerase is first phosphorylated with $[\gamma\text{-}^{32}P]ATP$ using purified yeast (or mammalian) TFIIH as the protein kinase. Typically, 0.2 μg of purified yeast RNA polymerase IIa is incubated with TFIIH and 2.5 μM $[\gamma\text{-}^{32}P]ATP$ (750 μCi) for 1 h at 30° in 20 mM Tris acetate, pH 7.8, 20 mM potassium acetate, 14% glycerol, 7 mM MgCl$_2$, 0.05 mM EDTA, 0.5 mM dithiothreitol (DTT), and 0.025% Tween 20 in a volume of 25 μl. The reaction is brought to 86 mM potassium acetate by adding 5 μl of exchange buffer (25 mM Tris acetate, pH 7.8, 20% glycerol, 5 mM MgCl$_2$, 0.5 mM DTT) containing 0.4 M potassium acetate. Next, the labeled polymerase is separated from residual $[^{32}P]ATP$ by chromatography at room temperature on DE52.[10] Specifically, the entire kinase reaction is loaded

[4] J. Archambault, R. S. Chambers, M. S. Kobor, Y. Ho, M. Cartier, D. Bolotin, B. Andrews, C. M. Kane, and J. Greenblatt, *Proc. Natl. Acad. Sci. USA* **94**, 14300 (1997).

[5] R. S. Chambers and M. E. Dahmus, *J. Biol. Chem.* **269**, 26243 (1994).

[6] R. S. Chambers and C. M. Kane, *J. Biol. Chem.* **271**, 24498 (1996).

[7] E. J. Cho, M. S. Kobor, M. Kim, J. Greenblatt, and S. Buratowski, *Genes Dev.* **15**, 3319 (2001).

[8] M. S. Kobor, J. Archambault, W. Lester, F. C. Holstege, O. Gileadi, D. B. Jansma, E. G. Jennings, F. Kouyoumdjian, A. R. Davidson, R. A. Young, J. Greenblatt, *Mol. Cell* **4**, 55 (1999).

[9] S. Hausmann and S. Shuman, *J. Biol. Chem.* **277**, 21213 (2002).

[10] J. D. Chesnut, J. H. Stephens, M. E. Dahmus, *J. Biol. Chem.* **267**, 10500 (1992).

by gravity into a small BioSpin column (Bio-Rad) containing 200 μl of DE52. Before loading, the column is washed with 5 column volumes of exchange buffer containing 2 M potassium acetate followed by 25 column volumes of exchange buffer containing 150 mM potassium acetate. Once the column has loaded, it is washed three times with 5 column volumes of exchange buffer containing 150 mM potassium acetate and 0.025% Tween 20. The polymerase is eluted with 4 column volumes (800 μl) of exchange buffer containing 0.8 M potassium acetate and 0.025% Tween 20. During the elution, one drop fractions are collected. Five to six fractions are pooled that incorporate the peak of the radioactivity. Buffer is exchanged by passing up to 100 μl of this pool over a BioSpin 30 column (Bio-Rad) preequilibrated by three sequential washes of 200 μl exchange buffer containing 150 mM potassium acetate. The loaded BioSpin 30 column is centrifuged for 2 min in a clinical centrifuge at 500 g. This spin column procedure is repeated once more, and the recovered sample is frozen in liquid nitrogen and stored at $-80°$. The specific activity of the polymerase is estimated by spotting a diluted aliquot on a Whatman GF/C filter followed by scintillation counting. Recovery from the DE52 column is assumed to be 50%, and thus the specific activity is an approximation.

The standard assay has 1.5 fmol of labeled RNA polymerase II incubated with 4 fmol of highly purified Fcp1p (or fractions from the purification) for 45 min at $30°$ in a 12-μl reaction. The reaction contains 50 mM Tris acetate, pH 7.8, 20% glycerol, 0.5 mM DTT, 10 mM MgCl$_2$, 0.025% Tween 20, 10 mM β-glycerophosphate, and 0.8 μM okadaic acid. Following incubation, reactions with Fcp1p are evaluated in one of two ways.

SDS–PAGE. The hyperphosphorylated and hypophosphorylated RNA polymerase II species of the largest subunit can be separated on a 5% SDS polyacrylamide gel using Laemmli loading buffer and otherwise standard electrophoresis conditions for SDS–PAGE. For the separation of these differentially phosphorylated forms of the largest subunit, 10 μl of the phosphatase reaction is mixed with 5 μg of yeast whole cell extract in Laemmli loading buffer; this addition results in better separation of the two forms of the largest subunit of the polymerase. The reason for this improvement in resolution is not clear. The amount of dephosphorylation can be evaluated using a phosphorimager to compare samples with and without the phosphatase. A unit of activity is defined as the amount of protein needed to release 1 pmol of phosphate per minute at $30°$. The position of the largest subunit can be confirmed by visualization with silver staining or by Western analysis.

Polyethyleneimine (PEI) Thin-Layer Chromatography. A more rapid and easily quantifiable assay utilizes thin-layer chromatography. In this assay, 2 μl from the 12-μl assay is spotted onto PEI-cellulose TLC plates

(Selecto Scientific) prewet with ddH$_2$O. The ascending chromatograph is developed with 1 M sodium formate, pH 3.5, until the solution runs about 2 in. up the chromatogram. Free phosphate ascends, whereas residual labeled RNA polymerase remains at the origin. The position of free phosphate is compared with the origin to determine the percentage of phosphate released. The specific activity of the incorporated ^{32}P on the date of assay is used to determine the moles of phosphate released in order to define a unit as the amount of protein needed to release 1 pmol of phosphate per minute at 30°.

p-Nitrophenylphosphate (PNPP) Substrate

In this assay, fractions containing Fcp1p are incubated in a 50-μl reaction containing 10% glycerol, 10 mM potassium acetate, 50 mM Tris acetate, pH 7.8, 0.5 mM DTT, 10 mg/ml PNPP (Sigma), and 10 mM MgCl$_2$. The reaction proceeds for 1 h at 30°, and is stopped by the addition of 50 μl 1 M Na$_2$CO$_3$. The released phosphate is determined by measuring the absorbance of the reaction at 410 nm using a millimolar extinction coefficient for p-nitrophenolate of 17.5 in a 1-cm cuvette. A unit of activity is defined as the amount of protein needed for the release of 1 pmol of phosphate in 1 h at 30°.[8]

Purification of *Saccharomyces cerevisiae* CTD Phosphatase

Overexpressed Protein from Bacterial Cells

The full-length *FCP1* gene (2199 bp) was cloned into the pET42a(+) expression vector (Novagen) using the *Bam*HI and *Hind*III restriction enzyme sites (E. Pierstorff and C. Kane, unpublished results). This produces a GST–*FCP1* fusion gene. Expression in *Eschericia coli* is carried out using a protocol modified from Hausmann and Shuman.[9] The pET42a(+)-*FCP1* plasmid is transformed into BL21(DE3) *E. coli*. Bacteria containing the plasmid are selected on LB media containing 100 μg/ml kanamycin. A 5-ml aliquot of LB + 100 μg/ml kanamycin is inoculated with a single colony of bacteria containing the plasmid. This culture is grown overnight (~16 h) with shaking at 37°, and the entire 5-ml culture is used to inoculate 250 ml of LB + 100 μg/ml kanamycin. This culture is grown with shaking at 37° to an optical density (OD)$_{600}$ of 0.6. Isopropyl-β-D-thiogalactoside (IPTG) and ethanol are added to final concentrations of 0.2 mM and 2%, respectively. Cultures are then grown with shaking at 17° for 20 h. Cells are harvested at 4° by centrifugation for 15 min at 2600 g, frozen in liquid nitrogen, and stored at −80° until used.

Cells are thawed on ice and resuspended in 20 ml of ice-cold buffer A (50 mM Tris–HCl, pH 7.5, 200 mM NaCl, 10% glycerol). All subsequent steps are done at 4°. Phenylmethylsulfonyl fluoride (PMSF) and lysozyme are added to concentrations of 500 μM and 100 μg/ml, respectively. This mixture is incubated on ice for 30 min for cell lysis. Triton X-100 is added to 0.1%, and extracts are sonicated to reduce viscosity and clarified by centrifugation for 45 min at 39,000 g.

In the meantime, 100 μl of glutathione-S-Sepharose slurry (Amersham Biosciences) is washed twice with 1 ml of buffer A + 0.1% Triton X-100. The soluble extract is added to buffer A-equilibrated glutathione-S-Sepharose beads. This mixture is incubated with gentle agitation for 1 h. The glutathione-S-Sepharose slurry is transferred to an empty 1.2-ml BioSpin column (Bio-Rad) and allowed to drain. The beads are washed twice with 1 ml of buffer A + 0.1% Triton X-100. The GST–Fcp1p fusion protein is eluted with two washes each with 100 μl of buffer A + 0.1% Triton X-100 + 50 mM reduced glutathione. The eluted protein is analyzed via SDS–PAGE and Coomassie blue staining. Protein is frozen in liquid nitrogen and stored at −80°. Aliquots are thawed on ice, and buffer exchange is carried out by passing the sample over a micro BioSpin P6 column (Bio-Rad) equilibrated in 50 mM Tris acetate, pH 7.8, 10% glycerol, 50 mM potassium acetate, 0.5 mM DTT, and 0.1 mM EDTA.

Recombinant Protein from Baculovirus-Infected Cells

Recombinant baculoviruses were produced using the GIBCO Bac-to-Bac system. Fcp1p is tagged at its amino terminus with a His6 tag that is separated from the Fcp1p open reading frame by a linker region. Viruses have been created for the production of wild-type Fcp1p, as well as two mutated versions that carry amino acid changes within the putative phosphatase catalytic site (D180E and D182E).[8] SF9 cells are used routinely for protein expression with cells harvested 48 h after infection with virus stock. It is recommended to first create a fresh virus stock and then perform titer and time course studies to determine optimal expression before commencing the purification.

The infected cell pellet is resuspended in 5 volumes of binding buffer [20 mM HEPES-KOH, pH 7.6, 500 mM NaCl, 5 mM imidazole, 10% glycerol, 1% NP-40, 5 mM 2-mercaptoethanol, protease inhibitors Boehringer Mannheim Complete–EDTA (1 tablet/25 ml)] per gram of cells (i.e., use 10 ml of binding buffer for 2 g of cells). The cell suspension is transferred to a metal beaker (50 ml) and kept on ice. Cells are lysed by sonication; e.g., use a Branson sonifier with a big horn (2 min, 50% cycle, setting 7), and the sample is kept cold. Foaming is also avoided. Insoluble material is removed by centrifugation (20', SS34 rotor, 18,000 rpm). The pellet will be relatively small. The supernate is recovered and kept on ice.

During the centrifugation, the Ni-agarose resin is prepared by taking 0.5 ml of Ni-agarose (the Qiagen resin has been used most frequently because it can be purchased precharged, but other resins may be used) in a 50-ml tube and equilibrating it with 2 × 25 ml of binding buffer. The amount of Ni-agarose depends on the amount of insect cells; 0.5 ml works for 20–30 plates (about 2 g of cells). Taking too much Ni-agarose increases the background binding of nonspecific proteins from the cell extract. The beads are collected by centrifugation (4 min at 2000 rpm in a tabletop centrifuge).

The clarified cell extract is added to the equilibrated beads, and the sample is rotated for 2 h at 4°. The beads are collected by centrifugation (4 min at 2000 rpm in a tabletop centrifuge), and the supernatant is removed carefully. The beads are washed three times for 10 min each with 20 ml of wash buffer [20 mM HEPES-KOH, pH 7.6, 500 mM NaCl, 10 mM Imidazole, 10% Glycerol, 5 mM β-mercaptoethanol, protease inhibitors Boehringer Mannheim Complete–EDTA (1 tablet/25 ml)] by rotating the tube in a cold room.

After the last wash, the remaining wash buffer is removed completely, and 2 ml of elution buffer (20 mM HEPES-KOH, pH 7.6, 500 mM NaCl, 400 mM imidazole, 10% glycerol, 1 mM EDTA, 1 mM DTT) is added. After the beads are resuspended in elution buffer, they are transferred to a 5-ml tube and rotated for 30 min at 4°. The beads are again collected by centrifugation, and the supernatant fluid is transferred to a fresh tube kept on ice. The elution procedure is repeated once again, and the two eluates are combined and dialyzed into phosphatase buffer (20 mM Tris–Cl, pH 7.9, 10% glycerol, 1 mM DTT, 0.1 mM EDTA) containing 100 mM NaCl. The dialysis continues for 3 h against 2 liters of this buffer, the buffer is changed and dialysis continues another 3 h. Longer dialysis times might reduce the activity.

All buffers should be kept at 4°, and 2-mercaptoethanol, DTT, and the protease inhibitor tablets are added freshly right before the purification. It is important to use HEPES buffer and 2-mercaptoethanol rather than Tris and DTT during the Ni-agarose purification.

The method outlined here generally yields Fcp1p that is at least 95% pure as judged by Coomassie blue staining. If higher purity is required, the protein can be purified further using ion-exchange chromatography. Chromatography on a Mono Q FPLC column works well for that purpose. The dialyzed sample is loaded onto a 10-ml Mono Q Sepharose column that has been equilibrated in phosphatase buffer containing 100 mM NaCl. When loading is complete, the column is washed with 10 volumes of phosphatase buffer containing 300 mM NaCl and eluted with a gradient from 300 to 550 mM NaCl.

Alternatively, if a FPLC system is not available, it is possible to further purify recombinant Fcp1p by affinity chromatography using a GST–RAP74 fusion protein. This method has been used successfully in the final purification step of recombinant human Fcp1p (M.S. Kobor, and J. Greenblatt, University of Toronto, unpublished results) and should be easily applicable to the purification of yeast Fcp1p. Fcp1p binds strongly to yRAP74 through regions located at its very carboxy terminus[11] and this can be exploited for purification, as truncated Fcp1p will not bind to this column and can be washed away. To do this, an affinity column (0.5 ml) is prepared containing GST–yRAP74 (amino acids 649–735)[4] at a concentration of 4 mg/ml. The column is equilibrated with 5 ml of phosphatase buffer containing 100 mM NaCl, and the protein sample containing Fcp1p is loaded slowly. The column is washed with 5 ml of the same buffer. Fcp1p is eluted by adding 2 ml of phosphatase buffer containing 1 M NaCl. The Fcp1p–Rap74 interaction is very salt sensitive, and this buffer will elute most of the Fcp1p. Finally, Fcp1p is dialyzed into phosphatase buffer containing 100 mM NaCl.

Native Protein from Yeast Cells Using the Tandem Affinity Purification Procedure

An alternative way to purify Fcp1p in its native form from yeast cells utilizes a yeast strain with chromosomal *FCP1* modified to produce an affinity-tagged protein. Toward this end, a yeast strain was constructed (YMK209) that carries a carboxy-terminal TAP tag on the *FCP1* gene in the chromosome. The published procedure[12] is then followed with some modifications for purification from 2 liters of yeast cells grown in YPD to midlog phase. Purifications are carried out routinely, with YMK209 and its isogenic derivative in which *FCP1* lacks the TAP tag.

A 2-liter culture of YMK209 is grown in 1% Bacto-yeast extract, 2% Bacto-peptone, and 2% glucose (YPD) until the cells reach midlog phase. The cells are collected by centrifugation at 4°, and the cell pellet is resuspended in 2 ml cold water per estimated gram of wet cells. The suspension is transferred to preweighed 50-ml Falcon tubes. The cells are collected by centrifugation in a clinical centrifuge at 4000 rpm for 4 min at 4°. The weight of the cell pellet is determined, and the pellet is resuspended in

---

[11] M. S. Kobor, L. D. Simon, J. Omichinski, G. Zhong, J. Archambault, J. Greenblatt, *Mol. Cell Biol.* **20**, 7438 (2000).

[12] O. Puig, F. Caspary, G. Rigaut, B. Rutz, E. Bouveret, E. Bragado-Nilsson, M. Wilm, and B. Seraphin, *Methods* **24**, 218 (2001).

1.3 volumes per gram of cells with yeast extraction buffer [YEB; 245 mM KCl, 100 mM HEPES-KOH, pH 7.9, 1 mM EDTA, 2.5 mM DTT, protease inhibitors Boehringer Mannheim Complete-EDTA (1 tablet/25 ml)] but without protease inhibitors. The cells are collected again in a clinical centrifuge at 4000 rpm for 4 min at 4°. This washed cell pellet is suspended in 1.3 volumes of YEB [now with protease inhibitors (Boehringer Mannheim Complete (1 tablet/25 ml)] per gram of cells. The cells are collected as before, and the cell pellet is frozen in liquid nitrogen and stored at −80° (usually, the cell pellet at this stage weighs around 7–10 g).

For the preparation of the starting yeast cell extract, the frozen plastic Falcon tube is cracked open to quickly recover the frozen cells. The cells are ground in a Krups coffee grinder (Model 203–70) in the presence of dry ice pellets for 2 min. Other coffee grinders may also be suitable, but they should be tested by examining the recovery of active Fcp1p after purification. Inefficient physical shearing and heating of the grinder should be avoided. The cell powder is transferred to a glass beaker and thawed evenly while at room temperature. The powder is resuspended in 0.8 volumes per gram of cells in YEB. The suspension is clarified by centrifugation for 2 h at 34000 rpm in a Beckman Ti70 rotor. The supernate is removed carefully, avoiding the overlaying lipid, and the clarified extract is dialyzed into 4 liters of IPP buffer (10 mM Tris–Cl, pH 7.9, 200 mM NaCl, 0.1% NP-40, 0.2 mM EDTA, 0.5 mM DTT) at 4° for 4 h. After dialysis, the extract is centrifuged briefly (0.5 h at 34000 rpm in a Beckman Ti70 rotor) to remove particulates.

The supernatant is added to 200 μl of IgG Sepharose 6 fast flow beads (Amersham Pharmacia) that have been equilibrated in 10 ml IPP buffer, and the suspension is incubated on a rotator for 2–4 h at 4°. This is best done in a 0.8 × 4-cm Poly Prep column (Bio-Rad). The column is drained by gravity flow and washed with 20 ml of IPP. The column is closed, and 1 ml of TEV cleavage buffer (50 mM Tris–Cl pH 7.9, 200 mM NaCl, 0.1% NP-40) and 100 units of TEV protease (Invitrogen) are added to the top; the column is sealed securely and rotated overnight at 4°. The eluate is recovered by gravity flow.

To the eluate is added 3 ml calmodulin-binding buffer and 3 μl 1 M CaCl$_2$. This solution is mixed with 100 μl of calmodulin beads (Stratagene) that have been equilibrated in 10 ml calmodulin-binding buffer (10 mM Tris–Cl, pH 7.9, 200 mM NaCl, 2 mM CaCl$_2$, 0.1% NP-40, 10 mM 2-mercaptoethanol) in a 0.8 × 4-cm Poly Prep column (Bio-Rad). The suspension is incubated at 4° for 1 to 2 h on a rotator. The column is drained by gravity flow and washed with 10 ml of calmodulin-binding buffer. Purified Fcp1p is eluted with a total of 1 ml of calmodulin elution buffer (10 mM Tris–Cl, pH 7.9, 200 mM NaCl, 0.1% NP-40, 3 mM EGTA, pH

8.0, 10 mM 2-mercaptoethanol) in 5 steps of 200 μl. The location of Fcp1p is determined by enzymatic assay and by SDS–PAGE.

Note that this protocol will yield native Fcp1p with the calmodulin-binding protein moiety attached to its carboxy terminus. Also, Fcp1p purified by this method appears as a single band on a silver-stained gel (M. Kobor and J. Greenblatt, University of Toronto, unpublished results), but the purification conditions (salt concentration, wash steps, etc.) can be modified easily if the goal is to identify Fcp1p-interacting proteins.

Native Protein from Yeast Cells

This procedure was worked out before the molecular identity of Fcp1p was known. The assay utilized RNA polymerase II radiolabeled *in vitro*, and the full procedure was first reported by Chambers and Kane.[6]

The yeast strain YPH/TFB1.6HIS is grown at 30° to midlog phase in 200 liters of YPD; harvested by centrifugation (1-kg yield); washed in cold distilled water; and resuspended in 0.5 volume of 3× lysis buffer (450 mM Tris acetate, pH 7.8, 150 mM potassium acetate, 60% glycerol, 3 mM EDTA, 3 mM DTT, 3 mM PMSF, 6 $\mu$$M$ pepstatin A, 6 μg/ml chymostatin, 1.8 $\mu$$M$ leupeptin). All steps are carried out at 4°. A whole cell extract is prepared as described previously,[13] with the following modifications. Poly-ethyleneimine is omitted, as this results in loss of CTD phosphatase activity, and cells are disrupted in a bead beater (Biospec Products, Inc.) with 8 × 1-min bursts interrupted with 2 min of cooling on ice. The extract is dialyzed against buffer A (50 mM Tris acetate, pH 7.8, 20% glycerol, 1 mM EDTA, 1 mM PMSF, 1 mM DTT) containing 0.05 M potassium acetate and stored as four equal aliquots at −80°. Each aliquot is diluted with buffer B (20 mM HEPES-KOH, pH 7.6, 20% glycerol, 1 mM EDTA, 1 mM PMSF, 1 mM DTT) to a protein concentration of 5 mg/ml, adjusted to 0.15 M potassium acetate, and loaded onto a DE52 column (25 × 5 cm) equilibrated in buffer B with 0.15 M potassium acetate. The DE52 column is washed with 1 column volume of buffer B with 0.15 M potassium acetate and step eluted with 2 column volumes of buffer B containing 0.25 M potassium acetate followed by 2 column volumes of buffer B with 0.45 M potassium acetate. CTD phosphatase activity elutes in the 0.45 M potassium acetate fraction and is dialyzed against buffer C (50 mM Tris acetate, pH 7.8, 20% glycerol, 0.1 mM EDTA, 0.5 mM DTT) containing 0.025 M potassium acetate, adjusted to 0.05 M potassium acetate, and loaded onto SP-Sepharose (15 × 2.5 cm) and Q-Sepharose (12 × 2.5 cm) columns (Amersham Pharmacia) connected in tandem and equilibrated in buffer

---

[13] W. J. Feaver, O. Gileadi, and R. D. Kornberg, *J. Biol. Chem.* **266,** 19000 (1991).

C containing 0.05 M potassium acetate. After the sample is loaded, columns are washed with 2 column volumes of buffer C with 0.05 M potassium acetate. The SP-Sepharose column is disconnected, and the Q-Sepharose column is developed with a 300-ml gradient of 0.05–1.2 M potassium acetate in buffer C. CTD phosphatase activity elutes at 0.86 M potassium acetate.

Q-Sepharose fractions containing CTD phosphatase activity are pooled; dialyzed against buffer C containing 10 mM potassium phosphate, pH 8.0, and 50 mM potassium acetate; and loaded onto blue Sepharose CL-6B (8 × 1.5 cm) (Amersham Pharmacia) and Bio-Gel HTP (8 × 2.5 cm) (Bio-Rad) columns connected in tandem and equilibrated in buffer C containing 10 mM potassium phosphate, pH 8.0, and 50 mM potassium acetate. After the sample is loaded and the columns are washed with 2 column volumes of buffer C with 10 mM potassium phosphate, pH 8.0, and 50 mM potassium acetate, the blue Sepharose column is disconnected, and the Bio-Gel HTP column is developed with a 200-ml gradient of 0.01–0.3 M potassium acetate, pH 8.0, in buffer C. CTD phosphatase activity elutes at 0.11 M potassium phosphate. Bio-Gel HTP fractions containing CTD phosphatase activity are pooled, adjusted to 0.75 M $(NH_4)_2SO_4$ and 0.025% Tween 20, and loaded onto a phenyl-Superose HR 5/5 column equilibrated in buffer D (same as buffer C but with 10% glycerol and 0.025% Tween 20) with 0.75 M $(NH_4)_2SO_4$. The column is developed with a 15-ml gradient of 0.75 to 0 M $(NH_4)_2SO_4$ in buffer D, and CTD phosphatase activity elutes at 0.55 M $(NH_4)_2SO_4$. Phenyl-Superose fractions containing CTD phosphatase activity are buffer exchanged to buffer D containing 0.3 M $(NH_4)_2SO_4$ in a microconcentrator (Centricon 30), concentrated, and loaded onto a Superose 12 HR 10/30 column (Amersham Pharmacia) equilibrated in buffer D containing 0.3 M $(NH_4)_2SO_4$. Superose 12 fractions containing CTD phosphatase activity are pooled, diluted with an equal volume of buffer C, and loaded onto a Mono Q HR 5/5 column (Amersham Pharmacia) equilibrated in buffer C containing 0.2 M potassium acetate. The Mono Q column is step eluted with 4 column volumes of buffer C containing 0.4 M potassium acetate and developed with a 30-ml gradient of 0.4–1.2 M potassium acetate in buffer C. CTD phosphatase activity elutes at 1.05 M potassium acetate. Fractions are stored at −80°.

[35] Functional Analysis of Transcription Factors CREB and CREM

By BETINA MACHO and PAOLO SASSONE-CORSI

This article describes some fundamental methods and reagents that have been applied successfully to the study of transcriptional regulation by factors of the CREB–CREM family. These proteins have been involved in a number of physiological responses and are thought to modulate gene expression by integrating signals originated from intracellular transduction pathways. We discuss experimental procedures to study CREB and CREM expression, their regulatory functions, the modulation of their activity by phosphorylation, the involvement of various kinases, and the identification and characterization of specific coactivators, which implicate tissue-specific mechanisms and chromatin remodeling events in the transcriptional response mediated by these factors.

The Transcriptional Response to Cyclic AMP

The finding that specific regulatory sequences and their binding factors are responsive to intracellular signals constituted an essential advance for a mechanistic and physiological understanding of gene expression. In eukaryotes, a family of cAMP-responsive nuclear factors mediates transcriptional regulation upon stimulation of the adenylate cyclase signaling pathway. In response to the activation of G-protein-coupled receptors by extracellular stimuli, cAMP levels increase intracellularly and mobilize the cAMP-dependent protein kinase A (PKA), which phosphorylates a number of cytoplasmic and nuclear substrates on serine residues in the context X-Arg-Arg-X-Ser-X. Among the known PKA substrates are transcription factors, which are thus activated and bind to distinct promoter sites to convey the signal.[1] Promoter elements that specifically mediate the transcriptional response to changes in the intracellular levels of cAMP have been identified by deletion analysis of a large number of regulatory regions.[2] The best characterized is the CRE (cAMP response element), whose consensus site is the 8-bp palindromic sequence TGACGTCA.[3] It

---

[1] G. M. Fimia and P. Sassone-Corsi, *J. Cell Sci.* **114**, 1971 (2001).

[2] E. Lalli and P. Sassone-Corsi, *J. Biol. Chem.* **269**, 17359 (1994).

[3] G. M. Fimia, D. De Cesare, and P. Sassone-Corsi, *Cold Spring Harb. Symp. Quant. Biol.* **63**, 631 (1998).

is interesting that this site is highly similar to the DNA-binding element for transcription factor AP-1, a regulator responsive to growth factor and phorbol esters.[4] Examples of cross-talk at the transcriptional level for CREs and AP-1 sites have been reported.[5–7]

The study of proteins recognizing the CRE has revealed a multigene family of transcription factors and has uncovered a striking versatility of transcriptional regulation in different physiological systems. Indeed, CRE-binding proteins have been found to play important roles in the physiology of the pituitary gland, in regulating spermatogenesis, in the response to circadian rhythms, and in the molecular basis of memory.[8] Within these physiological systems, CRE-binding proteins have been shown to display various mechanisms of action, e.g., through the involvement of different cofactors or kinases (see later).

A Family of Transcriptional Regulators

The CRE-binding protein (CREB) was first identified in DNAse I foot-printing assays as a nuclear protein in PC12 cells, which binds selectively to the CRE in the somatostatin promoter.[9] A year later, CREB cDNA was obtained through screening of a placental λgt11 library with the CRE sequence.[10] CREB is characterized by a leucine zipper, essential for dimerization, and an adjacent basic domain responsible for DNA binding. Thus, as Fos and Jun, CREB belongs to the bZip (basic domain-leucine zipper) class of transcription factors. Subsequent studies uncovered a multigene family of more than 10 members. Activating transcription factors (ATFs) were identified by a number of strategies, including low-stringency screening of either genomic or cDNA libraries and screening of various cDNA expression libraries with CRE and CRE-like sites.[11] Some ATF proteins, in particular ATF-2, are important players in the transcriptional program triggered during viral infections.[12] Interestingly, the DNA

[4] P. Sassone-Corsi, L. J. Ransone, and I. M. Verma, *Oncogene* **5**, 427 (1990).
[5] D. Masquilier and P. Sassone-Corsi, *J. Biol. Chem.* **267**, 22460 (1992).
[6] D. Monnier and J. P. Loeffler, *DNA Cell Biol.* **17**, 151 (1998).
[7] A. M. Ionescu, E. M. Schwarz, C. Vinson, J. E. Puzas, R. Rosier, P. R. Reynolds, and R. J. O'Keeefe, *J. Biol. Chem.* **276**, 11639 (2001).
[8] E. Lalli, J. S. Lee, M. Lamas, K. Tamai, E. Zazopoulos, F. Nantel, L. Penna, N. S. Foulkes, and P. Sassone-Corsi, *Phil. Trans. R. Soc. Lond. B Biol. Sci.* **351**, 201 (1996).
[9] M. R. Montminy and L. M. Bilezikjian, *Nature* **328**, 175 (1987).
[10] J. P. Hoeffler, T. E. Meyer, Y. Yun, J. L. Jameson, and J. F. Habener, *Science* **242**, 1430 (1988).
[11] T. W. Hai, F. Liu, W. J. Coukos, and M. R. Green, *Genes Dev.* **3**, 2083 (1989).
[12] M. G. Wathelet, C. H. Lin, B. S. Parekh, L. V. Ronco, P. M. Howley, and T. Maniatis, *Mol. Cell* **1**, 507 (1998).

sequence responding to the early adenoviral transactivator E1A is homologous to a CRE.[13]

An important member of the family, the CRE modulator (CREM), was identified during the search for negative regulators of CREB. CREM was isolated by screening a pituitary cDNA library with two probes spanning the CREB basic domain and the leucine zipper domain, respectively.[14] Several features of CREM have placed this transcription factors in a privileged position (see later). All members of the CREB–CREM family belong to the bZip class and act as dimers. In certain combinations they are also able to heterodimerize with each other, following a specific "dimerization code" that seems to be an inherent property of the leucine zipper structure of each factor. Outside the bZIP region, homology between the family members is quite poor. However, CREB shares extensive homology with ATF1 and CREM within the activation domain (AD, see later); thereby these three proteins compose a distinct subfamily.[15]

Phosphorylation-Dependent Activation

The structure of the AD in CREB and CREM is basically identical (Fig. 1). Two highly glutamine-rich domains, Q1 and Q2,[16] flank a middle region known as the phosphorylation (P) box, which contains a consensus phosphoacceptor site for PKA, corresponding to Ser133 in CREB and Ser117 in CREM, respectively. This site has been shown to act as a phosphoacceptor site for other kinases (see later). Phosphorylation at Ser133 is a prerequisite to turn these transcription factors into powerful activators, as it mediates the interaction with coactivators such as CBP (CREB-binding protein; see later).[17,18] The P box also contains an array of phosphoacceptor sites for various other kinases,[18,19] which, however, appear to be secondary to Ser133 and whose regulatory function awaits further elucidation. One of those, Ser142, appears to be particularly interesting: phosphorylation at Ser142 was first described as a negative event for Ser133 phosphorylation-mediated transcriptional activation,[20] while it

[13] D. M. Olive, W. al-Mulla, M. Simsek, S. Zarban, and W. al-Nakib, *Arch. Virol.* **112,** 67 (1990).

[14] N. S. Foulkes, E. Borrelli, and P. Sassone-Corsi, *Cell* **64,** 739 (1991).

[15] P. Sassone-Corsi, *Annu. Rev. Cell Dev. Biol.* **11,** 355 (1995).

[16] D. De Cesare, G. M. Fimia, and P. Sassone-Corsi, *Trends Biochem. Sci.* **24,** 281 (1999).

[17] G. A. Gonzalez and M. R. Montminy, *Cell* **59,** 675 (1989).

[18] R. P. de Groot, J. den Hertog, J. R. Vandenheede, J. Goris, and P. Sassone-Corsi, *EMBO J.* **12,** 3903 (1993).

[19] R. P. de Groot, R. Derua, J. Goris, and P. Sassone-Corsi, *Mol. Endocrinol.* **7,** 1495 (1993).

[20] P. Sun, H. Enslen, P. S. Myung, and R. A. Maurer, *Genes Dev.* **8,** 2527 (1994).

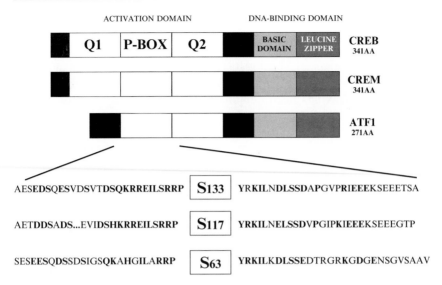

FIG. 1. Domain structure of the CREB family of proteins. The proteins have a highly similar organization. The glutamine-rich domains (Q1 and Q2) and the DNA-binding domain (basic domain and leucine zipper) are indicated in addition to the P box. The amino acid sequence of the P box of CREB, CREM, and ATF1 containing Ser133, Ser117, and Ser63 is shown. Phosphorylation at this residue turns CREB and CREM into activators through interaction with the coactivator CBP. This serine has been shown to be a phosphoacceptor site for various kinases.

was shown recently to be crucial for a calcium-mediated response.[21,22] While this observation is interesting, the relevance of this phosphorylation event with respect to recruiting the transcriptional machinery and within other physiological settings remains undetermined.

CRE-binding proteins can act as both activators and repressors of transcription. The activators mediate transcriptional induction upon their phosphorylation by PKA or other kinases.[17,18] Their expression is constitutive and widely distributed in various tissues in a housekeeping fashion, reminiscent of factors involved in homeostasis. CREM constitutes an exception, as it is highly abundant in adult testis, whereas expressed at very low levels in all other tissues.[15] CREM presents a number of remarkable characteristics. Through alternative splicing, several CREM isoforms originate, which encode activators (τ) as well as repressors of transcription

[21] J. M. Kornhauser, C. W. Cowan, A. J. Shaywitz, R. E. Dolmetsch, E. C. Griffith, L. S. Hu, C. Haddad, Z. Xia, and M. E. Greenberg, *Neuron* **34,** 221 (2002).
[22] D. Gau, T. Lemberger, C. von Gall, O. Kretz, N. Le Minh, P. Gass, U. Schmid, U. Schibler, H. W. Korf, and G. Schutz, *Neuron* **34,** 245 (2002).

(α, β, γ).[23,24] In testis, a developmental switch governs the various CREM isoforms, as in prepubertal animals CREM transcript levels are low and only repressor isoforms are detected.[25] However, during puberty, transcripts encoding the activator CREMτ accumulate to high levels in maturing germ cells, where CREM is thought to play a critical role in postmeiotic transcription. This has been demonstrated by the generation of mutant mice in which the CREM gene is disrupted by homologous recombination.[26,27] Homozygous CREM-null males are sterile and their germ cells cease differentiation at the spermatid stage and undergo apoptosis. This may be attributed to the lack of expression of several putative CREM target genes, i.e., protamines 1 and 2, transition proteins 1 and 2, and calspermin, all essential for the structuring of a mature spermatozoon.[26]

Among the repressors, the inducible cAMP early repressor (ICER) deserves special mention, as it is generated from a cAMP-inducible alternative promoter present in an intron of the CREM gene.[15] ICER is constituted of the sole leucine zipper and basic domain of CREM and thereby acts as a dominant-negative partner of either CREB or CREM. The induction of this powerful repressor is, at least in part, responsible for the transient nature of cAMP-induced gene expression.

Activation of Various Signaling Pathways Lead to CREB Phosphorylation

The same serine residue targeted by PKA can also be phosphorylated in response to other signals: growth factors, phorbol esters, or the Ca^{2+} ionophore A23187 leads to enhanced phosphorylation of either CREB at Ser133 or CREM Ser117; an increase in intracellular calcium by membrane depolarization, Epidermal growth factor (EGF), and Nerve growth factor (NGF), as well as ultra violet light, C (UVC), can increase phosphorylation of CREB Ser133.[18,28–31] Also, several kinases involved in signaling

[23] B. M. Laoide, N. S. Foulkes, F. Schlotter, and P. Sassone-Corsi, *EMBO J.* **12,** 1179 (1993).

[24] V. Delmas, B. M. Laoide, D. Masquilier, R. P. de Groot, N. S. Foulkes, and P. Sassone-Corsi, *Proc. Natl. Acad. Sci. USA* **89,** 4226 (1992).

[25] N. S. Foulkes, B. Mellstrom, E. Benusiglio, and P. Sassone-Corsi, *Nature* **355,** 80 (1992).

[26] F. Nantel, L. Monaco, N. S. Foulkes, D. Masquilier, M. LeMeur, K. Henriksen, A. Dierich, M. Parvinen, and P. Sassone-Corsi, *Nature* **380,** 159 (1996).

[27] J. A. Blendy, K. H. Kaestner, G. F. Weinbauer, E. Nieschlag, and G. Schutz, *Nature* **380,** 162 (1996).

[28] M. Sheng, M. A. Thompson, and M. E. Greenberg, *Science* **252,** 1427 (1991).

[29] D. De Cesare, S. Jacquot, A. Hanauer, and P. Sassone-Corsi, *Proc. Natl. Acad. Sci. USA* **95,** 12202 (1998).

[30] J. Xing, J. M. Kornhauser, Z. Xia, E. A. Thiele, and M. E. Greenberg, *Mol. Cell Biol.* **18,** 1946 (1998).

pathways stimulated by those factors have been demonstrated to target Ser 117/133: PKC, CamK (*in vitro*), p34^{cdc2}, and p70^{S6K} for CREM Ser117,[19,32] p90^{rsk2}, RSK-B, PKB (*in vitro*) and CamK (*in vitro*) for CREB Ser133.[33–37] Therefore, CREM/CREB not only modulate the cellular response to cAMP, but also promote gene expression in response to mitogenic and stress signaling pathways. They are thus thought to represent a convergence point for the integration of signals from diverse pathways.[15]

Investigating the Expression of CREB and CREM

The expression of both CREB and CREM is considered to be noninducible and somewhat weak in most tissues. One relevant exception is the high level of CREM expression in male postmeiotic germ cells, where it exerts an essential regulatory function.[3]

Riboprobes

The RNase protection assay (described in detail later) has been the main tool for studying CREM and CREB expression at the RNA level, as well as expression and induction of target genes. Although other methods, such as RT-PCR (especially useful for detecting extremely low-copy transcripts) and Northern analysis (essential to gather additional information on size and relative abundance of transcripts), can be utilized as well, RNase protection assays are still unparalleled in their robustness, reproducibility, sensitivity, and reliability. It has been shown that mRNAs with copy numbers as low as 10^5 copies per transcript can be detected.[38] To visualize CREB and CREM mRNA expression within animal tissues, *in situ* hybridization has been applied successfully. This methodology has been described in detail elsewhere.[39,39a]

---

[31] M. Iordanov, K. Bender, T. Ade, W. Schmid, C. Sachsenmaier, K. Engel, M. Gaestel, H. J. Rahmsdorf, and P. Herrlich, *EMBO J.* **16**, 1009 (1997).

[32] R. P. de Groot, L. M. Ballou, and P. Sassone-Corsi, *Cell* **79**, 81 (1994).

[33] J. Xing, D. D. Ginty, and M. E. Greenberg, *Science* **273**, 959 (1996).

[34] K. Du and M. Montminy, *J. Biol. Chem.* **273**, 32377 (1998).

[35] J. S. Arthur and P. Cohen, *FEBS Lett.* **482**, 44 (2000).

[36] P. K. Dash, K. A. Karl, M. A. Colicos, R. Prywes, and E. R. Kandel, *Proc. Natl. Acad. Sci. USA* **88**, 5061 (1991).

[37] B. Pierrat, J. S. Correia, J. L. Mary, M. Tomas-Zuber, and W. Lesslauer, *J. Biol. Chem.* **273**, 29661 (1998).

[38] J. Sambrook and D. W. Russel, *"Molecular Cloning,"* 3rd Ed. Cold Spring Harbor Laboratory Press, New York, 2001.

[39] L. Tessarollo and L. F. Parada, *Methods Enzymol.* **254**, 419 (1995).

[39a] D. Sassoon and N. Rosenthal, *Methods Enzymol* **225**, 384 (1993).

TABLE I
RIBOPROBES FOR EXPRESSION STUDIES OF CREM, ICER, AND ACT

	Fragment	From–to	Protected fragment	Ref.
CREMα	NcoI/HpaII	841–1156	316 bp	[14]
CREMβ/γ/τ	NcoI/HpaII	841–1156	218 bp	[14]
CREMτ	NcoI/PvuII	5–152	147 (CREMτ)	[40]
			109 (α/β/γ)	
CREMτ	PstI/SacI	290–590	300 (CREMτ)	[40]
			210 (α/β/γ)	
ACT	EcoRI	702–1072	370	[41]
ICER	XhoI	Exon 5	122	[42]

DNA sequence parts selected as probes for protection or *in situ* analyses are cloned into pBluescript to obtain the antisense probe by utilization of the T7 or T3 RNA polymerase. The sequence fragment should contain a suitable restriction site for linearization of the plasmid that would give rise to probes in the range of 100–400 bp. Alternatively, a suitable site on pBluescript can be utilized. Not any selected sequence fragment is likely to be a reliable riboprobe and often several fragments need to be tested to identify the ideal one. Successfully utilized riboprobes for CREM and CREB and related molecules are listed in Table I.

Antibodies

Various commercial and noncommercial antibodies have been utilized to study CREM and CREB expression at the protein level, both in immunofluorescence and in Western blot analyses. Coimmunoprecipitation experiments (see later) are also dependent on reliable and effective antibodies. Table II gives an overview of antibodies used in various studies.

Characterizing CREB Kinases

Several kinases have been found to phosphorylate CREB and CREM (see earlier discussion). As the function of CRE-binding proteins is modulated through phosphorylation, the identification of the kinases is an essential step toward the understanding of their physiological function. Various

[40] B. Mellstrom, J. R. Naranjo, N. S. Foulkes, M. Lafarga, and P. Sassone-Corsi, *Neuron* **10**, 655 (1993).

[41] G. M. Fimia, D. De Cesare, and P. Sassone-Corsi, *Nature* **398**, 165 (1999).

[42] J. H. Stehle, N. S. Foulkes, C. A. Molina, V. Simonneuax, P. Prevet, P. and Sassone-Corsi, *Nature* **365**, 314 (1993).

TABLE II
ANTIBODIES FOR EXPRESSION AND INTERACTION STUDIES OF CREM, CREB, AND COACTIVATORS,
UNPHOSPHORYLATED AND/OR PHOSPHORYLATED.[a]

Antibody		Description	IF	WB	IP	Ref.
Anti-CREB	R	TrpE–CREB fusion protein (Baculovirus)	Yes	Yes	Yes	43
Anti-pCREB	R	CREB peptide (CREBtide) phosphorylated *in vitro* with catalytic subunit of PKA	Yes	Yes	Yes	43
Anti-CREB	R	CREB peptide 128–141		Yes		44
Anti-CREB (220)	R	CREB peptide 136–150 recognizes CREB and pCREB		Yes	Yes	44
Anti-pCREB (5322)	R	CREB 128–141 phosphorylated by catalytic subunit of PKA, recognizes pCREB only		Yes		44
Anti-CREB (253)	R	Purified recombinant protein (1–341)		Yes	Yes	45
Anti-CREB	M	Santa Cruz Biotechnology C-21		Yes		
Anti-pCREB		Santa Cruz Biotechnology H-15		Yes		
Anti-CREM	R	CREM peptide (γ)		Yes	Yes	24
Anti-CREB	R	Against bacterial recombinant CREB		Yes		32
Anti-CREMτ	R	Purified bacterial recombinant CREMτ no cross-reaction with CREB	Yes	Yes		58
Anti-CREB	R	New England Biolabs		Yes		
Anti-pCREB	R	New England Biolabs		Yes		
Anti-CBP	R	Santa Cruz Biotechnology A-22		Yes	Yes	
Anti-CBP	R	N-terminal 22 amino acids of CBP	Yes			47
Anti-CBP (5614)	R	Recombinant CBP polypeptide (455–679)		Yes	Yes	45
Anti-pCBP	R	Peptide containing phosphorylated Ser301		Yes		46
Anti-ACT	R	ACT peptide 106–127	Yes	Yes	yes	41[b]
Anti-ACT	M	ACT peptide 106–127	Yes	Yes	Yes	
Anti-pCREB$_{Ser142}$	R	Against phosphopeptide 136–151	Yes	Yes		22

[a] IF, immunofluorescence; WB, Western blot; IP, immunoprecipitation; R, rabbit; M, Mouse; p, phosphorylated.
[b] Sassone-Corsi laboratory, not published yet.

[43] D. D. Ginty, J. M. Kornhauser, M. A. Thompson, H. Bading, K. E. Mayo, J. S. Takahashi, and M. E. Greenberg, *Science* **260,** 238 (1993).

[44] M. Hagiwara, P. Brindle, A. Harootunian. R. Armstrong, J. Rivier, W. Vale, R. Tsien, and M. R. Montminy, *Mol. Cell Biol.* **13,** 4852 (1993).

[45] P. Brindle, T. Nakajima, and M. Montminy, *Proc. Natl. Acad. Sci. USA* **92,** 10521 (1995).

[46] S. Impey, A. L. Fong, Y. Wang, J. R. Cardinaux, D. M. Fass, K. Obrietan, G. A. Wayman, D. R. Storm, T. R. Sonderling, and R. H. Goodman, *Neuron* **34,** 235 (2002).

approaches have been applied to determine CREB kinases, mostly using classical methods including in-gel kinase assays, phospo-mapping, and pharmacological treatments.

Characterizing the Coactivators of CREB and CREM

CBP

In somatic cells, phosphorylation of CREB at Ser133 facilitates the recruiting of CBP, a 265KD protein that interacts selectively with the phosphorylated P-box domain.[47] CBP was originally identified by screening a human thyroid λgt11 library with recombinant CREB labeled with [γ-^{32}P]ATP and using the obtained cDNA sequence to screen a mouse brain cDNA library to obtain the full sequence. Formation of the phospho–CREB–CBP complex promotes the interaction with TFIIB, a general transcription factor linked directly to RNA polymerase II.[48] Thus, CBP constitutes the link between CREB and the transcriptional preinitiation complex. This interaction is likely to require additional cofactors, of which the TBP-associated factor hTAF130, a subunit of the TFIID complex, constitutes an interesting example.[49] A functional homolog of CBP, p300, has a role in cell growth, differentiation, apoptosis, and DNA repair. In addition to contacting other elements of the transcription machinery, CBP/p300 have yet another function: through a histone acetyltransferase (HAT) activity they acetylate specific sites on histone tails and thereby promote transcription by contributing to the process of chromatin decondensation. Thus, CBP establishes a direct link between signaling and chromatin remodeling.[50]

ACT

In contrast to somatic cells, CBP/p300 do not seem to be involved in CREM-dependent activation in male germ cells. Indeed, the characterization of a novel transactivation mechanism came with the identification of the LIM-only protein ACT, a testis-specific member of the FHL (four and a half LIM domains) family of proteins, which is a powerful coactivator for CREM in postmeiotic male germ cells. The crucial finding in this

[47] J. C. Chrivia, R. P. Kwok, N. Lamb, M. Hagiwara, M. R. Montminy, and R. H. Goodman, *Nature* **365,** 855 (1993).

[48] R. P. Kwok, J. R. Lundblad, J. C. Chrivia, J. P. Richards, H. P. Bachinger, R. G. Brennan, S. G. Roberts, M. R. Green, and R. H. Goodman, *Nature* **370,** 223 (1994).

[49] K. Ferreri, G. Gill, and M. Montminy, *Proc. Natl. Acad. Sci. USA* **91,** 1210 (1994).

[50] P. Sassone-Corsi, *Science* **296,** 2176 (2002).

respect was that CREM, which in testis is unphosphorylated and thus unable to bind to CBP, bypasses its classical signaling-dependent mechanism by binding to ACT. This phosphorylation-independent activation seems to constitute a male germ cell-specific scenario.[3]

General Comments

The characterization of coactivators for CREB and CREM mainly relies on two steps: (1) demonstrating their interaction with either CREM and/or CREB and (2) demonstrating their coactivating properties in a CREM/CREB-dependent transcription assay. For interaction studies several *in vitro* and *in vivo* strategies are available, for example, the yeast two-hybrid system, GST pull-down assays, and coimmunoprecipitation experiments. This article reviews the coimmunoprecipitation approach, as it carries greatest *in vivo* significance. Coimmunoprecipitations can be performed using protein extracts from cultured cells or from animal tissue (method descriptions given later). In some circumstances, such as the favored protein is present in low amounts, cells can be transfected with the appropriate expression vectors in order to overexpress CREM or CREB and/or other relevant proteins. The limiting factor in this type of experiments is often the lack of a good-quality antibody for immunoprecipitation. In the case of CREB and CREM, some reliable tools exist and are listed in Table II. To reveal the coactivating properties of a potential cofactor, coexpression in transfected cells using a CRE-driven reporter has been used extensively and successfully (method description given later). Reporter genes classically used are the chloramphenicol acetyltransferase (CAT) or luciferase. Reporter constructs need to be designed to allow the measuring of reporter activity as a function of the transactivating potential of CREM/CREB without or with the respective coactivator(s) present. Some common reporters and expression constructs utilized in this context are listed in Table III.

Expression and Purification of CREB and CREM Proteins

The T7 expression system[51] provides a powerful and efficient method to produce large quantities of CREB and CREM proteins in bacterial cells. The expression and purification of the ectopically expressed proteins are remarkably easy, as CRE-binding proteins are soluble in bacteria and, due to their heat stability, can be purified easily from the lysate by a single

---

[51] F. W. Studier, A. H. Rosenberg, J. J. Dunn, and J. W. Dubendorff, *Methods Enzymol* **185,** 60 (1990).

TABLE III

REPORTER AND EXPRESSION PLASMIDS FOR CREM AND CREB TRANSACTIVATION ASSAYS
AND INTERACTION STUDIES

Construct	Description	Ref.
ptACE-CAT	Mouse testis angiotensin-converting enzyme (ACE) promoter from −688 to +17 in front of the CAT gene in pBLCAT$_3$	52
ptACE-CAT	Mouse testis ACE promoter from −91 to +17 in front of the CAT gene in pBLCAT$_3$	52
pSom-CAT	CRE from the rat somatostatin promoter in pSG5	14
pCαEV	Catalytic subunit of the mouse PKA gene in expression plasmid with mouse metallothionein 1 promoter	53
pSVCREB	Rat CREB cDNA in pSG5	25
pSVCREMα	Mouse CREMα in pSG5	14
pSVCREMβ	Mouse CREMβ in pSG5	14
pSVCREMγ	Mouse CREMγ in pSG5	14
pSVCREMτ	Mouse CREMτ in pSG5	25
pCALSP-CAT	Two CREs from calspermin promoter (−80 to +361) in pCAT basic	54
pG5E4CAT	Five Gal4 DNA-binding sites cloned upstream of theE4 promoter (−240/+38) in vector pSP72	55
pG4CREMτ	Mouse CREMτ in frame with Gal4DBD (1–147) in pG4MpolyII	23
pG4CREMτ1	Mouse CREMτ1 in frame with Gal4DBD (1–147) in pG4MpolyII	23
pG4CREMτ2	Mouse CREMτ2 in frame with Gal4DBD (1–147) in pG4MpolyII	23
pG4CREMτ117	Mouse CREMτ sequence with a Ser117 to Ala117 mutation in frame with Gal4DBD (1–147) in pG4MpolyII	23
pG4ACT	Mouse ACT in frame with Gal4DBD (1–147) in pG4MpolyII	41
pG4CREB	Rat CREB sequence with Gal4DBD (1–147) in pG4MpolyII	41
pG4-CBP-HAT	HAT domain of mouse CBP in pcDNA3 vector containing the Gal4DBD (1–147)	56
c-fos-CAT	Human c-fos promoter (−220/+42)	57

[52] Y. Zhou, Z. Sun, A. R. Means, P. Sassone-Corsi, and K. E. Bernstein, *Proc. Natl. Acad. Sci. USA* **93,** 12262 (1996).

[53] P. L. Mellon, C. H. Clegg, L. A. Correll, and G. S. McKnight, *Proc. Natl. Acad. Sci. USA* **86,** 4887 (1989).

[54] Z. Sun, P. Sassone-Corsi, and A. R. Means, *Mol. Cell Biol.* **15,** 561 (1995).

[55] F. Liu and M. R. Green, *Cell* **61,** 1217 (1990).

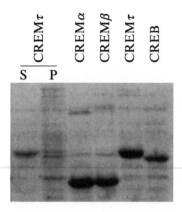

FIG. 2. Expression of CREB and CREM protein in bacteria. Coommassie brilliant blue staining of bacterial lysates. Lanes 1 and 2 show supernatant and pellet, respectively, of CREMτ producing bacteria; lanes 3–6 show supernatants of CREMα (33 kDa), CREMβ (33 kDa), CREMτ (42 kDa), and CREB (40 kDa) producing bacteria.

boiling step, which removes the majority of irrelevant bacterial proteins (Fig. 2). Proteins produced in this way have been utilized successfully for antibody production, DNA-binding analyses, and phosphorylation studies aimed at the identification of CREB and CREM kinases.[32,58]

For expression in bacteria, DNAs encoding CREM and CREB proteins are cloned into the T7 expression vector pET11d,[51] utilizing the NcoI–BamHI or NdeI–BamHI sites of pET11d. All constructs use the +3 ATG as the initiation codon. Transform Escherichia coli strain BL21(DE3) with these constructs and pick single colonies to inoculate 1-liter bacterial cultures in M9ZB medium[51] containing 0.1 mg/ml ampicillin. Grow cultures with shaking at 37° to an $OD_{600\ nm}$ of 0.6–0.8 and then induce by adding IPTG (0.4 mM) for 3 h. Spin down the cells at 4000 rpm for 10 min and resuspend the pellet in 0.1 volume of TE. Boil the cells in 1-ml aliquots for 8 min and then centrifuge for 10 min at full speed to remove denatured proteins. After this step, 90% of CREM or CREB protein is present in the supernatant and ready to use. For storage, add 30% glycerol and place at −70°. Protein yields obtained by this procedure are 5–20 mg/100 ml of bacteria.

[56] M. A. Martinez-Balbas, A. J. Bannister, K. Martin, P. Haus-Seuffert, M. Meisterernst, and T. Kouzarides, EMBO J. 17, 2886 (1998).
[57] P. Sassone-Corsi, Proc. Natl. Acad. Sci. USA 85, 7192 (1988).
[58] V. Delmas, F. van der Hoorn, B. Mellstrom, B. Jegou, and P. Sassone-Corsi, Mol. Endocrinol. 7, 1502 (1993).

Our "Microscale" RNase Protection Assay

Solutions

> 5× hybridization buffer: 200 mM PIPES, pH 6.4; 2 M NaCl; 5 mM
> EDTA; store at −20°
> Formamide: Take fresh stock bottle, aliquot, and store at −20°, thaw
> volume needed and refreeze
> 1× hybridization buffer: four parts formamide, one part 5×
> hybridization buffer
> RNase digestion buffer: 10 mM Tris–HCl, pH 7.5; 300 mM NaCl;
> 5 mM EDTA
> RNase T1 (Boehringer Mannheim), 100 U/ml
> RNase A, 10 μg/ml(27) (see comments)

Gel mix: 6% acrylamide, 19:1 bisacryl, 8 M urea, 1×TBE; dissolve little
by little, incubation at 37°C recommended

Preparation, Labeling, and Purification of the Antisense Riboprobe

A preparation of the linearized vector is phenol extracted and ethanol
precipitated. DNA is dissolved in distilled water at 1 μg/μl. Depending on
the orientation of the insert, T7 or T3 polymerase is used for the labeling
reaction. Riboprobes with a length between 120 and 400 bp have given best
results (see Table I).
Add in the following order:

> 4 μl 5× transcription buffer (Promega)
> 2 μl 100 mM dithiothreitol (DDT)
> 0.8 μl RNAse inhibitor (35 U/μl)
> 2 μl mix of ATP/CTP/GTP (5 mM each)
> 10 μl ^{32}P UTP (10 mCi/ml)
> 1 μl linearized template DNA
> 1 μl T7 or T3 RNA polymerase (Promega)

Keep the mix at room temperature always, as the spermidine in the
buffer can lead to DNA precipitation. Incubate the labeling reaction at
37° for 1 hour, remove the template by the addition of 1 μl RQ1 DNase
(Promega; 1 U/μl), and incubate at 37° for 15 min. The probe is purified
by phenol extraction and precipitation. Add 2 μl *E. coli* tRNA (Boehringer
Mannheim; 10 mg/ml) and 26.2 μl H$_2$O to a final volume of 50 μl. Extraction is performed with Tris (100 mM, pH 8.0)-equilibrated phenol/chloro-
form/isoamylalcohol (25:24:1). After centrifugation for 3 min at high
speed, transfer the aqueous phase into a fresh tube. Precipitate the RNA
with 200 μl 2.5 M ammonium acetate and 750 μl ethanol at −80° for

15 min. Spin at high speed for 10 min and then dissolve the pellet in 50 μl H$_2$O. Repeat the whole precipitation cycle two more times. Wash the final pellet with 75% ethanol/3 M sodium acetate, pH 5.2, dissolve in 100 μl 1× hybridization buffer, and utilize 1 μl for counting in a scintillation counter. Counts of over 1 million cpm/μl can be reached with good template preparations. For the protection assay, dilute the probe with 1× hybridization buffer to a concentration of 5×10^5 cpm in 30 μl. If a β-actin probe is used as an internal control, dilute it to 1.5×10^5 cpm in 30 μl for comparable signal strength.

RNA Preparation and Hybridization

The tissue under investigation is taken from the animal and put directly into an appropriate volume of RNASolve reagent (Omega Bio-Tek) or similar reagents. Follow the exact guidelines of the manufacturer for RNA preparation. Dry 1–4 μg of each RNA sample in a Speed-Vac in standard 0.5-μl Eppendorf tubes. Include one control sample with 2 μg *E. coli* tRNA. Add 10 μl of the diluted probe and place the tubes in a PCR machine preheated to 85° with the following program: 85° for 5 min and 45° indefinitely (overnight).

RNase Digestion and Purification

Vortex the RNase T1 stock bottle well and take out 16 μl. Centrifuge this suspension for 15 min at 4° at high speed. Carefully remove the supernatant and dissolve the pellet in 16 μl RNase digestion buffer. The enzyme pellet is translucent, thus extra care should be taken while removing the supernatant. Add 10 μl of the standard RNase A stock preparation and 24 μl H$_2$O to create a 50-μl stock mix. From this, 28 μl is taken and diluted in 1400 μl H$_2$O to obtain the proper working concentration. Take the RNA tubes from the machine and place them at room temperature. Add 117 μl of the working mix to each tube. Mix by pipetting vigorously and return the tubes to the PCR machine for an incubation of 1 h at 30°. Subsequently add to each tube 7 μl 10% SDS, as well as 1 μl of proteinase K. Vortex well and return to the PCR machine for an incubation of 20 min (or more) at 37°. Add 135 μl of phenol/chloroform to each tube, vortex vigorously, spin, and transfer the upper aqueous phase to a fresh 0.5-ml tube. Add 4 μl of 1 μg/μl *E. coli* tRNA as carrier RNA. Precipitate RNA with 350 μl ethanol. Mix well and incubate the precipitation at −80° for 25 min (the time is crucial) and then centrifuge at top speed at room temperature. Wash the pellet once with 70% ethanol. After removing the supernatant, spin the tubes briefly again to collect and remove all ethanol (also the fine droplets on the inside of the tube). Dissolve the pellet in 4 μl of formamide dye[38] and leave to dissolve for 10–15 min. Before loading the samples on the gel, heat them in the PCR machine at 85° for 3–4 min and then put them on ice immediately.

Gel

The 25-min precipitation time of the probe is ideal for preparation of the gel. Use a 6% acrylamide, 8 *M* urea, 1× TBE gel of the appropriate size using thin spacers (0.4 mm). Polymerization of the gel should continue for a maximum of 35 min. Before loading the samples, the wells should be rinsed very thoroughly with a syringe, as urea can precipitate in the wells and thus disturb loading and running samples into the gel. Run the gel at 300–500 V in 1× TBE buffer. Fix the gel in 10% acetic acid/10% methanol for 10 min and then dry it on a heated vacuum gel drier for 20 min at 80° before placing it into a cassette with a film for exposure. Exposure time can vary from overnight to several days depending on signal strength. An example of a protection assay is shown in Fig. 3.

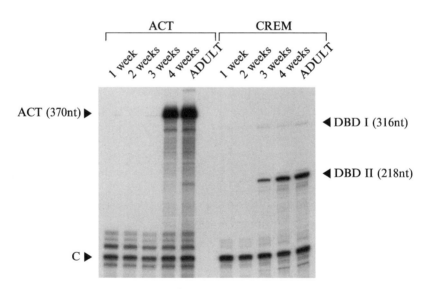

FIG. 3. RNase protection assay for analyses of CREM and ACT expression during testis development. RNAs were extracted from testes of mice at different ages (as indicated) and analyzed using specific riboprobes. Ten micrograms of total RNA from mouse testis was used in each lane. A β-actin riboprobe was used as an internal control (C).

Comments

RNase A Stock Preparation. The RNase A stock is prepared according to the standard procedure.[38] Even so the real activity of the enzyme is variable, thus testing of the new stock solution is recommended. Use your probe on a RNA sample likely to give you a specific signal (e.g., testis for CREMτ) and use a series of dilutions from your RNase A stock (1:1, 1:2, 1:4, 1:8). Judging the optimum digestion intensity from the clearness and specificity of obtained signals, you can chose one specific dilution as your working stock. Keep aliquots of the right dilution at $-20°$ to avoid multiple thawing/freezing. In some cases, sensitivity to RNase A activity can depend on the probe and an extra dilution curve might have to be performed for particularly sensitive probes.

Tubes. Nonautoclaved Eppendorf tubes are recommended to be used in all steps, as autoclaving often results in leaking lids. When vortexing radioactive material with phenol, tubes are required to be closed firmly and completely to avoid leakage and contamination.

Coimmunoprecipitation with CREM/CREB Proteins

Solutions

2× HBS buffer: 280 mM NaCl; 1.5 mM Na$_2$HPO$_4$ × 12H$_2$O; 50 mM HEPES; pH 7.05 with NaOH

2 M CaCl$_2$

EBC buffer: 50 mM Tris–HCl, pH 8.0; 170 mM NaCl; 0.5% NP-40; 50 mM NaF

Lysis buffer: EBC + 1 mM phenylmethyl sulfonyl fluoride + 10 μg/ml protease inhibitor cocktail (PIC)

Protein-Sepharose A (for polyclonal antibodies) or G (for monoclonal antibodies)

CREM in a Myc-tagged plasmid, e.g., pCS2MT[59] or pSV-CREM (see Table III)

The interactor to be tested in an expression plasmid with/without Myc-tag (e.g., pSG5/pCS2)

From Transfected Cells in Culture

Day 1 (Evening). Split COS cells 1:4 to 1:5 into four 10-cm dishes.

Day 2 (Evening). Bring all constructs to a concentration of 0.5 μg/μl and set up the transfections according to the following scheme:

[59] R. A. Rupp, L. Snider, and H. Weintraub, *Genes Dev.* **8,** 1311 (1994).

Sample	CREM/CREB	Factor to be tested	pBluescript	H_2O
1	—	—	20 μl	400 μl
2	10 μl	—	10 μl	400 μl
3	—	10 μl	10 μl	400 μl
4	10 μl	10 μl	—	400 μl

Add 60 μl $CaCl_2$ to each tube. Prepare 500 μl 2× HBS buffer in a 5-ml Falcon tube. Add the Ca^{2+}/DNA mix dropwise to the HBS buffer while vortexing the Falcon tube. Leave the transfection mix for 20 min at room temperature and then add dropwise to cells.

Day 3 (Morning). Wash cells twice with 1× phosphate-buffered saline (PBS) or growth medium without serum and replace full growth medium for 24 h.

Day 4/5 (Morning). Wash cells twice with 1× PBS and add 1 ml of cold (important!) lysis buffer per dish. Scrape cells and put them on ice immediately. Agitate tubes at 4° for 30 min and then spin for 10 min at full speed at 4°. In the meantime, wash four times 50 μl protein A/G-Sepharose twice with EBC buffer. Transfer the supernatant to a fresh tube and take an "input" sample of 50 μl. Immediately add 50 μl of Laemmli or SDS-loading buffer and boil sample for 5 min. Add the lysate to the prepared protein-Sepharose and agitate for about 1 h at 4°. This is the "preclearing" step, intended to deplete the lysate of all factors that bind to the Sepharose nonspecifically. After preclearing, spin tubes for 15 s (at 4°) and transfer the supernatants to fresh tubes. Add the respective antibody (e.g., 1 μl anti-Myc-tag antibody 9E10 or 1 μl anti-CREM antibody). Agitate tubes at 4° for a minimum of 3 h. This incubation period can be extended to 12 h if required by low binding of the respective factor, although in most cases 3–5 h should be sufficient. Again wash four times 50 μl protein A/G-Sepharose twice with EBC buffer and add the lysate–antibody mix. Agitate tubes for 1 more hour at 4°. Wash the Sepharose three times with EBC buffer (keep on ice all time) and add 50 μl Laemmli or SDS-loading buffer and boil samples. Run two parallel Western blots with all four samples and the four input samples, and probe the blots with antibodies against both factors tested.

From Mouse Tissue

This approach is important as it allows the study of native proteins. The procedure is very similar to the one described earlier, of course without the transfection. We recommend starting with a large quantity of tissue (e.g., ~4 mouse testes), which should be homogenized in cold lysis buffer and to extend the 3- to 5-h incubation for immunoprecipitation to a minimum of 12 h. Extra care should be taken to keep samples at $4°$ at all times. The ideal control for coimmunoprecipitation from tissue is a coimmunoprecipitation with the preimmune serum of the respective antibody/ies.

Transactivation Assays

Solutions

2× HBS
2 M $CaCl_2$
TEN buffer: 40 mM Tris–HCl, pH 7.4; 1 mM EDTA; 150 mM NaCl
0.25 M Tris–HCl, pH 8.0
1 M Tris–HCl, pH 7.4
β-Gal buffer: 60 mM Na_2HPO_4; 40 mM NaH_2PO_4; 10 mM KCl; 1 mM $MgCl_2$; 50 mM 2-mercaptoethanol (add before use)

COS cells are ideal for CREM/CREB transactivation assays utilizing a CAT (see earlier discussion) or a luciferase reporter. Split the cells on the day before the experiment 1:2 to 1:3 into 6-cm dishes so that they reach about 80% confluence for transfection the day after. Perform transfections with the Ca precipitation technique as described earlier, but with amounts of reagents like the following (example):

Sample	pSomCAT or G4CAT	pSVCREM or G4CREM	pSVLIM or ACT-AD	pCαEV (PKA)	lacZ	pSG5	H_2O
1	2 μl	1 μl	—	2 μl	1 μl	4 μl	to 250 μl
2	2 μl	1 μl	2 μl	—	1 μl	4 μl	to 250 μl
3	2 μl	1 μl	4 μl	—	1 μl	2 μl	to 250 μl
4	2 μl	1 μl	6 μl	—	1 μl	—	to 250 μl

30 μl 2 M $CaCl_2$; 250 μl 2 × HBS buffer; use 2-ml Eppendorf tubes.

After washing twice with 1 × PBS, harvest cells by scraping in 1 ml of TEN buffer. Collect cells by 1 min of centrifugation at 8000 rpm. Add 150 μl 0.25 M Tris, pH 8.0, to the pellet and lyse cells by vigorous vortexing and three cycles of freeze/thaw in liquid nitrogen and in a $37°$ water bath. Centrifuge at $4°$ for 10 min at maximum speed and transfer the supernatant (i.e., the extract) to a fresh tube. For normalization, determine the relative protein content of the samples with Bradford reagent: Mix 800 μl H_2O with 5 μl extract and add 200 μl Bradford reagent. Mix well and put into disposable cuvettes to measure the absorbance at 595 nm in a spectrophotometer. Normalize sample volumes for the CAT assay by taking 100 μl of the sample with the lowest $OD_{595\ nm}$ and adjust the other samples to the same volume and thus protein content by adding 0.25 M Tris, pH 8.0.

Example

Sample	$OD_{595\ nm}$	Sample	0.25 M Tris, pH 8.0
1	0.64	68.8 μl	31.2 μl
2	0.80	55.0 μl	45.0 μl
3	0.59	74.6 μl	25.4 μl
4*	0.44	100 μl	0 μl

CAT Assay

Prepare the following mix and add 50 μl to each sample (prepare a master mix): 60 MCi/mmol [^{14}C]-chloramphenicol, 2 μl 40 mM acetyl-CoA, and X μl 1 M Tris, pH 7.4, to reach 50 μl final volume.

Vortex samples, spin down briefly to collect liquids at the bottom of the tube, and incubate for 2 h in a 37° water bath. Spin samples briefly and add 500 μl ethylacetate per tube. Vortex vigorously and spin down at maximum speed for 3 min. Transfer the upper phase to a fresh tube and evaporate all liquid by placing tubes into a Speed-Vac for about 15 min. Meanwhile prepare a glass container lined with Whatman paper, add a fresh 200-ml mixture of 95:5 chloroform:methanol, and let it equilibrate for at least 30 min. When samples are dry, add 25 μl ethylacetate and vortex well. Spot samples on a Merck TLC plate (Silica gel 60 F_{254}), 2 cm from the bottom and at least 1.5 cm apart, by repeatedly applying 2–3 μl with a tip and letting the spot dry between applications. Put the plate in the glass container for 1 h and then dry on the bench for a few minutes. Expose a film to the TLC plate in a film cassette overnight at room temperature.

β-Galactosidase Assay

To normalize the results for transfection efficiency, we make use of the cotransfected lacZ plasmid to measure β-galactosidase activity in each sample. Mix 10 μl of extract with 0.6 ml of β-gal buffer and 60 μl o-Nitrophenyl-β-D galactopyranoside (ONPG) (freshly prepared at 4 mg/ml) and incubate samples in a 37° water bath for about 20 min. Stop the color reaction with 0.3 ml 1 M Na_2CO_3 and measure the $OD_{420\ nm}$ in a spectrophotometer.

Quantification of Results

The activity of each extract is calculated as the ratio of acetylated chloramphenicol to total (acetylated and nonacetylated) chloramphenicol.[38] Quantification of the spots can be achieved by scintillation counting of the respective areas on the TLC plate or, alternatively, by measurement of spot density on the autoradiograph with the phosphoimager.

Immunofluorescence Using CREB and CREM Antibodies

Our protocols are analogous to the excellent descriptions in Pines.[60]

Acknowledgments

We thank all members of the Sassone-Corsi laboratory for help and discussions. B.M. is supported by a Ph.D. fellowship of the Boehringer Ingelheim Funds. Work in our laboratory is supported by grants from CNRS, INSERM, CHUR, Human Frontier Science Program, Fondation pour la Recherche Médicale, and Association pour la Recherche sur le Cancer.

[60] J. Pines, *Methods Enzymol* **283,** 99 (1997).

[36] Functional Analysis of TFIID Components Using Conditional Mutants

By Joseph C. Reese and Michael R. Green

The general transcription factor TFIID plays a key role in the regulation of gene expression by RNA polymerase II. In yeast, the TFIID complex is composed of the TATA box-binding protein (TBP) plus 14 TBP-associated factors (TAFs). Roles of the TAF proteins in transcription have been examined by inactivating TAF function and monitoring the effects on gene expression. Seeing as all but one of the subunits of the TFIID complex are essential for viability in yeast, conditional mutants must be isolated and employed in their analysis.[1–5]

We have used both temperature-sensitive (ts) mutants and conditional depletion strategies to analyze TAF function.[6–8] Each strategy has its own

[1] J. C. Reese, L. Apone, S. S. Walker, L. A. Griffin, and M. R. Green, *Nature* **371,** 523 (1994).
[2] D. Poon, Y. Bai, A. M. Campbell, S. Bjorklund, Y. J. Kim, S. Zhou, R. D. Kornberg, and P. A. Weil, *Proc. Natl. Acad. Sci. USA* **92,** 8224 (1995).
[3] Z. Moqtaderi, J. D. Yale, K. Struhl, and S. Buratowski, *Proc. Natl. Acad. Sci. USA* **93,** 14654 (1996).
[4] N. L. Henry, A. M. Campbell, W. J. Feaver, D. Poon, P. A. Weil, and R. D. Kornberg, *Genes Dev.* **8,** 2868 (1994).
[5] M. R. Green, *Trends Biochem. Sci.* **25,** 59 (2000).
[6] S. S. Walker, J. C. Reese, L. M. Apone, and M. R. Green, *Nature* **383,** 185 (1996).
[7] L. M. Apone, C. M. Virbasius, J. C. Reese, and M. R. Green, *Genes Dev.* **10,** 2368 (1996).
[8] B. Li and J. C. Reese, *EMBO J.* **19,** 4091 (2000).

inherent advantages and disadvantages. Temperature-sensitive mutants allow for the rapid inactivation of a TAF protein under a variety of media conditions. This rapid inactivation allows one to analyze early time points, which minimizes secondary effects caused by the long-term loss of TAF function. However, transferring cells to 37° induces a heat shock that alters cell physiology and complicates analysis of the expression of certain classes of genes that are affected by stress. An additional disadvantage is that one cannot be assured that shifting a mutant to a nonpermissive temperature inactivates all of its functions. Conditional depletion/expression strategies use systems that place the TAF gene under the control of a tightly regulated promoter.[6,7,9] The advantage of this strategy is that the protein is depleted from the cell, alleviating concerns that the gene function is partially inactivated. However, a significant drawback of this strategy is it takes many hours (8 to 12) to deplete the protein, making it difficult to distinguish between primary and secondary effects. The length of time required to deplete the protein can be shortened somewhat by fusing a degradation signal to the protein, but hours are still required to deplete the protein.[9] Moreover, carbon source changes alter cell physiology and affect the expression of certain classes of genes. Whenever possible, both methodologies should be used so that the disadvantages of each individual strategy offset the other.

Characterization of Growth of Temperature-Sensitive TAF Mutants

Many procedures have been published on the generation of conditional mutants by *in vitro* chemical modification using hydroxylamine or by error-prone polymerase chain reaction (PCR), and replacement of the wild type using the plasmid shuffle technique.[10] We have used both strategies successfully to generate ts mutants of a number of TAFs. Instead of revisiting these methodologies, which are described in detail elsewhere,[10] this article discusses the characterization of ts mutants and their use in analyzing TFIID function.

Screens for ts mutants often yield a vast number of alleles. Several papers have demonstrated allele-specific transcriptional defects in different TAF mutants.[11,12] Thus, it is necessary to either analyze many alleles or

[9] Z. Moqtaderi, Y. Bai, D. Poon, P. A. Weil, and K. Struhl, *Nature* **383**, 188 (1996).

[10] C. Guthrie and G. R. Fink, "Guide to Yeast Genetics and Molecular Biology." Academic Press, New York, 1991.

[11] R. J. Durso, A. K. Fisher, T. J. Albright-Frey, and J. C. Reese, *Mol. Cell. Biol.* **21**, 7331 (2001).

[12] D. B. Kirschner, E. vom Baur, C. Thibault, S. L. Sanders, Y. G. Gangloff, I. Davidson, P. A. Weil, and L. Tora, *Mol. Cell. Biol.* **22**, 3178 (2002).

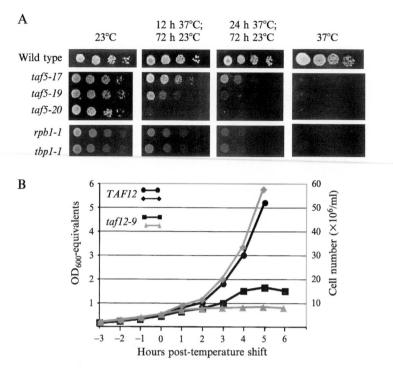

Fig. 1. Characterization of the growth of temperature-sensitive TAF mutants. (A) Analyzing recovery from a transient exposure to 37° distinguishes different mutant alleles of *TAF5*. Mutant alleles of RNA polymerase II (*rpb1-1*) and the TATA-binding protein (*tbp1-1*) were analyzed in parallel as controls. (B) Growth in liquid culture. Black lines indicate the OD_{600}, and gray lines represent cell number. Optical densities and cell numbers were corrected for to account for the dilution of the cultures during the experiment.

choose a few for detailed analysis. This is especially important when performing genome-wide based strategies such as DNA microarray studies, as the reagents used for this technique can be limiting, and analyzing multiple mutants can be cost prohibitive.[12-14] Therefore, it is advisable to characterize and compare mutants to select the most appropriate two or three to conduct such studies. In general, mutants that cause a more rapid loss of viability are more likely to severely disrupt the function of the

[13] F. C. Holstege, E. G. Jennings, J. J. Wyrick, T. I. Lee, C. J. Hengartner, M. R. Green, T. R. Golub, E. S. Lander, and R. A. Young, *Cell* **95,** 717 (1998).
[14] T. I. Lee, H. C. Causton, F. C. Holstege, W. C. Shen, N. Hannett, E. G. Jennings, F. Winston, M. R. Green, and R. A. Young, *Nature* **405,** 701 (2000).

protein. Therefore, the best mutants are usually those that display the "tightest" ts phenotype, a term that is subjective and open to interpretation. We prefer to use mutants that display the following growth characteristics: (1) rapid cessation of cell division at the nonpermissive temperature; (2) loss of viability after exposure to the nonpermissive temperature; and (3) normal phenotypes at the permissive temperature. These phenotypes can be assessed using the growth assays described in this section. In addition, we also analyze the effects of each mutant on steady-state mRNA levels and on the integrity of the TFIID complex; these methods are described in a later section.

Plating and Recovery Assay

This procedure is useful for comparing the growth of many mutants at one time. In this assay, each strain is grown in liquid culture, and the growth of each mutant is assessed by spotting serial dilutions on a plate, which provides a semiquantitative measure of growth and avoids the problems encountered by uneven streaking of cells across a plate. The plates are incubated at 37° for a period of time followed by a recovery step at the permissive temperature. This approach provides a measure of the loss of viability caused by the mutation. The example in Fig. 1A shows that analyzing mutants in this manner can reveal growth differences among different ts alleles that are not apparent when the strains are grown at 37° without a recovery period (far right panel). All of the mutants shown displayed the same lack of growth when exposed to only 37°; however, examination of their ability to recover from a transient exposure to 37° has revealed differences among the *TAF5* mutants.

1. Inoculate 5 ml of YPAD medium with three to four medium-sized colonies and grow at 23° to saturation (14–18 h).

2. Measure the OD_{600} of the saturated culture. It is critical to first dilute a small aliquot drawn from the overnight culture approximately 1/100 and 1/50 to achieve an optical density at 600 nm (OD_{600}) of 0.05–0.3 when reading the sample in the spectrophotometer because the linear relationship between OD_{600} and cell number breaks down when the OD_{600} of the sample exceeds 0.3 due to the light-scattering properties of yeast cell walls.

3. Using the original saturated culture, seed a 5-ml culture of fresh medium to an OD_{600} of 0.25 and grow at 23° to an OD_{600} of 1.0. Prepare four fivefold serial dilutions in media in sterile microfuge tubes. Alternatively, serial dilutions can be prepared from stationary cultures,

but we have found that this leads to either an overestimation or an underestimation of the viability of mutants.

4. Spot 2–3 μl of each dilution onto eight YPAD plates and let dry. It is important to vortex each tube regularly to prevent the cells from settling to the bottom of the tube. Incubate one plate at each of the following temperatures: $14°, 23°, 30°, 33°$, and $37°$. Incubate three additional plates at $37°$ and then transfer to $23°$ after 12, 24, and 48 h.

5. Photograph the plates at 24-h intervals over a period of 3 days.

Liquid Growth Curves

This procedure is the preferred method for quantitatively monitoring cell growth and the ability of each mutant to respond to the nonpermissive temperature. Many ts mutants arrest at a specific point in the cell cycle and/ or swell when incubated at $37°$. The swelling that occurs gives the misleading impression that the cells are dividing at a slower rate; thus, we prefer to simultaneously measure the OD_{600} and count the cells using a hemocytometer during the initial characterization of each mutant. Counting cells allows for visual inspection for a cell cycle arrest and gives a more accurate assessment of the severity of the growth phenotype of the mutant, as shown in Fig. 1B.

1. Inoculate 5 ml of YPAD medium with three to four colonies and grow at $23°$ until saturation (14–18 h).

2. Inoculate 50 ml of fresh medium contained in a 125-ml flask to a sufficient density so that 14–16 h later each culture will have an OD_{600} of approximately 0.3–0.6. The starting inoculum will differ for each strain based on its doubling time at $23°$.

3. The next morning, measure the OD_{600} of each culture. Adjust all cultures to an OD_{600} of 0.3. Continue incubating at $23°$ for an additional hour to allow the cells to adjust to the medium change. During this time the cultures will grow to an OD_{600} of 0.4–0.5.

4. Remove two 1-ml aliquots for the $t = 0$ time point. Use one to measure the OD_{600} and the other to count cells using a hemocytometer. Cells reserved for counting should be treated with either 0.1% sodium azide or 0.37% formaldehyde to suspend cell division and can be stored at $4°$ for a few days prior to counting.

5. For the temperature shift, place the flask into a shaking water bath set to $37°$. Alternatively, an equal volume of media warmed to $50°$ can be added to raise the temperature rapidly. We have observed little difference in results using either method as long as the culture volume is less than 100 ml and the culture-to-flask volume ratio is 1:4.

6. Remove 1-ml aliquots at 30-, 60-, 90-, 120-, 180-, 240-, and 360-min time points and monitor the OD_{600} of each sample. Fix the cells for counting as described earlier in step 4. It is very important to keep all the cultures at an OD_{600} between 0.4 and 1.2 throughout the entire time course. This will require constant dilution of the wild-type strain into fresh, prewarmed medium, whereas some ts mutants may need to be diluted only once. To minimize differences in culture conditions during dilution, use the same batch of media and dilute the culture by adding a portion of the starting culture to a parallel flask containing fresh media that was placed in the shaking water bath at least 30 min prior to dilution. Note that the dilution factor must be taken into account when constructing a growth curve.

Detection of Effects of the Loss of TAF Function on Steady-State mRNA Levels and Integrity of the TFIID Complex

One of the phenotypes associated with TAF inactivation is degradation of the mutant protein itself and the codegradation of other TFIID sub-units.[6,12–17] Therefore, when performing temperature shift experiments, we routinely collect a larger volume of culture and divide the cells into two aliquots at the washing step (step 5 in the following protocol): one aliquot is used to isolate total RNA to monitor changes in gene expression (Fig. 2A) and the other is used to prepare cell extracts for Western analysis (Fig. 2B). Changes in the expression of specific messages are measured by Northern blotting, S1 nuclease protection, or primer extension analysis. Analyzing changes in global gene expression is best done using DNA microarrays. However, it is important to first characterize the mutants prior to such analysis to select a few for these procedures. An effective way to compare mutants and select the most appropriate ones for further study is to monitor changes in total poly(A)$^+$ mRNA using poly(dT) as a probe. We routinely coprocess the RNA polymerase II mutant *rpb1-1* as a benchmark for rapidly ceased class II transcription.[14]

TAF Inactivation Protocol

1. Inoculate 150 ml of YPAD, contained in a 500-ml flask, with an aliquot of a saturated culture and grow overnight at $24°$ until an OD_{600} 0.4–0.5 is reached.

[15] J. C. Reese, Z. Zhang, and H. Kurpad, *J. Biol. Chem.* **275**, 17391 (2000).
[16] B. Michel, P. Komarnitsky, and S. Buratowski, *Mol. Cell* **2**, 663 (1998).
[17] S. L. Sanders, E. R. Klebanow, and P. A. Weil, *J. Biol. Chem.* **274**, 18847 (1999).

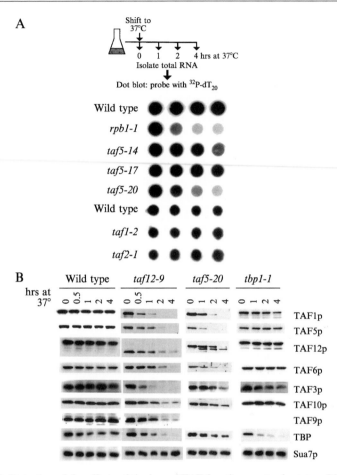

FIG. 2. Detection of the effects of the loss of TAF function on steady-state mRNA levels and the integrity of the TFIID complex. (A) Analysis of poly(A)$^+$ mRNA levels upon TAF inactivation. (B) Inactivation of a single TAF results in the codegradation of other TFIID subunits. The *tbp1-1* mutant was used as a positive control. Sua7p, a general transcription not part of TFIID, is used as a loading control.

2. The next morning, adjust all cultures to a similar OD_{600} (0.3–0.4) and continue growth for 1 h to allow the cells to acclimate to the medium change.

3. Remove an aliquot of the cell culture equal to 15 OD_{600} units (e.g., 30 ml of a culture with an OD_{600} of 0.5 equals 15 OD_{600} units) into a centrifuge tube and a smaller aliquot into a microfuge tube for measuring the OD_{600} of the culture. This will be the $t = 0$ time point.

4. Collect the cells by centrifugation at 3000 g for 5 min at room temperature. During this time, shift the culture to a 37° shaking water bath (Model G-76, New Brunswick Scientific or equivalent) and record the OD_{600}.

5. Pour off media and place the cell pellet on ice. Resuspend the cells in 2 ml of ice-cold wash solution (20 mM Tris–HCl, pH 7.5, 100 mM NaCl, 1 mM EDTA) and divide the cells into two microfuge tubes (one tube will be used for RNA analysis and the other for protein analysis).

6. Collect the cells by centrifugation in a microcentrifuge at 14,000 rpm for 2 min. Aspirate the supernatant and freeze the pellet on dry ice until all the time points are collected. The cells can be stored at −80° until use.

7. Repeat steps 3–6 for the remaining time points. Monitor the culture density throughout the course of the experiment and remove aliquots (approximately 15 OD_{600} units) at 30, 60, 120, 180, and 240 min posttemperature shift. Dilute the culture density, as required, to maintain an OD_{600} between 0.4 and 1.2 throughout the experiment. It is preferable to dilute a culture after removing an aliquot of cells for processing instead of just prior to removal because the dilution process causes a minor transient fluctuation in the expression of certain genes. We find it convenient to use the OD_{600} reading from the previous time point to estimate the volume of culture collected so that cells can be collected quickly. The small difference occasionally observed between the estimated OD_{600} and actual is corrected for during the washing step.

Isolation of Total RNA and Measurement of Total Poly(A)$^+$ mRNA

1. Heat an aliquot of acid phenol to 65° in a 50-ml centrifuge tube. The acid phenol should be prepared ahead of time. To do so, add 500 ml of diethylpyrocarbonate (DEPC)-treated water to solid phenol and heat at 60–65° until the phenol melts. Swirl to mix and then let stand to separate the phases. Remove the water, leaving an inch. Add 8-hydroxyquinilone to 0.1%. The solution should be stored at 4°.

2. Place the sample tubes on ice and immediately add 600 μl of room temperature TES (10 mM Tris–HCl, pH 7.4, 7.5 mM EDTA, 0.5% SDS). Vortex carefully to resuspend the cell pellet.

3. Add 600 μl of hot acid phenol to the first tube, cap, vortex for 10 s, and place into a heat block set at 65°. Repeat with all of the samples. Start timing after the last sample is added to the block. Vortex for 10 s every 5 min for 1 h.

4. Place the tubes on ice for 5–6 min. Spin at 14,000 rpm for 5 min at 4°. Carefully remove 550 μl of the aqueous phase, avoiding the interface, and transfer to a fresh tube.

5. Extract with 550 μl of room temperature acid phenol:chloroform: isoamyl alcohol (25:24:1).

6. Carefully remove only 450 μl of the aqueous layer and transfer to a fresh tube containing 25 μl of 1 M Tris–HCl, pH 8.0, and 35 μl of 5 M NaCl. Add 1 ml of absolute ethanol. Precipitate for 2 h at $-20°$ or 15 min on dry ice.

7. Recover nucleic acids by centrifugation at 14,000 rpm for 10 min at $4°$ and rinse the pellet with 75% ethanol.

8. Add 30 μl of DEPC-treated water and resuspend the RNA on ice. Measure the RNA concentration of each sample by recording the OD_{260}.

9. Add 1 μg of RNA (1 μl) to loading mix (20 μl formamide, 7 μl formaldehyde, 2 μl 20× SSC) and incubate for 15 min at $68°$. Place the samples on ice and add 78 μl of ice-cold 20× SSC (3 M NaCl, 1 M sodium citrate, pH to 7.0).

10. The manifold (Slot Blotter, Stratagene) is prepared using a charged nylon membrane (Zetabind, Cuno) and Whatmann 3MM paper as described by the manufacturer. Pass the samples through the membrane under gentle vacuum and wash each well twice with 10× SSC. Cross-link the RNA to the membrane by exposing it to 12 mJ/cm^2 UV (StrataLinker, Stratagene) and drying it overnight at room temperature.

11. Rehydrate the membrane in 1× SSC and incubate for 4–6 h at $37°$ in prehybridization solution [5× SSC, 5× Denhardt's solution (100× Denhardt's: 20 mg/ml each of Ficoll 400, polyvinylpyrrolidone, and bovine serum albumin), 0.5% SDS, and 0.1 mg/ml salmon sperm DNA]. Add 200,000 cpm/ml (determined in scintillation fluid) of ^{32}P end-labeled dT_{20} oligo (Pharmacia) to the prehybridization solution and incubate for 16–18 h at $37°$. Wash the blot once for 15 min at $37°$ in fresh prehybridization solution and twice for 15 min each in 2× SCC, 0.1% SDS at room temperature.

Notes. During the extraction procedure, do not exceed 15 OD units of cells using the specified cell-to-buffer ratios. The addition of 1 M Tris–HCl raises the pH of the solution and prevents the EDTA from precipitating out during the ethanol precipitation step. If RNA will be used for primer extension analysis, trace amounts of phenol should be removed by chloroform–isoamyl extraction prior to ethanol precipitation.

Analyzing Induced Gene Expression in Temperature-Sensitive Mutants

The protocol described earlier monitors the effects of TAF inactivation while genes are being activity transcribed and have preinitiation complexes formed at the time of the temperature shift. However, it is desirable and necessary to examine the effects of inactivating a TFIID subunit prior to the assembly of a preinitiation complex at the promoter, as it can provide some insight into the function of the TAF. This is achieved by preincubating the mutant at the restrictive condition, followed by an induction period. In designing an inactivation and induction scheme, it is advisable to keep the time period at 37° under 4 h and to choose a preincubation time that assures loss of TAF function. Typically a 1- to 2-hr preincubation time is sufficient.

Preparation of Mini Whole Cell Extract for Western Blotting

1. Place the sample tubes on ice and immediately add 50–75 μl of WCE buffer [50 mM Tris–HCl, pH 8.0, 5 mM EDTA, 0.6 M NaCl, 5% glycerol, 0.1% Triton X-100, 1 mM dithiothreitol (DTT)] containing the following protease inhibitors: 3 μg/ml each of pepstatin A, leupeptin, aprotinin, bestatin, and antipain; 1 μg/ml chymostatin; 2 mM benzamidine–HCl, and 0.5 mM phenylmethylsulfonyl fluoride (PMSF). The volume of the WCE buffer added is approximately 1–2 pellet volumes. Vortex to resuspend the cell pellet, and leave the samples on ice. Add 100 μl of glass beads (0.5 mm; Sigma).

2. Mix each sample at high speed on a vortex mixer six times for 20 s, cooling on ice for 2 min between bursts to avoid overheating the sample. Holding the tube at a slight angle (20°) while mixing is recommended.

3. Pellet the insoluble material by centrifuging at 14,000 rpm for 10 min at 4°.

4. Recover as much of the supernatant as possible without taking the debris pellet. The pipette tip can be placed into the beads. The recovery should be at least 40 μl. Transfer the supernatants to fresh tubes and spin again at 14,000 rpm for 15 min at 4°.

5. Recover a fixed amount of supernatant, usually 30 μl, and transfer to fresh tubes.

6. Remove a 3-μl aliquot, which will be used to determine the protein concentration, and then immediately add SDS–PAGE loading buffer to the remaining sample and boil for 3–4 min.

7. Determine the protein concentration for each sample: this procedure should yield a protein concentration of 5–10 mg/ml. Load an equal amount of protein (typically 20 μg) in each lane.

8. Fractionate the samples by SDS–PAGE, transfer to nitrocellulose, and detect TFIID components by immunoblotting using antibodies to TAFs and TBP.[6,11,15]

Notes. It is important to keep the pellet size constant between samples. Adjust the volume of culture collected according to its OD_{600} to achieve a constant pellet size. We have found that including NaCl to a concentration of 0.6 M helps to fully recover the TAF proteins and other yeast general transcription factors from the debris pellet.

Preparation of Midscale Whole Cell Extracts and Immunoprecipitation of TFIID

The aforementioned protocol analyzes for the presence of TFIID subunits in whole cell extracts, which, of course, does not indicate that TFIID is intact. Moreover, the stabilization of certain TAFs after the temperature shift in some mutants may result from their coexistence in the SAGA histone acetyltransferase complex.[18] Ideally, the structure of TFIID should be examined by purifying it to near homogeneity and analyzing the presence of all of its subunits. However, this is not practical when analyzing a number of different mutants. Instead, we have used a midscale whole cell extract procedure and coimmunoprecipitation using antibodies against TFIID subunits followed by Western blotting of the immunoprecipitates. The method is based on Wootner et al., with modifications.[1,19] In addition, these extracts can be used in in vitro transcription reactions to judge the competency of the mutant TFIID complexes to support transcription.[20]

Materials and Reagents

Buffer A. Filter a solution of 0.2 M Tris–acetate, pH 7.9, 0.39 M $(NH_4)_2SO_4$, 10 mM $MgSO_4$, 2 mM EDTA, and 20% glycerol. Prior to use, supplement with DTT to 2 mM and the following protease inhibitors: 3 μg/ml each of pepstatin A, leupeptin, aprotinin, bestatin, antipain, and chymostatin and 2 mM benzamidine–HCl.

Buffer B. Filter a solution of 20 mM HEPES–KOH, pH 7.5, 10 mM magnesium acetate, 150 mM potassium acetate, 10 mM EGTA, and 20% glycerol. Prior to use, supplement with DTT to 5 mM and the following protease inhibitors: 3 μg/ml each of pepstatin A, leupeptin, aprotinin, bestatin, antipain, and chymostatin and 2 mM benzamidine–HCl.

Buffer B'. Same as buffer B. Prior to use, supplement with DTT to 5 mM and the following protease inhibitors: 0.1 μg/ml each of pepstatin A, leupeptin, aprotinin, bestatin, antipain, and chymostatin and 1 mM benzamidine–HCl and PMSF.

---

[18] P. A. Grant, D. Schieltz, M. G. Pray-Grant, D. J. Steger, J. C. Reese, J. R. Yates, and J. L. Workman, Cell 94, 45 (1998).

[19] M. Wootner, P. A. Wade, J. Bonner, and J. A. Jaehning, Mol. Cell. Biol. 11, 4555 (1991).

[20] J. Reese, unpublished data.

IP(X) Buffer. Same as buffer B, except containing 0.1% NP-40 and DTT to 0.5 mM. X refers to the concentration (mM) of potassium acetate (KoAc).

1. For each mutant, inoculate four 2-liter flasks containing 500 ml YPAD from a starter culture and incubate at 23° until an OD$_{600}$ of 1.0–1.5 is reached. Add an equal volume of prewarmed (50°) media to two of the flasks such that the final temperature of the culture is 37°. Add an equal volume of ambient temperature media to the other two flasks. Incubate the cultures for an additional 3 h at either 23° or 37°, as appropriate.

2. Collect the cells by centrifugation at 5000 rpm for 5 min at 4° using a GSA (Sorvall) rotor. Pour off the supernatant and resuspend the cells in 25 ml of ice-cold distilled water per liter of culture (50 ml per treatment). Pool the samples grown at the same temperature into a single 50-ml tube and centrifuge at 3000 g for 10 min at 4°.

3. Resuspend the cells in 2 volumes of buffer A per wet gram of yeast. Add PMSF to a final concentration of 0.5 mM. Process the cells immediately or freeze in liquid nitrogen and store at −80° for later use.

4. In a cold room, thaw the cells in an ice-water bath and add 2 ml of chilled acid-washed glass beads per wet gram of yeast (~8-10 ml). Mix rapidly using a Vortex Genie-II six times for 30 s, with a 2-min cooling period in between bursts.

5. Transfer the cell lysate to prechilled 40-ml Oakridge tubes and centrifuge at 4000 rpm for 2 min at 4° using an SS-34 rotor (Sorvall).

6. Decant the supernatant into ultracentrifuge tubes, allowing the loose sediment to pour away from the pellet. Dilute to a final volume equivalent to 4 ml per gram of yeast using buffer A. Centrifuge at 44,500 rpm for 1.5 h at 4° in a 70Ti rotor.

7. Working in the cold room, carefully remove the supernatant using a pipette, avoiding the white lipid layer and the loose brown sediment at the bottom of the tube. Filter the solution through sterile cheesecloth into clean 50-ml screw-top centrifuge tubes.

8. Add solid ammonium sulfate [(NH$_4$)$_2$SO$_4$] to a final concentration of 0.337 g/ml extract. The (NH$_4$)$_2$SO$_4$ is added in 4 steps: one-third, one-third, one-sixth, and one-sixth of the total. After each addition, the (NH$_4$)$_2$SO$_4$ is allowed to dissolve while rocking gently at 4°. After each addition of (NH$_4$)$_2$SO$_4$, 10 μl of 1 M KOH is added per gram of (NH$_4$)$_2$SO$_4$.

9. After the final portion of (NH$_4$)$_2$SO$_4$ dissolves, rock gently for 30 min at 4°. Transfer to 40-ml prechilled Oakridge tubes. If strands of genomic DNA appear in the extract, pass it through cheesecloth prior to centrifugation.

10. Collect the precipitate by centrifugation at 12,500 rpm for 30 min at 4° using an SS-34 rotor.

11. Working in the cold room, resuspend the pellets in 50 μl of buffer B per gram of starting yeast cells. This will take some time and is facilitated by gently pipetting the dissolved portion over the pellet and gently dislodging the pellet from the side of the tube.

12. Transfer the extract into dialysis tubing (10–12 kDa cutoff, Spectrum Laboratories) that has been presoaked in buffer B. Dialyze against buffer B' for 1.5 h at 4°. Transfer to fresh buffer B' and continue to dialyze. Stop dialysis when the conductivity of the extract measures between 60 and 80 μS/cm when diluted into water 1/200 (about 2.5–3 h). Aliquot into microcentrifuge tubes, freeze in liquid nitrogen, and store at −80°. Do not centrifuge the extract after dialysis. Centrifuge to remove insoluble material immediately before use.

13. Dilute the extract to a final concentration of 3 mg/ml using IP(100) buffer in a final volume of 0.5 ml. Add an equal volume (0.5 ml) of IP(500) buffer to achieve a final KOAc concentration of 0.3 M. Mix well and incubate 10 min on ice.

14. Centrifuge the samples at 14,000 rpm for 20 min at 4° in a microcentrifuge. Transfer 950 μl of the clarified supernatant to a fresh tube containing 1–3 μl of rabbit serum.[6,11,15] For optimal results, the amount of antibody should be determined empirically, but in most cases 2 μl is sufficient.

15. Incubate the tubes for 2 h on ice and centrifuge at 14,000 rpm for 20 min at 4°. Transfer the supernatant to a fresh tube.

16. Add 30 μl of a 50/50 slurry of protein A-Sepharose beads (Pharmacia) prepared in IP(300) buffer. Incubate overnight at 4° mixing end over end.

17. Recover the protein A beads by centrifugation at 1000 rpm for 3 min at 4°. Save the supernatant.

18. Wash the immune complexes twice using 1 ml of ice-cold IP(300) buffer, mixing end over end for 5 min at 4°.

19. Resuspend the beads in 1 ml of IP(300) buffer and transfer to a fresh tube. This is an important step for reducing background.

20. Wash beads once more in 1 ml of ice-cold IP(300) buffer, mixing end over end for 5 min at 4°.

21. Wash beads briefly in IP(0) buffer. This is important to prevent the SDS from precipitating in the next step.

22. Elute the proteins by heating at 90° in 40 μl of 2 × SDS loading buffer.

23. Load one-third of the immunoprecipitate, 1% of the input whole cell extract, and 1% of the supernatant onto a 7.5% SDS–PAGE gel and 13% SDS–PAGE gel.

Notes. Higher KOAc concentrations can be used during the binding and washing steps. However, we find that TBP disassociation increases in buffers containing greater than 0.3 M KOAc.[15] Salts containing chloride ions should be avoided because the association of TBP with other TFIID complex components is much more sensitive to chloride than acetate.[20]

Conditional Depletion of TFIID Subunits

The most widely used conditional depletion strategy utilizes the *GAL1* promoter, which can be shut off by transferring cells from galactose- to dextrose-containing media. In this procedure, a strain containing the sole copy of a TAF gene under control of the *GAL1* promoter is constructed using one of two methods. In one, a construct containing the TAF gene of interest under control of the *GAL1* promoter is integrated into the genome of a haploid strain containing a deletion of the TAF gene, maintained by a plasmid-borne copy on a *URA3*-containing plasmid. The resulting strain is grown on rich medium containing galactose, and the loss of the plasmid copy is selected for on galactose plates containing 5-FOA.[6,7,10,15] In the other, the promoter of the TAF gene is replaced by a *GAL1* promoter using a PCR-generated cassette directly in a haploid strain.[21] We have found that it is essential to integrate the *GAL1*–TAF construct in order to achieve efficient depletion of the TAF protein.[20] TAF depletion and attenuation of cell growth are usually observed within 8 h (Fig. 3A). As observed with ts mutants, depleting one TAF subunit often results in the codegradation of other TFIID subunits so it is advisable to analyze the levels of all TAF proteins (Fig. 3B).

Depletion Methodology

1. Inoculate a 50-ml culture of YPA supplemented with 3% galactose and 0.2% sucrose and incubate the culture at 30° until the an OD_{600} of 0.8 is reached. Pellet the cells by centrifugation at 3000 g for 5 min.

2. Wash the cells in 50 ml of prewarmed YPAD (dextrose). This step is essential to achieve the most efficient depletion of protein. Pellet the cells by centrifugation, decant the supernatant, and resuspend in 25 ml of prewarmed YPAD. Set aside two 7-ml aliquots of cells for the $t = 0$ time point for protein extract preparation and RNA isolation (see earlier discussion). Add the remaining 7.5 ml to 100 ml of prewarmed YPAD and continue incubating at 30°.

[21] M. S. Longtine, A. McKenzie, D. J. Demarini, N. G. Shah, A. Wach, A. Brachat, P. Philippsen, and J. R. Pringle, *Yeast* **14**, 953 (1998).

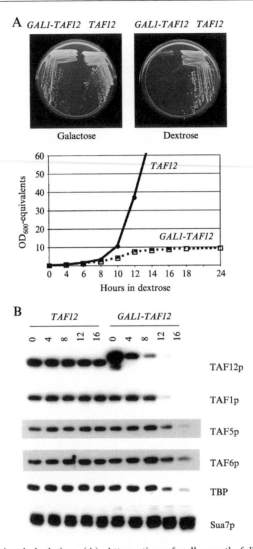

FIG. 3. Conditional depletion. (A) Attenuation of cell growth following conditional depletion of TAF12p. (B) Detection of the effects of conditional depletion of TAF12p on the integrity of the TFIID complex by immunoblot analysis. TAF12p in extracts prepared from the *GAL1–TAF12* strain migrates slower because of incorporation of a HA tag contained in the vector. Sua7p, a general transcription not part of TFIID, is used as a loading control.

3. Remove two aliquots of cells equal to 15 OD_{600} units at 4, 8, 12, 16, and 20 h after transferring the cells to dextrose-containing medium. Monitor the cell density every 2 h and dilute the cultures as needed into

fresh, prewarmed media to maintain the cell density between 0.25 and 1.5 throughout the entire experiment. This will require constantly diluting the wild-type strains and diluting the conditional strain through the first 8–10 h. It is preferable to dilute a culture after removing an aliquot of cells rather than just prior to removal because media changes caused by the dilution process cause a minor transient fluctuation in the expression of certain genes.

Notes. This protocol can be adapted to study induced gene expression. After a depletion period, typically at 12 h, the cells are treated with the inducing agents/conditions. We have also used cells depleted of TAFs using this protocol to prepare transcriptionally active yeast whole cell extracts.[20]

[37] Immunoaffinity Purification of Mammalian Protein Complexes

By YOSHIHIRO NAKATANI *and* VASILY OGRYZKO

Identification of the interaction partners of a particular protein became a valuable way to gain insight into the physiological role of the protein. Often, when studying a gene of unknown molecular function, the knowledge of the molecular complexes that include the encoded protein as a subunit can be a starting point for understanding the function of the protein. Many proteins appear to exist and function as stable multimeric protein complexes. For instance, the multimeric RNA polymerase II protein complex consists of 12 polypeptides.[1] It transcribes DNA templates as a complex, whereas individual subunits have no such activity, thus making it impossible to predict "bona fide" functions of an RNA polymerase II from a single subunit. Likewise, if the protein of interest functions as a complex, it is crucial to purify it as a complex to determine what it does in the cell.

The most widely used system to identify intermolecular interactive partners of a particular protein of interest is the yeast two-hybrid system.[2,3] However, given that this system is designed to detect mostly binary interactions, it is not suitable for investigating the intramolecular interaction partners of a protein if the latter is a component of a multimeric protein

[1] R. H. Ebright, *J. Mol. Biol.* **304,** 687 (2000).
[2] S. Fields and O. Song, *Nature* **340,** 245 (1989).
[3] A. Pandey and M. Mann, *Nature* **405,** 837 (2000).

complex. The other approach is affinity purification of interactive partners using the protein of interest as the ligand. Generally, the protein of interest is immobilized on a matrix and incubated with an extract *in vitro*, allowing one to purify intermolecular interaction partners.[3] An advantage of this method over the yeast two-hybrid system is that if the purification is performed under native conditions, intermolecular interaction partners can be purified as native forms. However, if the protein of interest exists in a stable multimeric protein complex, identification of intramolecular interaction partners of the immobilized protein would not be possible because the liganded protein would not be able to gain access and thus efficiently replace the corresponding endogenous protein in the complex.

For identification of intramolecular interaction partners, purification of the native protein complexes from the cells would be the best method. This would provide information on how the endogenous protein interacts with intramolecular interaction partners. However, in general, purification of native complexes is technically difficult and time-consuming, especially when the protein of interest is scarce. These limitations can be addressed with epitope tagging.[4–9] It has been shown that stably expressed exogenous proteins with epitope tags are integrated into the complexes when the complex is synthesized *de novo*. Such a protein complex can be purified rapidly by immunoprecipitation with antibodies against epitopes.

We took advantage of the epitope-tagging technique to develop a standardized approach to purifying multimeric protein complexes from mammalian cells. We put two different epitope tags tandemly (i.e., FLAG and HA) on either the N or the C terminus of the protein of interest. For rapid generation of cell lines that express the epitope-tagged protein of interest, we use retroviral transduction. The cells are sorted magnetically after transduction. Once stably transduced cell lines are established, the protein complex can be purified from cells in less than 24 h. This article describes both the generation of recombinant cells (molecular cloning and retroviral transduction) and the technique for purification of the complex (large-scale cell growth and affinity chromatography). Thus, the article is divided into two parts that correspond to these different steps in the procedure.

[4] C. M. Chiang *et al.*, *EMBO J.* **12**, 2749 (1993).
[5] J. Field *et al.*, *Mol. Cell. Biol.* **8**, 2159 (1988).
[6] T. Ikura *et al.*, *Cell* **102**, 463, (2000).
[7] H. Ogawa *et al.*, *Science* **296**, 1132 (2002).
[8] V. V. Ogryzko *et al.*, *Cell* **94**, 35 (1998).
[9] Q. Zhou, T. G. Boyer, and A. J. Berk, *Genes Dev.* **7**, 180 (1993).

Generation of Epitope-Tagged Protein–Expressing Cells

To generate stable cells that express double epitope–tagged versions of the proteins of interest, we constructed two retroviral vectors: pOZ-FH-N and pOZ-FH-C (Fig. 1). These vectors are derivatives of the MMLV-based pOZ vector, constructed by Bruce Howard and colleagues (N. Ziran, V. Ogryzko, T. Hirai, V. Russanova, and B. H. Howard, unpublished result). pOZ contains a bicistronic transcriptional unit that allows expression of two proteins from a single transcript. This design ensures tight coupling between expression of the gene of interest and the selection marker, the interleukin-2 receptor α chain (IL2Rα).[10] This approach is particularly advantageous when the expression of proteins of interest causes growth retardation.

pOZ-FH-N and pOZ-FH-C allow expression of the protein of interest in, respectively, the N-terminally and C-terminally tagged form. Generally speaking, the epitope tags are flexible and accessible to antibodies without disrupting the tagged complexes. Therefore, the position of the epitope tags has proved not to be an important issue in most proteins we have tested, except in those cases in which we found the nature of the complexes purified or the functional integrity of the protein to depend on the position of the epitope. Thus, it would be worth testing both an N-terminally and a C-terminally tagged protein.

Upstream of the *Xho*I site, pOZ-FH-N contains a Kozak sequence,[11] an initiation methionine, and FLAG[12] and HA[5] tags. The open reading frame (from the second codon to the stop codon) of protein of interest should be subcloned into the *Xho*I and *Not*I sites. We typically amplify the open reading frame by polymerase chain reaction (PCR) as an *Xho*I–*Not*I fragment and subclone it into pOZ-FH-N. If *Xho*I sites are present in the open reading frame, an alternative enzyme that produces an overhang comparable to *Xho*I (e.g., *Sal*I) can be used.

pOZ-FH-C contains the FLAG and HA tags and a stop codon downstream of the *Not*I site. If one amplifies the open reading frame by PCR, the 5′ PCR oligonucleotide should include a Kozak sequence[11] and an initiating methionine after the *Xho*I sequence. Note that the pOZ vector has a cryptic translation start site upstream of the cloning sites. To avoid formation of a fusion protein, the oligo nucleotide should be designed so that the protein of interest is out of frame from the *Xho*I (CTC–GAG) site.

---

[10] R. Padmanabhan *et al.*, *Anal. Biochem.* **170**, 341 (1988).
[11] M. Kozak, *J. Cell Biol.* **115**, 887 (1991).
[12] T. P. Hopp *et al.*, *Bio/Technology* **6**, 1204 (1988).

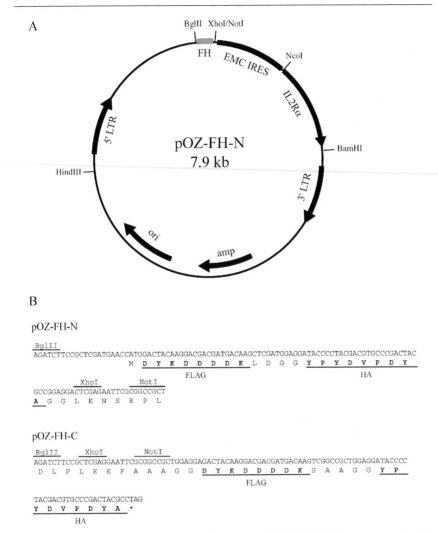

Fig. 1. (A) Restriction map of the retroviral vector pOZ-FH-N. (B) Cloning sites of pOZ-FH-N (top) and pOZ-FH-C (bottom), represented by the gray box (FH) for pOZ-FH-N in A.

A limitation of the retroviral system is the size of protein to be expressed. Retroviruses can package RNAs as large as 10 kb, measuring the distance from the start of the 5′ LTR to the end of the 3′ LTR. Thus, the maximum size of the insert in this system is ~6 kb. Although this limit has not been tested systematically, we have successfully expressed several proteins with molecular masses of about 100–130 kDa.[8]

Once the retroviral vector is constructed, we recommend checking expression of the epitope-tagged protein by transient transfection. We routinely transfect the retroviral plasmid into 293T fibroblasts. Forty-eight to 72 hours after transfection, the epitope-tagged protein is immunoprecipitated from cell lysates with anti-FLAG antibody–conjugated beads and is subsequently detected by immunoblotting with appropriate antibodies, e.g., antibodies against FLAG, HA, and/or the protein of interest.

Transfection of Retroviral Vector into Packaging Cells

In order to generate transduction-competent retrovirus, the retroviral vector is transfected into a cell line that expresses retroviral structural gene products (gag, pol, and env) necessary for the packaging of retroviral RNA into infectious particles.[13,14] Packaging cells used routinely for generation of amphotropic retrovirus, necessary to infect HeLa cells, are Bing or Phoenix A cells. Although the retroviral system employed is replication deficient, a safer alternative is to produce ecotropic retroviruses by transfecting the retroviral plasmid DNA into an ecotropic packaging cell line, such as Phoenix E. HeLa cells that stably express murine ecotropic receptors can be infected with the resulting viruses. In any event, work with amphotropic and ecotropic retroviruses should be done using appropriate safety precautions.

This section describes a calcium phosphate transfection method.[13,14] Although this method is widely employed for the introduction of retroviral vectors, lipid-mediated transfection methods can also be used. We found that transfection with FuGENE 6 (Roche) works as efficiently as the calcium phosphate method.

Reagents

2× HBS: 50 mM HEPES, 10 mM KCl, 270 mM NaCl, 11 mM dextrose (glucose), 1.5 mM Na_2HPO_4. Adjust pH to 7.05 at room temperature with NaOH, filtrate through a 0.45-μm filter, aliquot, and store at $-20°$

Procedure

1. Plate packaging cells (Bing or Phoenix A) at a density of 6×10^6 per 100-mm-diameter tissue culture dish in 10 ml of Dulbecco's modified Eagle's medium (DMEM) 10% fetal calf serum (FCS). Grow overnight in a 37° incubator (5% CO_2).

[13] F. Ausubel et al., "Current Protocols in Molecular Biology." Wiley, New York.
[14] J. Coffin, S. Hughes, and H. Varmus (eds.), "Retroviruses." Cold Spring Harbor Laboratory Press, Cold Spring Harbor, NY, 1997.

2a. Mix plasmids to be transfected and $CaCl_2$ solution as follows. First, dilute 2 M $CaCl_2$ with H_2O in a 1.5-ml Eppendorf tube. Then add the plasmids indicated to the calcium solution. Although packaging cells are supposed to express gag, pol, and env, these genes are relatively unstable. Thus, we cotransfect the retroviral vector along with expression plasmids for gag-pol and env.

24.8 μl	2 M $CaCl_2$
5 μg	retroviral plasmid
2.5 μg	gag-pol expression plasmid
2.5 μg	env expression plasmid
to the total volume of 200 μl	H_2O

2b. Place 200 μl of 2× HBS in a 14-ml round-bottom tube (Falcon 2059 or equivalent). Add the DNA/calcium mixture (from step 2a) into 2× HBS dropwise, mixing after each addition.

3. Add the transfection mixture (from step 2b) dropwise to the packaging cells in a dish (from step 1), stirring well after each addition, and incubate the plate in a 37° incubator (5% CO_2).

4. Twelve hours after transfection, replace the medium with fresh DMEM/10% FCS and incubate for an additional 48 h.

5. Centrifuge the culture supernatant at 2000 g for 5 min. To remove the packaging cells completely, filter the supernatant through a cellulose acetate membrane with a 0.45-μm pore size, which yields a retroviral stock. We recommend using the retroviral stock for transduction the same day.

Retroviral Transduction

The next step is the introduction of recombinant retrovirus into cells that are used for expression and purification of the epitope-tagged protein. We prefer to use HeLa S3 cells for large-scale protein purification because they grow as a suspension culture in a medium containing calf serum (CS), which is significantly less expensive than fetal calf serum. However, many other cell lines can be used if required. We have successfully purified complexes from NIH3T3 mouse fibroblasts and BJ-1 human fibroblasts.

We use HeLa S3 cells so that we can employ spinner culture. HeLa S3 cells can be grown in either DMEM/CS in a tissue culture flask or Joklik's medium/CS in a spinner bottle.[15] In general, we maintain HeLa cells in tissue culture flasks and grow them in spinner bottles for a large-scale culture. The amount of CS required for optimum cell growth depends on serum lots (ranging between 5 and 7%). Before purchasing, we carefully

---

[15] W. F. McLimans, *Methods Enzymol.* **58**, 149 (1979).

select a serum lot that supports optimum cell growth at less than 5%. When HeLa cells are grown in a tissue culture flask, close the cap loosely and grow them in a 37° tissue culture incubator with 5% CO_2. However, when using a spinner bottle, close the cap tightly and grow them on a magnetic stirrer in a normal 37° incubator. The density of HeLa cells should be kept between 2 and 6 $\times$ 10^5/ml.

Procedure

1. Grow HeLa cells in DMEM/5–7% CS in appropriate flasks; 5 $\times$ 10^5 cells per transduction would be required. When HeLa S3 cells are grown in DMEM/CS, they grow as a mixed population of floating and attached cells. When passaging the cells, tap the flask gently and transfer the cell suspension to a 50-ml conical tube (Falcon 2070 or equivalent). Cells that remain attached to the flask are washed with phosphate-buffered saline (PBS), trypsinized, and harvested. Combine these cells with the cell suspension in the Falcon tube.

2. Transfer the cell suspension containing 5 $\times$ 10^5 cells to a 15-ml conical tube and centrifuge at $\sim$2000 g for 5 min. While spinning, add 0.1 ml of 400 μg/ml polybrene to a 10-ml viral stock.

3. After centrifugation, discard the supernatant and resuspend the cell pellet into the 10-ml viral stock/polybrene mixture. Suspend thoroughly by pipetting up and down, transfer the suspension to a T75 flask, and place in a 37° incubator with 5% CO_2.

4. When the cell density reaches $\sim$5 $\times$ 10^5/ml (or 80% confluency), transfer the cells to 25 ml of the fresh medium in a T175 flask and allow them to continue to grow in a 37° incubator with 5% CO_2.

5. When the cell density reaches $\sim$5 $\times$ 10^5/ml (or 80% confluency) in the T175 flask, they are ready for sorting.

Sorting of Transduced Cells

The next step is selection of the transduced cells by magnetic affinity sorting with antibody against the surface selection marker.[10] Usually, two to three rounds of sorting are sufficient to achieve a pure population. However, the number of sorting rounds depends on protein of interest, especially when expression of the protein of interest causes growth retardation; given that small contaminants of nontransduced cells would, in such a case, eventually overtake the population, additional rounds of sorting may be required.

Materials

Dynabeads M-450 goat antimouse IgG (Dynabeads)
PBS/bovine serum albumin (BSA): 0.1% BSA in PBS

Storage solution: PBS/BSA in 50% glycerol
Mouse anti-IL2Rα antibody (Upstate Biotechnology, Inc.)
Magnets for separation Biomag flask separator (Polysciences)

Preparation of Anti-IL2Rα Antibody–Conjugated Magnetic Beads

1. Wash 100 mg of Dynabeads three times with 10 ml PBS/BSA in a T25 flask by magnetic separation. After the final wash, suspend the beads in 3.2 ml PBS/BSA.
2. Dissolve 200 μg anti-IL2Rα antibody in 0.8 ml PBS/BSA.
3. Transfer 0.8 ml anti-IL2Rα antibody (step 2) to the tube containing the bead suspension.
4. Rotate or shake the tube at 4° overnight.
5. Wash four times with 10 ml PBS/BSA by magnetic separation.
6. Resuspend the beads in 4 ml storage solution. Make aliquots and store at −20°.

Sorting Procedure

1. Add anti-IL2Rα antibody–conjugated magnetic beads (5 μl/ sample) to DMEM/5–7% CS (1 ml/sample) and suspend thoroughly by vortexing. If there are multiple samples, prepare enough suspension (e.g., 1.1 ml × sample number) for the last aliquot.

2. Add 1 ml bead suspension to the cell culture in the T175 flask. Incubate the flask in a 37° incubator (5% CO_2) for 30 min with occasional shaking. Observe binding of the beads to the cells under a microscope.

3. Transfer the suspension of cells and beads to a T25 flask as follows. Tap the T175 flask gently and transfer the suspension to a new T25 flask. To recover tightly attached cells from the T175 flask, wash them with PBS and then treat with 2 ml 0.05% trypsin solution. To avoid degradation of the surface marker, exposure to trypsin should be kept to a minimum. Thus, when cells start to be detached, tap the flask several times and add 5 ml of DMEM/CS immediately to neutralize the trypsin. Combine these cells with the suspension of those that have already been transferred to the T25 flask. Close the cap tightly.

4. Sandwich the T25 flask between magnetic plates (Biomag flask separator) and secure with rubber bands.

5. Stand the T25 flask/magnetic plate assembly vertically (the cap uppermost) in a tissue culture hood. Gently rock the flask occasionally to prevent the cells from accumulating at the bottom.

6. After 30 min, gently aspirate the medium containing the unbound cells.

7. To wash positive cells and remove negative cells, add 50 ml DMEM/5–7% CS to the T25 flask and close the cap tightly. Gently rock

the flask for a few minutes and aspirate the medium. Repeat this washing step two or three times.

8. After the final washing, remove the flask from the magnetic plates. Add 5 ml DMEM/5–7% CS and suspend the cells by pipetting up and down. Carefully wash the walls of the flask to recover the cells. Observe binding of the beads to the cells under a microscope and incubate the flask in a 37° incubator with 5% CO_2. If the cell number is less than ~1 × 10^5, transfer to a smaller flask.

9. When the cell density reaches ~5 × 10^5/ml (or 80% confluency), add 1 ml bead suspension (see step 1) to the flask. Repeat steps 2–8. For a further sorting, allow the sorted cells to grow in a 37° incubator with 5% CO_2.

10. Repeat cell sorting until ~100% cells bind to anti-IL2Rα–conjugated magnetic beads, as judged by viewing under a microscope.

Detection of Expression of the Epitope-Tagged Protein

Once ~100% of the cells become IL2Rα positive, propagate the cells in DMEM/5–7% CS in flasks. We routinely propagate cells to ~8 × 10^7 in four T175 flasks. These cells are used for (1) checking expression of the epitope-tagged protein; (2) preparing frozen stocks; and (3) large-scale cell culture.

We determine expression of the epitope-tagged protein by immunoblotting and *in situ* immunofluorescence microscopy with anti-FLAG and/or HA antibody. The former analysis provides information about the expression level and molecular weight of the epitope-tagged protein, whereas the latter shows expression level, subcellular localization, and variation of expression levels among various cells.

Epitope-tagged proteins may be detected by direct immunoblotting of cell lysates. However, given that the sensitivity and specificity of the FLAG and HA antibodies for immunoblotting are relatively low, we first concentrate epitope-tagged proteins by immunoprecipitation. In our experience, all tagged proteins tested so far are detected by immunoprecipitation followed by immunoblotting (see later). We also detect tagged proteins by *in situ* immunofluorescence microscopy.[16,17] However, if the copy number of a tagged protein is relatively low, they may not be detectable by the latter method.

[16] D. Spector, B. Goldman, and L. La (eds.), "Cells: A Laboratory Manual." Cold Spring Harbor Laboratory Press, Cold Spring Harbor, NY, 1998.
[17] E. Harlow and D. Lane, "Using Antibodies: A Laboratory Manual." Cold Spring Harbor Laboratory Press, Cold Spring Harbor, NY, 1988.

If a specific antibody that recognizes the endogenous protein is available, we also compare expression levels of the protein of interest in transduced (the endogenous and tagged protein) and nontransduced (the endogenous protein) HeLa cells. In our experience, transduced and nontransduced HeLa cells do not differ with regard to most proteins that we have tested. If the epitope-tagged protein can be separated from the endogenous one by SDS–PAGE, the endogenous and tagged protein levels in transduced cells can be analyzed.

Materials

RIPA buffer: 50 mM Tris–HCl, pH 8.0, 150 mM NaCl, 1% NP-40, 0.5% DOC, 0.1% SDS; store at 4°C

M2 anti-FLAG antibody-conjugated agarose (Sigma A 2220)

Procedure

1. Spin down 1×10^7 cells and wash with PBS.

2. Suspend the cell pellet in 0.8 ml RIPA buffer and transfer to a 1.5-ml Eppendorf tube.

3. Vortex for 1 min and rotate in a cold room for 30 min.

4. While extracting proteins, wash M2 anti-FLAG antibody–conjugated agarose beads to remove unconjugated antibody. Place ~10 μl (packed volume) antibody beads in a 0.5-ml Eppendorf tube and add 200 μl 100 mM glycine–HCl (pH 2.5). Vortex gently and leave for 2 min (do not expose to the acidic buffer for any longer than this). Spin down at 2000 g for 1 min and remove the supernatant with care. Suspend the beads in 200 μl 200 mM Tris–HCl (pH 8.0)/0.1% Tween 20. Spin down at 2000 g for 1 min and remove 180 μl supernatant.

5. Centrifuge the cell lysate (step 3) at 15,000 g for 5 min and transfer the supernatant to a new Eppendorf tube. Four hundred microliters of cell lysate is used for immunoprecipitation, whereas the rest is saved for other experiments, such as detection of protein of interest with a specific antibody.

6. Add 400 μl cell lysate to the washed beads (step 4). Rotate or shake the tube for 1 h in a cold room.

7. Wash three times with 200 μl PBS/0.1% Tween 20 by centrifugation at 2000 g for 1 min.

8. After the final wash, remove the supernatant. Spin again at 2000 g for 1 min and carefully remove residual buffer.

9. Resuspend in 10 μl elution buffer and incubate at room temperature for 30 min with occasional gentle vortexing (vortex very gently so that beads and buffer remain at the bottom of the tube).

10. Spin again at 2000 g for 1 min and carefully recover the supernatant, which can be used for immunoblotting with antibodies

against FLAG, HA, and/or the protein of interest. Alternatively recover the supernatant by passing through a microspin column (Bio-Rad Micro Bio-Spin or equivalent).

Large-Scale Cell Growth and Complex Purification

This section describes large-scale cell growth, cell fractionation, and affinity purification of the double-tagged protein. The scale of purification depends on the level of expression and purposes of the experiments. We purify complexes from as few as 2×10^7 cells (a single T175 flask) to as many as 8×10^{10} cells (100 liter culture). If identification of the complex by mass spectrometry is a goal and the complex is relatively abundant, purification of the complex from $\sim 6 \times 10^9$ cells (8 liter culture) would be sufficient. However, amounts required for the identification of proteins vary depending on equipment, methodologies, investigator's skill, character of the proteins, etc.

This section describes the protocol for tissue culture at the 8-liter scale. HeLa cells can be grown in Joklik medium as a suspension culture. For large-scale spinner culture, we place heavy-duty magnetic stirrers (Bellco or equivalent) in a large $37°$ incubator (Forma Scientific or equivalent). If a large incubator is not available, HeLa cells can be grown in a $37°$ warm room. Note that caps of the spinner flask should be closed tightly and the density of HeLa cells kept between 2 and 6×10^5/ml.

Growth of cells

Materials

1-liter spinner bottle (Bellco)
12-liter spinner bottle (Bellco)

Procedure

1. Spin down 4×10^7 cells (see *Detection of Expression of the Epitope-Tagged Protein*) and resuspend in 10 ml Joklik medium containing 5–7% CS. Transfer to 200 ml Joklik medium containing 5–7% CS in a 1-liter spinner bottle (2×10^5 cells/ml). Put the spinner bottle on a magnetic stirrer in a $37°$ incubator and stir at 60 rpm.

2. Monitor cell density every 24 h and add Joklik medium containing 5–7% CS prewarmed to $37°$ to a final density of 2×10^5 cells/ml.

3. After 72 h, the cells should have grown to yield $\sim 3.2 \times 10^8$ in ~ 800 ml. Transfer the whole culture to a 12-liter spinner bottle by decantation and add Joklik medium containing 5–7% CS prewarmed to $37°$ to a final density of 2×10^5 cells/ml.

4. Monitor cell density every 24 h and add Joklik medium as described earlier.

5. After 72 h, the cells should have grown to yield $\sim$2.5 $\times$ 10^9 in $\sim$6.4 liters. Add Joklik medium with 5–7% CS prewarmed to 37° to the final volume of 8 liters. Continue to grow until the density reaches 5$\sim$8 $\times$ 10^5 cells/ml.

6. Harvest cells by centrifugation.

Cell Fractionation and Complex Purification

A method of cell fractionation should be chosen that is appropriate for the subcellular localization and characterization of the protein of interest. We typically prepare cytoplasmic and nuclear extracts as well as nuclear pellet fractions from fresh HeLa cells.[18] These fractions can be stored at −80°.

Here we describe our purification method from nuclear extracts. At least in the beginning, we recommend performing parallel mock purification from nontransduced HeLa cells. HeLa nuclear extracts contain proteins that bind specifically to the M2 anti-FLAG antibody. Such "contaminants" can be detected in samples purified from both transduced and nontransduced cells. The "contaminants" can be removed by the second immunoaffinity purification, as they do not bind to the 12CA5 anti-HA antibody. Thus, complexes can be purified almost completely by the sequential immunoaffinity purification.

Anti-FLAG Antibody Immunoaffinity Purification

Materials

M2 anti-FLAG antibody-conjugated agarose (Sigma A 2220) 12CA5 anti-HA antibody-conjugated protein A–Sepharose: Immobilize 12CA5 anti-HA antibody to protein A–Sepharose 4 Fast Flow (Amersham Biosciences) at 6–10 mg/ml and cross-link with dimethyl pimelimidate using a standard protocol.[17]

Washing buffer: 20 mM Tris–HCl, pH 8.0, 100 mM KCl, 5 mM MgCl$_2$, 0.2 mM EDTA, 10% glycerol, 0.1% Tween. Just prior to use, add 10 mM 2-mercaptoethanol and 0.25 mM phenylmethylsulfonyl fluoride (PMSF)* Prepare 250 mM PMSF solution of PMSF in dimethyl sulfoxide (DMSO) and store at −20°.

FLAG elution buffer: 50 μg/ml FLAG peptide in washing buffer. To prepare the stock solution of FLAG peptide, reconstitute

---

[18] J. D. Dignam et al., Methods Enzymol. **101**, 582 (1983).

lyophilized FLAG peptide (Sigma) with 0.2 M Tris–HCl pH 8.0 to 5 mg/ml, make aliquots, and store at -20°.

HA elution buffer: 50 μg/ml HA peptide in washing buffer. Prepare the stock solution of HA peptide from reconstitute lyophilized HA peptide (Roche) as described for the FLAG peptide.

Procedure

1. Typically, an 8-liter culture yields ~15 ml (~8 mg protein/ml) nuclear extract. If a nuclear extract is stored at -80°, thaw it rapidly in a 37° water bath, shaking occasionally. Put the tube on ice just before thawing the extract completely.

2. Centrifuge the extract at 50,000 g for 30 min at 4°. Immediately after centrifugation, transfer the supernatant to a new centrifugation tube with a 10-ml pipette. Because the pellet is relatively soft and the lipid layer on the top diffuses rapidly, it is hard to obtain a clear supernatant, but this is not a cause for concern.

3. Centrifuge again at 50,000 g for 30 min at 4°.

4. While spinning the extract, wash M2 anti-FLAG antibody–conjugated agarose beads. The amount of beads required for purification depends on the level of the epitope-tagged protein in the extract and the accessibility of the epitope-tagged protein to the antibody. Thus, we recommend preliminary experiments with each protein to determine the optimum amount of beads. In our experience, 200 μl beads per 10 ml nuclear extract may be used for most proteins, although this may not be optimal in all cases. Pipette the beads onto a column (Bio-Rad Poly Prep or equivalent). Wash a sufficient amount of beads so that there is enough for the last aliquot, e.g., volume per sample × (sample number + 0.5). The packed volume of the beads can be measured by the scale on the column. Wash the beads with 10 volumes of 100 mM glycine–HCl (pH 2.5) to remove uncross-linked antibody.

5. Wash the beads with 10 bed volumes of 0.2 M Tris–HCl (pH 8) and then 10 bed volumes of washing buffer.

6. After the last drop from the column, close the bottom with the cap (supplied with columns) and add 1 volume of washing buffer to the column.

7. With an Eppendorf Pipetman P-1000 (or equivalent), transfer the bead suspension (50% slurry) to a 15-ml conical tube (Falcon 2096 or equivalent). To transfer the beads completely, cut the end of the tip with a clean razor blade. Suspend thoroughly by pipetting up and down with P-1000 (or equivalent), and transfer the aliquot to the tube. After the beads settle, read the packed volume of the beads off the scale on the tube.

If necessary, adjust the volume by adding or removing the beads. Keep the tube on ice.

8. Immediately after centrifugation (from step 3), transfer the supernatant to the tube containing M2 antibody beads (from step 7).

9. Close the cap tightly, and rotate or shake the tube for 3 h at 4°.

10. While rotating the tube, set up a column (Bio-Rad Poly-Prep column or equivalent) over a 50-ml conical tube using a plastic column holder (such as the one provided with Quiagen plasmid midi kit). Chill the column in a cold room.

11. Load the sample onto the column. Allow it to drain completely by gravity flow. Transfer the flow through to an appropriate tube and store it at −80°, if necessary.

12. Wash the beads by filling the column to the top with washing buffer and allow it to drain completely by gravity flow. Repeat the washing step for a total of three washes. Empty the 50-ml Falcon tube before each wash.

13. After the final wash, set up the column over a 15-ml conical tube. To remove liquid from the beads, spin the column at 1200 g for 5 min in a Beckman J6-M1 centrifuge with a JS-4.2 rotor or equivalent.

14. Immediately after centrifugation, close the bottom of the column with the cap. Load 200 μl (or volume equal to the packed volume of the beads) FLAG elution buffer to the beads and suspend well by vortexing gently. Incubate at room temperature for 30 min, mixing occasionally. Alternatively, incubate at 4° for 1 h. In most cases, we elute proteins at room temperature rather than at 4° because it is faster.

15. Remove the bottom cap and spin the column on a new 15-ml conical tube at 1200 g for 5 min. Transfer the purified sample to a 1.5-ml Eppendorf tube. Eighty to 90% of the complex can be recovered in the first elution. If necessary, repeat the elution step. When the sample is immediately further purified by anti-HA antibody immunoprecipitation, keep the sample on ice. Otherwise, freeze it on powdered dry ice and store at −80°.

Anti-HA Antibody Immunoaffinity Purification

1. Wash anti-HA 12CA5 antibody-conjugated beads as follows. Place ~20 μl (packed volume) antibody beads in a 0.5-ml Eppendorf tube and add 200 μl glycine–HCl (pH 2.5). Vortex mildly and leave for 2 min (do not expose to the acidic buffer any longer than this). Spin down at 2000 g for 1 min and remove supernatant with care. Suspend the beads in 200 μl 200 mM Tris–HCl (pH 8.0)/0.1% Tween 20. Spin down at 2000 g for 1 min, remove the supernatant, and resuspend in 200 μl of washing buffer.

2. Spin the beads at 2000 g for 1 min, remove the supernatant, and load ∼200 μl anti-FLAG antibody–immunopurified material. Save 10–20 μl anti-FLAG antibody–purified material for SDS–PAGE analysis. In general, we find that ∼20 μl of the antibody beads provides a large excess and is more than sufficient for ∼200 μl sample. However, we often find protein complexes with poor binding to anti-HA antibody, perhaps due to poor accessibility to the epitope. If a significant amount of the unbound complex is found, increase the amount of the antibody beads.

3. Spin down at 2000 g for 1 min and transfer the supernatant to a new 500-μl Eppendorf tube, saving the unbound material for SDS–PAGE analysis.

4. Resuspend in 200 μl washing buffer and transfer to a microspin column (Bio-Rad micro Bio-Spin or equivalent) set up over a 1.5-ml Eppendorf tube. Spin at 2000 g for 1 min.

5. Wash again with 200 μl washing buffer by centrifugation at 2000 g for 1 min.

6. After centrifugation, close the bottom with the cap (supplied with columns). Load 40 μl (or 2 bed volume) of the HA elution buffer to the packed beads. Incubate at room temperature for 1 h, mixing occasionally.

7. Recover the eluant by centrifugation at 2000 g for 2 min.

8. For the second elution, repeat steps 6 and 7.

Analysis of Affinity-Purified Materials

We typically analyze 2–10 μl anti-FLAG antibody–immunopurified material, 2–10 μl flow through of anti-HA antibody immunopurification, and 0.4–2 μl each of the first and second eluates from anti-HA antibody beads on an SDS–PAGE gel. Although silver staining is not quantitative, we analyze the samples along with a series dilution of a molecular weight marker for a rough estimate of the amount of the purified materials. We routinely run 4–20% Tris–glycine SDS–PAGE or 4–12% bis–Tris NuPAGE gels (Invitrogen), which are suitable for analysis of a wide range of molecular mass proteins ranging from less than 10 to over 500 kDa.

Acknowledgment

We thank members in the Nakatani and Ogryzko laboratories for valuable comments.

[38] Interaction of Gal4p with Components of Transcription Machinery *In Vivo*

By SUKESH R. BHAUMIK and MICHAEL R. GREEN

Transcriptional initiation of eukaryotic RNA polymerase II-dependent genes is triggered by the assembly of several dozen proteins on the promoter. These proteins include transcriptional activators, coactivators, chromatin-modifying proteins, general transcription factors (GTFs), and components of the RNA polymerase II holoenzyme complex.[1] Transcriptional activators bind with high affinity and specificity to upstream activating sequences (UASs) in the promoter and function by stimulating the assembly of the preinitiation complex (PIC) through a mechanism that is thought to involve a direct interaction with one or more components of the transcription machinery. Understanding the molecular mechanism by which activators function requires identifying the proteins, or "targets," with which it interacts.

In vitro protein–protein interaction experiments, such as GST pull-down assays, have been widely used to identify potential targets for a number of activators, including three well-characterized activators: yeast Gal4p, human p53, and viral VP16. The resulting picture is very complex: for a given activator, a confusingly large number of different targets have been identified *in vitro*. The identification of such a large number of targets may be due, in part, to the fact that GST pull-down assays can detect even very weak interactions, rendering this technique prone to artifacts. In addition, these *in vitro* assays have produced conflicting observations; for example, under some experimental conditions, VP16 interacts predominantly with TFIIB, whereas under other conditions, it interacts predominantly with TBP,[2,3] indicating that *in vitro* conditions can influence the ability of an activator to interact with its target. Although it is difficult to understand how an activator can interact with so many components of the transcription machinery during transcription activation, an alternative explanation is that the multiplicity of interactions between activators and components of the transcriptional machinery could point to a redundancy in recruitment options. These two opposing models underscore the need to examine activator–target interactions *in vivo* to identify bona fide target

[1] R. Tjian and T. Maniatis, *Cell* **77,** 5 (1994).
[2] K. F. Stringer, C. J. Ingles, and J. Greenblatt, *Nature* **345,** 783 (1990).
[3] Y. S. Lin and M. R. Green, *Cell* **64,** 971 (1991).

proteins and to establish an accurate mechanistic model of transcription activation.

Toward this goal, we have examined the yeast Gal4 protein, which has served as a paradigm for studying eukaryotic transcriptional activators. A number of *in vitro* targets of Gal4p have been identified, including general transcription factors TFIIA and TFIIB;[4–6] TATA box-binding protein (TBP);[7] putative coactivators Ada2p, Gal11p, and Sug1p;[7–10] the SRB/mediator complex component Srb4p;[8,11] and the chromatin modifying complexes SAGA and Swi/Snf.[12,13] To identify bona fide targets, we and others have used a formaldehyde-based *in vivo* cross-linking and chromatin immunoprecipitation (ChIP) assay to detect Gal4p–target interactions in living yeast cells.[14,15] These studies have shown that the yeast SAGA complex is an essential *in vivo* target of Gal4p.[14] The stage is now set to apply this technique toward identifying *in vivo* targets of other activator proteins, such as VP16 or p53.

Formaldehyde-Based *In Vivo* Cross-Linking and ChIP Assay

The formaldehyde-based *in vivo* cross-linking and ChIP assay was originally developed as a technique to monitor the association of a protein with its DNA-binding site.[16] The experimental strategy for this assay is outlined in Fig. 1. Cells are first treated with formaldehyde, which efficiently produces both protein–nucleic acid and protein–protein cross-links. Following whole extract preparation and sonication, the protein of interest (in this case, the putative target protein) is immunoprecipitated. The immunopurified complex contains not only the protein of interest, but also

[4] L. A. Stargell, Z. Moqtaderi, D. R. Dorris, R. C. Ogg, and K. Struhl, *J. Biol. Chem.* **275**, 12374 (2000).

[5] W. Wang, J. D. Gralla, and M. Carey, *Genes Dev.* **6**, 1716 (1992).

[6] Y. Wu, R. J. Reese, and M. Ptashne, *EMBO J.* **15**, 3951 (1996).

[7] K. Melcher and S. A. Johnston, *Mol. Cell. Biol.* **15**, 2839 (1995).

[8] C. J. Jeong, S. H. Yang, Y. Xie, L. Zhang, S. A. Johnston, and T. Kodadek, *Biochemistry* **40**, 9421 (2001).

[9] F. Gonzalez, A. Delahodde, T. Kodadek, and S. A. Johnston, *Science* **296**, 548 (2002).

[10] C. Chang, F. Gonzalez, B. Rothermel, L. Sun, S. A. Johnston, and T. Kodadek, *J. Biol. Chem.* **276**, 30956 (2001).

[11] S. S. Koh, A. Z. Ansari, M. Ptashne, and R. A. Young, *Mol. Cell* **1**, 895 (1998).

[12] N. Yudkovsky, C. Logie, S. Hahn, and C. L. Peterson, *Genes Dev.* **13**, 3269 (1999).

[13] C. E. Brown, L. Howe, K. Sousa, S. C. Alley, M. J. Carrozza, S. Tan, and J. L. Workman, *Science* **292**, 2333 (2001).

[14] S. R. Bhaumik and M. R. Green, *Genes Dev.* **15**, 1935 (2001).

[15] E. Larschan and F. Winston, *Genes Dev.* **15**, 1946 (2001).

[16] V. Orlando and R. Paro, *Cell* **75**, 1187 (1993).

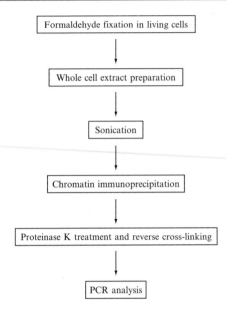

Fig. 1. Schematic outline for the formaldehyde-based *in vivo* cross-linking and chromatin immunoprecipitation (ChIP) assay.

other associated proteins and DNA. The premise behind this approach is that the target protein will be associated with a particular DNA sequence through its interaction with the activator bound there; in other words, the association of the target with DNA serves as a readout for its ability to interact with the activator. The ability of the target protein to associate with a DNA sequence is determined by polymerase chain reaction (PCR) analysis on the immunoprecipitated DNA. Prior to PCR analysis, the DNA is released from the cross-linked material by extensive digestion with proteinase K and mild heat treatment and is then purified by standard methods. A key advantage of using formaldehyde is that the cross-links are fully reversible. The immunoprecipitated DNA is then amplified by radioactive PCR using primer pairs flanking the activator binding site of interest. The PCR products are analyzed by native polyacrylamide gel electrophoresis followed by autoradiography.

A typical set of results is shown schematically in Fig. 2. The immunoprecipitated DNA should be PCR amplified in parallel with the control input (nonimmunoprecipitated) DNA so that the relative association of the target protein(s) with the DNA can be assessed quantitatively. The intensity of the signal is used to determine if a protein is a true *in vivo* target: a bona fide target should give a signal comparable to that obtained with the

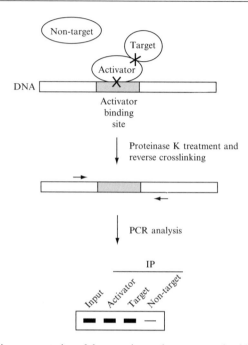

FIG. 2. Schematic representation of the experimental strategy used to identify the targets of activators *in vivo*. Formaldehyde treatment generates cross-links between the activator and its DNA-binding site and between the activator and its target protein. Immunoprecipitation is performed using antibodies against the activator itself and the putative target protein(s). Following immunoprecipitation, the immunoprecipitated DNA is analyzed by PCR using primer pairs flanking the activator-binding sites. As a control, input (nonimmunoprecipitated) DNA is also analyzed. A protein that is a bona fide *in vivo* target will be associated with the DNA via its interactions with the activator and will generate a strong PCR signal, whereas a protein that is not a true target (nontarget) will generate a signal of greatly reduced intensity.

activator itself, whereas a protein that is not a true target will associate with the promoter DNA at a background level. Typically, the association of the activator with an unrelated DNA sequence (e.g., the open reading frame of a gene) is used to monitor background binding levels.

Parameters to Consider when Designing the Experiment

DNA Substrate

The activator-binding site amplified in the PCR reaction can either be the endogenous binding site (i.e., the UAS in a promoter of a gene) or be an artificial binding site harbored on a plasmid. Monitoring the association of a target protein at the endogenous locus provides valuable information about the activator–target interaction at a physiologically relevant location. In this strategy, it is best to study the promoter of a highly expressed gene, as the promoter occupancy of transcriptional components will also be high, which will translate into a robust PCR signal. However, one disadvantage of using an endogenous promoter is that it is possible that other transcription components, such as GTFs, may mediate—or at least contribute to—the interaction between the activator and the putative target. Thus, it is not possible to ascertain whether the activator–target interaction is direct or indirect using an endogenous promoter.

One way to address this caveat is to monitor the ability of the target protein to be recruited by the activator to a plasmid bearing only the binding sites for the activator and no other promoter elements. This is done using only the activation domain of the activator of interest, which is typically fused with either the Gal4p or the LexA DNA-binding domain. The plasmid expressing the fusion protein is then transformed into cells carrying plasmid-borne Gal4p or LexA DNA-binding sites. The endogenous *GAL4* gene of the host cell should be deleted if the activation domain is fused with the Gal4p DNA-binding domain. The immunoprecipitated and input DNA are amplified using primer pairs flanking either the Gal4p or the LexA DNA-binding sites. Alternatively, the Gal4p or LexA DNA-binding sites can be integrated into the genome; however, they should be placed at a location far removed from any endogenous promoter elements (e.g., in an open reading frame).

Formaldehyde Concentration and Cross-Linking Time

The formaldehyde concentration and cross-linking time should be balanced for optimal fixation of proteins to chromatin. Generally, treatment with 1% formaldehyde for 15 min results in efficient cross-linking. Formaldehyde is a protein-denaturing agent that disrupts secondary and, in particular, tertiary structures, resulting in protein unfolding. The general sensitivity of proteins to formaldehyde *in vivo* should be determined empirically. Immunolocalization experiments can be used to determine a suitable concentration: formaldehyde concentrations that are too high will denature proteins and thus disrupt the interaction between the antibody and the protein of interest, resulting in loss of the fluorescence signal.

Overcross-linking can lead to excessive chemical modification of lysines and other preferred formaldehyde-reactive sites, resulting in partial or complete loss of antigen epitopes on the protein of interest. Furthermore, overfixed cells are refractory to sonication, which can result in the loss of up to 90% of the starting material. A time course experiment should be performed to obtain the optimal cross-linking time.

Antibodies

The use of affinity-purified antibodies is highly recommended, and polyclonal antibodies are preferred to monoclonals to avoid potential epitope-masking problems. Alternatively, the protein of interest can be epitope tagged at the C terminus in the genome by homologous recombination to avoid the time-consuming task of generating an antibody against the protein of interest. The experimental strategy and protocol for myc or HA epitope tagging the 3' end of a gene in the chromosome locus are described elsewhere.[17] The amount of antibody added to optimize the immunoprecipitation reaction should be determined empirically.

Immunoprecipitation Conditions

The ChIP assay described here uses highly stringent conditions: the immunoprecipitation and wash steps are carried out in a buffer of intermediate to high ionic strength and containing a combination of detergents. Under these conditions, some antibodies may work suboptimally or not at all, potentially leading to a negative result in the ChIP assay. Thus, the optimal immunoprecipitation conditions for the antibody should be determined experimentally. This can be achieved by changing the buffer and/or immunoprecipitation conditions and by analyzing the immunoprecipitate for the protein of interest by immunoblot analysis.

Protocol for Formaldehyde-Based In Vivo Cross-Linking and ChIP Assay

This protocol is used in our laboratory to study the association of various transcription factors with DNA in yeast. The protocol for the ChIP assay in mammalian cells can be found elsewhere.[18,19]

1. Grow a 50-ml yeast culture up to an OD_{600} of 1.0. Add 1.5 ml of 37% (v/v) formaldehyde (Sigma; to a final concentration of 1%) to the growing yeast culture. Fix the cells for 15 min at room temperature with occasional shaking.

[17] M. S. Longtine, A. McKenzie III, D. J. Demarini, N. G. Shah, A. Wach, A. Brachat, P. Philippsen, and J. R. Pringle, Yeast 14, 953 (1998).
[18] J. Wells and P. J. Farnham, Methods 26, 48 (2002).
[19] A. S. Weinmann and P. J. Farnham, Methods 26, 37 (2002).

2. Quench the cross-links by adding 2.5 ml of 2.5 M glycine (to a final concentration of 125 mM) to the culture. Incubate for 5 min at room temperature with occasional shaking.

3. Harvest the cells by centrifugation at 3500 rpm for 5 min at 4° (Beckman J-6B). Discard the supernatant and resuspend the cells in 2 ml ice-cold TBS (TE, pH 8.0, and 150 mM NaCl). Transfer equal amounts of the cell suspension into two 1.5-ml Eppendorf tubes. Spin down the cells at 13000 rpm for 2 min at 4° and discard the supernatant. Resuspend the cells in each tube in 1 ml ice-cold TBS and centrifuge again at 13,000 rpm for 2 min at 4° and then discard the supernatant. At this stage, the cell pellet can be frozen in liquid nitrogen and stored at −80°.

4. Resuspend the cell pellet in each tube in 200 μl ice-cold FA lysis buffer (50 mM HEPES, pH 7.5; 140 mM NaCl; 1 mM EDTA; 1% Triton X-100; and 0.1% Na-deoxycholate) containing a cocktail of protease inhibitors [1 mM phenylmethylsulfonyl fluoride (PMSF); 1 mM benzamidine; 25 μg/ml tosyl lysyl chloromethylketone (TLCK); 50 μg/ml tosyl phenylalanyl chloromethylketone (TPCK); 10 μg/ml aprotinin; 20 μg/ml antipain; 1 μg/ml leupeptin; and 1 μg/ml pepstatin). Add an equal volume of spherical-shaped acid-washed glass beads (425–600 μm, Sigma) and vortex the mixture at maximum speed for 30 min at 4° using the TOMY micro tube mixer MT-360 (Peninsula Laboratories, Inc.) or equivalent. Alternatively, samples can be vortexed six times (30 s each) on a vortexer at 4°; in between vortexes, let the samples sit on ice for at least 2 min to avoid overheating the samples.

5. Punch a small hole in the bottom of an Eppendorf tube with a 21G1$^{1/2}$ needle (Becton Dickinson) and collect the lysate in a fresh Eppendorf tube by brief centrifugation. Combine the lysates from both Eppendorf tubes into one tube. The final volume should be approximately 500 μl (400 μl FA lysis buffer + cell volume).

6. Sonicate the lysate using a microtip probe (output = 4) five times for 10 s in a Misonix sonicator (XL2020; Misonix Inc.). To avoid overheating the samples, let the samples sit on ice for at least 2 min in between 10-s pulses, and before each pulse, quickly dip the lower half of the Eppendorf tube in a dry ice–ethanol solution. This protocol should shear the chromatin to a final average size of 500 bp, which should be confirmed by agarose gel electrophoresis following the reverse cross-linking step (after step 14). The desired length of the sheared DNA can be obtained by changing the number of times the sample is sonicated.

7. Centrifuge the sonicated lysate at 13,000 rpm for 5 min at 4° and collect the supernatant (approximately 400 μl) in a fresh 1.5-ml Eppendorf tube. Set aside 5 μl of the lysate for analysis of the input DNA control sample (step 12).

8. Add 100 μl lysate to 300 μl FA lysis buffer containing the protease inhibitors. Add the primary antibody. Generally, we use 5 μl α-myc or α-HA antibodies, 2 μl α-RNA polymerase II antibody, and 3 μl of TBP and TAF polyclonal antibodies.[14,20,21]

9. Incubate with slow rotation the lysate–antibody mixture for 3 h at 4° and then add 50 μl of protein A or protein A/G plus agarose (according to the manufacturer's instructions) followed by incubation with slow rotation for 1 h at 4°.

10. Centrifuge the lysate briefly at 3000 rpm for 5 min to collect the agarose beads, remove the supernatant carefully, and wash the beads for 5 min at room temperature once with 1 ml of FA lysis buffer, twice with 1 ml of FA lysis buffer with 1 M NaCl, once with 1 ml of FA-W3 (10 mM Tris–HCl, pH 8.0; 0.25 M LiCl; 0.5% NP-40; 0.5% sodium deoxycholate; and 1 mM EDTA), and finally once with 1 ml of TE (pH 8.0). After the last wash, remove the last traces of TE using a 1-ml syringe affixed with a thin needle.

11. Resuspend the washed beads in 200 μl PK buffer (100 mM Tris–HCl, pH 7.5; 12.5 mM EDTA; 150 mM NaCl; and 1% SDS) and transfer them to a fresh Eppendorf tube containing 3 μl of 10 mg/ml proteinase K. Incubate for 5 h at 37°.

12. For the input DNA sample, add 5 μl of the starting sonicated extract (from step 7) to 180 μl PK buffer. Add 10 μl of 10 mg/ml proteinase K and incubate for 5 h at 55°.

13. To reverse the cross-links, incubate all samples for 6 h at 65°.

14. Purify the DNA by first extracting with 200 μl phenol:chloroform: isoamyl alcohol (25:24:1) and back extracting with 100 μl of a solution containing 50 mM Tris–HCl (pH 8.0) and 0.3 M sodium acetate. Add 10 μg glycogen to the extracted samples and precipitate the DNA by adding 750 μl ice-cold 95% ethanol. Dissolve the immunoprecipitated DNA in 20 μl TE (pH 8.0), and the input DNA in 100 μl TE (pH 8.0). At this point, the samples can be stored at $-20°$.

15. For PCR analysis, use 1 μl of the immunoprecipitated DNA sample or 1 μl of the input DNA sample in a total reaction volume of 25 μl containing 25 pmol primer, 0.2 mM each dCTP, dGTP, dATP and dTTP, and 0.25 μl [α^{32}-P]dATP (10 mCi/ml). Perform PCR amplifications as follows: 94° for 2 min; 23 cycles (94° for 30 s, 50° for 30 s, 72° for 1 min); 72° for 7 min, and hold at 10°. As with all PCR reactions, the annealing temperature can be variable and should be optimized for each primer pair

[20] X. Y. Li, S. R. Bhaumik, and M. R. Green, *Science* **288,** 1242 (2000).
[21] S. R. Bhaumik and M. R. Green, *Mol. Cell. Biol.* **22,** 7365 (2002).

(typically 5° below T_m). In addition, serial dilutions of input and IP DNA should be used to assess the linear range of DNA amplification.

16. To analyze the PCR products, add 8 μl of the PCR reaction to 4 μl DNA-loading buffer (0.25% bromphenol blue, 0.25% xylene cyanol, and 20% glycerol) and analyze on a 6% polyacrylamide gel.

17. Quantitate the immunoprecipitated DNA and input DNA signals by PhosphorImager analysis. The results are typically expressed as the ratio of the signal in the immunoprecipitated DNA sample relative to input.

Protocol for the ChIP Western blot

For the ChIP Western blot experiment, perform the ChIP assay up to the step 10 and then continue with the following steps.

1. Remove the immunoprecipitated material by incubating the beads in 40 μl of 50 mM Tris–HCl (pH 7.5), 10 mM EDTA, 1% SDS for 10 min at 65°.

2. Remove the beads by filtration and then add 10 μl of 5X SDS–PAGE buffer to the eluted material.

3. Boil the sample for 20 min. Analyze the sample by SDS–PAGE and immunoblot analysis using an antibody directed against the protein of interest.

Other Considerations

Although the ChIP technique can be used to identify proteins that associate with an activator, additional experiments are required to verify that the protein of interest is indeed a direct target of the activator. For example, several lines of evidence were used to conclude that SAGA is the direct target of Gal4p.[14] First, SAGA is required for Gal4p to stimulate PIC assembly and transcription. Second, the SAGA complex can be recruited to minimal Gal4p-binding sites in a plasmid by the Gal4p activation domain. Other proposed *in vitro* targets of Gal4p, RNA polymerase II, Srb4p, TFIIB, and TBP, are not recruited, suggesting that these factors are not required to bridge the Gal4p–SAGA interaction. Third, Gal4p interacts directly with a component of the SAGA complex *in vitro*. Finally, mutational and kinetic analysis was also consistent with the hypothesis that SAGA was the direct target of Gal4p. Therefore, the ChIP assay should be used in conjunction with transcriptional, biochemical, mutational, and/or kinetic analyses to identify the direct target(s) of an activator.

Conclusions

The formaldehyde-based *in vivo* cross-linking and ChIP assay has been used to identify *in vivo* targets of the yeast Gal4p activator protein. The identification of multiple *in vitro* targets is a phenomenon that is not unique to Gal4p. For example, the viral activator VP16 has been shown to interact *in vitro* with many factors, including the histone acetyltransferase complexes SAGA and NuA4,[22] the coactivator PC4,[23] components of the mediator complex,[24] TBP,[2,25] and the general transcription factors TFIIA,[26] TFIIB,[3] TFIIF,[27] and TFIIH.[28] Therefore, the formaldehyde-based *in vivo* cross-linking and ChIP assay may be a way to identify bona fide target proteins of other activators, such as VP16 or p53.

Acknowledgments

We thank Sara Evans for editorial assistance. This work was supported in part by an NIH grant to M.R.G. M.R.G. is an investigator and S.R.B. is an associate of the Howard Hughes Medical Institute.

[22] R. T. Utley, K. Ikeda, P. A. Grant, J. Cote, D. J. Steger, A. Eberharter, S. John, and J. L. Workman, *Nature* **394**, 498 (1998).
[23] H. Ge and R. G. Roeder, *Cell* **78**, 513 (1994).
[24] C. J. Hengartner, C. M. Thompson, J. Zhang, D. M. Chao, S.-M. Liao, A. J. Koleske, S. Okamura, and R. A. Young, *Genes Dev.* **9**, 897 (1995).
[25] C. J. Ingles, M. Sales, W. D. Cress, S. J. Triezenberg, and J. Greenblatt, *Nature* **351**, 588 (1991).
[26] N. Kobayashi, T. G. Boyer, and A. J. Berk, *Mol. Cell. Biol.* **15**, 6465 (1995).
[27] H. Zhu, V. Joliot, and R. Prywes, *J. Biol. Chem.* **269**, 3489 (1994).
[28] H. Xiao *et al.*, *Mol. Cell. Biol.* **14**, 7013 (1994).

[39] Dominant-Negative Mutants of Helix-Loop-Helix Proteins: Transcriptional Inhibition

By Vikas Rishi and Charles Vinson

The basic helix-loop-helix (B-HLH) family of transcription factors binds to sequence-specific DNA as homo- or heterodimers to modulate gene expression.[1,2] B-HLH proteins have been found in all eukaryotic

[1] T. D. Littlewood and G. I. Evan, *Protein Profile* **1**, 635 (1994).
[2] M. E. Massari and C. Murre, *Mol. Cell. Biol.* **20**, 429 (2000).

systems examined with over 125 members in the human genome.[3,4] B-HLH proteins regulate the expression of genes involved in cell growth[5] and differentiation of a variety of cell types, including muscle,[6] heart,[7] blood,[8] nerve,[9] and sex determination.[10]

Two general methods are used to determine the biological function of a gene. The first is to mutate the gene, either by changing the DNA sequence by classical mutagenesis or by deleting the gene directly. The gene can be deleted from the germ line, resulting in a gene "knockout,"[11] or from a particular tissue using the Cre/Lox system.[12] The second method, instead of altering the gene or deleting a gene, inhibits its activity. A variety of techniques can be used to achieve such inhibition. These include binding novel transcription factors to the promoter region of a gene to suppress its expression,[13] using RNAi-based strategies to inactivate mRNA function,[14] or using protein-based strategies that function by inactivating the protein product of a gene.[15]

What follows is a protein-based method that inhibits the DNA binding of B-HLH transcription factors.[16] B-HLH proteins contain a domain comprising both the DNA binding and dimerization regions and a transactivation domain that is critical for communicating with the basal transcriptional machinery. The transactivation domain can be either N- or C-terminal of the B-HLH domain. This modular structure facilitates the design of dominant negatives (DNs) to B-HLH proteins.

Traditionally, dominant negatives to dimeric transcription factors have been designed in one of two ways. The first is to produce the B-HLH domain without the transactivation domain.[17,18] Such a DN could bind DNA *in vivo* either as a homodimer or as a heterodimer with endogenous B-HLH proteins. The DN homodimers would bind to DNA but would

[3] J. C. Venter *et al., Science* **291,** 1304 (2001).
[4] V. Ledent, O. Paquet, and M. Vervoort, *Genome Biol.* **3** (2002).
[5] M. Henriksson and B. Luscher, *Adv. Cancer Res.* **68,** 109 (1996).
[6] H. Weintraub, *Cell* **75,** 1241 (1993).
[7] D. Srivastava, P. Cserjesi, and E. N. Olson, *Science* **270,** 1995 (1995).
[8] C. Murre, A. Voronova, and D. Baltimore, *Mol. Cell. Biol.* **11,** 1156 (1991).
[9] J. E. Lee, *Curr. Opin. Neurobiol.* **7,** 13 (1997).
[10] S. M. Parkhurst and P. M. Meneely, *Science* **264,** 924 (1994).
[11] S. L. Mansour, K. R. Thomas, and M. R. Capecchi, *Nature* **336,** 348 (1988).
[12] B. Sauer and N. Henderson, *New Biol.* **2,** 441 (1990).
[13] F. D. Urnov, E. J. Rebar, A. Reik, and P. P. Pandolfi, *EMBO Rep.* **3,** 610 (2002).
[14] A. P. McCaffrey *et al., Nature* **418,** 38 (2002).
[15] I. Herskowitz, *Nature (Lond.)* **329,** 219 (1987).
[16] D. Krylov *et al., Proc. Natl. Acad. Sci. USA* **94,** 12274 (1997).
[17] A. Clark and K. Docherty, *Biochem. J.* **296,** 521 (1993).
[18] S. P. Persengiev, L. R. Devireddy, and M. R. Green, *Genes Dev.* **16,** 1806 (2002).

not activate gene expression because they lack a transactivation domain. Heterodimers between the DN and the endogenous B-HLH protein would retain a transactivation domain from the endogenous B-HLH protein and thus some level of activity may be retained. Even under conditions where this class of DNs does inhibit the transactivation properties of a transcription factor efficiently, it will not inhibit any functions of the endogenous transcription factor that are a consequence of DNA binding but do not require the transactivation domain.

A second method for making a DN to B-HLH proteins is to mutate or delete the DNA–binding region.[19,20] Such DNs heterodimerize with endogenous factors and prevent DNA binding. To function, the DN molecules need to be overexpressed to overcome the significant stabilization that occurs when the native B-HLH dimer binds to DNA.

An alternative method to design DNs that abolish the DNA binding of B-HLH transcription factors in an equimolar competition is described.[16] Instead of mutating or deleting the basic region, it was replaced with an acidic protein sequence to produce the DN. The acidic sequence from the DN interacts with the basic region of B-HLH proteins and stabilizes heterodimerization between the two proteins. When bound to the DN, the B-HLH protein is unable to bind to DNA. The heterodimer between the DN and the B-HLH domain is more stable than the endogenous B-HLH dimer bound to DNA, overcoming the need to overexpress the DN. We refer to these DNs as A-HLH proteins with the A representing the acidic extension. A-HLH DNs heterodimerize with endogenous B-HLH proteins with the dimerization specificity of the HLH domain. A-HLH DNs have been shown to be active in a variety of biological systems.[16,21,22]

Structure of B-HLH Transcription Factors

The primary sequence of the B-HLH domain can be as short as 50 amino acids. Figure 1 presents the amino acid sequences of 12 B-HLH proteins and a consensus protein sequence that helps define the motif. There is a N-terminal cluster of basic amino acids that is critical for sequence-specific DNA binding followed immediately by an amphipathic α–helical sequence termed helix 1. Next comes a loop region of variable length (6 to 20 amino acids) followed by a second amphipathic α helix termed helix 2. Unlike a leucine zipper amphipathic α–helix where only the **a** and **d**

[19] D. MacGregor, L. Li, and E. Ziff, *J. Cell. Physiol.* **167**, 95 (1996).
[20] J. Kim and B. Spiegelman, *Genes Dev.* **10**, 1096 (1996).
[21] Y. Qyang et al., *Mol. Cell. Biol.* **19**, 1508 (1999).
[22] M. D. Galibert, S. Carreira, and C. R. Goding, *EMBO J.* **20**, 5022 (2001).

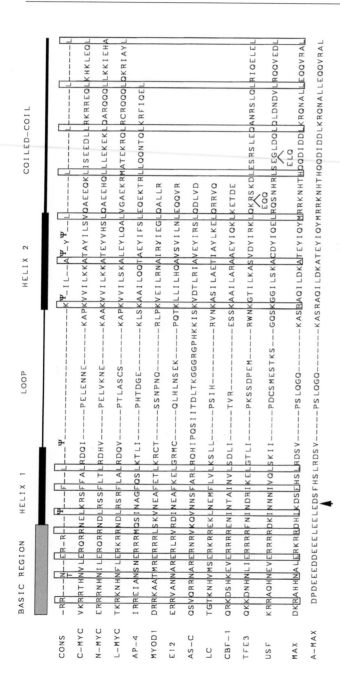

Fig. 1. Amino acid sequences corresponding to the DNA-binding and dimerization domains of seven B-HLH-ZIP and five B-HLH proteins. Polypeptide sequences designated by the one letter code were aligned to generate the most conservations. The top schematic presents the hypothesized boundaries of the different regions of the B-HLH-ZIP motif. The next line presents the consensus (CONS.). The amino acid sequences are from proteins found in vertebrates (MyoD, E12, AP-4, USF, TFE3, c-myc, n-myc, 1-myc, and Max), insects (AS-C), plants (LC), and yeast (CBF-1). The conserved amino acids in the basic helix 1 region are boxed to highlight the three–four repeat. This repeat brings these conserved amino acids to the same side of an α helix. Boxed amino acids in the helix 2–leucine zipper part of the figure correspond to the heptad repeat of hydrophobic amino acids that extends from the leucine zipper into helix 2. These two regions of the motif are separated by a loop region that contains a variable number of amino acids. Amino acid abbreviations are standard and include hydrophobic amino acids.

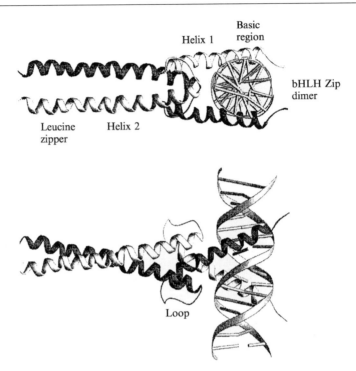

FIG. 2. The crystal structure of dimeric Max B-HLH-ZIP domain bound to DNA seen from two views. The two monomers are presented in different shades of gray.

positions are hydrophobic,[23] helix 2 also contains hydrophobic amino acids in the **e** and **g** positions.[23] In some B-HLH transcription factors, helix 2 continues C-terminally to form a leucine zipper coiled coil structure. These proteins are referred to as B-HLH-ZIP proteins. X-ray structures show that B-HLH and B-HLH-ZIP proteins bound to sequence-specific DNA have a similar overall structure.[24–26] For the purpose of this discussion, these two families are treated as having similar properties.

Figure 2 presents two views of the X-ray structure of the Max B-HLH-ZIP protein bound to DNA.[24] The top view is looking down the DNA and shows the two long α helices in each monomer of the dimer. One α helix is composed of the basic region and helix 1. The second α helix is composed

[23] C. Vinson and K. Garcia, *New Biol.* **4,** 396 (1992).
[24] A. Ferre-D'Amare, G. Prendergast, E. Ziff, and S. Burley, *Nature (Lond.)* **363,** 38 (1993).
[25] A. Ferre-D'Amare, P. Pognonnec, R. Roeder, and S. Burley, *EMBO J.* **13,** 180 (1994).
[26] T. Ellenberger, D. Fass, M. Arnaud, and S. Harrison, *Genes Dev.* **8,** 970 (1994).

of helix 2 and the leucine zipper. Helix 1 and helix 2 from each monomer form a parallel left-handed four-helix bundle in the dimer. Helix 1 interacts with the hydrophobic amino acids in the **e** and **g** positions from helix 2. Helix 2 forms a intermolecular coiled coil with the neighboring helix 2 using hydrophobic amino acids in the **a** and **d** positions to stabilize the interaction. Sequence-specific DNA binding is mediated by the basic region interacting with the major groove of DNA as an α-helical extension of helix 1. B-HLH proteins bind similar DNA sequences, which are variations of a sequence called the E box (5'-CANNTG-3') where N is any base, most commonly either CG or GC. Because of the symmetric nature of the structure, palindromic DNA sequences are often bound but this is not obligatory, even for homodimers.

DNA Binding Stabilizes B-HLH-ZIP Dimers

We have used circular dichroism (CD) spectroscopy, analytical ultracentrifugation, and the electrophoretic mobility shift assay (EMSA) to evaluate the biophysical properties of B-HLH-ZIP proteins and our designed dominant negatives. CD spectroscopy and analytical ultracentrifugation provide complementary data; CD spectroscopy monitors the presence of the α-helical structure, with more negative ellipticity representing a more α-helical structure, and analytical ultracentrifugation monitors the state of oligomerization. We have used CD and analytical ultracentrifugation to monitor structural changes that occur when B-HLH-ZIP proteins interact with each other or with DNA. CD spectra show a cooperative decrease in negative ellipticity with increasing temperature indicative of an α helix to random coil transition. Unlike most proteins, many B-HLH-ZIP proteins remain soluble when denatured and refold when cooled. Because of this reversible denaturation, we can use classical thermodynamic calculations to quantitate the stability of dimerization.

As an example, heterodimerization of the B-HLH-ZIP proteins Myc and Max was analyzed. At 6°, the Max B-HLH-ZIP domain sediments as a dimer in the analytical ultracentrifuge. The CD spectrum has minima at 208 and 222 nm, which are characteristic of an α-helical structure (Fig. 3A). The Myc B-HLH-ZIP domain alone sediments as a monomer and shows no α-helical structure. Mixing equimolar amounts of Myc and Max produces a CD spectrum with more negative ellipticity than the sum of the individual protein spectra, indicating that Myc interacts with Max (Fig. 3A). Analytical ultracentrifugation confirms that proteins in solution sediment dimers. The addition of DNA (a 22-bp double-stranded oligonucleotide containing a single E box) further increases the negative ellipticity in the

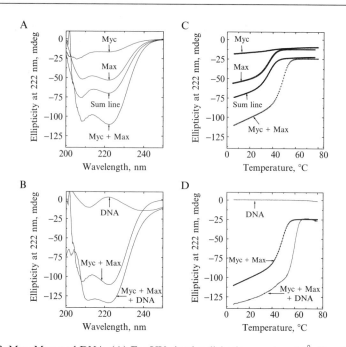

Fig. 3. Myc, Max, and DNA. (A) Far UV circular dichroism spectra at 6°: Myc B-HLH-ZIP domain, Max B-HLH-ZIP domain, and mixture of Myc and Max. The concentration of Myc, Max, and double-stranded DNA was 4.8 μM. The minima at 208 and 222 nm are indicative of an α-helical structure. (B) Far UV circular dichroism spectra at 6°: 22-bp double-stranded DNA containing E box, Myc + Max, and Myc + Max + DNA. The concentration of Myc, Max, and double-stranded DNA was 4.8 μM. The minima at 208 and 222 nm are indicative of an α-helical structure. The increase in ellipticity with the addition of DNA suggests that the Myc and Max basic regions are becoming α helical upon DNA binding. (C) CD thermal denaturation curves at 222 nm of the three samples presented in A. Lines through the Max and Myc + Max denaturation profiles are fitted curves assuming a two-state transition between α-helical dimers and nonhelical monomers. (D) CD thermal denaturation curves at 222 nm of the three samples presented in A. Note how the addition of DNA increases both ellipticity and stability.

CD spectrum, likely due to the basic region becoming α helical upon DNA binding (Fig. 3B).

The stability of the complexes presented in Figs. 3A and 3B can be determined using CD spectroscopy to monitor thermal denaturation at 222 nm (Figs. 3C and 3D). The melting temperature (the temperature at which 50% of the sample is not helical) provides a measure of stability of a particular B-HLH dimer. Max is α helical at low temperature and unfolds to a random coil with a T_m of 36°. In contrast, Myc is always unstructured

(Figs. 3A and 3C). The mixture of Myc and Max (T_m of 48°) is more stable than Max, again suggesting the formation of a Myc/Max heterodimer (Fig. 3C). The addition of DNA stabilizes the Myc/Max complex (Fig. 3D). Thus, to compete efficiently with Myc/Max binding to DNA, a DN to Myc should form a heterodimer with Myc that is more stable than Myc/Max bound to DNA.

Design of the Acidic Extension

Previous work showed that an amphipathic acidic sequence appended onto the N terminus of the dimerization domain of B-ZIP transcription factors produces robust DNs that inhibit B-ZIP DNA binding in an equimolar competition.[27–31] The amphipathic acid extension was designed to form a coiled coil structure with the B-ZIP basic region to stabilize the interaction between the DN and the B-ZIP domain. In applying the same concept to B-HLH proteins, a complication was that one did not know how to place the amphipathic acidic helix relative to helix 1. Next is appending the extension at each of three consecutive amino acid positions, which essentially rotated the amphipathic acidic helix through one helical turn. All three stabilized the interaction with Myc similarly, suggesting that the acidic, and not the amphipathic, property of the sequences was critical for DN properties. An acidic amino acid extension (EEEDDEEELEELE) proved to be a more efficient DN than any of the three orientations of the amphipathic acidic helix. The logic for the particular amino acid sequence is simple. Glutamic acid was used as the acidic residue, two aspartic acids were included to make the DNA sequence less repetitive and to prevent any potential cloning problems, and the LE sequence in bold is a *Xho*I site that was included for cloning purposes. Figure 1 shows the protein sequence of A-Max in which the acidic extension is appended onto the N terminus of the HLH domain of Max.

The acidic extension can interact with the basic regions from several B-HLH proteins, suggesting that it can be used in DNs designed against any B-HLH dimerization domain. Figure 4A presents CD–monitored thermal denaturations of Max, A-Max, and the mixture. A-Max is more stable than Max. More importantly, the mixture of A-Max and Max is more stable than the sum of their individual spectra, suggesting that they

[27] D. Krylov, M. Olive, and C. Vinson, *EMBO J.* **14**, 5329 (1995).
[28] M. Olive et al., *J. Biol. Chem.* **272**, 18586 (1997).
[29] S. Ahn et al., *Mol. Cell. Biol.* **18**, 967 (1998).
[30] C. Vinson et al., *Mol. Cell. Biol.* **22**, 6321 (2002).
[31] J. R. Moll, M. Olive, and C. Vinson, *J. Biol. Chem.* **275**, 34826 (2000).

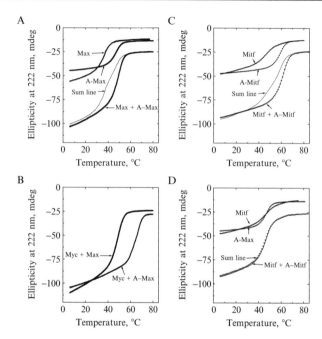

FIG. 4. The 12 amino acid acidic extension interacts with the basic regions from three B-HLH-ZIP domains. Circular dichroism (CD) thermal denaturations at 222 nm of Max, Mitf, and Myc. (A) Max, A–Max, and their mixture. The fourth curve is the sum of Max and A–Max curves, a curve that is expected if the two proteins do not interact. The fact that the mixture is more stable than the sum curve indicates that Max and A–Max heterodimerize. (B) Myc + Max and Myc + A-Max. (C) Mitf, A–Mitf, and their mixture. The fourth curve is the sum of Mitf and A–Mitf curves. (D) Mitf, A–Max, and their mixture. The fourth curve is the sum of Mitf and A–Max curves. The sum curve and the actual mixture melt superimpose, indicating that the two proteins do not interact.

form A-Max/Max heterodimers that are more stable than either homodimer. There is no increase in negative ellipticity in the mixture compared to the sum of the individual proteins, which suggests that the interaction between the acidic extension and the basic region does not result in the formation of an α-helical structure. Analytical ultracentrifugation confirms that the mixture of A-Max and Max forms a dimer in solution. Figure 4B presents the CD thermal denaturation of A-Max/Myc, which is more stable than Max/Myc. Thus the acidic extension from A-Max can also interact with the basic region of Myc. Again, there is no increase in negative ellipticity, indicating that the interacting region does not become α helical.

Interaction between acidic extension and basic region of Mitf, a member of the TFE subfamily of B-HLH-Zip protein, was also tested.

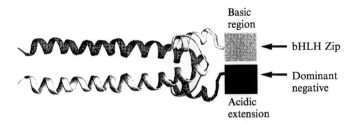

Basic
region

bHLH Zip

Dominant
negative

Acidic
extension

Fig. 5. Schematic of a heterodimer between a B-HLH-ZIP protein and an A-HLH-ZIP DN. The structure of the acidic extension and the basic region is not α helical and is presented as square shapes.

The acidic extension appended onto the N terminus of helix 1 of Mitf (A-Mitf) stabilized the interaction with the Mitf B-HLH-ZIP domain (Fig. 4C). Thus, the acidic extension interacts with all three basic regions examined, stabilizing the heterodimeric complex by between 3.5 and 6.3 kcal/mol. As a control for the specificity of the dimerization, Mitf was mixed with A-Max (Fig. 4D). The experimental thermal denaturation was superimposable with the sum of the separate denaturations of Mitf and A-Max, suggesting that these two proteins do not heterodimerize.[32] Thus the interaction between the acidic extension and the basic region cannot overcome the repulsion between the dimerization domains of Mitf and Max. This illustrates that the dimerization specificity of the A-HLH DNs is dominated by the HLH dimerization domain. This is important because it means that the DNs can heterodimerize with specific B-HLH proteins. A schematic of the DN interacting with the B-HLH-ZIP domain is shown in Fig. 5. Figure 6 presents CD thermal denaturations showing that Myc + A-Max is more stable than Myc + Max + DNA.

A-HLH DNs Inhibit DNA Binding of B-HLH Proteins in a
    Dimerization-Specific Manner

Electrophoretic mobility shift assays were used to examine whether the designed DNs would inhibit the DNA binding of B-HLH-ZIP proteins effectively, Figure 7 presents three B-HLH-ZIP proteins bound to an E box DNA sequence. Max, USF, and Mift at 10 nM all bind this DNA sequence. A-Max inhibits the DNA binding of Max at 1 molar equivalent, but does not inhibit the DNA binding of USF or Mitf even at 100 molar equivalents. Similarly, A-USF and A-Mitf specifically inhibit the DNA binding of USF and Mitf, respectively, at 1 molar equivalent but do not inhibit the DNA binding of other B-HLH proteins, even at 100

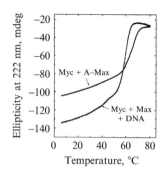

FIG. 6. CD thermal denaturation curves at 222 nm of Myc + Max + DNA and Myc + A-Max.

molar equivalents. These results demonstrate that A-HLH-ZIP DNs act in a dimerization-specific manner and highlight the importance of understanding the dimerization specificity of B-HLH proteins in order to predict which B-HLH proteins may be inhibited *in vivo*. This is particularly true for the heterodimerizing B-HLH proteins. For example, Max homodimerizes and heterodimerizes with the B-HLH families of Myc and Mad. Thus, A-Max is expected to inhibit the function of Myc, Mad, and Max proteins.

A-HLH DNs can be used in *in vitro* experiments, as shown in Fig. 7, to determine the dimerization specificity of different B-HLH domains, as has been done for B-ZIP proteins.[30] These experiments provide insight into expected partners *in vivo*. Phylogenies of B-HLH proteins that highlight amino acid relatedness and, by inference, suggest that related proteins have related dimerization properties can be used as a starting point to examine B-HLH dimerization specificity.[33,34] As one learns more about the dimerization properties of B-HLH proteins, one will be able to design DNs that inhibit different sets of B-HLH proteins.

Biological Uses of Dominant Negatives

Biological assays, examining either transactivation or transformation, have shown that A-HLH-ZIP proteins are more potent DNs than either B-HLH-ZIP or HLH-ZIP DNs.[16] The ability of the A-HLH to inhibit

[32] J. A. Zitzewitz, B. Ibarra-Molero, D. R. Fishel, K. L. Terry, and C. R. Matthews, *J. Mol. Biol.* **296**, 1105 (2000).

[33] W. R. Atchley, K. R. Wollenberg, W. M. Fitch, W. Terhalle, and A. W. Dress, *Mol. Biol. Evol.* **17**, 164 (2000).

[34] W. R. Atchley and W. M. Fitch, *Proc. Natl. Acad. Sci. USA* **94**, 5172 (1997).

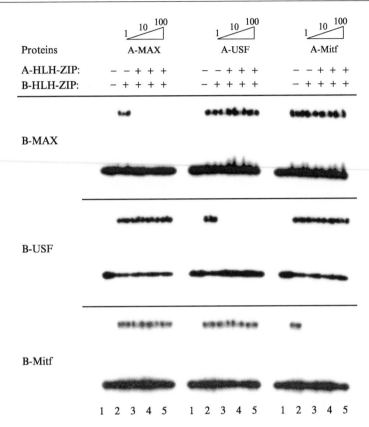

FIG. 7. B-HLH-ZIP DNA binding is inhibited by specific A-HLH-ZIP–dominant negatives. 10 nM of three B-HLH-ZIP dimers, the Max, USF, and Mitf homodimers are bound to a 28-bp DNA containing an E-box sequence. 1, 10, or 100 molar equivalents of the A-HLH-ZIP is added to the B-HLH-ZIP before the addition of DNA and the solution is electrophoresed on an EMSA. The sequence of the 28-mer DNA probes is shown with the consensus binding sites in bold: GTCAGTCAGGC**CACGTG**AGATCGGTCAG (E box).

DNA binding of B-HLH proteins in a dimerization-specific manner can be used to explore several biological issues. Families of B-HLH often have very similar B-HLH domains and might have redundant functions. Thus if a specific B-HLH gene is deleted, its function may be masked because related B-HLH proteins will compensate for its loss. Because the A-HLH DNs described here inhibit all B-HLH proteins that share a similar B-HLH domain, they are able to inhibit the gene products of several genes with overlapping functions. This broad activity is a particular advantage in vertebrate systems that have many families of related B-HLH proteins with potentially redundant functions.

B-HLH proteins bind similar DNA sequences with a consensus sequence known as the E box (5'-CANNTG-3'). However, it is not known whether different B-HLH dimers compete to bind the same DNA-binding site.[35] The A-HLH DNs described here should help unravel not only B-HLH-binding partners, but also the specificity of DNA binding. To determine DNA occupancy, chromatin immunoprecipitation assays can be used.[36–39] These assays cross-link proteins to DNA *in vivo* and then determine which DNA sequences are immunoprecipitated. By expressing A-HLH DNs *in vivo*, it is possible to sequester a particular B-HLH protein and prevent DNA binding. The CHiP assay could then be used to determine if a new B-HLH protein binds to DNA occupied previously by the target of the A-HLH DN. Alternative binding would suggest competition by B-HLH proteins for the same DNA sequence *in vivo*.

This dominant-negative approach combines genetics and pharmacologies to give a powerful research tool. One can use the paradigms of genetics to help unravel the functions of gene in biology but one can also use the tenets of pharmacology to understand if any of the phenotypes generated are clinically relevant. If so, we know that the target of the A-HLH DN is a potential molecular target. One can potentially design classical drugs with A-HLH-like properties to recapitulate the phenotype observed with A-HLH DNs.

[35] X. Luo and M. Sawadogo, *Proc. Natl. Acad. Sci. USA* **93,** 1308 (1996).
[36] V. Orlando and R. Paro, *Cell* **75,** 1187 (1993).
[37] B. Ren *et al., Genes Dev.* **16,** 245 (2002).
[38] A. S. Weinmann, P. S. Yan, M. J. Oberley, T. H. Huang, and P. J. Farnham, *Genes Dev.* **16,** 235 (2002).
[39] B. Ren *et al., Science* **290,** 2306 (2000).

[40]   Purification and Transcription Repression by Negative Cofactor 2

By S. Gilfillan, G. Stelzer, E. Kremmer, and M. Meisterernst

Negative cofactor 2 (NC2 also termed Dr1/DRAP) binds to promoters and inhibits transcription by RNA polymerase II by preventing formation of a *bona fide* preinitiation complex.[1,2] Biochemical and genetic data suggest that NC2 binds directly to TBP and to DNA and competes with the general transcription factors TFIIA and TFIIB.[3,4] The molecular details have been resolved at high resolution through determination of the crystal structure of the NC2–TBP–DNA complex.[5] NC2 dimerizes through histone fold domains of the H2A-H2B type. As shown in Fig. 1, NC2 is situated underneath the DNA surface contacting both DNA and TBP. The inhibitory complex contacts TBP from two sides via contacts with both subunits. Loop region 1 of NC2α in the helix 1–loop 1–helix 2–loop 2–helix 3 arrangement of the histone fold and helix 5 within the carboxy-terminal region of NC2β interact directly with TBP. The structure satisfactorily explains repression by the dimeric NC2 complex through sterical occlusion of both TFIIA and TFIIB.[1,2,4] It also provides a molecular model for a stimulatory function of NC2 through the histone fold in that it can, in principle, support binding of TBP to promoters.

NC2 is thought to be involved in the regulation of many genes. It is possible that it attenuates transcription in a global manner. However, evidence also shows that NC2 promoter occupancy is controlled by upstream regulatory pathways. This article focuses on the human NC2 complex, provides methods for its isolation from HeLa cells, its expression in *Escherichia coli*, and its analysis in DNA–TBP binding and in transcription, as well as its detection on endogenous genes through chromatin immunoprecipitation (ChIP). While there is no evidence for a related factor in prokaryotes, NC2 is conserved from yeast to humans. Thus, the methods are applicable to other mammalian cell lines and, at least in part, to other eukaryotic organisms. The ChIP techniques described here will allow tracking and characterization of NC2 effects on specific genes *in vivo*. They will further allow determination of the relative amount of NC2 on genes and its control by regulatory factors.

[1] M. Meisterernst and R. G. Roeder, *Cell* **67,** 557 (1991).
[2] J. A. Inostroza, F. H. Mermelstein, H. Ilho, W. S. Lane, and D. Reinberg, *Cell* **70,** 477 (1992).
[3] Y. Cang, D. T. Auble, and G. Prelich, *EMBO J.* **18,** 6662 (1999).
[4] J. Xie, M. Collart, G. Stelzer, M. Lemaire, and M. Meisterernst, *EMBO J.* **19,** 672 (2000).

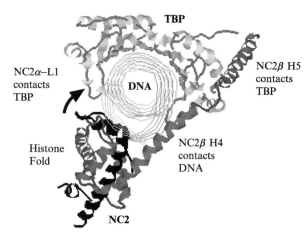

NC2α–L1 contacts TBP

DNA

NC2β H5 contacts TBP

Histone Fold

NC2β H4 contacts DNA

NC2

Fig. 1. NC2 dimerizes through histone fold domains and binds underneath the bent DNA–TBP complex.[5]

Purification of NC2 from HeLa Nuclear Extract

NC2 was originally defined as a factor that forms a stable complex with TBP in gel mobility shift experiments.[1] While NC2 is sometimes difficult to detect in nuclear extracts of mammalian cells, the complex is readily detectable upon fractionation on phosphocellulose (see Fig. 2). Both NC2α and NC2β subunits cofractionate on P11 and subsequent columns.[6] It has also been reported that Dr1 (NC2β) binds DNA–TBP in the absence of NC2α (DRAP).[2] This is usually not detected in crude fractions where carrier DNA is included in the gel shift experiments.

Materials

  Phosphocellulose (P11, Whatman)
  DE52 (Whatman)
  DEAE-Sephadex (A 25, Amersham Pharmacia Biotech)
  Mono Q (Amersham Pharmacia Biotech)

Solutions

  BC buffer: 20 mM Tris (pH 7.3 at 25°), 20% glycerol, 1 mM phenyl-methylsulfonyl fluoride (PMSF), 5 mM dithiothreitol (DTT), 1 mM EDTA, pH 8.0, 100 mM KCl (BC100), 200 mM KCl (BC200), etc.

[5] K. Kamada, F. Shu, H. Chen, S. Malik, G. Stelzer, R. G. Roeder, M. Meisterernst, and S. K. Burley, *Cell* **106,** 71 (2001).
[6] A. Goppelt, G. Stelzer, F. Lottspeich, and M. Meisterernst, *EMBO J.* **15,** 3105 (1996).

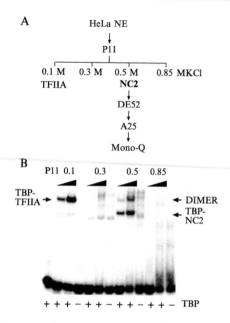

FIG. 2. (A) Purification chart for NC2 isolated from HeLa nuclear extracts and fractionated on phosphocellulose (P11, Whatman). (B) EMSA using a 35-bp adenovirus major late promoter fragment assaying HeLa extracts fractionated on P11 fractions where NC2 elutes exclusively in the 0.5 M KCl fraction.

Procedure

1. Prepare HeLa nuclear extract from 8×10^{10} cells.[7]
2. Dialyze the nuclear extract (280 ml: approximately 10 mg/ml) against BC100 buffer and load onto a preequilibrated P11 column (capacity 10 mg/ml column volume) with a flow rate of 1 column volume per hour. Approximately 50% of the protein will be in the flow-through fraction.
3. Wash column with BC100 (10 column volumes overnight).
4. Elute the column stepwise with BC300 (approximately 30% of total protein amount), BC500 (approximately 5% of total protein amount), and BC850 (approximately 2.5% of total protein amount).
5. NC2 elutes in the 0.5 M KCl fraction. Dialyze against BC100 and load on a DE52 column (2 mg/ml resin). Flow rate is approximately 1 column volume per hour. Wash with BC100 and elute with a linear KCl gradient (0.1 to 0.5 M KCl in BC buffer).

[7] M. Meisterernst, A. L. Roy, H. M. Lieu, and R. G. Roeder, Cell 66, 981 (1991).

6. NC2 elutes at 0.18 to 0.29 M KCl; pool fractions and dialyze against BC100. Load onto a DEAE-Sephadex (A25) column and elute with a linear gradient from 0.1 to 0.5 M KCl in BC buffer.
7. Pool the fractions in the range of 0.17 to 0.28 M KCl; dialyze against BC200 and load on a Mono Q column. Wash the column with 10 column volumes of BC200 and elute directly with BC400.
8. Analyze eluted fractions by SDS–PAGE.

Analysis of NC2 in the Electrophoretic Mobility Shift Assay (EMSA)

Annealed oligonucleotides or DNA fragments isolated from a vector can be used to analyze NC2–TBP and NC2 alone in EMSA. NC2 is a non-specific DNA-binding protein and will thus also bind DNA in the absence of TBP.[6] The addition of TBP shifts the complex to a lower mobility form that is more stable and more resistant to the addition of carrier DNA such as p[dGdC]. However, the amount of carrier must be determined carefully through titration. NC2 does not bind single-stranded DNA and RNA.

Solutions/Reagents

Annealing buffer: 200 mM NaCl, 10 mM Tris, pH 7.3, 1 mM MgCl$_2$
MicroSpin G-25 columns (Amersham Pharmacia)
Bovine serum albumin (BSA) (New England Biolabs)
Klenow fragment (MBI Fermentas)
[α-^{32}P]dNTP (3000 Ci/mmol, Amersham Pharmacia)
TGE: 25 mM Tris, pH 7.3, 248 mM glycine, 1 mM EDTA, pH 8.0

Annealing Oligonucleotides

1. Measure the concentration of oligonucleotides by A_{260}.
2. Mix oligonucleotides in annealing buffer (100 pmol of each oligonucleotide in a total volume of 20 μl) and transfer to a heat block at 95°. Turn the heat block off after 3 min and let the sample cool gradually to room temperature.
3. The DNA can be stored at 4° and labeled by Klenow fill-in by standard procedures.

Procedure for EMSA

The binding reactions for EMSA are 4 mM MgCl$_2$, 25 mM HEPES–KOH, pH 8.2, 0.4 mg/ml BSA, 5 mM DTT, 0.5 mM PMSF, 1–5 fmol/μl labeled DNA. The final concentration of glycerol is kept between 6 and 10%, and the final concentration of KCl is typically 70 mM. The binding reaction is carried out in a final volume of 20 μl.

1. Distribute the components for the binding reaction to each sample.
2. Add proteins; TBP is added last.
3. Mix gently by pipetting up and down.
4. Incubate at 27° for 30 min.
5. Load samples on a 5% gel (acrylamide:bisacrylamide, 50:1). The gel is prerun 1 h at 120 V in TGE, and the TGE buffer is changed after the prerun. The gel usually runs 3 h at 100 V with a 35-bp DNA.
6. Place the gel on 3MM Whatman paper and dry it on a gel dryer.
7. Expose the gel to a film (BioMax MR, Kodak), typically 1 h at −80° and/or quantify the bands using a phosphoimager.

Expression and Purification of Recombinant NC2 in *E. coli*

To test the influence of NC2 on transcription, the two subunits of NC2 can be expressed individually in *E.coli*. This usually results in low levels of NC2α. In contrast, NC2β is, under certain conditions, overexpressed greatly, but the protein is insoluble and found in inclusion bodies. When coexpressed, both subunits can be purified under native conditions. We used the pET11d expression vector and NC2 subunits that both harbor a hexa-histidine tag at their N termini (T7 system from Novagen).

Materials and Reagents

Escherichia coli BL21(DE3) pLys S (Novagen)
Isopropyl-β-D-thiogalactoside (IPTG, Roth)
Branson digital sonifier
Columns (0.5 ml, Bio-Rad)
Peristaltic pump P-1 (Amersham Pharmacia Biotech)
Ni^{2+}-NTA resin (Qiagen)
Proteins assay dye reagent concentrate (Bio-Rad)
Heparin Sepharose (Amersham Pharmacia Biotech)
BSA (Roche Diagnostics GmbH)

Solutions

Isopropyl-β-D-thiogalactoside, 0.5 M stock solution in water, sterile filtered
Imidazole hydrochloride: 1 M stock, adjust pH to 7.9 with NaOH
Lysis buffer: 500 mM NaCl, 10% glycerol, 20 mM Tris–HCl, pH 7.3. Add 2-mercaptoethanol (to 5 mM), IGEPAL CA-630 (NP-40) to 0.1%, leupeptin (to 2 μg/ml), pepstatin A (to 1 μg/ml), benzamidine (to 20 μg/ml), and PMSF (to 1 mM) freshly
HEPES–KOH 1 M stock (pH 7.9 at 25°)

Procedure

1. Transform *E. coli* BL21(DE3) pLysS with the vector-containing histidine-tagged NC2 and select on LB plates containing ampicillin and chloramphenicol.
2. Inoculate a 50-ml starter culture of LB/ampicillin/chloramphenicol and grow overnight.
3. Dilute from the starting culture into 500 ml LB/ampicillin/chloramphenicol (to OD_{600} 0.1) and grow with vigorous shaking at $37°$.
4. Add IPTG to 0.5 mM final concentration when OD_{600} is 0.6.
5. Grow the cells for 3 h at $30°$ with shaking.
6. Centrifuge the cells for 5 min at 5000 g.
7. Resuspend pellet in 20 ml lysis buffer (the resuspended bacteria can be frozen and stored at $-80°$). All remaining steps are performed at $4°$ with prechilled buffers.
8. Sonicate the sample on ice: amplitude 20% for 3 min; on 0.2 s, off 0.8 s (i.e., total elapsed time 15 min).
9. Spin for 30 min in SS34 rotor at 18,000 rpm (39,000 g).
10. Add imidazole to the supernatant to a final concentration of 5 mM and load onto a preequilibrated Ni^{2+} column (0.8 ml) using peristaltic pump P1 (flow rate: 5 ml/h).
11. Wash the column with 50 ml of BC400 including 10 mM imidazole (can be done overnight).
12. Elute the recombinant NC2 dimer with 8 ml 100 mM imidazole in BC400 (flow rate 4 ml/h) and collect 1-ml aliquots.
13. Analyze an aliquot from each fraction in a 1:10 dilution with a protein assay dye reagent concentrate from Bio-Rad (diluted 1:5 and filtered through 3MM Whatman paper).
14. Pool NC2-containing fractions and adjust salt concentrations to BC200 using BC0 buffer. Load with a peristaltic pump P1 onto a heparin column (0.5 ml, preequilibrated) at a flow rate of 4 ml/h.
15. Wash column with 30 ml BC200 (can be done overnight).
16. Elute NC2 with 6 ml BC600 (flow rate 4 ml/h) and collect 0.5-ml aliquots.
17. Measure protein concentration and adjust to at least 500 ng/μl with BSA. Usually, this protocol yields approximately 0.3 mg of recombinant NC2 proteins.
18. Add HEPES–KOH (pH 7.9 at $25°$) to 50 mM.
19. Aliquots are frozen in liquid nitrogen and stored at $-80°$.

Immunoprecipitation of NC2 Complexes

NC2 complexes are stable in solution and can be precipitated with antibodies that recognize either the α or the β subunit (see Fig. 3).[6] Antibodies are not yet available commercially. Hence, a stably expressed version of NC2 carrying a tag may be used instead. Here we used monoclonal antibodies directed against human NC2α. The antibody coimmunoprecipitates α and β both of recombinant and native protein complexes isolated from logarithmically growing HeLa cells. The complex is held together, at least in part, by hydrophobic interactions, which manifest in great salt resistance in the IPs (Fig. 3A) but dissociation by RIPA buffer.

Coupling of Antibody to Beads

All centrifugation steps are done for 4 min at 200 g.

1. Incubate 2 mg of NC2α monoclonal antibody (from rat 4G7) with 1 ml of protein G beads for 2 h at room temperature; collect beads by centrifugation.

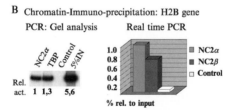

FIG. 3. (A) NC2 forms a stable complex in solution. (Top) IP and wash conditions. RIPA buffer contains 150 mM NaCl, 1% NP-40 (IGEPAL CA-630), 0.1% SDS, and 50 mM Tris–HCl (pH 8.0). BC buffer: 20 mM Tris–HCl (pH 7.3 at real time), 20% glycerol, 1 mM EDTA, and 150 mM KCl (BC150), 300 mM KCl (BC300), 500 mM KCl (BC500), or 850 mM KCl (BC850). (B) Chromatin IP using a monoclonal antibody against NC2α and polyclonal antibodies against TBP and NC2β, respectively. The control was performed with an isotype antibody of the rat monoclonal NC2α. The real-time PCR reaction is quantified by serial dilutions of input material. Radioactive PCR reaction is quantified by a phosphoimager.

2. Wash beads twice with 10 ml phosphate-buffered saline (PBS).
3. Wash beads twice with 0.2 M sodium borate, pH 9.0.
4. Resuspend beads in 10 ml of 0.2 M sodium borate, pH 9.0, and add 52 mg dimethyl pimelidate (DMP) (final concentration of 20 mM).
5. Incubate for 30 min at room temperature with gentle mixing.
6. Block excess DMP by washing the beads once in 0.2 M ethanolamine, pH 8.0, followed by an incubation for 2 h at room temperature in 0.2 M ethanolamine, pH 8.0, with gentle mixing.
7. Wash beads twice with 10 ml PBS.
8. Resuspend the beads in PBS and add sodium azide to a final concentration of 0.02% and store at 4°.

Procedure

1. Mix 20 μl protein G–NC2α antibody-coupled beads, 300 ng recombinant NC2 proteins, and 100 μl BC150. The BC150 buffer used in this method is supplemented with PMSF (final concentration 5 mM), DTT (final concentration 1 mM), and IGEPAL CA-630 (final concentration 0.1%).
2. Incubate for 3 h at 4° with gentle mixing.
3. Wash four times with 1 ml BC150. Recover the beads by centrifugation for 3 min at 300 g.
4. NC2 is eluted from antibody-coupled beads either with SDS buffer (4 min at 95°) or with a peptide corresponding to the epitope. Add 40 μl of the peptide (1 mg/ml in BC150) and incubate at room temperature for 1 h with shaking.
5. Analyze eluted samples by SDS–PAGE and Western blot.

Detection of NC2 on Genes by Chromatin IP[8,9]

NC2 can be found on many eukaryotic genes.[10] The amounts vary greatly depending on the microenvironment, the promoter structure, and signals (unpublished observations). However, the signals and factors that control NC2 occupancy are not fully understood. Hence, to understand the molecular processes on specific genes and their promoters, it will be interesting to monitor NC2 levels. The ChIP technique allows visualization of binding sites and levels of NC2 *in vivo*. As expected from the X-ray structure, NC2 is cross-linked readily to DNA and output signals are in the range of several percent of input comparable with histones.

[8] V. Orlando, H. Strutt, and R. Paro, *Methods* **11**, 205 (1997).
[9] V. Orlando, *Trends Biochem. Sci.* **25**, 99 (2000).
[10] J. V. Geisberg, F. C. Holstege, R. A. Young, and K. Struhl, *Mol Cell Biol.* **21**, 2736 (2001).

Materials

37% formaldehyde solution (Merck)
Branson sonifier
Glass beads (212–300 μm, Sigma)
Glycogen (Sigma)
Protein G Sepharose 4 fast flow (Amersham Pharmacia Biotech)
Bovine serum albumin (20 mg/ml, Roche Diagnostics GmbH)
Proteinase K solution (15.6 mg/ml, Roche Diagnostics GmbH)

Solutions

PBS: 137 mM NaCl, 2.7 mM KCl, 6.5 mM Na$_2$HPO$_4$, 1.5 mM KH$_2$PO$_4$, pH 7.2–7.4
RPMI 1640 medium with L-glutamine (GIBCO)
TBS: 10 mM Tris, pH 8.0, 150 mM NaCl
Buffer A (sonification buffer): 50 mM HEPES–KOH, pH 7.9, 140 mM NaCl, 1 mM EDTA, 1% Triton, 0.1% sodium deoxycholate, 1 mM PMSF
Buffer A with 500 mM NaCl
Buffer B: 10 mM Tris, pH 8.0, 250 mM LiCl, 0.5% IGEPAL CA-630 (Sigma), 0.5 % sodium deoxycholate, 1 mM EDTA
Buffer D (dialysis buffer): 5% glycerol, 1 mM EDTA, 10 mM Tris, pH 8.0, 1 mM PMSF
Buffer E (elution buffer): 1% SDS, 0.1 M NaHCO$_3$
10% sodium lauryl sarcosine

Cells

Jurkat J6 cells (human leukemic T-cell lymphoblast)

Procedure

1. Suspension cultures of human J6 Jurkat cells are used to prepare cross-linked extracts. Typically 2×10^8 cells are harvested at a density of 5×10^5/ml by centrifugation at 1000 rpm (204 g) in a benchtop centrifuge. Wash and resuspend in 50 ml PBS.

2. For cross-linking, add 1.35 ml of 37% formaldehyde to a 50-ml cell suspension (final concentration 1%) and incubate at room temperature for 9 min on a roller mixer. Stop the cross-linking reaction with 0.47 g of solid glycine to 50 ml (final concentration 125 mM).

3. Spin down the cells at 1000 rpm (204 g) and wash twice with cold TBS. The pellet can be stored at $-80°$.

4. Resuspend the cross-linked material in 10 ml buffer A, add glass beads, and sonicate in an ice-ethanol bath 10 times for 30 s with cooling phases of 1 min in between (duty cycle 70% and output control 3). The sample must not get warm!

5. Centrifuge for 10 min at 3500 rpm (2490 g) to remove glass beads. Collect the lysate in a new tube and adjust to 0.5% sarcosyl and 1.42 g/cm^3 CsCl.

6. Separate cross-linked material from free protein, DNA, and RNA by gradient centrifugation in a Beckman SW 41 swinging bucket rotor at 38,000 rpm for 36 to 72 h at 18°.

7. After centrifugation, puncture the tube with a syringe and collect 500-μl fractions dropwise. Check fractions on a 0.5 % agarose gel. Free DNA is usually in fractions 1–3, whereas cross-linked DNA (smear with a size of 2.0 to >30 kb) is found in fractions 6–12 corresponding to 1.38–1.39 g/cm^3 CsCl. Take an aliquot of the fractions containing cross-linked DNA, incubate for 2 h at 65° to remove cross-links, and check on a 1% agarose gel for DNA size. Pool fractions with an average size of 500 bp and dialyze overnight against buffer D.

8. Use 2.5 A_{260} for immunoprecipitation and adjust to buffer A in a total volume of 1 ml. As a preclearing step, 30 μl of protein G-Sepharose is added and incubated for 2 h at 4°.

9. Protein G-Sepharose is removed by centrifugation at 4000 rpm (1500 g) in a microcentrifuge. Add 15 μg of a monoclonal NC2α antibody to the supernatant and incubate overnight on a rotating instrument at 4°. As an ideal control, serve beads loaded with an isotype antibody.

10. Spin down at 13,000 rpm for 10 min to remove precipitates and transfer supernatant to fresh tubes.

11. Antibody-bound complexes are recovered by the addition of 30 μl protein G-Sepharose, 2 μg sonicated salmon sperm DNA, and 100 μg BSA to the sample and by incubation for 2 h at 4°.

12. Immunocomplexes are precipitated by centrifugation at 4000 rpm (1500 g). Wash twice with buffer A, twice with buffer A adjusted to 500 mM NaCl, twice with buffer B, and twice with TE (1 ml each time).

13. For elution, incubate the beads with 400 μl buffer E for 30 min at room temperature, spin down, and transfer the supernatant to a fresh tube.

14. Adjust to 500 mM NaCl and incubate for 5 h (or overnight) at 65° to reverse the cross-links. Add 6 μg proteinase K and adjust to 100 mM Tris, pH 6.8, and 50 mM EDTA. Incubate at 56° for 1 h.

15. Extract once with phenol–chloroform and once with chloroform–isoamylalcohol. Precipitate with 2.5 volumes ethanol and 10 μg glycogen.

16. Dissolve the pellet in 30 μl TE and use 1.5 μl per polymerase chain reaction (PCR). Ten percent of input material is processed like eluted samples (reverse cross-link and phenol/chloroform extract) and used as a standard in the PCR reaction. PCR reactions are performed either with *Taq* polymerase in the presence of 1 μCi [α-^{32}P]dCTP (10 μCi/μl, 3000 Ci/mmol) and separated on an polyacrylamide gel or by real time PCR according to the manufacturer's instructions. A serial dilution of input DNA serves as a standard.

Repression of Transcription by NC2

NC2 is a potent repressor of basal transcription in reconstituted *in vitro* systems. Efficient repression is dependent on the presence of both subunits and especially effective in reconstituted systems that contain TBP and lack TFIIA (Fig. 4) together with other recombinant and native partially purified basal factors (TFIIB, TFIIE, TFIIF and TFIIH, and RNA polymerase II, respectively). Repression is also seen, although to a lesser extent in the presence of TFIID (TBP and TAFs). In crude systems, excess NC2 also represses many (if not all) RNA polymerase II promoters. However, (and this discrepancy is not fully understood), stimulatory effects of NC2 were also reported for *yeast* and *drosophila* NC2 complexes.[10–14]

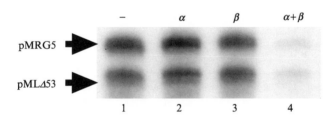

FIG. 4. Repression of transcription by NC2. Individual subunits and NC2 complex (approximately 50 ng) were added to a purified RNA polymerase II transcription system as indicated.

[11] H. Ge, E. Martinez, C.-M. Chiang, and R. G. Roeder, *Methods Enzymol.* **274,** 57 (1996).
[12] G. Prelich, *Mol Cell Biol.* **17,** 2057 (1997).
[13] M. Lemaire, J. Xie, M. Meisterernst, and M. A. Collart, *Mol Microbiol.* **36,** 163 (2000).
[14] P. J. Willy, R. Kobayashi, and J. T. Kadonaga, *Science* **290,** 982 (2000).

Materials

20 mg/ml bovine serum albumin (Roche Diagnostics GmbH)
40 U/μl RNase block (Stratagene)
Ultrapure NTP set (Amersham Pharmacia Biotech)
3'-O-Methylguanosine 5'-triphosphate (omG, Amersham Pharmacia)

Plasmids[6,11]

pMLΔ53, carrying a downstream G-less cassette
pMRG5, hybrid core promoter composed of HIV TATA box, the
  AdML initiator region, and five Gal4-binding sites upstream of the
  promoter upstream of G-less cassette

Solutions

Transcription stop solution: 7 *M* urea, 10 m*M* Tris, pH 7.8, 10 m*M*
  EDTA, 0.5% SDS, 100 m*M* LiCl, 100 μg/ml tRNA, 300 m*M*
  sodium acetate
Nucleotide mix: 10 m*M* ATP, 10 m*M* TTP, 0.5 m*M* CTP, 10 m*M*
  GTP, or 2 m*M* omG
Loading dye: 97% deionized formamide, 20 m*M* Tris, pH 7.3, 10 m*M*
  EDTA, 0.05% bromphenol blue
For reconstitution of basal transcription with a partially purified system,
see Fig. 4. The following transcription reactions are based on factors
purified as described elsewhere.[6,11]

Procedure

Prepare a DNA mix, per reaction add: 50 ng pMRG5, 50 ng pMLΔ53,
0.8 μl MgCl$_2$ (100 m*M* stock solution), 0.5 μl HEPES–KOH, pH 8.2 (1 *M*
stock solution), 0.1 μl DTT (1 *M* stock solution), 0.05 μl PMSF
(0.2 *M* stock solution), 0.2 μl omG mix, and 0.8 μl [α-^{32}P]CTP (10 μCi/μl,
3000 Ci/mmol); adjust the DNA mixture to a volume of 10 μl with H$_2$O
and BC buffer. The final salt concentration should be 60 m*M* KCl in a total
reaction volume of 20 μl.

1. In a separate microcentrifuge tube, mix 2 μl of a NC2-phospho-
   cellulose fraction (P11 0.5; 2 mg/ml) dialyzed to BC100, 10 ng
   recombinant TBP, 20 μg BSA, and 0.5 μl RNAse block to a total
   volume of 5 μl.
2. Mix 14 μl DNA mix and 5 μl protein mixture and add 1 μl either
   NC2α (40 ng/μl) or NC2β (100 ng/μl) or 1 μl of a premixed NC2α

(40 ng α and 100 ng $\beta/\mu l$) in BC100 buffer. Alternatively, the coexpressed NC2 complex can be added at similar concentrations.

3. Incubate at 28 ° for 60 min.

4. Stop with 400 μl stop solution, extract with phenol/chloroform, and precipitate with isopropanol for 1 h at −20°.

5. Spin down for 30 min at 13,500 rpm in a microfuge, wash with 80% ethanol, and dissolve the dried pellet in 10 μl loading dye.

6. Analyze the transcripts on a 5 % gel (acrylamide:bisacrylamide 19:1) containing 8 M urea.

Reconstitution of Transcription in Nuclear Extract

1. Prepare DNA mix as described for the purified system but without adding the DNA template. For runon transcription, use linear RNA polymerase II transcription templates (20 to 200 ng), appropriate nucleotide mix and 50 to 200 ng of purified recombinant NC2 complex. RNA polymerase III transcription is readily measurable, i.e., from the adenovirus VAI promoter, but is not decreased under these conditions.

2. Incubate DNA mix with NC2 premixed in HeLa nuclear extracts (25–50 μg of total protein) for 60 min at 28° and stop and process the reaction as described previously.

Acknowledgments

We thank Gregor Gilfillan and the members of the Meisterernst laboratory for critical reading of the manuscript. This work has been supported by grants of the DFG, the BMBF, and the EC to M.M.

[41] Hap1p Photofootprinting as an *In Vivo* Assay of Repression Mechanism in *Saccharomyces cerevisiae*

By Mitsuhiro Shimizu and Aaron P. Mitchell

Gene expression levels result from the dynamic interplay of activators and repressors. These factors may influence the basal transcriptional machinery directly or indirectly, through interactions that govern each other's activity or access to DNA. One fundamental question in analysis of a repressor is whether it acts by preventing transcriptional activators from binding to DNA target sites (i.e., UASs or enhancers). We have used a system to address this question that may be applicable to any repressor. The strategy employs *in vivo* ultraviolet (UV) photofootprinting to detect activator binding with one of the most widely used yeast reporter genes, *CYC1-lacZ*.

The *CYC1-lacZ* reporter gene, developed by Guarente and colleagues,[1] has been utilized for analysis of many different repressors and repression sites.[2-21] The *CYC1* upstream region in this reporter (plasmid pLGΔ 312S) includes binding sites for two different transcriptional activators: Hap1p, a Zn_2-Cys_6 zinc binuclear cluster protein that binds to UAS1 at -335 to -346 (numbered relative to the ATG initiation codon), and Hap2p/3p/4p/5p, the yeast CCAAT-binding complex that binds to UAS2 at -284 to -296. Our assay for activator binding focuses on the Hap 1p–UAS1 interaction.

UV photofootprinting has been widely used for analyzing the DNA structure and protein–DNA interaction *in vitro* and *in vivo*. This technique is based on the premise that changes in DNA structure or the formation of protein–DNA complexes will alter rates of dimerization of adjacent bases on a particular DNA strand.[22-26] The sites of UV photoproducts can be

[1] L. Guarente and M. Ptashne, *Proc. Natl. Acad. Sci. USA* **78,** 2199 (1981).
[2] S. R. Hepworth, L. K. Ebisuzaki, and J. Segall, *Mol. Cell. Biol.* **15,** 3934 (1995).
[3] K. S. Bowdish and A. P. Mitchell, *Mol. Cell. Biol.* **13,** 2172 (1993).
[4] H. S. Yoo and T. G. Cooper, *Mol. Cell. Biol.* **9,** 3231 (1989).
[5] M. Shimizu, W. Li, P. A. Covitz, M. Hara, H. Shindo, and A. P. Mitchell, *Nucleic Acids Res.* **26,** 2329 (1998).
[6] P. Herrero, M. Ramirez, C. Martinez-Campa, and F. Moreno, *Nucleic Acids Res.* **24,** 1822 (1996).
[7] K. D. Mehta and M. Smith, *J. Biol. Chem.* **264,** 8670 (1989).
[8] M. Vidal, A. M. Buckley, C. Yohn, D. J. Hoeppner, and R. F. Gaber, *Proc. Natl. Acad. Sci. USA* **92,** 2370 (1995).
[9] P. G. Siliciano and K. Tatchell, *Proc. Natl. Acad. Sci. USA* **83,** 2320 (1986).
[10] J. M. Lopes, K. L. Schulze, J. W. Yates, J. P. Hirsch, and S. A. Henry, *J. Bacteriol.* **175,** 4235 (1993).
[11] A. K. Vershon, N. M. Hollingsworth, and A. D. Johnson, *Mol. Cell. Biol.* **12,** 3706 (1992).
[12] L. W. Bergman, D. C. McClinton, S. L. Madden, and L. H. Preis, *Proc. Natl. Acad. Sci. USA* **83,** 6070 (1986).
[13] S. Sagee, A. Sherman, G. Shenhar, K. Robzyk, N. Ben-Doy, G. Simchen, and Y. Kassir, *Mol. Cell. Biol.* **18,** 1985 (1998).
[14] T. Wang, Y. Luo, and G. M. Small, *J. Biol. Chem.* **269,** 24480 (1994).
[15] P. A. Covitz and A. P. Mitchell, *Genes Dev.* **7,** 1598 (1993).
[16] A. D. Johnson and I. Herskowitz, *Cell* **42,** 237 (1985).
[17] P. A. Covitz, W. Song, and A. P. Mitchell, *Genetics* **138,** 577 (1994).
[18] C. A. Keleher, M. J. Redd, J. Schultz, M. Carlson, and A. D. Johnson, *Cell* **68,** 709 (1992).
[19] H. Friesen, S. R. Hepworth, and J. Segall, *Mol. Cell. Biol.* **17,** 123 (1997).
[20] R. Rodicio, J. J. Heinisch, and C. P. Hollenberg, *Gene* **125,** 125 (1993).
[21] R. A. Sumrada and T. G. Cooper, *Proc. Natl. Acad. Sci. USA* **84,** 3997 (1987).

detected by primer extension mapping using sequencing gel electrophoresis.[23] We have shown that binding of Hap1p to UAS1 is detectable by *in vivo* UV photofootprinting, following the protocol described in this article. The useful feature of the footprint is that Hap1p binding results in a *positive* signal and enhancement of formation of UV photoproducts.[27] Thus detection is extremely sensitive compared to a protection footprint because even limited occupancy of the UAS1 site yields a detectable signal. We have assigned the enhancement to Hap1p because enhancements of UV photoproducts were observed (i) within UAS1 (nucleotides −331 and −330), the known Hap1p-binding site, (ii) in cells from a *HAP1* wild-type strain and not in cells from an isogenic *hap1::LEU2* mutant, which lacks functional Hap1p, and (iii) only with DNA irradiated in whole cells and not with DNA that was purified prior to irradiation (see Fig. 1 and Shimizu *et al.*[27]). Thus the location and requirements for these enhancements demonstrate that they depend on Hap1p binding *in vivo*.

With this system, one can ask whether repressors affect binding of the Hap1p activator to DNA *in vivo*. For example, we used the system in characterization of the yeast repressor Rme1p, whose function is to inhibit meiosis in haploid yeast cells (Fig. 2). Analysis of the natural Rme1p repression target gene, *IME1*, indicated that Rme1p acts through a region far upstream of *IME1*, at −2146 to −1743 (relative to the AUG initiation codon). To determine whether repression by Rme1p causes inhibition of DNA binding by transcriptional activators, we placed the −2146 to −1743 region from the *IME1* locus upstream of the *CYC1* regulatory region in the *cyc1-lacZ* plasmid to create plasmid *RC-CYC1-lacZ* and confirmed by β-galactosidase assays that Rme1p represses *RC-CYC1-lacZ* expression. The following protocol describes assays of Hap1p-UAS1 binding in both repressed and derepressed strains.[27]

In Vivo Ultraviolet PhotoFootprinting Procedure

Sample Preparation

UV photofootprinting is performed as described by Axelrod and Majors[23] with minor modifications.[27,28]

[22] M. M. Becker and J. C. Wang, *Nature* **309**, 682 (1984).
[23] J. D. Axelrod and J. Majors, *Nucleic Acids Res.* **17**, 171 (1989).
[24] B. Suter, M. Livingstone-Zatchej, and F. Thoma, *Methods Enzymol.* **304**, 447 (1999).
[25] G. P. Pfeifer and S. Tornaletti, *Methods* **11**, 189 (1997).
[26] M. Shimizu, T. Mori, T. Sakurai, and H. Shindo, *EMBO J.* **19**, 3358 (2000).
[27] M. Shimizu, W. Li, H. Shindo, and A. P. Mitchell, *Proc. Natl. Acad. Sci. USA* **94**, 790 (1997).
[28] M. R. Murphy, M. Shimizu, S. Y. Roth, A. M. Dranginis, and R. T. Simpson, *Nucleic Acids Res.* **21**, 3295 (1993).

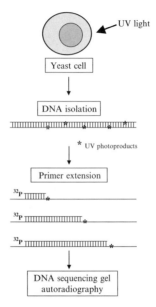

FIG. 1. Overview of *in vivo* photofootprinting. Yeast cells are UV irradiated, introducing UV photoproducts into DNA. These products block primer extension and may thus be mapped by the sizes of primer extension products, as detected by autoradiography. Presence of a protein bound to DNA alters local UV sensitivity and may either enhance or prevent production of a photoproduct, causing enhanced or reduced intensity of a primer extension reaction product.

1. Yeast cells are grown in 1000 ml YPAc (or SC-Ura) medium to early exponential phase (OD$_{600}$ 0.2–0.5) and harvested by vacuum filtration on a 90-mm filter.

2. Cells are resuspended in 15 ml of fresh medium.

3. Two 3.5-ml aliquots of the cell suspension are placed in petri dishes. One dish is for samples C1 and C2 and the other is for C3 and C4. In addition, four 1.5-ml samples are transferred to microfuge tubes (held on ice) for DNA preparations of unirradiated cells.

4. UV (254 nm) irradiations are carried out in a Stratalinker with open petri dishes.

Sample C1: 250 mJ (setting of 2500 with the Stratalinker)
Sample C2: 500 mJ (setting of 2500 for two irradiation intervals)
Sample C3: 750 mJ (setting of 2500 for three irradiation intervals)
Sample C4: 1000 mJ (setting of 2500 for four irradiation intervals)

CYC1-lacZ 5'region (−460 to +12)

```
ATAAGTAAAT GCATGTATAC TAAACTCACA AATTAGAGCT TCAATTTAAT
TATATCAGTT ATTACCCGGG AGCAAGATCA AGATGTTTTC ACCGATCTTT
-360                UAS1                           -311
CCGGTCTCTT TGGCCGGGGT TTACGGACGA TGGCAGAAGA CCAAAGCGCC
-310                UAS2                           -261
AGCTCTTTGG CGAGCGTTGG TTGGTGGATC AAGCCCACGC GTAGGCAATC
CTCGAGGTCG ACCTCGAGCA GATCCGCCAG GCGTGTATAT ATAGCGTGGA
GCCAGGCAAC TTTAGTGCTG ACACATACAG GCATATATAT ATGTGTGCGA
CGACACATGA TCATATGGCA TGCATGTGCT CTGTATGTAT ATAAAACTCT
TGTTTTCTTC TTTTCTCTAA ATATTCTTTC CTTATACATT AGGACCTTTG
TAGCATAAAT TACTATACTT CTATAGACAC ACAAACACAA ATACACACAC
-10        +1
TAAATTAATA ATGACCGGAT CC
```

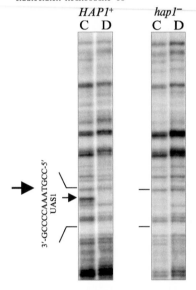

FIG. 2. Inhibition of Hap1p-UAS1 binding through Rme1p-dependent repression. *In vivo* UV photofootprints of the noncoding strand of the UAS1 region in a *CYC1-lacZ* plasmid that includes Rme1p-binding sites and the flanking region inserted upstream of UAS1. Expression of the *CYC1-lacZ* reporter is repressed in the *RME1* strain (left), but not in the *rme1-213* mutant strain (center) or in a strain with a functional *RME1* gene that carries *rgr1* and *sin4* mutations. For each strain, UV photoproducts in irradiated cells (lanes C) are compared with irradiated purified DNA (lanes D). The Hap1p-UAS1 photofootprint is detectable in the two derepressed strains (center and right) but not in the repressed strain (left). From Shimizu *et al.*[27]

After the each time of the irradiation, the cells should be resuspended by shaking or swirling the petri dish.

5. After the indicated doses, 1.5-ml samples are removed to microfuge tubes, and cells are pelleted for 1 min. The unirradiated samples (step 3) are also pelleted and treated in parallel through the following steps.

6. Cell pellets are resuspended in 0.5 ml of 1 M sorbitol, 0.1 M EDTA containing 0.1% 2-mercaptoethanol (added freshly).

7. Add 50 μl of Zymolase solution (10 mg/ml Zymolyase 100T in 40 mM potassium phosphate buffer, pH 7.2, 1 M sorbitol, 0.5 mM phenylmethylsulfonyl fluoride, 20 mM 2-mercaptoethanol) and incubate at 30° for 30 min.

8. Spin down the spheroplasts for 1 min in a microfuge and suspend in 0.5 ml of 50 mM Tris–HCl (pH 8.0), 20 mM EDTA.

9. Add 50 μl of 10% SDS and incubate at 68° for 30 min.

10. Add 10 μl of proteinase K (10 mg/ml), and incubate at 68° for 1 h.

11. Add 200 μl of 5 M potassium acetate and incubate on ice for 1 h.

12. Centrifuge for 15 min, and transfer the supernatants into new tubes.

13. Centrifuge for 5 min, and transfer the supernatants into new tubes.

14. Add 0.7 ml of isopropanol to each supernatant.

15. Centrifuge for 3 min, wash the pellet with 95% ethanol, and dry.

16. Dissolve the pellet in 300 μl of 10 mM Tris, 1 mM EDTA, pH 8.0.

17. Add 3 μl of RNase A (10 mg/ml). Incubate at 37° for at least 2 h.

18. Extract with phenol/CHCl$_3$ once, and with CHCl$_3$ once. Add 1/10 volume of 3 M sodium acetate, pH 5.2, and 3 volumes of ethanol. Centrifuge, wash the pellets with 70% ethanol, and dry.

19. For UV irradiation of the naked DNA, dissolve two of the un-irradiated DNA pellets in 300 μl of phosphate-buffered saline (PBS) (137 mM NaCl, 2.7 mM KCl, 10 mM, Na$_2$PO$_4$, 2 mM KH$_2$PO$_4$, pH 7.4), and place the entire sample in a petri dish. UV (254 nm) irradiations are carried out in a Stratalinker with open petri dishes:

> Sample D1: 120 mJ (setting of 1200 at the Stratalinker)
> Sample D2: 240 mJ (setting of 1200 for two irradiation intervals)

20. Ethanol precipitate, wash, and dry DNA pellets D1 and D2.

21. Dissolve all DNA samples for sequencing in 300 μl of 10 mM Tris–HCl, 0.1 mM EDTA (pH 8.0). Ethanol precipitate, wash the pellet with 70% ethanol, and dry the DNA pellets.

22. Dissolve each of the DNA pellets in 70 μl of TE.

23. Use 10 μl of each DNA sample for primer extension mapping.

Primer Extension Mapping

This primer extension protocol was adapted from Axelrod and Majors[23] with minor modifications.[27–30] The one-cycle primer extension reaction is performed to map UV photoproducts in multicopy plasmids, whereas multicycle primer extension is performed for those in the genome or on low-copy plasmids.

1. Primers are labeled radioactively as follows: 10 pmol of primer is combined with 10 μl of [γ-^{32}P]ATP ($\sim$6000 Ci/mmol, e.g., PB10218, Amersham Pharmacia Biotech) and 20 units of T4 polynucleotide kinase (total 25 μl) and incubated at 37° for 60 min. Then, 25 μl of water is added and the reaction is terminated by heating at 65° for 20 min. Excess [γ-^{32}P]ATP P]ATP is removed by passing through a Sephadex G-50 spin column (e.g., ProbeQuant G-50 microcolumns, Amersham Pharmacia Biotech).

2. Ten microliters of DNA (containing $\sim$10–50 μg of genomic DNA) is combined with 0.3 pmol of the ^{32}P end-labeled primer ($\sim$10^6 cpm), 5 μl of 5$\times$ *Taq* buffer, 1 μl of 5 mM dNTPs, 1.25 U of *Taq* polymerase, and H$_2$O (total 25 μl). In addition, 10 μl of unirradiated DNA is used for control DNA sequencing reactions (to map the sites of UV photoproducts precisely). For sequencing, the reactions contain 1 μl of 25$\times$ ddG, ddT, or ddC mix instead of the dNTP mix. If necessary, 25 μl of mineral oil may be layered on the reaction mixture.

3a. One-cycle primer extension: The sample is heated at 94° for 5 min (denaturation), cooled at 48° for 20 min (annealing), and incubated at 72° for 10 min (extension). Samples may then be held at 4°.

3b. Multicycle primer extension: The sample is subjected to several cycles of primer extension as follows: 94° for 1 min, 55° for 2 min, and 72° for 2 min, repeated 15 times. (If the signal is not strong enough, primer extension cycles may be increased up to around 30.) Samples may then be held at 4°.

4. Mineral oil is removed if necessary by the addition of 100 μl CHCl$_3$, and the DNA solution is transferred to a new tube and precipitated with ethanol.

5. The DNA pellet is dissolved in 5 μl of sequencing gel-loading buffer, heated at 95° for 5 min, and chilled on ice. Samples are electrophoresed on a 6% polyacrylamide sequencing gel containing 50% urea.

Recipes

Primer Design. Primers are designed as 33–38 nucleotides in length with 40–60% G+C content. If one is footprinting the *CYC1-lacZ* fusion template, it is important that primers anneal to the plasmid template and

[29] M. Shimizu, S. Y. Roth, C. Szent-Gyorgyi, and R. T. Simpson, *EMBO J.* **10,** 3033 (1991).
[30] S. Y. Roth, M. Shimizu, L. Johnson, M. Grunstein, and R. T. Simpson, *Genes Dev.* **6,** 411 (1992).

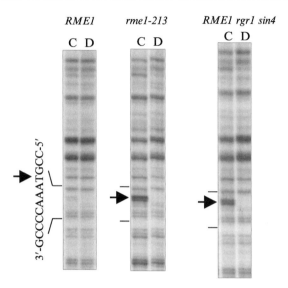

FIG. 3. UV photofootprinting of Hap1p-UAS1. (Top) Sequence of the *CYC1-lacZ* 5′ region from plasmid pLGΔ312S. Sequences from the 3′ end of *URA3* are italicized and end at a *Sma*I restriction site. Sequences of UAS1 and UAS2 in the CYC1 5′ region are underlined. *CYC1* sequences end at a *Bam*HI site, and the first few codons of the *CYC1-lacZ* fusion ORF are in bold-face type. (Bottom) *In vivo* UV photofootprints of the noncoding strand of the UAS1 region in a *CYC1-lacZ* plasmid.[27] The arrow indicates sites of enhancements of UV photoproducts in irradiated cells (lanes C) compared with irradiated purified DNA (lanes D). (Left) Reactions from a wild-type (*HAP1*) strain with functional Hap1p; (right) reactions from a mutant (*hap1::LEU2*) strain that lacks functional Hap1p. The major UAS1 UV photoproduct enhancements seen in intact cells depend on expression of functional Hap1p. From Shimizu *et al.*[27]

not to genomic *CYC1* sequences. For this purpose, we use primers that anneal to the junction sequences between the *CYC1* upstream region and its flanking region (*URA3* or an inserted repression region). For example, for footprinting of the noncoding strand of *CYC1-lacZ* fusion whose sequence is shown in Fig. 3, the primer TCAATTTAATTATATCAGT-TATTACCCGGGAGCA would be suitable.

5× *Taq* buffer: 50 m*M* Tris–HCl, pH 8.3, 250 m*M* KCl, 15 m*M* MgCl2, 0.15% NP-40, 0.15% Tween 20[31]

25× ddG mix: 1.25 m*M* ddGTP, 0.25 m*M* dGTP, 2.5 m*M* each of dATP, dTTP, and dCTP

[31] M. A. Innis, K. B. Myambo, D. H. Gelfand, and M. A. Brow, *Proc. Natl. Acad. Sci. USA* **85**, 9436 (1988).

25× ddA mix: 2.5 mM ddATP, 62.5 μM dATP, 2.5 mM each of dGTP, dTTP, and dCTP

25× ddT mix: 2.5 mM ddTTP, 0.125 mM dTTP, 2.5 mM each of dGTP, dATP, and dCTP

25× ddC mix: 2.5 mM ddCTP, 0.125 mM dCTP, 2.5 mM each of dGTP, dATP, and dTTP

Sequencing gel-loading buffer: 80% formamide, 1 × TBE buffer, 0.025% bromphenol blue, and 0.025% xylene cyanol FF

Troubleshooting

The purity of DNA samples is very important. If the sequencing gel is smeared, the DNA samples may be purified by passing through Sephadex G-50 spin columns or by using a DNA purification resin.

If the bands appear weak, the annealing temperature in the primer extension reaction may be lowered. In addition, an increase of cycles of primer extension may increase the signal intensity.

[42] Analysis of Activator-Dependent Transcription Reinitiation *In Vitro*

By RAPHAEL SANDALTZOPOULOS and PETER B. BECKER

The eternal struggle of Sisyphus, condemned by the Gods for unraveling their secrets, was to carry a heavy boulder to the top of a hill. Alas, every time he reached the top, the boulder would roll to the bottom of the hill and poor Sisyphus had to reinitiate his effort. Ancient Greeks perceived this seemingly unrewarded labor as the worst punishment and torture a conscious being could ever receive. In biology, many complex reactions have to occur repeatedly in order to be effective. In the case of gene transcription, where multiple initiation events are required in order to achieve appropriate mRNA levels, nature has made sure that "the boulder does not roll to the bottom of the hill, but is held up half-way": the reinitiation of transcription abbreviates the lengthy procedure required for the first successful initiation event.

The transcription output of a gene is a function of the efficiencies of transcription initiation, elongation, and the rate of transcription reinitiation. The overall number of mRNA molecules synthesized in a cell depends on the fraction of time in which a given promoter is active, as well as on the frequency with which polymerases initiate transcription from this site. In cell-free transcription systems, where usually a large population of identical promoters is analyzed at the same time, RNA synthesis is a function of the fraction of DNA molecules that are active as templates and on the average number of initiations from each promoter during the reaction time. Initiation necessitates the stepwise assembly of a preinitiation complex (PIC) on a promoter, involving dozens of polypeptides in a number of general transcription factors and RNA polymerase II itself. In contrast, reinitiation requires recruitment of only a subset of factors, as some factors (e.g., TFIID and TFIIA) remain associated with the promoter as a scaffold after promoter clearance by the polymerase.[1–4] Consequently, the kinetics of reinitiation are faster than the initial transcription cycle, allowing a fast and effective response to signals. Regulating the degree of reinitiation from a promoter thus contributes to adjusting gene expression levels to changing requirements. At this point a distinction between the terms "transcription reinitiation" and "polymerase recycling" should be noted: The term *reinitiation* refers to the reuse of a given promoter, but not necessarily by the same RNA polymerase molecule that catalyzed the first round of transcription. Polymerase *recycling* refers to the reuse of an RNA polymerase II complex after it has completed one round of transcription. The methodological approach described here focuses solely on transcription reinitiation without addressing the issue of polymerase recycling.

This article describes the technical details of a set of methods collectively capable of distinguishing between transcription initiation and reinitiation and measuring the initiation rate in cell-free transcription systems. These procedures were developed to study heat shock gene regulation. We discovered that one way by which *Drosophila* heat shock factor (HSF) activates transcription is by enhancing the reinitiation rate.[2] The focus of this article is to provide a detailed description of the methodology, which could not be described in detail in our original study due to space limitations. The study relied on the availability of a powerful cell-free transcription system, capable of supporting multiple initiation events.[5–8] Prior to induction, heat shock promoters are characterized by a paused polymerase: an RNA polymerase II complex has been recruited to the promoter, has initiated transcription, but is arrested to pause elongation some 25 nucleotides

[1] S. G. Roberts, B. Choy, S. S. Walker, Y. S. Lin, and M. R. Green, *Curr. Biol.* **5,** 508 (1995).
[2] R. Sandaltzopoulos and P. B. Becker, *Mol. Cell. Biol.* **18,** 361 (1998).
[3] L. Zawel, K. P. Kuman, and D. Reinberg, *Genes Dev.* **9,** 1479 (1995).
[4] N. Yudkovsky, J. A. Ranish, and S. Hahn, *Nature* **408,** 225 (2000).
[5] R. Sandaltzopoulos, C. Mitchelmore, E. Bonte, G. Wall, and P. B. Becker, *Nucleic Acids Res.* **23,** 2479 (1995).
[6] M. D. Biggin and R. Tjian, *Cell* **53,** 699 (1988).
[7] W. C. Soeller, S. J. Poole, and T. B. Kornberg, *Genes Dev.* **2,** 68 (1988).
[8] J. T. Kadonaga, *J. Biol. Chem.* **265,** 2624 (1990).

downstream of the transcription start site. Heat shock conditions induce the interaction of HSF with the promoter and a concomitant release of the polymerase to enter the elongation phase.[9–11] Pausing of the polymerase is controlled by upstream activators,[12] and the release of the first polymerase is followed by very rapid initiation cycles.[2,4,13] Results suggest that these subsequent initiation events are indeed *re*initiation cycles.

General Outline of the Experimental Strategy

Our experimental system consists of (i) a highly efficient nuclear extract, able to support multiple rounds of transcription *in vitro*; (ii) immobilized DNA templates, which allow to purify intermediates of the transcription process for detailed analysis; and (iii) a procedure suited to determine the absolute amounts of transcribed RNA precisely. The experimental procedures detailed here serve to establish whether the transcription templates are transcribed once or multiply, with the latter being a prerequisite for reinitiation, and whether these multiple rounds are due to several independent initiation events or to reinitiation. In order to achieve the first goal, we need to quantify the number of RNA molecules synthesized precisely and to relate it to the number of active DNA templates. Interpretation is unambiguous if the transcription system supports initiation from the majority of the templates. If, by contrast, only a small fraction of the templates is active, any observed increase in transcription could be due either to an increase in the fraction of active templates (i.e., initiation) or to multiple rounds of transcription from a constant number of active promoters.

Preinitiation complexes (PIC) may be assembled on the majority of template molecules during an incubation in nuclear extracts from *Drosophila* embryos in the absence of ribonucleotide triphosphates (NTPs). A subsequent addition of NTPs will lead to transcription from all templates. If the number of RNA molecules synthesized surpasses the number of DNA templates, some templates must have been used more than once. Reinitiation can then be distinguished from *de novo* initiation by its sensitivity to the detergent sarkosyl[14,15] and by monitoring template-bound transcription factors after the first initiation event.

[9] A. E. Rougvie and J. T. Lis, *Cell* **54,** 795 (1988).

[10] A. E. Rougvie and J. T. Lis, *Mol. Cell. Biol.* **10,** 6041 (1990).

[11] E. B. Rasmussen and J. T. Lis, *J. Mol. Biol.* **252,** 522 (1995).

[12] H. Tang, Y. Liu, L. Madabusi, and G. S. Gilmour, *Mol. Cell. Biol.* **20,** 2569 (2000).

[13] D. Yean and J. D. Gralla, *Nucleic Acids Res.* **27,** 831 (1999).

[14] D. K. Hawley and R. G. Roeder, *J. Biol. Chem.* **262,** 3452 (1987).

[15] M. N. Szentirmay and M. Sawadogo, *Nucleic Acids Res.* **22,** 5341 (1994).

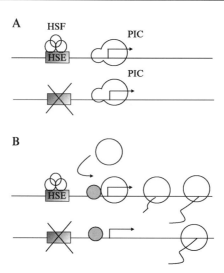

FIG. 1. Monitoring activator-dependent reinitiation of transcription: the case for heat shock control. (A) Complete preinitiation complexes (PIC) are assembled in the absence of NTPs on promoters containing or lacking heat shock elements (HSE), the binding sites for heat shock factor (HSF). (B) Upon addition of nucleotides, reinitiation cycles occur only if the promoter contains HSEs. Note that after the first transcription cycle, a scaffold containing some GTFs remains at the promoter (gray circle), facilitating rapid reinitiation.

In order to document the effect of a gene-specific transcription activator, such as HSF, on the reinitiation rate (Fig. 1), other phenomena, such as influences on the rate of PIC formation (i.e., initiation), must be excluded carefully. This requires monitoring the kinetics of PIC assembly in order to assure that the first initiation occurs equally well on all promoters.

RNA Quantification by Comparison to Reference RNA

RNA is quantified by reverse transcription, extending a radioactively labeled primer annealed some 100 nucleotides from the 5' end of the RNA. In order to distinguish the RNA transcribed in the cell-free system ("transcript") from endogenous RNA (nuclear extracts from embryos may contain significant amounts of RNA transcribed *in vivo* prior to cell homogenization), minigene templates containing a small deletion (e.g., 20 bp) between the transcription start and primer annealing sites are used (see Fig. 2).[5] RNA transcribed from the minigene is quantified by comparison to known amounts of a reference RNA added along with the transcription reaction stop mixture. This defined amount of RNA is copurified with

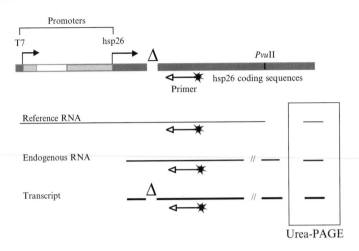

FIG. 2. Schematic representation of the *in vitro* transcription DNA template and the RNAs detected by reverse transcription. Hsp26 gene sequences are transcribed from the hsp26 promoter in the reaction or from an upstream T7 promoter (arrows with solid heads) in order to generate reference RNA. In order to distinguish the new transcript from hsp26 RNA endogenous to the crude nuclear extract, a minigene template is used from which some 20 bp have been deleted (Δ). All three RNAs are visualized by reverse transcription with a radioactive primer (arrows with empty heads) annealed some 100 nucleotides downstream of the start site.

the transcript and serves as a recovery and efficiency control during all subsequent steps.

Reference RNA differs from the transcript only by the fact that it is somewhat longer at the 5′ end due to its transcription from a T7 promoter in the plasmid vector as opposed to the native promoter (see Fig. 2). For our analysis of hsp26 transcription, we cloned the gene including 185 bp of the promoter sequence into pBluescript II. Transcription of the insert from the T7 promoter flanking the polycloning site resulted in an RNA containing the hsp26 promoter followed by coding sequences. Transcript, endogenous RNA, and reference RNA all contain the same hsp26-coding region. Thus, they can be reverse transcribed from the same radioactive primer and be distinguished by denaturing polyacrylamide gel electrophoresis based on their different sizes (see Fig. 2). Importantly, the unknown quantities of transcript can be compared to the reference RNA, whose absolute amount is known. The quantification of transcript is independent of the specific activity of the primer used for reverse transcription, its annealing efficiency, RNA recovery, and the efficiency of reverse transcription because any variability that may occur affects the reference RNA as much as the transcript. The preparation of reference RNA for the hsp26 gene is detailed in the following protocol.

1. Digest 5 μg template with PvuII to completion (e.g., overnight, using a five-fold excess of enzyme units). Monitor digestion by agarose gel electrophoresis. Complete digestion is important for the quantification of reference RNA, as all RNA molecules must have exactly the same length. PvuII cleaves in the coding region of the hsp26 gene, 638 bp downstream of the transcription start site.

2. Phenol/chloroform extract DNA. Precipitate with 1/10 volume sodium acetate and 2.5 volume ethanol.

3. Resuspend pellet in 10 μl H_2O. Use 1 μl the (500 ng) in a transcription reaction with T7 RNA polymerase (Stratagene) according to the supplier's instructions.

4. Extract reaction twice with acidic phenol: chloroform.

5. Precipitate RNA with 0.25 volume isopropanol and 0.25 volume 1.2 M NaCl, 0.8 M disodium citrate (without pH adjustment). Rinse pellet twice with 75% ethanol.

6. Air dry and resuspend in water.

7. Determine concentration of the RNA by reading the OD_{260}. One unit of optical density corresponds to 145 pmol/ml of a $\sim$835 bases long RNA molecule.

8. Dilute reference RNA to a final concentration of 100 fmol/μl in DEPC-treated H_2O containing 100 μg/ml yeast total RNA. This stock solution is diluted further as required before every use.

In order to evaluate accuracy and reproducibility, the entire procedure was repeated three times. Primer extension of calculated equal quantities of reference RNA from different stocks generated similar signal intensities. Different stocks were mixed to minimize random error. Typically, 1–5 fmol of reference RNA was added to each reaction with the stop mix. An even intensity in all lanes of the reference RNA signal indicates consistency of RNA recovery and reverse transcription. It also shows that the amount of labeled primer during reverse transcription is not limiting (no competition between reference RNA and transcript for annealing to labeled primer).

Optimization of Preinitiation Complex Formation

Transcription System

The following considerations may help optimize the *in vitro* transcription system for maximal template usage and PIC formation. The key parameter is the quality of the nuclear extract used as a source of transcription machinery. A detailed protocol for preparation of the nuclear extract from *Drosophila* embryos[6-8] can be found elsewhere.[5] We found that omission of the second wash of nuclei in the original protocols results in more efficient transcription. In general, we try to speed up extract preparation, especially the lengthy homogenization process, by avoiding very large batches of embryos and by using a continuous flow homogenizer (Yamato).

The main parameters to be optimized are the ratio between template DNA and extract and the time of incubation in the absence on NTPs. In order to ensure that transcription factors are in excess and that PICs are formed on most promoters, a fairly low ratio of template/extract is used. In order to sequester nonspecific transcription inhibitors present in the extract, the complete transcription premix is incubated for 5 min with carrier DNA (e.g., pUC plasmid) before the template is added.[5,16] The precise amount of carrier DNA required depends on the transcription extract and needs to be optimized, as addition of too much competitor DNA may have adverse effects. In our case, 1–3 μg carrier DNA were required per reaction containing 12.5 μl of extract.

Solid-Phase Transcription

Crucial to our experimental strategy is the immobilization of the template DNA on paramagnetic beads. Immobilization allows quantitative and nondisruptive purification of reaction intermediates in a single step. Complete PICs, assembled on the template in a nuclear extract devoid of NTPs, can be purified magnetically. The subsequent addition of NTPs, in the absence of extract, results in a single round of transcription.

Details of an exemplary immobilization procedure have been published.[17] In summary, a linear DNA fragment harboring the hsp26 gene promoter and downstream sequences is biotinylated at one end and bound to streptavidin-coated paramagnetic beads (Dynabeads). To facilitate immobilization and accurate quantification, one additional *Not*I and one *Afl*II restriction sites were introduced into the polylinker (between *Xho*I and *Apa*I sites) of the pBluescript II vector carrying the hsp26 gene promoter (see Fig. 3). Digestion with *Not*I and *Spe*I generates one short (13-bp) polylinker fragment and two long fragments (insert and vector). Incorporation of biotin-21-dUTP at the *Spe*I overhang by filling in with Klenow polymerase biotinylates only the hsp26-containing insert (in addition to the small

[16] L. A. Kerrigan, G. E. Croston, L. M. Lira, and J. T. Kadonaga, *J. Biol. Chem.* **266,** 574 (1991).

[17] R. Sandaltzopoulos and P. B. Becker, *in* "Chromatin Protocels," (P. B. Becker, ed.), p. 195, Humana Press, Totowa, NJ, 1998.

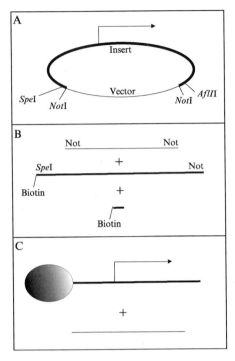

Fig. 3. Strategy for generation of an immobilized transcription template. (A) Salient features of the vector. Bold line: hsp26 gene sequences (insert) where the transcription start is indicated by an arrow. Strategic restriction sites in the polylinker are indicated. (B) DNA fragments after cleaving the plasmid under (A) with *Not*I and *Spe*I and filling in the ends with biotin-21-dUTP. Note that *Not*I overhangs are not biotinylated because they do not contain adenosines. (C) Immobilization of DNA fragments under (B) on streptavidin-coated beads. Note that the long vector fragment is not immobilized.

polylinker fragment, which can be disregarded as it is lost upon removal of free biotin by gel filtration). Prior to immobilization the DNA fragments are quantified by OD_{260} measurement. The fragment mixture is then incubated with Dynabeads (Dynal) under conditions (detailed by the supplier) where the biotinylated template fragment is immobilized quantitatively on the beads, but the nonbiotinylated vector fragment remains in solution (verified by agarose gel electrophoresis). This strategy allows one to calculate the amount of template DNA (in fmol) immobilized on a given amount of Dynabeads.

Digestion with *Afl*II and *Not*I (instead of *Spe*I and *Not*I) allows immobilization of the promoter in the reverse direction relative to the bead surface (transcription proceeding toward or away from the bead). We have not observed any significant difference in results as a function of template orientation.

In Vitro Transcription and Determination of Reinitiation Rate

Assembly of PICs on immobilized template DNA allows their purification from the crude extract. A subsequent addition of NTPs will trigger a single round of transcription from each PIC. The precise quantification of the RNA transcribed from a known amount of active template allows one to determine the average number of transcripts originating from each promoter. Transcription from preformed PICs is insensitive to up to 0.25% (w/v) of sarkosyl (*N*-lauryl sarcosine), a detergent known to prevent *de novo* assembly of PICs during initiation, but compatible with reinitiation events.[14,15] In the presence of sarkosyl, approximately one transcript per promoter is synthesized, which demonstrates that most of the templates are active.

This section describes how we determine the fraction of active templates using solid-phase methodology:

1. Resuspend stock of immobilized template (1 pmol of template/mg of bead suspension) by tapping. Pipette 8 μl of 30 ng/μl (9.5 fmol/μl) immobilized template into a fresh reaction tube.

2. Add 300 μl washing buffer [WB; 1× phosphate-buffered saline, pH 7.6, 0.1% (w/v) bovine serum albumin, 0.05% (v/v) Nonidet P-40].

3. Place on magnetic particle concentrator (MPC-6, Dynal, Oslo) for 1 min. Aspirate and discard supernatant.

4. Repeat steps 2–3.

5. Wash beads with 200 μl transcription reaction buffer [TB; 12.5 mM HEPES–KOH, pH 7.6, 6.25 mM MgCl$_2$, 0.05 mM EDTA, pH 8.0, 5% (v/v) glycerol, 50 mM KCl, 0.01% (v/v) Nonidet P-40].

6. Resuspend beads in 100 μl TB buffer.

7. Prepare the NTP premix (for eight reactions): 100 μl HEMG-100 [25 mM HEPEs–KOH, pH 7.6, 12.5 mM MgCl$_2$, 0.1 mM EDTA, pH 8.0, 10% (v/v) glycerol, 100 mM KCl], 40 μl 5 mM each of four NTPs, 4 μl 0.5 units/μl Inhibit Ace (5′ prime–3′ prime), 2 μl 1% (v/v) Nonidet P-40, 3.2 μl 0.1 M dithiothreitol, 8 μl 0.1 M creatine phosphate, 0.8 μl 100 ng/μl creatine phosphokinase (Roche), and 42 μl DECP-treated H$_2$O. Keep on ice.

8. Prepare PIC assembly mix by combining (for eight reactions): 100 μl extract, 0.8 μl 1 M HEPES–KOH, pH 7.6, 24 μl 1 μg/μl pUC, 2 μl 1 % (v/v) Nonidet P-40, 3.2 μl 0.1 M dithiothreitol, and 70 μl DEPC-treated H$_2$O.

9. Incubate both premixes at 26° for 5 min.
10. Concentrate beads, remove buffer, and resuspend in 200 μl PIC assembly mix.
11. Place on a thermomixer (Eppendorf) prewarmed at 26°. Set mixing to maintain beads in suspension with as mild agitation as possible. Long incubations (>20–30 min) may require additional agitation by tapping.
12. At various times (e.g., 1, 5, 10, 20, 30, 40, 50 min) take 10 μl into a fresh tube and place immediately on the MPC for 1 min.
13. Remove supernatant resuspend in 50 μl TB, and transfer to a fresh tube. Concentrate beads again and remove supernatant completely.
14. Resuspend beads in 25 μl of NTP premix.
15. Incubate for 20 min at 26° in a thermomixer.
16. Stop reactions with 250 μl stop mix [4 M guanidinium thiocyanate, 25 mM Na citrate, pH 7, 0.5 % (v/v) N-lauryl sarkosine, 0.1 M 2-mercaptoethanol] containing 10 μg $Saccharomyces$ $cerevisiae$ total RNA and 2 fmol reference RNA per reaction.
17. Add 25 μl 2 M Na acetate, pH 4.0, and vortex mix.

Proceed with RNA purification.

RNA Purification and Primer Extension Analysis

The RNA purification method is an adaptation of the method described by Chomczynski and Sacchi.[18] As a primer of reverse transcription, we use an oligonucleotide complementary to nucleotides 90–119 of hsp26 RNA. The short sizes of the resulting cDNA ensure efficient elongation by reverse transcriptase (RT). RNA purification and primer extension reactions are performed as follows.

1. Add 300 μl mix of 250 μl water-saturated phenol (not pH equilibrated) and 50 μl chloroform/isoamyl alcohol (49:1). Vortex mix.
2. Place all tubes on ice for 5 min. Then spin for 5 min.
3. Transfer aqueous phase to new tubes containing 250 μl isopropanol. Vortex mix and chill for 10 min at −20°.
4. Spin at full speed for 10 min in a bench-top centrifuge at 4°.
5. Discard supernatant completely (critical for high efficiency of the following steps) and dissolve white pellet in 100 μl stop mix (lacking additional carrier RNA).
6. Add an equal volume of isopropanol and mix thoroughly. Keep at −20° for at least 10 min.

---

[18] Chomczynski and N. Sacchi, $Anal.$ $Biochem.$ **162,** 156 (1987).

7. Centrifuge for 10 min. Discard supernatant and wash pellets with 70% (v/v) ethanol.

8. Dry pellets in a Speed-Vac concentrator for 2–3 min.

9. Dissolve pellets in 9 μl DEPC-treated H_2O containing 0.1 pmol of radioactively labeled oligonucleotide. Incubate for 5 min on a shaker.

10. Add 1 μl 10× primer annealing buffer (2.5 M KCl, 20 mM Tris, pH 7.9, 2 mM EDTA).

11. Denature for 5 min at 76°. Incubate for 30 min at annealing temperature. Place on ice. The optimal annealing temperature depends on the length and GC content of the oligonucleotide primer. Test different temperatures in pilot experiments. Primers with optimal annealing temperature between 50 and 55° give excellent results.

12. Prepare RT master mixture by combining (per reaction) 6 μl 5× MMLV-RT buffer (250 mM Tris, pH 8.3, at 37°, 15 mM $MgCl_2$), 8.6 μl DEPC-treated H_2O, 0.3 μl dithiothreitol, 3.5 μl 10 mM each dNTP in H_2O, 0.6 μl 0.5 U/μl Inhibit Ace, and 1 μl 200 units/μl MMLV reverse transcriptase (US Biochemicals).

13. Add 20 μl of master mixture to each tube. Incubate in a 37° water bath for 30 min.

14. Spin down briefly and add 300 μl cold ($-20°$) absolute ethanol.

15. Centrifuge for 10 min. Remove supernatant. With a radioactivity monitor, check that all radioactivity is in pellet. Wash with 700 μl 75% (v/v) ethanol (spin for 5 min). Dry pellets for 2–3 min in a Speed-Vac.

16. Dissolve pellets in 5 μl of a 2:1 fresh mix of formamide-loading buffer [94% (v/v) deionized formamide, 10 mM EDTA, 0.1% (w/v) xylene cyanol FF, 0.1% (w/v) bromphenol blue] and 0.1 M NaOH.

17. Denature for 5 min at 76° and chill on ice.

18. Separate on a denaturing 8% urea–polyacrylamide gel and prerun for about 30 min so that the gel temperature exceeds 50°. Apply 500 V (gel dimensions 200 × 200 × 1 mm). Fix and dry gel. Scan with a PhosphorImager.

19. Quantify radioactivity in bands corresponding to reference RNA and transcript. Subtract background values and determine transcription level relative to reference RNA in each lane. Calculate the ratio of RNA molecules per DNA template molecules at each time point. This ratio expressed percent equals to the percentage of active templates.

20. Plot the ratio of active templates as a function of time.

Under optimal conditions, the percentage of active templates must be very high, close to 100%. In our experience with heat shock genes, a plateau of PIC formation is reached quickly (10–15 min). However, minimal promoters may require a longer PIC assembly reaction. Duration of the

PIC formation reaction must be chosen such that all templates to be compared are fully loaded with PIC. Typically, an incubation of 30–40 min is sufficient.

Analysis of Proteins Assembled on the Template

Use of an immobilized DNA template allows visualizing proteins associated with the promoter before and after initiation of transcription. In contrast to *de novo* initiation, where all general transcription factors (GTFs) associate anew with a "virgin" promoter, TFIID and TFIIA remain associated with a promoter during reinitiation.[1–4] Therefore, detection of these factors at the promoter after the release of RNA polymerase into productive elongation provides additional evidence that subsequent transcription occurs by a reinitiation mechanism rather than by *de novo* initiation. Some proteins display significant binding to nonspecific DNA (e.g., the TATA-binding protein and RNA polymerase II), and therefore several antibodies against different GTFs should be used for conclusive results.[2,4] As a control for nonspecific "sticking" of proteins to DNA or the bead matrix, we use an equal amount of immobilized vector DNA lacking gene-specific sequences. This section describes a protocol for the visualization of promoter-associated proteins.

1. Pipette 10 μl of bead suspension containing immobilized template or control DNA (30 ng DNA/μl of bead suspension) into fresh reaction tubes. Prepare two samples each with specific and control DNA. In our case, approximately 0.3 μg of DNA was sufficient to detect general transcription factors by Western blotting with polyclonal antibodies. If the signal-to-noise ratio is not adequate, a shorter promoter fragment can be used (our specific template was 4.8 kbp long) as a large number of promoters should increase the signal without increase of background.

2. Concentrate magnetically and discard supernatant. Wash twice with 500 μl WB. Wash once with 300 μl TB.

3. Resuspend beads in 250 μl premix of 125 μl extract, 3 μl pUC (1 μg/μl), 2.5 μl 100 mM dithiothreitol, 12.5 μl 1% (v/v) Nonidet P-40 [0.05% (v/v) final concentration], and 107 μl DEPC-treated H$_2$O. The premix must equilibrate for 5 min at 26° to allow binding of inhibitors to pUC before combining it with the template beads.

4. Incubate at 26° with mild shaking for PIC assembly.

5. Concentrate on MPC for 1 min and discard supernatant.

6. Resuspend beads in 200 μl TB. Transfer into a fresh tube. Concentrate on MPC for 1 min and discard supernatant.

7. Repeat step 6.

8. Resuspend beads in one test and one control tube each in 10 μl of SDS-loading buffer and keep on ice. This sample monitors general factors within the PIC before initiation of transcription.

9. Resuspend beads in the remaining tubes in TB containing 0.5 mM of each NTP. Incubate at 26° with mild shaking for 20 min to allow transcription to proceed.

10. Repeat steps 5–7.

11. Resuspend beads in 10 μl of SDS-loading buffer. This sample monitors the promoter-bound factors after release of the polymerase.

12. Heat up all samples at 65° for 10 min. Do not boil. Concentrate for 2 min on MPC.

13. Transfer supernatant into a fresh tube and denature by incubating for 3 min at 95°.

14. Separate proteins by SDS–PAGE and analyze by Western blotting.

Activation of Reinitiation by Heat Shock Factor

Once the experimental system is well established, the effect of an activator on reinitiation may be demonstrated easily using templates in solution (i.e., not immobilized). Effects of upstream activators can be documented by comparison of reinitiation rates between templates that contain or lack the binding site for the factor or, alternatively, using transcription extracts devoid or depleted of the activator. This section provides a detailed protocol for analysis of the effect of *Drosophila* heat shock factor presented in detail elsewhere.[2,5] Three templates were used, namely a wild-type promoter, a minimal promoter, with sequences upstream of the TATA box deleted, and a promoter containing proximal heat shock elements directly upstream of the TATA box (see Fig. 4).

1. Adjust concentration of templates to 10 fmol/μl. Dispense 5.5 μl of each template into three different reaction tubes.

2. Mix 225 μl transcription extract from heat-shocked embryos (12.5 μl/ reaction), 54 μl 1 μg/μl pUC DNA, 7.2 μl 0.1 M dithiothreitol, 9 μl 5 U/μl Inhibit Ace, 4.5 μl 1% (v/v) Nonidet P-40, 18 μl 0.1 M creatine phosphate, 1.8 μl 100 ng/μl creatine phosphokinase, 40.5 μl DEPC-treated H$_2$O. Incubate premix for 5 min at 26°.

3. Add 110 μl of premix to each template. Mix gently and incubate for 40 min in a 26° water bath.

4. Prepare a set of five reaction tubes for each template. Add 5 μl of 5 mM each of all four NTPs. In the first tube of each set, add 1 μl 2.5 % (v/v) sarkosyl [final concentration 0.1 % (v/v)] to prevent reinitiation.

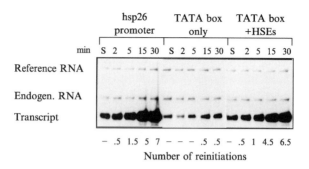

Fig. 4. An experimental result documenting HSE-dependent reinitiation from the hsp26 promoter. Experimental details are described in the text. Features of the three transcription templates are indicated above the panel. The duration of transcription (in minutes) after the formation of preinitiation complexes is indicated above the lanes. Reactions labeled "S" received sarkosyl and represent a single round of transcription. The average number of reinitiations (fold increase beyond the "S" reactions) is indicated at the bottom. High reinitiation efficiency required a promoter with intact HSEs (either in the wild-type hsp26 promoter or in a minimal promoter containing synthetic HSEs fused to a TATA box).

5. At time zero, add 20 μl of PIC assembly premix to each NTP-containing tube.

6. Add 250 μl denaturation mix to terminate transcription reactions in the second, third, fourth, and fifth tubes (see protocol for transcription on immobilized template, step 18) after 2, 5, 15, and 30 min at 26°, respectively. Terminate reaction containing N-lauryl sarkosine (first tube) 30 min after time zero. Continue with steps 16 and 17 describing the protocol for transcription on immobilized template.

7. Continue with RNA purification and primer extension analysis.

The transcription signal in the first lane of each set (i.e., containing N-lauryl sarkosine) corresponds to a single cycle of transcription (labeled "S" in Fig. 4). Any transcription signal above this level is due to reinitiation. The reinitiation rate as a function of time is determined for each template ($[signal_t\text{-}signal_0] / signal_0$). We found that there are over six rounds of reinitiation from the wild-type promoter during 30 min of transcription. In contrast, reinitiation from the minimal promoter was very poor (about 0.5 reinitiation per template molecule). Interestingly, a promoter containing just the proximal HSEs upstream of the TATA box was able to achieve a reinitiation rate similar to wild type, demonstrating that the activator is not only necessary, but also sufficient to confer maximal reinitiation rate.

Other Methodologies for Studying Reinitiation

The procedures detailed earlier described our experimental approach, which allows one to document an effect of HSF on the rate of reinitiation at the hsp26 promoter. Several other elaborate technical approaches have been employed by others to dissect transcription reinitiation, each offering certain advantages. Some rely on the prevention of *de novo* initiation, e.g., with sarkosyl,[13] or on the use of transcription extracts from conditional mutant strains.[4] Early studies used competition with reference templates in *trans* after PIC assembly.[8,14] Purified fractions have been used in multistep experiments after purification of intermediates.[1,3] The polymerase collision assay[19] uses an artificial impediment to the elongating polymerase (either a GTP-deficient reaction mix in combination with a G-less transcription cassette or specific interstrand cross-linking using a triplex-forming oligodeoxyribonucleotide with a psoralen group) such that the blocked polymerase pauses and remains associated with the template. Subsequent polymerases stall behind the paused one, creating an array of paused polymerases that can be visualized by a spectrum of shorter labeled transcripts. A rigorous comparison of all different approaches is beyond the scope of this article. Our approach, once the methodology is established, is straightforward, provided that a strong *in vitro* transcription system is at hand.

Albert Camus, through the prism of absurd logic, explained in his work "Le Mythe de Sisyphe" that Sisyphus was a passionate, optimistic, and fulfilled person, capable of comprehending the importance of his seemingly senseless repetition. Perhaps he was fulfilled because he did not need to consider speed or efficiency in order to master his task. However, when it comes to transcriptional responses of genes to the cellular environment, the possibility to adjust speed to changing needs is very important. In gene expression, repetition gives ample opportunity for regulation, as the frequency and efficiency of each cycle can be tuned. Reinitiation abbreviates the initiation process and hence leads to faster accumulation of mRNA. We are looking forward to the answer to the open question of how transcription regulators "prevent the boulder from rolling all the way down."

---

[19] M. N. Szentirmay, M. Musso, M. W. Van Dyke, and M. Sawadogo, *Nucleic Acids Res.* **26,** 2754 (1998).

Section IV

Transcription Initiation

[43] Molecular Analysis of Activator Engagement with RNA Polymerase

By LESTER J. LAMBERT, VIRGIL SCHIRF, BORRIES DEMELER, and MILTON H. WERNER

The interaction of a transactivating protein with the bacterial RNA polymerase (RNAP) is thought to be responsible for recruitment of RNAP to a specific promoter to be transcribed.[1] Numerous biochemical and genetic studies have identified the subunits of RNAP, which are targeted for protein–protein interactions by a transcription factor.[2,3] Two principal activator interaction sites within RNAP have been identified, namely the C-terminal domain of the α subunit (αCTD) and the C-terminal domain of the σ^{70} subunit. However, what an activator does to RNAP to promote transcription beyond recruitment has remained unclear due to the lack of a structural description of an activator bound to the holoenzyme. A molecular analysis of the interaction between activator and RNAP has been hampered by the difficulty in preparing large quantities of *Escherichia coli* RNAP or its subunits in complex with an activator in a form suitable for analysis by X-ray crystallography or nuclear magnetic resonance (NMR) spectroscopy. This problem has been overcome for RNAP itself, largely through the isolation and subcloning of RNAP subunits from thermophilic bacteria,[4–7] resulting in enzyme preparations that are amenable to X-ray and NMR analysis. The utilization of hyperstable enzyme components has led to the description of the three-dimensional structure of the RNAP catalytic subunits from *Thermus aquaticus*,[4] the σ^{70}-like conserved domains from the same organism,[5] and the RNA polymerase holoenzyme from *T. aquaticus*[6] and *T. thermophilus*.[7] An activator complex engaged with DNA and the *E. coli* αCTD has also been described[8] in which interactions with the activator and DNA helped to order the αCTD into the crystalline

[1] M. Ptashne and A. Gann, "*Genes and Signals.*" Cold Spring Laboratory Press, Cold Spring Harbor, NY, 2002.

[2] S. Busby and R. H. Ebright, *Cell* **79,** 743 (1994).

[3] A. Hochschild and S. L. Dove, *Cell* **92,** 597 (1998).

[4] G. Zhang, E. A. Campbell, L. Minakhin, C. Richter, K. Severinov, and S. A. Darst, *Cell* **98,** 811 (1999).

[5] E. A. Campbell, O. Muzzin, M. Chlenov, J. L. Sun, C. Anders Olson, O. Weinman, M. L. Trester-Zedlitz, and S. A. Darst, *Mol. Cell* **9,** 527 (2002).

[6] K. S. Murakami, S. Masuda, and S. A. Darst, *Science* **296,** 1280 (2002).

[7] D. G. Vassylyev, S. Sekine, O. Laptenko, J. Lee, M. N. Vassylyeva, S. Borukhov, and S. Yokoyama, *Nature* **417,** (2002).

array, resulting in suitably diffracting crystals for three-dimensional structure analysis. This article presents procedures for the preparation of free and activator-bound complexes with the second primary site of activator engagement within the bacterial RNAP, the C-terminal domain of σ^{70} known commonly as conserved region 4 (SR4). We further describe how the interactions between these proteins can be analyzed with NMR spectroscopy and protein–protein cross-linking. The methods described herein, in combination with fluorescence resonance energy transfer (FRET) of the free and activator-bound forms of RNAP[8a] can permit a detailed description of the structural context of an activator within the RNAP holoenzyme, providing first clues into how an activator may impose promoter choice on the enzyme. The interaction between SR4 and the transcriptional coactivator/repressor AsiA from T4 bacteriophage is utilized as an example.

Preparation of Proteins

SR4 from Escherichia coli σ^{70}

Structural analysis of E. coli σ^{70} has been limited by the inability, with one exception,[9] to obtain large quantities of the protein and/or its domain fragments in a form suitable for NMR or X-ray analysis. SR4 from σ^{70}, in particular, is overexpressed as a poorly behaved subdomain that does not form a discernible structure in solution as determined by circular dichroism or NMR spectroscopy. SR4 by itself falls out of solution as a "sticky" layer on regenerated cellulose membranes (such as those used in ultrafiltration) above 10 μM. However, in complex with an activator such as AsiA, SR4 forms a well-defined structure over much of its length. Thus, E. coli SR4 is prepared by isolation of the domain under denaturing conditions and is then refolded in the presence of AsiA to prepare the SR4/activator complex.

Cloning and Purification of SR4. A fragment encompassing residues 533–613 was chosen on the basis of breakpoints between predicted helix segments in the full-length protein using the PHDSec webserver.[10] The cDNA encompassing SR4 is inserted into pET30b (Novagen) and expressed without an affinity purification tag. The protein is expressed robustly in BL21(DE3) cells by 2-h induction with 1 mM

[8] B. Benoff, H. Yang, C. L. Lawson, G. Parkinson, J. Liu, E. Blatter, Y. W. Ebright, H. M. Berman, and R. H. Ebright, *Science* **297**, 1562 (2002).

[8a] T. R. Strick, A. Revyakin, J.-F. Allemand, V. Croquette, and R. H. Ebright, *Methods Enzymol.* **370** [49], this volume (2003).

[9] A. Malhotra, E. Severinova, and S. A. Darst, *Cell* **87**, 127 (1996).

[10] B. Rost and C. Sander, *Proteins* **19**, 55 (1994).

Isopropyl-β-D-thiogalactoside (IPTG) using adaptive control fermentation.[11] The culture is subsequently spun at 11,000 g for 20 min and the supernatant is discarded. The pellet is processed by homogenization to a smooth slurry in ice-cold 50 mM Tris–HCl, pH 8.0, 1 M NaCl buffer containing 10 mM benzamidine hydrochloride as a protease inhibitor. The cells are lysed on ice by a French press as described previously.[11] The lysis solution is centrifuged at 11,000 g and the supernatant discarded, as the desired product is exclusively in the insoluble fraction following lysis. Membranes and most proteins are then removed by homogenization of the pellet to a smooth slurry in ice-cold 50 mM Tris–HCl, pH 8.0, 250 mM NaCl buffer to which is added solid sodium deoxycholate to a final concentration of 5 mg/ml just before addition of the buffer to the lysis pellet. This suspension is incubated with continuous nutation for 18–24 hr at 4°. The incubation is complete when small white pellets are visible at the bottom of the nutating vessel and the pellet resulting from centrifugation of the suspension (23,000 g for 15 min) is off-white in color. The supernatant is discarded and the pellet is solubilized by homogenziation in ice-cold 50 mM Tris–HCl, pH 8.0, 50 mM NaCl, 6 M urea buffer. The dissolved pellet is then ultracentrifuged at 200,000 g for 1 h and the supernatant is recovered. The pellet following ultracentrifugation should be caramel colored and translucent. The desired protein is then isolated by cation-exchange chromatography at 4° (SP Sepharose, Amersham/Pharmacia Biotech) in 50 mM Tris–HCl, pH 7.4, 4 M urea using a salt gradient from 50 to 700 mM NaCl over 10 column volumes. Fractions containing the desired protein are identified readily by SDS–PAGE. These fractions are pooled and further purified using a 5 × 100-cm column of Sepharacryl S-100 equilibrated with 10 mM sodium phosphate, pH 7.2, 300 mM NaCl, and 4 M urea. Typically 10 to 20 mg of purified protein is obtained per gram of wet cell paste.

SR4 from Thermotoga maritima *sigA*

The conformation of SR4 alone in solution is best analyzed using the hyperstable forms of the domain, which can be subcloned from the sigA protein of thermophilic bacteria such as *T. maritima*. Unlike *E. coli* σ^{70}, free sigA in solution is quite well behaved for structural studies. SR4 from *T. maritima* sigA protein is cloned from genomic DNA isolated from 2 liters of culture grown on artificial seawater medium supplemented with lactose as described previously.[12] The culture is grown for 12 h to a density of 5.5×10^7 cells/ml and centrifuged at 11,000 g for 20 min. Genomic

[11] M. H. Werner, V. Gupta, L. J. Lambert, and T. Nagata, *Methods Enzymol.* **338,** 283 (2001).
[12] S. H. Brown, C. Sjøholm, and R. M. Kelly, *Biotech. Bioeng.* **41,** 878 (1993).

DNA is isolated using phenol/chloroform extraction as described[11] and resuspended in 50 mM Tris–HCl, pH 8.0, 1 mM Na$_2$EDTA. Standard polymerase chain reaction conditions are used to subclone a fragment encompassing residues 320–407 of sigA, equivalent to residues 533–613 of the *E. coli* protein. These two protein fragments are 65% identical in amino acid sequence. While the expression of sigA SR4 is about 10% of the level of the *E. coli* domain, sigA SR4 is expressed in folded form and isolated from the soluble fraction of a whole cell lysate. Expression is improved by subcloning into the pQE9 vector (Qiagen) in which the *lacI*^{q} gene is expressed in *trans* from a cotransformed vector (pREP4) as described by the manufacturer. Thus, HMS174 cells are grown by adaptive control fermentation[11] and induced with 0.5 mM IPTG for 2.5 h. The cells are harvested and lysed by a French press in the same buffer described for *E. coli* SR4, imidazole is added to 30 mM, and then the supernatant following centrifugation of the lysate is subsequently ultracentrifuged at 200,000 g for 1 h. This second supernatant is fractionated by Ni^{2+}-chelate chromatography using an imidazole gradient from 30 to 400 mM over 7 column volumes in 50 mM Tris–HCl, pH 7.4. A copurifying protease is removed by concentration of the desired protein fractions from Ni^{2+}-chelate chromatography to 50 ml and passage of this solution over a 5 × 100-cm Sephacryl S-100 column in PBS, pH 7.0.

T4 Bacteriophage AsiA

AsiA strongly represses gene expression from *E. coli* promoters harboring -10 and -35 consensus sequences recognized by σ^{70}. Thus, overexpression of this protein needs to be tightly controlled so that the toxicity builds up slowly as AsiA begins to block host gene expression due to its interaction with endogenous σ^{70}. This is achieved by using a salt-inducible promoter for T7 RNAP expression engineered into the BL21-SI cell line from GIBCO/BRL. Controlling the rate of polymerase expression in this cell line also regulates the rate at which the toxic protein AsiA builds up in the cell, enabling sufficient expression of the protein for structural studies by NMR. AsiA is expressed in BL21-SI cells by adaptive control fermentation[11] from the pET28b vector (Novagen) following induction with 0.5 mM IPTG for 2 h with concomitant addition of NaCl to a final concentration of 250 mM. In this case, the thrombin cleavable His$_6$ tag is retained from the cloning vector and used for Ni^{2+}-chelate chromatography purification of the SR4/AsiA complex. During growth and induction, care must be taken to maintain the concentration of NaCl in the growth medium below 30 mM prior to the induction period, as constitutive expression of toxic AsiA can result. The protein is isolated and purified under identical conditions to those described for sigA SR4.

Formation of E. coli *SR4 Complexes with AsiA*

Complexes of AsiA and *E. coli* SR4 are prepared in two ways with identical results. First, purified SR4 is diluted with phosphate-buffered saline (PBS) containing 50 mM imidazole and 6 M guanidine hydrochloride to a final concentration of 0.2–0.4 mg/ml. AsiA is added in a molar ratio of 0.8:1 AsiA:SR4, and the solution is dialyzed in a 1000 MWCO regenerated cellulose membrane at 4° against 4 liters of a buffer containing PBS and 50 mM imidazole. After 12–16 h, the dialysate is pumped over a 1-ml Ni^{2+}-chelate chromatography column, the column is washed with a further 5 column volumes and eluted with PBS containing 400 mM imidazole over 10 column volumes. In an alternative method, purified SR4 from *E. coli* is dialyzed at 4° against PBS containing 5% glycerol at a final protein concentration of 0.5 mg/ml to remove the denaturant. At low concentrations, SR4 remains in solution but is unfolded. Folding of SR4 is stimulated by the addition of AsiA at a molar ratio of 0.8:1 AsiA:SR4 and incubated an additional 20 h at 4°. Complexes of AsiA/SR4 from either method are then isolated by Ni^{2+}-chelate chromatography as described earlier. Regardless of the method, the His$_6$ affinity tag is removed by dialysis of the complex to lower the imidazole concentration to 50 mM followed by cleavage with thrombin (Amersham/Pharmacia Biotech) at 4° for 20 h at a final concentration of 15 units thrombin/mg of AsiA. Following cleavage, the reaction is quenched with 1 mM benzamidine hydrochloride, dialyzed to remove the imidazole, and concentrated to 0.7 mM in PBS (pH 7.0). Approximately 30% of the complex is lost as an off-white film on the ultrafiltration membrane used for concentration; no condition has been identified to suppress this effect.

Stoichiometry of Activator/RNAP Complexes

The stoichiometry of activator/RNAP complexes can be determined by a variety of analytical methods to include analytical ultracentrifugation and protein–protein cross-linking. To illustrate the application of each of these methods, analysis of the AsiA homodimer by analytical ultracentrifugation and application of protein–protein cross-linking to establish the stoichiometry of the AsiA/SR4 complex are described.

Analytical Ultracentrifugation of AsiA

All sedimentation velocity and equilibrium experiments are performed with a Beckman Optima XL-I analytical ultracentrifuge. Velocity, equilibrium, and Monte Carlo analyses are performed with the software package UltraScan 5.0.[13] Hydrodynamic corrections for buffer conditions are made according to data published by Laue et al.,[14] and the partial specific volume of AsiA is estimated by the method by Cohn and Edsall.[15] Monte Carlo analyses are calculated on a 40-processor Linux Beowulf cluster running Slackware Linux version 7.0. All samples are analyzed in a buffer containing 10 mM sodium phosphate, pH 6.2, 50 mM NaCl.

Sedimentation Velocity

Sedimentation velocity experiments are conducted to initially characterize AsiA with respect to composition, association state, and purity. Sedimentation velocity experiments are carried out at different concentrations[16] defined by the optical density (OD) of the protein solution (measured to be between 0.2 and 0.7) at either 230 or 280 nm. Samples are run in the AN-60-TI rotor at 20° spinning at 60,000 rpm. Double-sector aluminum centerpieces are employed, and scans are collected at either 230 or 280 nm in continuous radial mode with a 0.001-cm step size and no averaging. A van Holde–Weischet analysis of the data[17] revealed a mixture of monomer and dimer at a low concentration (230 nm) and mostly dimer at a high concentration (280 nm). Results of the velocity experiments are shown in Fig. 1. Figures 1A and 1B show experimental data at low and high concentration, respectively, and Figs. 1C and 1D show the van Holde–Weischet analysis of Figs. 1A and 1B. Figure 1E shows the combined integral distribution plot for both samples, clearly indicating a concentration-dependent, reversible association event.[18] The integral distributions of $S_{20,w}$ for the low concentration sample also display the typical half-parabola shape, indicative of self-association. The 280-nm data revealed a homogeneous composition, which made it suitable for finite element analysis, resulting in a calculated molecular mass of 19.64 ± 0.2 kDa, in good agreement with the theoretical molecular mass of the dimer (21,118 kDa). The frictional coefficient of 3.58×10^{-8} and frictional ratio, f/f_0, of 1.13

[13] B. Demeler, University of Texas Health Science Center at San Antonio, 2001. http://www.ultrascan.uthscsa.edu

[14] T. M. Laue, B. D. Shah, T. M. Ridgeway, and S. L. Pelletier, in "Analytical Ultracentrifugation in Biochemistry and Polymer Science" (S. E. Harding, A. J. Rowe, and J. C. Horton, eds.), p. 90. Cambridge University, United Kingdom, 1992.

[15] E. J. Cohn and J. T. Edsall, *Proteins, Amino Acids and Peptides as Ions and Dipolar Ions.* Reinhold, New York, 1943.

[16] L. M. Carruthers, V. R. Schirf, B. Demeler, and J. C. Hansen, *Methods Enzymol.* **321,** 66 (2000).

[17] K. E. vanHolde and W. O. Weischet, *Biopolymers* **17,** 1387 (1978).

[18] B. Demeler, H. Saber, and J. C. Hansen, *Biophys. J.* **72,** 397 (1997).

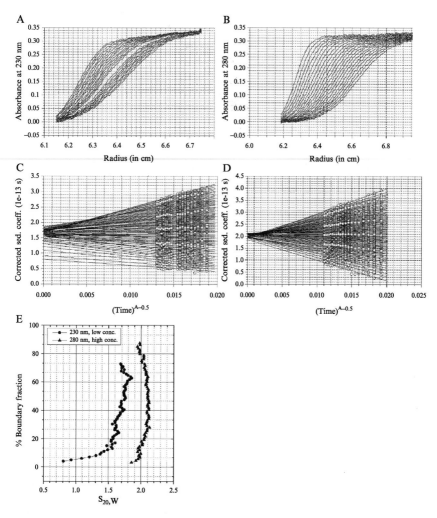

FIG. 1. Sedimentation velocity experiments used to characterize the AsiA homodimer. Sedimentation velocity absorbance scans at 60,000 rpm at low concentration (9.8 μM at 230 nm) (A) and at high concentration (68.4 μM at 280 nm) (B) and the respective van Holde–Weischet extrapolation plots (C, D) and overlaid integral distribution plots (E) are shown. Analysis results present a diffusion-corrected distribution of s values that is representative of composition. In this case, the van Holde – Weischet analysis shows evidence for the presence of both a monomer and a dimer in the low concentration sample and a completely dimerized sample at a high concentration. (E) The integral distribution plot for the low concentration sample displays the characteristic half-parabola shape indicative of a reversible association event. At a higher concentration, the distribution is nearly vertical, suggesting that the sample is homogeneous, sedimenting at ~2.0 S.

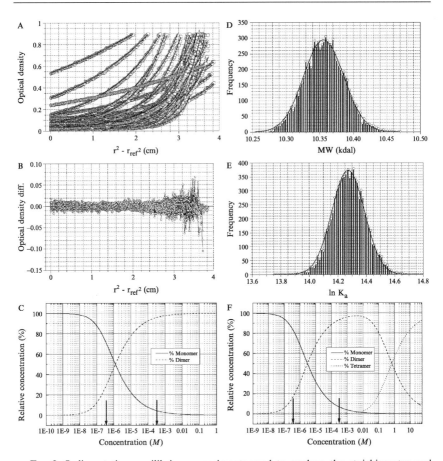

FIG. 2. Sedimentation equilibrium experiments used to analyze the stoichiometry and equilibrium association constant of AsiA. (A) Overlay of 38 wavelength scans and (B) residuals from the global fit of data as described in the text. The Monte Carlo distribution of the molecular weight (C) and the association constant (D) (ln K_a) are shown, which establish an experimental M_r 10.36 ± 0.077 kDa for the monomer and an association constant of 1.58 ± 0.404 × 10^6 M^{-1}. The equilibrium distribution plots for monomer–dimer (E) and monomer–dimer–tetramer (F) fits reveal that very little tetramer was present at the concentrations at which these measurements were taken (indicated by vertical arrows) and result in low confidence in the monomer–tetramer association constant. Thus, the system is best described by a monomer–dimer equilibrium.

suggest a mostly globular conformation of the dimer. Monte Carlo analysis of the finite element fit resulted in nearly Gaussian distributed parameters, whose distribution for the molecular weight is shown in Fig. 2D.

Sedimentation Equilibrium

Sedimentation equilibrium experiments are performed next to characterize the monomer–dimer equilibrium. Multiple loading concentrations ranging between 0.2 and 0.7 OD at each wavelength measured are analyzed simultaneously to ensure a good signal from both monomer and dimer species. Samples are run at 20° with speeds ranging between 15,000 and 50,000 rpm in double-sector epon/charcoal centerpieces using the AN-50-TI rotor. Scans are collected at equilibrium at either 230 or 280 nm in radial step mode with 0.001-cm steps and 50-point averaging. The simultaneous analysis of samples at multiple wavelengths and concentrations exploits the various absorption properties of the protein, enabling a higher degree of confidence in the fitted parameters.[19] In such a fit, parameters such as monomer molecular weight and association constants are considered global parameters and are forced to be the same for all included datasets. In order to compensate for the different absorption properties at different wavelengths, wavelength scans between 220 and 350 nm with 1-nm intervals are measured in triplicate, with 20 repetitions for each data point, all handled through the software packing running the XL-I centrifuge. Wavelength scans are fitted globally to a sum of Gaussian terms, whose width, amplitude, and offset are allowed to float but are considered global for all scans (Fig. 2A). Individual concentrations of the protein are adjusted by floating the amplitude of the sum for each scan. The resulting extinction profile is normalized by using an estimated extinction coefficient at 280 nm derived from the sequence of the denatured protein by the method of Gill and von Hippel.[20] For the global equilibrium analysis, 38 equilibrium scans from speeds ranging between 15,000 and 50,000 rpm and loading concentrations between 0.2 and 0.7 OD at both 230 and 280 nm are fitted to both monomer–dimer (Fig. 2C) and monomer–dimer–tetramer (Fig. 2F) models. Although the monomer–dimer–tetramer fit indicates the presence of small amounts of tetramer, the variance of the fit is reduced insignificantly by adding the additional parameter for the monomer–tetramer association. Thus, the system could be well described by a monomer–dimer equilibrium model, which results in random residuals and a monomer molecular mass of 10.36 ± 0.077 kDa (Fig. 2D), which is in excellent agreement with the molecular mass derived from the protein sequence (10,590). The association constant of $1.58 \pm 0.404 \times 10^6$ mol^{-1} suggests a fairly tight binding for the AsiA dimer (Fig. 2E). A distribution

[19] M. L. Johnson, J. J. Correia, D. A. Yphantis, and H. R. Halvorson, *Biophys. J.* **36,** 575 (1981).
[20] S. C. Gill and P. H. von Hippel, *Anal. Biochem.* **182,** 319 (1990).

of monomer and dimer species over the analyzed concentration range is shown in Fig. 2C. Monte Carlo analysis of the monomer–dimer–tetramer model results in a multimodal distribution for the monomer–tetramer association constant, indicating an ill-conditioned fit with multiple local minima, resulting from insufficient signal for the tetramer species. Monte Carlo distributions for molecular weight, monomer–dimer, and monomer–tetramer association constants from the monomer–dimer–tetramer fit are shown in Figs. 2C and 2F.

Protein–Protein Cross-Linking

An alternative, albeit less precise, method to determine the stoichiometry of an activator/RNAP complex is to perform protein–protein cross-linking and analyze the cross-linked species by SDS–PAGE or mass spectrometry. While mass spectrometry will undoubtedly provide the most precise measure of the complex mass and therefore the stoichiometry, mass spectrometry necessitates determination of the number of cross-links formed in order to control for the mass of the cross-linker and the number of cross-links in the interpretation of the complex composition. For this reason, a less laborious procedure, which can often result in a reasonable estimate of the apparent stoichiometry, can be achieved by SDS–PAGE of the cross-linked species. Cross-linking is particularly advantageous for AsiA/SR4, as the complexes tend to stick to many materials at modest concentrations, substantially degrading the accuracy of more precise measurements of mass such as analytical ultracentrifugation. Thus, AsiA/SR4 complexes are analyzed by cross-linking with short arm N-hydroxysuccimide (NHS) ester cross-linkers such as disuccinimidyl glutarate (DSG). DSG has a relatively short cross-linking length (7.7 Å), is water soluble, and conjugates primary amines. Because both AsiA and SR4 are rich in lysine residues, NHS ester cross-linkers are ideal for probing the interaction between these proteins. Fifteen micrograms of AsiA or the AsiA/ SR4 complex is cross-linked at a final DSG concentration of 67 μM in 10 mM sodium phosphate, pH 6.2, 50 mM NaCl and reacted for 30 min on ice in a final volume of 21 μl. The reaction is quenched with 1 μl 3 M Tris–HCl, pH 8.0, for 15 min followed by the addition of SDS loading dye. The cross-linked complexes are analyzed by 16% Tris–tricine PAGE, which reveals that the AsiA homodimer and the AsiA/SR4 complex possess nearly identical gel mobilities. Because the two proteins in the complex are nearly identical in mass, the similarity in gel mobilities suggests that one monomer of AsiA is replaced with SR4 in forming the heteromeric complex (Fig. 3). This indicates that the AsiA/SR4 complex is heterodimeric.

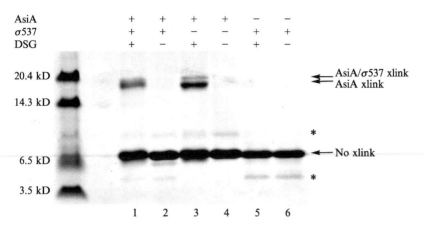

FIG. 3. Application of protein cross-linking to determine the stoichiometry of an activator/ RNAP complex. Denaturing SDS–tricine PAGE of homodimers and AsiA/SR4 complexes following cross-linking with DSG. SR4 fails to form a cross-link with DSG (lanes 5 and 6) in contrast to the AsiA homodimer (lanes 3 and 4) and the AsiA/SR4 complex (lanes 1 and 2). Cross-linking with the AsiA homodimer produces a major and minor species, which most likely result from mobility differences introduced by the number of cross-links formed between monomers (lane 3). Cross-linking the heterodimer (lane 1) produces a single species of slightly different mobility with apparently the same size as the cross-linked AsiA homodimer, suggesting that the heteromic complex with SR4 is approximately the same size as the AsiA homodimer. The percentage of the sample cross-linked in each case can be determined by quantitation of the cross-linked and uncross-linked bands indicated in the figure.

A conclusion regarding the stoichiometry of a protein–protein inter-action stabilized by a covalent crosslink can be further substantiated when subunits of a complex exchange with one another. Precross-linking the species in which a subunit exchange will take place can demonstrate that the inability for a homo-oligomer to dissociate could preclude the forma-tion of a heteromeric complex. This was indeed observed when purified AsiA (0.5 mg/ml) was cross-linked with 10 μl 2 mM DSG in dimethyl sulfoxide and incubated for 30 min at room temperature. The precross-linked AsiA failed to associate with SR4 in a Ni^{2+}-chelate pull-down assay employing a His_6-tagged SR4 molecule (Fig. 4).

The mass of the cross-linked species can be calibrated by comparison to a cross-linked, oligomeric protein(s) whose mass had been determined inde-pendently by analytical ultracentrifugation or mass spectrometry under native conditions. In this instance, the AsiA homodimer was an excellent choice given that a detailed analytical centrifugation analysis of this species had already been carried out. It is essential that the cross-linked standards used in the analysis of stoichiometry do not form artificial higher molecular

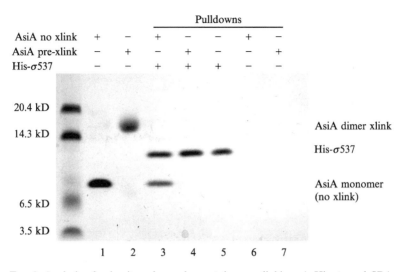

Fig. 4. Analysis of subunit exchange by protein cross-linking. A His$_6$-tagged SR4 was bound to Ni^{2+}-chelate Sepharose, and the bound suspension was used to probe the ability of SR4 to bind AsiA in the absence or presence of cross-linking. Precross-linked AsiA does not bind SR4 in this assay (lanes 2 and 4). In contrast, uncross-linked AsiA was bound efficiently by the SR4-loaded Sepharose beads (lanes 1 and 3). While the cross-linker may have simply blocked access to the binding surface for SR4 in lane 4, these data, in combination with the apparent size of the cross-linked AsiA/SR4 complex (Fig. 3), argue instead that the inability of the AsiA homodimer to dissociate following cross-linking precludes formation of the AsiA/ SR4 complex.

weight aggregates as a consequence of the cross-linking condition chosen. This can be largely prevented through the use of relatively short arm cross-linkers (less that 10 Å) and by performing a titration test for a cross-linker in which a fixed amount of protein is treated with a broad range of cross-linker concentrations and reaction times. The final choice for cross-linking condition is then made for the cross-linker concentration and time at which <10% of the interacting species is cross-linked and no "ladder" of higher-order cross-linked species is observed. Higher percentages of cross-linking are fine, as long as nonspecific aggregates are not visible in the gel.

Nuclear Magnetic Resonance Spectroscopy of Activator/RNAP Complexes

NMR spectroscopy is rapidly becoming one of the most powerful analytical tools for the analysis of macromolecular interactions. NMR can be used as an analytical tool to define structural parameters and characterize interactions between macromolecules or it can be used for complete structure determination. Thus, it is often possible to extract very detailed information from simple experiments, which can directly identify the amino acids participating in an interaction. Two methods are reviewed. In the first, a technique based on the analysis of chemical shift changes between free and bound states of a protein, recently dubbed NMR "footprinting," is described. In the second, an NMR experiment that displays interactions found only at the interface of a macromolecular complex is outlined. This experiment permits the direct identification of interacting residues at a protein–protein interface while removing the "background" of structural information that derives from either participant alone.

NMR "Footprinting"

An NMR "footprint" is measured as the change in chemical shift of a group of atoms that occurs when one molecule binds another. When the changes in chemical shift are mapped onto the three-dimensional structure of a molecule, they often cluster at a molecular surface, defining a "footprint" of the protein–protein interaction. This experiment is typically carried out with one protein labeled with ^{15}N and/or ^{13}C and all others unlabeled in a complex. An NMR experiment is employed that edits the spectrum such that only the chemical shifts of the labeled species are observed, thus the influence of a binding partner on the chemical shifts of the labeled protein are measured readily. A number of different approaches have been taken to the analysis of the chemical shift differences observed upon formation of a complex.[21–23] The most common monitor is the backbone amide proton and its attached nitrogen, as both atoms tend to be environmentally sensitive. Because the chemical shift differences are often larger for the proton than the nitrogen atom, some correction factors have been employed to attempt to give equal weight to the chemical shifts of the two atoms.[21] We prefer an alternative approach in which the chemical shift differences for proton and nitrogen represent the lengths of two sides of a right triangle. Plotting the calculated length of the hypotenuse then maps the aggregate change in chemical shift for a peak from the free to the interacting state.

The utility of such an experiment is not limited to interacting species, which are tightly bound to one another. For instance, the DNA-binding domain of the TFIIF subunit Rap30 interacts with DNA without a preference for

[21] S. Grzesiek, A. Bax, G. M. Clore, A. M. Gronenborn, J. S. Hu, J. Kaufman, I. Palmer, S. J. Stahl, and P. T. Wingfield, *Nature Struct. Biol.* **3,** 340 (1996).

[22] M. P. Foster, D. S. Wuttke, K. R. Clemens, W. Jahnke, I. Radhakrishnan, L. Tennant, M. Reymond, J. Chung, and P. E. Wright, *J. Biomol. NMR* **12,** 51 (1998).

[23] C. M. Groft, S. N. Uljon, R. Wang, and M. H. Werner, *Proc. Natl. Acad. Sci. USA* **95,** 9117 (1998).

sequence and binds only weakly to DNA ($K_d \approx 10 \ \mu M$) yet displays a clear NMR "footprint" (Fig. 5A).[23] In this case, the chemical shift differences defined a "footprint" that was in close agreement with the predicted protein/ DNA interaction surface based on the fold of the Rap30 DNA-binding domain (Fig. 5A, inset).[23] Tightly bound species that exchange subunits, such as the AsiA/SR4 complex can be more difficult to analyze, as the number of residues that experience significant chemical shift changes is likely to be high due to the exquisite sensitivity of the chemical shifts to the local environment. Moreover, the chemical shift changes are due, in this case, to subunit dissociation (AsiA dimer to AsiA monomer), as well as to the AsiA monomer binding SR4. For these reasons, the AsiA/SR4 complex displayed significant chemical shift changes across nearly one-half the length of the molecule (Fig. 5B). Thus, NMR "footprinting" failed to provide a definitive indication of the amino acids of AsiA, which may form contacts with SR4 in the complex.

The clustering of shift differences (which become the NMR "footprint") often overestimates the extent of a protein–protein interface. An example of this overestimate is the application of "footprinting" to the interaction between the phosphoprotein enriched in astrocytes, PEA-15, and ERK MAP kinase.[24] In this instance, the "footprint" of PEA-15 residues implicated in the protein interface with ERK (Fig. 5C) included a number of amino acids whose mutation did not affect binding to ERK (compare Fig. 5C with Fig. 5D).[24] The best interpretation of the "footprint," therefore, is to use it as a guide for targeted mutagenesis. A "footprint"–guided approach to choosing mutants identified the binding determinants at the PEA/ERK interface rapidly without having to test more than a fraction of the total number of residues in the molecule (Fig. 5D).

Interfacial NMR

The highest resolution approach to identifying the interacting amino acids at a protein interface is to distribute the chemical shifts from a labeled and unlabeled protein along different spectral axes in a multidimensional NMR experiment. Methods for achieving this specialized spectral editing are well established.[25–29] They rely on the ability to display the proton

[24] J. M. Hill, H. Vaidyanathan, J. W. Ramos, M. H. Ginsberg, and M. H. Werner, *EMBO J.* **21,** 6494 (2002).

[25] H. Kogler, O. W. Sorensen, G. Bodenhausen, and R. R. Ernst, *J. Magn. Reson.* **55,** 157 (1993).

[26] G. Otting, H. Senn, G. Wagner, and K. Wüthrich, *J. Magn. Reson.* **70,** 500 (1986).

[27] M. Ikura and A. Bax, *J. Am. Chem. Soc.* **114,** 2433 (1992).

[28] S. M. Pascal, A. U. Singer, G. Gish, T. Yamazaki, S. E. Shoelson, T. Pawson, L. E. Kay, and J. D. Forman-Kay, *Cell* **77,** 461 (1994).

[29] C. Zwahlen, P. Legault, S. J. F. Vincent, J. Greenblatt, R. Konrat, and L. E. Kay, *J. Am. Chem. Soc.* **119,** 6711 (1997).

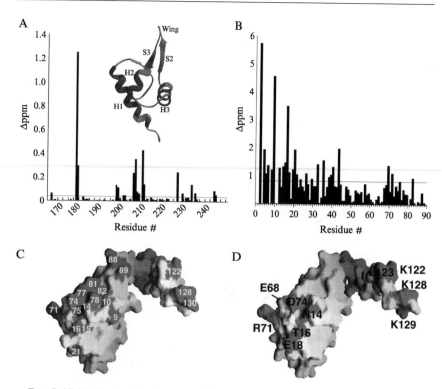

Fig. 5. NMR "footprinting" as a guide to targeted mutagenesis of a protein–protein interface. NMR "footprinting" can often lead to the identification of well-defined residues, which form a cluster on the surface of the molecule under study. (A) DNA titration of the DNA-binding domain of the TFIIF subunit Rap30 revealed a set of sharply defined shift changes that formed a specific cluster of perturbed residues on the surface of the protein.[23] In this example, the background of shift changes is minimal, indicated by the horizontal line across the plot. Mapping the shifted residues onto the three-dimensional structure of the domain (inset) defined a surface of the protein that is likely to be in contact with the DNA [red coloring on the indicated helices (H1–H3) and sheets (S1–S3)]. DNA-binding residues identified from the footprint in this instance are precisely those that would be predicted for a winged helix-turn-helix motif domain.[23] (B) A plot of chemical shift changes going from the AsiA homodimer to the AsiA/SR4 complex indicates a diffuse footprint accompanied by an intense background of shift changes (horizontal line in the plot). In this instance, one might conclude that nearly one-half of the 90 residue domain is a participant in the protein–protein interface, a conclusion that would be erroneous based on mutagenesis and NMR of the protein complex (see text).[34] (C and D) The best usage of the NMR "footprint" is as a guide for targeted mutagenesis of a protein interface, exemplified by the interaction between PEA-15 and ERK MAP kinase. (C) PEA-15 residues, which experience chemical shift changes and/or peak broadening in the presence of ERK, are indicated in red superimposed on a molecular surface representation of the structure of the protein.[24] Mutagenesis of these residues defined a subset of the footprint (shown in D), which most likely contacts ERK in the complex. (See color insert.)

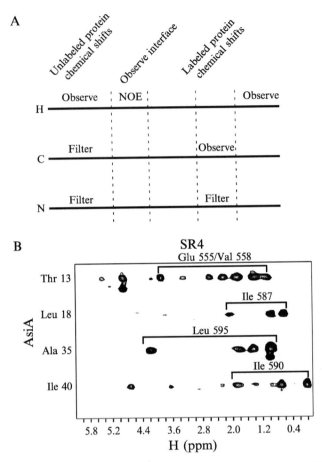

FIG. 6. NMR spectral editing techniques for interface analysis of an activator/SR4 complex. (A) Schematic of the NMR technique.[25-29] In the first half of the experiment, NMR signals of the labeled protein are suppressed, thus only the chemical shifts of the unlabeled protein are measured. The interaction between labeled and unlabeled components is then recorded as distance-dependent nuclear Overhauser effects (NOEs). Following the detection of NOEs, chemical shifts of the labeled component are recorded for both the side chain carbons and their attached hydrogens. The resultant spectrum displays only the chemical shifts of the labeled side chains, which transfer information via the NOE to the unlabeled side chains. (B) Subset of observed interfacial NOEs in the AsiA/SR4 complex. In each case, a particular CH group in the side chain of the indicated AsiA residue is shown to form multiple NOEs (bracketed) with hydrogens in SR4. Each of the NOEs represents a different hydrogen atom in the SR4 amino acid side chain (i.e., H_α, H_β) in close proximity (≤ 5 Å) to the indicated AsiA residue.

and carbon shifts along two spectral axes, while purging the carbon and attached proton chemical shifts along the third axis (Fig. 6). This third axis then displays the chemical shifts of only the unlabeled species (Fig. 6). This approach has led to identification of the residues in SR4 that are likely to be within 5 Å of residues in AsiA, a subset shown in Fig. 6B. Interestingly, many of these residues are implicated in the interaction with other activators that target SR4, suggestive of a common docking surface for SR4-interacting species.[30–33] As with NMR "footprinting," it is wise to follow-up the analysis of the edited/filtered NOE spectra with mutagenesis of the identified residues. This is particularly important for the identification of residues in the unlabeled species, as chemical shift degeneracies can make the assignment problematic.

Concluding Remarks

We have described procedures to prepare well-defined complexes of a transactivating protein with one of the primary targets in bacterial RNAP, the C-terminal conserved domain 4 (SR4) of σ^{70}. Through a combination of protein–protein cross-linking and NMR spectroscopy, the structure of the activator was described[34] and it was demonstrated that the dimeric activator dissociates to form a heterodimeric activator/RNAP complex. The stoichiometry of this complex was well described by cross-linking experiments using calibration standards whose stoichiometry was established by analytical ultracentrifugation. NMR analysis of the interactions between the activator and SR4 further identified a local surface of SR4, which is contacted by a hydrophobic surface of the activator to form the complex (Fig. 6B). These results are quite detailed and do not require a complete three-dimensional structure of the complex to be determined in order to characterize the binding surface. Thus, NMR is an immensely powerful tool for the analysis of the consequences of activator engagement with RNA polymerase.

Acknowledgment

The authors are grateful to Justine Hill for providing the images in Fig. 5C and 5D.

[30] N. Kuldell and A. Hocschild, *J. Bact.* **176,** 2991 (1994).
[31] M. Li, H. Moyle, and M. M. Susskind, *Science* **263,** 75 (1994).
[32] S.-K. Kim, K. Makino, M. Amemura, A. Nakata, and H. Shinagawa, *Mol. Gen. Genet.* **248,** 1 (1995).
[33] M. Lonetto, V. Rhodius, K. Lamberg, P. Kiley, S. Busby, and C. R. Gross, *J. Mol. Biol.* **284,** 1353 (1998).
[34] L. J. Lambert, V. Schirf, B. Demeler, M. Cadene, and M. H. Werner, *EMBO J.* **20,** 7149 (2001).

[44] Purification and Protein Interaction Assays of the VP16C Transcription Activation Domain

By Yuri A. Nedialkov, Dean D. Shooltz, and
Steven J. Triezenberg

Studies of eukaryotic transcription activator proteins that explore the mechanisms of transcription activation often address two kinds of questions. One important issue is to understand the structure of transcription activation domains (TADs), and specifically to identify those structural features most critical for their function. The second issue is to identify the interactions of transcription activators with other components of the transcription machinery.

Some structural features of TADs have been deduced from comparisons of primary structure and from mutational analyses. Many TADs can be grouped according to the predominance of particular amino acids, including acidic domains, glutamine-rich domains, and proline-rich domains.[1] In a number of cases, thorough mutational analyses have demonstrated the central importance of aromatic or bulky hydrophobic residues for the function of TADs.[2-6] Beyond these general characterizations, however, no consensus motifs have been identified as clear hallmarks of TADs. Greater insights into the secondary and tertiary structures of TADs have been hampered by the fact that many TADs, in isolation, seem disordered in aqueous solution.[7-9] Several studies have indicated that these domains assume a more ordered structure upon interaction with putative target or regulatory proteins.[10-12]

[1] P. F. Johnson, E. Sterneck, and S. C. Williams, *J. Nutr. Biochem.* **4,** 386 (1993).

[2] W. D. Cress and S. J. Triezenberg, *Science* **251,** 87 (1991).

[3] B. M. Jackson, C. M. Drysdale, K. Natarajan, and A. G. Hinnebusch, *Mol. Cell. Biol.* **16,** 5557 (1996).

[4] S. J. Triezenberg, *Curr. Opin. Genet. Dev.* **5,** 190 (1995).

[5] S. M. Sullivan, P. J. Horn, V. A. Olson, A. H. Koop, W. Nu, R. H. Ebright, and S. J. Triezenberg, *Nucleic Acids Res.* **26,** 4487 (1998).

[6] J. Lin, J. Chen, B. Elenbaas, and A. J. Levine, *Genes Dev.* **8,** 1235 (1994).

[7] F. Shen, S. J. Triezenberg, P. Hensley, D. Porter, and J. R. Knutson, *J. Biol. Chem.* **271,** 4819 (1996).

[8] P. O'Hare and G. Williams, *Biochemistry* **31,** 4150 (1992).

[9] L. Donaldson and J. P. Capone, *J. Biol. Chem.* **267,** 1411 (1992).

[10] P. H. Kussie, S. Gorina, V. Marechal, B. Elenbaas, J. Moreau, A. J. Levine, and N. P. Pavletich, *Science* **274,** 948 (1996).

[11] F. Shen, S. J. Triezenberg, P. Hensley, D. Porter, and J. R. Knutson, *J. Biol. Chem.* **271,** 4827 (1996).

Searches for proteins with which TADs interact to stimulate gene expression have embarked on both genetic and biochemical tacks and have revealed a surprising abundance of possibilities. Many of these putative targets are basal transcription factors, required to assist RNA polymerase II in establishing the preinitiation complex. The list of basal factors shown to bind TADs includes the TATA-binding protein (TBP),[13] TBP-associated factors (TAFs),[14,15] TFIIA,[16] TFIIB,[17] and TFIIH.[18] Components of the mediator protein complex, a subset of the RNA polymerase II holoenzyme, can also bind to TADs.[19,20] Other putative targets are known collectively as coactivator or adaptor proteins,[21] many of which are found in large multiprotein complexes that affect chromatin structure either by covalently modifying histones[22] or by using the energy of ATP hydrolysis in altering the interaction of nucleosomal proteins with DNA.[23,24]

This plethora of candidates raises important questions of specificity and function; with so many proteins potentially interacting with TADs, how can the biologically or mechanistically relevant interactions be distinguished? Preference will certainly be given to those interactions for which both convincing genetic and biochemical evidence can be marshalled.

VP16, a virion protein of herpes simplex viruses, is a potent transcriptional activator of viral immediate-early genes.[25] The TAD of VP16 resides in the carboxyl-terminal 78 amino acids (residues 413–490).[26] Within this domain, two distinct subregions have been identified,[5,27,28] each capable of stimulating

[12] M. Uesugi, O. Nyanguile, H. Lu, A. J. Levine, and G. L. Verdine, *Science* **277**, 1310 (1997).

[13] C. J. Ingles, M. Shales, W. D. Cress, S. J. Triezenberg, and J. Greenblatt, *Nature* **351**, 588 (1991).

[14] R. D. Klemm, J. A. Goodrich, S. L. Zhou, and R. Tjian, *Proc. Natl. Acad. Sci. USA* **92**, 5788 (1995).

[15] J. A. Goodrich, T. Hoey, C. J. Thut, A. Admon, and R. Tjian, *Cell* **75**, 519 (1993).

[16] N. Kobayashi, P. J. Horn, S. M. Sullivan, S. J. Triezenberg, T. G. Boyer, and A. J. Berk, *Mol. Cell. Biol.* **18**, 4023 (1998).

[17] Y. S. Lin, I. Ha, E. Maldonado, D. Reinberg, and M. R. Green, *Nature* **353**, 569 (1991).

[18] H. Xiao, A. Pearson, B. Coulombe, R. Truant, S. Zhang, J. L. Regier, S. J. Triezenberg, D. Reinberg, O. Flores, C. J. Ingles, and J. Greenblatt, *Mol. Cell. Biol.* **14**, 7013 (1994).

[19] S. S. Koh, A. Z. Ansari, M. Ptashne, and R. A. Young, *Mol. Cell* **1**, 895 (1998).

[20] S. Farrell, N. Simkovich, Y. Wu, A. Barberis, and M. Ptashne, *Genes Dev.* **10**, 2359 (1996).

[21] G. J. Narlikar, H. Y. Fan, and R. E. Kingston, *Cell* **108**, 475 (2002).

[22] S. L. Berger, *Curr. Opin. Genet. Dev.* **12**, 142 (2002).

[23] P. Sudarsanam and F. Winston, *Trends Genet.* **16**, 345 (2000).

[24] C. L. Peterson and J. L. Workman, *Curr. Opin. Genet. Dev.* **10**, 187 (2000).

[25] E. N. Campbell, J. W. Palfreyman, and C. M. Preston, *J. Mol. Biol.* **180**, 1 (1984).

[26] S. J. Triezenberg, R. C. Kingsbury, and S. L. McKnight, *Genes Dev.* **2**, 718 (1988).

[27] S. Walker, R. Greaves, and P. O'Hare, *Mol. Cell. Biol.* **13**, 5233 (1993).

[28] J. L. Regier, F. Shen, and S. J. Triezenberg, *Proc. Natl. Acad. Sci. USA* **90**, 883 (1993).

transcription and displaying both common and distinct activities and interacting partners. Although both subregions are rich in acidic amino acids and depend on specific aromatic residues for activity,[2,5] they have several distinguishing features. The VP16C subregion binds TBP more avidly than VP16N[11] (Y. A. Nedialkov and S. J. Triezenberg, manuscript in preparation). The VP16C fragment can also bind to human TAF32 or *Drosophila* TAF40[14,15] and can stimulate the formation of a complex comprising TFIID, TFIIA, and promoter DNA.[16] Both regions can activate reporter genes on transfected templates, but VP16N is much more effective on reporter genes integrated into a mammalian genome (P.J. Horn and S.J. Triezenberg, unpublished observations). In the more natural biological context of the herpesvirus genome, either region is sufficient to activate viral IE genes in normal patterns, although VP16C is less efficient.[29,30]

Many studies of transcriptional activation have employed the chimeric protein Gal4-VP16,[31] in which the DNA-binding domain of the yeast transcription factor Gal4 is fused to the TAD of VP16. Gal4-VP16 and deletion or substitution mutants thereof have often been used in genetic and biochemical screens for target proteins and activities.[32] This article describes a method for the purification of Gal4-VP16C that results in improved yield and quality compared with methods described previously for Gal4-VP16[32] or Gal4-VP16C.[7,16] A procedure for normalizing concentrations of Gal4-VP16C protein samples using a sandwich ELISA assay is also detailed. We describe the use of surface plasmon resonance (SPR) assays for quantitatively assessing the interactions of Gal4-VP16C and substitutions mutants thereof with target proteins. Finally, we provide a method for purifying the VP16C polypeptide, separate from any DNA-binding or purification tags, in yields and purity that may be appropriate for a range of structural analyses.

Purification of Gal4-VP16C

Rationale

Some of the activities of VP16C are distinct from those of VP16N or of the full-length activation domain. Exploration of those activities in *in vitro* experiments depends on the efficient preparation of purified protein. The following protocol offers a relatively simple method for obtaining sufficient

[29] R. Tal-Singer, R. Pichyangkura, E. Chung, T. M. Lasner, B. P. Randazzo, J. Q. Trojanowski, N. W. Fraser, and S. J. Triezenberg, *Virology* **259**, 20 (1999).
[30] W. C. Yang, G. V. Devi-Rao, P. Ghazal, E. Wagner, and S. J. Triezenberg, *J. Virol.* **76**, 12758 (2002).
[31] I. Sadowski, J. Ma, S. Triezenberg, and M. Ptashne, *Nature* **335**, 563 (1988).
[32] D. Tantin, T. Chi, R. Hori, S. Pyo, and M. Carey, *Methods Enzymol.* **274**, 133 (1996).

quantities of Gal4-VP16C protein for most biochemical and structural purposes and represents improvement in both yield and purity over previously described procedures.[7,16]

Methods

The bacterial expression plasmid pLA31ΔSma[7] encodes a fusion protein comprising the Gal4 DNA-binding domain (amino acids 1–147) fused to the carboxyl-terminal half of the VP16 activation domain (amino acids 451–490). Expression of the fusion protein is driven by the hybrid *tac* promoter[33] and thus is regulated by the lacI repressor protein in *Escherichia coli* strain Xa-90, which contains an episome bearing a *lacI^q* allele. The plasmid encodes ampicillin resistance. Comparable plasmids were constructed for expressing various substitution mutants of Gal4-VP16C,[5] and the following purification procedure was used for all such variants.

Overnight colonies plated on LB agar containing 100 μg/ml ampicillin are used to inoculate 35-ml overnight cultures in LB/Amp (100 μg/ml). This culture is then used to inoculate 3× 1-liter cultures grown at 37° to an A_{600} of 0.6 (around 2 h). Isopropyl-β-D-thiogalactoside (IPTG) is then added to a final concentration of 0.3 mM to induce expression of the fusion protein. After 3–4 h of vigorous shaking, the cells are harvested by centrifugation in 1-liter bottles (Sorvall H6000A) at 4000 rpm for 15 min. Each pellet is resuspended in 60 ml phosphate-buffered saline (PBS) by shaking at 400 rpm until the pellet is well suspended (approximately 5–10 min). The suspensions are then combined in a 250-ml centrifuge bottle and spun in a GSA rotor at 5000 rpm at 4° for 15 min. The supernatant is discarded; the pellet can be frozen at −80° if desired.

Frozen pellets are thawed at room temperature and then resuspended in 40 ml of cold PBS (10 mM Na_2HPO_4, 1.8 mM KH_2PO_4, 137 mM NaCl, pH 7.4) containing 10 μM zinc acetate, 20 mM 2-mercaptoethanol, and one tablet of complete protease inhibitor cocktail (Roche Diagnostics Corp., Indianapolis, IN). The cells are lysed by a French press (Thermo Spectronic, Rochester NY) until the optical density at 600 nm is reduced 10-fold. The lysate is centrifuged at 12,000 *g* at 4° for 15 min. To the supernatant, polyethylenimine is added dropwise while stirring at 4° to a final concentration of 0.22% (v/v), following by stirring for 10 more min. The suspension is centrifuged at 12,000 *g* for 10 min at 4°. To the cleared supernatant, dry ammonium sulfate is added gradually while stirring at 4° to a final concentration of 21% (w/v) followed by incubation at 4° for 1 h. Precipitated proteins are collected by centrifugation at 12,000 *g* for 30 min at 4°. Pellets are stored frozen at −80°.

[33] E. Amann, J. Brosius, and M. Ptashne, *Gene* **25,** 167 (1983).

Gal4-VP16C proteins are then purified by sequential chromatography using phosphocellulose and anion-exchange resins. Cellulose phosphate (P11, Whatman Specialty Products Inc., Fairfield, NJ) is activated according to the manufacturer's instruction and is then equilibrated with HEMG buffer (20 mM HEPES, pH 7.5, 1 mM EDTA, 5 mM MgCl$_2$, 10% glycerol) containing 200 mM NaCl, 10 μM zinc acetate, 0.1% NP-40, and 1 mM phenylmethylsulfonyl fluoride (HEMG$_{200}$). The precipitated proteins are resuspended in 60 ml HEMG$_{200}$ and loaded on a 90-ml bed of P11 in a 20 × 2-cm-diameter column. The P11 column is developed with a linear gradient of 200 to 1000 mM NaCl in HEMG. Fractions containing Gal4-VP16C are identified by a dot-blot immunoassay using a monoclonal antibody recognizing VP16C. Peak fractions are pooled and diluted in 10 volumes HEMG, pH 8.5 (without NaCl). This material is applied to a Spherogel TSK DEAE-5PW column (21.5 × 150 mm) using a Waters 600 E HPLC system. This column is developed with a 250-ml linear gradient of 0 to 750 mM NaCl in HEMG, pH 8.5, at a flow rate of 5 ml/min. Fractions containing the purified proteins (assessed by SDS–PAGE and silver staining) are stored as aliquots at −80°. Typical yields for Gal4-VP16C and substitutions mutants thereof range from 5 to 30 mg protein from each liter of bacterial culture medium.

Gal4-VP16C Sandwich Enzyme-Linked Immunosorbent Assay

Rationale

To obtain quantitative data in protein interaction assays, accurate measurements of the concentrations of purified Gal4-VP16C protein (wild type or mutant) are necessary. Total protein concentration assays were deemed insufficient, as they incorporate any contaminating proteins, including any proteolytic products of the desired protein. The Gal4-VP16C fusion proteins have no enzymatic activities to assay. Various DNA-binding assays (for the Gal4 domain) were deemed to be insufficiently quantitative and would not discriminate full-length product from truncated or degraded protein. Therefore, an ELISA assay was developed for normalizing concentration of the various Gal4-VP16C mutants using monoclonal antibodies that are now available commercially. This assay has the advantage of measuring intact proteins, using antibodies that recognize the Gal4 domain and the VP16C domain in a "sandwich" ELISA format[34–36] (see Fig. 1A).

[34] L. Belanger, C. Sylvestre, and D. Dufour, *Clin. Chim. Acta* **48**, 15 (1973).
[35] R. Maiolini and R. Masseyeff, *J. Immunol. Methods* **8**, 223 (1975).

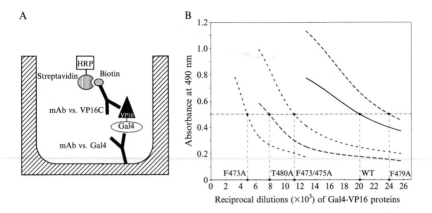

FIG. 1. (A) Schematic representation of the sandwich ELISA assay used to normalize concentrations of Gal4-VP16C fusion proteins. HRP, horseradish peroxidase; mAb, monoclonal antibody. (B) Results of ELISA assays for five different Gal4-VP16C preparations. Solid line, wild type (wt); broken lines, various substitution mutants. Horizontal line at 0.5 A_{490} is used to determine sample dilutions (vertical lines) yielding equivalent protein concentrations.

Methods

All of following steps are performed at room temperature. Immulon 4 HBX microtiter plates (Dynex Technologies, Inc., Chantilly, VA) are coated with 100 μl PBS containing 5 μg/ml of a monoclonal antibody (mAb) recognizing the Gal4 DNA-binding domain. The specific mAb employed for this purpose, designated Gal4–8, recognizes an undefined epitope within the first 92 amino acids of the Gal4 DNA-binding domain. This mAb was derived in our laboratory and is now available commercially from Zymed Laboratories, Inc., (South San Francisco, CA). After overnight incubation, a blocking solution is applied for 1 h (20 mM HEPES–HCl, pH 7.4, 137 mM NaCl, 20% normal goat serum, 0.5% Triton X-100, and 0.01% thimerosal). Wells are then washed twice with PBS containing 0.05% Tween 20 (T-PBS). Twofold serial dilutions of purified Gal4-VP16C proteins (wild-type or substitution mutants) are applied in blocking solution and plates are incubated for 1 h. Following three washes with T-PBS, plates are incubated for 40 min with biotinylated anti-VP16C mAb in blocking solution. Two different mAbs, both derived in our laboratory

[36] E. Harlow and D. Lane, "Antibodies: A Laboratory Manual." Cold Spring Harbor Laboratory Press, Cold Spring Harbor, NY, 1988.

and now licensed to Santa Cruz Biotechnology (Santa Cruz, CA), are appropriate for this purpose. The mAb designated VP16(1–21) recognizes an epitope that is disrupted by mutations of Phe residues at positions 473 and 475 of VP16C. The mAb designated VP16(7–9) recognizes an epitope that is disrupted by mutations of VP16C residues 483–486. These mAbs are labeled using ImmunoPure biotin-LC-hydrazide (Pierce, Rockford, IL) as specified by the supplier. The bound biotinylated mAb is detected by the addition of streptavidin conjugated with horseradish peroxidase (Pierce) in 1:10,000 dilution. The reaction is developed by the addition of O-phenylenediamine hydrochloride in stable peroxidase buffer (Pierce). The reaction is stopped by the addition of 1 M H_2SO_4, and the colored product is detected using a spectrophotometer (A_{490}) within 30 min.

Results

Representative results from ELISA assays of five different Gal4-VP16C protein preparations are shown in Fig. 1B. Traces shown represent smoothed fits to original data. Protein sample dilutions that contain equivalent concentrations of intact Gal4-VP16C fusion proteins are determined by identifying the dilutions that result in a given absorbance value (e.g., 0.5 as indicated by the horizontal dashed line in Fig. 1B). These values are then used to normalize Gal4-VP16C protein concentrations for subsequent use in transcription or protein interaction assays.

Biosensor Kinetic Measurements

Rationale

The binding constants describing the interaction of various basal transcription factors with wild-type or mutant Gal4-VP16C proteins can be determined using SPR. The SPR analysis described here employed a Biacore-2000 real-time kinetic interaction analysis system (Biacore, Inc., Piscataway, NJ). In this method, one of the two interacting macromolecules (designated the ligand) is tethered to a dextran-coated sensor chip. A solution bearing the second macromolecule (the "analyte") flows across the chip. Binding of the analyte to the ligand changes the refractive properties of the sensor chip, which are detected optically. These measurements are taken in real time during the flow of analyte across the chip (the binding phase) or during the subsequent flow of buffer lacking analyte (dissociation phase). Thus, this method can yield kinetic rate constants of the binding interaction. Others have reported the use of SPR assays to

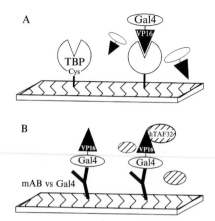

FIG. 2. Schematic representation of two alternative configurations for assessing interactions of transcription activation domains with basal transcription factors. (A) the yeast TBP protein (amino acids 61–240) is coupled covalently to Biacore CM5 sensor chips via Cys residues. Wild-type or mutant Gal4-VP16C proteins are used as analytes. (B) Gal4-VP16C is tethered by a monoclonal antibody (mAb) that is coupled covalently to the CM5 sensor chip. Samples containing putative target proteins (e.g., hTAF32) are used as analytes.

quantitate the interactions of RNA polymerase II and basal transcription factors[37] or for assessing the relative affinities of Gal4 activation domain mutants for TBP and TFIIB.[38]

Methods

TBP as Ligand in SPR Assays. In principle, either of the two partners in a binary interaction can be used as ligand in SPR assays. If a panel of mutant forms of one partner are to be compared, they are best used as analytes for binding to a uniform ligand immobilized on a sensor chip. In practice, this decision is influenced by any significant nonspecific binding of these proteins to the dextran-coated sensor chip. For example, we and others have noted a significant nonspecific binding of TBP to such chips when TBP is used as analyte,[37,38] perhaps due to a residual negative charge on unactivated portions of the sensor chip. Therefore, to permit analysis of the interaction of TBP with Gal4-VP16C and substitution mutants thereof, the TBP protein is tethered covalently to the sensor chip, and the Gal4-VP16C protein is used as an analyte (Fig. 2A).

[37] D. A. Bushnell, C. Bamdad, and R. D. Kornberg, *J. Biol. Chem.* **271,** 20170 (1996).
[38] Y. Wu, R. J. Reece, and M. Ptashne, *EMBO J.* **15,** 3951 (1996).

An amino-terminally truncated form of yeast TBP (amino acids 61–240) is purified from *E. coli* as described previously.[11] This form of TBP contains two cysteine residues at positions 78 and 164; Cys78 is more highly solvent exposed in the TBP crystal structures. The side chains of these Cys residues are used to immobilize TBP as the ligand on the sensor chip as follows. Carboxymethyl-coated sensor chips (CM5; Biacore Inc.) are activated by 0.05 MN-hydroxysuccinimide (NHS) and 0.2 M 1-ethyl-3-(3-di-methylaminopropyl)carbodiimide (EDC) as described in Biacore user manuals. A thiol-coupling reagent [2-(2-pyridinyldithio)ethaneamine hydrochloride] is then introduced (80 mM, 20 μl, 5 μl/min flow) in 0.1 M sodium borate buffer, pH 8.5. Subsequently, a solution of TBP (100 μg/ml) in 0.1 M sodium formate, pH 4.3, is passed across the sensor chip until approximately 300 response units (RU) are bound. Residual coupling groups are then blocked by washing with L-cysteine (50 mM in 1 M NaCl, 0.1 M sodium formate, pH 4.3; 20 μl, 5 μl/min flow rate). Wild-type or mutant Gal4-VP16C fusion proteins are then introduced at various concentrations (see Fig. 2) in TGEM buffer (20 mM Tris–Cl, pH 7.9, 10% glycerol, 1 mM EDTA, 5 mM MgCl$_2$, 1 mM DTT, 0.1% NP-40) containing 100 mM NaCl using a flow rate of 25 μl/min. The dissociation phase is performed using the same buffer without protein. Association rates (k_a) and dissociation rates (k_d) can be calculated using the 1:1 Langmuir association model in Biacore evaluation software version 3.0. The equilibrium dissociation constant (K_D) is calculated as k_d/k_a.

The sensor chips to which a given ligand is bound can be used repeatedly for assaying the binding parameters using various concentrations of analyte or using different analytes. However, the chips must be regenerated to remove the bound Gal4-VP16C fusion protein without denaturing the immobilized TBP. A variety of potential regeneration solutions suggested by the Biacore user manual were tested for their ability to dissociate Gal4-VP16 from TBP but to retain subsequent binding activity. The most suitable regeneration medium in our experience was 300 mM imidazole in 20% acetonitrile (data not shown). Regeneration efficiency was typically 90% or better, and sensor chips were used through three or more regeneration cycles.

Gal4-VP16C Proteins as Ligands in SPR Assays. In addition to comparing the binding of a set of Gal4-VP16C mutant proteins to a given ligand (e.g., TBP), it may be useful to compare the binding of various potential target proteins to the activator protein Gal4-VP16C. In such cases, the experiment is designed to immobilize Gal4-VP16C on the sensor chip and use the various interaction partners as analytes (Fig. 2B). For these experiments, the wild-type Gal4-VP16C fusion protein is immobilized on the detection chip as follows.

CM5 carboxymethyl sensor chips are activated with 0.2 M EDC and 0.05 M NHS according to the supplier's instructions. A monoclonal antibody recognizing the Gal4 DNA-binding domain (Gal4-8) is tethered covalently to the chip by injecting 100 μg/ml mAb diluted in 10 mM acetate buffer at pH 4.7 to yield approximately 1500–2000 redundant RU of covalently coupled protein. Unreacted ester groups are then blocked by washing with 1 M ethanolamine hydrochloride. An alternative procedure that may further reduce the residual negative charge on sensor chips has been described by others.[38] Purified Gal4-VP16C fusion proteins [or Gal4 (1-147) as a control] are then introduced at concentrations of approximately 50 μg/ml in HBS buffer (10 mM HEPES, 0.15 M NaCl, 3.4 mM EDTA, and 0.005% surfactant P20 at pH 7.4) to yield approximately 300 RU of fusion protein bound to the immobilized mAb. Sensor chips prepared in this manner can be used to assess the binding of several basal transcription factors, such as hTAF32, TFIIA, or TFIIB, presented as analytes in TGEM$_{100}$ buffer. Flow cells with immobilized Gal4-VP16C-specific mAb were regenerated using 100 mM HCl with greater than 95% efficiency. Fresh Gal4-VP16C protein was bound to the mAb following each regeneration cycle.

Results

Figure 3 presents the results of two experiments analyzing the interaction of Gal4-VP16C with TBP. In these experiments, TBP is immobilized on the sensor chip, and the Gal4-VP16C proteins are presented as analytes. The sensorgram shown in Fig. 3A describes the binding of wild-type Gal4-VP16C when presented at three different concentrations. From these association and dissociation curves, the software provided with the Biacore instrument can be used to calculate kinetic binding constants and thus an equilibrium binding constant. In this case, the calculated K_D value was 3.5×10^{-8} M. This value is quite consistent with results obtained by an equilibrium method using fluorescence spectroscopy,[11] in which the K_D was found to be 4×10^{-8} M. The convergence of the outcomes of these kinetic and equilibrium methods confers confidence in the parameters as determined under these conditions.

The sensorgram in Fig. 3B shows binding curves for wild-type and several mutant Gal4-VP16C proteins to the immobilized TBP. The reduced affinities of these mutants for TBP correlate well with their reduced abilities to activate transcription *in vitro*[16] and *in vivo*.[5] This correlation strongly supports the hypothesis that the association of this activator protein with TBP is a central feature of the mechanism of transcriptional activation.

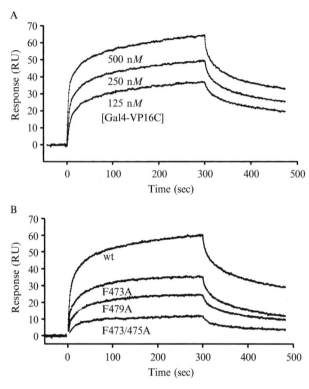

FIG. 3. Surface plasmon resonance (SPR) assays of the interaction between VP16C and yeast TBP. (A) Wild-type Gal4-VP16C binds to yTBP protein that is coupled covalently to the sensor chip. The binding phase and the dissociation phase extend from 0 to 300 sec and from 300 to 500 sec, respectively. Traces represent analyte solutions of Gal4-VP16C at three concentrations. (B) Mutations of VP16C that disrupt transcriptional activity also disrupt TBP interaction. Traces represent the binding of wild-type or mutant Gal4-VP16C proteins (at 500 nM) to yTBP.

Purification of an Isolated VP16 Activation Subdomain

Rationale

The interaction experiments described earlier explore the function of the VP16 transcriptional activation domain in an artificial context of fusion to the Gal4 DNA–binding domain. Other kinds of experiments, particularly those that focus on the physical structure of the activation domain, may be simpler to perform and interpret when the VP16 activation domain is isolated as a distinct entity. This section describes a simple and inexpensive *E. coli* expression system used to generate the isolated activation domain of VP16 (residues 413–490) or the individual subdomains VP16N (residues 413–456) or VP16C (residues 451–490). These polypeptides are obtained by proteolytic cleavage of GST fusion proteins followed by chromatographic purification.

Methods

The following procedures describe purification of the isolated C-terminal activation domain of VP16 from a 1-liter culture. Purification of either the full activation domain or VP16N will follow similar procedures.

The expression vector pGEX.VP16C (derived from pGEX2T, Amersham) encodes VP16 residues 452–490 as a thrombin-cleavable GST fusion protein under control of the *tac* promoter. pGEX.VP16C is transformed into *E. coli* strain BL21(DE3), and a 1-liter culture is grown at $37°$ to an OD_{600} of 0.6–0.9. IPTG is added to a 0.2 mM final concentration, and the culture is grown an additional 3 h at $30°$. Cells are harvested by centrifugation. Whole cell extracts can be analyzed by SDS–PAGE to monitor expression.

Cell pellets from a 1-liter growth are resuspended in 10 ml HEMGT-250 (24 mM HEPES, 0.1 mM EDTA, 12.5 mM MgCl$_2$, 10% glycerol, 0.1% Tween 20, 250 mM KCl, pH 7.9) with protease inhibitors, and lysed by two passes through a French press operating at 20,000 psi. The lysate is cleared by centrifugation at 10,000 g for 10 min at $4°$.

The lysate is diluted 10-fold with HEMGT-250 containing 5 mM DTT and is incubated with 1 ml of glutathione-Sepharose 4B (GSH) beads (Amersham) with gentle rotation for 3 h at $4°$. After incubation, the GSH resin is collected and washed sequentially with two 5-ml aliquots of HEMGT-250, two 5-ml aliquots of HEMGT-100, and two 5-ml aliquots of thrombin digestion buffer (20 mM Tris–Cl, 150 mM NaCl, 2.5 mM CaCl$_2$, pH 8.4).

While the fusion protein is still bound to the GSH beads, it is cleaved by thrombin digestion to liberate the VP16 activation domain. The digestion is performed in 2 ml of thrombin digestion buffer with 6 units of biotinylated thrombin (Novagen) and is incubated at $4°$ for 4 h with gentle rotation. Following digestion, the supernatant above the GSH resin and a subsequent 1-ml wash are collected and incubated with 0.2 ml streptavidin agarose beads (Novagen) at $4°$ for 3 h to trap the biotinylated thrombin. The supernatant above the beads and a subsequent 0.5-ml wash are collected and stored at $-80°$.

The VP16C polypeptide, once cleaved from the GST domain, can be subsequently purified through anion-exchange HPLC. The supernatant above the GSH beads, containing the products of the thrombin digestion,

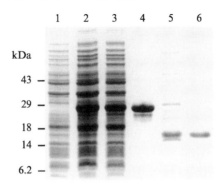

FIG. 4. Purification of VP16C polypeptide from a bacterially expressed GST fusion protein. Total cell lysates of uninduced (lane 1) and induced (lane 2) bacterial cultures containing a plasmid expressing a GST-VP16C fusion protein. Lane 3, cleared lysate of induced culture. Lane 4, proteins retained on glutathione-Sepharose beads. Lane 5, proteins released from glutathione-Sepharose by thrombin digestion. Lane 6, peak fraction of DEAE-5PW HPLC chromatography. Proteins were detected by Coomassie blue staining after SDS–polyacrylamide gel electrophoresis.

is diluted 10-fold in 20 mM Tris, pH 8.5, to reduce the salt concentration. The diluted VP16C is bound to a DEAE ion-exchange HPLC column (TSK DEAE-5PW, 2.15 × 15 cm, Beckman), and the column is developed with a 275-ml linear gradient from 0 to 750 mM NaCl in 20 mM Tris, pH 8.5 (flow rate 5 ml/min). The VP16C peptide elutes at about 435 mM NaCl.

The protein yield from this preparation is best estimated using either absorbance at 280 nm (based on a calculated molar extinction coefficient of 2560) or the bicinchoninic acid assay (BCA assay, Pierce), which provide comparable results. Coomassie dye-binding assays underrepresent the concentration of VP16C by a factor of about 3, probably due to the dearth of basic amino acids. The identity of the purified polypeptide can be verified by a Western blot with antibodies directed against the activation domain or by mass spectrometry. Polypeptide purity can be gauged by SDS–PAGE with Coomassie staining. Silver staining methods are inappropriate for VP16C, as this polypeptide stains very poorly with most silver-staining protocols. Due to its small size (4.2 kDa), the isolated VP16C activation domain migrates with the ion front in SDS–PAGE using traditional Laemmli buffers. To overcome this issue, we utilize a Tris–tricine buffer system[39,40] in which VP16C resolves well but migrates anomalously with

[39] H. Schagger and G. von Jagow, *Anal. Biochem.* **166,** 368 (1987).
[40] F. M. Ausubel, R. Brent, R. E. Kingston, D. D. Moore, J. G. Seidman, J. A. Smith, and K. Struhl, "Current Protocols in Molecular Biology." Wiley, New York, 2001.

an apparent molecular mass of about 18 kDa, probably due to its highly acidic character.

Results

Purification of the VP16C polypeptide as just described is documented in Fig. 4, with the final product showing only trace contaminants. Yields of approximately 1 mg per liter of bacterial culture are typical for this method. The resulting protein is sufficiently pure for many types of structural and interaction studies.

Acknowledgments

Our research on the VP16 transcriptional activator was supported by NIH Grant AI27323. This article builds on the efforts of former members of our laboratory, including Fan Shen, Susan Sullivan, Peter Horn, Lee Alexander, and Ryann Russell, whose contributions are greatly appreciated.

[45] Rapid Quench Mixing to Quantify Kinetics of Steps in Association of *Escherichia coli* RNA Polymerase with Promoter DNA

By Ruth M. Saecker, Oleg V. Tsodikov, Michael W. Capp, and M. Thomas Record, Jr.

Kinetic–mechanistic studies have characterized various steps in the process of forming the transcriptionally competent open complex between RNA polymerase (R) and promoter DNA (P). At the λP_R^{1-4} and *lac* $UV5^{5-8}$ promoters, investigations of the kinetics of association and dissociation as a function of temperature and (for λP_R) salt concentration led to the proposal that formation of the binary open complex (RP_o) involves at least two kinetically significant intermediates at each promoter (called I_1, I_2 at λP_R and RP_c, RP_i at *lac* UV5).

For λP_R:

$$R + P \underset{k_{-1}}{\overset{k_1}{\rightleftarrows}} I_1 \underset{k_{-2}}{\overset{k_2}{\rightleftarrows}} I_2 \underset{k_{-3}}{\overset{k_3}{\rightleftarrows}} RP_o \qquad (A)$$

Association kinetic studies are performed under pseudo first-order conditions where polymerase is in excess (at least fivefold) over promoter DNA (see Methods). At both the λP_R and the *lac* UV5 promoters, for

METHODS IN ENZYMOLOGY, VOL. 370

the conditions investigated, the step(s) involved in forming the first kinetically significant intermediate (I_1; RP_c) equilibrates rapidly on the time scale of its slow conversion to the second significant intermediate (I_2; RP_i): k_{-1} $\gg k_2$.[5,9] In both cases, the first kinetically significant intermediate (I_1; RP_c) is competitor sensitive [eliminated by a brief ($\sim$10 s) challenge with a polyanionic competitor such as heparin or poly (dAT)], whereas RP_o (as well as I_2 or RP_i) are competitor-resistant complexes under the conditions investigated.

Any investigation of the role of conformational changes in each of these steps requires a dissection of the kinetics of association to determine the equilibrium constant K_1 for the rapidly equilibrating first step in mechanism (A) and the rate constant k_2 for the subsequent conformational step. Such studies done over a broad range of temperatures and other solution variables allow structural transitions to be deduced.[1–3,5]

For the λP_R promoter in our assay buffer, conversion of I_1 to I_2 is too rapid above 15° to allow individual determination of k_2 and K_1 from analysis of kinetic data obtained by conventional manual mixing. This limitation applies to other manual-mixing studies of strong promoters, where a conventional "tau" analysis [see Eq. (9) later] typically requires a long extrapolation from data obtained at low polymerase concentrations where the kinetics are sufficiently slow. Such an extrapolation introduces a large uncertainty into the intercept ($1/k_2$) and therefore introduces a correspondingly large uncertainty in the determination of K_1. Use of a rapid mix–quench technique combined with nitrocellulose filter binding allows accurate determination of kinetic data in the millisecond to second time range.

Consequently, conditions can be investigated where the observed rate constant is determined more closely by k_2 (e.g., high polymerase concentrations).[1] The application of method described here allows a quantitative description of how the overall activation enthalpy and heat capacity of association are divided between formation of I_1 and isomerization of I_1 to I_2.[1] More generally, this technique can be exploited to dissect the kinetics

[1] R. M. Saecker, O. V. Tsodikov, K. L. McQuade, P. E. Schlax, Jr., M. W. Capp, and M. T. Record, Jr., *J. Mol. Biol.* **319**, 649 (2002).
[2] J.-H. Roe, R. R. Burgess, and M. T. Record, Jr., *J. Mol. Biol.* **184**, 441 (1985).
[3] J.-H. Roe and M. T. Record, Jr., *Biochemistry* **24**, 4721 (1985).
[4] M. L. Craig, O. V. Tsodikov, K. L. McQuade, P. E. Schlax, R. M. Saecker, M. W. Capp, and M. T. Record, Jr., *J. Mol. Biol.* **283**, 741 (1998).
[5] H. Buc and W. R. McClure, *Biochemistry* **24**, 2712 (1985).
[6] A. Spassky, K. Kirkegaard, and H. Buc, *Biochemistry* **24**, 2723 (1985).
[7] M. Buckle, I. K. Pemberton, J.-A. Jacquet, and H. Buc, *J. Mol. Biol.* **285**, 955 (1999).
[8] K. Brodolin and M. Buckle, *J. Mol. Biol.* **307**, 25 (2001).
[9] O. V. Tsodikov and M. T. Record, Jr., *Biophys. J.* **76**, 1320 (1999).

of formation of competitor-resistant complexes at other strong promoters, and thus should allow a comparison of these steps as a function of promoter DNA sequence and other solution variables.

Methods

Buffers

Our standard binding buffer (BB) contains 41 mM Tris–HCl (pH 8.0 at the temperature of the experiment), 120 mM KCl, 6.25 mM NaCl, 10 mM MgCl$_2$, 3.125% glycerol, 1 mM dithiothreitol (DTT), 100 μg/ml bovine serum albumin (Sigma). Storage buffer for RNA polymerase is 50% (v/v) glycerol, 10 mM Tris (pH 7.5 at 4°), 100 mM NaCl, 0.1 mM DTT, 0.1 mM EDTA. Wash buffer (WB) contains 10 mM Tris (pH 8.0 at room temperature), 100 mM NaCl, 0.1 mM Na$_2$EDTA.

Eσ^{70} RNA Polymerase

Escherichia coli K12 RNA polymerase holoenzyme (R) is purified by the modification of the procedure of Burgess and Jendrisak[10] described by Gonzalez et al.[11] In general, polymerase is >95% pure and >95% σ-saturated as judged by Coomassie brilliant blue R and silver-stained 6 and 12% SDS polyacrylamide gels. The physical concentration of RNA polymerase is determined by UV spectroscopy as described by Roe et al.[2] We store polymerase at −70° in storage buffer in 0.5- to 1.0-ml aliquots; working stocks are thawed and stored at −20°. Stocks of polymerase and all subsequent dilutions (see later) are made using low protein retention (e.g., silanized) polypropylene tubes.

It is invariably observed that not all polymerase molecules are active in promoter binding even at tight binding conditions. Consequently, the physical concentration of polymerase is often significantly larger than the active concentration. In doing any quantitative experiment as a function of [R]$_T$, it is of critical importance to determine the active concentration. We have noted that the binding activity of polymerase preparations increases from an initial value of ~15–40% to 60–80% over approximately a year when stored at −70° in storage buffer. However, the active concentration does decay very slowly in storage buffer at −20°. Thus it is important to determine binding activity at the time of the experiment. An activity assay can be done using forward titrations at constant [promoter] in the "complete

[10] R. R. Burgess and J. J. Jendrisak, Biochemistry 14, 4634 (1975).
[11] N. Gonzalez, J. Wiggs, and M. J. Chamberlin, Arch. Biochem. Biophys. 182, 404 (1977).

binding of limiting reagent regime" (e.g., [promoter] $\geq$5-fold greater than the observed dissociation constant, varying $[R]_T$). For λP_R, these assays are done at 1 nM promoter in BB at 28, 37, or 42° as described by Roe et al.[2]

Rapid Quench Association Kinetics Experiments

Association experiments described here were performed with a KinTek Chemical Quench Flow Model RQF-3 mixer (KinTek Co., Austin, TX), which uses only 10–20 μl volumes of sample per experiment and can measure reaction times as short as 2 ms. Formation of long-lived polymerase–DNA complexes that are stable to a challenge by the nonspecific polyanionic competitor heparin is measured as a function of time using nitrocellulose filter binding.

Use of an excess of active RNA polymerase over promoter DNA sites is necessary both to be able to neglect reductions in polymerase concentration from competitor-sensitive nonpromoter binding modes (e.g., end binding, interior nonspecific binding) and to be able to simplify the kinetic analysis in the pseudo first-order regime. Active polymerase concentrations typically range from 1 to 120 nM, whereas [DNA] is <1 nM. In the rapid mix machine used in Saecker et al.,[1] the sample loops were of equal volume (see Fig. 1). Thus the concentrations of polymerase and DNA in the injection syringes are twofold higher than the final macromolecular concentrations in the reaction loop. For example, for an experiment at 0.3 nM promoter, the λP_R [^{3}H]DNA fragment was diluted to 0.6 nM in binding buffer on ice.

Due to the sensitivity of protein nucleic acid interactions to salt concentration and to concentrations of buffer and other solutes, it is necessary to carefully design experiments in which the final concentrations of salts, buffer, and other solutes are constant at all polymerase concentrations. The combination of the small volume of polymerase used for each injection and the storage of polymerase in a buffer containing 50% glycerol required the following protocol to maintain a constant final concentration of glycerol. Serial dilutions of polymerase from the −20° stock (~5 μM) are made on ice in Tris–KCl buffer [41 mM Tris–HCl (pH 8.0 at the temperature of the experiment), 120 mM KCl, 1 mM DTT, 100 μg/ml bovine serum albumin] and adjusted to the final BB concentrations of $MgCl_2$, glycerol, and NaCl reported earlier using 40 mM $MgCl_2$, 20% (v/v) glycerol, 100 mM NaCl. These three concentrated stock solutions are made in 1× Tris–KCl buffer, pH 8.0, at the temperature investigated.

Before the start of a rapid mixing association experiment, RNA polymerase, DNA, and heparin solutions in BB are loaded into syringes,

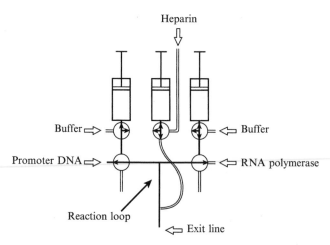

Fig. 1. A schematic of the KinTek Chemical Quench Flow Model RQF-3 mixer (KinTek Co., Austin, TX). This machine uses only ~15 μl of sample per time point and is capable of measuring time points as short as ~2 ms. RNA polymerase, promoter DNA, and the quench solution (containing a competitor such as heparin) are loaded in syringes as shown. Before mixing, samples equilibrate in the sample loops at the temperature set by the water bath (not shown). After a selected time of reaction, the sample is expelled from the sample loop using the quench solution and immediately filtered (see Methods). The reaction loop is flushed with double-distilled water and methanol, and dried before a new time point is taken.

connected to appropriate injection ports on the KinTek mixer machine, and equilibrated at the temperature of the experiment before mixing (see Fig. 1). A temperature-regulated water bath (e.g., Fisher Scientific Isotemp 1016D) is used to control the temperature of the sample and reaction loops in the KinTek apparatus, monitored using a Fluke 51k/j temperature probe. At time zero, DNA and polymerase (19 μl for each sample loop in the machine used in Saecker et al.[1]) are mixed rapidly (in less than 2 ms according to the manufacturer) in the reaction loop. At time t, the reaction is quenched and expelled from the reaction loop by mixing with ~220 μl of 100 μg/ml heparin. The entire sample is collected, filtered through nitrocellulose at room temperature, and washed with 0.5 ml of WB.

If the dissociation kinetics of preformed polymerase–promoter open complexes in the presence of heparin are sufficiently slow, as is the case for λP_R under the conditions investigated to date,[2,12] the amount of dissociation between the time from expulsion from the KinTek apparatus and filtering (30–60 s) will not be significant. If necessary, the salt concentration

[12] K. L. McQuade, Ph.D. Thesis, University of Wisconsin-Madison, Madison, WI, 1996.

of the quench solution can be adjusted to reduce the dissociation rate of competitor-resistant complexes (i.e., I_2 and RP_o) while allowing the competitor to trap free polymerase and polymerase bound originally in the competitor-sensitive complexes (i.e., I_1). Before the next sample is loaded, the reaction loop is flushed with water and then methanol and is dried by pulling air though the reaction loop. Before taking the next time point, the loop must be completely dry to ensure accurate sample volumes.

Nitrocellulose Filter-Binding Assays

Prior to use in binding assays, 0.45-μm nitrocellulose filters (BA85, Schleicher & Schuell, Inc., Keene, NH) are rinsed with either H_2O or WB and then soaked in WB for at least 30 min. Sample filtration and subsequent washing are done under a vacuum of 15–20 in of Hg using a 10-place filter manifold (Amersham Biosciences Corp., Piscataway, NJ). Total radiolabeled promoter DNA filtered (cpm_{TOT}, see later) is determined by spotting 10–100 μl from the reaction mix on a nitrocellulose filter. Following washing or spotting, each filter is transferred to a glass scintillation vial, dried under heat lamps, cooled, and dissolved for approximately 1 h in 0.75 ml of Cellosolve (ethylene glycol monoethyl ether, Sigma, St. Louis, MO). After filters are dissolved, 5 or 10 ml of ReadySafe liquid scintillation cocktail (Beckman Instruments, Fullerton, CA) and 0.1 ml of H_2O are added to each vial. Vials are vortexed vigorously until the solution is homogeneous. Counts per minute (cpm) in each vial are measured in a scintillation counter such as the Packard Tri-Carb 2100 TR (Perkin Elmer, Boston, MA).

Data Analysis to Obtain Observed Rate Constants and Fractional Occupancy of Promoter DNA

In the set of experiments reported in Saecker *et al.*[1] using a ^{3}H replacement-labeled 898-bp DNA fragment containing the λP_R promoter, we found that polymerase forms competitor-resistant complexes with a fraction of the DNA promoter molecules (10–20%) at very short times at all temperatures and polymerase concentrations. Because the fraction of DNA bound in these fast-forming complexes is not a function of polymerase concentration or temperature, it was concluded that these complexes are not bound at the promoter and that their formation is independent of formation of complexes at the promoter. We hypothesized that these fast-forming complexes result from polymerase binding to a subpopulation of DNA that is damaged by labeling and/or subsequent radioactive decay processes. For example, incomplete fill in by T4 DNA polymerase in the labeling reaction would produce a fragment with single-stranded 5' ends, which may mimic aspects of "tailed templates" and/or forks, which have been shown to bind holoenzyme tightly.[13] In general, because polymerase binds avidly to DNA ends, nicks, single-stranded DNA, and nonpromoter tight binding sites,[14] both the labeling method and the construction of the DNA fragment used in any experiment need to be evaluated carefully. If very high concentrations of protein are employed (≥ 0.3 μM), aggregation of both core and holoenzyme must also be considered, especially at low [salt].[15]

For both reversible- and irreversible-binding conditions, observed rate constants (β_{CR}) for forming competitor-resistant polymerase–DNA complexes at a given RNAP concentration were determined by nonlinear fitting of experimental data to the single-exponential dependence:

$$\text{cpm}_t = \text{cpm}_o + (\text{cpm}_{\text{plateau}} - \text{cpm}_o)(1 - e^{-\beta_{CR}t}) \tag{1a}$$

where counts per minute at time t (cpm_t) values were corrected for background retention of DNA on the filters in the absence of polymerase (typically $< 5\%$). In Eq. (1a), $\text{cpm}_{\text{plateau}}$ is the limiting value of cpm_{obs} as $t \to \infty$ (i.e., the plateau value at equilibrium) and cpm_o is the initial ($t = 0$) counts per minute. Because ~10–20% of the population of competitor-resistant complexes formed in the dead time of the experiments reported in Saecker et al.,[1] cpm_o was used as a fitting parameter. If this fast phase arises from damaged sites introduced by replacement labeling, it may not be significant for end-labeled DNA. In this case, cpm_o can be set equal to zero in Eq. (1a), yielding

$$\text{cpm}_t = (\text{cpm}_{\text{plateau}})(1 - e^{-\beta_{CR}t}) \tag{1b}$$

At all temperatures, the fraction (θ_{obs}^{CR}) of DNA in the form of competitor-resistant complexes at time t was determined from the ratio of the observed cpm to the total cpm (cpm_{TOT}), corrected both for background retention of DNA in the absence of polymerase and for filter efficiency E:

$$\theta_{\text{obs}}^{CR} = \frac{\text{cpm}_t}{\text{cpm}_{\text{TOT}}}\frac{1}{E} \quad \text{where } E \equiv \frac{\text{cpm}_{\text{plateau}}}{\text{cpm}_{\text{TOT}}} \tag{2}$$

The filter efficiency E is assumed to be independent of temperature. At $37°$ where 100% of λP_R promoter sites are occupied in excess RNA

[13] J. D. Helmann and P. L. deHaseth, Biochemistry 38, 5959 (1999).
[14] M. T. Record, Jr., W. S. Reznikoff, M. L. Craig, K. L. McQuade, and P. J. Schlax, in Escherichia coli and Salmonella typhimurium: Cellular and Molecular Biology (F. C. Neidhardt, ed.), 2nd Ed., Vol. I, p. 792. ASM Press, Washington, DC, 1996.
[15] S. L. Shaner, D. M. Piatt, C. G. Wensley, H. Yu, R. R. Burgess, and M. T. Record, Jr., Biochemistry 21, 5539 (1982).

polymerase,[4] E is determined directly. Values of E range from 70 to 100%, depending on DNA preparation and, in particular, on the extent to which unincorporated [^{3}H]ATP was removed. No significant difference in values of $cpm_{plateau}/cpm_{TOT} = E$ was observed in association experiments between 20 and 42°; therefore, we conclude that $\theta_{obs}^{CR} = 1$ in this temperature range and that E is independent of temperature.

On the 898-bp DNA fragment containing the λP_R promoter, experimental values of θ_{obs}^{CR} reflect both promoter– and nonpromoter–polymerase complexes. Because the formations of promoter ($\theta_{\lambda P_R}^{CR}$) and nonpromoter complexes [$\theta_{NP}^{CR} \equiv cpm_o/(cpm_{TOT}E)$] are independent events, we obtain

$$\theta_{obs}^{CR} = \theta_{\lambda P_R}^{CR} + \theta_{NP}^{CR} - \theta_{\lambda P_R}^{CR}\theta_{NP}^{CR} \tag{3}$$

The product in Eq. (3) corrects for the population of DNA fragments that have polymerase bound both at a nonpromoter site and at the promoter. At equilibrium, the fraction of DNA molecules with the promoter site occupied by polymerase in a competitor-resistant complex, $\theta_{\lambda P_R}^{eq,CR}$, is therefore

$$\theta_{\lambda P_R}^{CR,eq} = \frac{\theta_{obs}^{CR,eq} - \theta_{NP}^{CR}}{1 - \theta_{NP}^{CR}} \tag{4}$$

At temperatures in the range of 20–42°, where $\theta_{obs}^{CR,eq} = 1$ in excess $[R]_T$, Eq. (4) yields $\theta_{\lambda P_R}^{CR,\,eq} = 1$, as expected.

Kinetic Background

For all reaction conditions studied, the entire time course of formation of competitor-resistant complexes (RP_{CR}) at the λP_R promoter (P) in excess RNA polymerase is first order with observed rate constant β_{CR}:

$$-\frac{d\ln\Delta[RP_{CR}]}{dt} = \beta_{CR} \quad \text{or} \quad [RP_{CR}] = [RP_{CR}]_{eq}(1 - e^{-\beta_{CR}t}) \tag{5}$$

where $\Delta[RP_{CR}] \equiv [RP_{CR}]_{eq} - [RP_{CR}]$ and where the equilibrium concentration of competitor-resistant complexes at the promoter ($[RP_{CR}]_{eq}$) is related to the total promoter concentration $[P]_T$ by

$$\frac{[RP_{CR}]_{eq}}{[P]_T} \equiv \theta_{\lambda P_R}^{CR,eq} = \frac{[I_2] + [RP_o]}{[P]_T} = \frac{K_1 K_2(1 + K_3)[R]_T}{1 + K_1[1 + K_2(1 + K_3)][R]_T} \tag{8}$$

Where the formation of competitor-resistant complexes is irreversible, $\theta_{\lambda P_R}^{CR,eq} = 1$. More generally, where the formation of these complexes is reversible and the kinetics exhibit a decay to equilibrium, then $\theta_{\lambda P_R}^{CR,eq} < 1$. Kinetic data obtained under both reversible and irreversible conditions are analyzed using nonlinear fitting to Eq. (1) or (5) to obtain the rate constant β_{CR}.

Equation (5) can also be expressed in the form[9]:

$$\frac{d[RP_{CR}]}{dt} = \alpha_{CR}[P]_T - \beta_{CR}[RP_{CR}] \tag{7}$$

where the composite rate constant α_{CR} characteristic of the forward direction of the process in excess polymerase is related to the observed rate constant β_{CR} by the quantity $\theta_{\lambda P_R}^{CR,eq}$ [defined in Eq. (4)]:

$$\alpha_{CR} = \beta_{CR}\theta_{\lambda P_R}^{CR,eq} \tag{8}$$

In excess $[R]_T \gg [P]_T$, the rate constant α_{CR} is observed to be a hyperbolic function of polymerase concentration, and hence α_{CR}^{-1} is linear in $[R]_T^{-1}$ as represented by the empirical equation (commonly referred to as a tau analysis):

$$\frac{1}{\alpha_{CR}} = \frac{1}{k_a[R]_T} + \frac{1}{k_i} \tag{9}$$

where k_a designates a composite second-order association rate constant and k_i is a composite first-order isomerization rate constant. Equation (9) is a general result for mechanism (A) when the observed reversible or irreversible kinetics exhibit single exponential behavior. For this case, no approximations (e.g., steady state or rapid equilibrium assumptions) are required to derive the hyperbolic dependence of α_{CR} on $[R]_T$ represented in Eq. (9). Indeed, if single exponential behavior is observed experimentally for $[R]_T \sim k_a/k_i$, it follows that the initial binding step ($R + P \rightleftarrows I_1$) is in rapid equilibrium on the time scale of the second step ($I_1 \rightarrow I_2$). If the formation of competitor-resistant complexes is single exponential, then (without approximation)[9]:

$$k_a = K_1 k_2 \tag{10}$$

$$k_i = k_2 \tag{11}$$

Eqs. (9)–(11) yield the expression for α_{CR}:

$$\alpha_{CR} = \frac{k_a K_i [R]_T}{k_i + k_a[R]_T} = \frac{k_2 K_1 [R]_T}{1 + K_1[R]_T} \tag{12}$$

To obtain values of K_1 and k_2, the hyperbolic dependence of α_{CR} on polymerase concentration ($[R]_T$) should be analyzed by nonlinear fitting to Eq. (12).

From Eqs. (6), (8), and (12), the observed relaxation rate constant β_{CR} is

$$\beta_{CR} = \frac{k_2 K_1 [R]_T}{1 + K_1 [R]_T} + \frac{k_{-2}}{1 + K_3} \tag{13}$$

which is of the general form for a relaxation rate constant: $\beta_{CR} = \alpha_{CR} + k_d$, where α_{CR} is given by Eq. (12) and

$$k_d = \frac{k_{-2}}{1 + K_3} \tag{14}$$

Unlike α_{CR} [Eq. (12)], β_{CR} is not a hyperbolic function of $[R]_T$ and $(\beta_{CR})^{-1}$ is not linear in $([R]_T)^{-1}$. At low temperatures[1] or, more generally, for promoter sequences or polymerase variants where the formation of competitor-resistant complexes is significantly reversible (i.e., $\alpha_{CR} \neq \beta_{CR}$), the observed rate constant β_{CR} is therefore converted to α_{CR} using $\theta_{\lambda P_R}^{CR,eq}$ [Eq. (8) or k_d (Eq. (14)] before analysis to obtain K_1 and k_2. In reversible-binding situations where β_{CR} is determined, particular care is required to ensure that the quench procedure does not affect the determination of $[RP_{CR}]$.

Representative Rapid Quench Association Experiment at 37°

Figure 2 shows a representative example of raw data obtained using the method described here. The process of forming competitor-resistant complexes I_2 and RP_o at the λP_R promoter at 37° in BB at 123 nM RNA polymerase is fast and irreversible, going to completion in 20 s. (The value of cpm_t at 20 s is the same within uncertainty as that predicted from cpm_{TOT} and the filter efficiency E for $\theta_{\lambda P_R}^{CR,eq} = 1$.) Analysis of these data using Eq. (1a) yields a rate constant $\beta_{CR} = 0.25 \pm 0.02$ s^{-1}. Examples of kinetic data obtained using a combination of rapid and conventional mixing over a wide range of polymerase concentrations and reaction times are given in Saecker et al.[1] The combination of conventional and rapid quench mixing methods allows the accurate determination of the rate constant α_{CR} (equal to β_{CR} at 37° because $\theta_{\lambda P_R}^{CR,eq} = 1$) as a function of $[R]_T$ over two orders of magnitude in polymerase concentration (~1–125 nM).

Concluding Remarks

Rapid quench mixing of radiolabeled λP_R promoter DNA with E. coli RNA polymerase in a commercially available apparatus provides an accurate and efficient method of investigating the kinetics of association and subsequent conformational changes involved in forming long-lived complexes. Promoter binding as a function of time is assayed by nitrocellulose filter binding after quenching with a competitor (e.g., heparin). Under all conditions examined, the kinetics of formation of competitor-resistant complexes at the λP_R promoter (I_2, RP_o) are single exponential with

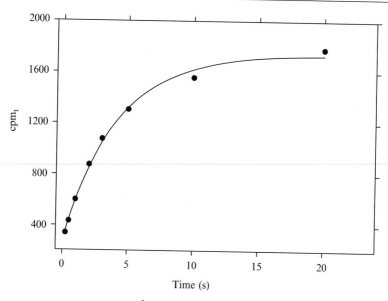

F$_{\text{IG}}$. 2. A representative time course for an association experiment with 123 nM RNA polymerase and 0.3 nM λP$_R$ [^{3}H]DNA at 37° in BB. Under these conditions, formation of competitor-resistant complexes is fast and irreversible ($\theta_{\lambda P_R}^{CR} = 1$), occurring on a timescale (<20 s) not accessible using conventional manual mixing. Fit of these data to Eq. (1a) yields the rate constant $\beta_{CR} = \alpha_{CR} = 0.25 \pm 0.02$ s^{-1} and background-corrected values of the initial (cpm$_o$) and plateau (cpm$_{plateau}$) counts per minute of 262 ± 33 and 1745 ± 38, respectively. For this experiment, the background-corrected total counts per minute (cpm$_{TOT}$) was 2450, and the filter efficiency was 0.71 ± 0.02, in the range (0.7–0.75) observed in this series of experiments.[1]

first-order rate constant β_{CR}. Interpretation of the polymerase concentration dependence of β_{CR} in terms of the three-step mechanism of open complex formation yields the equilibrium constant K_1 for formation of the first kinetically significant intermediate (I$_1$) and the forward rate constant (k_2) for the conformational change that converts I$_1$ to the second kinetically significant intermediate I$_2$:

$$R + P \underset{}{\overset{K_1}{\rightleftharpoons}} I_1 \overset{k_2}{\rightarrow} I_2.$$

Whereas previous manual mixing experiments at this promoter required conditions where association occurs on a time scale of minutes (e.g., low polymerase concentrations), fast mixing allows the association process to be investigated at high polymerase concentrations and solution conditions where the time scale is milliseconds or seconds. We exploited this method to determine K_1 and k_2 over a wide temperature range, which allowed a characterization of conformational changes occuring early in the process of open complex formation.[1] This method should be generally applicable to kinetic studies of polymerase–promoter interactions and thus should extend biophysical characterizations of transcription initiation.

Acknowledgments

Support of our research on the kinetics and mechanism of open complex formation is provided by NIH Grant GM23467 to M.T.R.

[46] Determination of RNA Polymerase Binding and Isomerization Parameters by Measuring Abortive Initiations

By Siddhartha Roy

Transcription initiation is a multistep process, not all of which has been well defined. Since the majority of transcription regulation occurs at these steps,[1] there is a need to obtain more information about these steps. To a first approximation, the first step in initiation is the formation of a closed complex from free promoter and RNA polymerase. This is followed by a series of isomerization steps culminating in an open complex. The final step is promoter clearance.[2] Each of these steps can be subdivided into substeps.

It would be ideal to have methods to assay each step separately and assess how regulatory molecules affect each of these steps. The abortive initiation method was developed as a kinetic method, which can separate the bimolecular closed complex formation step from the unimolecular isomerization step.[3,4] As mentioned, the latter step is composed of several resolved and unresolved unimolecular steps. The overall forward rate constant, obtained from the classical two-step scheme, probably largely reflects the slowest step in the pathway. Even today abortive initiation remains one of the best methods to estimate the rate constant for the isomerization and stability of the closed complex.[5] Previously, Gussin

[1] G. Lloyd, P. Landini, and S. Busby, *Essays Biochem.* **37**, 17 (2001).

[2] P. L. deHaseth, M. L. Zupancic, and M. T. Record, Jr., *J. Bacteriol.* **180**, 3019 (1998).

[3] C. L. Cech, J. Lichy, and W. R. McClure, *J. Biol. Chem.* **255**, 1763 (1980).

[4] W. R. McClure, *Proc. Natl. Acad. Sci. USA* **77**, 5634 (1980).

[5] G. N. Gussin, *Methods Enzymol.* **273**, 45 (1996).

discussed several aspects of abortive initiation assays.[5] This article focuses on those aspects that were either not treated or treated briefly in that article.

Principle of the Abortive Initiation Assay

The abortive initiation method hinges on the catalytic competence of the open complex to produce di- and trinucleotide products. The implicit assumption underlying the method is that the transcription initiation is a multistep sequential process without kinetically significant branches, although the generality of such an assumption has been challenged.[6,7] If the branched pathways are assumed to be kinetically insignificant, then the rates of formation of the di- or trinucleotide products are directly related to the concentration of the open complex. The rate of open complex formation is a function of the stability of the closed complex (K_s or K_B) and the rate of its isomerization (k_f) to open complex. Information about K_B can be obtained from the RNA polymerase concentration dependence of the rate of open complex formation, whereas information about k_f can be obtained from the RNA polymerase concentration-independent part. The scheme shown here is a two-step version of the transcription initiation and abortive initiation.

$$\text{R} + \text{P} \underset{k_{-1}}{\overset{k_1}{\rightleftharpoons}} \text{R·P}_c \xrightarrow{k_2} \text{R·P}_o \begin{cases} \text{trinucleotide + pyrophosphate} \\ \\ \text{dinucleotide + NTP} \end{cases}$$

Usually, a dinucleotide monophosphate corresponding to the first two nucleotides (at positions $+1$ and $+2$) is used as the initiating nucleotide and only one nucleoside triphosphate (corresponding to the third nucleotide in the template) is used.

Data Processing

Analysis of abortive initiation data is an involved process, and the true statistical significance of the obtained values of the kinetic constants can be uncertain in many cases. The most widely used technique today is to fit the *[product] versus time* profile at a single RNA polymerase concentration using nonlinear least-squares fit (with or without proper weighing) to

[6] R. Sen, H. Nagai, and N. Shimamoto, *J. Biol. Chem.* **275,** 10899 (2000).
[7] M. Susa, R. Sen, and N. Shimamoto, *J. Biol. Chem.* **277,** 15407 (2002).

obtain the value of lag time (τ_{obs}). A plot of τ_{obs} versus $RNAP^{-1}$ yields the K_B and the k_f.[5] Alternatively, a global-fit procedure can be used to extract the kinetic parameters directly. Global fitting of data sets is expected to be more robust against introduction of noise. In this procedure, data at different RNA polymerase concentrations are fitted simultaneously to obtain K_B and k_f directly. We estimated the robustness of the global fit procedure using simulated data. *[product] versus time* profile data were generated using a MATLAB (The Mathworks Inc., Natick, MA) program and the noise was added using a random number generator, which returns a random number whose values are normally distributed around a mean. Data generated in such a manner were then fitted globally using a MATLAB program, which uses a SIMPLEX optimization algorithm. To compare, the same data were analyzed individually using a nonlinear least-square fit to extract the value of lag time (τ_{obs}) and plotting τ_{obs} versus $RNAP^{-1}$ to extract the values of K_B and k_f. Figure 1 shows the global fit of simulated data at a certain noise level. When data were generated using a K_B value of $10^8 \, M^{-1}$ and a k_f value of $0.6 \, min^{-1}$ at this noise level, the global fit converged to a k_f value of $0.41 \, min^{-1}$ and a K_B value of $1.35 \times 10^8 \, M^{-1}$. When the same data were analyzed individually to extract the lag time and subsequently extract the values of the kinetic constants,

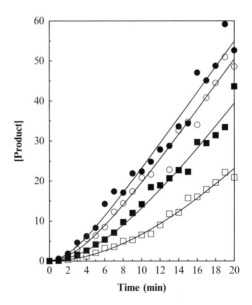

Fig. 1. Global fitting of simulated abortive initiation data. Data were generated using $k_f = 0.6$ and $K_B = 10^8$ with normally distributed noise as described in the text. The RNA polymerase concentrations chosen are 1 nM (□), 3 nM (■), 10 nM (○) and 30 nM (●). Data were generated by MATLAB using the method described in the text.

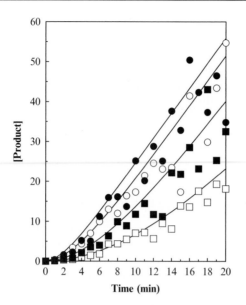

FIG. 2. Global fitting of simulated abortive initiation data generated with increased noise but with same k_f and K_B as in Fig. 1. Symbols are the same as in Fig. 1.

the values obtained were $k_f = 0.34$ min^{-1} and $K_B = 2.72 \times 10^8$ M^{-1}. With the noise level doubled, the global fit of data is shown in Fig. 2. The extracted values were $k_f = 0.57$ min^{-1} and $K_B = 9.68 \times 10^7$ M^{-1} by global fit and $k_f = 0.25$ min^{-1} and $K_B = 4.15 \times 10^8$ M^{-1} by individual fit. As expected, the global fit appears to be superior to individual fit, particularly with significant noise level.

An important dimension to processing abortive initiation data is regarding the steady-state assumption. In the original equations of McClure[4] which has been used widely, it was assumed that the closed complex is in steady state during progress to the open complex. Although this assumption is a good approximation for many promoters, it may introduce significant error for weak promoters having k_f less than 10^{-3} s^{-1}. In such a situation, it is appropriate to use the equation derived without the steady-state assumption[8]:

$$[\text{product}] = k_{cat} \cdot P_0[t - \{(k_1 \cdot R_0 + k_{-1} + k_2)/(k_1 \cdot k_2 \cdot R_0)\}$$
$$-\{2M/N(M-N) \exp (N_t/2)\} + \{2N/M(M-N) \exp (M_t/2)\}]$$

where $M = 0.5[-(k_1 R_0 + k_{-1} + k_2) - \sqrt{\{(k_1 R_0 + k_{-1} + k_2)^2 - 4 \cdot (k_1 k_2 R_0)\}}]$ and $N = 0.5[-(k_1 R_0 + k_{-1} + k_2) + \sqrt{\{(k_1 R_0 + k_{-1} + k_2)^2 - 4 \cdot (k_1 k_2 R_0)\}}]$

[8] S. Roy, *Anal. Biochem.* **271**, 86 (1999).

Thus, global fit using the analytical solution to abortive initiation rate equation offers a universal and superior method of obtaining the kinetic constants.

Although it is convenient to analyze abortive initiation data using a two-step model, useful information can be obtained for more complex kinetic schemes. A three-step model, consisting of an intermediate between the closed and the open complex, has been proposed for some promoters. In this model, the relationship between microscopic rate constants and kinetic parameters derived from abortive initiation data has been derived.[5]

Fluorescence-based Abortive Initiation Assays

In an earlier version of this method, radioactive nucleoside triphosphates were used to assay tri- or tetranucleotide formation. An abortive initiation assay system based on fluorescence has been developed.[9] One of the major advantages of this assay using hydrolysis of γ-ANS-UTP to ANS-PP$_i$ is its continuous nature. The quantum yield of γ-ANS-UTP is low, whereas it is many fold higher in the ANS-PP$_i$. Thus, hydrolysis of the α–β P-O-P bond of the γ-ANS-UTP during transcription leads to a significant fluorescence intensity increase. In the fluorescence-detected abortive initiation assay, usually γ-ANS-UTP and a dinucleoside monophosphate are used.

Synthesis and purification of γ-AmNS-UTP are straightforward and can be done using the protocol developed by Yarbrough et al.[10] γ-ANS-UTP (1-aminonaphthalene 5-sulfonate) is synthesized by water-soluble carbodiimide condensation of UTP and 1-amino naphthalene 5-sulfonate. The reaction is usually conducted in a large excess of 1-amino naphthalene 5-sulfonate.

The following solutions are prepared before initiating the reaction:

1. Ten milliliters of saturated solution of 1-amino naphthalene 5-sulfonate (approximately 0.2 M) after adjustment to pH 5.8 by NaOH solution
2. Four milliliters of UTP solution (12.5 mM), pH adjusted to 5.8
3. Two milliliters of 1.0 M water-soluble carbodiimide [1-ethyl-3-(dimethylaminopropyl) carbodiimide]

[9] P. J. Schlax, M. W. Capp, and M. T. Record, Jr., J. Mol. Biol 245, 331 (1995).
[10] L. R. Yarbrough, J. G. Schlageck, and M. Baughman, J. Biol. Chem 254, 12069 (1979).

Solutions 1 and 2 are mixed first followed by addition of the carbodiimide solution at room temperature. Reaction time is 2.5 h, and occasional adjustment of pH is necessary with a dilute NaOH solution. Completion of the reaction can be checked by thin-layer chromatography. For nucleotides, charge-based separation on polyethyleneimine cellulose plates is often useful.[11,12] The original purification was performed on DEAE-cellulose using a gradient of volatile buffer, which can be removed later by repeated lyophilization (0.05 to 0.4 M triethylamine bicarbonate, pH 7.5). An equally good alternative is reverse-phase chromatography. An acetonitrile gradient in 0.1 M triethylammonium acetate, pH 7.0, buffer on C-18 columns gives good separation and ease of separation.[13,14] Any excess buffer salts can be removed by a small Sephadex G-10 column after thorough drying.

Fluorescence-detected abortive initiation (FDAI) assays are typically performed at 37°. Fluorescence excitation and emission may be set at 360 and 460 nm, respectively. Buffers should be chosen such that there is no significant absorbance or fluorescence at the wavelengths of excitation and emission. Schlax and co-workers[9] have used 25 mM HEPES buffer, pH 7.5, containing 212 mM NaCl, 3 mM MgCl$_2$, 0.1 mM EDTA, 1 mM dithiothreitol, 6.2% glycerol, 0.1 mg/ml bovine serum albumin and a carryover of 1.0 mM Tris for the FDAI assay when studying the *Escherichia coli lac* promoter.[9] The initiating dinucleotide may be added at a concentration of 1.0 mM and γ-AmNS-UTP at 75 μM. It is preferable to use a reference standard to correct for lamp intensity fluctuations and drift. This is done with rhodamine B as a standard. In two-channel spectrofluorometers, such as SLM Aminco 8000C, the reference may be placed in the second channel whereas in single-channel instruments, such as Perkin-Elmer LS 50B, a motorized alternate sampling may be used to alternately collect data from reference and the reaction cuvette. Fluorescence intensity is expressed as the ratio of the reaction channel to the reference channel. Fluorescence data may be fitted directly to extract the τ_{obs} according to the following equation[15]:

$$F = F_0 + \beta t - \beta\tau_{obs}[1 - exp(-t/\tau_{obs})]$$

where F is the observed fluorescence intensity at time t, F_0 is the fluorescence intensity at time zero, and β is a constant in this formula, which is

[11] B. R. Bochner and B. N. Ames, *J. Biol. Chem.* **257**, 9759 (1982).
[12] R. C. Payne and T. W. Traut, *Anal. Biochem.* **121**, 49 (1982).
[13] A. Werner, *J Chromatogr.* **618**, 3 (1993).
[14] P. R. Brown, A. M. Krstulovic, and R. A. Hartwick, *Adv. Chromatogr.* **18**, 101 (1980).
[15] I. K. Kolasa, T. Lozinski, and K. L. Wierzchowski, *Acta* **49**, 659 (2002).

a function of quantum yields, extinction coefficients, promoter concentration, etc. For curve fitting, β may be treated as a fitting parameter. A similar equation may be used for global fitting. Schlax and co-workers[9] reported that a drift of unknown origin was present in their fluorescence assay. The drift did not originate from the spectrofluorometer. They reported that dilution of the assay mixture by even a small amount of distilled water results in significant reduction of the detected reaction velocity. The origin of such effects is not well understood and may even be variable from one assay to another. However, they should be accounted for, if present.

Caveat

An uncritical assumption in the analysis of abortive initiation data is that the reaction is a sequential pathway and not branched (Fig. 3). Therefore, the entity that catalyzes the formation of tri- or tetranucleotide products is on an intermediate pathway. This view has been challenged in recent papers. Shimamoto and co-workers suggested that the transcription initiation pathway is branched after initial formation of the closed complex.[6,7] One arm of the branch leads to a productive pathway, whereas the other leads to a moribund complex. Like the productive open complex, the moribund complex is also capable of catalyzing the abortive products. Because the concentration-dependent term is before branching, K_B derived from abortive initiation studies should not be significantly affected. Because the moribund complex is as capable of catalyzing abortive initiation products as the open complex in the productive pathway, the catalytic power of the moribund complex and kinetic constants of that pathway will

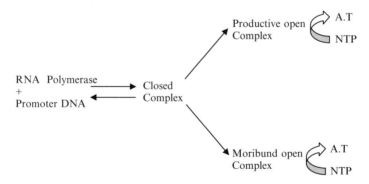

FIG. 3. Branched pathway of transcription initiation. The moribund complex is incapable of proceeding to elongation but capable of producing abortive transcripts.

determine the magnitude of the error in k_f determination. The magnitude may again vary from promoter to promoter. This raises the serious possibility that in some cases the kinetic constants derived from abortive initiation experiments may require significant correction.

A second problem comes from the fact that in most preparations of RNA polymerase, only a fraction is capable of binding to the promoter. This is generally true of DNA-binding proteins. The inactive fraction for RNA polymerase varies. Schlax and co-workers reported an active fraction of about 40%. Some commercial preparations showed about 20% active fractions (S. Roy, unpublished observation). Clearly, in such cases the kinetic parameters, particularly K_B, will be affected directly and need to be corrected for active fractions.

[47] Probing the Role of Region 2 of *Escherichia coli* σ⁷⁰ in Nucleation and Maintenance of the Single-Stranded DNA Bubble in RNA Polymerase-Promoter Open Complexes

By Pieter L. deHaseth and Laura Tsujikawa

The bacterial RNA polymerase (RNAP), referred to as the "holoenzyme," or Eσ, is a complex of the "core enzyme" (also core or E) and sigma factor (σ), a type of bacterial initiation factor.[1-4] The multisubunit core enzyme shows similarities in both sequence and structure to the multisubunit eukaryotic RNAP.[5-8] The role of the sigma factor is to recognize promoter DNA and to initiate the process of RNAP-induced strand separation to form the transcription-competent open promoter complex (see Refs. 3, 9, and 10 for reviews). The latter is the focus of this article.

[1] R. Burgess, A. Travers, J. J. Dunn, and E. K. F. Bautz, *Nature* **221,** 43 (1969).
[2] W. R. McClure, *Annu. Rev. Biochem.* **54,** 171 (1985).
[3] C. Gross, C. Chan, A. Dombroski, T. Gruber, M. Sharp, J. Tupy, and B. A. Young, *Cold Spring Harb. Symp. Quant. Biol.* **63,** 141 (1998).
[4] P. L. deHaseth, M. Zupancic, and M. T. Record, Jr., *J. Bacteriol.* **180,** 3019 (1998).
[5] L. A. Allison, M. Moyle, M. Shales, and C. J. Ingles, *Cell* **42,** 599 (1985).
[6] D. Sweetser, M. Nonet, and R. A. Young, *Proc. Natl. Acad. Sci. USA* **84,** 1192 (1987).
[7] G. Zhang, E. A. Campbell, L. Minakhin, C. Richter, K. Severinov, and S. A. Darst, *Cell* **98,** 811 (1999).
[8] E. Ebright, *J. Mol. Biol.* **304,** 687 (2000).

A major advance in our understanding of the transcriptional apparatus was achieved with the determination of the structure of the bacterial sigma factor,[11] core enzyme,[7] holoenzyme,[12,13] and a cocomplex of holoenzyme with a forked DNA[14] (see later) encompassing most of the sequence recognized by RNAP. These structures are of great importance to our understanding of the interaction of RNAP with promoter DNA[14,15] (see also Young et al.[16]). Relevant features of these structures include (1) the large total area of contact between core and σ^{70}, (2) DNA-binding groups of sigma factor that stick out from the body of the holoenzyme, and (3) the existence of separate tunnels for each strand of promoter DNA, leading the template strand deep into the hydrophobic core of the RNAP to the active site of the enzyme while keeping the nontemplate strand much closer to the surface.

RNAP–Promoter Interactions

In addition to the "primary" σ factor responsible for the recognition of promoters for housekeeping genes (σ^{70} in *Escherichia coli*), most bacteria have multiple minor sigma factors that help the cell deal with adverse conditions. Across eubacteria, the promoters recognized by holoenzymes containing the primary sigma factors are similar to the consensus σ^{70} promoter of *E. coli*. The −35 region (consensus TTGACA) is recognized by amino acids in σ^{70} region 4.2,[3,9] the "TG" ("extended −10") sequence by amino acids in region 3.0 (also referred to as region 2.5[17]), and the −12 T and −11 A of the −10 region (consensus TATAAT) by amino acids in region 2.4.[3,9] While it is likely that the TAAT sequence in the −10 region from −10 to −7 is recognized as single-stranded (SS) DNA,[18–21] it is not yet clear which amino acids are responsible.

[9] M. Lonetto, M. Gribskov, and C. A. Gross, *J. Bacteriol.* **174**, 3843 (1992).

[10] J. D. Helmann and J. Chamberlin, *Annu. Rev. Biochem.* **57**, 839 (1988).

[11] E. A. Campbell, O. Muzzin, M. Chlenov, J. L. Sun, C. A. Olson, O. Weinman, M. L. Trester-Zedlitz, and S. A. Darst, *Mol. Cell* **9**, 527 (2002).

[12] K. S. Murakami, S. Masuda, and S. A. Darst, *Science* **296**, 1280 (2002).

[13] D. G. Vassylyev, S.-I. Sekine, O. Laptenko, J. Lee, M. N. Vassylyeva, S. Borukhov, and S. Yokoyama, *Nature* **417**, 712 (2002).

[14] K. S. Murakami, S. Masuda, E. A. Campbell, O. Muzzin, and S. A. Darst, *Science* **296**, 1285 (2002).

[15] V. Mekler, E. Kortkhonjia, J. Mukhopadhyay, J. Knight, A. Revyakin, A. N. Kapanidis, W. Niu, Y. W. Ebright, R. Levy, and R. H. Ebright, *Cell* **108**, 599 (2002).

[16] B. A. Young, T. M. Gruber, and C. Gross, *Cell* **109**, 417 (2002).

[17] K. A. Barne, J. A. Bown, S. J. W. Busby, and S. D. Minchin, *EMBO J.* **16**, 4034 (1997).

[18] C. W. Roberts and J. W. Roberts, *Cell* **86**, 495 (1996).

[19] J. Qiu and J. D. Helmann, *Nucleic Acids Res.* **27**, 4541 (1999).

Formation of the open complex from free RNAP and promoter DNA requires conformational changes in both DNA and the RNAP.[22] McClure[23] showed that the RNAP concentration dependence of the kinetics of open complex formation allowed the process to be subdivided into a concentration-dependent step (closed complex formation) and a concentration-independent "isomerization" step (open complex formation). Subsequent work[24,25] revealed the presence of additional intermediates on the pathway to open complex formation. Steps in formation of an open complex are usually described as "bind, nucleate, melt."[26] A kinetic scheme with the currently recognized intermediates in the process is shown as[27,28]

$$R + P \rightleftharpoons I_0 \rightleftharpoons I_1 \rightleftharpoons I_2 \rightleftharpoons RP_o$$

SCHEME 1

where R is RNAP; P is promoter DNA; and I_0 ($= RP_c$) is the initial closed complex (RNAP–DNA contacts at the -35, TG and -10 sequences, but no strand separation). I_1 and I_2 are significant kinetic intermediates at the well-studied promoter P_R. I_2 but not I_1 was found to be resistant to a short challenge by the anionic competitor heparin.[27] RP_o is the open complex.

Nucleation ($=$ initiation) of strand separation is likely driven by binding free energy. The challenge is to correlate this scheme with structural studies.[29] As proposed by Saecker *et al.*, unfavorable conformational changes in I_1 have been introduced in both RNAP and promoter DNA. The distortion of DNA in I_1 by bending, twisting, or unwinding[30] would facilitate nucleation at $-11A$, possibly by flipping this base out of the helix[31–34] to generate the intermediate I_2. This would be similar to the structures

[20] D. L. Matlock and T. Heyduk, *Biochemistry* **39**, 12274 (2000).

[21] M. S. Fenton and J. D. Gralla, *Proc. Natl. Acad. Sci. USA* **98**, 9020 (2001).

[22] R. S. Spolar and M. T. Record, Jr., *Science* **263**, 777 (1994).

[23] W. R. McClure, *Proc. Natl. Acad. Sci. USA* **77**, 5634 (1980).

[24] H. Buc and W. R. McClure, *Biochemistry* **24**, 2712 (1985).

[25] J. H. Roe, R. R. Burgess, and M. T. Record, Jr., *J. Mol. Biol.* **176**, 495 (1984).

[26] M. T. Record, Jr., W. S. Reznikoff, M. L. Craig, K. L. McQuade, and P. J. Schlax, "*Escherichia coli* and *Salmonella*, Cellular and Molecular Biology" (F. C. Neidhardt, editor in chief), p. 792. ASM Press, Washington, DC, 1996.

[27] M. L. Craig, O. V. Tsodikov, K. L. McQuade, J. P. E. Schlax, M. W. Capp, R. M. Saecker, and M. T. Record, Jr., *J. Mol. Biol.* **283**, 741 (1998).

[28] R. M. Saecker, O. V. Tsodikov, K. L. McQuade, P. E. Schlax, M. W. Capp, and M. T. Record, Jr., *J. Mol. Biol.* **319**, 649 (2002).

[29] K. S. Murakami and S. A. Darst, *Curr. Opin. Struct. Biol.* **13**, 31 (2003).

[30] P. L. deHaseth and J. D. Helmann, *Mol. Microbiol.* **16**, 817 (1995).

[31] J. D. Helmann and P. L. deHaseth, *Biochemistry* **37**, 5959 (1999).

[32] M. S. Fenton, S. J. Lee, and J. D. Gralla, *EMBO J.* **19**, 1130 (2000).

characterized for, among others, the repair enzyme uracil N-glycosylase; accompanied by major changes in the conformation of the enzyme, the U is rotated completely out of the helix in a process called base flipping.[35–39] The nucleation step is likely rate limiting in open complex formation; for the P_R promoter, in both the forward and the reverse directions, the slowest step was found to be the interconversion between I_1 and I_2.[27] Several experimental approaches have established a crucial role for the $-11A$ in formation of an open complex.[20,31,34,40,41] Propagation of strand separation to yield RP_o involves unzipping of DNA from the $-11A$ in a downstream direction; concomitant conformational changes in the RNAP position each strand in its own tunnel.

Sigma Factor Region 2.3 and Nucleation of DNA Strand Separation

Sequence comparison of sigma factors has revealed the existence of four regions of significant sequence conservation.[9,42] This section focuses on region 2, which has been subdivided into 2.1, 2.2, 2.3, and 2.4. Some residues in region 2.1 and many in region 2.2 are involved in recognition of the core.[43] Region 2.4 carries out recognition of the -10 element (reviewed in Lonetto et al.[9]) and region 2.3 nucleation of strand separation, and likely also recognition of bases in the -10 element.[10,33,44–46]

In E. coli, region 2.3, extending from residues 417 to 434, has the sequence DKFEYRRGYKFSTYATWW. A comparison of the amino acid sequences of 53 primary sigma factors (supplementary material to

[33] M. Tomsic, L. Tsujikawa, G. Panaghie, Y. Wang, J. Azok, and P. L. deHaseth, J. Biol. Chem. 276, 31891 (2001).

[34] L. Tsujikawa, M. G. Strainic, H. Watrob, M. D. Barkley, and P. L. deHaseth, Biochemistry 41, 15334 (2002).

[35] Y. L. Jiang and J. T. Stivers, Biochemistry 41, 11236 (2002).

[36] P. M. Patel, M. Suzuki, E. Adman, A. Shinkai, and L. A. Loeb, J. Mol. Biol. 308, 823 (2001).

[37] G. Xiao, M. Tordova, J. Jagadeesh, A. C. Drohat, J. T. Stivers, and G. L. Gilliland, Proteins 35, 13 (1999).

[38] T. Hollis, Y. Ichikawa, and T. Ellenberger, EMBO J. 19, 758 (2000).

[39] D. R. Davies, I. Y. Goryshin, W. S. Reznikoff, and I. Rayment, Science 289, 77 (2000).

[40] Y. Guo and J. D. Gralla, Proc. Natl. Acad. Sci. USA 95, 11655 (1998).

[41] H. M. Lim, H. J. Lee, S. Roy, and S. Adhya, Proc. Natl. Acad. Sci USA 98, 14849 (2001).

[42] M. Gribskov and R. R. Burgess, Nucleic Acids Res 14, 6745 (1986).

[43] M. M. Sharp, C. L. Chan, C. Z. Lu, M. T. Marr, S. Nechaev, E. W. Merritt, K. Severinov, J. W. Roberts, and C. A. Gross, Genes Dev. 13, 3015 (1999).

[44] Y.-L. Juang and J. D. Helmann, J. Mol. Biol. 235, 1470 (1994).

[45] X. Huang, F. J. Lopez de Saro, and J. D. Helmann, Nucleic Acids Res. 25, 2603 (1997).

[46] G. Panaghie, S. E. Aiyar, K. L. Bobb, R. S. Hayward, and P. L. deHaseth, J. Mol. Biol. 299, 1217 (2000).

Campbell *et al.*[11]) confirmed the previously observed extensive sequence conservation in region 2.3.[9] Seven residues (F419, K426, F427, Y430, A431, W433, and W434 in the *E. coli* numbering used throughout this proposal) were invariant among these 53 sigma factors. Flanking the 2.3 region, in 2.2, residue K414, and in 2.4, residues Q446, R448, and R451 also are conserved for all sigma factors of the data set. The 2.3 regions of many minor sigma factors show much less conservation when compared to each other or to the primary sigma factors.

Amino acid residues Y430 and W433 are seen in the cocrystal structure[14] to be positioned close to the double-stranded (ds) – single-stranded junction of the forked DNA template, and thus are likely candidates involved in the nucleation of DNA melting. The following model[33] is supported by currently available data. In I_o (Scheme 1), both sequence-specific (H-bonding) and sequence-nonspecific (electrostatic) interactions serve to hold promoter DNA in a unique position with respect to aromatic amino acids Y430 and W433. These two residues likely cooperate to flip the −11A out of the DNA helix to nucleate the DNA-melting process. Several basic residues (K414, K418, R423, and K426) are likely involved in promoter DNA binding. The structure of the *E. coli* region 2.3[47] reveals that the side chains of these four residues and of Y425, Y430, W433, and W434 stick out on approximately the same face of the protein, where they can interact with promoter DNA.

Use of Forked DNAs to Dissect Steps in Formation of Open Complexes

Guo and Gralla[40] pioneered the use of forked DNAs as model templates for the study of DNA and protein sequences required for recognition by and stable complex formation with RNAP.[20,21,32–34,48] For these templates, the DNA from the −35 region through the first base pair of the −10 region is double stranded and the remainder is present as a single-stranded extension of the nontemplate strand. The interaction of RNAP with forked templates is a diagnostic tool for the ability to nucleate strand separation. Differentiation between binding and isomerization can be accomplished by subjecting RNAP–forked DNA complexes to a challenge with the poly anion, heparin (see later).

[47] A. Malhotra, E. Severinova, and S. A. Darst, *Cell* **87,** 127 (1996).
[48] L. Tsujikawa, O. V. Tsodikov, and P. L. deHaseth, *Proc. Natl. Acad. Sci USA* **99,** 3493 (2002).

Experimental Procedures

The experimental approach described here allows a quantitative evaluation of the extent to which a substitution in σ factor affects steps in the formation of an open complex or its stability. This can be achieved by determination of (1) the kinetics of open complex formation at promoters, (2) the affinity between RNAP and a forked DNA, (3) the stability of RNAP–forked DNA complexes, and (4) the extent of binding of ssDNA at a fixed concentration of RNAP. See Fig. 1 for sequences of the aforementioned DNAs.

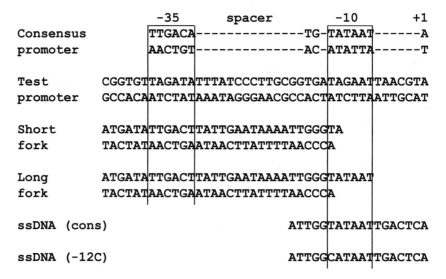

FIG. 1. Promoter, fork junction, and single-stranded DNA sequences. For the consensus *E. coli* σ^{70} promoter, only the sequence of the core promoter region, from the −35 region to the start site (+1), is shown. Lack of significant sequence conservation is indicated by hyphens. The test promoter is a variant of the P_{RM} promoter of bacteriophage λ. It forms open complexes at a rate that lends itself well to the use of manual methods. The two fork junction or forked templates have the identical bottom strand (template strand of the promoter from which the sequences were derived), but differ in the length of the 3′ extension of the top (nontemplate) strand. The short fork has an overhanging A, corresponding to the −11A of the −10 region of promoter DNA, whereas the overhang of the long fork extends to the 3′ end of the −10 region. The two single-stranded DNAs have the −10 sequence of the nontemplate strand of promoter DNA. The −12C oligodeoxyribonucleotide has a nonconsensus base at the −12 position, significantly weakening the specific interaction with the RNA polymerase.

Reagents and Techniques

RNAP. Purification of histidine-tagged σ factor is as described previously.[46] Reconstitution of holoenzyme from core polymerase (purchased from EpicentreTechnologies) and purified σ factor is accomplished by incubating the core (650 n*M*) and sigma factor (3.25 μ*M*) in 24 μl storage buffer [0.01 *M* Tris–HCl, pH 7.9, 0.1 *M* NaCl, 0.1 m*M* EDTA, 0.1 m*M* dithiothreitol (DTT), and 50% glycerol for 30 min on ice. The reconstituted holoenzyme is used without further purification. In this article, all RNAP concentrations refer to the total amount of core enzyme in solution (i.e., without correction for the amount of RNAP active in promoter binding).

Oligodeoxyribonucleotide (oligo) Purification and End Labeling. Synthetic oligos comprising each strand of a forked template (see Fig. 1) are purified by gel electrophoresis on a polyacrylamide gel containing 15% acrylamide, 7 *M* urea, and TBE buffer (100 m*M* Tris, pH 8.3, 100 m*M* boric acid, and 2 m*M* EDTA).[49] DNA bands are visualized by ultraviolet shadowing on a fluorescent TLC silica plate (Kodak) background. DNA is extracted by a crush-and-soak technique,[50] recovered by loading onto a C18 column, and eluted with 40% methanol. The eluate is dried in a speedvacuum concentrator, and DNA is redissolved in 50 μl of TE buffer (10 m*M* Tris, pH 7.9, and 1 m*M* EDTA). The concentration is determined by measuring absorbance at 260 nm. The extinction coefficient for each oligo is calculated using the biopolymer calculator program from the Schepartz Lab website (http://paris.chem.yale.edu/extinct.html). Oligos (10–30 pmol) are labeled at the 5′ end with ^{33}P in reactions containing T4 polynucleotide kinase (New England Biolabs) and [γ^{33}P]ATP (Dupont NEN).[49] The final concentration of labeled oligo is adjusted to 1 μ*M* by the addition of water.

Annealing Reactions. 5′ end-labeled top (long) strand oligos (100 n*M*) are annealed to unlabeled bottom (short) strand oligos (150 n*M*) in 40 μl buffer containing 50 m*M* NaCl and 25 m*M* Tris (pH 8.0). The mix is incubated in a temperature block at 90° for 5 min, after which the heat is turned off to allow slow cooling to room temperature.

Preparation of Labeled Promoter DNA by Polymerase Chain Reaction (PCR). The test promoter is a P_{RM} variant with an improved −10 region (TAGAAT instead of TAGATT) contained on a fragment also harboring

[49] J. Sambrook, E. F. Fritsch, and T. Maniatis, Molecular Cloning, 2nd Ed., pp. 13–45, *Cold Spring Harbor Laboratory Press, Plainview, NY*, 1989.

[50] J. Sambrook and D. W. Russell, Molecular Cloning, 3rd Ed., pp. 10–14, *Cold Spring Harbor Laboratory Press, Plainview, NY*, 2001.

a P_R promoter inactivated by mutation. For this promoter, the formation of open complexes is on the minute time scale, for an RNAP concentration of 200 nM. The labeled promoter is obtained by PCR using plasmid DNA as the template. PCR primers are designed to yield a promoter fragment of about 470 bp. Each 200-μl reaction contains 100 ng (30 fmol) of plasmid DNA and 50 nmol of each primer, at least one of which is 5' end labeled; amplification is for 30 cycles.

Buffer. All binding reactions are carried out in HEPES buffer: 30 mM HEPES, pH 7.5, 100 mM NaCl, 1 mM DTT, and 100 μg/ml bovine serum albumin.

Detection of RNAP–DNA Complexes by Electrophoretic Mobility Shift Assay (EMSA). Binding reactions are analyzed on 4 or 5% nondenaturing gels (Accugel 29:1, National Diagnostics) in 1 × TAE (40 mM Tris-acetate, 1 mM EDTA). The gels are run at 100 V for 2.5 h, dried, and exposed overnight to a phosphorimager screen (Molecular Dynamics). The percentages of bound and free DNA substrate are determined from band intensities using ImageQuant software (Molecular Dynamics).

KINETICS OF OPEN COMPLEX FORMATION. Open complex formation can be monitored by several methods, each exploiting different properties of the complex: appearance of $KMnO_4$ sensitivity due to strand separation, ability to make RNA, or stability as reflected by resistance to challenge by heparin. We have generally used the latter and have followed the kinetics of open complex formation, a sensitive method for detecting the effects of mutations in σ factor. Separate reactions were set up for each time point. Binding was initiated at zero time by the addition of RNAP (200 nM final) to promoter DNA (2 nM final, about 10^4 cpm) in 20 μl (final volume) of HEPES buffer prewarmed for 2 min at the desired temperature. Typically, eight time points are taken over the range of 30 s to 45 min. The reactions are challenged for 1 min with heparin (100 μg/ml final) prior to loading onto a running 4% nondenaturing gel. Band intensities are determined as indicated earlier. Binding data (percentage of DNA bound as a function of time) are fit to exponential functions, with the plateau level(s) and the rate constant(s) (k_{obs}) as variable parameters. A comparison of values of k_{obs} (normalized to $k_{obs} = 1$ for RNAP with wt σ^{70}) among the reconstituted mutant holoenzymes is shown in Fig. 3a.

Our fragment bears an additional promoter (the mutated P_R promoter retained some activity) at which open complex formation is much slower. In order to get a good fit, we use a double exponential equation: $y = A1\ [1\text{-}\exp(-k_{1obs}t)] + A2\ [1\text{-}\exp(-k_{2obs}t)]$, where y is the fraction of DNA bound, A1 and A2 are the plateaus for the two phases, and k_{1obs} and k_{2obs} are rate constants. The relevant rate constant is that for the faster of the two processes. With the promoter shown in Fig. 1, in HEPES

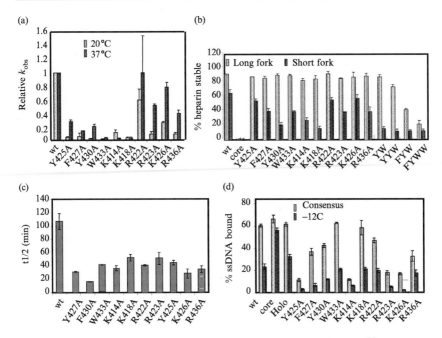

Fig. 2. Sample data on the interaction of RNAP bearing substitutions in σ^{70} with various DNAs. (a) Kinetics of open complex formation. Rate constants for the formation of stable complexes by mutant RNAP have been normalized to that for binding to wt RNAP. (b) Extent of formation of heparin-stable complexes with two different forked DNAs. YW: A substitutions at 430 and 433; YYW: A at 425, 430, and 433; FYW: A at 427, 430, and 433; FYWW: A at 427, 430, 433, and 434. (c) Stability of complexes with short fork DNA. Dissociation rate constants of heparin-stable complexes have been normalized to that for dissociation from wt RNAP. (d) Extent of complex formation with ssDNA. Two single-stranded oligodeoxyribonucleotides were used, differing in the sequence of the −12 base. See text for additional details.

buffer, formation of an open complex with 200 nM wt RNAP proceeds with a $k_{1obs} = 4 \pm 0.4$ min^{-1} at 37° (i.e., the reaction is half over in about 0.2 min).

The kinetics of open complex formation (Fig. 2a)[33,46] show that the alanine substitutions for basic and aromatic amino acids in region 2.3 of σ^{70} render the reconstituted RNAP cold sensitive[44,46,51,52] for open complex formation. This is as expected for residues important for the strand

[51] Y.-L. Juang and J. D. Helmann, *Biochemistry* **34**, 8465 (1995).
[52] S. E. Aiyar, Y.-L. Juang, J. D. Helmann, and P. L. deHaseth, *Biochemistry* **33**, 11501 (1994).

separation process: the effects of the substitutions become more prominent when less thermal energy is available to drive the melting process. Large effects of a single alanine substitution on the overall process of open complex formation are seen for Y425A, F427A, Y430A, W433A, K414A, and K418A. Runoff RNA synthesized after preincubation of promoter and (mutant) RNAP for a set amount of time gave similar results (shown elsewhere[46]). However, the effects of the Y430A and W433A substitutions at 37° were not evident; during the preincubation, the holoenzyme containing all but the most deleterious of σ^{70} substitutions is still able to bind a significant fraction of the promoter DNA.

BINDING OF FORKED DNAs. Labeled forked template (1 nM) is preincubated at room temperature for 1 min in HEPES buffer. RNAP is added to 65 nM, and the 10-μl reactions are incubated at 25° for 30 min, followed by a 10-min challenge with 100 μg/ml heparin. The mixtures are then subjected to EMSA, run at room temperature. Results are shown for both forked templates in Fig. 2b.[33] Single substitutions do not reduce the fraction of long fork DNA bound to RNAP, as the binding has "topped out," but they can be differentiated very well with short fork DNA. However, use of the long fork allows differentiation among the various holoenzymes reconstituted with multiply substituted σ^{70}, for which the binding to short fork has "bottomed out." Data in Fig. 2b[33] show that the single substitutions most deleterious to the formation of a heparin-resistant RNAP–forked DNA complex are Y430A, K414A, and K418A.

Additional information can be obtained by determining the extent of binding as a function of [RNAP], that is, a binding isotherm (Fig. 3). Individual binding reactions, with a final volume of 10 μl, are prepared, each with 1 nM forked DNA and RNA polymerase at concentrations ranging from 0 to 200 nM. The mixtures are incubated at room temperature for 30 min. To one set of reactions heparin is added to a final concentration of 100 μg/ml and to the other, the identical volume of ddH_2O. All reactions are incubated for an additional 10 min prior to the addition of 2 μl of a stock solution of loading dye (0.25% bromphenol blue, 0.25% xylene cyanol FF, and 30% glycerol). Fractions of DNA bound in the absence and presence of heparin, θ_{-hep}, and θ_{+hep}, respectively, are determined from band intensities, as described earlier.

Reactions, Equations, and Data Analysis. Complexes stable enough to survive a heparin challenge are assumed to contain RNAP in a conformation similar to that in an open promoter complex.[48] The interaction of RNAP with a forked template proceeds along the following scheme:

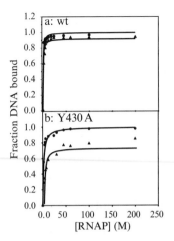

Fig. 3. RNAP–long fork interaction: Binding isotherms for formation of heparin-sensitive and -resistant complexes for wt holoenzyme and holoenzyme bearing the Y430A substitution in σ^{70}. Circles represent binding in the absence of a heparin challenge, and triangles represent binding following such a challenge. Data analysis and curve fitting were as described in the text. Best fits were obtained with $K_1 = 1.9 \times 10^8\ M^{-1}$ and $K_f = 11$ for RNAP containing wt sigma, and $K_1 = 1.8 \times 10^8\ M^{-1}$ and $K_f = 2.5$ for Y430A. See text and Table III.

$$
\begin{array}{cc}
K_1 & K_f \\
\end{array}
$$

$$
R + F \ \underset{\longleftarrow}{\overrightarrow{\hspace{1cm}}}\ RF_1 \ \underset{\longleftarrow}{\overrightarrow{\hspace{1cm}}}\ RF_{CR}
$$

SCHEME 2

where F is the forked DNA and RF_{CR} is the competitor (i.e., heparin) –resistant complex.[48] The equilibrium constants for the two steps are defined as follows:

$$
K_1 = \frac{k_1}{k_{-1}} = \frac{[RF_1]}{[R][F]} \tag{1}
$$

$$
K_f = \frac{k_f}{k_r} = \frac{[RF_{CR}]}{[RF_1]} \tag{2}
$$

where K_1 provides information about the affinity between RNAP and the forked DNA and K_f about the relative amounts of heparin-sensitive (RF_1) and heparin-resistant (RF_{CR}) complexes present at equilibrium. The experimental goal is to determine the values of K_1 and K_f. The relevant expression for θ_{+hep} is[53]

[53] O. V. Tsodikov and M. T. Record, Jr., *Biophys. J.* **76**, 1320 (1999).

$$\theta_{+\text{hep}} = \frac{K_1 K_f [R_f]}{(K_1 + K_1 K_f)[R_f]} \quad (3)$$

At the lowest concentrations of RNAP, the concentration of free RNAP, $[R_f]$, cannot be approximated adequately by the total concentration of the enzyme. Therefore, the following explicit expression for $[R_f]$ was used at all RNAP concentrations:

$$[R_f] = \frac{b + \sqrt{(b^2 + 4K_{\text{app}}[R])}}{2K_{\text{app}}} \quad (4)$$

where

$$b = -1 - (K_{\text{app}}[\text{forked DNA}] + K_{\text{app}}[R]) \quad (5)$$

For data obtained in the absence of a heparin challenge,

$$\theta_{-\text{hep}} = \frac{K_{\text{app}}[R_f]}{1 + K_{\text{app}}[R_f]} \quad (6)$$

$$\text{with} \quad K_{\text{app}} = K_1 + K_1 K_f \quad (7)$$

Values for K_1 and K_f are obtained by simultaneous fitting of data to the equations for the RNAP concentration dependence of $\theta_{+\text{hep}}$ [Eq. (3–5)] and $\theta_{-\text{hep}}$ [Eq. (4–7)]. To achieve this, the [RNAP] are entered twice in the same column and both sets of binding data (+ and − heparin) are entered in another column of the Sigma Plot spread sheet. In two adjacent columns, 0 or 1 is entered as multipliers to direct the use of one or the other equation. Sample pages containing data and equations are found in Tables I and II, respectively. See Fig. 3 for data and fits.

Data collected in Table III represent examples of the use of forked DNAs for determining the effects of amino acid substitution in σ^{70} on binding (K_1) and the subsequent steps (K_f). The major effects of substitutions for two aromatic amino acids, Y430A and W433A, are seen to be similar: reduction of the values of K_f. However, the K414A substitution primarily causes a reduction in K_1. In this assay, K418 has only a relatively small effect, mostly on K_f. The triple substitution YYW causes a drastic reduction in K_f as seen for both short and long forks. For the short, but not the long fork, an additional effect on K_1 is observed for as yet unknown reasons.

STABILITY OF RNAP–DNA COMPLEXES. In the aforementioned protocols, the added heparin serves to distinguish complexes with a very short half-life from those that have acquired significant stability. Upon addition of the heparin, it is observed that a fraction of the complexes dissociates

TABLE I
SAMPLE DATA ENTRY[a]

	1 [RNAP] (M)	2 frx. bound − or + hep	3 − heparin multiplier	4 + heparin multiplier
1	0	0	1	0
2	2E-09	0.556	1	0
3	4E-09	0.774	1	0
4	5E-09	0.808	1	0
5	6E-09	0.852	1	0
6	0	0	0	1
7	2E-09	0.288	0	1
8	4E-09	0.343	0	1
9	5E-09	–	0	1
10	6E-09	0.452	0	1

[a] For a subset of Y430A data shown in Fig. 3b. Lines 1–5, data without heparin challenge; lines 6–10, data with heparin challenge. Typically, each set of data contains 12 or more RNAP concentrations (2E-09 = 2 nM, etc.). To allow simultaneous fitting, Eq. 4–7 (for fitting data without heparin challenge) have column 3 as the multiplier, and Eq. 3–6 (for fitting data with heparin challenge) have column 4.

readily, with the remaining fraction dissociating much more slowly.[48] By following the time course of dissociation of "heparin-stable" complexes, any differential effects of the substitutions on the stabilities of these complexes can be determined. Dissociation rate constants, or complex half-lives, are determined from data obtained 5 min or more after the addition of heparin, using nonlinear fits $[y = \exp(-k_{diss}t)]$, or from the slopes of semilogarithmic plots of the fraction bound.

Binding data shown in Fig. 2c[33] indicate that of all single substitutions tested, Y430A causes the greatest decrease in the stability of the complex between a mutated RNAP and forked DNA: from a half-life of 100 min for complexes with wt RNAP[33] to 15 min.

INTERACTION OF RNAP WITH SINGLE-STRANDED DNA. End-labeled oligo (10 nM) is incubated with 65 nM RNAP for 30 min at 25°, followed by EMSA on a 5% nondenaturing gel run at 4°. Band intensities are determined as indicated earlier. If the gel is run at 25°, no bands other than that of the free DNA are observed.[32] It is likely that our protocol is not optimal, and that oligo binding mainly occurs upon loading the sample onto the gel at 4°.

Data in Fig. 2d[33] show that Y425, K414, R423, and K426 play a role in specific ssDNA binding. Specific recognition of the ssDNA is demonstrated

TABLE II
SAMPLE PAGE OF EQUATIONS USED FOR SIMULTANEOUS FITTING WITH SIGMA PLOT[a]

[Variables]
x = col(1)
y = col(2)

[Parameters]
k1 = 2e7
k2 = 1

[Equation]
dnat = 1e-9
k = k1 + k1*k2
b = -1 -k*dnat + k*x
rfree=(b + sqrt(b*b + 4*k*x))/(2*k)
f = col(3)*k*rfree/(1 + k*rfree) + col(4)*k1*k2*rfree/(1 + (k1 + k1*k2)*rfree)
fit f to y

[Constraints]
k1 > 0
k2 > 0

[Options]
tolerance = 0.0000100
stepsize = 100
iterations = 10000

[a] Here k1, K_1; k2, K_f; dnat, the total concentration of DNA (1 nM in this experiment); k, K_{app}; rfree, the concentration of free RNAP; x, column 1; and y, column 2, as in Table I. Locations of the multipliers in columns 3 and 4 of Table I are entered into the equation for f, under the header [Equation]. Values under the header [Parameters] are the initial estimates for k1 and k2.

by the observation that binding is reduced as a consequence of the T-12C substitution.[33]

Interpretation of Data

The following conclusions can be drawn from the aggregate of data: (1) K414 participates directly in double-stranded promoter DNA binding, while the role of K418 is not yet entirely clear. (2) Y430, crucial for interaction with the nucleation-modeling forked DNA, would be involved in the nucleation of DNA melting, possibly by stacking onto the −11 base after its flipping out of the helix. (3) K414, Y425, R423, and K426 interact with single-stranded DNA following nucleation, but perhaps concomitant with DNA strand separation. Apparently, K414 may engage in interactions with both dsDNA and ssDNA. W433 plays an important role, which was

TABLE III
COMPARISON OF BINDING EXPERIMENTS WITH SHORT AND LONG FORKS

	Short fork[a]		Long fork[a]	
RNAP[b]	K_1, 10^8 M^{-1}	K_f	K_1, 10^8 M^{-1}	K_f
wt σ^{70}	2.8 ± 0.1	1.1	1.9 ± 0.1	11.1 ± 0.1
Y430A	1.77 ± 0.04^c	0.41 ± 0.05^c	1.8 ± 0.1^c	2.5 ± 0.3^c
W433A	1.94 ± 0.03	0.44 ± 0.04	ND[d]	ND
YYW[e]	0.34 ± 0.05	0.21 ± 0.01	2.2 ± 0.1	1.7 ± 0.1
K414A	0.27 ± 0.01	0.65 ± 0.05	0.15 ± 0.03	3.8 ± 0.6
K418A	1.41 ± 0.02	0.35 ± 0.02	1.8 ± 0.1	3.1 ± 0.2

[a] See Fig. 1.
[b] with wt σ^{70} or σ^{70} bearing indicated substitution(s).
[c] Data from Tsujikawa *et al.*[48]
[d] Not done.
[e] Triple substitution: Y425A, Y430A, and W433A.

not captured by our model templates. It may interact (stack) with the downstream helix after base flipping. From the σ^{70} and holoenzyme structures it is evident that the F427 side chain is buried in the sigma structure so any effects of the F427A substitution are likely indirect. John Helmann's results with *Bacillus subtilis* Eσ^A RNAP[44,54] and Jay Gralla's with *E.coli* Eσ^{70} [32] are consistent with the aforementioned conclusions, particularly with respect to the roles of aromatic amino acids.

Acknowledgments

Research in our laboratory was supported by NIH Grant GM31808 to PLdH. We thank Drs. David Auble, Phil Rather, and David Samols for helpful comments and Dr. Oleg Tsodikov for his extensive prior help with analysis of the binding data.

[54] J. C. Rong and J. D. Helmann, *J. Bacteriol.* **176,** 5218 (1994).

[48]   On the Use of 2-Aminopurine as a Probe for Base Pair Opening During Transcription Initiation

By SIDDHARTHA ROY

Transcription initiation encompasses the steps from the initial interaction of RNA polymerase with the promoter to the promoter clearance. In prokaryotes, the protein activators or repressors that regulate transcription initiation act on many of these steps.[1–3] Thus, to have a complete understanding of the regulation process, it is important to estimate the steps of transcription initiation. It is also important to identify all the intermediates and elucidate their structural characteristics. Elucidation of the nature of the intervening transition states is also crucial.[4] During the transcription initiation process, the DNA goes from a base-paired duplex form in the closed complex to a base-pair disrupted form in the open complex and back to a base-paired duplex form after the RNA polymerase cleared the promoter. A real-time monitoring of kinetics of base pair disruption and reformation would be a valuable tool to investigate these steps.

Fluorescence spectroscopy is well suited for the study of such a process as the measurements can be performed in real time with high sensitivity. An ideal fluorescent probe for monitoring the base pair disruption process in the context of transcription initiation should have excitation wavelengths that are different from protein tyrosines or tryptophans and should have a minimal effect on the initial double helical structure of the DNA. In this respect, 2-aminopurine (2-AP) meets these criteria. 2-Aminopurine is an isomer of the natural base, adenine (Fig. 1). The amino group in the 6 position of adenine is shifted to the 2 position in 2-AP. This shifting does not change the number of hydrogen bonds but causes the NH .... OC hydrogen-bonded pair to shift from the major groove to the minor groove in a double-stranded DNA. Although this apparently does not cause any major distortion in the duplex that contains 2-aminopurine–thymine base pair,[5] a potential recognition group in the major groove is lost. This may result in a significant loss of interaction between DNA and proteins that interact

[1] G. Lloyd, P. Landini, and S. Busby, *Essays Biochem.* **37,** 17 (2001).
[2] S. Adhya and S. Garges, *J. Biol. Chem.* **265,** 10797 (1990).
[3] S. Adhya, M. Gottesman, S. Garges, and A. Oppenheim, *Gene* **132,** 1 (1993).
[4] S. Roy, S. Garges, and S. Adhya, *J. Biol. Chem.* **273,** 14059 (1998).
[5] T. M. Nordlund, S. Andersson, L. Nilsson, R. Rigler, A. Graslund, and L. W. McLaughlin, *Biochemistry* **28,** 9095 (1989).

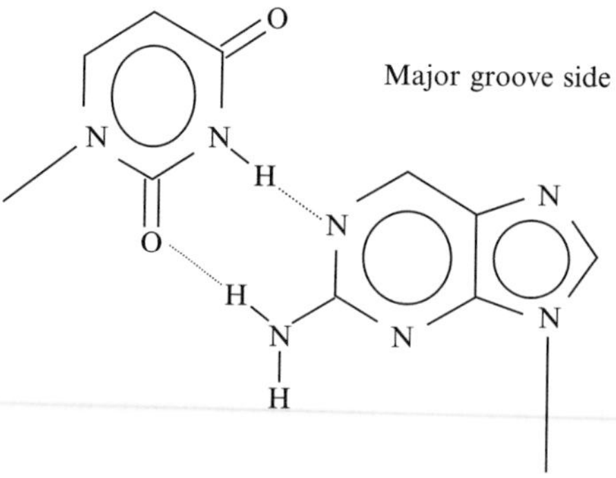

Major groove side

FIG. 1. Structure of 2-aminopurine–thymine base pair.

through the major groove.[6] The 2-AP absorption spectrum extends beyond that of the tyrosine and tryptophan (that is >300 nm), having an absorbance maximum around 310 nm, thus making it suitable for use in a protein–nucleic acid complex. The major advantage of 2-AP in a DNA-based reaction lies in the fact that quantum yield and lifetime are dependent on whether the base is stacked or not.[7,8] The fluorescence properties may also depend on its stacked neighbors, making this a suitable probe for studying any process involving base pair deformation and unstacking.[9]

Methods for Preparing 2-Aminopurine Containing Bacterial Promoters

2-Aminopurine containing oligonucleotides of shorter lengths are best obtained by chemical synthesis. Appropriately protected phosphoramidite (5′-dimethoxytrityl-N^2-(dimethylamino methylidene)-2′-deoxypurine riboside 3′-[(2-cyanoethyl)-(N,N-diisopropyl)]phosphoramidite) can be used directly in a standard DNA synthesizer. 2-AP containing oligonucleotides can now be obtained from many commercial organizations. For shorter templates, such as those used with bacteriophage T7 RNA polymerase, chemical synthesis followed by HPLC purification gives good results. For longer oligonucleotides, such as those used for *Escherichia coli* promoters with or without regulatory protein-binding sites, total chemical synthesis may be somewhat problematical, and assembly from shorter oligonucleotides by the ligase reaction is an alternative. In many of these

[6] H. M. Lim, H. J. Lee, S. Roy, and S. Adhya, *Proc. Natl. Acad. Sci. USA* **98,** 14849 (2001).
[7] J. M. Jean and K. B. Hall, *Proc. Natl. Acad. Sci. USA* **98,** 37 (2001).
[8] K. D. Raney, L. C. Sowers, D. P. Millar, and S. J. Benkovic, *Proc. Natl. Acad. Sci. USA* **91,** 6644 (1994).
[9] J. M. Jean and K. B. Hall, *Biochemistry* **41,** 13152 (2002).

promoters, the required length may exceed 70 bp, as the RNA polymerase footprint often extends from -35 to $+25$ or greater. If the regulatory sites are included, the required length may exceed 100 bp. Synthesis and purification of such long oligonucleotides pose considerable challenges because of low yields.

Annealing long oligonucleotides properly is a difficult problem. Proper annealing is of importance in 2-aminopurine assays as the presence of unpaired regions or mismatches is expected to completely alter the properties of the promoters and consequently the kinetics of transcription initiation. Slow annealing over a period of several hours works reasonably well. Repeated annealing with a gradually decreasing melting temperature may be tried to remove the partially mismatched structures. The quality of annealing is usually judged by running the duplex on a polyacrylamide gel alongside duplex markers (derived from restriction endonuclease cleavage) of similar size (see Fig. 2). If the two bands are of similar widths at similar loading concentrations, it may then be assumed that the oligonucleotides have annealed properly. Similar analysis can also be done using capillary electrophoresis.

A second method of preparation of 2-aminopurine containing templates is to use polymerase chain reaction. In this technique, two primers, one spanning from about -1 to about -80 and another from a downstream

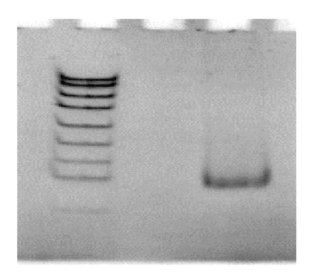

Fig. 2. Polyacrylamide gel electrophoresis of an annealed 90-bp 2-aminopurine containing DNA (right lane) and the *Msp*I digest of pUC19. Sizes of the fragments are 501, 489, 404, 331, 242, 190, 147, 110, 67, 34, and 26 bp. Suranjana Chattopadhyay and S. Roy, unpublished observation.

region several hundred bases away, are used to amplify the desired promoter region. The primer spanning the -1 to -80 region contains the 2-aminopurine substitution at the desired positions. Amplified 2-aminopurine containing DNA can then be purified by agarose gel electrophoresis or Fast Protein Liquid Chromatography (FPLC). It is usually convenient to include a transcription terminator in the amplified insert so that *in vitro* transcription can be done on the 2-AP containing template. This is a useful check on template quality and any down effect of the substitution.

A third method of preparing the templates is through use of DNA ligase. Assembly of long oligonucleotide duplexes through ligation of smaller oligonucleotides is an attractive proposition. One of the earliest attempts to assemble the long oligonucleotide duplex was through overlapping oligonucleotides. In this technique, several oligonucleotides, which have 5' or 3' overhangs, are assembled and ligated by DNA ligase. Such a technique has been used for 2-aminopurine containing oligonucleotides, and overall yields of 30–40% have been achieved after purification.[10]

Determination of Base Pair Opening Rates

For strong promoters, the k_f or the forward rate constant for isomerization (base pair disruption) is very fast; base pair opening rates can only be determined by stopped flow measurements (see Fig. 3 for example). This is true for many T7 promoters and some *E. coli* promoters. However, many *E. coli* promoters, particularly those activated by regulatory proteins, are weak at this step and their forward rates constants are such that the $t_{1/2}$ is in the range of several minutes. For these promoters, manual mixing experiments are adequate. There are several possible ways of measuring base pair opening rates under manual mixing conditions. One method is to use relatively high concentrations of template DNA and RNA polymerase (both considerably above the dissociation constant of the closed complex). Under such conditions upon initial mixing, there will be a rapid formation of closed complex followed by slow isomerization to the open complex. On rates for many promoters have been measured and generally fall around 10^7–10^8 M^{-1} s^{-1}.[11] Thus, at template and polymerase concentrations of 10^{-6} M each, the complex formation would be complete within the manual mixing time. The fluorescence increase observed after manual mixing would thus be entirely due to isomerization of the closed complex to open complex.

[10] J. J. Sullivan, K. P. Bjornson, L. C. Sowers, and P. L. deHaseth, *Biochemistry* **36,** 8005 (1997).
[11] M. Brunner and H. Bujard, *EMBO J.* **6,** 3139 (1987).

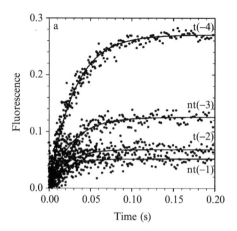

Time (s)

FIG. 3. Kinetics of open complex formation by the T7 RNA polymerase. Stopped flow kinetic traces show the time-dependent increase in the fluorescence of 2-AP at t(4), nt(3), t(2), and nt(1) positions in a 40-bp T7 promoter DNA upon binding to the T7 RNA polymerase. The promoter DNA (0.1 μM final) was mixed with the T7 RNA polymerase (0.4 μM, final concentrations) at $25°$ in a stopped flow instrument. The sample was excited at 315 nm, and the time-dependent fluorescence was measured above 360 nm. Each trace is an average of 10–15 measurements, and the corresponding *solid smooth line* is the fit to a single exponential. The "t" and "nt" refer to template and nontemplate strands.[12]

Method

One major problem in measuring fluorescence increase over a period of several minutes is machine drift. Machine drift can be corrected by a protocol in which the spectrometer alternately scans a reference cuvette and the sample cuvette throughout the observation period. The reference cuvette does not contain polymerase but otherwise is identical to the sample cuvette. The ratio of the two channels gives the kinetics of fluorescence increase corrected for machine drift. To reduce the consumption of template (due to relatively high concentrations used), use of fluorescence cells of 50 μl volume is desirable. Use of such cells requires capping of the cuvettes to minimize any solvent loss due to evaporation during the measurement process.

Equipment

Perkin-Elmer LS50B spectrofluorometer equipped with a motorized four-cell holder
One circulating water bath

[12] R. P. Bandwar and S. S. Patel, *J. Biol. Chem.* **276,** 14075 (2001).

One ordinary water bath with microtube holders

Two 50-μl capped cuvettes constructed for LS50B by Hellma Gmbh & Co.

One 100-μl and one 50-μl adjustable volume pipettes

RNA polymerase storage buffer

RNA polymerase holoenzyme (purchased from Epicetre Inc.) stored at 4° until use

Synthetic annealed 106-bp oligonucleotide duplex (concentration of 1 μM) containing *galP1* promoter labeled with 2-AP at the +3 position in 0.05 M Tris–HCl buffer, pH 7.5, containing 0.1 M NaCl and 10 mM MgCl$_2$

Circulate water through the four-cuvette holder cell at 25° for 1 h or more. Place a cuvette with sufficient water in it. Check the temperature in the sample chamber by dipping a thermometer bulb in the water or by a thermocouple. Readjust the bath temperature until the sample chamber temperature is 25°. Place the two dry capped cuvettes and wait for 30 min or more. Turn the other water bath on and set it to 25°. Turn the fluorometer on and allow it to stabilize. Set the excitation and emission wavelengths to 320 and 370 nm, respectively. Set the respective bandpasses to 10 and 15 nm. Initiate the alternate collection program and make it ready to initiate data collection. We use 15-s collections for each channel.

Pipette 50 μl of the oligonucleotide into two 0.5-ml autoclaved eppendorf tubes, cap them, and place them into the 25° water bath. In two separate tubes, incubate 10 μl of RNA polymerase (approximately 4–5 μM) and 10 μl of storage buffer at the same temperature. At 9 min, mix the storage buffer with one oligonucleotide sample and transfer to a capped cuvette. Mix the RNA polymerase at the 10th min with the other oligonucleotide and transfer to the other capped cuvette. The data collection is then initiated immediately. Data are usually collected for 60 min. The files are then saved and transferred to a spreadsheet program for further analysis. An example of the manual mixing experiment is shown in Fig. 4.

Data Analysis

The equation normally used to describe such a process is

$$\text{R} + \text{P} \underset{k_{-1}}{\overset{k_1}{\rightleftharpoons}} \text{R.P}_\text{c} \underset{k_{-2}}{\overset{k_2}{\rightleftharpoons}} \text{R.P}_\text{o}$$

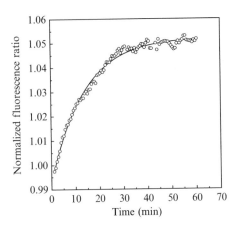

FIG. 4. Kinetics of fluorescence increase of a *galP1* promoter containing the DNA duplex when mixed with *E.coli* RNA polymerase at 25°. The two complementary oligonucleotides were chemically synthesized, purified, and annealed. The scan was recorded in a Perkin-Elmer LS 50B spectrofluorometer using the alternate scanning principle described in the text. The template and RNA polymerase concentration was 1 μM each in 0.05 M Tris–HCl buffer, pH 7.5, containing 0.1 M NaCl and 10 mM MgCl₂. The excitation wavelength was 320 nm and emission was set at 370 nm.

where R is the RNA polymerase, P is the promoter, suffixes c and o refer to closed and open complexes, respectively, and the k values are the respective rate constants. However, under conditions described earlier, the kinetic equation is a simple first-order equation and fluorescence growth can be described adequately by the following equation:

$$F_{obs} = F_0 + (F_\infty - F_0)[1 - \exp(-kt)]$$

where F_{obs} is the observed fluorescence (in this case the ratio of the two channels), F_0 is the zero time fluorescence, F_∞ is the fluorescence at infinite time, and k is the rate constant.

For other situations, where a more complete analysis is required, the following equation may be used. If the steady-state assumption is made and the pseudofirst-order condition $[R]_T \gg [P]_T$ is maintained, then the rate of open complex formation will be defined by the following rate equation:

$$[R \cdot P_o] = k_1[R]_T[P]_T/(k_{-2} + k_1[R]_T)$$
$$[1 - \exp\{-(k_2(k_{-2} + k_1[R]_T))t/(k_{-1} + k_2 + k_1[R]_T)\}] \qquad (1)$$

where $[R]_T$ is the total RNA polymerase concentration and $[P]_T$ is the total promoter concentration.

Converting them to fluorescence intensity,

$$F_{obs} = F_0 + \{(F_\infty - F_0)k_1[R]_T\}/(k_{-2} + k_1[R]_T)$$
$$[1 - \exp\{-(k_2(k_{-2} + k_1[R]_T))t/(k_{-1} + k_2 + k_1[R]_T)\}] \quad (2)$$

For slower promoters, k_{-2} is generally small and a modified equation can be used:

$$R + P \underset{k_{-1}}{\overset{k_1}{\rightleftarrows}} R.P_c \overset{k_2}{\rightarrow} R.P_o$$

In this case, the aforementioned rate equation simplifies to

$$[R.P_o] = [P]_T[1 - \exp\{-k_1k_2[R]_Tt/(k_{-1} + k_2 + k_1[R]_T)\}] \quad (3)$$

and the corresponding fluorescence version becomes

$$F_{obs} = F_0 + (F_\infty - F_0)[1 - \exp\{-k_1k_2[R]_Tt/(k_{-1} + k_2 + k_1[R]_T)\}] \quad (4)$$

For faster promoters, the rapid preequilibrium assumption has been used such that the equation reduces to

$$R + P \overset{K_s}{\rightleftarrows} R.P_c \underset{k_{-2}}{\overset{k_2}{\rightleftarrows}} R.P_o$$

where K_s is the equilibrium constant for the first step.

The kinetic parameters can also be obtained by measuring the fluorescence increase at several concentrations of RNA polymerase but still in stoichiometric excess over the DNA concentration. The fluorescence increase can then be fitted to a three-parameter exponential growth equation to obtain the observed rate constant (k_{obs}). The k_{obs} obtained at several different RNA polymerase concentrations can be fitted to the following equation to obtain K_s and k_f:

$$k_{obs} = k_f[RNAP]/(1 + K_s[RNAP])$$

Promoter Clearance Assay

After formation of the open complex, 12–14 bp are disrupted. In the presence of rNTPs, the polymerase initiates the synthesis of RNA, followed by clearance of the promoter by the polymerase and reformation of the disrupted bonds. The reformation step can be assayed directly by 2-aminopurine fluorescence. Sullivan et al.[10] have used multiple 2-amino-purine-labeled templates to assay the rate of promoter clearance of the P_{rm} promoter of bacteriophage λ. After formation of the open complex, rNTPs are added and the fluorescence of 2-aminopurine is followed imme-diately. Although the kinetics of loss of fluorescence intensity can be fitted to a single exponential, it may reflect a sum of several exponential decays if multiple 2-aminopurines are present in the template.

Concluding Remarks

A problem with working with short linear templates is the presence of ends, which the RNA polymerase is known to bind with high affinity.[13] The full kinetic scheme under such situation will be

$$
\begin{array}{ccccc}
 & k_1 & & k_2 & \\
R + P & \rightleftarrows & R \cdot P_c & \rightleftarrows & R \cdot P_o \\
K_e \uparrow\downarrow & k_{-1} & & k_{-2} & \\
R \cdot P_e & & & &
\end{array}
$$

where $R \cdot P_e$ is the end-bound RNA polymerase–oligonucleotide complex and K_e is the dissociation constant of the end-bound complex (making the reasonable assumption that end binding is fast compared to other pro-cesses). The result of end binding would be a decrease in free RNA poly-merase concentration. Whether the effective concentration of the template would be lower depends on whether $R \cdot P_e$ can bind a polymerase molecule at the promoter.

A second problem in determination of kinetic constants from 2-aminopurine assays is the determination of active fraction of the RNA polymerase. It is well known that only a fraction of many preparations of the RNA polymerase is competent to bind the template. This correction is required when rate constants are measured as a function of RNA polymerase concentrations. The correction is probably negligible when isomerization rates are being measured at relatively high RNA polymerase concentrations.

[13] P. Melancon, R. R. Burgess, and M. T. Record, Jr., Biochemistry 22, 5169 (1983).

[49] Single-Molecule DNA Nanomanipulation: Detection of Promoter-Unwinding Events by RNA Polymerase

By A. Revyakin, J.-F. Allemand, V. Croquette, R. H. Ebright, and T. R. Strick

This article describes a nanomanipulation technique that makes it possible to mechanically and quantitatively stretch and supercoil a single linear DNA molecule.[1-3] We show how this technique can be extended to the study of protein–DNA interactions that lead to DNA untwisting, particularly to the study of promoter unwinding by RNA polymerase during the initiation of transcription.

In our system, a linear 4-kb DNA molecule containing a single promoter is anchored at one end to a treated glass surface and at the other end to a small (1 μm) magnetic bead (Fig. 1A). Using a magnetic manipulator, the bead is pulled on and rotated, resulting in supercoiling of the DNA (one negative supercoil per each clockwise rotation; one positive supercoil for each counterclockwise rotation). The three-dimensional position of the bead is determined in real time using videomicroscopy and software analysis, yielding the DNA end-to-end extension, l. Changes in l as a function of force, F, and supercoiling, n, are calibrated.[4] The system is used to observe, in real time, protein–DNA interactions that affect supercoiling.[5] For example, when RNA polymerase is introduced and allowed to bind to and unwind promoter DNA, the corresponding changes in DNA supercoiling and end-to-end extension can be observed by monitoring the bead position in real time.

The system is prepared as follows. One end of the 4-kb DNA fragment containing a single promoter is ligated to a 1-kb multiply biotin-labeled DNA fragment, and the other end is ligated to a 1-kb multiply digoxigenin-labeled DNA fragment. Reaction of the resulting DNA fragment with a streptavidin-coated magnetic bead results in attachment of the DNA fragment to the bead, through multiple linkages, to the biotin-labeled end. Deposition of the resulting bead-attached DNA fragment onto an

[1] T. R. Strick, J. F. Allemand, D. Bensimon, A. Bensimon, and V. Croquette, *Science* **271**, 1835 (1996).

[2] J.-F. Allemand, D. Bensimon, R. Lavery, and V. Croquette, *Proc. Natl. Acad. Sci. USA* **95**, 14152 (1998).

[3] T. R. Strick, V. Croquette, and D. Bensimon, *Nature* **404**, 901 (2000).

[4] T. R. Strick, J.-F. Allemand, D. Bensimon, and V. Croquette, *Biophys. J.* **74**, 2016 (1998).

[5] M. Amouyal and H. Buc, *J. Mol. Biol.* **195**, 795 (1987).

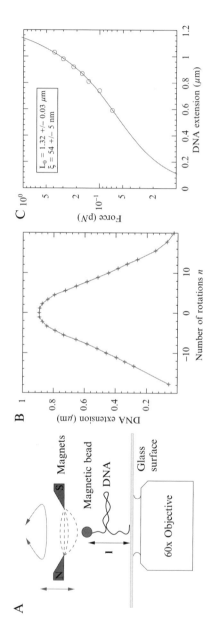

Fig. 1. Experimental system. (A) DNA is tethered between a magnetic bead and a glass surface as described in the text. Magnets located above the sample impose a vertical stretching force that can be changed by increasing or decreasing the distance between the magnets and the bead. Rotating the magnets causes the bead to rotate in a synchronous manner, supercoiling the tethered DNA, with each clockwise rotation introducing one negative supercoil and each counterclockwise rotation introducing one positive supercoil. The bead can be viewed under the microscope, and image analysis is used to determine the position of the bead above the surface and hence the extension of DNA, l. In an appropriate range of supercoiling, each rotation of the bead results in the introduction or removal of one plectonemic supercoil (plectoneme or loop; three plectonemes are shown) and a corresponding change in DNA extension, $\Delta l_{cal} \sim 60$ nm (because the contour length of a plectoneme is ~ 60 nm, but its contribution to vertical extension of the DNA is 0 nm). (B) Calibration curve showing DNA extension l vs rotation in RNA polymerase buffer (see text) at $34°$ for $F = 0.3$ pN. Rotation n is the number of turns applied to the 4-kb (1.3 μm) DNA via the magnets and the bead. The degree of supercoiling, $\sigma = n/Lk_0$, is the number of rotations of the bead, n, divided by the natural linking number, $Lk_0 = N/h$, where N is the number of base pairs and h is the number of base pairs per helical turn of the DNA ($h = 10.4$).[6] As the DNA is progressively over- or underwound at this low force, its extension drops regularly as additional plectonemes are formed. The dataset is collected at constant force, F, by rotating the magnets while their height above the sample is kept constant. In these conditions, each additional rotation (i.e., plectonemic supercoil) causes the extension to decrease by about 60 nm. As the force increases, this contraction rate drops according to $F^{-0.4}$.[7] lowering ionic strength causes an increase in Δl_{cal}.[4] (C) Force–extension data obtained on a single 4-kb DNA molecule at $\sigma = 0$. The solid line is a fit to the worm-like chain model of DNA elasticity (see text), giving a persistence length $\xi \sim 50$ nm and a crystallographic length of about 1.3 μm.

antidigoxigenin-coated glass surface results in attachment to the surface, through multiple linkages, at the digoxigenin-labeled end. Multiple linkages between the DNA fragment and the bead and the DNA fragment and the surface torsionally constrain the DNA with respect to the bead and surface and thereby couple supercoiling of the DNA to rotation of the bead relative to the surface.

Application of a magnetic field gradient above the surface makes it possible to pull on the bead, lifting the bead from the surface and stretching the DNA. The gradient is generated by a pair of magnets separated by a small gap (Fig. 2). Translating the magnets toward or away from the glass surface causes the force to increase or decrease, respectively. The stretching force can be measured by analyzing the Brownian motion of the bead and the end-to-end extension of the DNA, and forces from 10 fN to 100 pN can be applied and measured using this technique.

Rotating the bead by rotating the magnetic field mechanically and quantitatively twists the DNA. Control over the angular displacement of the bead is achieved by the magnetic field mentioned earlier. The bead becomes magnetized in a fixed orientation, causing it to behave effectively as a compass needle. Rotating the magnetic field causes the bead to rotate in a completely synchronous fashion. Each clockwise turn of the magnets unwinds the DNA by one turn, and each counterclockwise turn overwinds the DNA by one turn. A user-selected number of turns can be applied to the DNA in a fully controlled, fully reversible fashion.

In an appropriate range of supercoiling, each rotation of the bead causes a large, ~60-nm change in DNA end-to-end extension (see Fig. 1B), as it introduces or removes one plectonemic supercoil (which has a contour length of ~60 nm but which contributes ~0 nm to DNA end-to-end extension l). Systematically rotating the bead and monitoring the extension yields a calibration curve that relates extension to supercoiling (Fig. 1B). Thereafter, supercoiling can be determined simply by measuring extension and referring to the calibration curve. Continuous monitoring of extension gives real-time information on changes in DNA supercoiling, and thus makes possible real-time measurements of the topological effects of protein–DNA interactions.[3]

For a constant linking number, as is the case with this system when the magnets are not rotating, any change in DNA twist, Tw, must produce a corresponding change in DNA writhe, Wr (represented by plectonemic supercoils), according to the equation $\Delta Tw = -\Delta Wr$. We reasoned that

[6] W. Saenger, "Principle of Nucleic Acid Structure." Springer-Verlag, New York, 1988.
[7] T. R. Strick, G. Charvin, N. H. Dekker, J.-F. Allemand, D. Bensimon, and V. Croquette, *C. R. Phys.* **3**, 595 (2002).

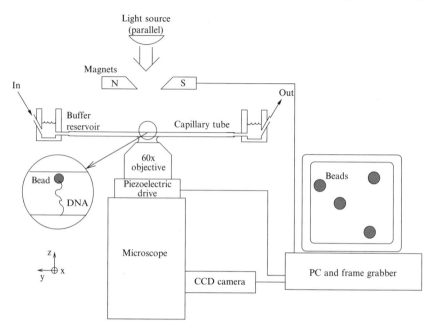

FIG. 2. Experimental setup. A treated glass capillary tube ($1 \times 1 \times 50$ mm, Vitrocom) is mounted on an inverted microscope (Olympus) and connected at its ends to buffer reservoirs. DNA is tethered at one end to the "floor" of the capillary and at the other to a micrometer-sized magnetic bead (Dynal or New England Biolabs, not shown to scale). The microscope is equipped with a high-NA 60 or 100× oil immersion lens (Olympus). The capillary and the objective are temperature controlled to within $0.1°$ using Peltier modules (Marlowe Electronics) and a temperature controller module (Wavelength Electronics, not shown). The fine focus of the objective is controlled by mounting it on a piezoelectric element (Physik Instrumente) operating in closed loop mode and driven with a PC-based interface. A pair of neodymium–iron–boron magnets ($1 \times 1 \times 0.5$ cm, Magnet Sales) is mounted north–south on motorized translation and rotation stages that are piloted using a PC-based controller (Physik Instrumente). A CCD camera (Jai CVM30) connected to the microscope relays images of the magnetic bead at video rates to the PC via a frame grabber (PCVision, Coreco Imaging). The computer uses protype image processing routines (XVIN[8]) to extract from this video the 3D position of the bead (to within about 10 nm). The vertical (z) position of the bead above the surface is determined by analysis of bead diffraction rings, which increase regularly in size as the bead moves away from the focal plane of the objective. The lateral (x and y) position of the bead is determined using two-dimensional particle tracking algorithms. From the position of the magnetic bead above the surface, the DNA end-to-end extension, l, is measured. Measurement of the bead mean-squared fluctuations perpendicular to the direction of stretching, $< \delta x^2 >$, yields the stretching force, $F = k_B T l / < \delta x^2 >$, where k_B is Boltzmann's constant and T is the absolute temperature.

[8] C. Gosse and V. Croquette, *Biophys J.* **82,** 3314 (2002).

promoter unwinding by RNA polymerase (or any other protein–DNA interaction that causes a local change in DNA twist) should be observable using this system. Similar to traditional topological assays on supercoiled DNA,[5] the approach exploits the fact that a change in twist upon promoter unwinding will result in a corresponding change in writhe (plectonemic supercoils). In this system the change in writhe is detected as a change in DNA end-to-end extension (Fig. 4).

Experimental Setup

A sketch and description of the experimental setup are provided Fig. 2. For the study of interactions between RNA polymerase and supercoiled DNA, this configuration has several advantages.

First, the magnetic manipulator is a user-friendly and robust tool for nanomanipulation experiments. The main magnets, made of high-grade neodymium–iron–boron and spaced by $\sim$1 mm, generate a magnetic field gradient that drops off exponentially over distances on the millimeter scale. As a result, the force imposed on the bead is not significantly affected by either changes in the DNA extension (no greater than 1 μm) or the surface position (due, for instance, to thermal drift).

Second, beads of different diameters can be employed to give the system different force and noise characteristics. With 1-μm-diameter magnetic beads, forces of about 1–2 pN and 1-s timescales—an important range in the study of protein–DNA interactions—can be accessed. Forces of up to about 100 pN can be reached using 4.5-μm-diameter beads, but for pN-scale forces, these beads do not achieve the temporal resolution of the smaller ones.

Third, reversible rotation of the bead, and corresponding supercoiling of the tethered DNA, is accomplished easily by rotating the magnets.

Methods and Protocols

For these DNA nanomanipulation experiments, a 4-kb linear DNA is labeled at one end (on both strands) with digoxigenin groups and at the other end (again on both strands) with biotin groups. Linear 4-kb DNA molecules are generated from a GC-rich DNA template by polymerase chain reaction (PCR). The template contains a unique promoter site of interest. One end of the 4-kb molecule is ligated to a DNA fragment labeled with biotin groups and the other end to a DNA fragment labeled with digoxigenin groups. The DNA is then attached to streptavidin-coated magnetic beads. This bead–DNA construct is then introduced into a glass capillary tube that has been functionalized with antidigoxigenin and to which the bead–DNA constructs bind at the remaining free DNA end.

The mechanical properties of the DNA are calibrated. This calibration is then used to measure RNA polymerase–DNA interactions via changes in overall DNA supercoiling.

Preparation of flow cell

In order to absorb antidigoxigenin to the glass surface, the surface must first be made hydrophobic. A simple procedure involving plasticization is presented, followed by an alternative procedure involving silanization. We then present procedures for functionalization and blocking of the resulting hydrophobic surface and assembly of the surface into a flow cell.

Preparing a Hydrophobic Surface by Plasticization

Materials

Square glass capillary, $1 \times 1 \times 50$ mm (Vitrocom)
Nitric acid (Sigma)
Polystyrene, $M_w \sim 280,000$, 0.1% (w/v) in toluene (Sigma)
Tygon tubing R3603, i.d. 0.0402 in., o.d. 0.1082 in. (Kimberly-Clarke)
Argon

1. Capillary tubes are cleaned in a nitric acid bath for 4 h and then rinsed extensively with water and dried with a stream of clean argon.
2. One end of the capillary is dipped into the polystyrene solution so as to draw in about 15 μl.
3. The capillary is slowly tipped at one end and then the other so as to coat the entire inner surface with the polystyrene solution. This should be done no more than two or three times, as a thick polystyrene layer will be more unstable than a thin one.
4. Excess solution is wicked out of the capillary, which then is dried with a stream of clean argon.
5. The ends of the capillary are fitted with a 2-cm length of Tygon tubing.

Preparing a Hydrophobic Surface by Silanization

Materials

Square glass capillary (as described earlier)
Phenyltrimethoxysilane (United Chemical)
Ethanol, anhydrous (Pharmco)
Acetic acid, pure (glacial) (Fisher)

1. Capillary tubes are cleaned in nitric acid and dried as described earlier.
2. A 95% ethanol solution in H_2O is adjusted to pH 5.0 with acetic acid.

3. Silane is added to the 95% ethanol solution to a final concentration of 2% and incubated with stirring for 5 min at room temperature.
4. Capillaries are incubated overnight at room temperature in the solution under gentle agitation.
5. Capillaries are rinsed briefly with ethanol and cured at $100°$ for 30 min. The ends of the capillary are then fitted with Tygon tubing.

Functionalization and Blocking of Hydrophobic Surface

Materials

Surface-modified capillary (see earlier description)
Phosphate-buffered saline (PBS, Fluke)
Polyclonal antidigoxigenin (Roche)
Bovine serum albumin fraction V (BSA; Roche); stock is 50 mg/ml in H_2O
Polyglutamic acid, M_w 1500–3000 (Sigma); stock is 10 mg/ml in PBS
Standard buffer (SB; 10 mM potassium–phosphate buffer, pH 8.0, 0.1 mg/ml BSA, 0.1% Tween 20)
NaN$_3$ (Sigma)

1. One hundred microliters of PBS containing 0.1 mg/ml polyclonal antidigoxigenin is injected into the capillary, which is then placed in a humid chamber and left to incubate overnight at $37°$.
2. A blocking solution of SB containing 10 mg/ml BSA, 3.3 mg/ml of polyglutamic acid, and 3 mM sodium azide is injected into the capillary, which is then placed in a humid chamber and left to incubate for at least 2 days at $37°$. (The resulting blocked capillary can be stored at $4°$ for up to 2 weeks.)

Assembly of Capillary into Flow Cell

Materials

Blocked surface-modified capillary (see earlier description)
Syringe pump (Fisher)
Microscope system (see Fig. 2)
Thermally regulated capillary holder (10 × 10-cm aluminum plate with 1.2 × 50-mm slot and attached Pelletier modules)
PBS
3-μm-diameter tosylactivated polymer beads (Bangs Laboratories)

1. To use a surface for experiments, the capillary is rinsed gently twice with 1 ml PBS to remove the blocking solution, inserted into a thermally regulated capillary holder, and connected to buffer-exchange reservoirs at each end (500-μl plastic wells, one connected to a syringe pump and the

other to a waste reservoir). The holder is attached to the microscope stage, and the microscope is focused on the "floor" of the capillary.

2. One hundred microliters of a 1:1000 dilution of 3-μm tosylactivated polymer beads in PBS is injected into the capillary. The beads are left to settle and bind to the surface for 30 min, or until they cease to move, at room temperature. (The beads later serve as references to correct for surface drift.)

3. A gentle (0.1–0.3 ml/min; typically 20 min) flow of PBS is established to rinse out unbound reference beads. (Each field of view as observed through the CCD camera should contain three to five beads.)

4. One to 2 ml of buffer SB is flowed (0.2–0.3 ml/min, typically 10 min) through the capillary.

Preparation of DNA Constructs for Nanomanipulation

Preparation of DNA Fragment

As mentioned previously, the DNA fragment under study should be ~4 kb in length, corresponding to an end-to-end extension under stretching of up to ~1.3 μm. (For longer DNA fragments, real-time measurements of extension show higher noise relative to expected signal.) The DNA fragment should contain a centrally located promoter and should otherwise be G/C- rich (to avoid nonspecific melting of A/T-rich regions). For this construct, the *rrnB* P1 promoter was amplified from plasmid pTZ19rrnBP1 (S. Nechaev and K. Severinov, unpublished results) using the add-on PCR primers GAGAGAGGTACCGGTTGAATGTTGCGCGGTCAG and GAGAGAGGTACCGTTGTTCCGTGTCAGTGGTG. The resulting 94-bp DNA fragment (spanning positions −78 to +16, where +1 is the transcription start site) was then cloned into the unique *Kpn*I site of the *Thermus aquaticus rpoC* gene (located at position 2889 downstream of the *rpoC* ORF), yielding plasmid pART$_{aq}$RpoC/rrnBP1.[9] The same procedure can be used to prepare constructs for the study of other promoter sequences.

Materials

Add-on PCR primer "*Mlu*I": gagagaacgcgtgaccttctggatctcgtccaccagg (10 μM)

Add-on PCR primer "*Not*I": gagagagcggccgcgagaagatccgctcctggagc-tacg (10 μM)

pARTaqRPOC/rrnBP1 plasmid, 1:200 dilution of DNA miniprep

Dimethyl sulfoxide (DMSO; Sigma)

dNTP mix, 2.5 mM each (Roche)

---

[9] L. Minakhin, S. Nechaev, E. A. Campbell, and K. Severinov, *J. Bacteriol.* **183**, 71 (2001).

Thermostable DNA polymerase, high fidelity, 3 U/μl, and 10× PCR
   buffer (Roche)
*Mlu*I and *Not*I restriction enzymes (New England Biolabs)
Qiaquick gel extraction kit (Qiagen)
Spin column, S400HR (Amersham)
Thermal cycler
Lyophilizer
PCR solutions and temperature cycles are set up as follows.

Ingredient	Volume (μl)	Step	Temperature (°)	Time (min)
Template DNA	1	1	60	0.5
Add-on primer "*Not*I"	1.5	2	68	4
Add-on primer "*Mlu*I"	1.5	3	94	2
DMSO	2.5	4	60	0.5
dNTPs	4	5	68	4
10× PCR buffer	5	6	Goto 3, 30×	
H$_2$O	34	7	68	7

1. Prepare five tubes of the aforementioned mixture (250 μl total).
2. Place in thermal cycler and incubate for 5 min at 95°.
3. To each tube add 0.5 μl of DNA polymerase and pipette up and down.
   (Tubes should be left in the thermal cycler at 95° during this step.)
4. Run the PCR program shown above.
5. Purify the 4-kb PCR product on 0.7% agarose gel containing 0.5 μg/
   ml ethidium bromide and extract the 4-kb product. Typical yields
   are on the order of 2 μg (∼1 pmol).
6. Digest product with restriction enzymes *Mlu*I and *Not*I per
   instructions of the manufacturer for 1 h at 37°. Heat inactivate for
   20 min at 65°.
7. Purify product using the spin column as per instructions of the
   manufacturer.
8. Adjust the concentration of the product to 50 n*M*.

Preparation of 1-kb Labeled DNA Fragments

Two 1-kb DNA fragments—one labeled with dUTP-biotin and the
other labeled with dUTP-digoxigenin—are generated by PCR using the
same template. One primer for the biotin–labeling PCR incorporates a
*Not*I restriction site for subsequent cleavage and ligation with the 4-kb

DNA fragment. One primer for the digoxigenin-labeling PCR incorporates a *Mlu*I restriction site for subsequent cleavage and ligation with the 4-kb DNA fragment.

Materials

Add-on PCR primer "*Mlu*I", 10 μM
Add-on PCR primer "*Not*I", 10 μM
Primer "TaqRpoC-1" TCCTGGCGCAGGTAGATGAG, 10 μM
Primer "TaqRpoC-2" CTGATGCAAAAGCCCTCGGG, 10 μM
Plasmid pARTaqRpoC/rrnBP1, 1:200 dilution of miniprep DNA (see earlier description)
Thermostable
Thermostable DNA polymerase, high fidelity, 3U/μl, and 10× PCR buffer (Roche)
DMSO (Sigma)
dNTP mix, 2.5 mM each (Roche)
Biotin-16-dUTP, 1 mM (Roche)
Digoxigenin-11-dUTP, alkali stable, 1 mM (Roche)
Qiaquick PCR cleanup kit (Qiagen)
Spin column S400HR (Amersham)
Thermal cycler
Lyophilizer
PCR volumes and temperature cycles are as follows.

| | Volume | | | | |
| | Biotin labeling | Dig labeling | Step | Temperature (°) | Time (min) |
Ingredient					
Template DNA	1	1	1	58	0.5
Add-on primer "*Not*I"	1.5		2	72	0.75
Primer TaqRpoC-1	1.5		3	95	2
Add-on primer "*Mlu*I"		1.5	4	58	0.5
Primer TaqRpoC-2		1.5	5	72	0.75
DMSO	2.5	2.5	6	Goto 3, 30x	
dNTPs	4	4	7	72	5
Biotin-16-dUTP	1				
Dig-11-dUTP		1			
10× PCR buffer	5	5			
H$_2$O	34	34			

1. Prepare five tubes of 50 μl of the aforementioned mixture (250 μl total per labeling reaction).

2. Place in thermal cycler and incubate for 5 min at 95°.
3. To each tube, add 0.5 μl of DNA polymerase and pipette up and down. (Tubes should be left in the thermal cycler at 95° during this step.)
4. Run the PCR program shown above.
5. Purify the 1-kb PCR products on 0.7% agarose gel containing 0.5 μg/ml ethidium bromide and extract the 1-kb product. Typical yields are ~2 μg (1 pmol).
6. Digest purified, biotin-labeled DNA with NotI for 1 h at 37° as per manufacturer's instructions. Heat inactivate enzyme for 20 min at 65°. Digest purified, digoxigenin-labeled DNA with MluI for 1 h at 37° as per manufacturer's instructions. Heat inactivate enzyme for 20 min at 65°.
7. Purify product using the spin column as per instructions of the manufacturer.
8. Adjust the concentration of the product to 200 nM.

Ligation of 4-kb DNA Fragment to 1-kb-Labeled DNA Fragments

Materials

Purified DNA fragments (see earlier description)
T4 DNA ligase, 10 U/μl and 10× reaction buffer (New England Biolabs)
EDTA (Sigma)
Tris–HCl, pH 8 (Sigma)

Ingredient	Volume (μl)
4 kb DNA (50 nM)	4
1 kb Dig-labeled fragment (200 nM)	2.5
1 kb Biotin-labeled fragment (200 nM)	2.5
10× T4 DNA ligase buffer	2
H₂O	8
T4 DNA ligase (10 U/μl)	1

1. Mix items listed and incubate for 3 h at room temperature.
2. Inactivate ligase by the addition of EDTA to 10 mM followed by a 10-min incubation at 65°.
3. Dispense into 5-μl aliquots and store at −20°.
4. For binding to magnetic beads, the DNA construct is diluted to ~50 pM with 10 mM Tris, pH 8.0, 10 mM EDTA. Dilute solution can be stored at 4° for 3–6 months before the proportion of DNA

molecules observed to be supercoiled under the microscope decays by about half.

Anchoring of DNA to Beads

Materials

SB (see above)
PBS (see above)
BSA (see above)
1-μm-diameter magnetic beads, 10 mg/ml (New England Biolabs or Dynal) dig- and bio-labelled 4-kb.
DNA fragment (see above)

1. Wash 10 μl of 1-μm magnetic beads in 200 μl PBS supplemented with 1 mg/ml BSA.
2. Resuspend beads in 10 μl PBS supplemented with 1 mg/ml BSA.
3. Deposit a 0.5-μl drop of DNA fragment (50 pM) at the bottom of a small microfuge tube.
4. Load a wide-bore pipette tip with 90 μl of SB.
5. Deposit the 10 μl of beads onto the drop of DNA.
6. Immediately dilute the reaction with the 90 μl SB. This should be done by gently depositing the SB onto the bead + DNA solution.
7. Resuspend the beads to homogeneity by tipping the tube upside down (without causing the liquid to drop) or spinning the tube between thumb and forefinger.

Anchoring of DNA to Flow Cell

Materials

Flow cell (see above)
DNA fragment tethered to beads (see above)
SB (see above)

1. Before injecting the bead–DNA mixture into the flow cell, the magnets should be moved at least 2 cm away from the flow cell.
2. Inject 15 μl of the bead–DNA mixture into one of the plastic reservoirs connected to the capillary tube.
3. Inject another 20–50 μl of SB into the reservoir to ensure that the majority of beads enter the capillary and are distributed evenly along the length of the capillary.
4. Allow magnetic beads to sediment and incubate for ~15 min at room temperature. Magnetic beads should be seen to move about on the surface and not appear immobile. If beads appear immobile, reblock surface for

1–2 h by incubating with BSA and polyglutamic acid as described. [Other possible ways of reducing nonspecific interactions include exposing the surface for 5–10 min to a 1 M solution of NaN_3 or a 10% solution of sodium dodecyl sulfate, the surface can also be exposed to a 10-mg/ml solution of polyglutamic acid in PBS.]

5. Establish a gentle flow (150–200 μl/min, typically 20 min) of SB to remove unbound beads from the surface. (Buffer is injected into the input reservoir using a motorized syringe pump.) The output reservoir is drained by gravity feed. Approximately every 2 min a rod with a small (2-mm diameter) magnet at the end is slowly passed just over the capillary to lift unbound beads off the surface and into the flow field.)

6. Turn off flow and move magnets as close as possible to the sample without coming into contact, causing DNA molecules to extend away from the surface. (Forces range from $\sim$0.5 to 1 pN, depending on the bead.)

Rapid Selection of a Single Supercoilable DNA

At forces between 0.5 and 1 pN, a magnetic bead tethered by a single 4-kb DNA molecule hovers $\sim$1 μm above the surface and displays rapid, constrained Brownian motion. Beads tethered by a single supercoilable molecule can be identified rapidly by observing the behavior of beads as the magnets are rotated. (Changes in the vertical position can be judged by eye by focusing the image slighly above the bead; the bead will appear to grow larger as it moves away from the focal plane toward the surface.)

The magnets first are rotated 15–20 turns counterclockwise (in the direction of positive supercoiling). For a bead associated with a single supercoilable DNA molecule, the DNA molecule will form $\sim$15 plectonemic supercoils, and the bead will be observed to move toward the surface.[4] After returning to the initial position, the magnets are then rotated 15–20 turns clockwise (in the direction of negative supercoiling). For a bead associated with a single supercoilable DNA molecule, the bead will not be observed to move toward the surface upon rotation 15–20 turns clockwise, because under such levels of negative supercoiling DNA denatures and extends.[4] (If the bead does descend in this case, it is probably tethered to the surface by two DNA molecules.)

Calibration of Bead Image, Torsional Zero of DNA, and Force

Calibration of Bead Image

The bead is tracked in x, y, and z at 30 Hz to a resolution of $\sim$10 nm using the software package XVIN, which contains image treatment routines.[8] A reference bead attached to the surface is tracked simultaneously to correct for surface drift. Parallel illumination of the sample permits observation of diffraction rings around the beads, thus allowing for measurement of the z position.

The x and y positions of a bead from one frame to the next are determined by correlating the bead profile with its mirror image obtained by symmetry about the position of the bead at the previous frame[8] (see Fig. 3). To measure the z position of the bead, an initial calibration of its diffraction ring pattern as a function of distance from the objective must be performed. Then, when the objective position is fixed, changes in the ring pattern of the bead can be converted into vertical displacements of the bead relative to the focal plane.

To perform this calibration, the DNA is first extended by increasing the force to ~ 1 pN. (This reduces the bead's Brownian motion.) The image is then focused just above the bead, and the objective is stepped away (i.e., focusing deeper into the sample) in 0.3-μm, ~ 1-s increments. At each step, the radial intensity profile of the diffraction rings is measured and added to a calibration file. The process is then repeated on one of the tosylactivated polymer "reference" beads immobilized on the surface during preparation of the flow cell. The magnetic bead of interest is then made to recoil to the surface (by rotating the magnets ~ 25 turns counterclockwise and moving

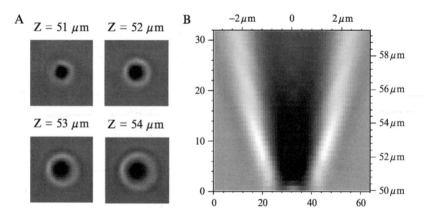

FIG. 3. Determination of vertical (z) displacement of bead via calibration of bead diffraction rings. (A) Images of a 1-μm bead obtained at different focus positions. Diffraction rings formed around the bead grow as the image is focused farther above the bead. [The value above each image represents the position (in μm) of the piezoelectric objective driver at which the image was taken.] (B) Calibration file consisting of a "stack" of bead images, where for each focus position (y axis, 0.3-μm increments) the intensity profile of the bead (gray scale) has been averaged over $360°$ and plotted (x axis) as a function of distance from the bead center. This calibration of the ring pattern as a function of focus position is then used to convert fluctuations in apparent ring size into vertical (z) displacement.

the magnets ~ 1.5 mm from the sample), and the magnetic bead of interest and the immobilized reference bead are tracked simultaneously by comparing their diffraction patterns to the respective calibration files. This provides a reference for the position of the magnetic bead of interest when the DNA end-to-end extension is zero. Furthermore, tracking of both the bead of interest and the reference bead during experiments makes it possible to measure the position of the magnetic bead while correcting for thermal drift of the surface.

Determining the Rotational Zero of the DNA

The DNA end-to-end extension is measured as a function of supercoiling at a low stretching force ($F < 0.3$ pN) in the standard buffer (see Fig. 1B). The extension of DNA is greatest when the molecule is torsionally relaxed ($\sigma = 0$). The magnet rotation is adjusted so as to satisfy this condition before proceeding to the force calibration. This position is the "rotational zero" of the molecule.

In the regime where DNA extension changes linearly with rotation, an extra rotation of the bead generates an additional plectonemic supercoil (loop) along the DNA and its extension decreases by about 80 nm per turn (for experiments performed in the standard buffer and at forces of about 0.1 pN) (Fig. 1B). Calibration data must be acquired whenever the environmental conditions (ionic strength, temperature, or force) are changed. (Increasing ionic strength reduces the slope of the extension vs supercoiling curves, as the radius of plectonemic loops decreases when DNA charge is screened out.[4] Increasing temperature decreases the natural linking number of the DNA, as the helical pitch increases with temperature.[10] In our hands, the rotational zero of the molecule moves to lower n by ~ 0.01 degrees/°/bp (T.R.S., unpublished results). In addition, increasing force reduces the per-turn contraction rate of DNA.[4,7])

Force Calibration

The force acting on the magnetic bead is measured by determining DNA end-to-end extension, l, and also the mean-square Brownian fluctuations of the bead perpendicular to the direction of stretching, $< \delta x^2 >$. These parameters are related to the stretching force, F, as $F = k_B T l / < \delta x^2 >$, where k_B is Boltzmann's constant and T is absolute temperature.

The applied force and resulting DNA extension (see Fig. 1C) are measured for different positions of the magnets above the sample. The DNA must be torsionally relaxed during these measurements, and points should

---

[10] M. Duguet, *Nucleic Acids Res.* **21,** 463 (1993).

be taken in the 0.05 to 1 pN range. Fitting force–extension data to the worm-like chain model of polymer elasticity[11,12] gives a persistence length of about 53 nm and a crystallographic length of about 1.3 μm. A numerical solution is provided[12]:

$$F = \frac{k_B T}{\xi}\left[\frac{1}{4(1-l/l_0)^2} - \frac{1}{4} + \frac{l}{l_0} + \sum_{i=2}^{i=7} a_i \left(\frac{l}{l_0}\right)^i\right] \tag{1}$$

where ξ is the persistence length ($\xi = 53$ nm for DNA), l_0 is the crystallographic length ($l_0 = 1.3$ μm for 4 kb DNA) and $a_2 = -0.5164228$, $a_3 = -2.737418$, $a_4 = 16.07497$, $a_5 = -38.87607$, $a_6 = 39.49944$, $a_7 = -14.17718$.

Force Calibrations Based on Stretching Transitions in Supercoiled DNA

The following procedure yields an estimate of the stretching force exerted by the magnetic field gradient:

1. Rotate magnets clockwise $\sim$25 turns (in the direction of negative supercoiling) and move magnets $\sim$1–2 mm away from sample, allowing bead to recoil to surface.

2. Increase the force progressively by moving magnets closer to the sample (in $\sim$0.2-mm increments) until an abrupt increase in DNA extension occurs. The force at which this occurs, F_c^-, is 0.3 pN in SB at room temperature.[4] (The abrupt increase in extension that occurs for negatively supercoiled DNA at F_c^- is due to the formation of denatured regions in the DNA[13] and the concomitant disappearance of plectonemic supercoils with negative topology.)

A similar procedure also holds for positively supercoiled DNA:

1. Rotate magnets counterclockwise $\sim$25 turns (in the direction of positive supercoiling) and move magnets $\sim$1–2 mm away from sample, allowing bead to recoil to surface.

2. Increase the force progressively by moving magnets closer to the sample (in $\sim$0.2-mm increments) until an abrupt increase in DNA extension occurs. The force at which this occurs, F_c^+, is 3 pN in SB at room temperature.[4] (The abrupt increase in extension that occurs for negatively supercoiled DNA at F_c^+ is due to the formation of hypertwisted domains in

[11] C. Bustamante, J. F. Marko, E. D. Siggia, and S. Smith, *Science* **265,** 1599 (1994).
[12] C. Bouchiat, M. D. Wang, S. M. Block, J.-F. Allemand, T. R. Strick, and V. Croquette, *Biophys. J.* **76,** 409 (1999).
[13] T. R. Strick, V. Croquette, and D. Bensimon, *Proc. Natl. Acad. Sci. USA* **95,** 10579 (1998).

the DNA[2] and the concomitant disappearance of plectonemic supercoils with positive topology.)

These transitions at limiting forces F_c^- and F_c^+ can serve to estimate the force exerted on the bead for a given distance between the magnets and the sample. Increasing the salt concentration delays the onset of these secondary structural transitions and increases the limiting forces; thus in SB supplemented with 150 mM NaCl $F_c^- = 1$ pN.[4]

It is important to keep in mind that when a negatively supercoiled and structurally homogeneous DNA is required (i.e., with no alternative DNA structures, such as denaturation bubbles, and forks), it must not be stretched by forces greater than F_c^-.

Application to the Study of Promoter Unwinding

Single-molecule techniques make it possible to observe in real time the interactions between a single protein and a single DNA molecule.[3,7,14,15] This provides unique information on the catalytic step size of the enzyme, rate-limiting steps in the cycle, and the influence of torsion and force on different steps of the enzymatic cycle. Unwinding of a promoter site by RNA polymerase should affect the overall topology of the DNA molecule bearing the promoter[5]; this may be detected in real time in a single-molecule assay using supercoiled DNA.

Detecting Promoter Unwinding

Materials

Escherichia coli RNA polymerase holoenzyme (Epicentre)
RNA polymerase reaction buffer (25 mM HEPES–NaOH, pH 7.9, 100 mM NaCl, 10 mM MgCl$_2$, 3 mM 2-mercaptoethanol, and 200 μg/ml BSA)
Flow cell and calibrated DNA (see earlier description)

1. Rinse flow cell with RNA polymerase reaction buffer (0.1–0.3 ml/min, 3 ml total).
2. Set temperature to 34° and DNA stretching force to $F\sim0.3$ pN.
3. Calibrate DNA extension as a function of supercoiling as described earlier. In these conditions, changing the DNA rotation by one turn causes a change in DNA extension $\Delta l_{cal} \sim 60$ nm.

---

[14] M. D. Wang, M. J. Schnitzer, H. Yin, R. Landick, J. Gelles, and S. Block, *Science* **282,** 902 (1998).

[15] B. Maier, D. Bensimon, and V. Croquette, *Proc. Natl. Acad. Sci. USA* **97,** 12002 (2000).

4. Rotate magnets to supercoil DNA by the desired amount (typically the DNA is supercoiled by seven turns to give $|\sigma| = 0.018$); direction of rotation depends on the sign of supercoiling one wishes to study.
5. Begin monitoring DNA extension by tracking bead position.
6. Rinse flow cell with 3×100 μl of reaction buffer containing the desired concentration of RNA polymerase holoenzyme (typically $\sim$ nM). After these preliminary washes, add another 200 μl of reaction buffer containing RNA polymerase holoenzyme at the same concentration to the flow cell.
7. Terminate flow and verify that the DNA extension returns to position prior to injection (to ensure that the flow due to injection has not affected the system).
8. Monitor DNA extension as a function of time to observe RNA polymerase/DNA interactions.

Spatial Analysis of Unwinding Signal

Because promoter unwinding disrupts approximately one turn of the double helix (14 bp[16]), in experiments using negatively supercoiled DNA, a single promoter unwinding event should result in the loss of approximately one plectonemic supercoil and a Δl_{obs} $\sim$60-nm *increase* in the distance between the bead and the surface (Fig. 4A). Reversal of promoter unwinding should be accompanied by reappearance of the plectonemic supercoil and a return of the bead to its initial position. Analogously, in experiments with positively supercoiled DNA, a single reversible promoter unwinding event should result in the transient gain of approximately one plectonemic supercoil and a $\sim$60-nm *decrease* in end-to-end extension (Fig. 4B). The loss and appearance of the plectonemic supercoil amplify the signal, as a unit change in local DNA twist is converted into a large ($\sim$60 nm) change in overall DNA extension. It is important to point out that, for the forces considered here, the contribution of the bubble to extension is essentially zero.[15]

Temporal Analysis of Unwinding Signal

Temporal analysis of the unwinding signals may also be performed on such time traces (Fig. 4). One can measure the time interval t_{closed} between consecutive promoter unwinding events (related to the rate of formation of

[16] R. H. Ebright, *J. Mol. Biol.* **304,** 687 (2000).

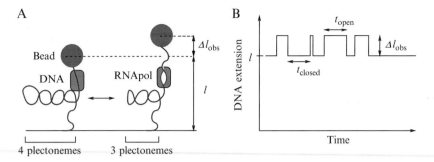

Fig. 4. Detection of promoter unwinding by RNA polymerase. (A) Approach (with negatively supercoiled DNA). (Left) Supercoiled DNA forms negative ($-$) supercoils, storing the torsional constraint as negative writhe ($Wr < 0$). RNA polymerase (shaded box) is shown bound to DNA. (Right) Promoter unwinding by RNA polymerase will cause the DNA to lose $\sim$1 turn of the double helix, causing the twist to decrease by one: $\Delta Tw = -\Delta Wr$. Conservation of DNA linking number according to White's theorem ($\Delta Tw + \Delta Wr = 0$[17]) implies that this decrease in twist must be compensated by an increase in writhe; this translates into the disappearance of one ($-$) plectonemic supercoil and a subsequent increase in DNA extension $\Delta l_{obs} \sim$60 nm. (B) Idealized data. Promoter unwinding events are detected as increases in DNA extension. Promoter rewinding events are detected as decreases, to the initial state, in DNA extension. The amplitude of unwinding in helical turns can be estimated by dividing the observed extension change, Δl_{obs}, by the extension change per turn, Δl_{cal}, obtained from the extension vs supercoiling calibration curve taken in reaction conditions. In addition, the lifetime, t_{open}, of the unwound state, as well as the time between unwinding events, t_{closed}, can be measured directly. Analyzing the lifetime distributions of these events can provide useful new information on the reaction pathway.

the unwound complex) and the duration t_{open} of each promoter unwinding event (i.e., the lifetime of the unwound complex). Statistical analysis of the distribution of both intervals and lifetimes of unwound complexes can provide information on the kinetics of the reaction pathway. The dependence of lifetime distributions on DNA supercoiling, protein concentration, temperature, nucleotides, effectors, and activators (such as CAP protein) should then allow for a better understanding of the mechanisms involved in promoter unwinding by bacterial RNA polymerase.

Signal-to-Noise Analysis

We now consider the temporal and spatial resolution afforded by this system.

[17] Alberts, Bray, Lewis, Raff, Roberts, and Watson, "Molecular Biology of the Cell." Garland, 1989.

Thermal agitation of the magnetic bead generates noise of amplitude δz in the real-time measurement of the DNA extension. This noise precludes measurements on the subsecond scale. However, this noise can be reduced to yield a robust signal-to-noise ratio (signal of $\sim$60 nm and noise of $\sim$15 nm give a signal-to-noise ratio of $\sim$4) by averaging the real-time signal over a $\sim$1-s timescale.

The rationale for the aforementioned conclusion is as follows: fluctuations (noise), δz, in bead z position are related to the random thermal (Langevin) force, F_L, acting on the bead and the stiffness, k_z, of plectonemically supercoiled DNA: $\delta z = F_L/k_z$. The rms Langevin force is given by $F_L = \sqrt{4k_BT \cdot 6\pi\eta r \Delta f}$, where Δf is the bandwidth in Hertz and $6\pi\eta r$ is the viscous drag coefficient of the bead. With the viscosity of water $\eta = 10^{-3}$ poise, the bead radius $r = 0.5$ μm, and $k_z \sim 7 \times 10^{-7}$ N/m (for negatively supercoiled 4-kb DNA at a force of $\sim$0.3 pN), we obtain $F_L \sim 10$ fN. $\sqrt{\Delta f}$ and $\delta z \sim 15$ nm $\sqrt{\Delta f}$. Thus as the bandwidth increases (i.e., as temporal resolution is improved), there is less time over which to average out thermal fluctuations and the noise level increases (i.e., the accuracy of spatial measurements worsens). With a bandwidth of 1 Hz (achieved by signal averaging over 1 s), we have an acceptable compromise between temporal and spatial resolution with a signal-to-noise ratio of 4.

Detecting DNA Bending/Compaction

Formation of the RNA polymerase–promoter open complex is known to be accompanied by apparent compaction of DNA of 5–30 nm due to introduction of a net bend in the DNA and wrapping of the DNA on and around the surface of RNA polymerase (C. Rivetti, N. Naryshkin, and R. H. Ebright, unpublished results). Apparent compaction of DNA will result in different signals—and therefore will be detectable and quantifiable—in experiments with positively supercoiled DNA or negatively supercoiled DNA (Fig. 5A). Apparent DNA compaction will cause a net decrease in the end-to-end extension of the DNA regardless of the sign of the supercoiling. Thus, in experiments with negatively supercoiled DNA, promoter unwinding will cause DNA extension to increase, but associated DNA compaction will partly offset the increase in extension (Fig. 5A). (The two effects will subtract from one another, yielding a smaller signal than if no compaction occurs.) Conversely, in experiments with positively supercoiled DNA, promoter unwinding will cause DNA extension to decrease, and associated DNA compaction will cause DNA extension to decrease further (Fig. 5B). (The two effects will add.) Thus compaction will be observable as a difference in signal amplitude observed upon promoter

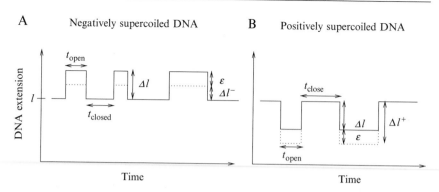

FIG. 5. Detection of DNA compaction by DNA polymerase. Idealized data showing the effects of DNA topology and DNA compaction on the expected signals. Solid lines represent signals expected for open complex formation not accompanied by DNA compaction; dotted lines represent the expected signals for open complex formation accompanied by DNA compaction. (A) For negatively supercoiled DNA in the absence of compaction, promoter unwinding will cause an *increase* in extension of amplitude Δl. Compaction by length ε will reduce the signal amplitude to $\Delta l^- = \Delta l - \varepsilon$. (B) For positively supercoiled DNA in the absence of compaction, promoter unwinding will cause a *decrease* in extension of amplitude Δl. Compaction by length ε will increase the signal amplitude to $\Delta l^+ = \Delta l + \varepsilon$. The extent of compaction, ε, can be determined according to $\varepsilon = (\Delta l^+ - \Delta l^-)/2$.

unwinding on positively vs negatively supercoiled DNA and will be quantifiable as one-half the difference in signal amplitude (see legend to Fig. 5). We note that DNA bending may produce a similar effect. In principle, it should therefore be possible to distinguish between protein–DNA interactions that result solely in a change of local DNA twist and those that result in full or in part from a local change in DNA compaction and/or bending.

Concluding Remarks

We described a nanomanipulation technique that is used to quantitatively supercoil a single DNA molecule. Once the response of the DNA to mechanical changes in supercoiling has been calibrated, the system can be used to observe in real-time interactions between a single enzyme molecule and its supercoiled DNA substrate. We discussed applying this method to the study of RNA polymerase-induced promoter unwinding, as well as bending/compaction of the DNA by RNA polymerase. With this assay it will be possible to quantitatively study the role of DNA sequence and supercoiling, temperature, nucleotides, effectors, and activators in formation of the transcription bubble. This technique may be applicable to the real-time study of a variety of protein–DNA interactions that cause deformation of the DNA, such as the unwinding of origins of replication or the binding of transcription factors.

Acknowledgments

T.R.S is grateful to the Cold Spring Harbor Laboratory Fellow program for support. R.H.E is supported by NIH Grant GM41376 and a Howard Hughes Medical Institute Investigatorship. V.C. and J.-F.A. thank the Ecole Normale Superieure (Paris), the CNRS, and the Universities of Paris VI and VII, as well as the ARC (V.C.) and the French Ministry of Research ACI Program (J.-F.A.) for continued support.

[50] Simple Fluorescence Assays Probing Conformational Changes of *Escherichia coli* RNA Polymerase During Transcription Initiation

By RANJAN SEN and DIPAK DASGUPTA

The promoter recognition by *Escherichia coli* RNA polymerase (RNAP) σ^{70} holoenzyme is a multistep process involving complicated conformational changes in both DNA and RNA polymerases. From a detailed thermodynamic, kinetic, and biochemical characterization[1] of this process, it has been established that RNAP first forms a closed complex (RP_c) in which the promoter DNA is in a closed duplex state. Then RP_c converts into an open complex (RP_0) through multiple intermediate states (RP_i). In open complex, the promoter is melted and ready to start transcription. This conversion from closed to open complex not only changes the state of DNA, but major conformational changes in RNAP also occur, which involve the "jaw closure" on the downstream part of the DNA.[1,2] Each of these conformational states during the pathway of open complex formation could be potential targets of activator and repressor molecules. Therefore, it is important to probe each of these states structurally to understand the basic mechanism of regulation of gene expression (Fig. 1).

[1] P. L. deHaseth, M. L. Zupanic, and M. T. Record, Jr., *J. Bact.* **180**, 3019 (1998).
[2] K. S. Murakami, S. Masuda, E. A. Campbell, O. Muzzin, and S. A. Darst, *Science* **296**, 1285 (2002).

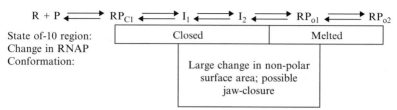

FIG. 1. Summary of the intermediates in the process of open complex formation. R and P are RNAP and promoter, respectively. RP_c, I_1, and I_2 are the intermediates with duplex DNA, whereas DNA is melted in RP_{o1} and RP_{o2}. From Sen and Dasgupta.[6]

These intermediate states can be trapped at lower temperatures.[3,4] This article describes a simple temperature-dependent fluorescence assay of the gradual conformation changes of RNAP in these intermediates. Assays involve a wide range of temperature from 4° to 37°C so that we can probe as many intermediate states as possible.[5,6]

Methods

Enzymes, DNA, and Chemicals

Escherichia coli RNA polymerase holoenzyme is from Pharmacia. Specific activity of the enzyme is 950 units/mg. We use the strong *trp* promoter cloned in expression vector pDR720 in all our assays. Fluorescence assays of RNA polymerase are done in buffer T (50 mM Tris–Cl, pH 8.0, 10 mM MgCl$_2$, 100 mM KCl), and those with TbCl$_3$ are done in similar buffer containing 20 mM Tris–Cl, pH 7.5. Tb(III) tends to precipitate at a higher pH. The reagents are of analytical grade, and the buffers are free from fluorescence impurities.

Fluorescence Spectroscopy

Fluorescence assays involve two types of fluorophores: (1) the intrinsic fluorescence of RNA polymerase originating from the tryptophan residues, which is measured by exciting the enzyme at 295 nm and following the

[3] A. Spassky, *J. Mol. Biol.* **188**, 99 (1986).
[4] D. C. Straney and D. M. Crothers, *Cell* **43**, 449 (1985).
[5] R. Sen and D. Dasgupta, *Biochem. Biophys. Res. Commun.* **201**, 820 (1994).
[6] R. Sen and D. Dasgupta, *Biophys. Chem.* **57**, 269 (1996).

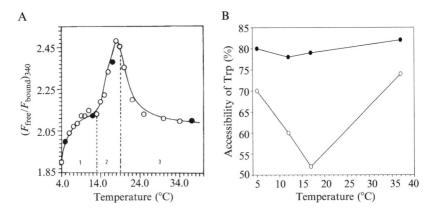

FIG. 2. (A) Relative emission intensity $[F_{free}/F_{bound}$, where F_{free} and F_{bound} denote emission intensities of free RNAP (10 units) and in the presence of 12 nM pDR720, respectively, at 340 nm] of RNAP as a function of temperature. Temperature ranges for different equilibria are demarcated by broken lines and are indicated by numerals 1, 2, and 3. Temperatures corresponding to the closed circles are chosen for further conformational analysis. (B) Changes in the accessibility of tryptophan residues of RNAP (10 units) alone (●) and in the presence of 12 nM pDR720 (○) as a function of temperature. From Sen and Dasgupta.[6]

emission at 340 nm, and (2) a fluorescent substrate Tb–GTP as an extrinsic fluorophore,[7] which is measured by excitation at 295 nm and emission at 545 nm.

All fluorescence spectra are recorded with a Hitachi 4010 spectrofluorimeter in a CAT mode with a multiple scan attached to a constant temperature water bath. The excitation wavelength is 295 nm, and emission and excitation slit widths are 5 and 10 nm, respectively. Correction due to the innerfilter effect is not considered because sample absorbance is below 0.04. All spectra are recorded in a temperature-controlled cuvette holder.

The RNA polymerase holoenzyme (10 units) is incubated with 12 nM of DNA at 4°, and the temperature is increased at intervals of 1 to 2° up to 37°. Fluorescence spectra are recorded at each temperature. Similar recordings are done for free enzyme under identical conditions. Relative intensity (F_{free}/F_{bound}) is plotted against temperature to generate the curve in Fig. 2A.

The Tb–GTP complex is made by mixing the two components, TbCl$_3$ and GTP.[7] RNAP is added to Tb–GTP. To the preformed RNAP–TbGTP complex, a saturating amount of DNA is added. Fluorescence polarization

[7] D. Chatterjee and V. Gopal, *Methods Enzymol.* **274**, 456 (1996).

anisotropy (FPA) and acrylamide quenching of Tb–GTP fluorescence at 545 nm are performed at four different temperatures (5, 12, 17, and 37°).

Fluorescence Polarization Anisotropy

FPA is a measure of the rotational freedom of a fluorophore in the solution. In general, a small fluorophore is depolarized in solution due to random Brownian motion and gives rise to a low FPA value. When it is bound to a macromolecule, its rotational motion is restricted and it becomes more polarized, resulting in a higher FPA value. The extent of polarization (also FPA value) or the orientation of the dipole moment associated with the fluorophore depends on the local environment of the binding site of the macromolecule.[8] Therefore, changes in the local conformational of the binding site will give different FPA values of the fluorophore bound at that site. These FPA values can be obtained by using emission and excitation polarizers, which split the excitation and emission beams into vertical and horizontal planes so that orientation of the dipole moments in these two planes can be measured. This polarizer is a common accessory of any standard fluoremeter.

One can measure the changes in FPA values of the fluorophore Tb–GTP bound to the RNA polymerase that acts as a reporter of different conformational states of the substrate-binding site. FPA (r) is measured from the following equation:

$$< r > = (I_{VV} - GI_{VH})/(I_{VV} - 2GI_{VH}) \tag{1}$$

where I denotes the intensity and subscripts refer to vertical or horizontal positioning of excitation and emission polarizers, respectively.[8]

Fluorescence Quenching

Fluorescence quenching is a process that decreases the intensity of the fluorescence emission. Quenching by small molecules either in solvent or bound to the protein in close proximity to the fluorophore can greatly decrease the quantum yield of a protein. Quenching may occur by several mechanisms. Most common ones are (1) dynamic or collisional quenching, where a random noninteractive collision of a small molecule deactivates the excited state of the fluorophore, and (2) static quenching, where a small molecule makes a ground state complex with the fluorophore so that it becomes nonfluorescent. The extent of dynamic quenching depends on the accessibility of the fluorophore to the quencher. Therefore, quenching data

---

[8] J. R. Lakowicz, "Principles of Fluorescence Spectroscopy." Plenum Press, New York, 1983.

give an idea of the solvent accessibility of the fluorophore and report the tertiary or quaternary structural changes of the macromolecule.

We used a neutral quencher, acrylamide, to measure the accessibility of tryptophan residues in RNA polymerase to get an idea of changes in tertiary structure. We probed the accessibility of the Tb–GTP fluorophore bound at the active site of the enzyme in a similar manner. Quenching data were analyzed according to Stern–Volmer method using the following equation:

$$F_O/F = 1 + K_{sv}[Q] \tag{2}$$

where F_O and F are fluorescence intensities of the fluorophore in the absence and presence of the quencher, respectively. K_{SV} and Q denote the quenching constant and concentration of the quencher, respectively. The fraction of accessibility (f_e) of the tryptophan residues was calculated from the equation:

$$F_O/(F_O - F) = 1/(K_{SV}f_e[Q]) + 1/f_e \tag{3}$$

Results

Tryptophan Fluorescence of RNA Polymerase as a Reporter of the Existence of Multiple Intermediates in the Pathway of Open Complex (RP_O) Formation

Figure 2A is the ratio of fluorescence intensity of free and promoter bound RNAP as a function of temperature. The ratio was plotted to normalize the effect of temperature on fluorescence. This plot could be described as a sum of three hyperbolic curves and the multiphasic nature had originated from the transition of one type of complex to another with a different fluorescence quantum yield. We ascribe this multiphasic curves to following equilibria involving four types of complexes in the pathway of open complex formation.

$RP_c \leftrightarrow RP_{i1}$ equilibrium 1 (between 4 and 13°)

$RP_{i1} \leftrightarrow RP_{i2}$ equilibrium 2 (between 13 and 19°)

$RP_{i2} \leftrightarrow RP_o$ equilibrium 3 (between 19 and 24°)

This trend agreed well with the earlier temperature-dependenent formation of intermediates during this process.[3,9,10] We chose RNAP–promoter

[9] B. Hoefer, D. Muller, and H. Koster, *Nucleic Acids Res.* **80**, 2544 (1985).
[10] P. Schikor, W. Metzger, W. Werel, H. Lederer, and H. Heumann, *EMBO J.* **8**, 2745 (1990).

complexes at 5, 12, 17, and 37 °, which represent homogeneous population of RP_c, RP_{i1}, RP_{i2}, and RP_o, respectively. We probed the tertiary structures and active site accessibility of RNAP in these complexes.

Probing Changes in Tertiary Structure of RNAP by Fluorescence Quenching of Tryptophan Residues

At first, RP_c was made by incubating RNAP and promoter DNA at 5 ° and then other complexes were made from RP_c by increasing the temperature slowly to the desired value (12, 17, and 37 °). At each of these temperatures, sufficient time was allowed to reach the equilibrium. To each of these complexes we added increasing concentration of acrylamide and followed the quenching of tryptophan fluorescence at 340 nm. Fluorescence data were analyzed according to Eqs. (2) and (3) to obtain the accessibility of trytophan residues in each of these complexes (Fig. 2B). Similar data were also obtained for free RNAP at these temperatures to rule out any artifact due to temperature change. It was observed that about 80% of tryptophan residues of the free RNAP remain accessible to the solution at all these temperatures, which indicated that the tertiary structure of RNAP over this temperature range remained the same. Results from promoter-bound RNAP clearly indicated that the overall accessibility of tryptophan residues changed significantly in different intermediate complexes. This suggests that a major conformational change in the tertiary structure of RNAP occurs during the RP_c to RP_o conversion, which was originally proposed by Roe et al.[11] from thermodynamic analysis.

A Fluorescent Substrate, Tb–GTP, as a Probe for Active Site Conformation of RNAP during RP_c to RP_o Conversion

Tb–GTP, a fluorescent analogue of GTP, has been used successfully to probe the active site of RNAP.[7] Tb–GTP binds to the I site of RNAP and was shown to be incorporated in the RNA chain by abortive initiation assays. The Tb–GTP complex was formed by mixing $TbCl_3$ with GTP in T buffer, pH 7.5, and the complex formation was monitored by fluorescence. Tb–GTP, upon excitation at 295 nm, gives rise to two emission peaks at 488 and 545 nm.[6,7] Stability of the Tb–GTP complex (K_d 0.2 μM) was greater than that of the Tb–GTP–RNAP complex. Under the experimental conditions (see legend to Fig. 3), it is unlikely that Tb(III) will dissociate from the complex.

The fluorescence polarization anisotropy (FPA, $<r>$) of Tb–GTP bound to RNAP was measured both in the presence and in the absence of promoter

[11] J. H. Roe, R. R. Burgess, and M. T. Record, Jr., J. Mol. Biol. 184, 441 (1985).

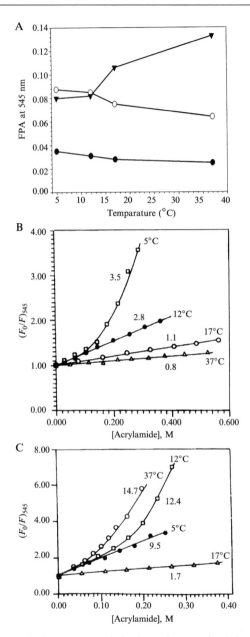

Fig. 3. (A) Changes in fluorescence polarization anisotropy (at 545 nm) of Tb–GTP as a function of temperature under different conditions: (i) alone (o), (ii) in the presence of RNAP (●), and (iii) in the presence of pDR720 (▼). Concentrations of Tb, GTP, RNAP, and pDR720 were 77 μM, 12 μM, 10 units, and 12 nM, respectively. Stern–Volmer plots for quenching of Tb–GTP under different conditions: (B) in the presence of RNAP alone and (C) in the presence of both RNAP and pDR720. Quenching constants (M^{-1}) are shown above or below each curve. Temperatures are also indicated. From Sen and Dasgupta.[6]

DNA at specified temperatures (Fig. 3A). $<r>$ values were obtained according to Eq. (1). In general, FPA decreases with temperature due to enhanced Brownian motion at higher temperatures. This was observed for both TbGTP alone and the Tb–GTP–RNAP complex (Fig. 3A). However, a marked deviation of this trend was observed in the ternary complex of the Tb–GTP–RNAP–promoter. FPA values were found to increase upon formation of RP_{i2} ($\sim 14°$) and reached the maximum in the RP_o complex ($37°$). Increasing rigidity of the Tb–GTP molecule bound to RNAP along the pathway of formation of RP_o could suggest an active site rearrangement.

Acrylamide quenches the fluorescence of the Tb–GTP complex. This helped us monitor the extent of accessibility of the active site to the solution in different RNAP–promoter complexes. Complexes were formed at different temperature as described earlier and an increasing concentration of acrylamide was added. A decrease in fluorescence intensity was analyzed according to Eq. (2) to obtain the Stern–Volmer plots (Figs. 3B and 3C). A decrease in the quenching constant with temperature in the presence of RNAP alone is a typical feature of static quenching (Fig. 3B), where a ground state nonfluorescent complex is formed between Tb–GTP and acrylamide. However, curves obtained in the presence of RNAP and promoter did not follow any particular trend, rather having a characteristic quenching constant. This suggests that each of these intermediates has its own active site conformations.

Concluding Remarks

Simple fluorescence assays to monitor the conformational changes of RNAP during the formation of an open complex at *trp* promoter are described here. These assays can be applied easily to other promoters, such as LacUV5, λP_R, and T7A1. It will be interesting to see what type of fluorescence profile (like Fig. 2A) will be generated on other promoters. These assays will be more informative if RNAP with different σ subunits as well as well as different promoter mutants with known biochemical phenotypes are used. Another interesting application of these assays will be to study the effect of repressor and activator molecules on RNAP–promoter combinations. It is expected that the fluorescence properties of either the active site or the whole RNAP will alter in response to the repressor and activator molecules.

Finally, as fluorescence signals are indirect measurements of conformational changes, it is important to use these techniques on biochemically characterized RNAP–promoter complexes. If these were to be applied to a new system, it is advisable to perform parallel biochemical assays such as DNA footprinting to confirm the existence of intermediate states.

[51] Measuring Control of Transcription Initiation by Changing Concentrations of Nucleotides and Their Derivatives

By David A. Schneider, Heath D. Murray, and Richard L. Gourse

Nucleotides and their derivatives participate in the regulation of gene expression in a large number of biological networks, many of which have been explored in detail in prokaryotic systems.[1-3] With the recent discovery that small molecules can also interact with mRNAs, thereby regulating transcription termination and translation initiation,[4,5] there is increasing interest in the measurement of the concentrations of small molecules. This article focuses on measurements of nucleoside triphosphates (NTPs) and of guanosine 5'-diphosphate 3'-disphosphate (ppGpp), as well as on assays for the responses of promoters to changes in the concentrations of these molecules in *Escherichia coli*.

An unusual kinetic characteristic of rRNA promoters, namely the intrinsically short-lived open complex these promoters form with RNA polymerase (RNAP), results in a requirement for unusually high concentrations of the initiating nucleoside triphosphate (iNTP) for initiation of transcription.[3,6] Changing concentrations of the iNTPs thereby result in the differential regulation of rRNA promoters relative to other promoters. Similarly, ppGpp differentially regulates promoters because of their unique kinetic characteristics. ppGpp is produced by the cell in response to starvation for amino acids and other nutrients (sometimes referred to as the stringent response[2]). ppGpp binds directly to RNAP, negatively regulating rRNA transcription initiation directly[7] and positively regulating some other promoters indirectly.[8] ppGpp decreases the lifetimes of the

---

[1] R. Landick, C. L. Turnbough, Jr., and C. Yanofsky, *in* "*Escherichia coli* and *Salmonella*: Cellular and Molecular Biology" (F. C. Neidhardt *et al.*, eds.), p. 1263. ASM Press, Washington, DC, 1996.

[2] M. Cashel, D. R. Gentry, V. J. Hernandez, and D. Vinella, *in* "*Escherichia coli* and *Salmonella*: Cellular and molecular biology" (F. C. Neidhardt *et al.*, eds.), p. 1458. ASM Press, Washington, DC, 1996.

[3] D. A. Schneider, T. Gaal, and R. L. Gourse, *Proc. Natl. Acad. Sci. USA* **99**, 8602 (2002).

[4] W. Winkler, A. Nahvi, and R. Breaker, *Nature* **419**, 952 (2002).

[5] B. A. McDaniel, F. J. Grundy, I. Artsimovitch, and T. M. Henkin, *Proc. Natl. Acad. Sci. USA* **100**, 3083 (2003).

[6] T. Gaal, M. S. Bartlett, W. Ross, C. L. Turnbough, Jr., and R. L. Gourse, *Science* **278**, 2092 (1997).

[7] M. M. Barker, T. Gaal, C. A. Josaitis, and R. L. Gourse, *J. Mol. Biol.* **305**, 673 (2001).

open complexes that all promoters form with RNAP, but it inhibits rRNA promoters specifically because these promoters are rate limited by this step in the transcription mechanism.

Studies have determined the times in bacterial growth when ppGpp and NTPs actually change, thus defining the physiological conditions when each effector is responsible for the regulation of transcription.[9] This article provides protocols for characterizing the effects of ppGpp and iNTPs on promoter activity *in vitro* and *in vivo* and for measuring the concentrations of the nucleotides themselves.

Characterization of the Effects of ppGpp and iNTP Concentration on Transcription Initiation *in Vitro*

Rationale

Because active prokaryotic RNAP can be purified readily,[10,11] and it is often difficult to design experiments that differentiate direct from indirect effects *in vivo*, whenever possible we examine potential transcription factors for direct effects on promoter activity *in vitro*. Although it is feasible to dissect effects of regulatory molecules into individual steps in the initiation pathway,[7,12] for simplicity here we discuss effects of ppGpp and iNTPs on promoter activity using multiple round transcription assays that do not focus on individual kinetic intermediates.

Stepwise Procedure for Measuring Effects of ppGpp or iNTPs on Transcription Initiation

Transcription reactions are generally performed in small volumes (10–25 μl for most applications). Each reaction contains 0.6 nM supercoiled plasmid template in the transcription buffer described later. With linear templates, higher concentrations of DNA are used (e.g., 2–4 nM). The concentration of salt (generally KCl or NaCl) must be optimized for the specific promoter used. Effects of ppGpp or iNTPs are more pronounced at high salt concentrations. For exploring effects of changing

[8] M. M. Barker, T. Gaal, and R. L. Gourse, *J. Mol. Biol.* **305**, 689 (2001).

[9] H. D. Murray, D. A. Schneider, and R. L. Gourse, *Mol. Cell* **12**, 125–134 (2003).

[10] R. R. Burgess and J. J. Jendrisak, *Biochemistry* **14**, 4634 (1975).

[11] H. Tang, K. Severinov, A. Goldfarb, and R. H. Ebright, *Proc. Natl. Acad. Sci. USA* **92**, 4902 (1995).

[12] M. T. Record, Jr., W. S. Reznikoff, M. L. Craig, K. L. McQuade, and P. J. Schlax, In "*Escherichia coli* and *Salmonella*: Cellular and Molecular Biology" (F. C. Neidhardt *et al.*, eds.), p. 792. ASM Press, Washington, DC, 1996.

iNTP or ppGpp concentrations on supercoiled templates containing *rrn* P1 promoters, 150 mM NaCl or 170 mM KCl is appropriate. To examine the effects of the initiating NTP concentration on promoter activity, the concentration of the iNTP is varied while the other three NTPs are kept constant. For example, for a promoter initiating with ATP, we use 10–2000 μM ATP, 100 μM GTP, 10 μM CTP, 10 μM UTP, and 5 μCi[α-^{32}P] P] UTP for a 25-μl reaction.

To test effects of ppGpp on transcription, we generally perform reactions using 500 μM ATP, 250 μM GTP, 100 μM CTP, and 10 μM UTP (and 5 μCi[α-^{32}P] UTP) in the presence of 100–200 μM ppGpp (available commercially from TriLink, Inc. or purified according to Cashel[13]). Control reactions should be included containing buffer instead of ppGpp. Reactions are set up to contain everything except RNAP, brought to 30°, and then reactions are initiated by the addition (with mixing) of RNAP ($\sim$1–4 nM for supercoiled templates, $\sim$20 nM for linear templates) and stopped after 15 min with an equal volume of loading dye containing formamide (see later). Products are placed on ice immediately and can be stored at $-20°$ until separation by denaturing polyacrylamide gel electrophoresis and visualization by phosphorimaging. Conditions for electrophoresis depend on the size of the RNA product and resolution required. Band intensity can be quantified. using ImageQuant software (Molecular Dynamics).

Reagents

All solutions are made from RNase-free dry stocks (Sigma) and filter sterilized.

Transcription buffer (made as a 10× stock and diluted into the reaction accordingly): final concentration (1×) is 170 μM KCl, 10 mM MgCl$_2$, 40 mM Tris–HCl, pH 8.0, and 1 mM dithiothreitol (DTT). Bovine serum albumin (BSA; 100 μg/ml final), DNA template, NTPs, and ppGpp (as required) are added to the reactions at the time of the experiment.

RNAP dilution buffer: Same as 1× transcription buffer, but without NTPs and template

Formamide loading buffer: 95% formamide, 20 mM EDTA, pH 8.0, 0.5% bromophenol blue, 0.5% xylene cyanol FF

Polyacrylamide gel: 6% polyacrylamide (19:1 polyacrylamide:bisacrylamide), 7 M urea, 1× TBE buffer (90 mM Trizma base, 90 mM boric acid, 2.5 mM EDTA). Polymerize with TEMED (70 μl per

[13] M. Cashel, *Anal. Biochem.* **57**, 100 (1974).

100 ml gel solution) and ammonium persulfate (APS) (850 μl of 10% APS per 100 ml gel solution), electrophorese at ~12 V/cm using 1× TBE running buffer.

Other Comments

When examining the effects of NTP concentration on promoter activity, we generally plot total transcription (y axis) versus NTP concentration (x axis). All promoters require higher concentrations of the iNTP than the elongating NTPs for transcription. We use the term "NTP sensing" to describe only promoters, such as *rrnB* P1, with a requirement for an unusually high concentration of the initiating NTP.[6] To determine whether a promoter fits into this category, it is useful to use well-characterized non-NTP-sensing promoters such as *lacUV5* or λP_R, as well as *rrnB* P1, for comparison with the test promoter.

Some promoters utilize alternative start sites when the concentration of the iNTP is low relative to the concentration of the other NTPs. For example, when CTP concentrations are high (~100 μM) and ATP concentrations are low (10–50 μM), a CTP:ATP ratio never present *in vivo*, *rrnB* P1 will initiate with CTP at position -1 (numbering relative to the normal start site, $+1$). This switch would result in an underestimate of the ATP requirement for initiation at *rrnB* P1. Therefore, it is sometimes necessary to work with low concentrations of the other 3 NTPs (~10 μM) when varying the concentration of the initiating NTP in order to avoid physiologically irrelevant switches in the start site.

We generally observe no more than a two- to three-fold inhibition of *rrnB* P1 promoter activity by ppGpp *in vitro*,[7] a weaker inhibition than generally obtained under stringent conditions *in vivo*. Therefore, as with measurements of NTP sensing *in vitro*, it is important to include ppGpp-insensitive control promoters for comparison. Potential explanations for the difference in inhibition by ppGpp *in vitro* versus *in vivo* are discussed by Barker *et al.*[7]

Measurement of Promoter Activity *in Vivo* Using Reporters

Rationale

Promoter fusions have been employed to measure transcription activity in many different systems. In each case, a relatively straightforward enzymatic assay (e.g., β-galactosidase derived from the *lacZ* gene, chloramphenicol acetyltransferase from the *cat* gene, or firefly luciferase from the *luc* gene) is employed using commercially available reagents. Although reporter systems have limitations, they have proven extremely useful for many applications. We generally employ single-copy promoter–*lacZ* fusions to measure responses of rRNA promoters to changes in steady-state growth rate, as described here.

Stepwise Procedure

λphage-based systems allow for stable insertion of promoter–*lacZ* fusions in single copy into the chromosome at the λ attachment site. Details of the design and construction of these reporter systems have been published previously.[14–16] Further details of the β-galactosidase assay and relevant reagents are provided by Miller.[17]

The culture is grown for at least three to four generations to an OD_{600} of ∼0.3. After recording the OD_{600}, the culture is placed on ice for at least 20 min and lysed by sonication. Other methods are available for lysis or permeabilization, but we have found that these methods are affected by media conditions, probably because the properties of the cell wall and/or membranes vary with growth conditions.[18] We add 1 ml of chilled cells to 4 ml of chilled Z buffer[17] and sonicate in a plastic tube twice for 20 s each using a microtip at the maximum power setting, chilling the tube on ice between the 20-s sonications.

Z buffer and sonicated cells are added to a total volume of 1 ml in a clear test tube (the amount of lysate is adjusted to the strength of the promoter, as required). The reaction is initiated by the addition of 200 μl *O*-nitrophenylgalactopyranoside (ONPG, 4 mg/ml), and the start time and stop time of the reaction are recorded. When the color of the reaction turns from clear to light yellow, the reaction is terminated by the addition of 500 μl of 1 *M* Na_2CO_3. A control reaction containing only Z buffer, ONPG, and stop solution is used to blank the spectrophotometer, the optical density of each sample is measured at 420 and 550 nm, and the promoter activity is calculated according to the equation provided by Miller.[17]

Other Comments

Measuring transcription using promoter–*lacZ* fusions is fast, the required reagents are inexpensive, and only standard laboratory equipment is needed. However, because the output of the assay is a protein, potential

[14] K. P. Bertrand, K. Postle, L. V. Wray, Jr., and W. S. Reznikoff, *J. Bacteriol.* **158**, 910 (1984).
[15] R. W. Simons, F. Houman, and N. Kleckner, *Gene* **53**, 85 (1987).
[16] L. Rao, W. Ross, J. A. Appleman, T. Gaal, S. Leirmo, P. J. Schlax, M. T. Record, Jr., and R. L. Gourse, *J. Mol. Biol.* **235**, 1421 (1994).
[17] J. H. Miller, *In* "Experiments in Molecular Genetics." Cold Spring Harbor Laboratory, Cold Spring Harbor, NY, 1972.
[18] B. L. Wanner, R. Kodaira, and F. C. Neidhart, *J. Bacteriol.* **130**, 212 (1977).

effects on steps other than transcription initiation (e.g., transcription elongation, mRNA decay, translation initiation, or translation elongation) can confound the analysis. For this reason, it is important to compare the activities of only fusions making the same RNA transcript. Furthermore, *lacZ* fusions report accumulation, not synthesis rate, as β-galactosidase is a stable protein. This limits the utility of β-galactosidase activity as a reporter of rapid changes in promoter activity in nonsteady-state situations, such as after shifts in growth phase, upshifts, downshifts, or after inactivation of mutant gene products by temperature shifts.

Measuring Promoter Activity *In Vivo* by Primer Extension

Rationale

In many cases, regulation of transcription by small molecules is most apparent in nonsteady-state situations such as those mentioned earlier when responses to changes in nutritional or environmental conditions are rapid and/or transient. Reporter-based systems for measuring promoter activity are not well suited to these types of experiments. The most direct way to quantify promoter activity in these situations is by direct analysis of the short-lived RNA products themselves using primer extension. For this purpose, we generally measure the transcripts generated by the same promoter–*lacZ* fusions described here.

Stepwise Procedure

RNA Extraction. Our RNA extraction method is derived from previously described boiling lysis procedures. Six milliliters of culture is added directly to 1.5 ml of hot lysis solution (1 M NaCl, 2.5% SDS, 0.05 M EDTA). The lysis solution is in a 50-ml conical tube positioned in a beaker partially filled with boiling water. After 90 s of continued heating, the lysed culture is extracted at least twice with an equal volume of phenol:chloroform (1:1) to remove cell debris. A predetermined quantity of extracted RNA containing an mRNA that hybridizes to the same labeled primer as the test mRNA is added to the lysis mix at the time of the first addition of phenol:chloroform. This mRNA should be at least 30 nucleotides longer or shorter at the 5' end than the test RNA, allowing this transcript to serve as a recovery marker.[3] Separation of aqueous and organic phases is accomplished by centrifugation for 10 min at 3200 rpm in a swinging bucket centrifuge. After centrifugation, the purified RNA from the aqueous phase is transferred to another 50-ml tube containing 0.4 ml 3 M sodium acetate (pH 5.3) and 8.5 ml 100% ethanol, precipitated overnight at $-20°$, and collected by centrifugation at 3200 rpm for 30 min at $4°$. The resulting RNA pellet is allowed to dry, resuspended in 400 μl 10 mM Tris–HCl (pH 8.0), and transferred to a 1.5-ml microcentrifuge tube containing 800 μl 100% ethanol and 40 μl 3 M NaOAc. This mixture is then reprecipitated at $-80°$ for 20 min, centrifuged at $4°$ in a microcentrifuge for 30 min at 14,000 rpm, washed with 70% ethanol, and resuspended in 50 μl 10 mM Tris–HCl (pH 8.0). This procedure, although somewhat lengthy, yields concentrated RNA suitable for primer extension analysis that is stable at $-20°$ for at least several weeks.

Primer Extension. The primer used for reverse transcription should be ~20–25 nucleotides long and hybridize specifically to the mRNA of interest near (within ~80–150 nucleotides) the 5′ end of the transcript. The primer is labeled at its 5′ end using [γ-^{32}P]ATP and polynucleotide kinase according to the manufacturer's specifications (Promega). If single nucleotide resolution is desired, the primer should be gel purified to ensure that only the correct species is present.

An aliquot of the sample RNA is added to a mixture of excess labeled primer and M-MLV reverse transcription buffer as described by the manufacturer (Promega). The sample is heated to $80°$ for 10 min to denature the nucleic acids, cooled slowly to room temperature, and incubated for 5 min at the desired temperature for the extension reaction ($40–50°$, depending on the melting temperature of the primer). To begin the reaction, dithiothreitol (3.3 mM), dNTPs (400 μM each), and M-MLV reverse transcriptase (4 units/μl, Promega) are added to each prewarmed sample and incubated for 30 min. The reaction is stopped by the addition of an equal volume of formamide loading solution (see earlier discussion). Alternatively, the reaction can be stopped by the addition of 2 volumes of cold ethanol and 1/10 volume of 3 M NaOAc, precipitated, and resuspended in a small volume of formamide loading dye.

Radioactive cDNAs are heated to $90°$ for 5 min before electrophoresis to denature the cDNA from the RNA transcript and then are analyzed on denaturing polyacrylamide gels. To determine the size of cDNA products, a sequencing ladder should be generated with the same labeled primer used for RNA hybridization.

Other Comments

The protocol presented earlier is one of many that have been described for the extraction of nucleic acids from bacterial cells. It has a major advantage over commercially available kits for RNA purification in that it allows extraction from more cells, increasing signal strength. The method also includes steps where the RNA can be concentrated to increase signal strength even further. Perhaps most importantly, there is little or no elapsed time between termination of cell growth, cell lysis, and organic extraction, reducing chances for degradation from endogenous RNases. Nevertheless, for some applications, we have had success with commercial RNA extraction kits (e.g., the BioRad AquaPure RNA isolation kit).

Extraction of NTPs and ppGpp from *Escherichia coli*

Rationale

The most important requirements for small molecule extraction methods are speed, reproducibility, and sensitivity. This section discusses two methods used to analyze fluctuations in NTP and/or ppGpp concentrations and briefly describes a third method that has been developed for the bioassay of ATP concentrations.

Stepwise Procedure

Formic Acid Extraction. This protocol, described by Jensen and colleagues many years ago,[19] in principle leads to immediate cell lysis and protein precipitation by the addition of total cell culture directly to cold acid. Cultures are grown under the desired conditions to an $OD_{600}>0.1$ (below this cell density, detection is difficult). The method can be used to extract either unlabeled or radioactively labeled nucleotides (see later). Before extraction, 40 μl of 2 M formic acid is dispensed to an appropriate number of 1.5-ml microcentrifuge tubes on ice. After recording the OD_{600}, 200 μl of culture is added directly to the cold formic acid, mixed briefly, and allowed to stand on ice for at least 20 min. Cell debris is pelleted in a microcentrifuge for 5 min at 14000 rpm, and the supernatant is transferred to a clean tube for storage at $-20°$ where the extracted NTPs are stable for several days. If high-pressure liquid chromatography (HPLC) is employed for detection, the extract should be mixed with an equal volume of solvent A (see later) and passed through a 0.2-μm nitrocellulose filter prior to loading. In most cases, even for high-abundance molecules, this extraction is performed after growth of cells in ^{32}P phosphate, after which detection is by thin-layer chromatography (TLC).

Formaldehyde/Alkaline Extraction. This method has been used primarily because it allows for the concentration of cells prior to extraction, facilitating visualization of low-abundance molecules. The following

[19] K. F. Jensen, U. Houlberg, and P. Nygaard, *Anal. Biochem.* **98,** 254 (1979).

protocol was adapted from that of Little and Bremer[20] and can be scaled up or down as required. After recording the optical density (OD_{600}>0.1), culture growth is terminated (and NTPases are inactivated) by the addition of formaldehyde (10 ml of culture is added to 1 ml of 1.9% formaldehyde in a glass centrifuge tube and allowed to stand on ice for up to 20 min). The formaldehyde-fixed cells are then pelleted by centrifugation (5 min at 6000 rpm), the supernatant and excess medium are removed carefully, the pellet is resuspended in 0.5 ml, 0.1 M KOH, and the sample is allowed to stand on ice for 30 min. The extract is then transferred to a 1.5-ml microcentrifuge tube containing 0.5 ml of chromatography solvent (solvent A for HPLC or phosphate buffer for TLC, see later) and 5 μl 88% H_3PO_4. Debris is removed by centrifugation in a microcentrifuge at 14,000 rpm for 5 min at 4°, and the supernatant can then be stored at -20° for no more than 2 days. Extracts must be passed through a 0.2-μm nitrocellulose filter prior to examination by HPLC.

Other Comments

Each extraction method has limitations. Formic acid extraction liberates more ATP per cell than formaldehyde treatment and extraction with alkali, but because the extract is not made from cells that have been concentrated, even relatively abundant nucleotides are difficult to detect without radiolabeling. Radiolabeling (with ^{32}P) allows detection of high- and low-abundance molecules such as NTPs and ppGpp by TLC (although ppGpp is only two- to three-fold above background when extracted during log phase). For issues associated with radiolabeling, see later.

Some laboratories routinely include a filtration step in which cultures are passed through glass fiber filters before extraction and then these filters are added directly to cold formic acid for extraction. In theory, this could increase detection of low-abundance molecules, but differential filtration could increase error.

The formaldehyde/alkali method has been used for examining both NTPs and ppGpp, but we have observed large effects of the growth medium on the efficiency of ATP extraction.[21] For this reason, we suggest use of the formic acid method whenever possible, and the formaldehyde/alkali method should not be employed, especially when comparing NTP concentrations from cultures grown in different media.

[20] R. Little, and H. Bremer, *Anal. Biochem.* **126,** 381 (1982).
[21] D. A. Schneider, and R. L. Gourse, unpublished results (2003).

Chromatographic Separation of NTPs and ppGpp from *E. coli* Extracts

Rationale

We have used two methods for the separation of nucleotides from each other and other small molecules: ion-paired, reverse-phase HPLC and TLC. For most applications, either method can be adapted for the quantitation of most NTPs. The TLC method described here requires radiolabeling of nucleotides. Separation of NTPs from dNTPs requires a second chromatographic dimension but is usually unnecessary. While the HPLC method described here does not require radiolabeling, it does not separate CTP from cellular contaminants and is more time-consuming than one-dimensional TLC.

Detection of ppGpp is challenging during logarithmic growth, especially at higher growth rates, because of its low abundance. HPLC methods have been used previously for the detection of ppGpp under these conditions, but we have found that two small peaks comigrate with the ppGpp peak under these conditions, confounding quantitation. Therefore, we feel that ^{32}P labeling of cells followed by TLC is preferable for accurate quantitation of ppGpp during steady-state growth.

Stepwise Procedure

PEI-Cellulose TLC. This method is adapted directly from Jensen *et al.*[19] Detection of isotopically labeled molecules on PEI-cellulose TLC plates (J. T. Baker) is now performed by phosphorimaging, however, greatly facilitating quantitation and data analysis. Cells are grown in MOPS minimal medium[22] containing 1 mM KH$_2$PO$_4$ and ~20 μCi ^{32}P KH$_2$PO$_4$ (NEN DuPont) per 1 ml of culture. We allow the phosphate pools to equilibrate for at least two to three generations of growth. Before loading samples on TLC plates, each plate is prepared by sequential immersion in distilled H$_2$O, air drying, immersion in methanol, and a second air-drying step. Twenty microliters of extract (see earlier discussion) from cells grown to an OD$_{600}$ of at least 0.1 is sufficient for visualization of NTPs and ppGpp. Ten microliters of extract can be spotted on the plate at one time, followed by air drying before an additional sample is added. The running buffer is 0.85 M KH$_2$PO$_4$ (pH 3.4). When the solvent front approaches the top of the plate, the plate is removed from the chromatography tank and air dried. The region of the plate containing the unincorporated label (the solvent front) can either be cut off or washed out of the plate by immersion

[22] F. C. Neidhardt, C. A. Bloch, and D. F. Smith, *J. Bacteriol.* **119**, 736 (1974).

in 10% citric acid. To isolate NTPs from dNTPs (which are of much lower abundance), the dried plate can be run in a second dimension (for protocol, see Jensen et al.[19]). As with one-dimensional TLC, spots are quantified by phosphorimaging.

Ion-Paired, Reverse-Phase HPLC. Many methods have been developed for the analysis of both NTPs and ppGpp by HPLC. Many of these methods work well with purified nucleotides but are not suitable for detection in complex extracts. The following protocol is effective for separation and quantitation of NTPs from *E. coli* cells, although it is not sufficient for ppGpp.

Prior to preparation of extracts, purified standards should be tested using the same equipment and solvent preparations that are to be used with extracts. This will establish when individual NTPs should elute. Individual columns and solvents have slightly different characteristics, potentially causing peaks to shift and cause misidentification. The following method has been optimized for the analysis of GTP, ATP, and UTP. With this protocol, CTP elutes earlier, is more affected by slight solvent variations, and is often masked by extract contaminants. In general, the simplest way to adjust conditions so as to shift NTPs away from other peaks is to vary the pH of solvent A.

Column: Supelco LC-18T
Solvent A: 0.06 M $NH_4H_2PO_4$, 5 mM picA (Waters), pH 4.2
Solvent B: 100% acetonitrile; both solvents should be filtered and degassed before use

Method

Initial conditions: 100% solvent A, 0% solvent B, flow rate 1 ml/min (same flow rate in all steps)
Step 1: 20-min gradient from 0 to 15% solvent B (starts immediately upon loading sample)
Step 2: 10-min gradient from 15 to 20% solvent B
Step 3: 1-min gradient from 20 to 100% solvent B
Step 4: 5-min isocratic 100% solvent B
Step 5: return to initial conditions (100% solvent A until baseline stabilizes)

To verify which peak is the molecule of interest, extracts can be spiked with pure samples and compared with chromatographs generated without spiking. NTPs are quantified by integration of the area under the appropriate peak (most HPLC software is capable of performing this task) and comparison with standard curves.

Other Comments

The methods described here cannot be used with certain culture media. Most important, because the TLC procedure utilizes labeling of NTPs with radioactive phosphate, it is inappropriate for use with media containing high (or varying) levels of phosphate, such as complex media and phosphate-buffered minimal media. Additionally, we have found that some common components of bacterial media, such as casamino acids, can contain varying amounts of phosphate, making quantitative comparisons impossible. Although HPLC does not require radiolabeling, NTPs often coelute with medium components present in acid extracts. The formaldehyde/alkali extracts are not subject to this problem, as the medium is purified away during the initial centrifugation step.

Other issues to be considered include differences in cost, technical expertise, and reliability. Analysis of NTPs by HPLC can be costly and time-consuming, although it can be more versatile than TLC.

A recurring question with all methods for the measurement of nucleotide concentrations is whether the observed value is representative of the free, unbound pool of nucleotide or whether the method employed liberates the total nucleotide pool. It seems logical that the fraction of interest for regulatory studies would be the "free" pool, i.e., that available to the enzyme of interest. We have developed a bioassay for measuring available ATP *in vivo* by expression of an enzymatic reporter, a variant of firefly luciferase with an apparent K_{ATP} in the range present in growing bacterial cells. This bioassay has proven useful for comparing free ATP concentrations present in cells growing in a wide range of media.[21]

Concluding Remarks

There is an increasing appreciation for the role(s) of small molecules in complex regulatory networks. The control of rRNA expression in *E. coli* is a system in which two types of small molecules, ppGpp and iNTPs, play critical roles in controlling transcription initiation. The methods described here allow for correlation of transcription initiation with the concentrations of these effector molecules *in vitro* and *in vivo*. Such analyses are leading to a better understanding of how small molecules contribute to the regulation of gene expression.

[52] Identification of Promoters of *Esherichia coli* and Phage in Transcription Section Plasmid pSA850

By Dale E. A. Lewis

Transcriptional regulation of structural genes is an important aspect by which *Escherichia coli* cells control the catabolism and anabolism of genes. In the past, "minicircle" DNA templates were isolated *in vivo* to study transcriptional regulation of the *gal* promoters *in vitro* in the presence of DNA-binding proteins such as galactose repressor (GalR), lactose repressor (LacI), histone-like protein (HU), and cyclic AMP and its receptor protein (cAMP-CRP).[1] One of the advantages of this system is that *gal* promoters or any promoter of interest can be studied *in vitro* under different physiological conditions in the absence of any competing promoter for RNA polymerase. However, the minicircles do not contain any control promoter, which can be used to normalize the amount of transcript originating from the promoter of interest. Because the identities and origins of the other transcripts from promoters on the plasmid were unknown, it was necessary to isolate minicircles. This article identifies several major promoters that have enabled us to compare physiological effects on these promoters and to better understand certain features of transcriptional regulation, which were not obvious using minicircles. We have also identified a promoter in the bacterial attachment site (*att*B'OB). This promoter may be involved in site-specific recombination between the phage attachment site (*att*P'OP) and *att*B'OB. This article presents a detailed account of how the 5' and 3' ends, as well as the direction of unknown transcripts, can be identified by using *in vitro* transcription and primer extension assays.

Reagents and Enzymes

All restriction endonucleases, including alkaline phosphatase, are from New England Biolabs. *E. coli* RNA polymerase (specific activity: 885 U/mg, 7.6 U/μl; 2500 U/mg, 1 U/μl) is from Amersham Life Science and USB. The recombinant RNasin ribonuclease inhibitor (40 U/μl) is from Promega. Denaturing polyacrylamide gel solutions (Gel-mix 8) and T4 DNA ligase are from Invitrogen. Primers are from BioServe Biotechnologies. XL polymerase chain reaction (PCR) and DNA sequencing (dRhodamine terminator

---

[1] H. E. Choy and S. Adhya, *Proc. Natl. Acad. Sci. USA* **90,** 472 (1993).

METHODS IN ENZYMOLOGY, VOL. 370

cycle sequencing ready reaction) kits are from Applied Biosystems. [α-^{32}P]UTP (specific activity = 3000 Ci/mmol, 10 μCi/μl) and [γ-^{32}P]dATP (specific activity = 7000 Ci/mmol, 167 μCi/μl) are from ICN.

Plasmids

The parental plasmid (pSA508) is a derivative of pIBI24 (International Biotechnologies) and contains the following features: the phage attachment site (*att*P'OP), the corresponding bacterial attachment site (*att*B'OB), the ρ-independent transcription terminator site of *rpoC*, and multiple cloning sites (mcs)[2,3] (Fig. 1). A ρ-independent transcription terminator, *rpoC*, is located downstream of O_I to generate transcripts of 125 and 130-nucleotides from *P1* and *P2*, respectively, as described here (Fig. 1). A 166-bp *Eco*RI–*Pst*I fragment containing the wild-type *gal* regulatory region from −75 to +91 is cloned into pSA508 to generate pSA850 (Fig. 1). The recombinant plasmids are transformed into *E. coli* DH5α-competent cells (Invitrogen). Plasmids used in this study are listed in Table I. Plasmid pSA850 is used as a template for all PCR amplifications unless specified. Plasmid pSA913 contains a deletion of a 147-bp fragment (−222 to −76), which is generated by PCR with the appropriate primers on pSA850. The fragment is digested with AflIII and *Eco*RI and is cloned into pSA850, which is digested with the same enzymes. Similarly, pSA914 and pSA915 contain a 56-bp deletion fragment (−213 to −158 and −157 to −102, respectively). Plasmid pSA951 contains a 50-bp deletion fragment (+144 to +194), and pSA952 contains a 99-bp deletion fragment (+144 to +143). Plasmid pSA950 is generated by PCR using pI24 as a template to generate a 229-bp (−119 to +110) *Bam*HI–*Hin*DIII fragment, which contains the *gal* operator/promoter region. The fragment is cloned into pSA508 such that the initiation of *P1* and *P2* starts downstream of the *rpoC* terminator and transcribes toward the terminator. In addition, *att*P'OP is deleted in pSA950.

All digested fragments are purified by a QIAquick PCR purification kit (Qiagen) and separated from the parental plasmid on a 1% agarose gel, which is electrophoresed in 40 mM Tris base, 40 mM acetic acid, 1 mM EDTA, pH 7.8 (1× TAE), buffer at 100 V. A vertical gel slice is removed from the gel and stained in 0.5 μg/ml ethidium bromide solution to localize the digested fragments. The stained gel slice is realigned with the unstained gel, and the unstained fragments are sliced from the

[2] H. E. Choy and S. Adhya, *Proc. Natl. Acad. Sci. USA* **89**, 11264 (1992).
[3] C. Squires, A. Krainer, G. Barry, W.-F. Shen, and C. L. Squires, *Nucleic Acids Res.* **9**, 6829 (1981).

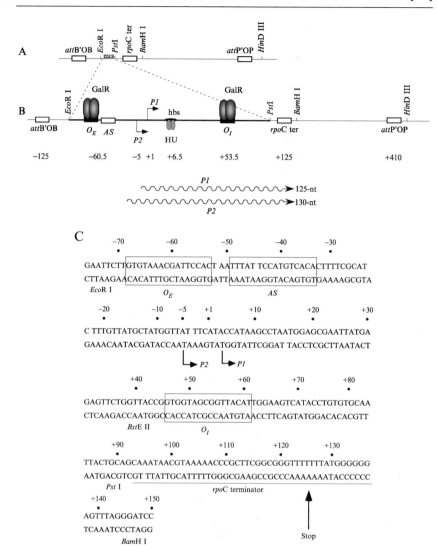

FIG. 1. Schematic diagrams of the regions of interest on the plasmid used in this study. (A) The vector (pSA508) contains the bacterial attachment site for λ DNA integration (attB'OB), the phage λ attachment site (attP'OP), multiple cloning sites (mcs), and the ρ-independent terminator (rpoC ter). (B) The gal operator/promoter region was cloned into the mcs of pSA508. The transcription start site of P1 is referred to as +1; numbers upstream and downstream of +1 are indicated by negative (−) and positive (+) signs. P2 is located at −5. O_E (−60.5) and O_I (+53.5) represent gal operators, AS, activating site for the cAMP–CRP complex. GalR dimers bound to O_E and O_I are shown as an oval shape. Histone-like protein (HU) binds to its binding site (hbs) at +6.5. The rpoC ter is located at +125 to generate

TABLE I
LIST OF PLASMIDS

Plasmid	Features	Source
pI24	*gal*ETKM'	Sankar Adhya
pSA508	mcs, *att*B'OB, *att*P'OP, *rpo*C terminator	Choy and Adhya, 1993
pSA850	Insert 166-bp *gal* DNA (−75 to +91)	Lewis and Adhya, 2002
pSA889	Δ O_E (−53 to −68)	This work
pSA896	Δ O_I (+46 to +61)	This work
pSA913	Δ 147-bp (−222 to −76)	This work
pSA914	Δ 56-bp (−213 to −158)	This work
pSA915	Δ 56-bp (−157 to −102)	This work
pSA921	Δ 356-bp (+195 to +551)	This work
pSA950	Insert 229-bp *gal* DNA (−119 to +110)	This work
pSA951	Δ 50-bp (+144 to +194)	This work
pSA952	Δ 99-bp (+144 to +243)	This work
pDL88	GC to CG @+121, ΔTA between +121 & +127	This work
pDL89	GC to TA @+120	This work
PDL274	Δ 123-bp (+428 to +550)	This work
PDL275	Δ 130-bp (+421 to +550)	This work
PDL276	Δ 190-bp (+361 to +550)	This work

gel. The DNA is eluted from the gel slice according to the protocol outlined in the QIAquick gel extraction kit (Qiagen). The fragments are ligated to the pSA850 or pSA508 vector, which is also digested with the same enzymes and dephosphorylated with alkaline phosphatase. The recombinant plasmids are transformed into maximum efficiency cells (DH5α). Purification of the plasmids is performed according to the protocol outlined in the Qiagen plasmid Midi kit. The pellet is suspended in 1× TE (10 mM Tris base, 1 mM EDTA, pH 8.0) and stored at 4°. This method of purification yields high-quality RNase-free plasmids for *in vitro* transcription assays. The plasmids are prepared for sequence using the dRhodamine terminator cycle sequencing reaction kit. Briefly, the reaction contains 8 μl terminator dye, 1 μl primer (6–8 pmol/μl), 1 μl DNA template (0.5–0.8 μg/μl), and 10 μl H$_2$O. Cycle sequencing is performed on a Gene-Amp PCR System 2400 (Applied Biosystems) with the following program: step 1, 25 cycles at 96° for 10 s, 50° for 5 s, and 60° for 4 min and step 2, hold at 4°. Sequencing products are precipitated with 74 μl 70% EtOH/0.5 mM MgCl$_2$. The dry pellets are suspended in 14 μl template

transcripts length of 125 nucleotides for *P1* and 130 nucleotides for *P2*. Arrows show the direction of transcription from *P1* and *P2*. (C) The DNA sequence of the *gal* region. O_E, O_I, and AS are boxed. The *rpo*C terminator sequence is underlined, and the arrow indicates the termination of transcription from *P1* to *P2*.

suspension reagent, heated to 95° for 2–4 min, chilled on ice, and analyzed on an ABI Prism 310 genetic analyzer to verify the necessary base pair changes that are inserted or deleted and to detect if the DNA contains other mutations as a result of PCR and cloning errors.

In Vitro Transcription Assays

Preparation of Gel Plates

The preparation of gel plates is an important procedure in achieving good results. Wash the plates with warm water and rinse them with deionized water (Millipore water is preferable). Wipe the plates with 70% (v/v) EtOH and coat them with Photo-Flo 2000 solution (1/25 dilution) so that the plates can be separated easily after gel electrophoresis. Coat the comb with Photo-Flo 2000 so that it can be removed easily after gel polymerization without tearing the well. Allow the gel to polymerize for at least 2 h to ensure good well formation. The gel can be left at room temperature overnight or at 4° for a few days. Care should be taken in removing the comb to avoid damaging the wells. After the comb is removed, rinse the wells with deionized water or 1× TBE (89 mM Tris base, 89 mM boric acid, 2 mM EDTA, pH 8.3) buffer. Wash the plates with deionized water to remove any polymerized gel on the outside of the glass. Prerun the gel at constant power of 65 W for 45–60 min to reach an operating temperature of approximately 45–50°. Rinse wells with buffer before loading the samples. Miniflex flat tips work well in loading the samples.

Preparation of Reagents

0.4 M Tris acetate, 0.4 M magnesium acetate, 2.0 M potassium glutamate, 100 mM dithiothreitol (DTT) and 10 mM ATP

RNA polymerase buffer: 10 mM Tris, pH 8.0, 100 mM NaCl, 0.1 mM EDTA, 0.1 mM DTT, and 50% (v/v) glycerol

GalR and CRP buffer: 50 mM Tris, pH 8.0, and 0.75 M KCl

HU buffer: 1/10 GalR buffer

The loading dye contains 90% (v/v) formamide (distilled), 10 mM EDTA, 0.1% (w/v) xylene cyanol (XC), and 0.1% (w/v) bromphenol blue (BPB). It is advisable to filter the loading dye solution to remove any particles of XC and BPB. The nucleotide mixture (20×) contains 2 mM GTP, 2 mM CTP, and 0.2 mM UTP. To prepare the radioactive nucleotide mixture (10×), use 2.5 μl 20× NTPs, 0.5 μl [α-^{32}P]UTP (10 μCi/μl), and 2.0 μl H$_2$O. These reagents can be stored at 4° for several months. 10× NTPs must be prepared fresh because it contains radioactive UTP.

Procedure for setting up Reactions

Step 1. Preparing a master mixture for six reactions, use the volume of reagents in Table II. This setup is suitable for experiments to study GalR, HU, and CRP effects on transcription. Alternatively, cAMP can be added to individual reactions with CRP. This is done in the experiment described in this article. If GalR and HU are being studied, cAMP would be eliminated and 4.75 μl H_2O would be added to make a total of 25 μl. Because only six reactions are needed, a mixture for seven reactions is set up to ensure an equal volume (25 μl) of mixture is transferred into each reaction.

Step 2. Assembly of reactions with GalR, HU, and/or CRP. Reagents and proteins are added in the order presented in Table III. It is important to prepare the reactions on ice (4°). Eppendorf RNase-free tubes (0.65 ml) are preferable, as the total reaction volume is 50 μl. However, 1.7-ml tubes can also be used. The incubation of the reactions is done in a heating block with water at 37°.

Mix the reactions gently by tapping the tube with the tip of the finger to mix the RNAP before centrifuging the tubes for 2–3 s in a microcentrifuge (Eppendorf 5415C) at room temperature to collect the residues from the walls of the tube. The reactions are incubated at 37° for 5 min in a heating block. As soon as rxn-1 is added to the block, the timer is started and then at 0.5 min, rxn-2 is added to the block, at 1.5 min, rxn-3 is added, at 2.0 min, rxn-4 is added, at 2.5 min, rxn-5 is added, and at 3.0 min, rxn-6 is added. To start the transcription reaction, 5.0 μl NTPs (10×) is added to rxn-1 at 5.0 min, rxn-2 at 5.5 min, rxn-3 at 6.0 min, rxn-4 at 6.5 min, rxn-5 at 7.0 min, and rxn-6 at 7.5 min. Pipette the NTPs up and down two to three times and mix the reaction by gently tapping the tube with the tip of the finger. The reactions are incubated for an additional 10 min at 37° before they are terminated by the addition of an equal volume (50 μl) of

TABLE II
REAGENTS USED

Reagents Presence	for 1 reaction	for 7 reactions
Tris acetate (0.4 M)	2.5 μl	17.5 μl
Magnesium acetate (0.4 M)	1.25 μl	8.75 μl
Potassium glutamate (2 M)	5.0 μl	35.0 μl
ATP (10 mM)	5.0 μl	35.0 μl
DTT (100 mM)	0.5 μl	3.5 μl
rRNasin (40 U/μl)	1.0 μl	7.0 μl
cAMP (100 μM)	5.0 μl	35.0 μl
DNA (20 nM)	5.0 μl	35.0 μl
H_2O (ultrapure)	0 μl	0 μl
Total volume	25.25 μl	176.75 μl

TABLE III
REAGENTS AND PROTEINS

Reaction	1	2	3	4	5	6
Proteins/rxn	Control	GalR	GalR + HU	CRP	GalR + CRP	GalR + HU + CRP
Mixture	25 μl	25 μl	25 μl	25 μl	25 μl	25 μl
GalR buffer	10 μl	5 μl	5 μl	5 μl	0 μl	0 μl
HU buffer	5 μl	5 μl	0 μl	5 μl	5 μl	0 μl
GalR (0.8 μM)	0 μl	5 μl	5 μl	0 μl	5 μl	5 μl
HU (0.8 μM)	0 μl	0 μl	5 μl	0 μl	0 μl	5 μl
CRP (0.5 μM)	0 μl	0 μl	0 μl	5 μl	5 μl	5 μl
RNAP (0.2 μM)	5 μl	5 μl	5 μl	5 μl	5 μl	5 μl
Incubation				37° for 5 min		
NTP (10×)	5 μl	5 μl	5 μl	5 μl	5 μl	5 μl
Total volume	50 μl	50 μl	50 μl	50 μl	50 μl	50 μl
Incubation				37° for 10 min		
Loading dye				50 μl		

loading dye and placed on ice. The dye is added to rxn-1 at 15 min, rxn-2 at 15.5 min, rxn-3 at 16.0 min, rxn-4 at 16.5 min, rxn-5 at 17.0 min, and rxn-6 at 17.5 min. It is recommended that not more than 20 reactions be done at the same time. The samples are centrifuged for 2–3 s and heated to 90° for 2–3 min to denature the mRNA. The samples are centrifuged again and chilled on ice before being loaded (6 μl) onto an 8% sequencing gel and electrophoresed at constant power of 65 W in 1× TBE buffer. The remaining samples can be stored at −80° for at least 3 days, but the mRNA will degrade slowly over time. The electrophoresis run should take 1.5–2 h depending on the expected size of the transcription product. If the length of the transcript is known, the migration of the dyes (XC and BPB) can be used as indicators of when the gel should be stopped.

After electrophoresis, the plates are separated and the gel is attached to one plate. The gel is transferred to 3MM chromatography paper (Whatman) and covered with Saran wrap. A second 3MM paper is placed under the first paper to avoid contaminating the gel dryer with radioactivity. The gel is dried at 80° for 1–2 h depending on the concentration of the gel. The dried gel is placed in a cassette with one intensifying screen and exposed to Kodax X-Omat AR film. The cassette is enclosed in an X-ray film security bag (PGC Scientific) to prevent light from entering the cassette and damaging the film. The cassette is stored at −80° for 12–15 h. Allow the cassette to reach room temperature before developing the film. Depending on the intensity of the transcripts, the exposure time of the gel to film can be increased or decreased. At the end of the experiment, monitor the working area for radioactivity contamination.

Quantitation of Transcript Product

RNAI transcripts (108 nucleotides) are used as an internal control to quantitate the relative amount of *gal* transcripts.[4,5] If desired, the gel can be scanned using the ImageQuant program (Molecular Dynamics) and the ratio of the transcripts area to that of *RNAI* can be compared to determine the effect of GalR, HU, or CRP on the transcript of interest.

5' Mapping of Transcripts

The 5' end of transcripts on pSA850 are mapped using primers PEgus–3 [5' CCAATGTAACCGCTACCACC 3' (+64 to +45)], 508–1 [5' TGCAT GCTCGGTACCCGGGTCGACG 3' (−78 to −92)], LacZ-2 [5' GGTC TA-GAAATTCCCTATAGTGAGTCG 3' (−151 to −177)], and POP-1 [5' CA-CAGTATCGTGATGACAGAGGCAGGGAGTGGGAC   3'(+257   to

[4] J. Tomizawa, T. Itoh, G. Selzer, and T. Som, *Proc. Natl. Acad. Sci. USA* **78,** 1421 (1981).
[5] D. E. A. Lewis and S. Adhya, *J. Biol. Chem.* **277,** 2498 (2002).

+291)]. T4 polynucleotide kinase and [γ-^{32}P]ATP are used to label the primers. Briefly, a mixture of 5 μl primer (10 pmol/μl), 5 μl T4 polynucleotide kinase buffer (10×), 2 μl T4 polynucleotide kinase (10U/μl), 2 μl [γ-^{32}P]ATP, and 36 μl H$_2$O is incubated at 37° for 20–30 min. The reactions are heated at 90° for 2–3 min to inactivate the enzyme. The unincorporated label is removed by using ProbeQuant G-50 microcolumns (Amersham Pharmacia Biotech). The columns are centrifuged at 2000 rpm for 2 min using an Eppendorf microcentifuge (Model 5415C) to remove interstitial fluid. The columns are transferred to new 1.5-ml Eppendorf tubes, and each radioactive primer is loaded on to a column, which is centrifuged again at the aforementioned speed to remove unincorporated label from the primers. Primer extension assays are performed as described in the protocol of AMV reverse transcriptase by Promega. *In vitro* transcription assays are performed as described earlier to obtain total transcribed RNA, which are placed at 90° for 2–3 min in the absence of loading dye. To 5 μl of total RNA transcript is added 1 μl of γ-^{32}P-labeled primer (1 pmol/μl) and 5 μl AMV primer extension 2× buffer (100 mM Tris–HCl, pH 8.3, 100 mM KCl, 20 mM MgCl$_2$, 1 mM spermidine, 2 mM GTP, CTP, ATP, and TTP). Reactions are incubated at 58° for 20 min and cooled to room temperature for 10 min using GeneAmp PCR system 2400. An aliquot of 9 μl mixture [5 μl AMV primer extension 2× buffer, 1.4 μl sodium pyrophosphate (40 mM), 1 μl AMV reverse transcriptase (0.05 U) and 1.6 μl H$_2$O] is added to the reactions and incubated at 42° for 30 min. The reactions are terminated by the addition of an equal volume (20 μl) of loading dye [98% formamide, 10 mM EDTA, 0.1% (w/v) XC, and 0.1% (w/v) BPB], The reactions (6 μl) are loaded on to an 8% sequencing gel as described previously.

DNA Sequencing

DNA sequencing is carried out according to the double-stranded (ds)DNA cycle sequencing system protocol from Invitrogen. The DNA template (1 μg/μl) is heated to 95° for 5 min and chilled on ice. The reaction mixture contains 5 μl labeled primer (10 pmol/μl), 1 μl DNA, 4.5 μl *Taq* sequencing buffer (10×), 0.5 μl *Taq* DNA polymerase (2.5 U/μl), and 25 μl H$_2$O. An 8-μl of aliquot mixture is added to four tubes containing 2 μl termination mix A, 2 μl termination mix C, 2 μl termination mix G, or 2 μl termination mix T. The sequencing of DNA is performed on GeneAmp PCR System 2400. The program includes the following: step 1, 95° for 5 min; min; step 2, 20 cycles at 95° for 30 s, 55° for 30 s, and 70° for 1 min; step 3, 10 cycles at 95° for 30 s and 70° for 30 s; and step 4, hold at 4°. An aliquot of 5 μl loading dye (5×) is added to each reaction. The sequencing cycles should take approximately 90 min. The labeled primer is used for both primer extension and DNA sequencing reactions. DNA sequencing reactions (2–3 μl) are loaded on to an 8% sequencing gel as described earlier.

3′ Mapping of Transcripts

To map the 3′ end of the *gal* transcripts, a linear DNA template (10 nM) containing the promoter of interest is used to reduce interference from other promoters. *In vitro* transcription is performed in the presence of cAMP–CRP complex and 3′-*O*-methylguanosine 5′-triphosphate, an analog of guanosine 5′-triphosphate (Amersham Biosciences). We found that the amount of *gal* transcripts on the linear template is reduced drastically than on supercoiled DNA templates. Therefore, the cAMP-CRP complex is used to enhance *P1* expression. During transcription, the 3′-*O*-methylguanosine 5′-triphosphate is incorporated at every guanosine position in the transcript, resulting in transcription termination. Thus, the position of every guanosine residue will be mapped and the 3′ end of the transcript will be determined by this method.

Transcription from Promoters on Plasmid

A typical *in vitro* transcription result is shown in Fig. 2. The transcription start sites of *P1* and *P2* were 5 bp apart, which account for the difference in the length of their RNA products (Fig. 1). As expected, in the presence of GalR, *P1* was repressed whereas *P2* was activated (Fig. 2, lane 2). In the presence of GalR and HU, both *P1* and *P2* transcripts were repressed (Fig. 2, lanes 3 and 4). The cAMP–CRP complex repressed *P2* and activated *P1* (Fig. 2, lane 5). However, activation of *P1* by the cAMP–CRP complex was eliminated by the presence of GalR and HU due to two mechanisms of repression. First, the binding of GalR to O_E repressed *P1* by contacting and inhibiting RNA polymerase at the *P1* promoter (contact inhibition).[6] Second, the binding of GalR and HU to their respective sites prevented *P1* and *P2* expression due to DNA loop formation (looping repression). GalR, HU, or the cAMP–CRP complex did not affect *RNAI* transcripts. In addition to *P1*, *P2*, and *RNAI* transcripts, other transcripts were noted that are referred to as *PA*, *PB*, and *PC* for simplicity in this study. The *PA* transcript was of interest because it was repressed slightly by DNA looping (Fig. 2, lanes 3 and 4). Is DNA looping affecting other promoters outside of the loop? It was also observed that DNA looping or the cAMP–CRP complex did not affect the *PB*

---

[6] S. Adhya, M. Geanacopoulos, D. E. A. Lewis, S. Roy, and T. Aki, Cold Spring Harbor Symposia on Quantitative Biology, Vol. LXIII (1998).

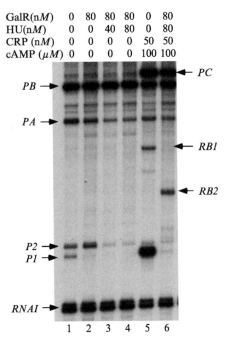

GalR(nM)	0	80	80	80	0	80
HU(nM)	0	0	40	80	0	80
CRP (nM)	0	0	0	0	50	50
cAMP (μM)	0	0	0	0	100	100

FIG. 2. RNA transcripts made from promoters on pSA850. RNAI transcripts are transcribed from the *rep* promoter on the plasmid and used as an internal control. *P2* and *P1* are *gal* promoters. P_A, P_B, and P_C transcripts are from unknown promoters. RB1 and RB2 indicate "roadblock" transcripts. The transcription products were resolved on an 8% denaturing polyacrylamide gel. The amount of proteins (GalR, HU, and CRP) and cAMP present in each reaction is indicated above the lanes.

transcript, although the *PB* transcript was reduced slightly in lane 5 (Fig. 2). A simple explanation could be due to the recruitment of RNA polymerase, as *P1* and *PC* were enhanced by the cAMP–CRP complex. The *PC* transcript was barely detectable in the absence of the cAMP–CRP complex. However, it was enhanced greatly by this complex and was not affected directly by GalR or DNA looping (Fig. 2, lanes 5 and 6). Minor transcript products appeared in lanes 5 and 6, which appear to be cAMP–CRP complex dependent. The two bands, as shown later, were probably caused by the collision of elongating RNA polymerase into a protein that lay in its path. The two bands are referred to as *RB1* and *RB2* to indicate roadblock of RNA polymerase. How certain can it be that the two bands were not initiated from their own promoter? When the *PC* promoter was deleted, the bands disappeared (data not shown). Therefore, it is likely that a fraction of the transcripts originating from *PC* was blocked by the cAMP–CRP complex bound to its site at −40.5 to create *RB1* (Fig. 2, lane 5) and GalR bound to O_E (-60.5) to create *RB2* (Fig. 2, lane 6) based on the expected length of *RB1* and *RB2* transcripts. In addition, when *AS* and O_E were not present, the bands were not observed (see later).

Mapping PA Transcript

Deletion of O_E *operator*

The origin and direction of the *PA* transcript were investigated because DNA looping repressed it. Perhaps this information could shed more light on the mechanism of DNA looping in the *gal* system. Two plasmids containing an operator deletion of O_E (pSA886) or O_I (pSA896) were used to investigate whether the length of the *PA* transcript would change if it were transcribing through the operators and terminating at the *rpoC* terminator (Fig. 3). The result on the wild-type DNA template was as expected (Fig. 3, lanes 1–4). On the mutant template with O_E deletion (Fig. 3, lanes 5–8), the *P1* and *P2* transcripts were unchanged as expected, as the promoter starting points were downstream of O_E (Fig. 3AII). DNA looping repression of *P1* and *P2* requires both operators. In addition, the repression of *P1* alone requires the binding of GalR to O_E. However, the length of the *PA* transcript was decreased by 16 bp, size of O_E, indicating that this transcript was transcribing through O_E. The length of the *PB* transcript was not affected by O_E deletion, indicating that this promoter was not transcribing through the *gal* region. Therefore, *PB* must be located somewhere else on the plasmid.

Deletion of O_I *Operator*

On the mutant template with O_I deletion (Fig. 3, lanes 9–12), the length of *P1* and *P2* transcripts was decreased by 16 bp because O_I was located downstream of the promoters (Fig. 3AIII). The binding of GalR to O_E activated the *P2* transcript and repressed that from *P1*. The length of the *PA* transcript was reduced by 16 bp, indicating that this transcript was being transcribed through O_I and may be stopping at the *rpoC* terminator. The length of the *PB* transcript was not affected by O_I deletion, suggesting that it was not transcribing through the *gal* region.

Mapping PA Orientation

The aforementioned result suggests that the *PA* transcript should be present in the original plasmid (pSA508), which lacks the *gal* DNA (166 bp). Therefore, transcription products from the wild-type plasmid (pSA850) and pSA508 were compared to investigate whether *PA*, *PB*,

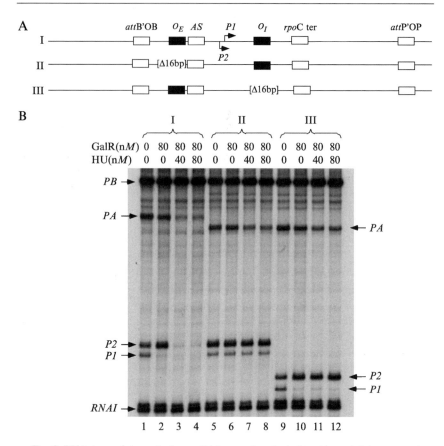

Fig. 3. RNA transcripts made from wild-type and mutant plasmids containing one or two *gal* operators. (A) Schematic of the DNA region of interest: (I) DNA contains two operators (lanes 1–4), (II) O_E (16 bp) is deleted (lanes 5–8), and (III) O_I (16 bp) is deleted (lanes 9–12). (B) RNA transcripts made from the wild-type and mutant plasmids described (see Figs. 1 and 2).

and *PC* products were altered (Figs. 1 and 4). The result with pSA850 was as expected (Fig. 4, lanes 1–4). The result on plasmid pSA508 without *gal* DNA showed that the length of *PB* was unchanged and that the *PA* transcript was migrating to a new level below the *RNAI* transcripts. This is in agreement with an expected length of 166 bp less than that with *gal* DNA. This hints that the *PA* transcript is probably initiating upstream of *rpoC* and is transcribing toward it. In addition, the *PC* transcript migrated to a lower level in pSA508, suggesting that this transcript is initiating upstream of *rpoC*. Note that *RB1* and *RB2* were not observed in pSA508 as the O_E and *AS* sites of *gal* DNA were not present where GalR and the cAMP–CRP complex bind to block transcripts from *PC*.

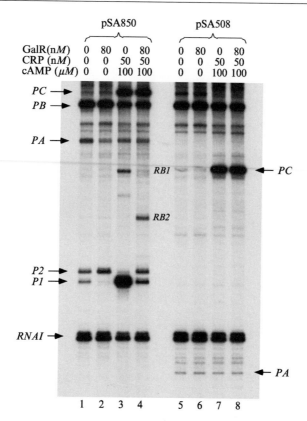

FIG. 4. RNA transcripts made from pSA850 (wild type) and pSA508 (vector) plasmids. pSA850 contains *gal* DNA (lanes 1–4), and pSA508 contains no *gal* DNA (lanes 5–8). See the legend of Fig. 2 for details.

Mapping 5′ Start Sites of PA and PC

Based on the data just described, plasmid pSA850 was used to map the transcription start point (tsp) of *PA* by primer extension and DNA sequencing (Fig. 5). Lanes 1–4 showed sequencing reactions of pSA850 and lanes 5–9 showed primer extension data in the presence and absence of GalR, HU, and/or the cAMP–CRP complex by using primer PEgus3 (+64 to +45). Primer extension data are consistent with *in vitro* data observed earlier. *PC* was observed in the presence of the cAMP–CRP complex (lanes 8 and 9), whereas *P2*, *P1*, and *PA* were observed in lanes 5–9 depending on whether GalR and HU were present. *P1* was mapped to +1 and *P2* to −5 as expected. The *AS*, O_E, and *Eco*RI sites were identified.

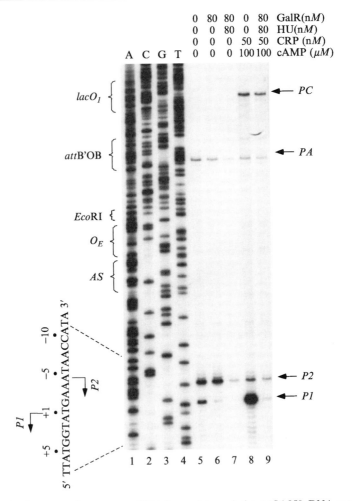

Fig. 5. Primer extension assays on RNA transcripts made from pSA850. DNA sequencing of the plasmid with primer PEgus-3 is from lanes 1–4. Primer extension products from RNAs made in *in vitro* transcription and extended with Pegus-3 are from lanes 5 to 9. The DNA sequence of the *gal* promoter region is at the left of the figure for reference. Other regions identified are located at the left.

In addition, the *att*B'OB site was identified encircling the *PA* transcript. This suggested that *PA* was originating from the *att*B'OB site. Moreover, *PC* was identified to be initiating from the *lacO*$_1$ region.

Primer 508-1 (−78 to −92) was used to identify the tsp of *PA*. Figure 6 shows that *PA* started with a G residue (complement to C) and mapped at −120 to generate a transcript of 245 nucleotides through the *gal* DNA.

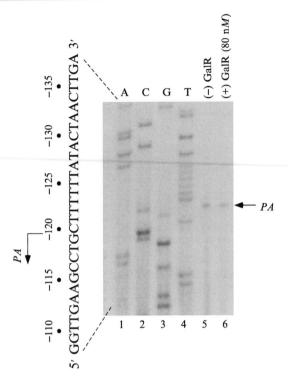

Fig. 6. Primer extension assays on RNA transcript made from the *att*B'OB region. DNA sequencing of the plasmid with primer 508-1 is from lanes 1 to 4. Primer extension products from RNAs made in *in vitro* transcription and extended with 508-1 are from lanes 5 to 6. The DNA sequence of the *att*B'OB region is at the left of the figure for reference. PA indicates the transcript originating from the *att*B'OB.

This correlates well with migration of the *PA* transcript. The *PA* transcript was not labeled in the presence of $[\gamma\text{-}^{32}P]ATP$, thus supporting the idea that it started with another nucleotide instead of an adenosine residue (data not shown).

To map the *PC* transcript, primer lacZ-2 (-151 to -177) was used on *in vitro* transcription in the presence and absence of the cAMP–CRP complex (Fig. 7, lanes 5 and 6). The start site of *PC* mapped to -239 and started with an A residue. This generates a *PC* transcript of 364 nucleotides. In fact, *PC* was identical to the *lac* promoter, which depends on the cAMP–CRP complex for its activity. The plasmid contains the lac operators (O_1 and O_2); therefore, the plasmid can also be used to study the *lac* promoter.

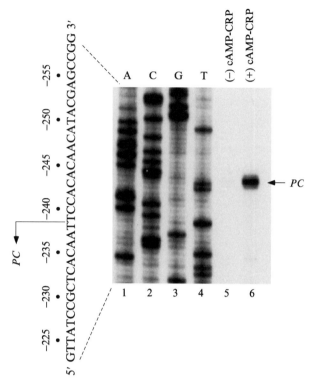

FIG. 7. Primer extension assays on RNA transcript made from the $lacO_I$ region. DNA sequencing of the plasmid with primer lacZ-2 is from lanes 1 to 4. Primer extension products from RNAs made in *in vitro* transcription and extended with lacZ-2 are from lanes 5 and 6. The DNA sequence of the $lacO_I$ region is at the left of the figure for reference. *PC* indicates the transcript originating from the $lacO_I$.

Deletion Analysis of the PA Start Site

Deletion of T7 and attB'OB Regions

To be certain that the *PA* transcript was transcribing from the *att*B'OB site, a series of deletion was done. First, a 147-bp (−222 to −76) fragment containing a *T7* promoter (tsp -164) and the *att*B'OB sites was deleted (Fig. 8AII). Next, a 56-bp (−213 to −158) fragment containing the *T7* promoter site was deleted (Fig. 8AIII), and finally, a 56-bp (−157 to −102) fragment containing the *att*B'OB site was deleted (Fig. 8AIV). The wild-type DNA template result was as expected (lanes 1–6). Note that GalR failed to repress *P1* and activate *P2* in the presence of the cAMP–CRP complex (lane 5). In general, the cAMP–CRP complex activates *P1*

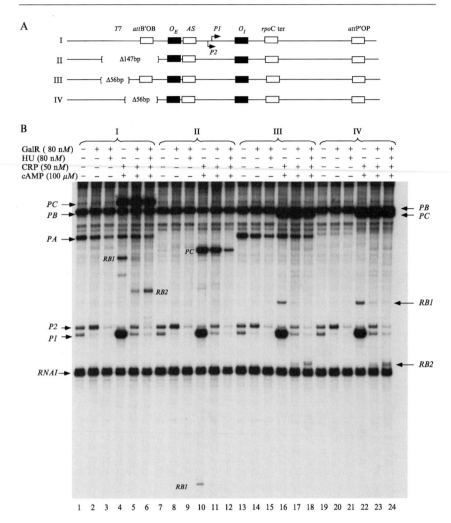

FIG. 8. RNA transcripts made from wild-type and mutant plasmids. (A) Schematic of the DNA region of interest: (I) wild-type plasmid showing the location of the T7 transcription start site region. Mutant plasmids with a deletion of 147 bp (II), 56 bp (III), and 56 bp (IV). (B) RNA transcripts made from wild-type (lanes 1–6) and mutant (lanes 7–24) plasmids.

and represses *P2*. Similarly, GalR represses *P1* and activates *P2*. Hence, the presence of both proteins nullified the effect of each protein. GalR and HU repressed both promoters even in the presence of the cAMP–CRP complex (lane 6). Deletion of *T7* and the *att*B'OB region (Fig. 8AII) removed the *PA* transcript, confirming that this promoter was located in that region (lanes 7–12). The *PB* transcript was not affected, implying that it was located somewhere else on the plasmid. The length of the cAMP–CRP-dependent promoter *PC* was decreased by 147 nucleotides, confirming that this promoter was located to the left of the *att*B'OB region and transcribing through the *gal* DNA toward the *rpoC* terminator. Note that the *RB1* band was observed in lane 10. The presence of GalR (lane 11) reduced *PC* expression, which was reduced further by DNA looping (lane 12). There was no roadblock transcript observed in lanes 11 and 12 to account for the drastic reduction of the *PC* transcript. Because the *PC* transcript was transcribing from P_{lac} at -239, the deletion of 147 bp brought the tsp of P_{lac} closer to O_E. Therefore, RNA polymerase was unable to initiate transcription efficiently from *PC* due to the steric hindrance of DNA looping created by GalR bound to O_E and O_I.

Delection of T7 Region

Deletion of the *T7* region (Fig. 8AIII) does not remove the *PA* transcript, indicating that this promoter start site was located downstream of the *T7* region and transcribing rightward toward the *rpoC* terminator (lanes 13–18). The *PB* transcript was unaffected as expected, but *PC* was reduced by 56 nucleotides. Note that *RB1* and *RB2* were reduced in length and migrated faster than that in wild type (compare lanes 4 and 16, 6 and 18). Transcription was also done with *T7* RNA polymerase and only *T7* transcripts (289 nucleotides) were transcribed (data not shown). When the *T7* region was deleted, no transcript was observed.

Delection of att*B'OB Region*

Deletion of the *att*B'OB region (Fig. 8AIV) removed *PA*, indicating that this promoter mapped to the *att*B'OB site as described earlier (lanes 19–24). The *PB* transcript was unaffected as expected, but the *PC* transcript was reduced by 56 nucleotides.

Analysis of the PB Transcript

Delection of the att*P'OP Region*

From the results given earlier, it is reasonable to speculate that *PB* was initiating from the right of *rpoC* and transcribing leftward toward the terminator or perhaps it was located somewhere else on the plasmid. To address this, 50- and 99-bp fragments downstream of the *rpoC* terminator were deleted and the length of the *PB* transcript was analyzed (Figs. 9AII, 9BII, 9AIII, and 9BIII). As can be seen, the *PB* transcript was reduced by 50 bp (lanes 5–8) and 99 bp (lanes 9–12), indicating that this promoter was

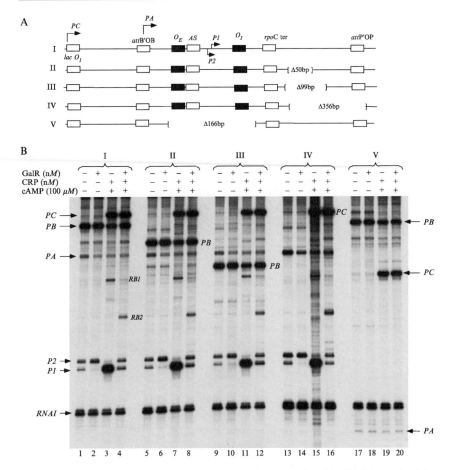

FIG. 9. RNA transcripts made from wild-type and mutant plasmids. (A) Schematic of the DNA region of interest: (I) wild-type plasmid. Mutant plasmids with a deletion of 50 bp (II), 99 bp (III), 356 bp (IV), and pSA508 (V) without the 166-bp *gal* DNA. (B) RNA transcripts made from wild-type (lanes 1–6) and mutant (lanes 7–16) plasmids and vector pSA508 (lanes 17–24).

transcribing toward the *rpoC* terminator. The *PA* and *PC* transcripts were unaffected, as these promoters start upstream of the *rpoC* terminator. Also, *RB1* and *RB2* were unaffected by these deletions. Next, the *att*P'OP region (356 bp) was deleted to investigate whether the *PB* promoter was transcribing from this region (Fig. 9AIII). Results show that the *PB* transcript was absent, confirming that it started in the *att*P'OP region (lanes 13–16). To map the 5' start site of *PB*, primer extension experiments were performed using a labeled primer POP-1 (+257 to +291) on the total RNA transcript and isolated *PB* transcript (Fig. 10). A preparative sequencing gel (8%)

was run with total transcripts from pSA850. The gel was exposed to film to visualize the location of the *PB* transcript, which was excised from the gel, and the RNA was eluted by electroelution and precipitated. The total length of *PB* (364 nucleotides) is shown in lane 6. Primer extension from total RNA and isolated PB transcripts are shown in lanes 5 and 7, respectively. Results showed that both RNAs generated extension products, which mapped to the adenosine residue (underlined) in the *att*P'OP sequence (5'TATTGAT**A**TTT 3'). The *PB* promoter correlates to the *RIII* promoter identified in the phage attachment site.[7] Other products were identified on total RNA as *PX*, *PY*, *PW*, and *PZ*. The *PZ* promoter correlates to the *RII* promoter.[7] Attempts were made to assign these transcripts to bands between *PA* and *PB* (Fig. 9BI lane 1). To this end, constructs were made with deletions of 123 and 130 bp, which remove *PB*, and a deletion of 190 bp, which remove *PB*, *PX*, *PY*, and *PZ* (Fig. 10B). This was confirmed by *in vitro* transcription, which showed that the *PB* transcript was absent from lanes 3–8 and that two of the bands below *PB* in lane 1 were absent when 190 bp was deleted (Fig. 10C, lanes 7 and 8). This result allows the mapping of those transcripts to the 190-bp DNA fragment, but the author was unable to assign them correctly.

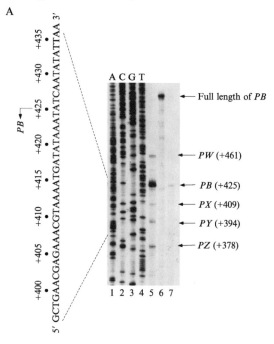

FIG. 10. (*Continues*)

[7] J. Kur, N. Hasan, and W. Szybalski, *Virology* **168,** 236 (1989).

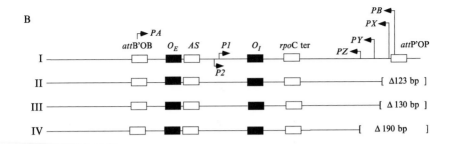

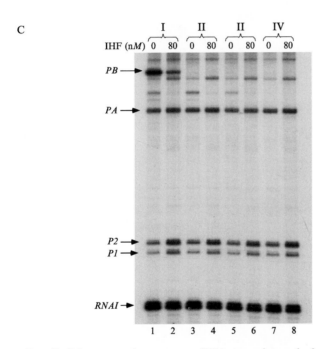

FIG. 10. Primer extension assays on RNA transcript made from the *att*P'OP region. (A) DNA sequencing of the plasmid with primer POP-1 is from lanes 1 to 4. Primer extension products from RNAs made in *in vitro* transcription and extended with POP-1 are in lane 5. Full length of *PB* isolated and purified from an 8% denaturing polyacrylamide gel (lane 6). Full length of *PB* from lane 6 extended with POP-2 (lane 7). The DNA sequence of the *att*P'OP region is at the left of the figure for reference. *PB* indicates the transcript originating from the *att*P'OP. *PW, PX, PY,* and *PZ* are other transcripts identified by POP-2 in lane 5. (B) Schematic of the DNA region of interest: (I) wild-type plasmid. Mutant plasmids with a deletion of 123 bp (II), 130 bp (III), and 190 bp (IV). (C) RNA transcripts made from wild-type (lanes 1–2) and mutant (lanes 3–8) plasmids in the presence and absence of 1Hf.

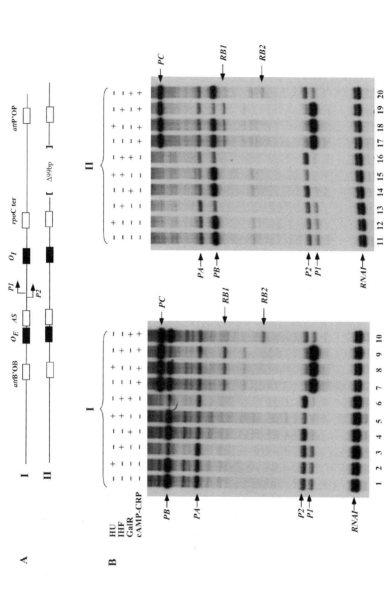

FIG. 11. RNA transcripts made from wild-type and mutant plasmids. (A) Schematic of the DNA region of interest: (I) wild-type plasmid and (II) mutant plasmids with a 99-bp deletion. (B) RNAI transcripts made from wild-type (lanes 1–10) and mutant (lanes 11–20) plasmids, in the presence of HU, IHF, GalR, and the cAMP–CRP complex (see Fig. 2 legend).

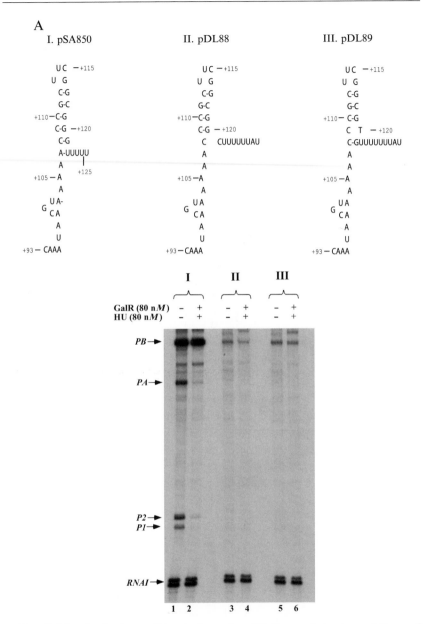

FIG. 12. Mutation in the *rpoC* terminator region. (A) Proposed structures of the *rpoC* terminator region with wild-type RNA (I) and mutation RNAs (II and III). (B) RNA transcripts made from wild-type (lanes 1–2) and mutant (lanes 3–6) plasmids in the presence of GalR and HU.

Effect of IHF on PB Transcript

Because the *att*P'OP region contains binding sites for the integration host factor (IHF), it was tested whether IHF would repress the *PB* transcript. Results showed that IHF does not affect *P1*, *P2*, *PA*, *PC*, or *RNAI* transcripts (Fig. 11); however, it repressed the *PB* transcript, confirming that *PB* was transcribing from IHF region (lanes 3, 6, and 9 (pSA850) and regions 13, 16, and 19 (pSA952). It was observed that IHF repressed two transcripts migrating between *PA* and *PB* (lanes 3, 6, and 9). This

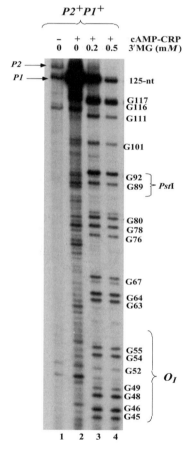

FIG. 13. RNA transcripts made from a linearized DNA fragment containing both *gal* promoters, *P2* and *P1*, in the presence of the cAMP–CRP complex and 3'-O-methylguanosine 5' triphosphate (3'MG). The position of guanine residues in the transcript is indicated on the right of the 8% sequencing gel.

result was also observed (Fig. 10c lane 2) where 1Hf repressed PB and the two transcripts between PA and PB.

RpoC Terminator is Bidirectional

From this study, *PA, PC, P1*, and *P2* were shown to be transcribing from the right of the *rpoC* terminator, whereas *PB* was transcribing from the left of the *rpoC* terminator. This indicates that the *rpoC* terminator was acting

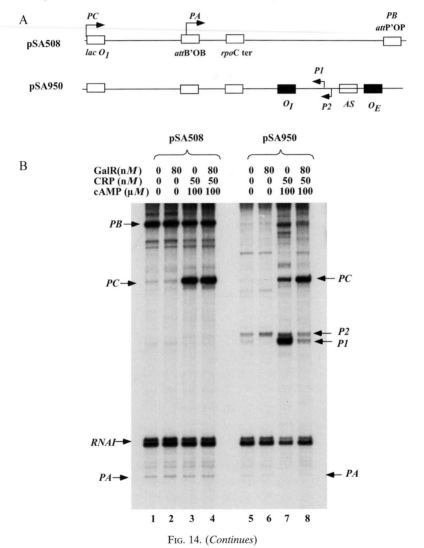

FIG. 14. (*Continues*)

C

+90   +100   +110   +120   +130   +140
 •   •   •   •   •   •
pSA950 5' CTGCAAAACGGTCATGATG<u>GGGATCCC</u>TAAACTCCCCCCATAAAAAAACCCGCCGAA

+150   +160
 •   •
GCGGGTTTT<u>ACGTTATTTGCTGCAG</u> 3'
         ↑
        STOP

pSA950

D

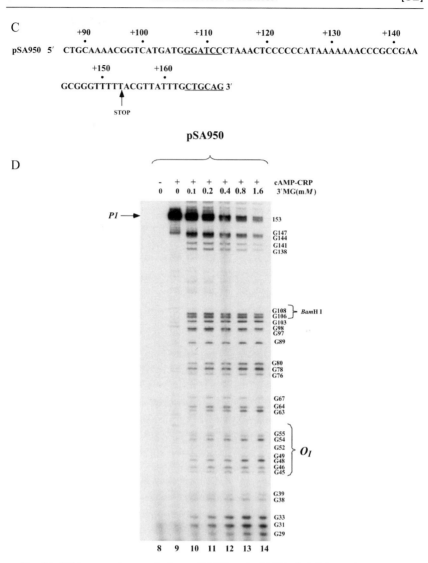

Fig. 14. RNA transcripts made from pSA508 and pSA950. (A) Schematic of the DNA region of interest: (I) pSA508 and (II) pSA950 containing the *gal* sequence downstream of the *rpoC* terminator with the promoters transcribing toward the terminator. (B) RNA transcripts made from pSA508 (lanes 1–4) and pSA950 (lanes 5–8) in the presence of GalR and the cAMP–CRP complex. (C) DNA sequence of pSA950 around the *rpoC* terminator with the arrow marking the end of the transcript. (D) RNA transcripts made from a linearized DNA fragment isolated from pSA950 in the presence of the cAMP–CRP complex and 3'-*O*-methylguanosine 5' triphosphate. The position of guanine residues in the transcript is indicated on the right of the 8% sequencing gels (see Fig. 2 legend).

bidirectionally. If the *rpoC* terminator were disrupted, it would affect the termination of these transcripts. Two DNA mutations in the *rpoC* region were isolated as a result of PCR or cloning error (Fig. 12A), which disrupted the terminator loop. One of the *rpoC* terminator mutants (pDL88) contains two mutations: a GC to CG at +121 and a TA deletion between +122 and +127. The other terminator mutant (pDL89) contains a GC to TA at +120. The result shows that *P1*, *P2*, and *PA* transcripts were not detected, as the *rpoC* terminator was defective (Fig. 12B, lanes 3–6). Also, the *PB* transcript was reduced drastically due to ineffective termination (lanes 3–6).

Determination of the 3′ End of Transcripts

It was determined previously that the *P1* and *P2* transcripts were 120 and 125 nucleotides, respectively. Upon examining the *gal* DNA sequence in Fig. 1, it appears that the transcripts should terminate between +122 and +130 in the stretch of U's instead of at +120 G. To determine the length of the transcripts, mapping of the 3′ end of the *gal* transcripts was repeated using 3′-*O*-methylguanosine 5′-triphosphate. A 260-bp PCR fragment containing wild-type promoters $P2^+P1^+$ (pSA850) was used to prevent the interference of any other promoter. Results showed that this method mapped the 3′ end of the *gal P1* transcript by identifying the guanosine residues in the transcript (Fig. 13, lanes 3 and 4). The guanosine residue at +117 can be accounted for accurately. Then, by extrapolating to the 3′ end, it can be deduced that *P1* terminated at +125 nucleotides. This means that the length of *P2*, *PA*, and *PC* is 130, 245, and 364 nucleotides, respectively.

To map the 3′ end of the *rpoC* terminator in the opposite directions, pSA950 was used. Because *PB* is independent on the cAMP–CRP complex, the *gal* DNA was cloned downstream of the *rpoC* terminator and *PB* was deleted. *In vitro* transcription results showed that *P2* and *P1* were terminated at the *rpoC* terminator (Fig. 14B lanes 5–8). The PC transcript was as expected (lanes 3, 4, 7, and 8). A 287-bp PCR fragment on pSA950 was used to map the 3′ end of the *P1* transcript in the opposition directions (Figs 14C and 14D, lanes 10–14). Guanosine residues at +147 can be accounted for accurately. By extrapolating to the 3′ end, it can be deduced that *P1* terminated at +153 nucleotides. This means that the length of *PB* is 322 nucleotides.

This article illustrated the power of *in vitro* transcription and primer extension to map the 5′ and 3′ ends of transcripts from *att*B'OB and *att*P'OP regions. This same principle can be applied to unknown transcripts. During DNA manipulation (deletion, insertion, or substitution), new transcripts can be generated and it is important to understand the origin of the transcript to aid in the interpretation of data.

[53] Enhancer-Dependent Transcription by Bacterial RNA Polymerase: The β Subunit Downstream Lobe Is Used by σ^{54} During Open Promoter Complex Formation

By Siva R. Wigneshweraraj, Sergei Nechaev, Patricia Bordes, Susan Jones, Wendy Cannon, Konstantin Severinov, and Martin Buck

The DNA-dependent RNA polymerase (RNAP; EC 2.7.7.6) of *Escherichia coli*, the best-characterized multisubunit RNAP, is composed of a core enzyme (E, subunit composition $\alpha_2\beta\beta'\omega$) and one of seven identified molecular species of the σ subunit (Eσ, subunit composition $\alpha_2\beta\beta'\omega\sigma$). Advances in high-resolution structural analysis of the bacterial RNAP have opened up opportunities to study the functional role that each structural module of the RNAP plays in transcription.[1,2] A mobile structural module of *E. coli* RNAP, known as the β downstream lobe (residues 186–433), was shown to contribute to stable open promoter complex formation during transcription directed by RNAP containing the σ^{70} factor.[3] This article describes experimental systems used to probe the function of the β subunit downstream lobe in the context of RNAP containing the major variant σ subunit, the enhancer-dependent σ factor, σ^{54}.[4] Both enhancer-dependent RNAP (σ^{54}–RNAP) and enhancer-independent RNAP (σ^{70}–RNAP) are capable of promoter recognition that results in the formation of the closed promoter complex. σ^{70}–RNAP closed promoter complexes can isomerize rapidly into transcriptionally active open promoter complexes in the absence of additional activators or energy sources. In contrast, σ^{54}–RNAP complexes remain closed unless an enhancer DNA-bound activator and an energy source in the form of ATP or GTP hydrolysis is provided. The ATPase activity of the activator induces the propagation of initial DNA melting or distortion in the closed σ^{54}–RNAP promoter complexes toward the transcription initiation start point and allows open promoter complex formation.[5,6]

[1] K. S. Murakami, S. Masuda, and S. A. Darst, *Science* **296**, 1280 (2002).

[2] K. S. Murakami, S. Masuda, E. A. Cambell, O. Muzzin, and S. A. Darst, *Science* **296**, 1285 (2002).

[3] S. Nechaev, M. Chlenov, and K. Severinov, *J. Biol. Chem.* **275**, 25516 (2000).

[4] M. Buck, M. T. Gallegos, D. J. Studholme, Y. Guo, and J. D. Gralla, *J. Bacteriol.* **182**, 4129 (2000).

[5] X. Zhang, M. Chaney, S. R. Wigneshweraraj, J. Schumacher, P. Bordes, and M. Buck. *Mol. Microbiol.* **45**, 895 (2002).

[6] Y. Guo, C. M. Lee, and J. D. Gralla, *Genes Dev.* **14**, 2242 (2000).

Preparation of Proteins

Overexpression and Purification of Recombinant σ^{54}

For historical reasons, we used the σ^{54} protein from *Klebsiella pneumoniae*. The σ^{54} from *K. pneumoniae* and *E. coli* are essentially interchangeable and have no significantly different properties. *K. pneumoniae* σ^{54} can be expressed to high levels in *E. coli* cells using the T7 RNAP–T7 promoter system. Plasmid pSRWσ^{54}, which contains the *K. pneumoniae* σ^{54} under the control of the T7 promoter, is transformed into *E. coli* strain B834(DE3) (Novagen). Freshly transformed cells are used to inoculate 1 liter of LB medium (50 cfu/liter) containing kanamycin (50 μg/ml) and are grown at 37° with vigorous shaking in a 2-liter flask. At $OD_{600} = 0.5$, the culture is shifted to a 25° water bath and shaken for an additional 30 min before induction by the addition of isopropylthio-β-D-galactosidase (IPTG) to 1 mM. After 3-h induction at 25°, cells are harvested by centrifugation (10,000 rpm; 10 min at 4° in a Beckman JA-14 rotor), resulting in 1–2 g of a wet cell pellet, which can be stored at −80°. For lysis, the cell pellet is resuspended in 20 ml of ice-cold lysis buffer [10 mM Tris–HCl, pH 8.0, 50 mM NaCl, 1 mM dithiothreitol (DTT), 0.1 mM ethylenediaminetetraacetic acid (EDTA), and 5% (v/v) glycerol] containing a cocktail tablet of protease inhibitors (Complete, Roche Diagnostics) and lysed in a cell disrupter (Constant Systems Ltd., UK). The cell lysate is centrifuged at 18,000 rpm in a Beckman JA-20 rotor at 4° for 30 min to remove cell debris and inclusion bodies. The overexpressed σ^{54} is usually in the soluble cell fraction, and the supernatant can be used directly for the purification steps. The supernatant is transferred to a fresh centrifuge tube, and streptomycin sulfate (Sigma) [from a 20% (v/v) stock] is added gradually to a final concentration of 2.0% (v/v) at 4° and incubated while stirring gently on ice for 30 min. The streptomycin sulfate precipitate is then recovered by centrifuging at 18,000 rpm in a Beckman JA-20 rotor for 30 min at 4°. The resulting supernatant is transferred to a fresh centrifuge bottle, and solid ammonium sulfate is added slowly to final concentration of 70% (w/v) and incubated while stirring gently at 4° on ice for 30 min. The ammonium sulfate precipitate is centrifuged (18,000 rpm; 30 min at 4° in a Beckman JA-20 rotor). The resulting pellet is then resuspended in 5 ml of Sepharose buffer A [20 mM imidazole, pH 7.0, 100 mM NaCl, and 5% (v/v) glycerol] and dialyzed overnight at 4° in Sepharose buffer A. The dialysate is centrifuged (18,000 rpm; 30 min at 4° in a Beckman JA-20 rotor) and loaded onto a preequilibrated (according to manufacturers' instructions) 5 ml Hi-Trap Q Sepharose column (Amersham Biosciences) using a fast protein liquid chromatography (FPLC) machine (Amersham Biosciences). The column is washed with 100 ml of Sepharose buffer A. The σ^{54} is eluted (in 1-ml fractions) with a 200-ml linear gradient of 0 to 1 M NaCl in Sepharose buffer A. Fractions are analyzed by sodium dodecyl sulfate–polyacrylamide gel electrophoresis (SDS–PAGE) and fractions containing the σ^{54} are pooled and dialyzed overnight against heparin buffer A [10 mM Tris–HCl, pH 8.0, 50 mM NaCl, 1 mM DTT, 0.1 mM EDTA, and 5% (v/v) glycerol] at 4°. The dialysate is centrifuged (18,000 rpm; 30 min at 4° in a Beckman JA-20 rotor). The resulting supernatant is transferred to a fresh centrifuge tube, and solid MgCl$_2$ is added to a final concentration of 10 mM and stirred gently until all the MgCl$_2$ crystals are dissolved. The supernatant is then loaded onto a preequilibrated (according to manufacturers' instructions) 5-ml Hi-Trap heparin column (Amersham Biosciences) using an FPLC (Amersham Biosciences) to separate σ^{54} from the core RNAP. The column is washed with 100 ml of heparin buffer A. The σ^{54} is eluted (in 1-ml fractions) with a 200-ml gradient of 0 to 1 M NaCl in heparin buffer A containing 10 mM MgCl$_2$. Fractions are analyzed by SDS–PAGE and those containing pure σ^{54} are pooled and dialyzed overnight against storage buffer [10 mM Tris–HCl, pH 8.0, 50 mM NaCl, 0.1 mM EDTA, 1 mM DTT and 50% (v/v) glycerol] at 4°. Aliquots of σ^{54} are stored either at −20° (short-term) or at −80° (long term). The typical yield is about 5–10 mg (protein estimated by Bradford assay[7] using bovine serum albumin as standard).

Overexpression, Purification, and In Vivo Reconstitution of Hexahistidine-Tagged Wild-Type and β(Δ186–433) Core RNAP

Plasmids expressing the hexahistidine-tagged (6His-tagged) β subunit, either wild-type or lacking residues 186–433, a nonessential, evolutionarily variable domain of E. coli RNAP that forms the RNAP downstream lobe and is required for the action of termination factor Alc,[8] are transformed into E. coli strain XL-1 blue (Stratagene). Freshly transformed cells are used to set up a 200-ml starter culture that is grown overnight at 37° in LB medium containing ampicillin (200 µg/ml) in a 1-liter flask. For overexpression, four 1-liter batches of LB medium are inoculated with 50 ml of starter culture and grown at 37° with ampicillin (200 µg/ml) with vigorous shaking in 2.5-liter flasks. At OD$_{600}$ = 0.4, expression of the β subunit gene is induced by the addition of IPTG at a final concentration of 1 mM and growth is continued for another 3–4 h. Cells are harvested by centrifugation (5000 rpm; 10 min at 4° in a Sorvall GS-3 rotor). If not used

[7] M. Bradford, Anal Biochem. **72**, 248 (1976).
[8] K. Severinov, M. Kashlev, E. Severinova, I. Bass, K. McWilliams, E. Kutter, V. Nikiforov, L. Snyder, and A. Goldfarb, J. Biol. Chem. **269**, 14254 (1994).

immediately, the cell pellet can be stored at $-80°$. The cell pellet is resuspended in 30 ml of grinding buffer [50 mM Tris–HCl, pH 8.0, 200 mM NaCl, 5 mM β-mercaptoethanol, 1 mM EDTA, 1 mM phenylmethylsulfonyl fluoride (PMSF) and 5% (v/v) glycerol] and disrupted by sonication, and the lysate cleared by centrifugation (15,000 rpm; 15 min at 4° in a Sorvall SS-34 rotor). The supernatant is transferred to a centrifuge tube, polymin P (Kodak) [from a 10% (v/v) stock made in H_2O, pH adjusted to 8.0 with HCl] is added gradually to a final concentration of 0.8% (v/v) at 4°, and the mixture is incubated on ice for 10 min. The precipitate, which is formed upon polymin P addition, is collected by centrifugation (5000 rpm; 5 min at 4° in a Sorvall SS-34 rotor), and the pellet is washed in 20 ml of buffer A [10 mM Tris–HCl, pH 8.0, 500 mM NaCl, 5 mM β-mercaptoethanol, and 5% (v/v) glycerol] by brief sonication followed by centrifugation (10,000 rpm; 15 min at 4° in a Sorvall SS-34 rotor). The washed pellet is resuspended carefully in ~0.5 ml of buffer A containing 1 M NaCl using a glass rod, followed by further addition of buffer A containing 1 M NaCl to a 20-ml final volume and brief sonication to ensure that a homogeneous suspension is obtained. Following centrifugation (10,000 rpm; 5 min at 4° in a Sorvall SS-34 rotor), the pellet is resuspended carefully as before. Supernatants from the two 1 M NaCl wash steps (~40 ml total volume) are transferred into a fresh centrifuge bottle, and proteins are precipitated by the addition of ammonium sulfate powder to 70% saturation (0.45 g/ml of supernatant). Proteins, which include RNAP, are allowed to precipitate for 1 h or, preferably, overnight at 4°. Precipitated proteins are collected by centrifugation (15,000 rpm; 30 min at 4° in a Beckman JA-20 rotor) and resuspended in 30 ml of buffer A (containing no NaCl). After centrifugation (15,000 rpm; 30 min at 4° in a Sorvall SS-34 rotor), the supernatant is loaded onto a 5-ml Heparin Hi-Trap column (Amersham Biosciences) equilibrated with buffer A (containing 100 mM NaCl) and connected to an FPLC system. After loading, the column is washed with 15 ml of the same buffer. Bound proteins are eluted with three steps of increased concentrations of NaCl in buffer A (0.3, 0.6, and 1.0 M). Eluting proteins are monitored by UV absorbance (OD_{260}), and entire peaks are collected in a single vessel. Fractions are analyzed by SDS–PAGE, and the RNAP-containing fraction (usually, the 0.6 M NaCl fraction) is supplemented with 2 mM imidazole and loaded onto a precharged (with $NiCl_2$) and preequilibrated (according to manufacturers' instructions) 1-ml Hi-Trap chelating HP column (Amersham Biosciences) connected to an FPLC system (Amersham Biosciences). Nonspecifically bound proteins are removed by washing the column with 3 ml of buffer A containing 500 mM NaCl and 2 mM imidazole, followed by washing the column with 10 ml buffer A containing 500 mM NaCl and 20 mM imidazole. RNAP is eluted with buffer A containing 500 mM NaCl and 200 mM imidazole. RNAP-containing fractions are identified by SDS–PAGE. The Ni chromatography step allows the separation of RNAP containing the plasmid-borne β subunit from chromosomally encoded, wild-type RNAP and is critical for preparation of mutant RNAP free of contaminating wild-type enzyme. Because the mutant β subunit is substantially smaller than the wild-type β subunit, a 6–8% (w/v) SDS–PAGE is sufficient to screen for fractions eluted by 200 mM imidazole buffer containing equimolar amounts of RNAP β' and shortened β subunit, which do not contain the full-length β subunit. In contrast, the wash fractions should contain equimolar amounts of full-sized β and β' and no shortened β. The 200 mM imidazole fraction contains highly pure RNAP, which is a mixture of core RNAP and σ^{70} RNAP holoenzyme. To obtain pure core RNAP enzyme, the 200 mM imidazole fraction is diluted threefold in buffer A and loaded onto a 1-ml ProteinPak HQ column (Waters) preequilibrated (with buffer A containing 200 mM NaCl) and attached to an FPLC. After loading, the column is washed with 2–3 ml of buffer A containing 200 mM NaCl, and RNAP is eluted with a 10-ml linear gradient (from 200 to 400 mM NaCl) in buffer A. RNAP core and RNAP holoenzyme are eluted at about 320 and 340 mM NaCl, respectively. Fractions are analyzed by SDS–PAGE, and peak fractions containing pure core RNAP are pooled and concentrated by ultrafiltration through a Centricon-100 centrifugal concentrator (Amicon) to approximately 1 mg/ml (estimated by Bradford assay[7] using bovine serum albumin as standard). Glycerol is added to 50% (v/v), and core RNAP is aliquoted and stored at $-20°$. The typical yield is approximately 200 μg of mutant core RNAP from a 4-liter cell culture.

Overexpression and Purification of 6His-Tagged E. coli σ^{54} Activator Phage Shock Response Protein F (PspF)

To assess the activities of wild-type and mutant Eσ^{54}, we use two forms of the E. coli σ^{54} activator PspF, namely PspF and PspF$_{1-275}$. PspF$_{1-275}$ lacks the carboxyl-terminal DNA-binding domain and essentially represents the central catalytic domain of σ^{54} activators and has the advantage of being able to activate the Eσ^{54} without the requirement for an enhancer sequence, hence simplifying the design of activity assays. pspF and pspF1–275 genes are cloned into pET28b$^+$ (Novagen) and expressed using the T7 RNAP–T7 promoter system. The encoded proteins (PspF and PspF$_{1-275}$, respectively) contain an amino-terminal 6His tag and a site for thrombin-mediated cleavage for the removal of the 6His tag if desired. The culturing procedure used to overexpress PspF$_{1-275}$ is identical to that used for σ^{54} (see earlier discussion). The cell pellet is resuspended in 25 ml of ice-cold

Ni buffer A [25 mM NaH$_2$PO$_4$, pH 7.0, 500 mM NaCl, and 5% (v/v) glycerol] containing a cocktail tablet of protease inhibitors (Complete, Roche Diagnostics) and lysed in a cell disrupter (Constant Systems Ltd., UK). The cell lysate is centrifuged at 18,000 rpm in a Beckman JA-20 rotor at 4° for 30 min. The overexpressed 6His-tagged PspF$_{1-275}$ is usually in the soluble cell fraction, and the supernatant can be used directly for the affinity purification step. For affinity purification, 25 ml of the supernatant is loaded onto a precharged (with NiCl$_2$), preequilibrated (according to manufacturers' instructions) 5-ml Hi-Trap chelating HP column (Amersham Biosciences) using an FPLC machine (Amersham Biosciences). Nonspecifically bound proteins are removed by washing the column with 50 ml of 0.05 M imidazole in Ni buffer A. The 6His-tagged PspF$_{1-275}$ is eluted (in 1-ml fractions) with a 40-ml linear gradient of 0.05 to 0.8 M imidazole in Ni buffer A. Fractions are analyzed by SDS–PAGE, and fractions containing the 6His-tagged PspF$_{1-275}$ are pooled and dialyzed overnight against storage buffer [10 mM Tris–HCl, pH 8.0, 50 mM NaCl, 0.1 mM EDTA, 1 mM DTT, and 50% (v/v) glycerol] at 4°. Aliquots of 6His-tagged PspF$_{1-275}$ are stored at −80°. The typical yield is about 5–10 mg (protein estimated by Bradford assay[7] using bovine serum albumin as standard), and typical purity is >98%.

Full-length PspF can be overproduced using E. coli strain B834(DE3), but in less than optimal quantities. Further, using E. coli strain B834(DE3) as the expression host limits the solubility of the overexpressed 6His-tagged PspF. In order to maximize the amount of soluble over expressed 6His-tagged PspF, a different expression host was used, E. coli strain C41,[9] which has uncharacterized mutations that allow higher-level expression of proteins that are toxic. For overexpression, a freshly transformed colony of E. coli strain C41 harboring plasmid pET28b$^+$–pspF is used to inoculate 10 ml of LB medium containing kanamycin (25 μg/ml) (starter culture). Following overnight growth at 37°, two 1-liter aliquots of LB medium in 2.5-L flasks are inoculated with 1 ml of the starter culture and grown at 37° with vigorous shaking. At OD$_{600}$ = 0.4, the culture is shifted to a 16° water bath and shaken for an additional 30 min before induction by the addition of IPTG to 0.5 mM. After 12–16 h at 16°, the culture is harvested by centrifugation (10,000 rpm; 10 min at 4° in a Beckman JA-14 rotor), resulting in ∼10 g of a wet cell pellet, which can be stored at −80°. For lysis, the cell pellet is resuspended in 10 ml of ice-cold Ni buffer A containing a cocktail tablet of protease inhibitors (Complete, Roche Diagnostics), processed, and purified as described for PspF$_{1-275}$ by FPLC chromatography using a Hi-Trap chelating HP column (Amersham Biosciences).

[9] B. Miroux and J. E. Walker, J. Mol. Biol. 260, 289 (1996).

Peak fractions are pooled and dialyzed overnight against PspF storage buffer [10 mM Tris–HCl pH 8.0, 500 mM NaCl, 0.1 mM EDTA, 1 mM DTT, and 50% (v/v) glycerol] at 4°. Aliquots of PspF are stored at −80°. The typical yield is about 5–6 mg (protein estimated by Bradford assay[7] using bovine serum albumin as standard), and typical purity is >95%.

Characterization of σ^{54}-RNAP

In Vitro *Reconstitution of σ^{54}-RNAP*

While many methods exist for assessing RNAP holoenzyme formation *in vitro*, the native gel assembly assay, a semiquantitative assay, has often been the method of choice for convenient and quick analysis of σ^{54}–RNAP formation and the effect of the $\beta(\Delta186$–433) mutation on holoenzyme formation. In this assay, σ^{54}, core RNAP, and holoenzyme are resolved on a native gel, and holoenzyme formation is judged by depletion of the bands corresponding to the core RNAP and the appearance of a new, faster, migrating band corresponding to the holoenzyme. Binding reactions are conducted in core binding buffer [40 mM Tris–HCl, pH 8.0, 100 mM NaCl, 1 mM DTT, 0.1 mM EDTA, and 10% (v/v) glycerol]. First, core RNAP (250 nM) and different amounts of σ^{54} (at typical molar ratios of 1:1 to 1:4 of core RNAP to σ^{54}) are mixed together at 4° in a final reaction volume of 10 μl. Following incubation at 30° for 10 min, 2 μl of native-loading dye [from a 5 × stock: core binding buffer containing 50% (v/v) glycerol and 0.05% (w/v) bromphenol blue] is added to each reaction. For electrophoresis, samples are loaded onto 4.5% (w/v) native polyacrylamide Bio-Rad Mini-Protean gels. The gels are run in Tris–glycine buffer (25 mM Tris and 200 mM glycine, pH 8.6) at 50 V for 2 h at room temperature. The electrophoresed proteins are visualized by staining and destaining the gels with Coomassie blue dye [0.5% (w/v) Coomassie blue (Sigma), 50% (v/v) methanol, and 7.5% (v/v) acetic acid] and destain solution [45% (v/v) methanol and 9% (v/v) acetic acid], respectively. A representative native gel is shown in Fig. 1 in which the binding activity of wild-type and $\beta(\Delta186$–433) core RNAP to σ^{54} is compared.

Promoter Complex Formation by σ^{54}-RNAP

To monitor the contribution of RNAP surfaces to σ^{54}-dependent promoter complex formation, an electrophoretic mobility shift assay (EMSA) is used. The promoter DNA probe is prepared using two fully complementary (for homoduplex probe; Fig. 2) 88-bp oligonucleotides that comprise the *Sinorhizobium meliloti nifH* promoter sequence (from positions −60

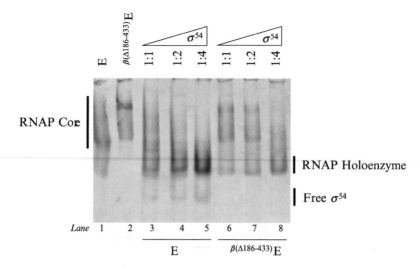

FIG. 1. Nativel gel [4.5% (w/v)] showing the binding of increasing amounts of σ^{54} (250–1000 nM) to free wild-type (E) and $\beta(\Delta186–433)$ core RNAP (lanes 1 and 2, respectively) to form the RNAP holoenzyme (lanes 3–5 and 6–8, respectively). The migration positions of core RNAP, free σ^{54}, and holoenzyme are indicated.

to +28 with respect to the transcription start site at +1). Heteroduplex variants of promoters are used routinely to study the interactions made by the RNAP during open complex formation and promoter escape.[10] Two heteroduplex promoter probes often used to assess and dissect the contribution of σ^{54} and RNAP to promoter complex formation by the σ^{54}–RNAP *en route* to transcription initiation are the early melted and late melted promoter probes (Fig. 2). Early melted and late melted promoter probes represent the conformation of promoter DNA as it is thought to exist within the closed and open σ^{54}–RNAP promoter complexes, respectively. Like the homoduplex promoter probe, heteroduplex promoter probes are made using two 88-bp oligonucleotides comprising the *S. meliloti nifH* promoter, with one of the oligonucleotides (nontemplate strand) containing the region of mismatched sequences to generate the heteroduplex segment (Fig. 2). Thus, in the case of early melted and late melted promoter probes, positions -12 and -11 and positions −10 to −1 are mismatched, respectively.

Preparation of Homoduplex and Heteroduplex Promoter Probes. Oligonucleotides are purchased from MWG-Biotech AG, Germany. For heteroduplex formation, sequences are chosen in order to prevent

[10] W. Cannon, S. R. Wigneshweraraj, and M. Buck, *Nucleic Acids Res.* **30,** 886 (2002).

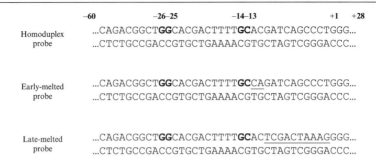

Fig. 2. *Sinorhizobium meliloti nifH* homoduplex and heteroduplex (early melted and late melted) promoter probes used for EMSAs. The consensus GG and GC elements of enhancer-dependent promoters are in bold, and their positions with respect to the transcription start site at +1 are given. Mutant sequences introduced in the nontemplate strand to create the heteroduplex segments are underlined.

purine–pyrimidine mismatched base pairing. Prior to duplex formation, one of the oligonucleotides is ^{32}P end labeled at the 5′ end. End labeling is conducted in 20-μl reactions in T4 polynucleotide kinase buffer (One-Phor-All buffer, Amersham Biosciences) containing 20 pmol of oligonucleotide, 35 μCi of [γ-^{32}P]ATP (10 mCi/ml), and 10 U of T4 polynucleotide kinase (Amersham Biosciences). Following incubation at 37° for 30 min, the reaction is incubated further at 70° for 10 min to inactivate the T4 polynucleotide kinase. For duplex formation, respective pairs of oligonucleotides with the unlabeled strand present at a twofold molar excess (10 pmol in 20 μl) are incubated at 95° for 3 min in TM buffer (10 mM Tris–HCl, pH 8.0, and 10 mM MgCl$_2$) and then chilled rapidly in iced water for 5 min to allow duplex formation.

Assays to Measure Closed and Open Promoter Complex Formation by σ^{54}-*RNAP.* Reactions are conducted in STA buffer [25 mM Tris–acetate, pH 8.0, 8 mM Mg-acetate, 10 mM KCl, 1 mM DTT, and 3.5% (w/v) polyethylene glycol (PEG-6000, Sigma)]. Simple binding assays for σ^{54}-RNAP closed promoter complex formation contained a 16 nM ^{32}P-labeled homo- or heteroduplex probe and 100 nM σ^{54}-RNAP (assembled using a 1:4 molar ratio of core RNAP to σ^{54}) in a final volume of 10 μl. Reactions are incubated at 37° for 5 min and stopped by adding 2 μl of native-loading dye. For electrophoresis, 4 μl of the reaction is loaded onto 4.5% (w/v) native polyacrylamide Bio-Rad Mini-Protean gels. The gels are run in TBE (89 mM Tris, 89 mM *ortho*-boric acid, and 2 mM EDTA) buffer at 60 V for 80 min at room temperature. To assess open promoter complex formation, reactions are composed essentially as described earlier using the homoduplex promoter probe. Following incubation at 37° for 5 min,

4 μM PspF$_{1-275}$ and 1 mM ATP or GTP are added to the reaction and incubated further for 5 min at 37° to activate open or initiated promoter complex formation, respectively. (*Note.* GTP allows the synthesis of a 3-bp transcript from the *S. meliloti nifH* promoter; Fig. 2.) Closed σ^{54}–RNAP promoter complexes are sensitive to heparin, whereas open or initiated σ^{54}–RNAP promoter complexes are resistant. Therefore, to destroy closed and unstable promoter complexes, the reactions are stopped by adding native-loading dye containing heparin to a final concentration of 100 $\mu g/$ ml and, following incubation at 37° for 5 min, electrophoresed as described previously. The gels are dried and quantified using a PhosphorImager (BAS-1500, Fuji, Japan).

Assays with Heteroduplex Promoter Probes. Wild-type σ^{54}–RNAP promoter complexes formed on the early melted promoter probe are resistant to heparin, in contrast to those formed on homoduplex promoter probes in the absence of activation or initiation.[10] This stability is mainly attributed by the ability of the σ^{54}–RNAP to recognize and bind tightly to the DNA fork junction structure present within the early melted promoter probe; this interaction is of regulatory significance for transcription activation by σ^{54} activators.[6,10] Wild-type σ^{54}–RNAP promoter complexes formed on the late melted promoter probe are sensitive to heparin and only become heparin resistant following activation (with PspF$_{1-275}$ and ATP) or initiation (using PspF$_{1-275}$ and GTP).[11] This suggests that conformational changes must occur within the σ^{54}–RNAP for it to acquire heparin stability on the late melted promoter probe.[11] Therefore, EMSAs are designed and conducted (exactly as described earlier) using these heteroduplex promoter probes to assess the contribution the β downstream lobe makes toward (i) early DNA interactions made by the σ^{54}–RNAP and (ii) the acquisition of heparin stability in response to activation.

In Vitro *Transcription Assays*

One of the most common *in vitro* assays used to assess open complex formation is the single-round transcription assay. In this assay a plasmid harbors the promoter, to which the RNAP binds, initiates transcription, and transcribes until it encounters a factor-independent transcription termination sequence. Transcription elongation is conducted in the presence of heparin, which prevents reinitiation and thus results in a single round of transcription from the promoter. In the case of σ^{54}-dependent transcription, the single-round transcription assay is used mainly to check whether mutant forms of σ^{54}–RNAP are able to form activator-dependent open

[11] W. Cannon, M. T. Gallegos, P. Casaz, and M. Buck, *Genes Dev.* **13,** 357 (1999).

promoter complexes (regulated transcription assay) and whether the mutations have resulted in an activator-independent σ^{54}–RNAP that transcribes via an unstable open complex (deregulated transcription assay).

Regulated Transcription Assay. Standard reaction mixtures contain 100 nM σ^{54}–RNAP (reconstituted with 1:4 ratio RNAP to σ^{54}) and 20 U of RNase inhibitor (RNasin from Promega, UK) in STA buffer in a final reaction volume of 10 μl. Reactions are incubated at 37° for 5 min, and template DNA is added to a final concentration of 20 nM. For activation from solution using PspF$_{1-275}$, plasmid pMKC28,[12] which harbors the *S. meliloti nifH* promoter, is used as the template. For activation from an enhancer-bound activator using full-length PspF, plasmid pSLE1,[13] which contains the σ^{54}-dependent *pspA* promoter and a PspF-specific upstream enhancer region, is used as the template. [*Note.* Plasmid pTE103,[14] the parent vector for pMKC28 and pSLE1, contains the phage T7 early transcriptional terminator sequence downstream of the multiple cloning site region (MCS). Promoters inserted into the MCS of pTE103 direct transcription clockwise (5′ to 3′ direction) to generate a discrete transcript of ∼470 bp.] Following incubation at 37° for 5 min, 2 mM ATP and 4 μM PspF$_{1-275}$ or 100 nM PspF are added to initiate open complex formation for 5 min at 37°. For transcript production, 2 μl of a master mix containing final concentrations of 1 mM ATP, CTP, and GTP, 0.5 mM UTP, 3 μCi of [α-^{32}P]UTP (20 mCi/ml), and 100 μg/ml heparin are added and incubated further for 5 min at 37°. The reaction is stopped by adding 4 μl of formamide-loading dye [3% (w/v) xylene, 3% (w/v) bromophenol blue, and 800 μl of 250 mM EDTA in 10 ml of deionized formamide]. The reaction is heated at 90° for 1 min and put on ice. Half (7 μl) of the reaction is used for electrophoresis on a 6% (w/v) polyacrylamide–6 M urea gel. Following electrophoresis, the discrete ^{32}P-labeled RNA product can be visualized and quantified by PhosphorImager (BAS-1500, Fuji, Japan) analysis of the dried gel. A representation of a 6% (w/v) denaturing gel used for analyzing activator-dependent transcription by the wild-type and $\beta(\Delta186-433)$ RNAP is shown in Fig. 3.

Deregulated Transcription Assay. For activator-independent transcription by σ^{54}–RNAP, the reactions are composed as described earlier for the regulated transcription assay, but pMKC28 is used as the preferred template, the activator is omitted, and a form of σ^{54} altered in its regulatory properties is used.[4] Following incubations to reconstitute the σ^{54}–RNAP

[12] M. Chaney and M. Buck, *Mol. Microbiol.* **33,** 1016 (1200).
[13] S. Elderkin, S. Jones, J. Schumacher, D. Studholme, and M. Buck, *J. Mol. Biol.* **320,** 23 (2002).
[14] T. Elliot and E. P. Geiduschek, *Cell* **36,** 211 (1984).

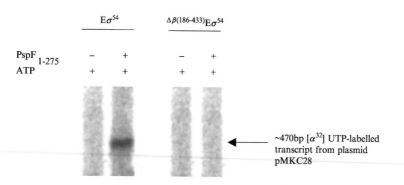

FIG. 3. Section of a 6% (w/v) denaturing gel showing that β subunit residues 186–433 are required for regulated (activator-dependent) transcript formation from the plasmid pMKC28 containing the *S. meliloti nifH* promoter.

and allow closed complexes to form, 4 mM GTP is added to allow activator-independent initiation to occur from the *S. meliloti nifH* promoter (Fig. 2). The reaction is then processed and the transcript is analyzed exactly as described for the regulated transcription assay.

Concluding Remarks

The protocols described in this article provide a simple step-by-step guide to obtain a relatively pure preparation of proteins required to study enhancer-dependent transcription. The experimental assays are designed to measure *in vitro* reconstitution of the σ^{54}–RNAP and to assess the extent of DNA opening (open complex formation) by the σ^{54}–RNAP in response to activation. We have used these assays in conjunction with potassium permanganate probing[15] of wild-type and $\beta(\Delta186$–$433)$Eσ^{54}–RNAP promoter complexes to show that RNAP β subunit residues 186–433 are used commonly by the enhancer-independent Eσ^{70} and enhancer-dependent Eσ^{54} for open promoter complex formation *en route* to transcription initiation.[3,16]

[15] J. D. Gralla, M. Hsieh, and C. Wong, *in* "Footprinting Techniques for Studying Nucleic Acid-Protein Complexes" (A. Revzin, ed.), p. 107. Academic Press, Orlando, FL, 1993.
[16] S. R. Wigneshweraraj, S. Nechaev, K. Severinov, and M. Buck, *J. Mol. Biol.* **319**, 1067 (2002).

[54] Mutational Analysis and Structure of the Phage SP6 Promoter

By INKYUNG SHIN and CHANGWON KANG

Substantial parts of a genome can regulate the functions of DNA and RNA. Regulation of gene replication and expression critically depends on sequence-specific binding proteins or ribonucleoprotein complexes in processes such as DNA synthesis, repair and transcription, and RNA editing, splicing, translation, and degradation.

The exploding availability of DNA sequence data provides a challenge to devise ways of displaying data in a manner that conveys comprehensive clues about function. The most frequently available data are comparative sequence information, and such data have most often been dealt with by constructing consensus sequences. Multiple sequences with a common specific function can be collected from natural genomes or from pools of artificial sequences, and a consensus sequence(s) can be identified. However, consensus sequences convey only part of the information available from comparative sequence data. Although sequence logos, based on information theory, quantitatively display information at each position in a functional site, they do not reflect different activities of individual sequences.[1] However, comprehensive mutational analysis and measurements of activities of all mutant sequences would provide maximal information. When each residue is mutated or modified and the activity of each variant is measured quantitatively, the results would provide much greater information than the consensus sequences.

A typical example of a functional DNA sequence is a promoter. Extensive information on *Escherichia coli* and phage T7 promoters has been accumulated through many independent approaches, but none of these covered the entire length of the promoters. This article provides a method for making all possible substitutions into every position of the phage SP6 promoter and measuring their effects on promoter activity *in vivo* and *in vitro*.[2] The relative *in vivo* activity of promoter variants can be determined by measuring the interference to replication of promoter-bearing plasmids by transcription.[2] This assay has an advantage over direct assays of RNA measurement or indirect assays for a reporter protein. Also, a graphic method called "activity logos" can be used to display saturation mutagenesis data.

---

[1] T. D. Schneider and R. M. Stephens, *Nucleic Acids Res.* **18**, 6097 (1990).
[2] I. Shin, J. Kim, C. R. Cantor, and C. Kang, *Proc. Natl. Acad. Sci. USA* **97**, 3890 (2000).

Activity logos, based on energetics, provide a quantitative view of the energetic contribution of each residue to promoter function.[2]

The Phage SP6 Promoter

Phage SP6 promoters consist of a highly conserved 20-bp region from −17 to +3. The consensus sequence is ATTTAGGTGACACTATAGAA, where the underlined G is the transcription start site +1. SP6 promoters also share high degrees of sequence homology with those of phages T7, T3, and K11. All these phage promoters share a core sequence that extends from −7 to +1, whereas the region −12 to −8 varies.[3,4] Mutational analyses of the phage T7 promoter suggest that a specific RNA polymerase-binding domain extends from −17 to −5 and that a transcription initiation domain is located from −4 to +5.[5] However, previous analyses on a limited set of phage SP6 promoter mutants suggested that polymerase–DNA contacts are distributed in a broader region (to approximately position −3) in the SP6 promoter than in the T7 promoter.[3,6] *In vitro* transcription experiments using −9 and −8 mutants in various salt conditions suggested that SP6 RNA polymerase uses more nonionic forces for promoter interaction than T7 polymerase.[3]

Analysis of 11 known SP6 promoter sequences showed that only positions −17 to +3 contain significant sequence information (Table I). Also, escape from abortive initiation cycling occurs after the formation of 6 nucleotide RNA in SP6 transcription.[7] An *in vitro* evolution study showed no sequence requirements at positions −22 to −18.[8] Thus, the coverage from −17 to +5 would include all residues of the SP6 promoter.

Saturation Mutagenesis of the SP6 Promoter

During chemical synthesis of promoter oligonucleotides, random mutations are introduced into the region corresponding to −25 to +24 under conditions allowing the maximum yield of single residue substitutions in the region from −17 to +5. Oligonucleotides for mutagenesis are synthesized on a 0.2 μmol scale with a Gene Assembler Plus DNA synthesizer

[3] S. S. Lee and C. Kang, *J. Biol. Chem.* **268,** 19299 (1993).
[4] K. G. Han, D. H. Kim, E. Junn, S. S. Lee, and C. Kang, *J. Biochem. Mol. Biol.* **35,** 637 (2002).
[5] K. A. Chapman, S. I. Gunderson, M. Anello, R. D. Wells, and R. R. Burgess, *Nucleic Acids Res.* **16,** 4511 (1988).
[6] S. S. Kim, Y. Hong, and C. Kang, *Biochem. Mol. Biol. Int.* **31,** 153 (1993).
[7] S. C. Nam and C. Kang, *J. Biol. Chem.* **263,** 18123 (1988).
[8] R. R. Breaker, A. Banerji, and G. F. Joyce, *Biochemistry* **33,** 11980 (1994).

TABLE I

OCCURRENCE OF EACH NUCLEOTIDE IN NONTEMPLATE, UPPER STRANDS OF THE
11 KNOWN SP6 PROMOTERS[a]

Position[b]	-20	-19	-18	-17	-16	-15	-14	-13	-12	-11	-10	-9	-8	-7	-6	-5	-4	-3	-2	-1	+1	+2	+3	+4	+5	+6	+7	+8
Consensus[c]				A	t	T	T	A	g	G	t	G	A	C	A	C	T	A	T	a	G	a	a					
G	1	4	6						7	11	3	11									11	3		6	4	4	1	6
A	5	1		10	4			11	2				11		11			11		10		8	10		5	6	7	3
T	1	4	5	1	7	11	11				8						11		11	1			1	1	2		1	1
C	4	2							2					11		11								4		1	2	1

[a] J. Rush and C. C. Richardson, personal communication.
[b] Base pair positions are numbered from transcription initiation site +1.
[c] In the "Consensus" row, bases occurring in more than 60% of the noncoding strand sequences are shown by lowercase letters, whereas 100% conserved bases are shown by capital letters.

(Amersham Pharmacia): MutAC (5'-CTGGATCCaTTTaGGTGacacTa-TaGaaGaAGTGATCAGTCTAGATGCG-3') and MutGT (5'-CTGGA TCCAtttAggtgACACtAtAgAAgAAGTGATCAGTCTAGATGCG-3'). The sequences are of upper, nontemplate strands from −25 to +24. They carry mixtures of bases in the positions designated by lowercase letters; in these positions, the indicated base is 91%, whereas the other three bases are 3% each. A 16 nucleotide primer is synthesized, annealed to the 3' ends of MutAC and MutGT, and extended by the Klenow fragment of *E. coli* DNA polymerase I to make double-stranded DNA. The resulting 49-bp duplex is then cleaved with *Xba*I and inserted into the *Xba*I–*Sma*I site of pSV2 that is described later. Competent *E. coli* JM109 cells are transformed with the ligation mixtures, and transformants are grown on LB plates containing 100 μg/ml ampicillin. Then, individual promoter variants cloned in pSV2 are sequenced in order to isolate most of the 66 possible mutants. When 132 clones (twofold of the number of possible mutations) are sequenced, about two-thirds of all possible mutants are isolated from these random pools. The remaining one-third of mutants can be obtained from other oligonucleotides in the same way.

The promoter assay vector pSV2 was derived from plasmid pGEM4Z (Promega). Digestion of pGEN4Z with *Nde*I and *Eco*RI removes the SP6 promoter sequence. The 2.5-kb *Nde*I–*Eco*RI fragment of pGEM4Z was self-ligated after the ends were filled using the Klenow enzyme. The 1-kb *Bam*HI–*Bgl*II fragment of plasmid pGEMEX-2 (Promega) was inserted at the *Bam*HI site of the SP6 promoterless plasmid, resulting in pSV2 (Fig. 1).

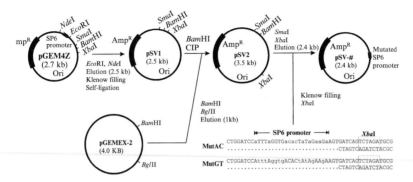

FIG. 1. Construction of the SP6 promoter assay vector pSV2. This promoterless plasmid is used to clone all SP6 promoter variants. The method of its construction is described in the text.

Relative Promoter Strength *In Vivo*

Construction of a Two-Vector System

Replication of ColE1-type plasmids is mainly regulated by two complementary RNA species, RNA I and RNA II, encoded by upstream regions of the replication origin.[9] Part of the RNA II transcript forms a stable hybrid with the template DNA near the origin. Then, RNase H-mediated cleavage of hybrids at the origin yields a primer for the initiation of unidirectional DNA synthesis. RNA I can form a complex with RNA II and prevent it from serving as a primer. Excessive transcription into the replication origin results in enhanced levels of RNA I, thereby decreasing the plasmid copy number. Such interference of plasmid replication by strong SP6 transcription has been observed in a group of SP6 promoter-bearing plasmids, derived from the ColE1 replicon, in the presence of the phage RNA polymerase. This interference results in curing or reduction in the copy number of the promoter-bearing plasmids.[10]

The *in vivo* transcription activity of each cloned variant can be quantified by measuring the copy number of an SP6 promoter-bearing plasmid, rather than by measuring expression of a reporter gene. The mutagenized SP6 promoter is ligated upstream of RNA I. Thus, if plasmid pSV-# is transformed in *E. coli* JM109 containing pACSP6R, which carries the gene encoding the SP6 RNA polymerase, RNA transcripts that extend to RNA I

---

[9] H. Masukata and J. Tomizawa, *Cell* **44**, 125 (1986).
[10] Y. S. Kwon, J. Kim, and C. Kang, *Genet. Anal. Biomol. Eng.* **14**, 133 (1998).

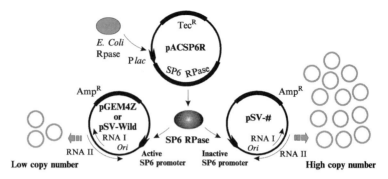

Fig. 2. *In vivo* transcription assay based on interference of the replication of pSV-# plasmids by SP6 transcription. SP6 RNA polymerase is produced from the gene under control of the *lac* promoter in plasmid pACSP6R. It recognizes an SP6 promoter variant on a pSV-# plasmid and produces a transcript that contains a 3' RNA I sequence. This RNA interacts with RNA II and reduces the pSV-# copy number. Thus, the copy number depends on the strength of the promoter variant.

are synthesized from the SP6 promoter by SP6 polymerase. When a promoter variant is active, an extended form of RNA I is synthesized that acts as an inhibitor of the replication primer (RNA II), resulting in reduction in the copy number of pSV-# (Fig. 2). However, if a variant pSV-# is inactive, the copy number increases extensively compared to pSV wild type.

The SP6 RNA polymerase gene under control of the *lac* promoter was cloned at the *Pvu*II/*Sca*I site of plasmid pACYC184 and the resulting plasmid was named pACSP6R.[11] Plasmid pACSP6R has a p15A replication origin and a tetracycline resistance gene. The second plasmid, pSV-# containing an ampicillin resistance gene, has a ColE1 replication origin and is compatible with pACSP6R.

Determination of Plasmid Copy Number

The copy number of pACSP6R is invariantly 220 per cell in the presence of strong or weak promoters, measured against the chromosome copy number as follows.

1. *Escherichia coli* JM109 cells are cultured at $37°$ with vigorous shaking until 0.7 OD_{600}.
2. One milliliter of cell culture is mixed with 1/4 volume of stop solution (5% phenol in ethanol) and centrifuged at 12,000 rpm for 3 min.

[11] W. Jeong and C. Kang, *Biochem. Mol. Biol. Int.* **42,** 711 (1997).

3. Pellets are washed with 0.5 ml Mg^{2+}-free H1 minimal salt solution [10 mM K_2HPO_4, 10 mM KH_2PO_4, 1 mM $(NH_4)_2SO_4$] and recentrifuged.
4. Incubate at $-20°$ for 30 min.
5. Add 50 μl RNase–lysozyme solution (10 mM Tris–HCl, pH 8.0, 20 μg/ml RNase, 1.5 mg/ml lysozyme) and vortex. Incubate at room temperature for 15 min.
6. Add an equal volume of 0.6% SDS and vortex for 10–15 min. Incubate at 37 ° for 15 min.
7. Mix 8 μl of incubation solution with 2 μl loading buffer (40% sucrose, 0.5% bromphenol blue) and subject to electrophoresis on a 0.8% agarose gel.
8. Stain the gel with ethidium bromide and destain in water. Photograph the gel using Polaroid film type 667. Quantify the densities of DNA bands by Imagequant version 3.3 (Molecular Dynamics).

The variable copy number of each pSV-# is determined by comparison to pACSP6R. *E. coli* JM109 cells grown in LB medium containing 100 μg/ml ampicillin and 25 μg/ml tetracycline are harvested and plasmids are prepared from 3-ml aliquots by the alkaline lysis method. After electrophoresis, DNA band densities are analyzed as described earlier. The copy number of each pSV-#, N_i, is calculated as

$$N_i = \frac{\text{band intensity of each pSV-\#per bp}}{\text{band intensity of pACSP6R per bp}} \times 220$$

Each copy number value is obtained by averaging more than three experimental measurements.

Determination of In Vivo *Promoter Activity*

A kinetic model for replication control of ColE1-type plasmid, which was developed previously by Brendel and Perelson,[12] is used to establish the relationship between promoter activity and plasmid copy number. In our system, the Rom protein is not involved, and the interaction between RNA I and RNA II is weakened. (The fourth base upstream of the replication start site of pGEM4Z is G, not C. The base difference in transcripts produces a different secondary structure, resulting in impairment of the inhibitory antisense property of RNA I.) Thus, all the kinetic parameters can be adopted from the original model except for those involving the RNA I–RNA II complex. Values of these parameters are adjusted to set the copy

[12] V. Brendel and A. S. Perelson, *J. Mol. Biol.* **229**, 860 (1993).

number at 400 per cell (for pGEM4Z derivatives). The rate constant of synthesis of RNA I, k_I, is calculated from the measured plasmid copy number, N_i, of that variant.

$$k_I^i = 2330 \times \left(\frac{1}{N_i} + 8.4 \times 10^{-5} \right)$$

The relative activity of each promoter variant, a_i, compared with the consensus promoter is calculated as

$$a_i = \frac{k_I^i - k_I^{no}}{k_I^{con} - k_I^{no}}$$

where k_I^{con} is for the plasmid containing the consensus SP6 promoter and k_I^{no} is for pSV2 that does not contain an SP6 promoter sequence.

Relative Promoter Strength *in Vitro*

The strengths of SP6 promoter variants are determined individually *in vitro* by measuring the production of runoff transcripts from a linearized plasmid containing each variant. For template linearization, the plasmids are digested with a suitable restriction endonuclease. It is important that restriction digestion be performed to completion, as a small amount of undigested plasmid DNA can give rise to very long transcripts. When templates contain 3' protruding ends, extraneous transcripts appear in addition to the expected transcripts.[13] After restriction digestion, the linearized plasmid is extracted with phenol/chloroform and precipitated with ethanol. Transcription reactions are carried out as follows.

1. Add the following components at room temperature in the order listed.

2 μl	10$\times$ transcription buffer
2 μl	100 mM dithiothreitol
4 units	RNasin ribonuclease inhibitor
4 μl	2.5 mM ribonucleotide mixture
0.6 μg	linearized template DNA
0.2 μl	[α-^{32}P]CTP (400 Ci/mmol; 1 Ci = 37 GBq)
5 units	SP6 RNA polymerase
Nuclease-free H$_2$O, if necessary,	to final volume of 20 μl

(10$\times$ transcription buffer consists of 400 mM Tris–HCl, pH 7.5, 60 mM MgCl$_2$, and 20mM spermidine. The mixture should be kept at room

---

[13] E. T. Schenborn and R. C. Mierendorf, Jr., *Nucleic Acids Res.* **13**, 6223 (1985).

temperature, as DNA can precipitate in the presence of spermidine if kept at 4°.)

2. Incubate at 37° for 30 min.
3. Add an equal volume of gel-loading buffer [80% (v/v) deionized formamide, 10 mM EDTA, pH 8.0, 0.025% xylene cyanole, 0.025% bromphenol blue].
4. Heat the mixture at 75° for 5 min and then apply 10 μl to an 8 M urea/polyacrylamide gel in TBE buffer. Electrophoresis is carried out at 500 V for 2–3 h.

After electrophoresis, gels are exposed to a PhosphorImager screen and analyzed using a Storm 860 scanner (Molecular Dynamics).

When circular plasmid DNA instead of linearized DNA is used as a template, transcripts are precipitated by the addition of 100 μl ice-cold 10% trichloroacetic acid. After standing on ice for 10 min, the precipitates are collected on glass fiber filters (Boehringer Mannheim Biochemicals) and washed twice with 200 μl ice-cold 5% trichloroacetic acid. [32]P radioactivity is then measured using a Beckman 7500 liquid scintillation counter and a toluene-based fluor. The templates presumably form negatively supercoiled structures *in vivo*, thus *in vivo* activities of promoters correlate more closely with their *in vitro* activities when the promoters are assayed on naturally supercoiled templates.

The variants can also be analyzed for gel retardation to assess polymerase binding or formation of closed complexes. SP6 RNA polymerase–DNA complexes are produced by combining a 0.01-pmol ^{32}P-labeled 24-bp DNA probe (the upper strand sequence is 5'-CGATTTAGGTGACACTATA-GAAGA-3') containing an SP6 promoter sequence with 0.01–0.1 pmol purified SP6 RNA polymerase in the standard transcription buffer plus a nonspecific competitor (2 μg salmon sperm DNA) in a 10-μl reaction volume. After incubation at 37° for 10 min, the mixture is loaded carefully onto an 8% nondenaturing polyacrylamide (1:29 bis ratio) gel in TBE buffer. A constant temperature gel apparatus (Hoefer) is used to maintain the gel running temperature within 1° of the reaction incubation temperature. After running, the gel is exposed to a PhosphorImager screen.

Purification of SP6 RNA Polymerase

We constructed a plasmid that contained the fragment of the phage SP6 genome encoding SP6 RNA polymerase. The SP6 RNA polymerase gene under the control of the *lac* promoter was cloned at the *Pvu*II/*Sca*I site of plasmid pACYC184 having the replication origin of p15A, and the resulting plasmid was named pACSP6R. In plasmid pACSP6R, the SP6

RNA polymerase gene is expressed under the control of the inducible *lac* promoter by *E. coli* RNA polymerase. An *E. coli* cell usually contains approximately 10 copies of the repressor, which are able to reduce *lac* expression by a factor of 1000 as compared to the fully induced state. This value pertains to one copy of the *lac* operon per cell. However, if the *lac* operator is carried on a high copy number plasmid, the repressor molecules are titrated out by the high copy number of the operator sequence. Thus, the regulated promoter becomes constitutive. Because the plasmid carrying the SP6 enzyme gene does not contain *lacI*, the use of a repressor-over-producing, i^q but $ompT^+$ strain of *E. coli*, JM109, was adopted to attain controlled expression of the gene.

SP6 RNA polymerase is purified as follows.

1. Shake a cell culture of JM109/pACSP6R (3–5 ml) at 37° in LB medium containing 25 μg/ml of tetracycline.
2. At 0.4 OD_{600}, add isopropyl-β-D-thiogalactoside to a final concentration of 0.1 mM and allow further growth for 4 h.
3. Harvest cells and wash them twice with buffer A (40 mM $KHPO_4$, pH 7.7, 1 mM EDTA, 1 mM dithiothreitol, 5% glycerol) containing 50 mM NaCl.
4. Resuspend cells in a 1.5-ml tube in 0.5 ml buffer A containing 50 mM NaCl and 10 mM phenylmethylsulfonyl fluoride (PMSF) and immediately sonicate on an ice bed. (The PMSF protease inhibitor solution should be prepared at 0.1 M in ethanol and stored at −20° because of its short life in aqueous media.)
5. Centrifuge lysates at 12,000 rpm for 10 min to remove cell debris.
6. Load supernatant onto a phosphocellulose column (0.5–0.6 ml bed volume) preequilibrated with buffer A containing 50 mM NaCl.
7. Wash with 5 volumes buffer A containing 100 mM NaCl.
8. Elute with 3 volumes buffer A containing 400 mM NaCl.
9. Add glycerol to samples to a final concentration of 50%.

Graphic Presentation of Saturation Mutagenesis Data

The methods described in this article allow for all possible residues at each position to be assayed individually *in vivo* and *in vitro* for their effects on promoter strength. All of these data can be displayed graphically by the activity logos we developed based on energetics.[2] Transcription activity depends on the product of the equilibrium of polymerase–promoter binding and the rates of isomerization and promoter clearance. The binding constants and rates are exponential functions of free energy and activation energy, respectively, in the same form. Activity of a sequence would be a sum of individual residue contributions for a first approximation. The relative energetic importance of a position can be estimated by adding the effects of each possible base on activity. At each position, the energetic contribution of each of the four bases can be estimated from the relative activity of each base.

Our extensive set of *in vivo* and *in vitro* activity data on all the point substitutions of the SP6 promoter is displayed by activity logos in Fig. 3. They are also compared with sequence logos that reflect the consensus sequence of 11 natural promoters. The height of the logo for a position

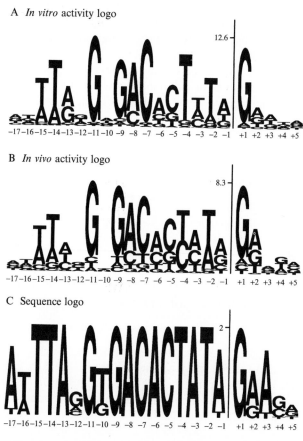

Fig. 3. Activity and sequence logos of the phage SP6 promoter. The two activity logos were constructed based on the energetic contribution to promoter activity of each base pair in every position.[2] Sequence logos were generated from 11 known SP6 promoter sequences (Table I) using the program found at http://www.bio.cam.ac.uk/seqlogo/logo.cgi.[1] Copyright 2000 National Academy of Sciences, U.S.A.

indicates the relative importance of that position for promoter activity, or relative dependency of activity on the kinds of base in that position. According to the activity logos, positions -11, -7, and $+1$ appear to be the most critical for SP6 promoter strength. However, sequence logos suggest that many more positions are highly important. Sequence logos appear to be high-contrast, low-resolution versions of activity logos.

Here, the relative *in vivo* activity of promoter variants is measured using the interference to plasmid replication by transcription. This indirect assay has an advantage over direct assays of RNA measurement or indirect assays for activity of a reporter protein. In this *in vivo* assay, strong mutational effects result in high, rather than low, copy numbers of plasmid, which can be measured accurately. This advantage is significant in the construction of activity logos because strong mutational effects are weighted more logarithmically than weak ones.

Unlike sequence logos based on information theory, our activity logos derived from measurements of individual activities are based on energetics and provide a quantitative view of the energetic contribution of each residue to promoter function.

Acknowledgments

We thank Professor Charles R. Cantor at Boston University for his contribution to the development of activity logos. This work was partially supported by grants from Korea Science and Engineering Foundation (R01-1999-00111), from National Research Laboratory Program (M1-0104-00-0179), and from Brain Korea 21 Project.

[55] Fluorescence Methods for Studying the Kinetics and Thermodynamics of Transcription Initiation

By Smita S. Patel and Rajiv P. Bandwar

DNA-dependent RNA polymerases (RNAP) are key enzymes that catalyze RNA synthesis using the sequence information of the genome in a process known as transcription.[1,2] Transcription consists of several stages: (i) initiation, (ii) abortive RNA synthesis, (iii) promoter clearance,

[1] M. T. Record, Jr., W. S. Reznikoff, M. L. Craig, K. L. McQuade, and P. J. Schlax, *in* "*Escherichia coli* and the *Salmonella typhimurium*: Cellular and Molecular Biology," 792, American Society for Microbiology, Washington, DC, 1996.

[2] P. H. von Hippel, D. G. Bear, W. D. Morgan, and J. A. McSwiggen, *Annu. Rev. Biochem.* **53**, 389 (1984).

(iv) elongation, and (v) termination. The RNAP binds to the promoter DNA during initiation to form a closed complex, which is converted to an open complex by melting of a short stretch of DNA in the vicinity and including the RNA synthesis start site (+1). Two nucleoside triphosphate (rNTPs) complementary to the +1 and +2 bases of the template bind in the RNAP active site to form the initial transcribing complex. Phosphodiester bond formation reaction between the two rNTPs results in pppNpN, the first RNA product, which is elongated by the sequential addition of nucleoside monophosphate (rNMP) to the 3'-OH group of the last nucleotide added. During initiation, short RNAs (up to 12-mer) tend to dissociate from the RNAP–DNA complex as abortive products. When RNAs longer than 12-mer are made, RNAP clears the promoter and enters the elongation phase where RNA synthesis is efficient and processive.

During transcription, RNAs that code for proteins, structural RNAs such as ribosomal RNA, and tRNAs are synthesized. Thus, transcription is a key process that needs to be regulated for controlled cell growth and development. Each stage of transcription is regulated by mechanisms involving interactions of the RNAP with the DNA, accessory proteins, and ligands. The primary targets for regulation are the rate-limiting step(s) and the reversible ones (equilibrium constant close to one) because small perturbation in these steps is expected to have the greatest effect on the overall efficiency of transcription. To understand the mechanism of transcription and its regulation, first it is necessary to elucidate the kinetic pathway by dissecting the elementary steps and determining their intrinsic rate constants and then determining which steps are affected by either promoter sequence variation and/or accessory factors.

This article focuses on transient-state kinetic approaches to elucidate the pathway of transcription initiation. We discuss these methods for T7 RNAP, a 99-kDa single subunit phage polymerase that has been characterized extensively both structurally and mechanistically.[3–6] However, the methods described here are general and applicable to studies of other enzymes and polymerases. Transient-state kinetic experiments are designed to follow the formation and decay of reacting species as a function of time.[7,8] The concentrations are determined directly by radiometric methods or indirectly through optical changes associated with the formation of intermediates

[3] W. T. McAllister, *Cell. Mol. Biol. Res.* **39,** 385 (1993).

[4] G. M. Cheetham and T. A. Steitz, *Curr. Opin. Struct. Biol.* **10,** 117 (2000).

[5] W. T. McAllister, *Nucleic Acids Mol. Biol.* **11,** 15 (1997).

[6] R. Sousa, *Uirusu* **51,** 81 (2001).

[7] K. A. Johnson, *in* "The Enzymes" (D. S. Sigman, ed.), p. 1. Academic Press, San Diego 1992.

[8] K. A. Johnson, *Methods Enzymol.* **249,** 38 (1995).

and products. The kinetics are measured as a function of enzyme or substrate concentration, and rate versus concentration dependencies are fit to a reaction pathway. Kinetic data in most cases are too complex and hence are best analyzed by computational methods.[9–11]

We describe fluorescence methods to determine the equilibrium-binding constants and both fluorescence and radiometric methods to determine the intrinsic rate constants of the transcription initiation pathway. The initial steps of RNAP binding to the promoter DNA and formation of the open complex were measured by the stopped-flow fluorescence method. The steps of nucleotide binding and RNA synthesis were measured by both stopped-flow fluorescence and radiometric rapid chemical-quenched flow methods. We emphasize that two or more types of experiments should be used in conjunction and data should be fit globally by numerical methods. The greater the number of experiments, the better the chance of determining the correct mechanism and accurately deriving the equilibrium and rate constants of the reaction pathway.

2-Aminopurine as a Probe for RNAP–DNA Interactions

To probe structural changes in DNA, normal bases can be substituted with fluorescent base analogs such as 2-aminopurine (2-AP; Fig. 1B) or pyrrolo-dC that form Watson–Crick-type base pairs. The fluorescence intensity of these base analogs is highly sensitive to base-stacking interactions in DNA, and placement of these analogs at chosen positions in the DNA helix allows measurement of local events, such as binding of proteins, conformational changes in DNA, and melting transitions.[12–24] The fluorescence

[9] C. Frieden, *Trends Biochem. Sci.* **18**, 58 (1993).
[10] S. S. Patel, R. P. Bandwar, and M. K. Levin, *in* "Kinetic Analysis of Macromolecules: A Practical Approach" No. 267 (K. A. Johnson, ed.). Oxford Univ. Press, Oxford, 2003.
[11] C. Frieden, *Trends Biochem. Sci.* **19**, 181 (1994).
[12] D. C. Ward, E. Reich, and L. Stryer, *J. Biol. Chem.* **244**, 1228 (1969).
[13] J. M. Jean and K. B. Hall, *Proc. Natl. Acad. Sci. USA* **98**, 37 (2001).
[14] E. L. Rachofsky, R. Osman, and J. B. A. Ross, *Biochemistry* **40**, 946 (2001).
[15] S. K. Mishra, M. K. Shukla, and P. C. Mishra, *Spectrochim. Acta A Mol. Biomol. Spectrosc.* **56A**, 1355 (2000).
[16] B. W. Allan, J. M. Beechem, W. M. Lindstrom, and N. O. Reich, *J. Biol. Chem.* **273**, 2368 (1998).
[17] R. P. Bandwar and S. S. Patel, *J. Biol. Chem.* **276**, 14075 (2001).
[18] Y. Jia and S. S. Patel, *J. Biol. Chem.* **272**, 30147 (1997).
[19] T. M. Nordlund, S. Andersson, L. Nilsson, R. Rigler, A. Graslund, and L. W. McLaughlin, *Biochemistry* **28**, 9095 (1989).
[20] K. D. Raney, L. C. Sowers, D. P. Millar, and S. J. Benkovic, *Proc. Natl. Acad. Sci. USA* **91**, 6644 (1994).

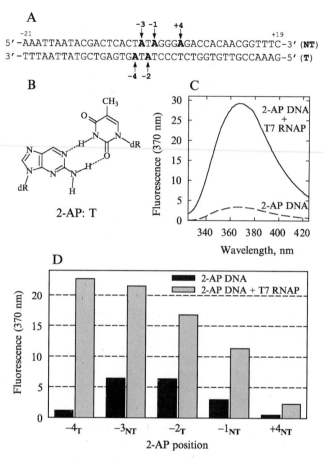

FIG. 1. Fluorescence changes in 2-AP DNA on binding to T7 RNAP. (A) A 40-bp ds DNA containing a consensus T7 promoter sequence substituted individually with 2-AP as indicated. (B) 2-AP forms a Watson–Crick base pair with T. (C) Typical emission spectra of a singly substituted 2-AP DNA in the absence (broken line) and in the presence (continuous line) of T7 RNAP measured upon excitation at $\lambda = 315$ nm. (D) Fluorescence intensities of DNAs (0.5 μM) substituted singly with 2-AP (as indicated in (A); in the absence (black bars) and in the presence (gray bars) of T7 RNAP (4.0 μM) after subtracting the fluorescence intensity of the nonfluorescent T7 RNAP–DNA complex.

[21] J. T. Stivers, *Nucleic Acids Res.* **26,** 3837 (1998).
[22] J. J. Sullivan, K. P. Bjornson, L. C. Sowers, and P. L. deHaseth, *Biochemistry* **36,** 8005 (1997).
[23] A. Ujvari and C. T. Martin, *Biochemistry* **35,** 14574 (1996).
[24] C. Liu and C. T. Martin, *J. Mol. Biol.* **308,** 465 (2001).

intensity changes of 2-AP in DNA has been widely used to measure the equilibrium-binding constant of RNAP to promoter DNA or to measure the kinetics of DNA binding as well as promoter strand separation in real time using stopped-flow methods.[17,18,23–25]

To measure open complex formation, the preferred position of 2-AP incorporation in the promoter DNA is the region that gets unwound during initiation. This region in most promoters is A/T rich and hence several adenines are available for substitution. In the T7 promoter, the -4 to $+2$ region (numbering relative to transcription start site at $+1$) is unwound during open complex formation.[26] Synthetic DNAs (40-mer) are made with 2-AP at -1_{NT}, -3_{NT}, $+4_{NT}$, -2_T, or -4_T (Fig. 1A). The oligodeoxynucleotides are purified by standard methods and their concentrations are determined by absorbance measurement at 260 nm and the calculated extinction coefficient of the DNA. The double-stranded (ds)DNA is prepared by mixing single-stranded (ss)DNAs in a 1:1 ratio, which is determined by titrating the two strands and resolving ssDNA from dsDNA by native PAGE.

The absorption spectrum of a 2-AP DNA shows a strong peak at 260 nm corresponding to the absorption by normal bases and a shoulder in the region 305–315 nm corresponding to 2-AP absorption. Upon excitation at 315 nm, the 2-AP DNA shows a broad fluorescence spectrum, with a peak at 370 nm (Fig. 1C). The fluorescence intensity of 2-AP ssDNA is quenched when the ssDNA is annealed to its complementary strand. On an average, the fluorescence intensity of the free 2-AP base is quenched $\sim$95 % upon incorporation into ssDNA and a further 30–90% quenching is observed when the ssDNA is converted to dsDNA. The neighboring base sequence influences the 2-AP fluorescence intensity, which increases when the 2-AP base unstacks from the neighboring bases.[14]

The fluorescence intensity of 2-AP dsDNA promoter increases when it binds to T7 RNAP as shown in Fig. 1C.[17,23] dsDNA promoters containing 2-AP in the -4 to -1 region show a significantly higher intensity increase than the DNA with 2-AP at $+4_{NT}$ (Fig. 1D). This indicates that the $+4$ bp is not significantly perturbed (unpaired or unstacked) when T7 RNAP binds the promoter. A peculiarly large increase in fluorescence intensity is observed when -4_T 2-AP dsDNA binds to T7 RNAP (about 20-fold from free dsDNA fluorescence compared to 3- to 4-fold increase at the other three positions).[17] This indicates that 2-AP at -4_T unstacks from the neighboring bases during open complex formation, which is evident from the crystal structure of the T7 RNAP–DNA complex.[27]

[25] Y. Jia, A. Kumar, and S. S. Patel, *J. Biol. Chem.* **271,** 30451 (1996).
[26] G. M. T. Cheetham and T. A. Steitz, *Science* **286,** 2305 (1999).

Equilibrium Dissociation Constant

The mechanism by which RNAPs melt a specific region of the promoter DNA is not known. No external energy source, such as ATP hydrolysis, is required by T7 RNAP to melt the dsDNA. Hence, the energy for DNA melting and stabilizing the open complex is obtained from specific interactions with the promoter DNA sequence. The free energy of specific interactions between RNAP and DNA can be calculated from the measured equilibrium dissociation constant, K_d, using the relationship $\Delta G = -RT \ln K$, where R is the gas constant, T is temperature in $^{\circ}$K, and K is $1/K_d$.

The observed free energy of promoter DNA binding to RNAP, ΔG_{obs}, is the sum of the intrinsic binding energy ΔG_{int} (which is the potential energy available from specific interactions of RNAP with the promoter) and the energy utilized in causing conformational changes in RNAP and/or DNA denoted by ΔG_{open}. Thus, ΔG_{obs} is smaller than ΔG_{int} by the amount ΔG_{open}. To determine the free energy utilized in open complex formation, one needs to know the values of ΔG_{obs} and ΔG_{int}. Several structurally different DNAs containing the consensus T7 promoter sequence were synthesized to estimate the free energy of promoter DNA melting (Table I). The dsDNA contains the T7 promoter consensus sequence from -21 to $+19$. The bulge-8 DNA contains a noncomplementary region from -4 to $+4$ and is considered a mimic of the melted DNA bound in the open complex; hence, for the bulge-8 DNA $\Delta G_{open} \approx 0$ and ΔG_{obs} is a close estimate of ΔG_{int}. It is believed that some form of DNA distortion, such as DNA bending and/or twisting, may play a role in initiating DNA melting during open complex formation.[28,29] Bulge-1 DNAs containing a single mismatch at each of the four TATA positions were used to estimate the energetics of TATA deformation.

The equilibrium dissociation constant of the reaction between T7 RNAP and DNA is determined by monitoring the fluorescence changes of 2-AP DNA. Of the various methods that have been used to measure the K_d of the RNAP–DNA complex, such as membrane-binding assay or gel-shift assay, fluorescence-based titration is a true equilibrium method that allows measurement of K_d values in the nanomolar to micromolar range. Fluorimetric titrations are carried out at 25° in a 3-ml quartz cuvette. A defined concentration (5–50 nM) of T7 RNAP in 2.5 ml buffer is titrated with -4_T 2-AP DNA added in small aliquots (2–10 μl) at regular intervals.

[27] G. M. T. Cheetham, D. Jeruzalmi, and T. A. Steitz, *Nature* **399**, 80 (1999).
[28] P. L. deHaseth, M. L. Zupancic, and M. T. Record, Jr., *J. Bacteriol.* **180**, 3019 (1998).
[29] J. D. Helmann and P. L. deHaseth, *Biochemistry* **38**, 5959 (1999).

TABLE I
VARIOUS FORMS OF THE CONSENSUS T7 PROMOTER DNA USED FOR ENERGETICS OF T7 RNAP BINDING

DNA	Description	Sequence[a]	K_d (±SE)	ΔG_{obs} (kcal/mol)
Double Stranded	40-bp fully duplex	AAATTAATACGACTCACTATAGGGAGACCACAACGGTTTC TTTAATTATGCTGAGTG**A**TAT<u>CC</u>CTCTGTGTTGCCAAAG	150–480 nM^b	-8.6–9.3^b
Bulge-8	8 base bubble from −4 to +4	AAATTAATACGACTCAC^{CCGCATAC}GACCACAACGGTTTC TTTAATTATGCTGAGTG**A**_{TATCCCT}CTGGTGTTGCCAAAG	1.3±0.1 pM	-16.2 ± 0.1
Bulge-1[c]	AA mismatch at −2	AAATTAATACGACTCACTA^AA<u>G</u>GGAGACCACAACGGTTTC TTTAATTATGCTGAGTG**A**T_AT<u>C</u>CCTCTGGTGTTGCCAAAG	0.29±0.09 nM	-13.0 ± 0.2^c

[a] Upper row, 5'-3' nontemplate strand; lower row, 3'-5' template strand; G:C base pair, transcription start site, +1; **A**, 2-AP.
[b] A range of K_d and ΔG_{obs} is provided from four different experiments.
[c] Other bulge-1 DNAs with mismatches at −4(AA), −3(TT), and −1(TT) provided ΔG_{obs} in the range of −12 to −13 kcal/mol.[31]

The solution is stirred in the cuvette for 2–3 min to allow the reaction to reach equilibrium. The absorbance at 315 and 370 nm, and the fluorescence intensity at 370 nm after excitation at 315 nm, is measured. Titration is continued until a linear increase in fluorescence intensity is observed. Fluorescence intensity values at each [DNA] are corrected for innerfilter effect using[30]

$$F_{i,c} = F_{i,\text{obs}} \times 10^{0.5[Abs_{i,\text{ex}} + Abs_{i,\text{em}}]} \qquad (1)$$

where $F_{i,c}$ is the corrected fluorescence intensity, $F_{i,\text{obs}}$ is the observed fluorescence intensity, $Abs_{i,\text{ex}}$ is the absorbance at 315 nm, and $Abs_{i,\text{em}}$ is the absorbance at 370 nm at point i in the titration. The concentration of total T7 RNAP, $[E]_i$ and DNA, $[D]_i$ at each point is determined after accounting for dilution:

$$[E]_i = \frac{v_0}{v_{i,f}} \times [E]_0 \qquad (2)$$

$$[D]_i = \frac{v_0}{v_{i,f}} \times [D]_0 \qquad (3)$$

where $[E]_0$ and $[D]_0$ are the initial concentrations, $v_{i,f}$ is the volume up to point i, and v_0 is the initial volume. The corrected fluorescence $F_{i,c}$ is plotted as a function of total $[D]_i$ and fit to Eqs. (4) and (5) to obtain the equilibrium dissociation constant, K_d:

$$F_{i,c} = C + [D]_i \times f_D + [ED] \times (f_{ED} - f_D) \qquad (4)$$

where C is a constant, f_D, and f_{ED} are the fluorescence coefficients for 2-AP DNA (D) and T7 RNAP–DNA complex (ED), respectively. The $[ED]$ is defined by quadratic Eq. (5):

$$[ED] = \frac{K_d + [E]_i + [D]_i - \sqrt{\left(K_d + [E]_i + [D]_i\right)^2 - 4 \times [E]_i \times [D]_i}}{2} \qquad (5)$$

The fluorescence intensity change in 2-AP dsDNA upon binding to T7 RNAP is shown in Fig. 2A. Titration data were fit to Eqs. (4) and (5) to obtain the K_d of dsDNA (Table I). As shown in Fig. 2B, the bulge-8 DNA showed stoichiometric binding and provided a K_d in the picomolar range. Because of the stoichiometric binding of the bulge-8 DNA, the measured K_d value is associated with a large error. In order to obtain a more accurate K_d value, titration must be carried out with RNAP

[30] T. M. Lohman and D. P. Mascotti, *Methods Enzymol.* **212**, 424 (1992).

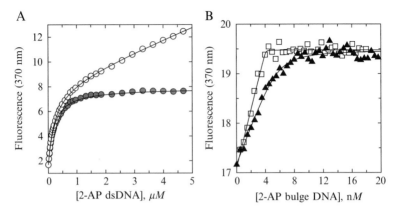

F$_{IG}$. 2. Equilibrium binding of T7 RNAP to dsDNA and bulge DNAs. DNAs are described in Table I. The -4_T adenine was substituted with 2-AP in each DNA, and the 2-AP fluorescence at 370 nm was measured upon excitation at 315 nm during titration of T7 RNAP at 25°. (A) T7 RNAP (50 nM) was titrated with -4_T 2-AP dsDNA. The observed fluorescence intensity was corrected for the innerfilter effect [Eq. (1)], and the plot of corrected fluorescence (○) was fit to Eqs. (4) and (5) with a K_d of 246 ± 10 nM. The linear fluorescence increase due to free 2-AP dsDNA was subtracted (●) using the value of f_D [Eq. (4)]. (B) Bulge-8 titration at 5 nM T7 RNAP (□) provided a K_d of 2 ± 10 pM. The bulge-1 (-2 AA mismatch) titration at 7.5 nM T7 RNAP (▲) provided a K_d of 0.29 ± 0.09 nM.[31]

and DNA in the picomolar range. Measurements at such low protein and DNA concentrations are not reliable partly because protein and DNA tend to adsorb on the solid surfaces during the experiment. Therefore, the K_d of bulge-8 DNA was determined from k_{on} and k_{off} rate constants using the relationship $K_d = k_{off}/k_{on}$ (discussed in the next section).

The K_d for the dsDNA promoter provided the ΔG_{obs} to be around -9 kcal/mol (Table I). Similarly, the K_d for the bulge-8 DNA provided the ΔG_{int} of -16 kcal/mol. The 7-kcal/mol energy difference in the binding of dsDNA relative to premelted bulge-8 DNA is the energy utilized in opening the dsDNA. The ΔG_{obs} of bulge-1 DNAs were in the range of -12 to -13 kcal/mol, which is 3–4 kcal/mol different than either bulge-8 or dsDNA. Introducing a mismatch in the TATA region resulted in a net favorable interaction of the DNA with T7 RNAP and saved about 3 to 4 kcal/mol. The gain in free energy from the mismatch is greater than the energy needed to melt a single TA bp (about 2 kcal/mol in an AT-rich sequence). The mismatch in the TATA region enhances dsDNA flexibility,

[31] R. P. Bandwar and S. S. Patel, *J. Mol. Biol.* **324**, 63 (2002).

which in turn allows the DNA to maximize interactions with the RNAP. These results indicate that a bent DNA conformation might be an intermediate in the pathway to open complex formation.

Stopped-Flow Kinetics of DNA Binding

Association Rate Constant

The kinetics of complex formation are most straightforward for the premelted DNAs. To measure the association rate constant (k_{on}) of T7 RNAP and bulge-8 DNA, the -4_T 2-AP bulge-8 DNA is mixed with T7 RNAP in a stopped-flow instrument and the time-dependent increase in fluorescence intensity is measured at $\lambda \geq 360$ nm with continuous excitation at 315 nm (Fig. 3A).[31] The kinetics is measured under pseudo first-order conditions (one reacting species in excess of the other) to simplify data analysis. The stopped-flow kinetics of bulge-8 DNA binding was measured at a constant [T7 RNAP] and increasing [2-AP bulge-8 DNA] and is shown in Fig. 3B. The increase in fluorescence intensity with time at each concentration was fit to a single exponential:

$$y = A \times \left(1 - e^{-kt}\right) + C \qquad (6)$$

where y is observed fluorescence, A is the amplitude of fluorescence change, t is time, k is the rate constant, and C is the y intercept. The observed rates increases linearly with increasing [2-AP bulge-8 DNA] (Fig. 3C), indicating that bulge-8 DNA binding is a one-step reaction:

$$E + D \underset{k_{off}}{\overset{k_{on}}{\rightleftharpoons}} ED$$

Reaction 1

The linear plot was fit to Eq. (7) to obtain the association rate constant of bulge-8 DNA binding to T7 RNAP.

$$k_{obs} = k_{on} \times [D] + k_{off} \qquad (7)$$

The k_{off} was estimated to be close to zero from the y intercept. Note that the value of k_{off} is obtained by extrapolation to the y axis and hence not determined accurately. For practical reasons, data cannot be collected at concentrations where the observed rates would be close to k_{off}. The value of k_{off} or the dissociation rate constant of DNA from T7 RNAP can be measured directly.

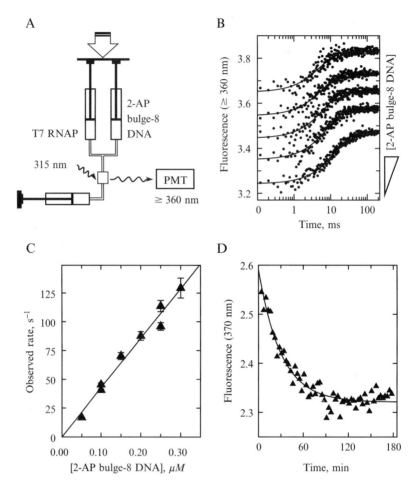

FIG. 3. Association and dissociation kinetics of bulge-8 DNA. (A) Schematic of the stopped-flow experiment. T7 RNAP and 2-AP bulge-8 DNA are mixed rapidly at 25°, and fluorescence intensity at $\lambda \geq 360$ nm is measured upon excitation at $\lambda = 315$ nm. (B) Time-dependent fluorescence increase at various concentrations of 2-AP bulge-8 DNA and at a fixed [T7 RNAP] (25 nM, final). Each kinetic trace (dots) is an average of about 5–10 traces and fit to the single exponential Eq. (6) (continuous line). (C) The observed rate versus [2-AP bulge-8 DNA] provided a slope ($=k_{\text{on}}$) of 432 ± 27 s^{-1} μM^{-1} and a y intercept ($=k_{\text{off}}$) ~ 0 [Eq. (7)]. (D) The kinetics of 2-AP bulge-8 DNA dissociation from T7 RNAP. At time zero, 2 μM nonfluorescent bulge-8 DNA was mixed with a complex of 2-AP bulge-8 DNA and T7 RNAP (100 nM 2-AP bulge-8 DNA + 25 nM T7 RNAP), and the fluorescence intensity was measured versus time. The observed fluorescence intensity decay was fit to Eq. (8) with a $k_{\text{off}} = 5.57 \pm 0.45 \times 10^{-4}$ s^{-1}. The bulge-8 DNA K_{d} ($=k_{\text{off}}/k_{\text{on}}$) was calculated to be 1.3 ± 0.1 pM.[31]

Dissociation Rate Constant

The dissociation rate constant, k_{off}, is measured directly by mixing a complex of T7 RNAP and -4_T 2-AP bulge-8 DNA with an excess of nonfluorescent bulge-8 DNA.[31] When the fluorescent 2-AP bulge-8 DNA dissociates from T7 RNAP, it gets diluted in a pool of excess nonfluorescent bulge-8 DNA and the probability of its rebinding to T7 RNAP is decreased significantly. The fluorescence intensity of 2-AP DNA in the complex is higher relative to free DNA and therefore decreases over time as the 2-AP DNA dissociates irreversibly from the complex (Fig. 3D). The kinetics of DNA dissociation were fit to Eq. (8) to obtain k_{off}:

$$y = A \times \left(e^{-kt}\right) + C \qquad (8)$$

Generally, this experiment is carried out in a stopped-flow instrument if the dissociation rates are fast (> 0.01 s^{-1}, half-life < 60 s) or in a standard spectrofluorometer if the rates are slow (< 0.01 s^{-1}). When slow dissociation rates are measured, it is important that the sample be exposed to light intermittently to minimize photobleaching.

Kinetics of Open Complex Formation in the Absence of Initiating Nucleotide

The kinetics of dsDNA binding to T7 RNAP and open complex formation can be determined by monitoring in real time the fluorescence changes of 2-AP in a stopped-flow instrument. T7 RNAP is mixed with 2-AP dsDNA and the fluorescence intensity at $\lambda \geq 360$ nm is monitored as a function of time with continuous excitation at 315 nm. A representative kinetic trace is shown in Fig. 4A, which was fit to the single exponential equation, Eq. (6). The observed rate was the same regardless of the position of 2-AP in the promoter within the -4 to -1 TATA sequence, indicating that the TATA region of the T7 promoter opens in a concerted manner.

To dissect the steps of DNA binding and to determine the rate of open complex formation, the kinetics of dsDNA binding were measured (under pseudo first-order conditions of excess T7 RNAP over DNA). The observed rates increased in a hyperbolic manner with increasing [T7 RNAP] (Fig. 4B, black circles), which indicates that unlike the bulge-8 DNA that binds in one step, the dsDNA binds to T7 RNAP with a minimum of two steps as shown in Reaction 2:

$$E + D \underset{k_2}{\overset{k_1}{\rightleftharpoons}} ED_c \underset{k_4}{\overset{k_3}{\rightleftharpoons}} ED_o$$

Reaction 2

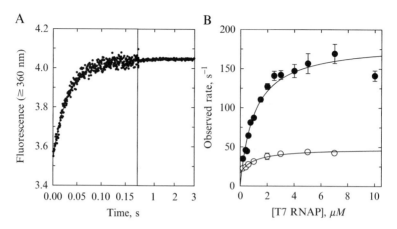

Fig. 4. Stopped-flow kinetics of dsDNA binding with and without 3'-dGTP. (A) A representative kinetic trace of the reaction between T7 RNAP (0.5 μM, final) and -4_T 2-AP dsDNA (0.15 μM, final). The kinetics were measured at increasing [T7 RNAP] in the absence and in the presence of 3'-dGTP (500 μM, final), which was premixed with T7 RNAP. Kinetic data were fit to Eq. (6) (continuous line). (B) Observed rates in the absence (●) and in the presence (○) of 3'-dGTP. Continuous lines are the fit to hyperbolic Eq. (9), which provided $K_{1/2} = 1.0 \pm 0.4\ \mu M$, $k_{max} = 183 \pm 14\ \mathrm{s}^{-1}$, and $y0 = 0$ in the absence of 3'-dGTP, and $K_{1/2} = 1.0 \pm 0.3\ \mu M$, $k_{max} = 32 \pm 2\ \mathrm{s}^{-1}$, and $y0 = 16 \pm 3\ \mathrm{s}^{-1}$ in the presence of 3'-dGTP.[32]

The association of T7 RNAP (E) and DNA (D) results in an intermediate denoted as ED_c, a closed complex, which precedes the formation of the open complex, ED_o. Although the reaction consists of two steps, a single kinetic phase was observed, which is possible if $E + D$ to ED_c is a rapid equilibrium step (i.e., $k_2 \gg k_3$). In this case, the rate dependency can be fit to

$$k_{obs} = \frac{k_{max} \times [E]}{K_{1/2} + [E]} + y_0 \qquad (9)$$

The aforementioned treatment provided a $K_{1/2}$ of 1.0 μM and a k_{max} of 183 s^{-1}. If $E + D$ to ED_c is not a rapid equilibrium step, then the meaning of $K_{1/2}$, k_{max}, and $y0$ is complex and analysis of kinetics is more difficult. In such a case, additional experiments are necessary to constrain the fit and to determine the intrinsic rate constants with more certainty.

[32] N. M. Stano, M. K. Levin, and S. S. Patel, J. Biol. Chem. 277, 37292 (2002).

Kinetics of Open Complex Formation in the Presence of
  Initiating Nucleotide

The RNA initiation sequence of the T7 consensus promoter is
(+1)GGG, hence in the presence of GTP, G-ladder synthesis is observed.
The 3'-deoxyguanosine triphosphate (3'-dGTP) lacks the 3'-OH group,
and use of this analog allows one to examine the steps up to GTP binding
without the complexity of RNA synthesis.[32] dsDNA-binding experiments
in the presence of 3'-dGTP are carried out by mixing T7 RNAP and 3'-
dGTP with the -4_T 2-AP dsDNA. A time-dependent increase in the
fluorescence of 2-AP DNA was observed, and the kinetics is measured at
increasing [T7 RNAP]. The observed rate increases in a hyperbolic manner
with increasing [T7 RNAP] (Fig. 4B, white circles). When data were fit
to Eq. (9), it showed that the k_{max} is about six-fold slower (32 s^{-1}) and
$K_{1/2}$ is 1.0 μM in the presence of 500 μM 3'-dGTP. Thus, the observed
rate of open complex formation is slower in the presence of the initiating
nucleotide.

In order to understand why the observed rate of open complex formation in the presence of 3'-dGTP (32 s^{-1}) is slower than the rate in the
absence of nucleotide (183 s^{-1}), the kinetics of dsDNA binding were measured at increasing [3'-dGTP] (Fig. 5A). Interestingly, the observed rates
decreased with increasing [3'-dGTP] and the corresponding fluorescence

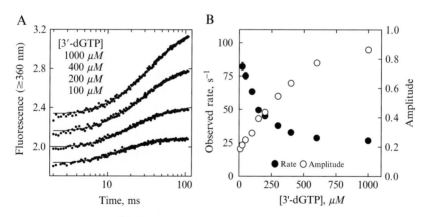

Fig. 5. Effect of [3'-dGTP] on the kinetics of dsDNA binding. (A) Kinetics of 2-AP
dsDNA (0.15 μM, final) binding to T7 RNAP (1.0 μM, final) mixed with increasing
[3'-dGTP]. The kinetic traces were fit to Eq. (6). (B) Observed rates (●) and amplitudes (○)
are plotted versus [3'-dGTP].[32]

amplitude increased (Fig. 5B). This dependency is characteristic of a reaction in which steps of DNA binding and promoter opening precede 3′-dGTP binding, as shown in Reaction 3:

$$E + D \underset{k_2}{\overset{k_1}{\rightleftharpoons}} ED_c \underset{k_4}{\overset{k_3}{\rightleftharpoons}} ED_o \underset{K_{d,3'-dGTP}}{\overset{3'-dGTP}{\rightleftharpoons}} ED_oG$$

Reaction 3

Kinetic data from three experiments shown in Fig. 4 and 5 were fit globally to Reaction 3 to obtain the intrinsic rate constants (shown in Fig. 8).[10,32] The derived rate constants reveal that T7 RNAP binds the dsDNA promoter to form a closed complex ED_c with a close to diffusion-limited association rate constant. The ED_c to ED_o conversion is reversible and occurs with an unfavorable equilibrium constant of 0.13. The initiating rNTP binds to ED_o with a K_d close to 70 μM, which corresponds to the +2 NTP K_d. Hence, data support a model in which ED_c to ED_o conversion is driven by the binding of the GTP that base pairs with the template at the +2 position.[32]

Kinetics of NTP Binding during Initiation

The fluorescence intensity of 2-AP introduced in the −4 to −1 region or at $+4_{NT}$ of the promoter DNA increases when the T7 RNAP–DNA complex binds the +1 and +2 nucleotides, which is GTP.[18,33] This fluorescence change provides the necessary signal to measure the kinetics of GTP binding by the stopped-flow method. The kinetics are measured by mixing the T7 RNAP-2-AP DNA complex with GTP and measuring the fluorescence intensity at $\lambda \geq 360$ nm with continuous excitation at 315 nm. A representative kinetic trace of GTP binding is shown in Fig. 6A. The kinetics were measured at increasing [GTP] to dissect the steps of GTP binding. The observed rates increased in a hyperbolic manner with increasing [GTP] (Fig. 6B), indicating that GTP binding is not a simple one-step process. The observed kinetics indicate that GTP binding occurs with a minimal two-step mechanism shown in Reaction 4:

$$ED_o + GTP \underset{K_d}{\rightleftharpoons} ED_oGG \underset{k_3}{\overset{k_2}{\rightleftharpoons}} ED'_oGG$$

Reaction 4

The GTP dependency was fit to Eq. (10) to obtain the apparent K_d or the cumulative K_d of +1 and +2 GTPs equal to 400 μM and the maximum rate of the conformational change (k_2) following GTP binding equal to 14 s^{-1} with k_3 close to zero:

[33] R. P. Bandwar, Y. Jia, N. M. Stano, and S. S. Patel, *Biochemistry* **41,** 3586 (2002).

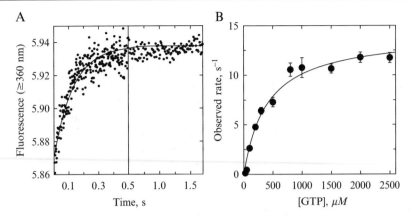

FIG. 6. Kinetics of initiating GTP binding by the stopped-flow method. (A) A representative kinetic trace of a reaction between GTP (2500 μM, final) and a preformed T7 RNAP–DNA complex (-4_T 2-AP dsDNA, 0.15 μM + T7 RNAP, 0.45 μM, final). Kinetics were measured at increasing [GTP] and data were fit to Eq. (6) (continuous line). (B) The observed rate versus [GTP] was fit to Eq. (10) with an apparent GTP $K_d = 400 \pm 70 \ \mu M$, $k_2 = 14.2 \pm 0.8 \ s^{-1}$, and $k_3 = 0$.

$$k_{obs} = \frac{k_2 \times [GTP]}{k_d + [GTP]} + k_3 \qquad (10)$$

To dissect the K_d values of +1 and +2 GTP, for example, to determine the K_d of +2 GTP, one can carry out the same experiment in the presence of GMP, which binds to the +1 position only.[33] In the presence of saturating concentration of GMP, GTP-binding kinetics provide the K_d of +2 GTP. Such experiments in the presence of 600 μM GMP provided a GTP K_d of 80 μM,[33] which is the upper limit for the K_d of +2 GTP. Another way to obtain K_d values of +1 and +2 NTPs is to alter the initiation sequence of the consensus promoter to (+1)GAC. With this promoter, GTP is the +1 nucleotide and ATP is the +2 nucleotide. The binding of ATP to the -2_T 2-AP DNA–T7 RNAP complex showed an increase in fluorescence intensity with time.[32] This signal was used to measure the K_d of ATP equal to 40 μM. Similarly, the K_d of +1 GTP was measured as 400 μM in the presence of saturating ATP.[32] Thus, +1 NTP binding is about 10 times weaker than binding of the +2 NTP.

Kinetics of Initial RNA Synthesis

The binding of +1 and +2 NTP is followed by the synthesis of pppNpN, the first RNA product. In the consensus T7 promoter, the initiation sequence is (+1)GGG; hence in the presence of GTP alone, 2-mer, 3-mer, and longer G-ladders (up to 6- to 8-mer) from slippage reactions are synthesized. The steady-state rate of G-ladder synthesis is close to 0.1 mol of RNA synthesized s^{-1} mol^{-1} of RNAP–DNA complex at $25°$.[34] This rate may not represent the rate of pppGpG synthesis because the steady-state rate in general is dictated by the rate of the slowest step in the pathway. To determine the rate of pppGpG synthesis, it is necessary to measure the presteady-state kinetics of RNA synthesis during initiation. In the presteady-state experimental setup, the concentrations of T7 RNAP and promoter DNA are high in order to measure RNA synthesis accurately in the first turnover, and the reactions are carried out in a rapid chemical-quenched flow instrument. A preformed complex of T7 RNAP and dsDNA promoter is mixed with GTP and $[\gamma\text{-}^{32}P]$GTP from separate syringes of a rapid chemical-quenched flow instrument (Fig. 7A), and the reactions are quenched after various time intervals with 1 N HCl. A typical time course of pppGpG (2-mer) and longer G-ladder synthesis is shown in Fig. 7B. The presteady-state kinetics of RNA synthesis shows a burst; i.e., a fast exponential phase followed by a linear phase (Fig. 7C). The kinetics were fit to:

$$y = A \times \left(1 - e^{-kt}\right) + b \times t + C \tag{11}$$

where A is burst amplitude, k is burst rate, b is steady-state rate, and C is the y intercept. Burst kinetics indicate that the synthesis of 2-mer and 3-mer RNA is rapid at the T7 RNAP active site. The steady-state rate (b) of G-ladder synthesis is slower and limited either by the rate of product dissociation or the rate of T7 RNAP recycling on the promoter. The steady-state rate is therefore a complex parameter that cannot be interpreted meaningfully in terms of the mechanism of transcription initiation. The burst amplitude (A) provides the concentration of active T7 RNAP–DNA complex and the burst rate (k) the rate of pppGpG synthesis. It is clear from Fig. 7B that 2-mer RNA does not accumulate in the presteady-state time scale. Thus, the 2-mer to 3-mer conversion rate is fast relative to the rate of 2-mer formation. Hence, the burst rate provides an estimate of 2-mer formation.

Presteady-state burst kinetics were measured at increasing [GTP] (Fig. 7C). The burst rate increased hyperbolically with increasing [GTP] (Fig. 7D). The hyperbolic dependency was fit to Eq. (10), which provided the apparent K_d of +1 and +2 GTPs and the rate of pppGpG synthesis. The K_d of initiating GTPs from this radiometric assay (330 ± 90 μM) is very close to the K_d (400 ± 70 μM) obtained from fluorescence stopped-flow experiments described earlier. Similarly, the rate of pppGpG synthesis

[34] Y. Jia and S. S. Patel, *Biochemistry* **36,** 4223 (1997).

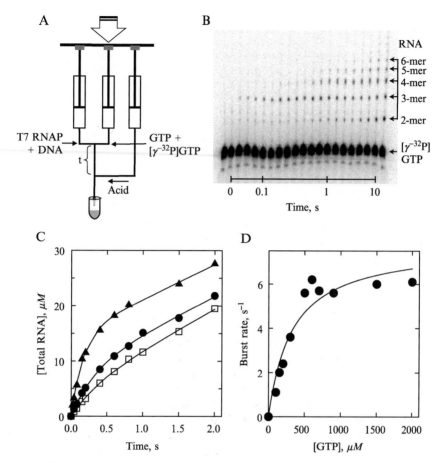

FIG. 7. Kinetics of GTP binding and initial RNA synthesis by the chemical quenched flow method. (A) T7 RNAP (15 μM, final) and dsDNA (10 μM, final) from one syringe were mixed with GTP ($+$ [γ-^{32}P]GTP) from a second syringe and allowed to react for a predetermined time, t, before quenching with 1 N HCl from a third syringe. (B) Typical sequencing gel (23% denaturing PAGE) shows the RNA products. (C) Representative time courses of total RNA synthesis at [GTP] = 150 μM ($\square$), 200 μM ($\bullet$), and 600 μM ($\blacktriangle$). Biphasic kinetics were fit to Eq. (11) (continuous lines). (D) The burst rate versus [GTP] was fit to Eq. (10) with an apparent GTP K_d = 330 $\pm$ 90 μM and k_2 = 7.8 $\pm$ 0.7 s^{-1}.[33]

(7.8 $\pm$ 0.7 s^{-1}) is about two times slower than the rate of the conformational change upon GTP binding obtained from the stopped-flow experiments (14.2 $\pm$ 0.8 s^{-1}). Thus, a combination of radiometric rapid chemical-quenched flow and fluorescence stopped-flow methods provides

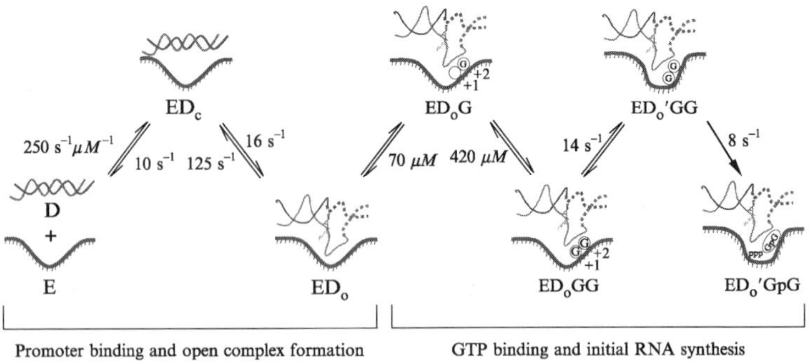

FIG. 8. Kinetic pathway of transcription initiation by T7 RNAP with a consensus promoter. Rate constants for steps up to ED_oG were obtained from global analysis of the kinetic experiments described in Figs. 4 and 5. The equilibrium constant of the ED_oG to ED_oGG step was measured as described.[32] Rates of the ED_oGG to $ED_o'GG$ and $ED_o'GG$ to $ED_o'GpG$ steps were measured as described in Figs. 6B and 7D, respectively.[32,33] (See color insert.)

values of initiating GTP K_d, the rates of conformational change, and pppGpG synthesis.

A complete kinetic pathway of transcription initiation depicted in Fig. 8 is built from the information obtained from the various experiments discussed. The basic knowledge of the mechanism of transcription for a consensus promoter without the effector allows one to then understand how variations in the promoter sequence and the presence of accessory factors regulate transcription. Such analysis with several T7 promoters indicated that multiple steps in the pathway of transcription initiation are affected by the promoter sequence[33] and by the repressor T7 lysozyme. Some of these steps include the stability of the closed and open complexes, the rate of first phosphodiester bond formation, and the affinity for the +1 initiating nucleotide. The rates of the individual steps are affected to a small degree, but the cumulative effect is significant and regulates the yield of the final RNA transcript. A complete understanding of the mechanism of transcription will require not only knowledge of the kinetic pathway, but also structures of the intermediates in the pathway.

Acknowledgment

This work was supported by National Institutes of Health Grant GM51966.

[56] In Vitro Studies of the Early Steps of RNA Synthesis by Human RNA Polymerase II

By Jennifer F. Kugel and James A. Goodrich

We have experimentally isolated five steps in the RNA polymerase II transcription reaction (Fig. 1). These steps include preinitiation complex formation, initiation, escape commitment, promoter escape, and transcript elongation.[1,2] Preinitiation complex formation minimally involves RNA polymerase II and the general transcription factors binding to promoter DNA. When preinitiation complexes are provided with nucleoside triphosphates, transcription initiates and RNA polymerase II produces short 2 and 3 nucleotide RNAs. These complexes are unstable and the RNA products are frequently released in an abortive manner. When the RNA transcript is lengthened to 4 nucleotides, RNA polymerase II undergoes a transition referred to as escape commitment in which the ternary transcribing complex becomes stable and committed to proceeding forward through the reaction. Promoter escape then occurs as escape-committed complexes transform into elongation complexes, which subsequently complete synthesis of full-length RNA.

The protocols discussed here focus on methods to study the early steps in transcript synthesis (i.e., initiation, escape commitment, and promoter escape). These approaches can be used to determine how different factors affect specific steps in early RNA synthesis. Each of the steps in early transcription has the potential to be influenced by transcriptional activators and/or general transcription factors such as TFIIE and TFIIH. For example, initiation and postinitiation events,[3] promoter proximal pausing,[4] promoter clearance,[5] and transcript elongation (as reviewed refs. in[6-8]) have been shown to be targets of transcriptional regulators. In addition,

[1] J. F. Kugel and J. A. Goodrich, J. Biol. Chem. 275, 40483 (2000).
[2] J. F. Kugel and J. A. Goodrich, Mol. Cell. Biol. 22, 762 (2002).
[3] J. Liu, S. Akoulitchev, A. Weber, H. Ge, S. Chuikov, D. Libutti, X. W. Wang, J. W. Conaway, C. C. Harris, R. C. Conaway, D. Reinberg, and D. Levens, Cell 104, 353 (2001).
[4] K. P. Kumar, S. Akoulitchev, and D. Reinberg, Proc. Natl. Acad. Sci. USA 95, 9767 (1998).
[5] S. Narayan, S. G. Widen, W. A. Beard, and S. H. Wilson, J. Biol. Chem. 269, 12755 (1994).
[6] J. W. Conaway, A. Shilatifard, A. Dvir, and R. C. Conaway, Trends Biochem. Sci. 25, 375 (2000).
[7] D. Reines, R. Conaway, and J. Conaway, Curr. Opin. Cell Biol. 11, 342 (1999).
[8] A. Shilatifard, J. W. Conaway, and R. C. Conaway, Curr. Opin. Gen. Dev. 7, 199 (1997).

FIG. 1. Model depicting the steps in the RNA polymerase II transcription reaction. See text for description. R, general transcription factors (e.g., TBP, TFIIB, TFIIF, and RNA polymerase II); P, promoter DNA; PIC, preinitiation complex; RP_I · (3nt RNA), initiated complex containing the 3 nucleotide RNA; RP_{EC} · (4nt RNA), escape committed complex containing the 4 nucleotide RNA; R_E · (15nt RNA), elongation complex containing the 15 nucleotide RNA; R_E · (FL RNA), elongation complex containing full-length RNA. Adapted from Kugel and Goodrich.[2]

TFIIH contains two ATP-dependent helicases[9,10] thought to function during early steps in transcription.[11] We have found that TFIIE and TFIIH increase the fraction of complexes that produce full-length RNA by acting at a specific step in early transcription, namely escape commitment.[1]

Forming Preinitiation Complexes

Theory

Preinitiation complex formation involves the assembly of general transcription factors and RNA polymerase II on the promoter DNA. The following protocols for studying promoter escape, initiation, and escape commitment all require that functional preinitiation complexes first be assembled at the promoter. The minimal components needed to form functional preinitiation complexes include recombinant TBP, recombinant TFIIB, recombinant TFIIF, purified RNA polymerase II core enzyme, and negatively supercoiled promoter DNA. The DNA template must be negatively supercoiled to detect long RNA products in the absence of TFIIE and TFIIH.[12–15] If linear DNA templates are to be used, then TFIIE and TFIIH must be added to preinitiation complexes prior to initiating

[9] L. Schaeffer, R. Roy, S. Humbert, V. Moncolli, W. Vermeulen, J. H. J. Hoeijmakers, P. Chambon, and J.-M. Egly *Science* **260**, 58 (1993).

[10] R. Drapkin, J. T. Reardon, A. Ansari, J.-C. Huang, L. Zawel, K. Ahn, A. Sancar, and D. Reinberg. *Nature* **368**, 769 (1994).

[11] A. Dvir, J. W. Conaway, and R. C. Conaway, *Curr. Opin. Genet. Dev.* **11**, 209 (2001).

[12] J. D. Parvin, H. T. M. Timmers, and P. A. Sharp, *Cell* **68**, 1135 (1992).

[13] C. M. Tyree, C. P. George, L. M. Lira-DeVito, S. L. Wampler, M. E. Dahmus, L. Zawel, and J. T. Kadonaga, *Genes Dev.* **7**, 1254 (1993).

[14] J. D. Parvin and P. A. Sharp, *Cell* **73**, 533 (1993).

[15] J. A. Goodrich and R. Tjian, *Cell* **77**, 145 (1994).

transcription, and the reactions must include hydrolysable ATP (in the form of either rATP or dATP). Additional factors can be added during preinitiation complex formation. For example, in this purified transcription system, TBP can be replaced by purified TFIID and recombinant TFIIA to study transcriptional activation.[16] Similar conditions for preinitiation complex formation can be used if a nuclear extract is the source of the general transcription machinery. In addition, activators or other regulatory proteins can be included during preinitiation complex formation.

Preparation of Transcription Factors and DNA Templates

Methods for preparing recombinant (TBP, TFIIB, TFIIE, TFIIF, and TFIIA) and native (core RNA polymerase II, TFIIH, and TFIID) human transcription factors have been described in detail elsewhere.[16,17] The DNA template should be purified from *Escherichia coli* strains under conditions that maintain the negative superhelicity. Qiagen kits can be used for the preparation of plasmids as long as RNA Guard (Amersham Pharmacia) is used in transcription reactions as described later.

Method

Preinitiation complexes are assembled in 20-μl reactions. The protein components are preincubated in a volume of 10 μl (reaction tube A), and the DNA template is preincubated in a volume of 10 μl (reaction tube B). The protein and DNA components are then combined to allow preinitiation complex assembly (add tube B to tube A). The schematic in Fig. 2 provides a timeline of the method for assembling preinitiation complexes and testing activity by *in vitro* transcription. Table I shows the recipes for reaction tubes A and B for a single 20-μl reaction; the constituents are explained in detail later.

In vitro transcription reactions are assembled such that half the reaction volume consists of DB(100) buffer and proteins [which are dialyzed against DB(100)]. Consequently, components of the DB(100) buffer are twice as concentrated as needed for preinitiation complex assembly and *in vitro* transcription. For example, the glycerol concentration in DB(100) is 20%; the final concentration in the reaction is 10%. Therefore, in the complete 20-μl reaction, the correct concentrations of all buffer components are achieved by adding 10 μl composed of DB(100) and proteins and 10 μl composed of RM4X, DNA, and water. When titrating components

[16] S. K. Galasinski, T. N. Lively, A. Grebe de Barron, and J. A. Goodrich, *Mol. Cell. Biol.* **20**, 1923 (2000).

[17] J. F. Kugel and J. A. Goodrich, *Proc. Natl. Acad. Sci. USA* **95**, 9232 (1998).

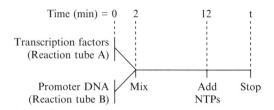

Fig. 2. Time course for assembly of preinitiation complexes *in vitro*. General transcription factors and DNA template are incubated separately at 30° for 2 min prior to mixing. Preinitiation complexes form during the subsequent 10-min incubation at 30°. Nucleotides are then added to initiate transcription.

TABLE I
COMPONENTS OF TRANSCRIPTION REACTIONS

Component	Reaction tube A	Reaction tube B
DB(100)	2.0 μl	5.0 μl
RM4X	2.5 μl	2.5 μl
Water	2.5 μl	1.5 μl
DNA template	—	1.0 μl
Protein mix	3.0 μl	—
Total volume	10.0 μl	10.0 μl

or adding factors to transcription reactions, it is important to maintain this buffer ratio. The following protocol explains preinitiation complex formation using a minimal transcription system consisting of negatively supercoiled promoter DNA, TBP, TFIIB, TFIIF, and RNA polymerase II. To determine the volumes of solutions to make, refer to Table I and consider the total number of reactions to be performed.

1. DB(100) buffer. Make an appropriate amount of DB(100) buffer containing 20 mM Tris, pH 7.9, 20% glycerol, 1 mM dithiothreitol (DTT), 100 mM KCl, and 100 μg/ml bovine serum albumin (BSA) (New England Biolabs). Store on ice. This is the same buffer that all of the general transcription factors are dialyzed against prior to storage, with two exceptions: (1) the dialysis buffer does not contain BSA and (2) RNA polymerase II is unstable in KCl; therefore the KCl is replaced with 100 mM ammonium sulfate as described.[17]

2. RM4X buffer. Make an appropriate amount of buffer RM4X containing 40 mM HEPES pH 8.0, 2 mM DTT, 16 mM MgCl$_2$, and 3 units/μl of RNA Guard (Amersham Pharmacia). Store on ice.

3. DNA template. Prepare a DNA solution by diluting stock plasmid DNA in water to a concentration of 20 nM. One microliter will be added to each reaction to obtain a final concentration of 1 nM in a 20-μl reaction. Store on ice.

4. Reaction tube A. Assemble reactions in 1.5-ml tubes. Each reaction will require two tubes (A and B) because the protein and DNA components are preincubated separately. To each tube add DB(100), RM4X, and water, as indicated in Table I. The protein mix will be added later.

5. Reaction tube B. Add DB(100), RM4X, DNA template, and water, as indicated in Table I.

6. Protein mix. Prepare a protein mix containing all the general transcription factors diluted into DB(100) to the following concentrations: 1 ng/μl TBP, 3 ng/μl TFIIB, 2 ng/μl TFIIF, and 9 ng/μl RNA polymerase II. Three microliters of the protein mix will be added to each reaction. Assemble the protein mix on ice in the following order: DB(100), TBP, TFIIB, TFIIF, and RNA polymerase II. Mix by flicking with finger (do not vortex). Special consideration must be taken when working with transcription factor proteins. All protein factors should be frozen in liquid nitrogen in small aliquots and stored at $-80°$ after purification. Thaw the small aliquots of proteins by warming between fingers, placing quickly on ice. Many factors can be refrozen in liquid nitrogen and thawed rapidly three or four times before loss of activity is noticed (e.g., TBP, TFIIB, and TFIIF). RNA polymerase II should never be frozen and thawed more than twice.

7. Preincubation of proteins and DNA. Add the protein mix to reaction tube A and place at $30°$ in a staggered manner such that each reaction is incubated for the same amount of time. For example, at time point 0 min, add 3 μl of protein mix to reaction tube 1A, mix by flicking with finger, and place at $30°$. At the same time, place reaction tube 1B at $30°$. Repeat this for each pair of reaction tubes at staggered time intervals.

8. Formation of preinitiation complexes. Allow tubes A and B to incubate 2 min at $30°$. At this point, add tube B to tube A to attain a total reaction volume of 20 μl. Mix by flicking with finger and return to $30°$. Repeat this for each pair of reaction tubes at the correct time.

9. Incubate at $30°$ to allow preinitiation complexes to form. It is important to determine the amount of time required to complete preinitiation complex formation at the promoter of interest to decide on the optimal time to use. We allow 10 min under the conditions described here. At the completion of preinitiation complex formation, nucleotides are added to initiate transcription, as described in the following sections.

Additional Comments

If TFIIE and TFIIH are to be included in reactions, they can be added directly to the protein mix. We have also found that TFIIE and TFIIH can be added after other factors are assembled into preinitiation complexes and immediately prior to nucleotides.[18] TFIID and TFIIA can be added in place of TBP. In this case, the final concentration of $MgCl_2$ in reactions should be increased to 6 mM. If transcriptional activation is to be studied, the activator(s) can be prebound to the DNA template by adding it to reaction tube B and removing an equivalent amount of DB(100). If a nuclear extract is to be used, replace the protein mix with nuclear extract and adjust the volume of DB(100). If the contents of reaction tube B are the same for every reaction performed, then a single large reaction tube B can be assembled and preincubated at 30°. Aliquots of 10 μl can then be added to the A tubes. By doing so, the preincubation time for tube B will vary for each reaction. This is acceptable when reaction tube B contains just DNA and buffer components (i.e., activators are not included).

Monitoring the Transition from Preinitiation Complexes
    to Elongation Complexes

Theory

The transition from preinitiation complexes to elongation complexes includes at least three steps: initiation, escape commitment, and promoter escape (see Fig. 1). To isolate these early steps in transcript synthesis from downstream steps, such as elongation, RNA synthesis must be paused after promoter escape. At the AdMLP, promoter escape is complete by synthesis of a 15 nucleotide RNA. This point was defined through rate measurements in which we found synthesis of a 15 nucleotide RNA to be rate limiting and kinetically distinct from elongation of a 15 nucleotide RNA to full-length RNA.[1,17] The point at which promoter escape is complete may vary from promoter to promoter and can be determined using kinetic experiments as described previously.[17]

Initiating transcription with a limited set of nucleotides is the simplest technique for pausing RNA polymerase II at the completion of promoter escape. For example, at the AdMLP (in a G-less cassette template), transcription can be paused after synthesis of a 15 nucleotide RNA by initiating transcription with the dinucleotide ApC (many dinucelotides can be purchased from Sigma), CTP, and UTP (Fig. 3A). RNA polymerase II will pause at the thymidine at +16 (template strand) due to the lack of ATP

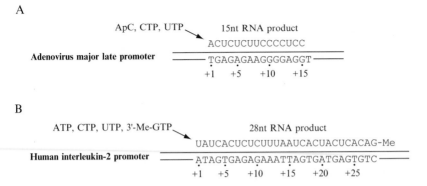

FIG. 3. Methods to pause RNA polymerase II soon after completion of promoter escape. (A) Schematic of the AdMLP. The template strand sequence in the initial transcribed region is shown. The sequence of the 15 nucleotide RNA produced *in vitro* in the presence of ApC, CTP, and UTP is shown above the DNA. (B) Schematic of the human IL-2 promoter. The template strand sequence in the initial transcribed region is shown. Shown above the DNA is the sequence of the 28 nucleotide RNA produced *in vitro* in the presence of ATP, CTP, UTP, and the chain terminator 3'-*O*-methylguanosine 5'-triphosphate (Amersham Pharmacia) (3'-Me-GTP), which ensures that RNA polymerase II does not read through the G at position + 28.

in the reaction. These paused, ternary complexes are stable over the course of the reaction. As another example, the steps of early transcription at the human interleukin-2 (IL-2) promoter can be studied easily because the IL-2 gene does not contain any cytosines in the template strand until +28 (Fig. 3B). Therefore, transcription will stop after synthesis of a 28 nucleotide RNA in the presence of ATP, CTP, UTP, and the chain terminator 3'-*O*-methylguanosine 5'-triphosphate (3'-Me-GTP), which ensures that RNA polymerase II does not read through the guanosine at position +28.[18]

If the promoter and gene of interest do not consist of sequences easily amenable to pausing RNA polymerase II, manipulations of the DNA template can be considered. First, the DNA can be mutated to allow pausing of RNA polymerase II at +16. Second, the promoter can be isolated and fused to a downstream sequence that is amenable to study. Third, a linear DNA template can be used such that the downstream region ends between +15 and +30; therefore, RNA polymerase II will "run off" the end and limit transcript synthesis to a short RNA. When performing runoff transcription, two additional experimental parameters must be considered. Because the DNA template is linear, TFIIE, TFIIH, and

[18] H. A. Ferguson, J. F. Kugel, and J. A. Goodrich, *J. Mol. Biol.* **314,** 993 (2001).

ATP (or dATP) must be included in the reaction.[12–15,17] In addition, to ensure that new preinitiation complex formation does not influence studies of the early steps of transcript synthesis, transcription must be limited to a single round of synthesis. Single round transcription can be obtained with the use of a nonspecific competitor DNA such as poly(dI-dC)·poly(dI-dC).[18]

Method

The following experiment describes analyzing early transcription at the AdMLP by pausing RNA polymerase II in elongation complexes containing a 15 nucleotide RNA.

1. Nucleotide premix. Prepare a solution of 11× nucleotides by diluting stock solutions of nucleotides into TE buffer (10 mM Tris, pH 7.9, 0.1 mM EDTA) to the following concentrations: 11 mM ApC, 7 mM UTP, 275 μM [α-^{32}P]CTP (2.5 μCi/μl). Each reaction will get 2 μl of the nucleotide premix. Both nonradioactive and radioactive CTP are added to the nucleotide premix to obtain the final concentration and specific activity.

2. Add 2 μl of nucleotide premix to each reaction containing preinitiation complexes (assembled as described earlier). Mix by flicking with finger. Incubate at 30° for 30 min. This is sufficient time to complete promoter escape at the AdMLP; however, this should be determined for the promoter of interest.

3. To stop transcription, add 100 μl of a stop solution containing 3.1 M ammonium acetate, 10 μg of carrier yeast RNA, and 15 μg of proteinase K.

4. Ethanol precipitate the transcription reactions by adding 300 μl of cold 100% ethanol. Vortex. Spin at 16,000 g for 20 min at 4°. Remove the ethanol. The ethanol will have a significant number of radioactive counts and should be disposed of appropriately.

5. Add 200 μl of cold 80% ethanol. Spin at 16,000 g for 5 min at 4°. Remove the ethanol.

6. Dry the pellets in a Speed-Vac.

7. Resuspend the pellets in 4 μl of formamide loading buffer (95% formamide, 15 mM EDTA, 0.025% bromphenol blue).

8. Load the entire reaction on a 14% polyacrylamide (19:1) gel (20 × 20 cm; 0.4-mm spacers) containing 7 M urea and 0.5× TBE (45 mM Tris, 45 mM boric acid, 1 mM EDTA).

9. Run the gel in 0.5× TBE at 17 W until the bromphenol blue is two-thirds of the way down the gel.

10. Dry gel and visualize transcript RNA by phosphorimagery.

Additional Considerations

Kinetic differences between the early steps of transcription make promoter escape experimentally discernible from initiation and escape commitment. Both initiation and escape commitment are complete within seconds at the AdMLP and at the IL-2 promoter.[1,18] Because promoter escape is the rate-limiting step in early transcription at these two promoters, any factor or experimental parameter that affects the rate at which a short RNA is produced must do so by affecting promoter escape. It remains to be determined whether promoter escape is rate limiting at most promoters.

Monitoring the Formation of Functional Initiation Complexes by Abortive Initiation

Theory

Abortive initiation is a steady-state assay that monitors functional initiation complexes. RNA polymerases synthesize short (2 to 3 nucleotides) aborted RNA products when preinitiation complexes are provided with limited nucleotide substrates specific to the transcription start site sequence of a given promoter.[19–23] These short RNA products are released by RNA polymerase II, which remains stably bound to the promoter (Fig. 4A).

The choice of nucleotide substrate to use depends on the sequence around the transcriptional start site of the promoter to be studied. At many promoters, multiple different combinations of nucleotides will allow abortive initiation because RNA polymerase II can initiate at a range of sites surrounding +1. For example, at the AdMLP the following nucleotide combinations all produce 2 or 3 nucleotide abortive products: the dinucleotide ApC and [α-^{32}P]UTP produce ApCpU; CpA and [α-^{32}P]CTP produce CpApC; and ATP and [α-^{32}P]CTP produce pppApC.

Short RNAs migrate aberrantly in polyacrylamide gels. The number of phosphates on the 5′ end of the RNA affects migration dramatically; therefore, to simplify analysis, we treat abortive RNAs with a phosphatase to remove all 5′ phosphates. This also better separates small RNA products

[19] W. R. McClure, C. L. Cech, and D. E. Johnston, *J. Biol. Chem.* **253,** 8941 (1978).

[20] W. R. McClure, *Proc. Natl. Acad. Sci. USA* **77,** 5634 (1980).

[21] J. A. Goodrich and W. R. McClure, *J. Mol. Biol.* **224,** 15 (1992).

[22] D. S. Luse and G. A. Jacob, *J. Biol. Chem.* **262,** 14990 (1987).

[23] D. S. Luse, T. Kochel, E. D. Kuempel, J. A. Copploa, and H. Cai, *J. Biol. Chem.* **262,** 289 (1987).

A

Abortive initiation assay

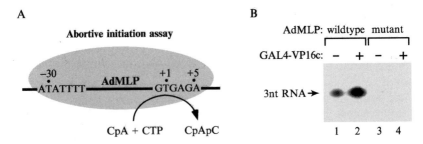

B

Fig. 4. GAL4-VP16c stimulates abortive initiation. (A) Schematic depicting the abortive initiation assay used at the AdMLP. The template strand sequence is shown along with the RNA transcript produced. See text for details. (B) The amount of abortive product increases in the presence of GAL4-VP16c. Transcription reactions contained TFIIA, TFIIB, TFIID, TFIIE, TFIIF, TFIIH, core RNA polymerase II, and 200 μM dATP in addition to the CpA and CTP, as discussed in the text. Activator was included or omitted as indicated. Reactions in lanes 1 and 2 received a plasmid containing the wild-type AdMLP. Reactions in lanes 3 and 4 received a plasmid containing the AdMLP with a mutation at the transcription start site: +1 changed from T to C on the template strand.[15] The 3 nucleotide RNA product is shown.

from unincorporated ^{32}P-labeled nucleotides. The positions at which short RNA products migrate do not correlate with size in the predicted manner. For example, a 2 nucleotide RNA migrates more slowly than a 3 nucleotide RNA. Moreover, sequence affects migration. For these reasons, it is important to compare all transcripts with RNA standards of identical size and sequence to the expected products, as described in the protocol.

For each nucleotide combination tested, a parallel reaction should be performed with a mutant promoter in which the start site position is changed such that the combination of nucleotides used will not produce a transcript. Any transcript produced from the mutant template originates from nonspecific start sites on the plasmid DNA and can be subtracted as background from the transcript produced from the wild-type template.

Method

The following protocol uses the example of the AdMLP and the nucleotides CpA and CTP, but can be used for other nucleotide combinations and promoters.

1. Nucleotide premix. Prepare a solution of 11× nucleotides by diluting stock solutions of nucleotides into TE buffer to the following concentrations: 11 mM CpA and 5.5 μM [α-^{32}P]CTP (2.5 μCi/μl). Each reaction will get 2 μl of the nucleotide premix. Both nonradioactive and radioactive CTP are added to the nucleotide premix to obtain the final concentration and specific activity.

2. Add 2 μl of nucleotide mix to each reaction containing preinitiation complexes (assembled as described earlier). Mix by flicking with finger. Incubate at 30° for 30 min.

3. Stop reactions by transferring to 70° for 3 min.

4. Add calf intestinal alkaline phosphatase (10 units) to each reaction and incubate at 37° for 20 min.

5. Add 3 μl of a stop solution containing 200 mM EDTA, 20% glycerol, and 0.025% bromphenol blue to each reaction. Vortex. Reactions can be stored at $-20°$ prior to electrophoresis.

6. Load 6 μl of each reaction on a 20% polyacrylamide (19:1) gel (20 × 20 cm; 0.4-mm spacers) containing 7 M urea and 0.5× TBE. Also load RNA size standards. Size standards are 2,3,4, and 5 nucleotide RNA oligonucleotides (Dharmacon Research) with the same sequence as the expected RNA transcripts produced from the promoter of interest. Sufficient quantities should be loaded such that the standards can be visualized by UV shadowing (1 nmol is ample).

7. Run the gel in 0.5× TBE at 17 W until the bromphenol blue is half-way down the gel.

8. The gel can be subjected to phosphorimagery, and RNA size standards can be visualized by UV shadowing with a 254-nm light after drying the gel or while it is still wet.

Interpretation of Results

In the experiment shown in Fig. 4B, we tested whether the fusion protein GAL4-VP16c, which has been shown to activate transcription *in vitro*,[24] could stimulate abortive initiation. When GAL4-VP16c was added to abortive initiation reactions, a fivefold increase in the amount of 3 nucleotide RNA product was observed (compare lane 2 to lane 1). As a control, a mutant AdMLP in which the +1 was changed from U to C (template strand) was included in lanes 3 and 4. No abortive product was observed, indicating that the CpApC transcript observed with the wild-type template was produced from the AdMLP start site. Because abortive initiation only monitors those steps in transcription that occur during production of a 3 nucleotide RNA (i.e., preinitiation complex formation and initiation), we conclude that GAL4-VP16c can activate transcription at one or both of these steps.

---

[24] J. A. Goodrich, T. Hoey, C. J. Thut, A. Admon, and R. Tjian, *Cell* **75**, 519 (1993).

Monitoring Escape Commitment

Theory

Escape commitment occurs during the formation of ternary complexes that are stable and committed to proceeding forward through the remainder of the transcription reaction.[1,2] This occurs as RNA polymerase II leaves the abortive mode of transcript synthesis, as discussed in the previous section, and synthesizes a short 4 nucleotide RNA that remains stably associated with the transcribing preinitiation complex (see Fig. 1). Consequently, one method to experimentally monitor escape commitment is to isolate ternary complexes containing 4 nucleotide RNAs. The use of gel filtration spin columns allows the separation of 4 nucleotide RNAs contained in stable ternary complexes, which pass through the column, from short RNAs that are abortively synthesized and released, which are retained in the column.[2,23] In this way, transcripts made during escape commitment can be monitored independently from those made during initiation.

The primary consideration in monitoring escape commitment is how to limit transcript synthesis to a 4 nucleotide RNA. The sequence of some promoters will naturally halt transcription at +5 if only a subset of nucleotides is added, as was the case with the IL-2 promoter.[18] When characterizing escape commitment at the AdMLP, we made a mutation at +5 (+5mt AdMLP) such that the longest RNA that could be made was 4 nucleotide when transcription was initiated with ApC, UTP, and CTP (Fig. 5A).[1] Of

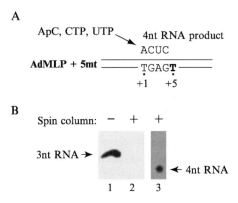

Fig. 5. Escape committed complexes contain 4 nucleotide RNAs stably bound. (A) The sequence of the template strand of the AdMLP +5mt is shown along with the 4 nucleotide RNA produced. (B) Escape commitment results in stable ternary complexes with 4 nucleotide RNA bound. Reactions were passed over G25 spin columns as indicated. Lane 3 is a longer exposure of lane 2. The positions of 3 and 4 nucleotide RNA products are indicated. Adapted from Kugel and Goodrich.[2]

the promoters studied to date, escape commitment depends on the length of the transcribed RNA (4 nucleotides) and not on the sequence of the promoter.[2] Therefore, mutations that limit transcript synthesis to 4 nucleotide do not affect the characteristics of escape commitment.

Method

The following experiment describes analyzing escape commitment at the +5mt AdMLP.

1. Nucleotide premix. Prepare a solution of $11\times$ nucleotides by diluting stock solutions of nucleotides into TE buffer to the following concentrations: 11 mM ApC, 1.1 mM UTP, and 5.5 μM [α-^{32}P]CTP (2.5 μCi/μl). Each reaction will get 2 μl of the nucleotide premix. Both nonradioactive and radioactive CTP are added to the nucleotide premix to obtain the final concentration and specific activity.

2. Add 2 μl of nucleotide premix to each reaction containing preinitiation complexes (assembled as described in an earlier section). Mix by flicking with finger. Incubate at 30° for 30 min.

3. While transcription is occurring, equilibrate G25 gel filtration spin columns (Amersham Pharmacia). To do so, snap off the bottom of the column and place in a 1.5-ml tube. Spin at 735 g for exactly 1 min. Remove the column to a new 1.5-ml tube. Add 100 μl of a solution consisting of 50 μl DB(100), 25 μl RM4X, and 25 μl ddH$_2$O to each column. Spin at 735 g for exactly 1 min. Place the equilibrated column into a new 1.5-ml tube in which to collect the eluate.

4. Pipette each transcription reaction on the top of an equilibrated G25 spin column. Spin at 735 g for exactly 2 min.

5. Heat the eluates at 70° for 3 min.

6. Treat the eluates with calf intestinal alkaline phosphatase (10 U) at 37° for 20 min.

7. Add 3 μl of stop mix containing 200 mM EDTA, 20% glycerol, and 0.025% bromphenol blue. Vortex.

8. Load 8–10 μl of reaction on a 20% polyacrylamide (19:1) gel (20 × 20 cm; 0.4-mm spacers) containing 7 M urea and 0.5× TBE. Also load RNA size standards, as described in the abortive initiation section.

9. Run the gel in 0.5× TBE at 17 W until the bromphenol blue is half-way down the gel.

10. Dry gel and visualize transcript RNA by phosphorimagery and size standard RNAs by UV shadowing with 254-nm light.

Interpretation of Results

Data illustrating the aforementioned technique are shown in Fig. 5B. As shown in lane 1, large quantities of 3 nucleotide RNA are visible when the reaction was not passed through a spin column; however, 3 nucleotide RNA is not detected in the eluate from a spin column, as shown in lane 2. This indicates that the 3 nucleotide RNA was abortively produced and released and not part of stable ternary complexes. When lane 2 was exposed longer, 4 nucleotide RNA that eluted from the spin column as part of ternary complexes could be seen. This RNA was produced during escape commitment. The longer exposure was necessary because the 4 nucleotide RNA was produced only once per actively transcribing complex, whereas the 3 nucleotide RNA was produced and released multiple times per complex.

Quantitation of Template Usage

Theory

It is common that during *in vitro* transcription in mammalian systems only a fraction of preinitiation complexes give rise to full-length RNA products. An estimation of this fraction is obtained from calculating template usage under single round transcription conditions. Template usage is defined as the moles of transcript RNA that are produced per mole of template DNA added to the reaction. Under multiple round conditions and when monitoring abortive products, it is possible for more than one RNA molecule to be produced per template DNA. Therefore, under these conditions, the template usage calculation merely provides insight into the efficiency of the transcription reaction over time.

Template usage calculations are useful because they can reveal how specific factors and conditions affect the *in vitro* transcription reaction. For example, we have shown that TFIIE and TFIIH increase the fraction of complexes that produce a full-length RNA (as determined by template usage) under single round conditions.[17]

Method

The following protocol used to calculate template usage assumes the RNA transcript is labeled with $[\alpha\text{-}^{32}P]CTP$. It is important to do this for every experiment performed to account for decay of the radioactivity over time.

1. Save a portion of the nucleotide premix used to initiate transcription. Make serial dilutions of the nucleotide premix in the range of 1:1000 to 1:100,000.

2. Spot 10 μl of each dilution of Whatman paper and expose to a phosphorimager screen along with the transcription experiment.
3. Calculate the moles of CTP in each spot. Remember that both radioactive and nonradioactive CTP are included in the nucleotide premix.
4. Quantitate the spots using phosphorimagery and calculate the phosphorimager units per mole of CTP.
5. Quantitate the RNA band using phosphorimagery.
6. Calculate the moles of CTP incorporated into transcript RNA. For this calculation, divide the phosphorimager units in the RNA band (value from step 5) by the phosphorimager units per mole of CTP (value from step 4).
7. Calculate the moles of RNA produced. Divide the value from step 6 (moles of CTP in the transcript) by the number of cytosines in the transcript RNA.
8. To calculate template usage, divide the moles of RNA produced in the transcription reaction (value from step 7) by the moles of DNA template added to the reaction.

Interpretation of Results

It has been widely observed that template usage values for *in vitro* transcription by RNA polymerase II are quite low. Values for a single round of transcription are usually below 10% and often approximate 1%. This indicates that a large fraction of the preinitiation complexes abort prior to the production of a full-length RNA and that the transcription reaction branches between functional and aborted pathways during at least one point in the reaction.[1,17]

[57] Site-Specific Protein-DNA Photocross-Linking
of Purified Complexes: Topology of the RNA
Polymerase II Transcription Initiation Complex

By Diane Forget and Benoit Coulombe

The initiation of mRNA synthesis requires the formation of a preinitiation complex containing RNA polymerase II (RNAPII) and the general initiation factors TFIID (or TBP), TFIIB, TFIIE, TFIIF, and TFIIH on promoter DNA.[1] Because many general initiation factors and RNAPII are multisubunit proteins, the preinitiation complex can comprise up to

50 polypeptides. Some of these are targeted to the promoter through direct interactions with specific sequence elements (such as the interaction of TBP with the TATA box), whereas other factors are recruited mostly through protein–protein interactions with these DNA-bound factors. In the presence of ATP, the preinitiation complex melts promoter DNA in the region of the transcriptional initiation site, thereby making the template strand available for the initiation of RNA synthesis.

Overview of the Procedure

Protein–DNA photocross-linking has proved to be the method of choice to analyze the molecular organization and topology of large nucleo-protein complexes such as those involved in the transcription reaction. In the past, our laboratory used cross-linking probes carrying a photoreactive nucleotide at specific positions along promoter DNA to localize the components of the transcription machinery in the preinitation complex.[2–5] A possible limitation of protein–DNA photocross-linking relates to the specificity of preinitiation complex assembly. Because most general initiation factors and RNAPII have an affinity for any DNA (although lower than that for promoter DNA), it is often difficult to set up conditions that will systematically and exclusively allow for the formation of specific complexes on our various photoprobes. In order to circumvent this problem, we developed a method for cross-linking proteins to DNA in purified complexes. The overall procedure is summarized in Fig. 1. Briefly, complexes are first assembled by mixing transcription factors with a photoprobe that juxtaposes one (or a few) photoreactive nucleotide with one (or a few) radiolabeled nucleotide at a specific location in the promoter. The complexes are submitted to an electrophoretic mobility shift assay (EMSA) in a native gel that is then irradiated with ultraviolet (UV) light in order to cross-link the proteins to DNA. The specific complexes are then localized on the gel, purified by cutting out the gel slices, and processed for the identification of cross-linked polypeptides. Because this procedure helps reduce to a minimum the nonspecific cross-linking signals due to aggregation, it has

[1] B. Coulombe and Z. F. Burton, *Microbiol. Mol. Biol. Rev.* **63**, 457 (1999).
[2] F. Robert, D. Forget, J. Li, J. Greenblatt, and B. Coulombe, *J. Biol. Chem.* **271**, 8517 (1996).
[3] D. Forget, F. Robert, G. Grondin, Z. F. Burton, J. Greenblatt, and B. Coulombe, *Proc. Natl. Acad. Sci. USA* **94**, 7150 (1997).
[4] F. Robert, M. Douziech, D. Forget, J. M. Egly, J. Greenblatt, Z. F. Burton, and B. Coulombe, *Mol. Cell* **2**, 341 (1998).
[5] M. Douziech, F. Coin, J. M. Chipoulet, Y. Arai, Y. Ohkuma, J. M. Egly, and B. Coulombe, *Mol. Cell Biol.* **20**, 8168 (2000).

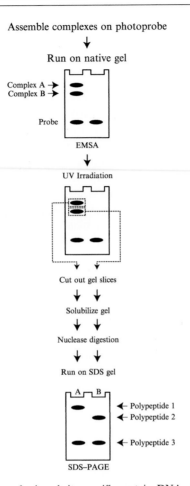

FIG. 1. Overall scheme for in-gel site-specific protein–DNA photocross-linking.

allowed us to define with more precision the topological organization of the RNAPII preinitiation complex.

Preparation of Photoprobes

General Considerations

The nucleotide derivative 5-(N-(p-azidobenzoyl)-3-aminoallyl)-dUTP (AB-dUTP or N_3R-dUTP) possesses a side chain that places a photoreactive nitrene in the major groove of the DNA helix 10 Å away from the

DNA backbone.[6] For this reason, the cross-linking of a polypeptide to the photoprobe does not require a direct interaction of the polypeptide with the DNA helix.

Chemical Synthesis of Photoreactive Nucleotide AB-dUTP

A volume of 100 μl of a 100 mM 4-azidobenzoic acid N-hydroxysuccinimide solution (ABA-NSH) in dimethylformamide (DMF) is added to 1 ml of 2 mM 5-(N-(3-aminoallyl))-dUTP (5-aa-dUTP) in 100 mM sodium borate, pH 8.5, and the reaction is incubated at room temperature for 3 h. During this time the reaction mixture frequently forms a precipitate that can be discarded. The reaction mixture is then applied to 0.2 ml DEAE-Sephadex A-25 in a 10-ml disposable column equilibrated in 100 mM TEAB, pH 8.5 (triethylammonium bicarbonate buffer). The column is washed six times with 1 ml of the same buffer and eluted with a step gradient of 0.2 to 1.0 M TEAB, pH 8.5, at intervals of 0.02 M (1 ml per step fraction). Fractions of 1 ml are collected, evaporated to dryness by vacuum centrifugation, and dissolved in 100 μl of deionized water. A volume of 2 μl of each fraction is analyzed by thin layer chromatography on polyethyleneimine (PEI)-cellulose plates (J. T. Baker). The chromatogram is run in 1 M LiCl, and the dried plates are visualized with UV (254 nm). R_f values for 5-aa-dUTP and AB-dUTP are 0.54 and 0.098, respectively. Concentrated stocks of AB-dUTP are stable for several years at $-70°$ when protected from ambient light.

Enzymatic Synthesis of Photoprobes

A schematic representation of the procedure for the synthesis of photoprobes is shown in Fig. 2. In this example, one photoreactive nucleotide is placed at position +1 and three radiolabeled guanosines at positions −1, −3, and −4 of the adenovirus major late promoter. Annealing of both specific and upstream primers to single-stranded templates is performed as follows: 500 ng of single-stranded (ss) DNA (approximately 0.5 pmol) is mixed with 40 ng (approximately 5 pmol each) of both specific and upstream primers, 1 μl of 10 $\times$ buffer A (300 mM Tris–HCl, pH 8.0, 500 mM KCl, and 70 mM MgCl$_2$; freshly prepared) is added, and the volume made up to 10 μl with deionized water. The reaction is mixed, incubated for 5 min at 90°, and then for 30 min at room temperature.

Incorporation of the photoreactive AB-dUTP and the radiolabeled nucleotide is achieved via primer extension with T4 DNA polymerase. From this point on, all manipulations must be carried out under reduced light

---

[6] B. Bartholomew, G. A. Kassavetis, B. R. Braun, and E. P. Geiduschek, *EMBO J.* **9,** 2197 (1990).

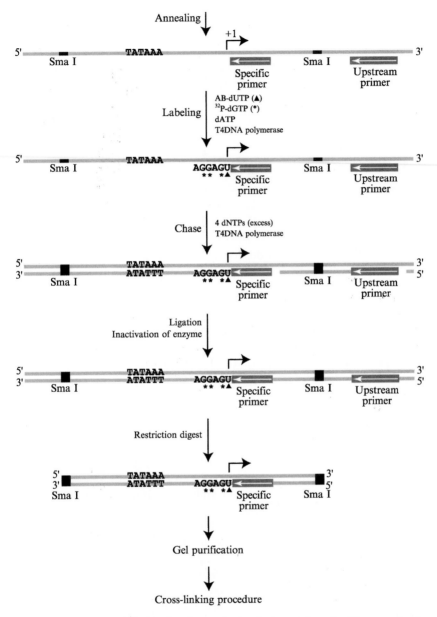

FIG. 2. Experimental design for the synthesis of photoprobes. In this example, the photoprobe contains one photonucleotide (U) at position +1 and three radiolabeled guanosines (G) at positions −1, −3, and −4 of the adenovirus major late promoter.

conditions. Bovine serum albumin (BSA) (10 mg/ml, 0.5 μl), AB-dUTP (1 μl; the amount of the nucleotide derivative to be added is determined empirically for each preparation and is generally between 0.5 and 2 μl), 20 μCi of the appropriate [α^{32}P]dNTP (3000 Ci/mmol) ([α^{32}P]dGTP for the example in Fig. 2), 5–10 U of T4 DNA polymerase, and 10 × buffer A (1 μl) are then added and the volume is made up to 20 μl with deionized water. The reaction is incubated for 30 min at room temperature and is then chased with all four dNTP by adding 5 μl of dNTP mix (10 mM each dATP, dCTP, dGTP, and dTTP in 1 × buffer A) and incubating for 5 min at room temperature followed by 20 min at 37°. The addition of dNTP in large excess is crucial because it is necessary to limit the incorporation of radiolabeled and photoreactive nucleotides during the extension of the photoprobes. Following probe synthesis, nicks at the 5' end of the primers are repaired by the addition of T4 DNA ligase (5 U) and ATP (1 mM final concentration) and incubating at room temperature for 1 h. The reaction is then heated at 65° for 20 min in order to inactivate the ligase.

The synthesized DNA is digested with restriction endonucleases (10–20 U) for 90 min in order to release the photoprobe (Sma 1 in the example shown in Fig. 2). The probe is purified on a native gel by adding 5 μl of a 6X gel loading solution (0.25% bromophenol blue, 0.25% xylene cyanol, and 30% glycerol in deionized water) to the digest and electrophoresing through a native 8% polyacrylamide gel (40:1 acrylamide:bis) in 1X TBE buffer. The TBE buffer is prepared as a 5X stock by mixing 54 g Tris-base, 27.5 g boric acid, and 20 ml 0.5 M EDTA, pH 8.0, in 1 liter of deionized water. The gel is run at 150 V for about 1 h in 1X TBE buffer.

After the run the gel sandwich is removed from the gel box, and plates are separated such that the gel remains on one of them. The gel/glass plate is then wrapped in plastic and aluminum foil and moved to a dark room (a conventional red light can be used). Kodax X-OMAT AR film is placed on a clean bench, the foil is removed, and the gel is placed on the film with the glass plate facing up (e.g., gel side down) for 2 to 5 min. The gel is removed and rewrapped with the foil and the film is developed (an example of the autoradiogram of a gel used for photoprobe purification is shown in Fig. 3). The band corresponding to the photoprobe can be identified easily because the size of the fragment generated by the restriction enzyme is known. Using a scalpel blade, the film is cut so that the square piece containing the band corresponding to the photoprobe is removed. This operation leaves the film with a window at the position of the photoprobe. The film is superimposed on the gel and an ink marker is used to mark the square corresponding to the photoprobe. The gel slice is then cut out with a clean scalpel blade and chopped into small pieces (six to

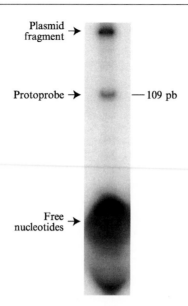

Plasmid fragment →

Protoprobe → — 109 pb

Free nucleotides →

FIG. 3. An autoradiogram of a gel used for the purification of photoprobes. The position of the DNA fragment carrying both photoreactive and radiolabeled nucleotides at a specific location is shown.

eight fragments). These are placed in a 1.5-ml microcentrifuge tube and sufficient 10 mM Tris, pH 7.9, added to completely submerge the gel (about 125–200 μl). The probe is eluted by incubating overnight at room temperature, and the liquid containing the probe is collected and purified on a microspin S-200 HR column (Amersham) in order to remove any salts and other putative contaminants. A 1-μl aliquot of the photoprobe solution is then counted by liquid scintillation, and the probe is diluted to the appropriate count number (see later) with deionized water. The probe is now ready for use and can be stored in the dark at 4° for 1–2 weeks. Fresh probes (less than a week old) give the best results.

Precautions

i. AB-NSH and AB-dUTP are manipulated under indirect lighting conditions using a 40-W incandescent lamp. The use of a standard dark room is not necessary. As a rule, we find that conditions providing just enough light to be able to work are acceptable.

ii. The specific primer must be designed in such a manner that T4 DNA polymerase only adds a few nucleotides. In the example shown in Fig. 2, incorporation during the site-specific labeling is restricted to positions −4 to +1 by omitting dCTP from the reaction. The success of this step can be monitored by analysis of the reaction products on a sequencing gel.

In-Gel Protein–DNA Photocross-Linking

Synthesis of N,N'-Bisacryloylcysamine (Bac)

Bac is a disulfide-containing analog of bis-acrylamide.[7–9] Its chemical synthesis and use in native gels are essentially as described by Naryshkin et al.[10] with minor modifications. Polyacrylamide:bac gels can be dissolved by the use of reducing agents. Acryloyl choride is highly toxic and all manipulations in this section must be performed in a fume hood. Cystamine dihydrochloride (20 g) is dissolved in 200 ml of 3 M NaOH, whereas acryloyl chloride (21.5 ml) is dissolved in 200 ml chloroform. The resulting solutions are then mixed in a 2-L flask (as acryloyl cholride reacts with water, it is added drop-by-drop). Two phases will form: an upper, aqueous phase and a lower, organic phase. The reaction is stirred magnetically for 3 min at room temperature, followed by 15 min at 50°. Immediately after the latter incubation the reaction mixture is transferred to a 2-L separating funnel and phases are allowed to separate. The lower, organic phase is then transferred to a 1-L beaker. Impure bac is precipitated by chilling on ice for 10 min and is collected by filtration in a Büchner funnel. The crude precipitate is dissolved in 150 ml chloroform in a 1-L beaker and is recrystallized by placing the beaker on a stirrer plate and stirring for 1 min at room temperature, followed by 5 min at 50°, and then placing the beaker on ice for 10 min. The precipitated bac is collected by filtration in a Büchner funnel and transferred to a 250-ml flask. The flask is sealed with Parafilm, piercing the seal several times with a syringe needle, and the flask is placed in a vacuum dessicator and dried under vacuum for 16 h at room temperature. Expect a yield of 7.0–9.0 g.

Preparation of Polyacrylamide:Bac Gels

Polyacrylamide:bac gels for the EMSA are prepared as follows. A 20% acrylamide:bac (19:1) stock solution is prepared by dissolving 19 g of acrylamide and 1 g of bac in 80 ml of water in a 200-ml beaker and stirring

[7] J. N. Hansen, Anal. Biochem. **76,** 37 (1976).
[8] J. N. Hansen, B. H. Pheiffer, and J. A. Boehnert, Anal. Biochem. **105,** 192 (1980).
[9] J. N. Hansen, Anal. Biochem. **116,** 146 (1981).
[10] N. Naryshkin, Y. Kim, Q. Dong, and R. H. Ebright, Methods Mol. Biol. **148,** 337 (2001).

for 30 min at 60° (the solubility of bac in water is increased by adding the acrylamide before adding the bac and by performing the addition at 60°). The volume is then adjusted to 100 ml with water and the solution is allowed to cool down to room temperature prior to filtering through a 0.22-μm filter unit and storing at 4° in the dark (stable for a few weeks). The gel is assembled using one glass plate that has been siliconized by applying 100 μl of Surfacil siliconizing agent and spreading evenly with a Kimwipe. A 4.5% polyacrylamide:bac gel in 0.5× TBE buffer is prepared by mixing 11.25 ml of an acrylamide:bac (19:1) stock solution, 5 ml of 5× TBE, 100 μl MgCl$_2$, and 33.7 ml water and preheating the slab gel assembly and the gel mixture at 50° in an incubator for 30 min. Polymerization is initiated by adding 250 μl TEMED and 125 μl freshly prepared 10% ammonium persulfate, and the gel is poured immediately into the slab gel. Allow 20 min for polymerization at 50° (the TEMED and ammonium persulfate concentrations are critical variables in the preparation of polyacrylamide:bac gels). Finally, the gel and the 0.5× TBE reservoir containing 2 mM MgCl$_2$ are prechilled by placing them in a 4° cabinet for 3 h (the polyacrylamide:bac is stable for up to 72 h at 4°, but it is better to use it fresh).

Electrophoresis of Protein–DNA Complexes in Native Gels

The reactions are assembled by mixing recombinant TBP (80 ng), recombinant TFIIB (80 ng), RAP30 (300 ng), RAP74 (600 ng), calf thymus RNA polymerase II (300 ng), TFIIE34 (160 ng), and TFIIE56 (380 ng) in a total of 20 μl of buffer G (12 mM HEPES, pH 7.9, 60 mM KCl, 0.12 mM EDTA, 8 mM MgCl$_2$, 50 ng/ml BSA, 5 mM 2-mercaptoethanol, and 12% glycerol) and then adding 2.5 mg/ml poly(dIdC.dIdC) (1 μl) and 6000 cpm of the photobrobe per [α^{32}P]dNTP residue incorporated. All work with the photoprobe should be performed under reduced lighting conditions. The binding reactions (21 μl) are incubated for 30 min at 30°. The EMSA are performed as described previously.[11] The 4.5% acrylamide:bac gels are run in a 4° cabinet for 60 min at 400 V.

UV Irradiation of Gels

UV irradiation of protein–DNA complexes is performed in the gel. One glass plate is removed and the gel is irradiated immediately with a 254-nm UV light for 10 min using a Hoefer UVC 500 ultraviolet cross-linker.

---

[11] B. S. Wolner and J. D. Gralla, *Mol. Cell Biol.* **20**, 3608 (2000).

Localization and Isolation of Protein–DNA Complexes

In order to localize and isolate the complexes of interest, one of the glass plates is removed and the gel is covered with Whatman paper. The other glass plate is then removed and the gel is covered with plastic wrap. The gel is exposed to a phosphorimager screen overnight and the image printed. Using a scalpel blade, the print is then cut so that the square piece containing the band corresponding to the complex is removed. This operation leaves the print with a window at the position of the complex of interest. The print is then placed on the gel, and the gel slice corresponding to the square cut is excised with a clean scalpel blade. The excised gel is transferred to a 1.5-ml microcentrifuge tube, taking care to avoid carrying any pieces of Whatman paper and plastic wrap.

Solubilization of Complexes

Gel slices are solubilized by the addition of 10 μl of 1 M dithiothreitol (DTT) (about 0.4 M final) and heating for 10 min at 37° (2–4 M 2-mercaptoethanol can be substituted for 1 M DTT). Then 40 μl of ND buffer (20 mM HEPES, pH 7.9, 100 mM KCl, 0.2 mM EDTA, 0.2 mM EGTA, and 20% glycerol) containing 0.4 M DTT is added and the sample is heated at 37° for 25 min.

Nuclease Digestion and SDS–PAGE Analysis

The gel slice is solubilized by adding 13 μl of 0.5 U/μl DNase in 30 mM CaCl$_2$ and incubating at 37° for 20 min. A volume of 3.9 μl 10% SDS is then added and the solution is boiled for 3 min. Finally, 2.5% acetic acid/15 mM ZnCl$_2$ (5.2 μl) and 80,000 U/ml S1 nuclease (2.6 μl) are added and the reaction is incubated at 37° for 20 min. The reaction is stopped by adding 15 μl 5× loading buffer (80 mM Tris, pH 6.8, 12.5% glycerol, 2.5% SDS, 0.9 M 2-mercaptoethanol, and 0.2% bromophenol blue) and the sample is boiled for 5 min prior to loading on a standard SDS–PAGE gel along with prestained molecular weight markers. The gel is run at 30 mA while the proteins are in the stacking gel and at 55 mA when they are in the separating gel. Detailed procedures for SDS–PAGE electrophoresis have been described.[12] Finally, the gel is transferred to Whatman paper and dried under vacuum. The dried gel is exposed to Biomax screens and films.

---

[12] J. Sambrook, E. F. Fritsch, and T. Maniatis, in "Molecular Cloning: A Laboratory Manual." Cold Spring Harbor Laboratory, Cold Spring Harbor, NY, 1989.

Analysis of RNA Polymerase II Preinitiation Complex

The power of the in-gel photocross-linking procedure to analyze the RNAPII preinitiation complex is illustrated in the example shown in Figs. 4 and 5. Using calf thymus RNAPII and recombinant TBP, TFIIB, RAP74, RAP30, TFIIE56, and TFIIE34 for preinitiation complex assembly on photoprobes −39/−40 (Fig. 4) and +1 (not shown), EMSA analysis revealed the formation of two distinct complexes in the native gels. Assembly of these complexes is promoter specific because a mutation in the TATA box (TATAAA to TAGAGA; not shown) or the omission of TBP in the assembly mixture (Fig. 4, compare + and −) abolishes the formation of the two complexes. The band corresponding to each complex assembled on photoprobes −39/−40 and +1 is excised from the gels and analyzed as described earlier. RPB2, RAP74, TFIIE34, and RAP30 crosslink to positions −39/−40, and RPB1 and RPB2 to position +1 using both

FIG. 4. Electrophoresis of protein–DNA complexes in a native gel. Complexes were assembled with calf thymus RNAPII, TFIIB, TFIIF, and TFIIE in either the presence (+) or the absence (−) of TBP on photoprobe −39/−40. Two complexes, A and B, are resolved using the electrophoretic mobility shift assay (EMSA).

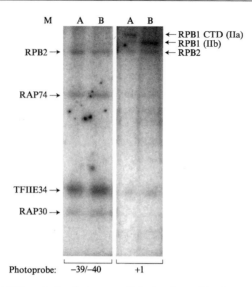

FIG. 5. SDS–PAGE analysis of the cross-linked polypeptides in complexes A and B. Complexes A and B were assembled on photoprobes −39/−40 and +1. Gels were irradiated with UV light, and gel slices containing each complex were excised and processed as described in the text. Although no difference is observed in polypeptides that cross-link to photoprobe −39/−40, the form of RPB1 that cross-linked to position +1 varies when complex A is compared to complex B. The difference in the molecular weight of RPB1 indicates that the IIa form of RNAPII (e.g., with a hypophosphorylated CTD) is found in complex A, whereas the IIb form (e.g., without the CTD due to its proteolysis during purification) is present in complex B.

complexes A and B. Interestingly, the IIa form of RPB1 with an intact hypophosphorylated CTD (about 220 kDa) is found in the complex of lower mobility (complex A), whereas the IIb form of RPB1 with a proteolyzed CTD (about 180 kDa) is found in the complex of higher mobility (complex B). This result indicates that one major difference between complexes A and B is the form of RNAPII that is associated with the general initiation factors that form the preinitiation complex. In support of this conclusion, the preparation of calf thymus RNAPII used in our experiments contains approximately equal amounts of the IIa and IIb forms of RNAPII.

Acknowledgments

We are grateful to our colleagues who encouraged us to develop an in-gel photocross-linking method for the analysis of RNA polymerase II complexes. We also thank the members of our laboratory for helpful discussions, Diane Bourque for art work, and Will Home for critical reading of the manuscript.

[58]   Assay of Promoter Melting and Extension of mRNA: Role of TFIIH Subunits

By FRÉDÉRIC COIN and JEAN MARC EGLY

In order to gain access to the nucleotides of the template strand, DNA-dependent RNA polymerase II (RNA pol II) requires DNA strand melting prior to initiation, promoter escape, and elongation. The general transcription factor TFIIH plays an active role in mRNA extension. Interestingly, further studies demonstrated its involvement in nucleotide excision repair but also recently in RNA pol I transcription. The study of the role of TFIIH in transcription was hampered by the absence of functional assays and by the difficulty in obtaining large quantities of a highly purified factor. Human TFIIH was reconstitued in insect cells, allowing the study of the role of each enzymatic subunit of this factor in transcription. This article describes new procedures developed to facilitate the purification of TFIIH from a low amount of cells, as well as several assays to analyze each step of transcription.

The transcription of eukaryotic protein-coding genes involves complex regulation of RNA pol II activity in response to physiological and developmental conditions. TFIIH is one of the multisubunit complexes involved in this regulation. The identification of the components of this factor was a breakthrough in the understanding of its function; it plays a central role in protein coding and ribosomal genes transcription, DNA repair, and possibly cell cycle control.[1,2] Human TFIIH contains nine subunits ranging from 89 to 32 kDa in size (Fig. 1A),[3,4] with some of them having enzymatic activities: the two largest subunits, XPB and XPD, exhibit DNA unwinding (helicase) and DNA-dependent ATPase activities. Cdk7, cyclin H, and MAT1 form a kinase complex found either free or associated with TFIIH. This complex, called CAK (for cdk activating kinase), is capable of phosphorylating the carboxyl-terminal domain (CTD) of the largest subunit of the RNA pol II as well as cyclin-dependent kinases such as cdk1/cdc2, cdk2, and cdk4.[5] Three human genetic disorders are connected with mutations in XPB and XPD: xeroderma pigmentosum (XP), Cockayne syndrome

[1] J. H. J. Hoeijmakers, J. M. Egly, and W. Vermeulen, *Curr. Opin. Genet. Dev.* **6**(1), 26 (1996).
[2] S. Iben *et al.*, *Cell* **109**, 297 (2002).
[3] J. Q. Svejstrup, P. Vichi, and J. M. Egly, *Trends Biochem. Sci.* **20**(10), 346 (1996).
[4] P. Frit, E. Bergmann, and J. M. Egly, *Biochimie* **81**, 27 (1999).
[5] T. Riedl and J. M. Egly, *Gene Expr.* **9**, 3 (2000).

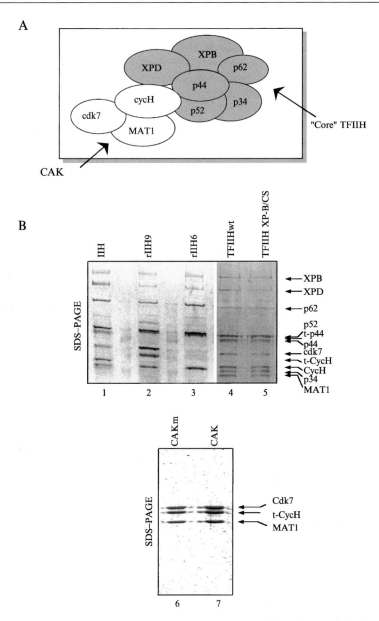

FIG. 1. Purification of TFIIH through cobalt or immunoaffinity columns. (A) Schematic view of TFIIH with its two subcomplexes: the core (gray) and the CAK (white). (B) Five hundred nanograms of rIIH9 (lane 2), rIIH6 (lane 3), and immunopurified TFIIHwt or from XP-B/CS patient cell lines (lanes 4 and 5, respectively) was loaded on a 11% polyacrylamide gel and analyzed by silver staining. Two hundred nanograms of either mutant (CAKm, lane 6)

(CS), and trichothiodystrophy (TTD). Patients are characterized by photosensitivity (XP, CS, and TTD), pigmentation abnormalities of the skin in sun-exposed areas (XP), high incidence of skin cancer (XP), neurological and developmental problems (XP, CS), dwarfism (TTD), and brittle hairs and nails (TTD).[6] Interestingly, several clinical features, such as developmental and neurological abnormalities, are difficult to explain on the basis of the NER defect. Indeed, the involvement of a transcription factor in a repair pathway was the first indication that XP, TTD, and CS could result from a defect in both NER and transcription.[7]

mRNA synthesis has been dissected into several distinct steps, including preinitiation complex assembly, transcription initiation, promoter escape, elongation, and termination. Studies indicate that TFIIH and its ATP-consuming enzymatic subunits play a role in at least the three first steps. mRNA synthesis requires ATP at a number of transcriptional steps,[8,9] including formation of an open initiation transcription complex, synthesis of an RNA transcript,[10] and phosphorylation of the largest RNA pol II subunit.[11] In 1989, Sawadogo and Roeder[9] demonstrated that AMP-PNP, an analog of ATP, cannot be used to initiate RNA synthesis. A few years later, the Conaways[12] showed that the rat transcription factor δ, which is similar to human TFIIH, possesses an ATPase activity, strongly stimulated in the presence of a DNA fragment encompassing the adenovirus major late promoter (Ad$_2$MLP) TATA box.

The ability to reconstitute and produce TFIIH in the baculovirus/insect cells system[13] and to purify this factor rapidly from various normal and patients-derived cell lines[14–16] allows us to systematically investigate the role of each TFIIH subunit in transcription. This article describes the

[6] D. Bootsma et al., in "The Genetic Basis of Human Cancer" (B. Vogelstein and K. W. Kinzler, eds.), p. 245. McGraw-Hill, New York, 1998.
[7] D. Bootsma and J. H. J. Hoeijmakers, Nature 363, 114 (1993).
[8] D. Bunick et al., Cell 29, 877 (1982).
[9] M. Sawadogo and R. G. Roeder, J. Biol. Chem. 259, 5321 (1984).
[10] Y. Jiang, M. Yan, and J. D. Gralla, Mol. Cell. Biol. 16, 1614 (1996).
[11] H. Lu et al., Nature, 358, 641 (1992).
[12] R. C. Conaway and J. W. Conaway, Proc. Natl. Acad. Sci. USA 86, 7356 (1989).
[13] F. Tirode et al., Mol. Cell. 3, 87 (1999).
[14] G. LeRoy et al., J. Biol. Chem. 273, 7134 (1998).
[15] G. S. Winkler et al., J. Biol. Chem. 273, 1092 (1998).
[16] F. Coin et al., EMBO J. 18, 1357 (1999).

or wild-type (CAK, lane 7) kinase complexes containing, in addition to a tagged cyclin H (t-CycH) and MAT1, either mutant or wild-type cdk7, respectively, was loaded on a 15% polyacrylamide gel and Coomassie stained (SDS–PAGE). HeLa TFIIH (IIH) is used as a qualitative control lane 1.[17]

methodology developed to study the role of TFIIH in gene expression *in vitro* and exposes the main conclusions of these experiments. The authors point out that this manuscript is not a review on transcription initiation and promoter escape, but a review on methods. As thus, experiments described here are illustrated by results obtained in our laboratory. Readers will find general and detailed information on the topic in Dvir *et al.*[18]

TFIIH Purification

The availability of wild-type and mutants TFIIH is of prime importance for studying the involvement of TFIIH in transcription initiation and promoter escape. Such TFIIH can be purified from patient cell lines harboring XP, TTD, or CS phenotypes or from baculovirus-infected insect cells. The purpose is to set up fast and efficient methods to purify TFIIH from a reduced amount of cells.

Cell Extracts

Preparation of Human Whole Cell Extracts (WCE)

To allow purification of TFIIH, extracts are prepared by mechanical lysis of the cells. This section describes the preparation of HeLa whole cells extracts, while a similar protocol has been adapted in our laboratory for transformed patient cell lines. A list of the buffers used is given in Table I.

a. All steps should be carried out on ice or at 4°. Cells (0.9×10^9) are pelleted (5000 rpm, 5 min) and resuspended in 20 ml cold $1\times$ phosphate-buffered saline (PBS, Sigma). The volume of the pellet (V1) is measured and the cells are washed with 100 ml of $1\times$ PBS and pelleted (5000 rpm, 5 min).

b. The pellet is resuspended in $4 \times$ V1 volumes of TE buffer supplemented with 5 mM dithiothreitol (DTT) and kept on ice for 20 min. min. Cells are lysed with 20 strokes in a 25-ml homogenizer (Kontes Glass Co., USA), and the lysate is diluted with $4 \times$ V1 volumes of buffer A under gentle stirring, which prevents genomic DNA fragmentation and allows efficient precipitation in the following steps. One V1 volume of $4 M$ (saturated) ammonium sulfate (AS) is added slowly to the lysate and stirred for 30 min before being ultracentrifuged (50,000 rpm, 2h30, in a R60Ti Beckman rotor).

[17] M. Gerard *et al.*, *J. Biol. Chem.* **266**, 20940 (1991).

[18] A. Dvir, J. W. Conaway, and R. C. Conaway, *Curr. Opin. Genet. Dev.* **11**, 209 (2001).

c. The supernatant (volume, V2) is clarified through gauze, and proteins are precipitated with 0.33 g AS per milliliter supplemented by 1 μl 10 M NaOH/g AS. The mixture is stirred gently for 30 min and centrifuged (30,000 rpm, 20 min, in a R45Ti rotor). The pellet is resuspended in 1/10 V2 of buffer B and dialyzed (dialysis tubing-visking 7-30/32″, Medicell Intl) 8 h against 1 liter of buffer B. WCE is then centrifuged (11,000 rpm, 10 min in a R45Ti rotor) to remove aggregates, snapfrozen, and stored at $-80°$ or used immediately for purification of TFIIH. The total protein content is measured by the Bradford method (Bio-Rad).

Preparation of Baculovirus-Infected Insect Cell Extracts

Typically, 1×10^8 High 5 (H5) cells are used for one preparation of 10 μg of TFIIH. Insect cells are infected with combinations of baculoviruses expressing XPB, XPD, p62, p52, His p44, p34, cdk7, cyclinH, or MAT1 (the P44 subunit is histidine tagged for further purifications on a metal chelate affinity column). To study the role of CAK on TFIIH transcription, recombinant TFIIH containing the six subunits of the core (XPB, XPD, p62, p52, His-p44, p34) and the free CAK complex (cdk7, His-cyclinH, or MAT1) are produced separately. The cells are collected 48 h after infection, washed once with 1× PBS 30% glycerol, dounced in buffer C supplemented with 5 mM 2-mercaptoethanol, 0.5 mM phenyl-methyl sulfonyl fluoride (PMSF), and 1X protease inhibitor cocktail [anti-pain, chymostatin, pepstatin (Sigma) and aprotinin, leupeptin (Roche)],

TABLE I
BUFFERS

A	50 mM Tris–HCl, pH 7.9, 50% glycerol, 10 mM MgCl$_2$, 25% sucrose, 2 mM DTT
B	50 mM Tris–HCl, pH 7.9, 15% glycerol, 5 mM MgCl$_2$, 40 mM (NH$_4$)$_2$SO$_4$, 0.2 mM EDTA, 1 mM DTT
C	20 mM Tris–HCl, pH 7.9, 20% glycerol, 0.1% Nonidet P-40, 150 mM NaCl
D	10 mM Tris–HCl, pH 7.8, 17% glycerol (v/v), 0.5 mM DTT, 5 mM MgCl$_2$
E	50 mM Tris–HCl, pH 7.9, 15% glycerol, 5 mM MgCl$_2$, 0.2 mM EDTA, 1 mM DTT, 50 mM KCl
F	50 mM Tris–HCl, pH 7.9, 20% glycerol, 300 mM KCl
G	25 mM Tris–HCl, pH 7.9, 17% glycerol (v/v), 0.5 mM EDTA, 0.2 mM DTT, 5 mM MgCl$_2$
H	20 mM Tris–HCl, pH 7.9, 250 mM NaCl
I	50 mM Tris–HCl, pH 7.9, 10% glycerol, 0.1 mM EDTA, 50 mM KCl, 0.5 mM DTT
J	20 mM Tris–HCl, pH 7.9, 2 m EDTA, 200 mM NaCl
K	10 mM Tris, HCl pH 7.9, 1 mM EDTA, 10 mM NaCl
TE	10 mM Tris–HCl, pH 7.9, 1 mM EDTA

and then centrifuged at 11,000 rpm for 30 min. The supernatant is used directly for TFIIH purification or frozen at $-80°$.

Chromatography

Principle

TFIIH was originally purified through classical protocols starting from HeLa WCE and involving several chromatographic steps.[17] Such technology, very useful for the preparation of large quantities of pure TFIIH from HeLa cell lines, is not adapted for the purification from patients cell lines. These cells are usually very difficult to grow in culture and/or necessitate expensive mediums to obtain a small amount of cells. Then, a procedure to immunopurify TFIIH was developed using a monoclonal antibody and a peptide capable of displacing TFIIH from this antibody.[14–16] In parallel, we succeeded in expressing the nine (or less) subunits of human TFIIH in insect cells followed by subsequent purification with an antibody or metal chelate affinity column.[13]

Purification

Heparin Column

HeLa or Patient Cell Lines. WCE (1×10^9 cells) is centrifuged (5000 rpm, 15 min, $4°$) to remove aggregates that could obstruct the column and is applied onto a heparin Ultrogel (hep-UG A4R Sepracor) column to eliminate cellular proteins and DNA [1 ml of gel in 10 ml Poly-Prep chromatography columns (Bio-Rad), gravitational flow] equilibrated with buffer D containing 100 mM KCl. After washing with 2 resin volumes of buffer D, the elution is performed sequentially with 2 resin volumes of buffer D containing 0.22, 0.40, and 1.00 M KCl. The 0.40 M KCl hep–UG eluate is dialyzed to equilibrium 2×2 h against 1 liter of buffer E.

H5 Insect Cells. Insect cell crude extracts are first loaded onto a hep–UG column [1 ml of gel in 10 ml Poly-Prep chromatography columns (Bio-Rad) gravitational flow] preequilibrated in buffer C containing 0.1 M KCl. H5 extracts are fractionated onto the hep–UG column to remove part of the endogeneous insect cell proteins and incomplete TFIIH subcomplexes (considering the interactions essential to a six-subunit complex, it is possible statistically to generate up to 63 subcomplexes, each containing from one to six subunits). The column is washed three times with 1.5 resin volume of buffer C containing 0.40 M NaCl, and the proteins are eluted with 3 × 1.5 resin volume of the same buffer containing 0.50 M NaCl. Typically, 0.50 M NaCl fractions are pooled and dialyzed 2 h against either buffer E for immunoprecipitation or buffer F for cobalt affinity purification.

IMMUNOPRECIPITATION. Typically, 10–12 ml (10 mg of protein) of the dialyzed hep–UG 0.4 M (WCE) or hep–UG 0.5 M (insect cells) fractions are incubated overnight at 4° with 400 μg of the purified 1H5 anti-p44 monoclonal antibody[19] cross-linked to 200 μl of protein A–Sepharose CL-4B (Amersham Pharmacia Biotech AB). After spinning 5 min at 1500 rpm at 4°, the supernatant is removed and the resin is washed three times with 10 bead volumes of buffer G containing 0.4 M KCl and 0.1% Nonidet P-40 and twice with buffer G containing 0.05 M KCl and 0.01% Nonidet P-40. The absorbed material is eluted for 6 h at 4° in 400 μl of buffer G containing 0.05 M KCl, 0.01% Nonidet P-40, 0.2 mg/ml insulin, and 2.0 mg/ml of the peptide competitor representing the first N-terminal 17 amino acids of p44 (MDEEPERTKRWEGGYER). This elution step is generally not repeated, as nearly all TFIIH is present in the first eluate. We obtain 10 μg (25 ng/μl) of highly purified TFIIH starting with 10 mg of proteins from the hep–UG 0.40 M (HeLa) or 0.50 M (H5) fractions. TFIIH fractions are aliquoted (freezing and thawing TFIIH fractions could affect the activity of the complex) and stored at −80°.

COBALT AFFINITY COLUMN. Recombinant TFIIH from infected insect cells containing nine or six subunits can be purified by a metal chelate affinity column. Collected hep-UG 0.50 M fractions dialyzed against buffer F are incubated for 1 h at 4° with 1/40th fraction volume of Co^{2+} chelate His Bind resin (Novagen, USA). After packing the column, the resin is washed extensively with 3 × 1.5 column volume of buffer F containing 0.01 M imidazole to eliminate complexes lacking the tagged p44 subunit. Proteins are then eluted in two column volumes of buffer F containing 0.1 M EDTA. Although the nine polypeptides can be detected in the 0.01 M imidazole washes, assays of transcription have demonstrated that these fractions are inactive. This most probably reflects the presence of rIIH subcomplexes. The 0.10 M EDTA fraction is then dialyzed 2 × 2 h against buffer G. The yield is higher than for p44 immunopurification (approximately 16 μg of TFIIH at 80 ng/μl). TFIIH fractions are aliquoted and stored at −80°.

FREE CAK. CAK contains three subunits: cdk7, cyclin H, and MAT1. For purification of the free CAK complex, cyclin H has been histidine tagged.[20] The H5 cell extract is loaded onto 1 ml of Co^{2+} chelate His Bind

[19] S. Humbert et al., EMBO J. **13**, 2393 (1994).
[20] M. Rossignol et al., J. Biol. Chem. **274**, 22387 (1999).

resin (Novagen, USA). The column is then washed with 6 column volumes of buffer H containing 0.04 M imidazole, and elution is performed with 6 column volumes of buffer H containing 0.1 M imidazole. The fractions are pooled, dialyzed against buffer E containing 0.05 M KCl, and incubated for 3 h at 4° with 250 μg of antibody against cdk7 cross-linked to 250 μl of protein A–Sepharose. The beads are washed three times in buffer E containing 0.50 M KCl, and elution is performed for 12 h at 4° in 250 μl of buffer B containing 0.05 M KCl and 2.0 mg/ml of peptide competitor. rCAK is aliquoted and stored at −80°. Typically, 1 liter of culture (10^9 cells) yields 25–35 μg of kinase complex. Alternatively, a CAK complex containing an inactive cdk7 subunit (rCAKm)[21] can be purified following the same protocol.

ANALYSIS AND OBSERVATIONS. The presence of TFIIH in hep–UG fractions can be verified by SDS–PAGE and Western blotting. As mercaptoethanol is a potent inhibitor of TFIIH fixation to the cobalt column, dialysis of hep–UG fractions should be performed carefully. TFIIH purified from classical chromatographic steps (Fig. 1B, lane 1[18]), the recombinant TFIIH purified through cobalt (lanes 2 and 3) or through immunoaffinity columns (lanes 4 and 5), and the rCAK (lanes 6 and 7) are resolved by SDS–PAGE following silver staining.

Enzymatic Assays

As TFIIH was first described as a transcription factor, a runoff transcription assay is used to follow its activity during purification. The same assay was used to measure the effects of mutations in TFIIH on its transcriptional activity. The synthesis of a 309 nucleotide transcript results from a multistep reaction, including the formation of a stable preinitiation complex followed by promoter opening, first phosphodiester bond formation, promoter escape, and elongation. To further investigate the role of TFIIH and its subunits in these different steps, several assays have been developed to test abortive initiation, promoter opening, and promoter escape.

In Vitro Runoff Transcription Assay

Principle

Ad$_2$MLP is currently used as template DNA to evaluate the transcription activity of TFIIH contained in WCE preparations or purified from human or insects cells (reconstituted transcription system). In both

---

[21] J. P. Tassan *et al.*, *EMBO J.* **14,** 5608 (1995).

instances, the transcription reactions are performed in two steps: a preincubation of the DNA template with either the WCE or the various components of the basal transcription machinery (i.e., TFIIA, TFIIB, TFIID/TBP, TFIIE, TFIIF, and TFIIH, as well as RNA pol II) to favor the assembly of the transcription initiation complex and an RNA synthesis step, which starts upon the addition of ribonucleotides.

Preparation of the Ad₂MLP template

a. The DNA template is obtained by ligation of an *Eco*RI/*Bam*HI fragment, corresponding to nucleotides −677 to +33 of the Ad₂MLP in which upstream regulatory sequences are deleted (Δ−372/−34), to the *Bam*HI/*Sal*I fragment from pBR322 (New England Biolabs). The resulting fragment (648 bp) is subcloned into the *Eco*RI/*Sal*I sites of pUC19 (New England Biolabs), generating the pUC309 plasmid.[18] Bacterial cells (DH5α strain of *Escherichia coli*) are transformed with pUC309 for the large-scale preparation of plasmid DNA by the alkaline lysis method.[22]

b. After *Eco*RI/*Sal*I digestion of 1 mg of pUC309, the DNA template is purified by a sucrose density gradient [5 to 20% gradient in buffer J containing 10% Sarkosyl poured into 40-ml polyallomer tubes (Beckman)]. After the addition of EtBr (100 μl of a 10-mg/ml stock) to the solution, ultracentrifugation is carried out (30,000 rpm, 15 h, 20°) in a SW28 rotor.

c. The band corresponding to the 648-bp fragment is collected, EtBr is removed with butanol, proteins are extracted with phenol:chloroform, and DNA is precipitated with 1/10 volumes 3 M sodium acetate and 2.5 volumes ethanol. After centrifugation, the pellet is resuspended in ultrapure water.

Transcription Reaction with WCE

a. Fifty to 75 μg of WCE is incubated for 15 min at 25° with 50–70 ng of Ad₂MLP template in buffer 1 with 5 mM MgCl₂ in a 20-μl final reaction volume.

b. ATP, GTP, and UTP (250 μM each), 10 μM CTP, 4 μCi [α-³²P]CTP (400 Ci/mmol, Amersham), and buffer I up to 25 μl in 6.5 mM MgCl₂ final concentration are added to the preincubation reaction before incubation for 45 min at 25°. As radiolabeled nucleotides are now incorporated, all subsequent steps should be carried out behind a Plexiglas shield.

---

[22] J. Sambrook, E. Fritsch, and T. Maniatis, *"Molecular Cloning: A Laboratory Manual."* Cold Spring Harbor Laboratory Press, Cold Spring Harbor, NY, 1989.

c. Transcription stop buffer (400 μl) containing 50 mM sodium acetate, pH 5.2, 0.5% SDS, and 50 μg/ml tRNA is added to the reaction. Extraction is performed by adding 300 μl phenol, vortexing, adding 300 μl chloroform, vortexing again, and spinning in a bench centrifuge for 5 min. After collecting the aqueous phase, nucleic acids are precipitated with 1/20 volume 5 M ammonium acetate and 2.5 volumes ethanol and recovered by centrifugation.

d. The pellet is dried and resuspended with sequencing gel-loading buffer. The mixture is boiled for 5 min and kept on ice before loading on a 5% denaturing polyacrylamide gel [acrylamide/bisacrylamide (19:1), urea 7 M, 1 $\times$ Tris-borate electrophoresis buffer].[22] After electrophoresis, the gel is autoradiographed overnight at $-80°$ with an intensifying screen.

Reconstituted Transcription System

This protocol is essentially used to follow the activity of the TFIIH-containing fractions. By introducing mutations in TFIIH subunits or by using TFIIH purified from patient cell lines, we determined the functions of the XPB and XPBhelicases, as well as that the CAK complex in the transcription reaction.

Assay. Purification of the general transcription factors has been already described.[23] Usually, 30 ng of TBP, 15 ng of TFIIB, 100 ng of TFIIEα, 60 ng of TFIIEβ, 4 μg of partially purified human TFIIA, 80 ng of TFIIF, and 10 μg of partially purified RNA pol II are preincubated with the DNA template for 15 min before starting RNA synthesis upon the addition of NTPs.

Observations. The optimal amount of each factor to be added to the *in vitro* transcription reaction should be determined for each new batch of proteins. As the MgCl$_2$ concentration is critical in runoff transcription, it should be of 5 mM in the preincubation reaction and reach 6.5 mM in the elongation step. Phosphorylation of the large subunit of RNA pol II by cdk7 is not required when using Ad$_2$MLP as a template. To study the effect of the phosphorylation of the large subunit of RNA pol II by cdk7, the dihydrofolate redutase promoter and TFIID (and not TBP) should be used under the same condition.

Analysis of Results. The reconstituted transcription system using recombinant TFIIH was a very useful tool to analyze the function of each helicase in the first steps of transcription initiation. When an artificial mutation in the ATPase domain (Fig. 2A, lane 2) or an natural mutation found in patients (lane 7) is introduced in XPB, the resulting TFIIH fails

[23] J. C. Marinoni, M. Rossignol, and J. M. Egly, *Methods* 12, 235 (1997).

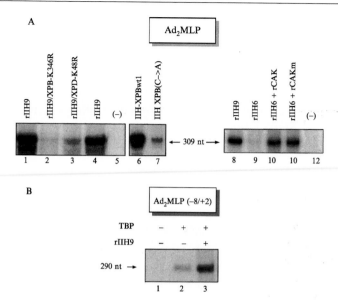

FIG. 2. *In vitro* runoff transcription. (A) *In vitro* runoff transcription is performed in the presence of all the basal transcription factors and recombinant IIH9, IIH9/XPD-K48R, IIH9/XPB-K346R mutated in their consensus ATP-binding site (lanes 1 to 4), or immunopurified TFIIH from XP-B/CS patient cell lines (lanes 6 and 7) or in the presence of IIH6 with or without wild-type or mutant CAK (lanes 8 to 11). Runoff transcription results in a RNA product of 309 nucleotides. −, without TFIIH. (B) *In vitro* runoff transcription is performed with an opened template (−8/+2) in the presence or absence of rIIH9. Runoff transcription results in a RNA product of 290 nucleotides. As a negative control, TBP is omitted in the reaction.

to promote transcription.[13,16,24] This makes XPB a strong candidate for playing a role in the formation of the open initiation complex. In contrast to XPB, XPD is dispensable for the initiation of transcription and RNA synthesis, as a TFIIH complex containing a mutation in the ATP-binding domain of XPD is still active in transcription (lane 3[13]). *In vitro*-reconstituted transcription also uncovers the role of cdk7 as a component of TFIIH in the transcription reaction. Cdk7 (and CAK) is not necessary for transcription initiation of the Ad₂MLP. Indeed, TFIIH lacking the CAK subcomplex or containing an inactive cdk7 subunit (rCAKm) is able to initiate transcription (Fig. 2, lanes 9 and 11, respectively). However, when present, CAK stimulates *in vitro* basal transcription from minimal promoters, even if cdk7 is inactive (Fig. 2, compare lane 10 with lanes 9 and 11).

[24] G. Weeda *et al.*, *Cell* **62**, 777 (1990).

Runoff Transcription without TFIIH

Principle

To investigate whether XPB is involved in promoter opening, we set up an assay in which we take advantage of the observation that the Ad_2MLP promoter becomes single stranded between nucleotide -8 and $+2$ during the initiation step. Using such a partially opened Ad_2MLP $(-8/+2)$ template, we and others[25] analyzed the role of TFIIH in promoter opening.

Production of Heteroduplex

For production of the heteroduplex Ad_2MLP $(-8/+2)$ template, two different double-stranded oligonucleotides (5'AATTCCCTATAAAGG-GGGTGGGCGCGCGT**AGCAGGAGTG**TCTCTTCCTCG3' and 5'A-ATTCCCTATAAAAGGGGGTGGGCGCGC**TTCGTCCTCAC**TCTC-TTCCTCG3') with an inverted region (bold sequence), from -8 to $+2$ nucleotides including the transcription start site, are cloned in the *Eco*RI/*Bam*HI sites of the pUC309 plasmid,[18] taking the place of the wild-type Ad_2MLP promoter and generating new parental pUC309A and pUC309B plasmids, respectively.

  a. One hundred micrograms of pUC309A and pUC309B is digested with 250 U of restriction enzymes *Eco*RI or *Hin*dIII (New England Biolabs) for 2 h at $37°$.

  b. After phenol/chloroform extraction and ethanol precipitation, these two linear molecules are then mixed (at a concentration of 20 ng/μl) in buffer K and denatured for 15 min at $95°$.

  c. The reaction is placed on ice and, after the addition of 150 mM, heated at $55°$ for 30 min. The reaction is then allowed to slowly cool down at room temperature for rehybridization.

  d. ATP is added to a final concentration of 2 mM and ligation is initiated by the addition of 1600 U of T4 DNA ligase (New England Biolabs) for 5 min at $20°$. This short time reaction favors the ligation of circular opened plasmids.

  e. After purification on the CsCl gradient, circular Ad_2MLP $(-8/+2)$ plasmids are linearized with *Sal*I and purified by sucrose density gradient as described earlier. Runoff transcription of these templates gives rise to a 290 nucleotides transcript.

[25] F. C. Holstege, P. C. van der Vliet, and H. T. Timmers, *EMBO J.* **15,** 1666 (1996).

Assay

In vitro-reconstituted transcription using Ad$_2$MLP ($-8/+2$) can be performed as described earlier. Removing TBP from the reaction serves as a negative control.

Analysis of Results

By creating an opened region around the start site of the Ad$_2$MLP, we were able to circumvent the requirement for TFIIH in transcription, indicating that TFIIH is critical in creating a single-stranded region around the promoter. Indeed, when TFIIH is omitted from the reaction, transcription still occurs (Fig. 2, lane 2). Interestingly, the addition of TFIIH stimulates runoff transcription (Fig. 2, lane 3), revealing a potential additional role for TFIIH in transcription initiation or promoter escape. Removing TBP can serve as a negative control (Fig. 2, lane 1).

Abortive Initiation Assay

Principle

To investigate the requirement of the various TFIIH subunits for synthesis of the first phosphodiester bond, we use the abortive initiation reaction in which a priming CpA dinucleotide replaces the NTP specified at the predominant start site of the Ad$_2$MLP. Such a reaction generates trinucleotides (CpApC) that have a radioactive cytidylyl extension on the 3' end.

Assay

a. Preinitiation complexes are assembled using 10 ng of Ad$_2$MLP, 4 ng of TBP, 2 ng of TFIIB, 160 ng of TFIIF, 26 ng of TFIIE, 200 ng of TFIIH, and 10 μg of partially purified RNA pol II in buffer I containing 0.4 mg/ml BSA and 5 mM MgCl$_2$ for 30 min at 25°.

b. Diphosphodiester bond synthesis is initiated by the addition of priming dinucleotides CpA (0.5 mM; Sigma) and MgCl$_2$ to 6.5 mM, dATP to 4 μM, and 4 μCi of [α-^{32}P]CTP (400 Ci/mmol, Amersham). The reaction is then allowed to proceed for 30 min.

c. The reactions are stopped with 0.10 M EDTA and 0.5 mg/ml proteinase K and are supplemented further with 10 μl of sequencing gel-loading buffer.

d. Routinely 10 μl of the mixture is applied to a 20% acrylamide gel (19:1 acryl/bisacryl), 8.3 M urea gel and run at 20 W constant power for 2–2.5 h. After migration, the gel is autoradiographed overnight at $-80°$.

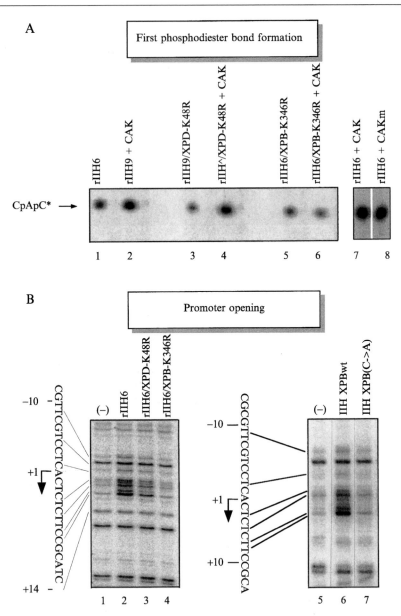

FIG. 3. First phosphodiester bond formation and promoter opening. The abortive assay is performed in the presence of all the basal transcription factors and recombinant IIH6, IIH6/XPD-K48R, or IIH6/XPB-K346R mutated in their consensus ATP-binding site and in the presence or absence of CAK as indicated at the top of the figure. CpApC is indicated by an arrow. Promoter opening experiments using KMnO₄ footprinting were done with either

Observations

Because free $[\alpha\text{-}^{32}P]CTP$ is not removed by ethanol precipitation, the gel is highly radioactive and should be manipulated with care behind a Plexiglas shield.

Analysis of Results

We found that the entire TFIIH complex in which the XPB helicase is mutated in its ATP-binding site cannot initiate synthesis of the first phosphodiester bond (Fig. 3A, compare lanes 2 and 6). In contrast, TFIIH with a XPD containing a mutation in its ATP-binding site allows significant formation of the first phosphodiester bond (Fig. 3A, compare lanes 2 and 4). The abortive synthesis assay also allows us to investigate the role of CAK on the initiation reaction. Indeed, the IIH9 complex containing an inactive cdk7 mutant is as efficient as a complex containing wild-type cdk7 in forming the first phosphodiester bond (Fig. 3A, compare lanes 7 and 8). Taken together, these results suggest that the ternary CAK complex improves the efficiency of the initiation reaction from cdk7 kinase-independent promoters such as Ad_2MLP.

Promoter Opening and Permanganate Assay

Several experiments suggest that TFIH is involved in promoter opening. Indeed, opening of the DNA around the transcription start site requires ATP as a source of energy.[26] The need for TFIH to allow transcription initiation from the Ad_2MLP depends on the topology of the promoter: a negatively supercoiled but not linearized DNA template circumvents the requirement for TFIIH.[27,28] Transcription from supercoiled templates does not require energy, as AMP–PNP can be used instead of ATP. The presence of two helicases in TFIIH, as well as the inhibition of runoff transcription and initiation when XPB ATPase is absent, makes the subunit a strong candidate for playing a role in promoter opening. To study the involvement

---

[26] W. Wang, M. Carey, and J. D. Gralla, *Science* **255,** 450 (1992).
[27] J. D. Parvin and P. A. Sharp, *Cell* **73,** 533 (1993).
[28] F. C. P. Holstege *et al., EMBO J.* **14,** 810 (1995).

---

recombinant wild-type rIIH6 (lane 2) or ATPase site mutated rIIH6 complexes (lanes 3 and 4), in addition to 50 ng of rCAK (left), or TFIIH immunopurified from XP-B/CS patients (right), –, experiment without TFIIH. The promoter sequence, as well as the opened region, is indicated.

of TFIIH in this crucial step, we developed the probing of single-stranded regions by the permanganate assay.

Principle

Potassium permanganate (KMnO$_4$) is a chemical probe that reacts preferentially with single-stranded thymines. This oxydation modifies the 5,6 bond of pyrimidines to 5,6 cis-diols, which constitute a strong block for DNA polymerases. Here we used this property to detect modified thymines by a primer extension assay.

Observations

We noticed that the presence of TFIIA has no influence on KMnO$_4$ assay and can be removed. The amount of protein must be determined experimentally; in our hands, only the concentration of TBP was critical and must be adjusted. Modifications of experimental conditions can lead to the absence of the signal or an increased background. Therefore, several precautions should be taken to obtain interpretable results. (a) Because DTT or 2-mercaptoethanol is a strong scavenger of KMnO$_4$, and therefore can result in misinterpretation of data, we recommend the use of highly concentrated protein preparations to reduce the concentration of these inhibitors. (b) Strong vortexing of DNA or freezing/thawing of the stock solution should be avoided to prevent denaturation. (c) KMnO$_4$ stock solution can be stored at 4° for less than 1 month. (d) The KMnO$_4$ concentration and reaction time must also be optimized experimentally. (e) We suggest omitting ATP or one of the crucial transcription factor (such as TBP or TFIIH) to serve as controls.

Assay

a. The reaction mixture containing 15 ng of TBP, 15 ng of TFIIB, 100 ng of TFIIE α, 60 ng of TFIIE β, 80 ng of TFIIF, and 10 μg of a partially purified fraction of RNA pol II in addition to 20 ng of Ad$_2$MLP template is incubated for 30 min at 25° with 100 ng of TFIIH in a 20-μl reaction with buffer I and 5 mM MgCl$_2$.

b. ATP and CTP (200 μM) are added and the reaction is allowed to proceed for 5 min. ATP alone allows promoter opening, but the addition of CTP increases the signal by stabilizing the preinitiation complex.

c. Two microliters of the 160 mM KmnO$_4$ solution (Janssen Chemika) is incubated with the reaction for 2 min at room temperature, after which the reaction is stopped by the addition of 2 μl of 14.4 M 2-mercaptoethanol. The mixtures are then put on ice. The 160 mM KMnO$_4$ solution is prepared from a 10× stock kept at 4°.

d. After the addition of 200 μl of transcription stop, the DNA is extracted by phenol–chloroform. The mixture is vortexed briefly and the aqueous phase is recovered by centrifugation for 5 min at room temperature. After ethanol precipitation, the DNA is dissolved in 35 μl of H_2O by vigorous pipetting (avoid vortexing to protect DNA integrity).

e. Mix 49.5 μl of H_2O, 10 μl of 10x *Taq* polymerase buffer [166 mM $(NH_4)_2SO_4$, 670 mM Tris–HCl, pH 8.8), 67 mM $MgCl_2$, 100 mM 2-mercaptoethanol, 1 mg/ml bovine serum albumin], 4 μl of 5 mM dNTP mix, 1 μl of 32P 5' end-labeled primer (4.10^7 cpm/μl) and 0.5 μl (2.5 U/ml) of *Taq* polymerase together with DNA and cover with 100 μl of mineral oil. The DNA is denatured at 94° for 1.5 min, hybridized for 2 min at 57°, and extended for 3 min at 72°. Polymerase chain reaction cycles are repeated 10 times. Samples are then extracted with phenol–chloroform and precipitated with ethanol. The pellet is dried and resuspended in sequencing gel-loading buffer, boiled for 5 min, and kept on ice before loading on a 8% denaturating polyacrylamide gel and autoradiographed at −80°.

Analysis of Results

The $KMnO_4$ assay was used to investigate the role of the XPB and XPD helicases of TFIIH on promoter opening. In the presence of wild-type TFIIH, the *Taq* polymerase stops at positions +8, +7, +5, +3, −1, and −7, indicating modified thymines and promoter opening around the start site[13,16,25] (Fig. 3B, lanes 2 and 6). TFIIH complexes mutated in the ATPase domain of XPB or purified from XP-B/CS patients are inactive (Fig. 3B, lanes 4 and 7, respectively), whereas XPD mutated rIIH6 displays an intermediate footprinting pattern (lane 3).[13] These data highlight the crucial role of XPB in transcription initiation and fully support *in vivo* data.[29–31]

Promoter Escape

TFIIH also plays a role in the subsequent elongation step, when RNA pol II escapes the promoter.[32,33] Promoter escape takes place from the synthesis of the first phosphodiester bond of nascent transcript through the point when transcripts reach the size of 15 nucleotides. After transcription

[29] S. N. Guzder *et al.*, *Nature* **369,** 578 (1994).
[30] S. N. Guzder *et al.*, *Nature* **367,** 91 (1994).
[31] P. Sung *et al.*, *EMBO J.* **7,** 3263 (1988).
[32] A. Dvir, R. C. Conaway, and J. W. Conaway, *Proc. Natl. Acad. Sci. USA* **94,** 9006 (1997).
[33] J. Bradsher, F. Coin, and J. M. Egly, *J. Biol. Chem.* **275,** 2532 (2000).

initiation, RNA pol II is not yet committed to transcript elongation and has a tendency to pause, resulting in some cases in so-called abortive initiation transcripts. To rescue transcription, RNA pol II needs additional factors, including TFIIH. Although previous studies have shown that TFIIH plays a critical role in this step (see, e.g., Sung et al.[32]) the contribution of each of the helicases and CAK subunits has emerged only recently.

Principle

To discriminate between the roles of TFIIH in promoter opening and elongation, we have designed a premelted heteroduplex DNA template. RNA termination is obtained when RNA pol II incorporates the chain-terminating ATP-analog cordycepin, allowing the accumulation of products that escape the promoter as a group of oligomers of 17 and 31 nucleotides in length, whereas shorter oligonucleotides (12 nucleotides) indicative of RNA pol II that fails to clear the promoter, are observed in the absence of promoter escape. In order to initiate transcription in the absence of ATP, we have used dinucleotide priming with ApG as the initiating substrate (Fig. 4, bottom).

Assay

a. Preinitiation complexes are assembled using 10 ng of the Ad_2MLP $(-8/+2)$ heteroduplex and 4 ng of TBP, 4 ng of TFIIB, 160 ng of TFIIF, 26 ng of TFIIE, 200 ng of TFIIH, and 10 μg of partially purified fraction RNA pol II in buffer I containing 200 mM ApG (Sigma), 0.4 mg/ml BSA, and 5 mM $MgCl_2$ for 15 min at $25°$.

b. Transcription is started by the addition of 5 mM CTP and GTP, 4 μCi of $[\alpha\text{-}^{32}P]$ UTP (400 Ci/mmol, Amersham), 100 mM cordycepin (TriLink, San Diego, Ca), 4 mM dATP, and 6.5 mM $MgCl_2$ final concentration and incubated for 30 min at $30°$.

c. The reactions are stopped in the presence of 30 mM EDTA and 0.1 mg/ml proteinase K and are supplemented with 10 μl of sequencing gel-loading buffer.

d. Routinely 10 μl of such reactions is applied to a 20% acrylamide gel (19:1 acryl/bisacryl), 8.3 M urea gels, and run at 20 W for 2–2.5 h. The gel is dried and autoradiographed overnight at $-80°$.

Analysis of Results

We observed that CAK (and XPD, data not shown) plays no apparent role in promoter escape, as the ability of wild-type TFIIH to rescue transcription is unchanged in its presence (Fig. 4, compare lanes 3 and 4).

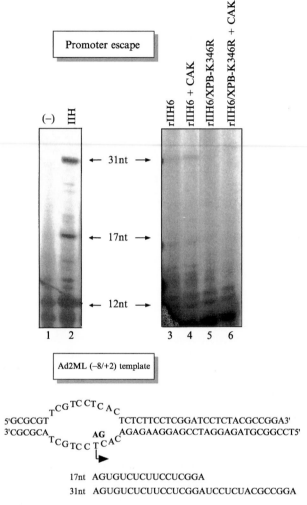

FIG. 4. Promoter escape with opened Ad2ML promoter. Transcription reactions employed a premelted heteroduplex DNA as the template, which is single stranded between nucleotide −8 and +2 as indicated. In order to initiate transcription efficiently, we used the dinucleotide ApG in a 40-fold excess over the concentration of GTP in the reactions. TFIIH (HeLa) is a TFIIH fraction purified through classical chromatographic steps.[18] (Bottom) A diagram of the heteroduplex Ad2ML template used in these studies and the expected transcripts generated.

In contrast, the ability of TFIIH that contains a XPB mutated in the AT-Pase domain to rescue transcription is strongly diminished (Fig. 4, lanes 5 and 6). This observation supports the general idea that the primary effect of CAK on cdk7 kinase-independent promoters is to increase the efficiency of the initiation reaction by the preinitiation complex and shows that XPB helicase activity is not only required for promoter opening, but also for the subsequent steps, such as promoter escape.

Conclusion

The reconstitution of TFIIH in insect cells, as well as the development of methods to investigate each reaction involved in RNA synthesis, led to the dissection of the role played by each enzymatic subunits of TFIIH in transcription initiation and promoter escape. XPB is responsible for both ATP-dependent formation of the open complex prior to transcription and ATP-dependent suppression of the arrest of early RNA pol II elongation products during promoter escape. While the function of XPB is well defined, the role of XPD in transcription seems unclear. For structural reasons, its presence stimulates initiation and promoter escape. Future work will help define the nature of such stimulatory function and the role played by XPD helicase activity in transcription. Finally, CAK plays a role in stabilization of the preinitiation complex largely attributed to a nonenzymatic function of this complex.

A detailed model for the topology of the RNA pol II initiation complex has been proposed on the basis of cross-linking experiments. This technique provided an extensive map of the possible interactions of both RNA pol II and general factors with DNA and evidence for the existence of the so-called "wrapping preinitiation complex."[34] In this model, promoter DNA is tightly wrapped around RNA pol II and the general initiation factors. Evidence that the XPB helicase makes contacts with promoter DNA upstream and downstream of the transcriptional start site strengthen data obtained with classical methods described in this article.[35]

Finally, impairment of transcription initiation or promoter escape could lead to dramatic effects on cell physiology. In this respect, the finding that mutations found in XP-B/CS patients can block the regulation of TFIIH function by the FUSE-binding protein (FBP) and FBP-interacting repressor (FIR) illustrates how activators and repressors can act on initiation and promoter escape through TFIIH. Indeed FBP and FIR activate or inhibit, respectively, transcription by modulating initiation and promoter escape through TFIIH.[36,37] Further work on promoter initiation and escape will most probably integrate activators and will explain, at the molecular level, the function of these proteins.

[34] F. Robert et al., Mol. Cell 2, 341 (1998).
[35] M. Douziech et al., Mol. Cell. Biol. 20, 8168 (2000).
[36] J. Liu et al., Mol. Cell 5, 331 (2000).
[37] J. Liu et al., Cell 104, 353 (2001).

Acknowledgments

This review is a compilation of techniques that were set up and optimized by people from our laboratory. We are thankful to all the members of our group and particularly to M. Chipoulet, A. Frey, and C. Braun for sharing protocols and technical advice and to S. Larochelle for critical reading of this work. This work was supported by grants from the INSERM, the CNRS, the Ministère de la Recherche et de l'Enseignement Supérieur, the Association pour la Recherche sur le Cancer, and EEC.

[59] Assays for Investigating the Mechanism of Promoter Escape by RNA Polymerase II

By Arik Dvir, Joan Weliky Conaway, and Ronald C. Conaway

Transcription initiation by RNA polymerase II from its promoters is a complex process that requires at minimum the five general initiation factors TFIIB, TBP, TFIIE, TFIIF, and TFIIH and an ATP(dATP) cofactor. Biochemical studies of transcription initiation by RNA polymerase II in this minimal enzyme system have revealed that initiation occurs by a multistep mechanism that begins with the assembly of polymerase and all five general initiation factors into a stable preinitiation complex at the promoter and proceeds with ATP(dATP)-dependent unwinding of promoter DNA surrounding the transcriptional start site by the TFIIH XPB DNA helicase to form the open complex, synthesis of the first few phosphodiester bonds of nascent transcripts, and escape of polymerase from the promoter.

Although most research on the mechanism of RNA polymerase II transcription initiation has focused on delineating the steps in assembly of the preinitiation complex and on elucidating the mechanics of ATP(dATP)-dependent formation of the open complex, evidence from a series of studies from our laboratory and elsewhere has identified promoter escape by RNA polymerase II as a distinct stage of transcription initiation and has brought to light roles for the general initiation factors and an ATP(dATP) cofactor in this process.[1-6] In summary, these studies suggest

[1] A. Dvir, R. C. Conaway, and J. W. Conaway, *J. Biol. Chem.* **271**, 23352 (1996).

[2] A. Dvir, R. C. Conaway, and J. W. Conaway, *Proc. Natl. Acad. Sci. USA* **94**, 9006 (1997).

[3] A. Dvir, S. Tan, J. W. Conaway, and R. C. Conaway, *J. Biol. Chem.* **272**, 28175 (1997).

[4] R. J. Moreland, F. Tirode, Q. Yan, J. W. Conaway, J. M. Egly, and R. C. Conaway, *J. Biol. Chem.* **274**, 22127 (1999).

[5] K. P. Kumar, S. Akoulitchev, and D. Reinberg, *Proc. Natl. Acad. Sci. USA* **95**, 9767 (1998).

[6] J. F. Kugel and J. A. Goodrich, *Proc. Natl. Acad. Sci. USA* **95**, 9232 (1998).

that efficient promoter escape by RNA polymerase II requires conversion of the early elongation complex to an "escape-competent" intermediate in a step that exhibits a transient requirement for the general initiation factors TFIIE and TFIIF, the TFIIH XPB DNA helicase, an ATP(dATP) cofactor, and template DNA extending 40 to 50 bp downstream of the transcriptional start site. Failure of early RNA polymerase II elongation complexes to undergo conversion to an escape-competent intermediate results either in abortive transcription with the release of the nascent transcript and dissociation of polymerase from the DNA or in transcriptional arrest by polymerase at promoter-proximal sites 9 to 13 bp downstream of the transcriptional start. Thus, efficient promoter escape by RNA polymerase II requires that the early elongation complex undergoes a critical ATP(dATP)-dependent structural transition that most likely depends on the interaction of polymerase and/or one or more of the general initiation factors TFIIE, TFIIF, and TFIIH with template DNA extending 40 to 50 bp downstream of the transcriptional start site.

With findings indicating that DNA-binding transcriptional activators can dramatically increase the efficiency of promoter escape by RNA polymerase II[5,7,8] in a process that may depend on the multiprotein, polymerase-associated mediator complex, it is likely that future investigations of promoter escape will shed new light not only on the basic mechanisms of transcription initiation by RNA polymerase II and general initiation factors, but also on the mechanisms by which DNA-binding transcriptional activators regulate gene expression. This article describes methods and approaches that have proven useful in our own investigations of the mechanism of promoter escape by RNA polymerase II and should be valuable for future efforts to define the mechanism of this important transcriptional stage in greater detail.

Materials

Unlabeled ultrapure ribonucleoside 5'-triphosphates, deoxyribonucleoside 5'-triphosphates, and 3'-O-methylguanosine 5'-triphosphate (3'-O-MeGTP) are from Pharmacia Biotech Inc. ATPγS is from Boehringer Mannheim. Polyvinyl alcohol (type II), α-amanitin, hexokinase-agarose, and the dinucleotides CpA, CpU, ApC, ApG, GpA, UpU, and UpG are from Sigma. Bovine serum albumin (Pentex fraction V) is from ICN Immunobiologicals. The human placental ribonuclease inhibitor (RNasin) and Klenow fragment are from Promega. The Ultrogel AcA 34 gel

[7] A. Fukuda, Y. Nogi, and K. Histake, *Proc. Natl. Acad. Sci. USA* **99**, 1206 (2002).
[8] J. Kim, J. Lu, and P. G. Quinn, *Proc. Natl. Acad. Sci. USA* **97**, 11292 (2000).

filtration resin is from BioSepra Inc. $[\alpha\text{-}^{32}P]$rNTPs (>400 Ci/mmol) is from Amersham Corp.

Methods

Preparation of RNA Polymerase II and General Initiation Factors

RNA polymerase II[9] and TFIIH[10,11] are purified from rat liver nuclear extracts as described. Recombinant TFIIB (rat α[12]) and yeast TBP (AcA 44 fraction[13,14]) are expressed in *Escherichia coli* and purified as described. Recombinant TFIIF is expressed in *E. coli* strain JM109(DE3) by coinfection with M13 bacteriophage vectors M13mpET-RAP30 and M13mpET-RAP74 and purified as described.[15] Recombinant TFIIE is expressed in *E. coli* and purified as described,[16] except that the TFIIE 56-kDa subunit is expressed in *E. coli* strain BL21(DE3)-pLysS.

Reconstitution of Promoter-Specific Transcription by RNA Polymerase II and General Initiation Factors

In our laboratories, the adenovirus 2 major late (AdML) core promoter in pDN-AdML,[17] which includes AdML DNA sequences from positions −50 to +10 with respect to the transcriptional start site, has served as a valuable model for investigating the mechanism of promoter escape by RNA polymerase II (Fig. 1). One advantage of this AdML core promoter derivative is that it directs synthesis by RNA polymerase II of a 15 nucleotide G-less transcript that is of sufficient length for polymerase to have escaped the promoter and that can be visualized easily as a discrete transcript in reactions containing the RNA chain-terminating nucleotide 3′-O-MeGTP. As our DNA template, we typically use an ~300 bp *Kpn*I to *Nde*I DNA fragment derived from pDN-AdML. This DNA fragment includes AdML core promoter sequences from −50 to +10, as well as

[9] H. Serizawa, R. C. Conaway, and J. W. Conaway, *Proc. Natl. Acad. Sci. USA* **89,** 7476 (1992).

[10] R. C. Conaway and J. W. Conaway, *Proc. Natl. Acad. Sci. USA* **86,** 7356 (1989).

[11] R. C. Conaway, D. Reines, K. P. Garrett, W. Powell, and J. W. Conaway, *Methods Enzymol.* **273,** 194 (1996).

[12] A. Tsuboi, K. Conger, K. P. Garrett, R. C. Conaway, J. W. Conaway, and N. Arai, *Nucleic Acids Res.* **20,** 3250 (1992).

[13] M. C. Schmidt, C. C. Kao, R. Pei, and A. J. Berk, *Proc. Natl. Acad. Sci. USA* **86,** 7785 (1989).

[14] J. W. Conaway, J. P. Hanley, K. P. Garrett, and R. C. Conaway, *J. Biol. Chem.* **266,** 7804 (1991).

[15] S. Tan, R. C. Conaway, and J. W. Conaway, *BioTechniques* **16,** 824 (1994).

[16] M. G. Peterson, J. Inostroza, M. E. Maxon, O. Flores, A. Admon, D. Reinberg, and R. Tjian, *Nature* **354,** 369 (1991).

[17] R. C. Conaway and J. W. Conaway, *J. Biol. Chem.* **263,** 2962 (1988).

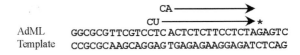

AdML   GGCGCG TTCGTCCTC ACTCTCTTCCTCTAGAGTC
Template   CCGCGC AAGCAGGAG TGAGAGAAGGAGATCTCAG

Ad(-9/-1)   GGCGCG AAGTAGAAG ACTCTCTTCCTCTAGAGTC
Template   CCGCGC AAGCAGGAG TGAGAGAAGGAGATCTCAG

Fig. 1. Structures of the AdML core promoter in pDN-AdML and the Ad(-9/-1) bubble template. (Top) Sequence of the AdML core promoter in pDN-AdML. The position of the first G residue of the transcript is marked with an asterisk. Locations at which the dinucleotides CpU and CpA prime synthesis of transcripts are indicated. (Bottom) Structure of the Ad(-9/-1) bubble DNA template.

vector DNA sequences extending ~250 bp downstream of the AdML transcriptional start site.

To reconstitute transcription initiation, RNA polymerase II preinitiation complexes are assembled at 28° at the AdML core promoter by a preincubation of at least 15 min of 60 μl reaction mixtures containing 20 mM HEPES–NaOH (pH 7.9), 20 mM Tris–HCl (pH 7.9), 60 mM KCl, 7 mM MgCl$_2$, 0.1 mM EDTA, 1 mM dithiothreitol (DTT), 0.5 mg/ml bovine serum albumin, 2% (w/v) polyvinyl alcohol, 7% (v/v) glycerol, 6 units RNasin, ~100 ng of DNA template, ~10 ng of recombinant TFIIB, ~20 ng of recombinant TFIIF, ~20 ng of recombinant TFIIE, ~50 ng of recombinant yeast TBP, ~40 ng of purified TFIIH, and 0.01 unit of RNA polymerase II. Depending on the experiment, transcription reaction mixtures can be scaled up or down proportionately in volume.

Use of Dinucleotide-Primed Transcription Assays to Investigate the Mechanism of Promoter Escape by RNA Polymerase II

It is difficult to resolve very short transcripts (<8 to 10 nucleotides in length) from the radioactive ribonucleoside triphosphates used as substrates for RNA synthesis. For this reason, dinucleotide-primed transcription assays, in which the initiating nucleotides are dinucleotides lacking 5'-phosphates, have proven valuable for investigating the requirements for both transcription initiation and synthesis of the first few phosphodiester bonds of nascent transcripts by RNA polymerase II. Dinucleotide-primed transcription initiation by RNA polymerase II can occur over an approximately 9 bp region surrounding the transcriptional start site.[18] Thus, at the

[18] M. Samuels, A. Fire, and P. A. Sharp, *J. Biol. Chem.* **259,** 2517 (1984).

AdML core promoter, for example, the dinucleotides CpA and CpU will support synthesis of radioactively labeled transcripts when reactions are carried out in the presence of appropriate mixtures of ribonucleoside triphosphates (Fig. 1). Short dinucleotide-primed transcripts are conveniently analyzed in 25% (w/v) acrylamide, 3% (w/v) bisacrylamide, 7 M urea gels, which effectively separate radioactively labeled trinucleotide and larger transcripts from unincorporated $[\alpha\text{-}^{32}P]$rNTPs and which are capable of resolving short transcripts that differ by only a single nucleotide.

Preparation and Analysis of Early RNA Polymerase II Elongation Intermediates

To prepare early RNA polymerase II elongation intermediates that have not yet escaped the promoter, preinitiation complexes assembled at the AdML core promoter are incubated at 28° for 10 to 20 min in the presence of 200 μM CpU or CpA, 1 nM UTP, 5 μCi $[\alpha\text{-}^{32}P]$CTP, and 5 μM ATP or dATP. In the presence of severely limiting concentrations of UTP, RNA polymerase II initiates and synthesizes 4 to 8 nucleotide transcripts at the rate of ~1 to 2 nucleotides per minute (Fig. 2, lanes 1 and 4), allowing preparation of early, "preescaped" transcription complexes. These early elongation complexes can be chased into 16 to 18 nucleotide long, 3′-O-MeG-terminated RNAs by the addition of ATP, CTP, UTP, and 3′-O-MeGTP (Fig. 2, lane 2).

Early RNA polymerase II elongation complexes are prone to transcriptional arrest after synthesizing 9 to 13 nucleotide long transcripts; lane 3 of Fig. 2 shows the accumulation of arrested 9 to 13 nucleotide long transcripts synthesized when RNA polymerase II elongation complexes containing 4 to 8 nucleotide transcripts are chased in the presence of ATPγS, an inhibitor of the TFIIH DNA helicase. RNA polymerase II elongation complexes that have synthesized 16 to 18 nucleotide long, 3′-O-MeG-terminated transcripts are considered to have escaped the promoter, as further elongation of these transcripts is independent of TFIIH and an ATP cofactor with a hydrolyzable β-γ phosphoanhydride bond. We therefore calculate the efficiency of promoter escape by RNA polymerase II by measuring the ratio of 3′-O-MeG-terminated RNAs to total transcripts.

In our investigations of the mechanism of promoter escape by RNA polymerase II, it has proven valuable to isolate transcriptionally active early RNA polymerase II elongation intermediates that have not yet escaped the promoter in a form that is free of ribonucleoside triphosphates and abortive transcripts. To accomplish this, AcA 34 gel filtration is used to isolate early RNA polymerase II elongation complexes free of unicorporated nucleotides and abortive transcripts. Transcription reaction mixtures

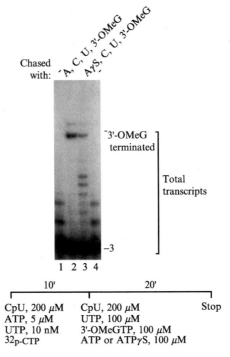

Fig. 2. Transcripts synthesized by early RNA polymerase II elongation complexes. Preinitiation complexes were assembled as described in the text. Early elongation complexes were formed in the presence of the indicated ribonucleoside triphosphates (*lanes 1 and 4*) and chased into longer products in the presence of ATP (*lane 2*) or ATPγS (*lane 3*).

scaled up to ~300 μl are applied at room temperature to ~2-ml AcA 34 gel filtration columns packed into Pasteur pipettes fitted with glass wool plugs and preequilibrated in 20 mM HEPES–NaOH, pH 7.9, 20 mM Tris–HCl, pH 7.9, 60 mM KCl, 7 mM MgCl₂, 0.1 mM EDTA, 1 mM DTT, 0.5 mg/ml bovine serum albumin, 2% (w/v) polyvinyl alcohol, and 7% (v/v) glycerol. Fractions of ~100 μl are collected. The elution of stable RNA polymerase II elongation intermediates can be monitored using a handheld survey meter. The RNA polymerase II elongation complex elutes in the void volume of the AcA 34 gel filtration column, whereas unincorporated nucleotides partition into the included volume. Elongation complexes isolated by this method can be used immediately or stored for several days at 4° without significant loss of activity. Isolated RNA polymerase II elongation complexes containing radioactively labeled 4 to 8 nucleotide long transcripts can be chased into 3'-*O*-MeG-terminated RNAs by the addition of ATP, CTP, UTP, and 3'-*O*-MeGTP.

In a complementary approach, we have investigated the requirement for an ATP(dATP) cofactor in promoter escape by RNA polymerase II by removing ATP from preparations of early elongation complexes with immobilized hexokinase. To accomplish this, RNA polymerase II preinitiation complexes assembled at the AdML core promoter are incubated 10 to 20 min at 28° in 60-μl reaction mixtures containing 200 μM CpU, 1 nM UTP, 5 μM ATP, and 5 μCi [α-^{32}P]CTP. Early RNA polymerase II elongation complexes are then incubated for 15 min at 28° with 3 mM glucose and 30 μl of a 1:1 slurry of hexokinase–agarose beads preequilibrated in 20 mM HEPES–NaOH, pH 7.9, 20 mM Tris–HCl, pH 7.9, 60 mM KCl, 7 mM MgCl$_2$, 0.1 mM EDTA, 1 mM DTT, 0.5 mg/ml bovine serum albumin, 2% (w/v) polyvinyl alcohol, and 7% (v/v) glycerol ($\sim$ 5 units of hexokinase). The hexokinase–agarose beads are removed by centrifugation for 30 s at 16,000 g. The removal of ATP can be monitored using a luciferase-based bioluminescence assay (luciferase–luciferin preparation from firefly, Sigma Chemical Co., St. Louis, MO).

Use of Bubble DNA Templates to Investigate the Mechanism of Promoter Escape by RNA Polymerase II

Premelted, "bubble" DNA templates have proven valuable for our investigations of the roles of TFIIE, TFIIH, and an ATP(dATP) cofactor in promoter escape by RNA polymerase II because none of these reaction components (which are needed to unwind the promoter at the transcriptional start site during initiation from normal duplex templates) are required for promoter-specific initiation from these templates.[19–21] The bubble DNA template used most commonly in our laboratories is the AdML core promoter derivative designated Ad(−9/−1), which contains a premelted region from positions −9 to −1 with respect to the transcriptional start site (Fig. 1).[2]

Although bubble DNA templates can be prepared by annealing synthetic oligonucleotides, we have found it useful to use bubble templates longer than those made easily from synthetic oligonucleotides. As a consequence, we have developed a method for preparing bubble templates from single-stranded M13 DNA. To prepare M13 phage for preparation of the Ad(−9/−1) DNA template, we cloned the EcoRI to HindIII fragment from pDN-AdML into the polylinker of double-stranded M13mp19 linearized with the same restriction enzymes to generate M13mp19-AdML.[2]

[19] D. Tantin and M. Carey, J. Biol. Chem. **269,** 17397 (1994).
[20] G. Pan and J. Greenblatt, J. Biol. Chem. **269,** 30101 (1994).
[21] F. Holstege, D. Tantin, M. Carey, P. C. van der Vliet, and H. Th. M. Timmers, EMBO J. **14,** 810 (1995).

M13mp19-AdML bacteriophage are grown in XL-1 Blue bacterial cultures, and single-stranded phage DNA, which corresponds to the template strand of the AdML promoter, is prepared as described.[22] Two single-stranded synthetic oligonucleotides are used as primers to direct the synthesis of double-stranded DNA using single-stranded M13mp19-AdML DNA as the template. One oligonucleotide contains the nontemplate strand of AdML promoter sequences from positions −59 to +17 with respect to the transcriptional start site, and the other oligonucleotide contains the same DNA sequence except for the 9 base noncomplementary sequence AAGTAGAAG from positions −9 to −1.

To prepare the wild-type AdML control DNA template and the Ad(-9 to −1) bubble template, approximately 150 fmol of the appropriate oligonucleotide is mixed with ∼ 50 fmol of single-stranded M13mp19-AdML DNA in a hybridization solution containing 25 mM Tris–HCl, pH 7.6, 50 mM KCl, and 8 mM MgCl$_2$. The mixtures are heated for 5 min at 94° and then cooled to room temperature for 2 h. The annealed oligonucleotide primers are then extended for 3 h at 37° by the Klenow fragment of *E. coli* DNA polymerase I in reaction mixtures containing 10 units of Klenow fragment, 100 μM dATP, 100 μM dCTP, 100 μM dGTP, and 100 μM dTTP. The DNA templates are digested with *Kpn*I, which cleaves the DNA ∼50 bp upstream of the transcriptional start site, extracted once with phenol/chloroform, ethanol precipitated, and then digested with *Ava*II to generate ∼410-bp DNA fragments.

Template fragments are fractionated on a low percent agarose gel containing 5 to 10 ng/ml ethidium bromide. Bands containing the DNA fragments are cut from the gel with a razor blade and diced into ∼1-mm cubes. The agarose cubes are transferred into a microcentrifugal ultrafiltration device (Ultrafree-MC, Millipore), which is then placed in liquid nitrogen for 1 to 2 min. After a 10-min spin at maximum speed in a microcentrifuge, the filtrate, which contains the DNA fragments, is precipitated with ethanol, washed with 70% ethanol, and resuspended in 20 mM Tris–HCl, pH 7.5). The purified DNA fragments can then be used as templates in RNA polymerase II transcription reactions.

Acknowledgments

Research in the authors' laboratories is supported by National Institutes of Health Grant R37 GM41628 (R.C.C. and J.W.C.) and by National Science Foundation Grant MCB-0215992 and funds provided by the Oakland University Research Excellence Program in Biotechnology (A.D.).

[22] J. Sambrook, E. F. Fritsch, and T. Maniatis, "Molecular Cloning, a Laboratory Manual." Cold Spring Harbor Laboratory Press, Cold Spring Harbor, NY, 1989.

Author Index

Numbers in parentheses are footnote reference numbers and indicate that an author's work is referred to although the name is not cited in the text.

A

Abrahams, J. P., 50
Achberger, E. C., 11
Ackers, G. K., 321
Acosta, D., 207
Adams, P. D., 51
Adams, S. R., 204
Ade, T., 400
Adhya, S., 262, 556, 568, 569, 618, 619, 625, 627
Adman, E., 556
Admon, A., 523, 524(15), 697, 735
Adrian, M., 29(23), 30
Agalioti, T., 379
Aggarwal, A. K., 46, 251
Ahn, B. E., 73
Ahn, K., 687(10), 688
Ahn, S. H., 143, 461
Aiyar, A., 40
Aiyar, S. E., 556, 559(46), 561, 561(46), 562(46)
Aki, T., 627
Akman, L., 339
Akoulitchev, S., 687, 733, 734(5)
Aksoy, S., 339
Albert, A. C., 121, 122, 132, 137(17)
Alberts, B., 15, 595
Albright-Frey, T. J., 416(11), 417, 424(11), 427(11)
Alekshun, M. N., 279
Allan, B. W., 670
Allemand, J.-F., 369, 506, 577, 578(4), 579, 589(4), 591(4; 6), 592, 592(4)
Allen, N. S., 377
Alley, S. C., 446
Allis, C. D., 378
Allison, L. A., 553
al-Mulla, W., 398

al-Nakib, W., 398
Alper, S., 12
Amann, E., 525
Amemura, M., 521
Ames, B. N., 551
Amit, R., 371
Anders Olson, C., 505
Anderson, A. L., 368
Anderson, K., 374
Andersson, S., 568, 670
Andrews, B., 387
Anello, M., 659
Angerer, A., 182(20), 183
Anikin, M., 84
Ansari, A. Z., 446, 523, 687(10), 688
Anthony, J. R., 54, 55(8), 61(7; 8), 62(8), 63(8)
Anthony, L. C., 13, 14(20), 54, 55(7), 62(7), 63(7), 181, 195, 209
Aplin, R. T., 175
Apone, L. M., 415, 415(6; 7), 416, 424(6), 425(1), 427(6), 428(6; 7)
Appleman, J. A., 610
Arai, N., 735
Arai, Y., 702
Arakawa, T., 315
Aravind, L., 340
Archambault, J., 42, 387, 392
Argos, P., 255
Armitage, J. P., 54, 59(2)
Armstrong, R., 403
Arnaud, M., 458
Arnosti, D. N., 182(11), 183
Arthur, J. S., 401
Arthur, T. M., 174, 192, 207, 208(11), 209, 209(11), 211(11)
Artsimovitch, I., 13, 14(20), 139, 174, 192, 207, 208(11), 209(11), 211(11), 606
Atchley, W. R., 464

Jacobsen, M. P., 367
Jacobson, B. A., 338
Jacquot, S., 400
Jaehning, J. A., 15, 425
Jaffar Ali, B. M., 371
Jaffe, H., 81
Jagadeesh, J., 556
Jäger, J., 37
Jahnke, W., 517
Jair, K.-W., 280
Jakimowicz, D., 339, 341(25)
Jameson, G. B., 49
Jameson, J. L., 397
Jankins, N. A., 6
Jansma, D. B., 387, 389(8), 390(8)
Jaquet, J.-A., 535(7), 536
Jean, D., 379
Jean, J. M., 569, 670
Jegou, B., 403(58), 407
Jendrisak, J. J., 15, 16, 17(25), 27, 57(19), 58, 283, 291, 327, 537, 607
Jennings, E. G., 387, 389(8), 390(8), 417, 420(14)
Jensen, D. B., 182(17), 183, 184
Jensen, G. J., 50, 138, 154(5)
Jensen, K. F., 613, 615(19), 616(19)
Jenuwein, T., 378
Jeon, Y. H., 40
Jeong, C. J., 446
Jeong, W., 662
Jerebtsova, M., 386
Jeruzalmi, D., 672(27), 673
Jia, Y., 670, 672, 672(18), 682, 682(18), 683(33), 684, 685(33), 686(33)
Jiang, Y. L., 556, 715
Jin, D. J., 3, 6, 9, 18, 55, 57(12; 16), 59(12; 16), 63(12), 76, 174, 179, 217, 283, 284, 288(9), 290(1), 291, 292, 295
Jin, H., 278
John, S., 454
Johnson, A. D., 480
Johnson, K. A., 669
Johnson, K. S., 208
Johnson, L., 484(30), 485
Johnson, M. L., 364, 366, 367, 368(67; 68), 513
Johnson, P. F., 238, 522
Johnston, D. E., 695
Johnston, S. A., 446
Joliot, V., 454
Jonassen, I., 256

Jones, D. P., 174, 184
Jones, E. W., 142
Jones, S., 28, 646, 656
Jones, T. A., 39, 51, 52(25)
Jordan, P. A., 301, 302(6)
Josaitis, C. A., 110, 111(5), 118(5), 606, 607(7), 609(7)
Joyce, C. M., 84
Joyce, G. F., 659
Juang, Y.-L., 11, 12(39), 19(11), 20(11), 21, 22(39), 556, 561, 561(44), 567(44)
Junn, E., 659

K

Kadonaga, J. T., 283, 324, 325(1), 477, 488, 493, 493(8), 501(8), 688, 694(13)
Kaestner, K. H., 400
Kaganman, I., 89, 94
Kaguni, J. M., 338, 339, 343, 345
Kaidow, A., 295
Kain, S., 3
Kamada, K., 467(5), 468
Kanazawa, S., 379
Kandel, E. R., 401
Kandler, O., 66
Kane, C. M., 386, 387, 394(6)
Kang, C. M., 658, 659, 659(2), 661, 662, 666(2), 667(2)
Kang, J. G., 74, 79, 82(6)
Kang, J. K., 80
Kang, M. E., 158, 159(11)
Kao, C. C., 735
Kapanidis, A. N., 95, 205, 554
Kaplan, S., 54, 55, 59(2), 65, 65(17)
Karl, K. A., 401
Karlin, S., 239, 239(33), 245(33)
Karls, R. K., 54, 55, 57(12), 59(4; 12), 63(4; 12; 20), 64, 65(20)
Karp, M., 148
Karplus, M., 40
Karsch-Mizrachi, I., 244
Kashlev, M., 96, 138, 138(4), 139, 141(4), 142(4), 143(4; 24), 144, 208, 648
Kassavetis, G. A., 121, 165, 704
Kassir, Y., 480
Katayama, T., 339, 343, 345
Kato, J., 339
Katzameyer, M. J., 85
Kaufman, J., 517, 519(21)

Subject Index

A

2-Aminopurine, transcription
  initiation studies
  equilbrium dissociation constant
    determination with T7 RNA
    polymerase, 673–677
  kinetic analysis of base pair opening
    approaches, 571–572
    data collection and analysis, 573–576
    equipment, 573
    labeled promoter preparation, 569–571
    machine drift concerns, 572
    principles, 568–569
    promoter clearance assay, 575–576
  stopped-flow assays with T7 RNA
    polymerase
    binding to 2-aminopurine-labeled
      promoters
      association rate constant
        determination, 677
      binding in absence of initiating
        nucleotide, 679–680
      binding in presence of initiating
        nucleotide, 681–682
      dissociation rate constant
        determination, 679
    initial RNA synthesis, 683–686
    nucleotide binding during transcription
      initiation, 682–683
    principles, 668–672
ANN, *see* Artificial neural network
Artificial neural network
  Dragon Promoter Finder 1.3
    advantages, 249–250
    artificial neural network training and
      tuning, 244–245
    model testing and validation, 245–249
    sensitivity and specificity, 240, 249
    structure of program, 241–244
    Web access, 240–241
  protein-binding DNA sequence
    identification

analysis of sequences, 235
  ranking function preprocessing, 236–237
AsiA
  homodimer analysis by analytical
    ultracentrifugation
    overview, 509–510
    sedimentation equilibrium, 513–514
    sedimentation velocity, 510, 512
  purification of T4 protein, 508
  s^{70} SR4 complexes
    cross-linking, 514–516
    nuclear magnetic resonance studies
      footprinting, 517–518
      interfacial nuclear magnetic
        resonance, 518, 521
    preparation, 509
    transcriptional repression, 508

B

Biotin acceptor peptide tag, *see* RNA
  polymerase II

C

CIITA
  acetyltransferase activity
    assay, 385–306
    functions, 379
    GTP stimulation, 379
  purification from baculovirus–Sf9 insect
    cell system
    anti-FLAG bead activation and
      immunoprecipitation, 382–383
    cell growth, 380
    cell lysis, 382
    infection of cells, 382
    materials, 380
    overview, 379
    transfection, 380–381
    viral assay, 381
    viral stock preparation, 381–382
  purification from transfected HeLa cells,
    383–384

ISBN: 0-12-182273-7

90000

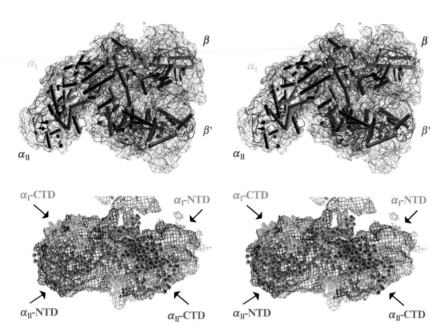

RAY *ET AL.*, CHAPTER 3, FIG. 3. (Top) Stereo view of the global fit of the *E. coli* homology model of RNAP into the three-dimensional map derived from cryo-EM studies. This orientation of RNAP—looking into the active site of the enzyme—illustrates its typical "crab claw" shape. The Cα backbone is represented as cylinders for α helices and arrows for β sheets. (Bottom) Stereo views of the refined fit illustrate localization of the α_I and α_{II} C-terminal domains relative to the corresponding N-terminal domains based on their individual crystal structures. Relative to the global view above, these domains have been rotated anticlockwise by 90° about an axis perpendicular to the plane of the paper and flipped 180° around a horizontal axis in the plane of the paper.

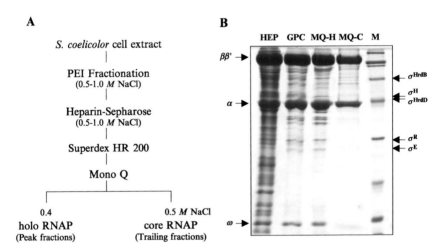

HAHN *ET AL.*, CHAPTER 7, FIG. 2. Purification steps of RNA polymerase from *S. coelicolor*. (A) Schematic procedure for the purification of *S. coelicolor* RNA polymerase. (B) SDS–PAGE profile of eluates from Heparin-Sepharose 6B (lane HEP), Superdex 200 (lane GPC), and Mono Q columns (MQ-H and MQ-C). Peak holoenzyme activity is eluted from the Mono Q column at 0.4 *M* NaCl (lane MQ-H) and is followed by core RNAP fractions (lane MQ-C). Eluates are analyzed on a 0.1% SDS–13% polyacrylamide gel and stained with Coomassie brilliant blue R-250. Positions of RNA polymerase subunits and several associated sigma factors are indicated along with molecular mass markers (lane M; 14.4, 21.5, 31.0, 45.0, 66.2, 97.4, 116.3, and 200 kDa).

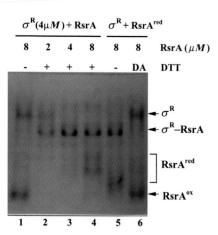

HAHN *ET AL.*, CHAPTER 7, FIG. 4. Redox-dependent formation of the RsrA–σ^R complex monitored by native PAGE. σ^R (4 μM) and RsrA (2–8 μM, lanes 2–4) were incubated in 25 μl of N_2-saturated binding buffer. In a control reaction, DTT was omitted from the binding mixture (lane 1). Reduced RsrA (RsrA^{red}, 8 μM) was incubated with σ^R (4 μM) in the binding buffer without DTT (lane 5) or with added diamide (DA; 1 mM) (lane 6). Samples were separated by electrophoresis on a native 10% polyacrylamide gel and visualized by Coomassie blue staining. Positions of σ^R, the reduced and oxidized forms of RsrA (RsrA^{red}, RsrA^{ox}), and the σ^R–RsrA complex are indicated. Data are taken from Kang *et al.*

C

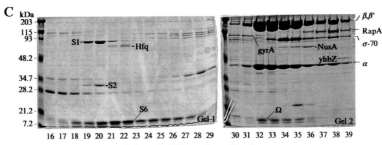

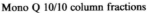

Mono Q 10/10 column fractions

D

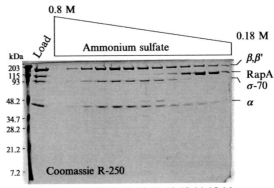

Phenyl superose column fractions

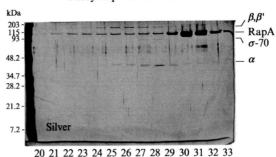

Superose 6 HR 10/30 column fractions

SUKHODOLETS *ET AL.*, CHAPTER 25, FIG. 1. (A) Schematic for the purification procedure of native RapA from the RNAP holoenzyme–RapA complex. (B) Mono Q column protein profile. (C) Mono Q column fractions stained with Coomassie brilliant blue R-250; 10 μl from each fraction was mixed with an equal volume of $2\times$ Laemmli sample buffer and loaded per lane of a 10% SDS–polyacrylamide gel. Prestained protein markers (Bio-Rad, broad range) were loaded in the first lane of each gel. The S1, S2, S6, Hfq, GyrA, NusA, RNAP Ω subunit, and yhbZ proteins were identified based on the following N-terminal protein sequences: MTESFAQLFE (S1), XTVSMRDMLK (S2), MRHYEIVFM (S6), AKGQSLQDPFL (Hfq), XDLAREITPVNI (GyrA), MNKEILAVVEAVSNE (NusA), ARVTVQDAVEKIGNR (Ω), and MKFVDEASILVVA (yhbZ). (D) Phenyl-Superose 5/5 column fractions stained with Coomassie brilliant blue R-250 (top) and Superose 6 HR 10/30 column fractions stained with silver (bottom). Protein samples for denaturing electrophoresis were made as described (C), and the Superose 6 fraction containing the purified RapA protein used in the enzymatic assays is indicated.

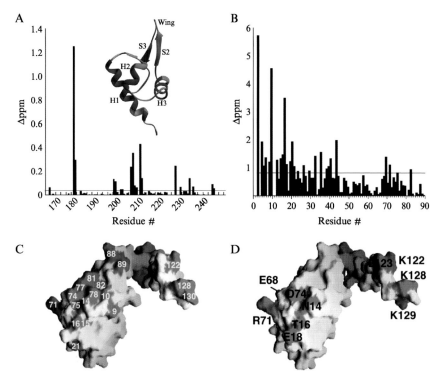

LAMBERT *ET AL.*, CHAPTER 43, FIG. 5. NMR "footprinting" as a guide to targeted mutagenesis of a protein–protein interface. NMR "footprinting" can often lead to the identification of well-defined residues, which form a cluster on the surface of the molecule under study. (A) DNA titration of the DNA-binding domain of the TFIIF subunit Rap30 revealed a set of sharply defined shift changes that formed a specific cluster of perturbed residues on the surface of the protein. In this example, the background of shift changes is minimal, indicated by the horizontal line across the plot. Mapping the shifted residues onto the three-dimensional structure of the domain (inset) defined a surface of the protein that is likely to be in contact with the DNA [red coloring on the indicated helices (H1–H3) and sheets (S1–S3)]. DNA-binding residues identified from the footprint in this instance are precisely those that would be predicted for a winged helix-turn-helix motif domain. (B) A plot of chemical shift changes going from the AsiA homodimer to the AsiA/SR4 complex indicates a diffuse footprint accompanied by an intense background of shift changes (horizontal line in the plot). In this instance, one might conclude that nearly one-half of the 90 residue domain is a participant in the protein–protein interface, a conclusion that would be erroneous based on mutagenesis and NMR of the protein complex (see text). (C and D) The best usage of the NMR "footprint" is as a guide for targeted mutagenesis of a protein interface, exemplified by the interaction between PEA-15 and ERK MAP kinase. (C) PEA-15 residues, which experience chemical shift changes and/or peak broadening in the presence of ERK, are indicated in red superimposed on a molecular surface representation of the structure of the protein. Mutagenesis of these residues defined a subset of the footprint (shown in D), which most likely contacts ERK in the complex.

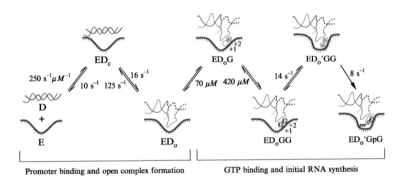

| Promoter binding and open complex formation | GTP binding and initial RNA synthesis |

PATEL AND BANDWAR, CHAPTER 55, FIG. 8. Kinetic pathway of transcription initiation by T7 RNAP with a consensus promoter. Rate constants for steps up to ED_oG were obtained from global analysis of the kinetic experiments described in Figs. 4 and 5. The equilibrium constant of the ED_oG to ED_oGG step was measured as described. Rates of the ED_oGG to $ED_o'GG$ and $ED_o'GG$ to $ED_o'GpG$ steps were measured as described in Figs. 6B and 7D, respectively.